Practical Guide for Biomedical Signals Analysis Using Machine Learning Techniques

Practical Guide for Biomedical Signals Analysis Using Machine Learning Techniques

A MATLAB® Based Approach

Abdulhamit Subasi

ELSEVIER

ACADEMIC PRESS

An imprint of Elsevier

Academic Press is an imprint of Elsevier
125 London Wall, London EC2Y 5AS, United Kingdom
525 B Street, Suite 1650, San Diego, CA 92101, United States
50 Hampshire Street, 5th Floor, Cambridge, MA 02139, United States
The Boulevard, Langford Lane, Kidlington, Oxford OX5 1GB, United Kingdom

Library of Congress Cataloging-in-Publication Data
A catalog record for this book is available from the Library of Congress

British Library Cataloguing-in-Publication Data
A catalogue record for this book is available from the British Library

ISBN 978-0-12-817444-9

For information on all Academic Press publications
visit our website at https://www.elsevier.com/books-and-journals

Publisher: Masucci Stacy
Acquisition Editor: Rafael E. Teixeira
Editorial Project Manager: Sara Pianavilla
Production Project Manager: Kiruthika Govindaraju
Cover Designer: Mark Rogers

Typeset by SPi Global, India

Dedication

A huge thank to my parents for always expecting me to do my best and telling me I could accomplish anything, no matter what it was.

To my wife, Rahime, for her patience and support.

To my wonderful children, Seyma Nur, Tuba Nur, and Muhammed Enes. You are always in my heart and the joys in my life.

To those who read this book and appreciate the work that goes into it, thank you. If you have any feedback, please let me know.

Abdulhamit Subasi

Contents

Preface

Biomedical signals have been recorded and analyzed for several years. New signal-processing methods have uncovered information that entirely altered the previous approaches for diagnosing different diseases. Actually, it is not easy to visualize a situation when diseases related to the heart, brain, or muscles were diagnosed without including certain information derived from biomedical signals. This kind of information is crucial for therapeutic devices in cardiac, neurological, and neuromuscular applications. The goal of this book is to present a comprehensive overview of methods related to biomedical signal processing. Because biomedical signal processing has mostly been synonymous with the processing of electrocardiograms (ECG), electroencephalograms (EEG), and electromyograms (EMG), we have focused on problems relevant to these three types of signals.

Biomedical signal processing has reached certain maturity as an academic subject, and there are many useful books related to this subject. Because biomedical signal processing is an interdisciplinary subject, it has to be taught in different ways depending on the educational program. It is evident that a master course on biomedical signal processing may embrace widely different contents such as choice of biomedical signals, processing methodologies, and important machine learning and pattern recognition techniques. Furthermore, this book may be used as a reference for practicing engineers, physicians, and researchers working with biomedical signals.

Nowadays, there is great interest in machine learning applications in health, biomedicine, and biomedical engineering. This book presents innovative developments in applied machine learning and biomedical signal processing. We describe different machine learning applications in biomedical signal analysis, including cardiology, electroencephalography, and electromyography. Biomedical signal analysis is the application of engineering techniques to problems in medicine and health care. Machine learning techniques are computing algorithms, including artificial neural networks, fuzzy logic, and support vector machines. Most of the machine learning techniques is nonlinear mapping between the inputs and outputs of a model representing the operation of a biomedical system.

This book consists of five chapters, each with an overview of signal processing and machine learning techniques. Chapter 1 is an overview of biomedical signals and machine learning techniques. Chapter 2 focuses on biomedical signals such as EEG, EMG, ECG, etc. Chapter 3 presents different signal-processing techniques commonly used in the analysis of biomedical signals. Chapter 4 presents some feature extraction and dimension reduction techniques. Chapter 5 provides an overview of machine learning techniques widely used for the detection and classification of biomedical signals.

Various biomedical signals are discussed with a particular focus on machine learning methods' potential as a tool for diagnostics. Bioelectric potentials representing brain activity are outlined based on EEG recordings. Here we focused on several applications including identification of abnormal EEG activity in subjects with neurological diseases (e.g., epilepsy) and in the control of external devices using EEG waveforms, known as brain-computer interfaces (BCI). It also discusses the analysis of neuromuscular pathologies that affect EMG-related signal characteristics. Electrical potentials produced by the heart are recorded on the body's surface and characterized using an ECG. The ECG can be used to identify many cardiac abnormalities. Techniques that have been applied to the analysis of ECG signals for the diagnosis of cardiovascular diseases are also presented with several examples.

This book should be of considerable help to a broad readership, including researchers, professionals, academics, and graduate students from a wide range of disciplines who are beginning to look for applications in biomedical signal analysis. The text provides a comprehensive account of recent research in this emerging field, and we anticipate that the concepts presented here will generate further research in this multidisciplinary field.

Acknowledgments

I would like to thank Sara Pianavilla and Rafael E. Teixeira who provided great support and did a lot of work for this book.

Also, I would like to thank Kiruthika Govindaraju for being patient in getting everything necessary completed for this book.

As well, my sincere thanks to Mark Rogers and other staff who worked on the design of the book.

Abdulhamit Subasi

Chapter 1

Introduction and Background

In last decade, there has been vast improvement in the analysis of biomedical signals, especially in the applications of signal processing and machine leaning techniques. Some of the applications that have been successfully applied toward different biomedical signals include parametric and nonparametric spectral estimation, higher order spectral estimation, time-frequency (TF) methods, discrete wavelet transform (DWT), wavelet packet transform (WPT), tunable Q wavelet transform (TQWT), dual tree complex wavelet transform (DTCWT), empirical wavelet transform (EWT), and empirical mode decomposition (EMD). Also, biomedical signals, such as electroencephalograms (EEG), evoked potentials (EP), electromyograms (EMG), electrocardiograms (ECG), phonocardiograms (PCG), photoplethysmography (PPG), and electrohysterograms (EHG), have been used for the diagnosis of different disorders, significantly promoting developments in signal processing and pattern recognition (Gath & Inbar, 2013).

Biomedical signals carry information that might not be directly observed but instead is hidden in the signal's structure. This hidden information can be converted into meaningful interpretations. Furthermore, biomedical signals show the properties of biological systems, and they are helpful for identifying various pathological conditions. Hence, by using different signal processing tools and machine learning methods, biomedical signal analysis has become a crucial tool for extracting clinically substantial information hidden in biomedical signals. Biomedical signal analysis is an interdisciplinary topic used for developing automated diagnostic systems for decision support. Generally, expert physicians manually or visually evaluate biomedical signals, but this might sometimes result in an unreliable or inefficient diagnostic. Therefore, the aim of automated biomedical signal analysis is to diminish the subjectivity of manual evaluation of biomedical signals. Using computer-aided biomedical signal analysis to evaluate different signal characteristics helps in making an objective decision and improves measurement accuracy. Furthermore, to reduce subjectivity of measurement, biomedical signal analysis is used for feature extraction and dimension reduction to support characterizing and understanding of the information that exists in a physiological signal. Feature extraction methods can be used to extract valuable information that cannot be seen easily in the signal via visual assessment (Sörnmo & Laguna, 2005).

1.1 ELECTROENCEPHALOGRAPHY

Control signals originating from the brain are sent through the body to perform numerous functions. EEG is the brain's electrical activity recorded by means of noninvasive electrodes positioned on the scalp. The EEG signal characterizes the brain's complete electrical activity coming from many neuronal activities. Hans Berger recorded the first human EEG signal in 1924, which prompted substantial improvements in human brain research. Berger used two electrodes to acquire a weak current from a human being's head by means of a galvanometer. Today, clinical EEG equipment is used to support physicians' decision-making processes for monitoring routine brain electrical activity. Drowsiness, excitement, relaxation etc. affect EEG signals. In clinical tests, recording electrodes are positioned on the scalp using international 10–20 standards (Klem, Lüders, Jasper, & Elger, 1999). It is crucial for good EEG recording to have a clean connection between the skin and the electrode. To decrease the impedance between the electrode and the scalp, a conductive gel is generally used. EEG signals are then processed by low- and high-pass filters to filter the slow electrogalvanic signals and to eliminate any electromyographic activity from the EEG signals. EEG signals are the most important source of information for analyzing brain activity and brain status. Because EEG recordings contain a large amount of data, analysis and processing of these signals are crucial. The mathematical modeling of EEG signals and the widespread use of machine learning methods play a crucial role in the analysis of this data and to support decision making. EEG analysis therefore enables the treatment and diagnosis of neurological diseases in a clinical environment. Machine learning tools also assist in the recognition of patterns inside the EEG signals, enhancing the identification and detection of brain disorders and pathologies (Begg, Lai, & Palaniswami, 2008).

The EEG signal from the scalp has a time duration of 0.01 to 2s and an amplitude of around 100 μV (Kerem & Geva, 2017). The frequency components of the EEG signals are generally employed for analysis, taking into account of

Practical Guide for Biomedical Signals Analysis Using Machine Learning Techniques. https://doi.org/10.1016/B978-0-12-817444-9.00001-5

following frequency bands: delta (up to 4Hz), theta (4–8Hz), alpha (8–12Hz), beta (12–26Hz), and gamma (26–100Hz). Waveform activities differ from each other according to the brain function related to the specific mental and physical tasks. For instance, low-frequency waves (delta and theta) dominate during sleep times, whereas an EEG signal acquired during awake times includes a higher percentage of high-frequency waves (alpha and beta). Also, transitioning from an eyes-open state to an eyes-closed state changes the EEG frequency spectrum; the state with closed eyes has a diverse peak in the spectrum of around 10Hz (Felzer & Freisieben, 2003). The EEG signals are semistationary, time-dependent, and nonstationary in the waveforms. Hence, these characteristics cannot be detected easily. Power spectrum achieves a quantitative measure of the frequency distribution of the EEG at a cost of the amplitude distribution and information related to the EEG pattern existence. Even though these primary efforts are unsatisfactory, they allow the use of frequency analysis in the analysis of brainwave activity. Therefore, TF methods, such as wavelets, are applied for feature extraction from EEG signals (Bigan, 1998). Furthermore, TF methods, such as DWT, WPT, TQWT, DTCWT, EWT, and EMD, are essential to explain the different behavior of the EEG, expressing it in both the time and frequency domain. It should also be highlighted that the wavelet is appropriate for the analysis of nonstationary signals, such as EEG, ECG, and EMG. Therefore, the wavelet is appropriate for detecting transient events, such as spikes, that occur during epileptic seizures (Adeli, Ghosh-Dastidar, & Dadmehr, 2007; Bronzino, 2000; Subasi, 2007; Subasi & Gursoy, 2010).

There are around 2 million epilepsy patients in the United States alone, and responsive therapeutic intervention facilitated by seizure detection algorithms will increases the efficiency of the method. Raghunathan, Jaitli, and Irazoqui (2011) devised a two-stage cascaded seizure detection solution, with full detection efficiency. Their proposed solution is based on the usage of features that results in unique patterns during the seizure. The proposed technique shows a high sensitivity rate and a low detection duration. Yuan, Zhou, Liu, and Wang (2012) proposed a new method for multichannel long-term EEG. Novel nonlinear features of EEG signals are derived from the fractal geometry, as the linear feature comes from the relative fluctuation index. The vector of the feature is then merged into an extreme learning machine for classification. For more stable results, postprocessing techniques, such as smoothing and channel fusion, are employed, which was tested on 21 subjects with segment-based and event-based analysis.

The lack of a clear difference in EEG activities makes visual detection of different disorders from EEG signals challenging (Bigan, 1998). Therefore, computer-aided decision-support systems were developed to allow for more precise and quicker detection of disorders using EEG recordings. Artificial neural network (ANN)-based classifiers' performance was compared by Pang, Upton, Shine, and Kamath (2003). They trained an ANN with features selected from a raw EEG signal. A system for the automatic analysis and detection of epileptic seizures using wavelet transform to extract features from EEG signals was developed by Bigan (1998). An ANN model was used for the automated analysis of EEG recordings. A DWT and a mixture of expert models were employed for EEG signal classification by Subasi (2007). In Subasi and Gursoy (2010), DWT was used for feature extraction; principal component analysis (PCA), independent component analysis (ICA), and linear discriminant analysis (LDA) were used for dimension reduction; and support vector machines (SVM) were used for the classification of EEG signals. Subasi, Kevric, and Canbaz (2017) proposed an epileptic seizure detection model to determine the optimum parameters of SVMs by employing particle swarm optimization (PSO) and genetic algorithm (GA). Alickovic, Kevric, and Subasi (2018) compared wavelet packet decomposition (WPD), DWT, and EMD for epileptic seizure detection and seizure prediction. Soleimani-B, Lucas, Araabi, and Schwabe (2012) proposed a robust technique to elaborate and evolve a neuro-fuzzy model that works as an online adaptive method with a patient-independent parameter for seizure prediction. Furthermore, prediction is improved by using multiple features to detect the preictal patterns. Liu, Zhou, Yuan, and Chen (2012) proposed wavelet decomposition of multichannel intracranial EEG (iEEG) within five scales by selecting three frequency bands. They extracted effective features, such as relative amplitude, relative energy, and coefficient of variation, and fluctuation index at particular scales. SVM was employed for classification, achieving low false detection rate and high sensitivity for seizure detection in long-term iEEG. Williamson, Bliss, Browne, and Narayanan (2012) combined multivariate EEG features with patient-specific machine learning for seizure prediction. The proposed algorithm calculates the covariance matrices and eigenspectra of space-delay correlation from EEG data, and classifies the data using SVM. Aarabi and He (2012) implemented a rule-based, patient-specific, seizure prediction framework for focal epilepsy. They used five univariate measures, including largest Lyapunov exponent, Lempel-Ziv complexity, correlation entropy, correlation dimension, and noise level as well as one bivariate measure, nonlinear interdependence, which were extracted from electrodes implanted deep in the brain.

One of the most disturbing results of injury or disease to the nervous system is the loss of motor function. The development of EEG-controlled assistive devices during the last few decades enable control of artificial prostheses, wheelchairs, and computers by using the electrical activity of the brain. But the performance of these artificial devices is low compared with the accuracy and speed of natural limb movements. In a practical motor, imagery-based, brain-computer interface (BCI) framework, the user imagines moving different body parts such the right hand or leg movements, and the computer learns how to distinguish different patterns of the concurrently acquired EEG activity. Actually, the main objective of the BCI is to employ the brain's signals and produce the necessary commands to control external devices. The most significant application is to bypass the disabled parts of the body or energize paralyzed organs. BCI systems are employed as an exceptional communication approach for people with different neuromuscular disorders, such as spinal cord injury, amyotrophic lateral sclerosis (ALS), cerebral palsy, and stroke (Sanei, 2013).

Furthermore, brain signals are employed by a BCI to interpret EEG signals into commands for external device control, such as computers, tablets, mobile phones, wheelchairs, and prosthesis. A BCI framework offers a new connection between the brain and the world that circumvents the human body. Hence, this framework can be employed to improve the life-quality of people with disabilities (Gao, Guan, Gao, & Zhou, 2015). BCI systems are conventionally implemented to support people who have different anomalies, such as a locked-in state. Because locked-in patients are reliant on others, BCI can offer them better quality of life with greater autonomy. Recently, researchers were interested in different applications of BCI technology to enhance neurorehabilitation for people with strokes and other chronic disorders (McFarland & Wolpaw, 2017). Noninvasive BCI systems employ EEG signals to interpret a subject's thoughts or intentions into a device-control signal, which enables a disabled person to interact with a computer, wheelchair, or neuroprosthesis. The mental practice of a hand or foot movement can be achieved by motor imagery (MI) without executing it (Zarei, He, Siuly, & Zhang, 2017). P300 event-related potential (ERP) (da Silva-Sauer, Valero-Aguayo, de la Torre-Luque, Ron-Angevin, & Varona-Moya, 2016; Gao et al., 2015; Lafuente, Gorriz, Ramirez, & Gonzalez, 2017; Lee & Kim, 2018; Sellers, Krusienski, McFarland, Vaughan, & Wolpaw, 2006; Shahriari & Erfanian, 2013), slow cortical potentials, oscillations in alpha and beta range (Guger et al., 2001; Pfurtscheller et al., 2006), and steady-state visual-evoked potentials (SSVEP) (Zhang, Guan, & Wang, 2008) have been employed in BCI systems. The common BCI technique is the P300-based ERP, which happens about 300 ms after an occasional or substantial stimulus is received (da Silva-Sauer et al., 2016; Wolpaw, Birbaumer, McFarland, Pfurtscheller, & Vaughan, 2002). Most of the time, MI approaches are used for BCI systems based on oscillations to produce event-related synchronization (ERS) and event-related desynchronization (ERD) in the α and β frequency bands of the EEG. These kinds of BCI systems are employed for navigation of wheelchairs or cursor control on computer screens. Consequently, many MI-based BCI methods, such as foot movement, left/right hand movement, tongue movement, and/or mental counting, are implemented. Flickering lights that carry EEG oscillations with the same frequency as the stimulation source are employed in SSVEP-based BCI systems. SSVEP schemes employ different stimulation frequencies to control robots or mobile phones. On the other hand, the P300-based BCI systems employ the P300 component in the EEG generally to enhance the communication speed of spelling devices (Guger et al., 2009). The computer actions can be controlled with subject's thoughts after the training phase. Hence, BCI systems mainly employ EEG signals to control a wheelchair, a robotic arm, or a computer cursor (Sanei, 2013). Farwell and Donchin (1986) initially implemented a BCI system that employed an oddball paradigm to allow the subject to spell words on a computer by successively selecting alphabet letters. In this model, all the alphabet letters and other useful symbols are displayed on the computer screen, allowing the subject to select a required character by concentrating visually on the relevant cell for a short time; meanwhile, the column or the row representing the focused-upon cell flashes to produce a P300 component in the EEG (Shahriari & Erfanian, 2013).

MI is the interpretation of the subject's motor intention into control signals. Several applications can be employed according to the MI rhythms, such as playing a computer game or controlling a mouse (Ramadan & Vasilakos, 2017). MI is useful in the treatment of neurological motor disabilities caused by stroke, spinal cord injuries, and Parkinson's disease (Vasilyev, Liburkina, Yakovlev, Perepelkina, & Kaplan, 2017). Because most stroke patients suffer from complete paralysis of a limb, this physiological signal modality offers some benefits through which brain activity related to the motor control can be monitored. The BCI under the MI paradigm is employed to give feedback to patients on their engagement with MI tasks (Leamy et al., 2014).

Kayikcioglu and Aydemir (2010) used a two-dimensional feature vector extracted from the second-order polynomial coefficients of EEG signals. Then, the k-nearest neighbor (k-NN) algorithm was employed for classification. Significant improvement is achieved for the accuracy and speed of the classification for data set Ia of BCI competition II (http://www.bbci.de/competition/ii/#datasets). Leamy et al. (2014) compared the stability and reliability of motor-related EEG features from a machine learning per spective in stroke subjects. Li, Zhang, Khan, and Mi (2018) proposed a novel framework that combines the fre quency band selection of common spatial pattern algorithms and the least squares twin SVM optimized with PSO for rec ognition of MI patterns. Baali, Khorshidtalab, Mesbah, and Salami (2015) proposed linear prediction singular value decomposition in MI pattern recognition for EEG-based BCI. Kevric and Subasi (2017) compared three popular signal-processing methods—DWT, EMD, and WPD—for EEG signal decomposition in BCI. The BCI competition III dataset IVa, 2-class motor-imagery dataset, was employed. They achieved 92.8% classification accuracy using k-NN classifier with WPD. Rao and Scherer (2010) used two different BCI systems to control a humanoid robot navigating in a virtual world. Zarei et al. (2017) combined the PCA and the cross-covariance technique (CCOV) for dis criminatory information extraction from mental states based on EEG in BCI implementations. They achieved high perfor mance for the proposed feature set. A supervised filter method was employed by Martín-Smith, Ortega, Asensio-Cubero, Gan, and Ortiz (2017) for evolutionary multi objective feature selection to classify high-dimensional data. The features were extracted from the EEG signals using multiresolution analysis in different domains, and then linear discriminant analysis was used for the BCI clas sification task. Miao, Wang, and Liu (2017) proposed linear discriminant criteria to select the channels with high discrim inative powers. Moreover, an artificial bee colony algorithm was used to determine the optimal global combination of time intervals and frequency bands at the same time without prior knowledge of common spatial pattern feature extraction and classification.

Nijboer et al. (2008) studied the effectiveness of a P300-based BCI interpretation device for people with ALS. A stepwise linear discriminant function was employed to classify the signal after each set of flashes. They performed one experiment with six participants using a 6 × 6 matrix and another experiment with four participants using either a 6 × 6 or a 7 × 7 matrix to create novel and spontaneous statements. Guger et al. (2009) implemented the P300 speller using two different models: (i) a single character (SC) speller that flashes every character separately, and (ii) a row/column (RC) speller that represents multiple characters at once. So, more reliable control with a higher P300 amplitude was expected with the SC flasher in which the target character appears. Shahriari and Erfanian (2013) developed a P300-based framework that enables individuals to interact with 1 of 36 symbols represented on a 6 × 6 matrix. A unique ERP characterized by P300 is produced once the subject realizes a target symbol. Moreover, to improve the recognition of the P300 in the EEG, a subspace-based spatial filter was employed in the BCI framework.

Ikegami, Takano, Kondo, Saeki, and Kansaku (2014) employed both the two-step speller and the conventional RC speller to assess the performance of a P300 BCI system. The performance in the control subjects was significantly higher than ALS patients, but the accuracy of ALS patients was increased significantly with new visual stimuli. Hence, the proposed method achieved better accuracy for ALS patients as compared with the conventional RC speller. Akram, Han, and Kim (2014) integrated a dictionary with two model interfaces to provide word proposals to users, making typing both easier and faster. Moreover, in another study, Akram, Han, and Kim (2015) also suggested a novel P300 classifier to increase the classification performance and word typing speed by employing initial character typing with word proposals. The T9 interface is modified as a mobile phone keypad utilized for text messaging. A random forest (RF) classifier with multiple decision trees significantly improved P300 classification accuracy. da Silva-Sauer et al. (2016) investigated how the user's concentration affected the performance of a spelling task using three different P300-based BCI schemes: one with a 7 × 6 symbol matrix, another with a word predictor, and a third using a P300-based speller displaying a reduced 4 × 3 symbol matrix, similar to the T9 mobile phone interface.

Guy et al. (2018) used a visual P300 speller BCI system to investigate if disabled people with ALS could use a BCI in a daily environment. The framework composed of EEG acquisition connected to a separate keyboard-display control software. Ryan, Townsend, Gates, Colwell, and Sellers (2017) combined both a color model and a checkerboard model. They found that the checkerboard color matrices achieve faster communication, higher personal preference, and higher accuracy than the traditional gray-to-white checkerboard model. Additionally, the color stimuli improved ERPs in the parietal/occipital electrodes and achieved better classification accuracy. Wang, Gu, Yu, and Li (2017) employed a self-training, regularized, weighted online sequential extreme learning machine (WOS-ELM) classifier to train a robust P300 classifier with lower computational complexity. After a short training session, the P300 speller system was converted into an input mode to improve spelling accuracy after a few characters were input. Lee and Kim (2018) proposed an ensemble SVM to improve a P300 speller where the different classifiers are trained by diverse datasets to increase the diversity of the ensemble. Hence, increasing diversity increases the classification performance of the ensemble algorithm. Liu et al. (2018) used a deep convolutional neural network, termed BN3, to automatically detect the discriminative spatiotemporal features of the P300 signal. They showed that BN3 achieved better character recognition performance than present detection algorithms.

Human beings spend about one-third of their lives asleep. Internal brain activities during sleep cannot be easily explained or understood, as a set of highly complex patterns happen in the human brain during sleep. Recently, an excessive number of studies has been devoted to sleep analysis and its relationship to other psychological situations. Unfortunately, the number of subjects suffering from sleep disorders is noteworthy, and it makes it harder for subjects to participate in normal events; it also affects a subject's health condition in several ways. Recent research assumes that sleep may have a substantial mission in memory consolidation, in which certain memories are established whereas other less important memories vanish (Sejnowski & Destexhe, 2000; Stickgold, 2005). Hence, it is imperative to have a precise scheme for sleep analysis and monitoring. The goal of sleep analysis and monitoring is usually done by expert(s) distinguishing several sleep stages during the entire night. Polysomnography (PSG) is a multivariate system that records several biomedical signals, such as EEG, ECG, EMG, and electrooculogram (EOG). Once biomedical signals are recorded, Rechtschaffen and Kales (R & K) (Rechtschaffen & Kales, 1968) standard is utilized for classification of sleep and used as a tool for classification of sleep stages in different studies worldwide despite its weaknesses (Himanen & Hasan, 2000). The R & K standard was enhanced by the American Academy of Sleep Medicine (AASM) (Berry et al., 2012). Consequently, sleep scoring discriminates wakefulness (W) and one of two sleep stages: rapid eye movement (REM) and nonrapid eye movement (NREM). NREM can be further separated into four different stages numbered as 1, 2, 3, and 4.

Numerous researches have been implemented with the aim to define and distinguish different sleep stages (Acharya et al., 2015; Fraiwan, Lweesy, Khasawneh, Wenz, & Dickhaus, 2012; Hassan & Subasi, 2017; Rodríguez-Sotelo et al., 2014; Zhu, Li, & Wen, 2014). In the literature, several methods have been developed for feature extraction from EEG signals, which can consistently detect sleep dynamics. These contain time-domain statistic approaches (Gunes, Polat, Dursun, & Yosunkaya, 2009), spectral analysis (Li, Wong, & de Bruin, 2012), TF analysis such as Wigner-Ville distribution (Bajaj & Pachori, 2013) and wavelet analysis (Fraiwan et al., 2012; Subasi, 2005), dynamic warping (Long, Fonseca, Foussier, Haakma, & Aarts, 2014), graph domain analysis (Zhu et al., 2014), coherence (Duckrow & Zaveri, 2005), etc. After the feature extraction, to classify different sleep stages (such as wakeful vs. sleep, or wakeful vs. REM vs. NREM, etc.). different machine learning algorithms are applied. They range from simple LDA to nonlinear and highly complex Gaussian mixture model-based classifiers (Acharya, Chua, Chua, Min, & Tamura, 2010). For the classification of different sleep stages, several traditional methods, namely LDA (Liang, Kuo, Hu, Pan, & Wang, 2012), neural networks (Shimada, Shiina, & Saito, 2000), SVM (Enshaeifar, Kouchaki, Took, & Sanei, 2016; Wu, Talmon, & Lo, 2015), k-NN (Li et al., 2012), hidden Markov model (Doroshenkov, Konyshev, & Selishchev, 2007), fuzzy systems (Held et al., 2006; Jo, Park, Lee, An, & Yoo, 2010), etc., are implemented to discriminate between several sleep stages. Recently, ensemble machine learning classifiers were developed to enhance the performance of a single classifier. Alickovic and Subasi (2018) developed a robust sleep stage classification framework for automated classification of sleep stages from a single-channel EEG. In the developed framework, signals taken from a Pz-Oz electrode were denoised using multiscale principal component analysis, then the most useful features were extracted by utilizing DWT. Then statistical values of DWT subbands were calculated. In the last step, extracted features were classified by rotational support vector machine (RotSVM).

1.2 ELECTROMYOGRAPHY

Scientists have divided the human body into many different systems. One of them is the neuromuscular system, which is composed of two subsystems: the skeletal-muscular system and the nervous system. Human action is provided by the neuromuscular system. The central nervous system (CNS) controls the muscles through nerve signals. Nerves act as a wire, carrying electrical current from the CNS to the muscles. When the signals from the CNS arrive at the muscle, it produces movement by contraction or relaxation, which causes the pulling or releasing of the bone. EMG can be defined as a technique of working on neuromuscular conditions based on cell action potential while the muscle is active. The amplitude (0.01-10 mV on average) and the frequency (10-2000 Hz on average) of EMG signals depend on location, etiology, and type of pathology. For example, although pulse duration of an EMG signal shows the location and metabolic condition of the muscle, odd spikes may point to myopathy. The neuromuscular system is also under threat of disease, like all other systems in the human body. These diseases can originate in the nervous system, the neuromuscular junction, and the muscle fibers. Although some neuromuscular diseases cause weakness in the muscles, some cause neuron or muscle death. Accurate diagnosis is very important for successful recovery of the patient who is suffering with neuromuscular diseases. In many cases, clinical inspections are not enough to identify the disease and its location (Preston & Shapiro, 2012). Sometimes the symptom's location and disease location are far from each other. Therefore, correct localization of disease has high importance for accurate diagnosis and full recovery. Electrodiagnostic (EDX) studies can be used to determine and diagnose a patient with neuromuscular disorders. EDX consists of two methods: nerve conduction studies (NCS) and EMG (Begg et al., 2008).

Usually, the EMG signal is interpreted by an expert neurologist who can analyze EMG waves using needle conduction, and who has knowledge about generic muscle waves. In many cases, it is crucial to find such an expert neurologist for patient requisition. Therefore, designing an automated diagnosis system using EMG signals is becoming more important each day This requirement can be served by machine learning techniques. It is possible to implement an automated diagnosis system to help the neurologist determine neuromuscular disorders based on classifying EMG signals using machine learning techniques. To design an automated diagnosis system, first the EMG signal is preprocessed, and features are extracted that contain time domain and/or frequency domain information by using Fourier, wavelet, or autoregressive (AR) coefficients, or other signal-processing techniques. After the feature extraction from the EMG signals, the data is used as input to a classifier such as ANNs, k-NN, SVM, or decision tree to detect and identify the disease (Begg et al., 2008).

The challenges of developing a useful automated system that can successfully diagnose neuromuscular diseases are still continuing. The EMG signal features vary in a wide scale among the patients. For example, amplitude and duration depends on patient ages. We need to design a signal processing technique that can save or find the distinctive data in raw EMG signals to overcome this problem. The quality of the features affects the accuracy of the classifiers (Begg et al., 2008).

EMG is a device that records the motor unit action potential (MUAP). The needle or fine wire EMG and surface EMG (sEMG) are the main types of EMG. The required instruments to record EMG signals are electrodes, a signal acquisition system with signal filters, amplifiers, and visual displays. To record a useful EMG signal, the EMG must be sensitive to noise and artifacts. Therefore, the instruments used for acquiring an EMG signal are produced with some standards such as filter bandwidth, gain, and input impedance (Preedy & Peters, 2002; Begg et al., 2008).

The wire electrode or needle electrode can reach an individual motor unit and acquire the action potential more precisely than the sEMG. Simply, a needle electrode is a wire connected to a stainless steel electrode. The most common type is a concentric ring needle with a monopolar single electrode. Single fiber EMG (sFEMG) and macro EMG are two other types of needle electrodes (Togawa, Tamura, & Oberg, 1997). Another electrode type is a wire electrode, which is more flexible than a needle electrode and made of silver, nickel, platinum, or chromium. It is preferred for kinesiology studies. The cannula is used to insert the wire electrode into the muscle, and later it is removed. The EMG measures and records the voltage with respect to the ground, so a ground electrode is used. Generally, it is inserted in a place where there is no muscle activity, such as over the elbow (Begg et al., 2008).

It is difficult to detect minor aberrations or different patterns of anomalies by MUAP assessment (Preston & Shapiro, 2012). A number of studies have been performed to design an accurate automated diagnosis system by employing different classification algorithms to classify EMG signals (Richfield, Cohen, & Albers, 1981; Subasi, Yilmaz, & Ozcalik, 2006). It is possible to find commercial versions of some algorithms on the market, but almost none of them are used at clinics to diagnose neuromuscular disorders (Bozkurt, Subasi, Koklukaya, & Yilmaz, 2016). AR analysis and time domain analysis were used together to classify the EMG signals by Pattichis and Elia (1999). De Michele, Sello, Carboncini, Rossi, and Strambi (2003) applied wavelet cross-correlation analysis on two different muscles and discovered that it is possible to make detailed classifications. MUAP parameters were used as input to parametric sequential pattern recognition classifiers by Pattichis, Schizas, and Middleton (1995). Loudon, Jones, and Sehmi (1992) used eight different MUAP parameters as input to the statistical pattern recognition techniques to classify the EMG data. Hassoun, Wang, and Spitzer (1994a, 1994b) applied time domain parameters as input to a tree-layered ANN and used a "pseudo-unsupervised" algorithm as a classifier in their study. Two different classification methods were used by Christodoulou and Pattichis (1999) at the same time to classify the EMG signals. They proposed to use unsupervised machine learning algorithms, including self-organized feature maps, learning vectors, and Euclidian distance. GAs were used to classify EMG signals by Schizas and Pattichis (1997). In recent years, wavelet neural network (WNN) was used to analyze EMG signals. Subasi et al. (2006) used AR signal processing with WNN to classify EMG data. They stated that they classified EMG signals with an accuracy of 90.07%, and that it was possible to develop a simple, accurate, and reliable automated classification system for routine clinical usage. Katsis, Goletsis, Likas, Fotiadis, and Sarmas (2006) used SVM to classify EMG data with an accuracy of 93%, 95%, and 92% for normal, myopathy, and neuropathy, respectively. The features of EMG signals were extracted by wavelet method and used as input to ANN algorithm and learning vector quantization (LVQ) by Guo, Yang, Chen, Wang, and Li (2006). Jiang, Wang, Wang, and Jin (2006) performed wavelet transformation of EMG signals, then used the statistical characteristic of wavelet coefficients as an input to an ANN. By using similar techniques, Cai, Wang, and Zhang (1999) extracted feature vectors by using wavelet transform and used them as input to an ANN, which was trained by a standard backpropagation algorithm to classify the EMG signals.

EMG presents comprehensive information to describe neuromuscular activity and muscular morphology. The EMG signals must be decomposed, classified, and analyzed to describe a muscle using quantitative EMG (QEMG) data. To diagnose neuromuscular disorders, an EMG signal must be classified for the detection of abnormalities (Yousefi & Hamilton-Wright, 2014). Recently, Rasheed, Stashuk, and Kamel (2008) developed a model to distinguish individual MUAP waveforms from a raw EMG signal to extract relevant features and classify the MUAPs. The adaptive fuzzy k-NN classifier with time domain features with 93.5% accuracy and wavelet domain features with 92.6% accuracy were achieved. The EMG signal is segmented utilizing a threshold technique to identify possible MUAPs. A statistical pattern recognition technique is utilized for clustering the identified MUAPs. After employing an AR method for feature extraction, MUAPs are classified utilizing a binary SVM classifier. The classification accuracy of binary SVM with AR features is 100% (Kaur, Arora, & Jain, 2010).

Subasi (2012a) utilized several feature extraction methods and machine learning techniques, such as multilayer perceptron neural network (MLPNN), dynamic fuzzy neural network (DFNN), and adaptive neuro-fuzzy inference system (ANFIS), to classify EMG signals and compare them according to their accuracy. In the proposed framework, neuro-fuzzy classification methods were able to classify EMG signals with high accuracy; the ANFIS with AR + DWT feature extraction methods was 95% accurate. Moreover, Subasi (2012b) developed an effective combination of classifier and features to classify EMG signals. LDA, radial basis function neural network (RBFNN), MLPNN, C4.5 DT, SVM, and fuzzy SVM (FSVM) classifiers were used with statistical features extracted from DWT subbands. It is reported that the FSVM and the statistical features extracted from DWT subbands utilizing the internal cross-validation method achieved better performance than other classifiers. Subasi (2013) proposed a framework in which DWT is utilized to decompose the EMG signals into the frequency subbands, and then a set of statistical features are extracted from these subbands. Substantial improvements in terms of classification accuracy were realized by the developed PSO-SVM classification system with an overall accuracy of 97.41%. Furthermore, Subasi (2015) used an evolutionary approach to classify EMG signals utilizing an SVM classifier. In the developed framework, the EMG signals were decomposed into the frequency subbands utilizing DWT, then a set of statistical features was extracted from the subbands; evolutionary SVM achieved an accuracy of 97% using ten-fold cross-validation. A comparison research was done between combined neural network (CNN) and feedforward error backpropagation ANN (FEBANN) classifiers by Bozkurt et al. (2016). CNN was more successful than FEBANN with an accuracy of 92%, but it was not fast enough for practical usage.

Gokgoz and Subasi (2014) studied the effects of multiscale principal component analysis (MSPCA) denoising in EMG signal classification. A multiple single classification (MUSIC) signal processing technique was utilized for feature extraction to classify EMG signals into normal, ALS, or myopathic. It was discovered that MSPCA denoising improved the classification accuracy. After denoising EMG signals with MSPCA, the classification accuracy was 92.55% for SVM, 90.02% for ANN, and 82.11% for k-NN, respectively. The same researchers (Gokgoz & Subasi, 2015) presented a framework for classification of EMG signals utilizing MSPCA for denoising, DWT for feature extraction, and decision tree algorithms for classification. The presented framework achieved the best performance utilizing k-fold cross-validation with 96.67% total classification accuracy by utilizing DWT and RF.

Parsaei and Stashuk (2013)employed K-means clustering, and the supervised classification is realized by utilizing a certainty-based algorithm. The proposed framework had an average accuracy of 86.4% for simulated and 96.4% for real data, respectively. Dobrowolski, Wierzbowski, and Tomczykiewicz (2012) used the wavelet index to classify myogenic, neurogenic, or normal EMG signals. The proposed method achieved 99.4% accuracy with the fast-DWT and SVM-training algorithm called "sequential minimal optimization" (SMO). Kamali, Boostani, and Parsaei (2013) proposed a scheme that utilizes both time and time-frequency features of an MUAP along with an ensemble of SVM classifiers. Time-frequency features are DWT coefficients of the MUAP. Time-domain features consist of peak-to-peak amplitude, turn, area, duration, and phase of the MUAP. Evaluation results of the proposed framework utilizing EMG signals (myopathic, neuropathic, and control) showed that the system achieved an average of accuracy of 91%. Kamali, Boostani, and Parsaei (2014) employed an ensemble of SVMs classifiers to determine the class label (myopathic, neuropathic, or normal) using both time domain and TF domain features of the EMG signal. The performance of the proposed model has an accuracy of 97%.

Artameeyanant, Sultornsanee, and Chamnongthai (2016) proposed a normalized weight, vertical visibility algorithm as a feature extraction method for ALS and myopathy detection. In the proposed method, the features are extracted by utilizing selective statistical mechanics and measurements, and the extracted features are used as a feature matrix for classifier input. Finally, powerful classifiers such as MLPNN, SVM, and k-NN classifiers are employed to categorize signals into healthy, ALS, and myopathy. They achieved 98.36% accuracy with Nikolic Dataset and 99.17% accuracy with Physionet EMG dataset. Naik, Selvan, and Nguyen (2016) presented a classification technique for neuromuscular disorders (myopathic and ALS) that utilized a single-channel EMG sensor. The single-channel EMG signals are decomposed by employing ensemble EMD algorithm, then the Fast ICA algorithm is used for dimension reduction. A reduced set of features is classified utilizing linear discriminant analysis, and the classification results are fine-tuned with a majority voting scheme. The proposed method achieved higher classification accuracy (98%) with a clinical EMG database. Khan, Singh, and Tiwari (2016) proposed a framework that utilizes both time domain and TF domain features extracted from the EMG signals. Several classification approaches, including single classifier and multiple classifiers with time domain and time frequency domain features, were examined. SVM and k-NN classifiers are employed to predict class label (normal, neuropathic, or myopathic). A DWT-based feature extraction scheme with a multiclassifier model achieved 97% accuracy. Sengur, Akbulut, Guo, and

Bajaj (2017) proposed a deep learning-based classifier for effective categorization of normal and ALS EMG signals. They used different TF methods combined with a convolutional neural network for EMG signal classification. They employed ALS and normal EMG signals, and achieved 96.80% accuracy with CWT and CNN. Mishra, Bajaj, Kumar, Sharma, and Singh (2017) employed improved empirical mode decomposition (IEMD) in combination with the least squares support vector machine (LS-SVM) classifier for the analysis of ALS and normal EMG signals. The proposed technique achieved 96.33% accuracy. Hazarika, Dutta, Boro, Barthakur, and Bhuyan (2018) presented a real-time feature extraction and fusion model for an automated classification of EMG signals as ALS, myopathy (MYO), or normal (NOR) using DWT and canonical correlation analysis (CCA). The obtained discriminant features are fed to the k-NN classifier, and 98.80% accuracy is achieved with twofold cross-validation.

The sEMG electrode is made up of Ag/AgCl in the shape of a disk or bar. It is more beneficial than wire or needle electrodes, because it is used to attach to the body instead of being inserted into it. Before attaching the electrode, the skin must be shaved, cleaned by alcohol or medical napkins, and gel applied to get good conduction between skin and the electrode. The human skeletal-muscular system's primary duty is to provide the forces needed to do a variety of activities. The system consists of the muscular and the nervous systems to create the neuromuscular system. The skeletal-muscular system is composed of muscle sets connected to bones, and a motion is performed once nerve signals produce muscle contractions and relaxations (Begg et al., 2008). EMG is the recording and understanding of these muscle activities. Initially, EMGs were recorded mainly for investigative or diagnostic aims, but with the progression of bioelectric technology, EMGs have also become a basic instrument in accomplishing artificial control of limb motion, such as functional electrical stimulation and rehabilitation (Bronzino, 2000). sEMG signals produced by the electrical activity of muscle fibers during contraction are taken from the skin surface. Multichannel sEMG signals, acquired by electrodes placed on the muscles, can be employed to recognize the user's activity intent, as every muscle activity corresponds to a specific pattern of activation. Therefore, sEMG signals have a substantial control input for exoskeleton, prostheses, and rehabilitation robots. Current prosthetic devices mainly use sEMG signals from the body of the patients mounted at different positions along the body. EMG signals from the body of the patients correspond to a local, area-specific level, and play an important role in pattern classification (Zhang & Sup, 2014).

A biosignal refers to any signal in a living organism that can be measured and monitored on a frequent basis. The nervous system is in charge of controlling muscle movement, which entails of contraction and relaxation of the muscle via signals obtained by the muscles, which are detectable by the use of biosensors. EMG is a neuromuscular diagnosis method used in medicine and health care, and can be used as a man-machine interaction (MMI) (Merletti & Di Torino, 1999). It records and analyses the electrical signals produced by the skeletal muscles (Robertson, Caldwell, Hamill, Kamen, & Whittlesey, 2014). One EMG application is the treatment of stroke patients where intensive therapy can help in rehabilitation (Lum, Burgar, Shor, Majmundar, & Van der Loos, 2002; Riener, Nef, & Colombo, 2005). Another EMG application area is the development of EMG-controlled assistive devices for disabled individuals such as control of prosthesis, electrical wheelchairs, virtual mouse, virtual world, etc. (Abdullah, Subasi, & Qaisar, 2017; Oskoei & Hu, 2007; Podrug & Subasi, 2015). EMG signals are a feasible solution for hand movement recognition (Xing, Yang, Huang, Wang, & Zhu, 2014) because every movement on the transmitted signal has a definite signature. Hence, EMG signals can be involved in the decision-making process for effective control of a robotic exoskeleton. It can also be applied toward leisure activities such as video games. The crossroads for the use of EMG is the that its comfort level is higher for an arm amputee who uses a glove with EMG electrodes as opposed to using EEG electrodes that show potential in the head's region. The biomedical signals are of a multidimensional nature. Hence, it is hard to find a robust feature extraction and machine learning algorithm for the MMI (AbdelMaseeh, Chen, & Stashuk, 2016; Ju, Ouyang, Wilamowska-Korsak, & Liu, 2013; Vidovic et al., 2016; Yan, Cempini, Oddo, & Vitiello, 2015; Young, Smith, Rouse, & Hargrove, 2013). Machine learning algorithms have the ability to distinguish between different types of signals for EMG-controlled assistive devices.

sEMG signals have been extensively used for rehabilitation purposes and prosthetic device control, aimed toward amputees or those with congenitally deficient limbs, as well as for improving the interface between humans and computers (Coelho & Lima, 2014). Enhanced sEMG signals can be employed to classify different kinds of myoelectric signals. The variation of EMG signals affects the classification performance of algorithms due to the different types of contraction. To eliminate such variations, several biomedical signal-processing methods are utilized along with their signal denoising and feature extraction capabilities for EMG signals. Then, a set of feature reduction techniques is employed to reduce the high-dimensional features. Finally, a variety of machine learning techniques are used for robust classification (Al-Faiz & Miry, 2012; Chowdhury et al., 2013; Phinyomark et al., 2013; Tsai, Hsieh, Luh, & Lin, 2014; Xing et al., 2014). The hands of a human being play a crucial role for grasping and manipulating different objects. Even the loss of a single hand affects activity, and a prosthetic hand can be a solution to equip the armless subject (Kurzynski, Krysmann, Trajdos, & Wolczowski, 2016). An upper limb prosthesis is mainly based on myoelectric control employed to recognize sEMG signals that are realized during muscle contraction. Because a majority of the muscles that produce finger motion is left in the stump after a hand amputation, they can be used for myoelectric control; the activity of these muscles can be employed for prosthesis control. The control of the prosthesis is solely based on a set of repeatable muscle contractions, which can be differentiated from ordinary arm functions. Some valuable features can be extracted from the EMG signals that deliver information about muscle activity for prosthesis control. EMG signals can be investigated with its amplitudes, and phases in time domain and frequency. Feature extraction can be realized using several methods including signal amplitude, AR model coefficients, frequency characteristic and power spectrum, and TF characteristic of the EMG (Wojtczak, Amaral, Dias, Wolczowski, & Kurzynski, 2009).

An sEMG signal contains substantial information on muscle activity and is widely used for myoelectric control (AlOmari & Liu, 2014). The application of sEMG differs from rehabilitation devices to human-machine interaction systems. Several intelligent models were developed to identify the neuromuscular activities in prosthetic device control (AlOmari & Liu, 2015; Rechy-Ramirez & Hu, 2015). Because rehabilitation devices and assistive robotic systems have joysticks and keyboards, many disabled people struggle to access them, and innovative human-machine interfaces are needed (Rechy-Ramirez & Hu, 2015). Karimi, Pourghassem, and Shahgholian (2011) accomplished a high classification rate for 10 hand motions using an ANN with GAs. Rafiee, Rafiee, Yavari, and Schoen (2011) proposed a new method for the recognition of intramuscular and sEMG signals acquired from the upper forearm for six hand motions to select the appropriate sensors for each motion. Different statistical features were extracted from the WPT of the EMG signals as well. Yan, Wang, and Xie (2008) used a new approach to train the artificial limbs of amputees with four different motions. They employed wavelet packet energy (WPE) for feature extraction from the sEMG signal and mutual information for dimension reduction. Furthermore, they used fuzzy multilayer perceptron and LS-SVM for classification. Huang, Xie, Guo, and Chen (2012) proposed an ant colony optimization (ACO)-based feature selection to distinguish eight hand motions. Xing et al. (2014) proposed a real-time sEMG classification system to identify six types of wrist motions. They used WPT and nonparametric-weighted feature extraction for feature extraction and SVMs for classification. Kurzynski et al. (2016) introduced the concept of a multiclassifier system to control a bioprosthetic hand with a dynamic ensemble selection structure. Moreover, feedback information coming from bioprosthesis sensors is applied to adjust the classification task by employing adaptive tuning capabilities of base classifiers.

Boyali and Hashimoto (2016) proposed a myoelectric pattern recognition system using spectral collaborative representation-based classification to yield high classification accuracy for hand gestures. They built a training dictionary for recognizing signal patterns, and manual feature selection was automatically eliminated. Tsai et al. (2014) proposed short-time Fourier transform (STFT)-based features to compare classification results from training and validation motion patterns produced from different muscle contractions. To characterize the relationships between the muscles in multichannel EMG signal studies, straightforward features are directly combined into a feature vector. The performance of straightforward features and STFT-ranking features, with the same experimental protocol and the same conditions, were discussed. AlOmari and Liu (2015) classified eight different hand motions using GA and PSO. They integrated these optimization algorithms with wavelet transform. In addition, four different dimensionality reduction methods were used with an SVM classifier. Wołczowski and Zdunek (2017) analyzed numerous computational approaches for dimensionality reduction using four multifactor decomposition methods for classification of 11 grasping movements. The motion classification accuracy can be increased by employing multichannel sEMG signals. The classification accuracy is directly proportional to the number of acquisition channels. However, it also increases the system complexity because the system has to handle more datasets. To eliminate the limitations of the multichannel sEMG schemes, single-channel sEMGs are also employed Tavakoli, Benussi, and Lourenco (2017). However, the major issue of these arrangements is poor reliability, and the challenge in such arrangements is very low signal-to-noise ratio (SNR) and crosstalk Rasool, Bouaynaya, Iqbal, and White (2014).

In the last decade, the upper-limb prosthesis field faced rapid development because of an improvement in sensors, actuators, controllers, and digital manufacturing. Moreover, numerous complex prosthetic terminals were implemented with multiple degrees of freedom to achieve the capabilities of the human hand in terms of grasping (Khezri & Jahed, 2009; Tsai et al., 2014). For instance, recent commercial hands such as i-limb (TouchBionics), Vincent (Vincent Systems), and Bebionic (Rechy-Ramirez & Hu, 2015) utilize five or six actuators to develop quite precise grasps. Conversely, several amputees prefer a simple and lightweight prosthetic hand. In spite of improvements on the complex prosthetic terminals, the control of such systems requires many control inputs. Existing human-machine interfaces created by a small number of sEMG electrodes does not allow for adequate inputs for controlling the hands. It is apparent that full control of a high-quality prosthesis hand needs a larger number of EMG sensors, which increases the complexity and cost of the system. Naik and Kumar (2010) and Ajiboye and Weir (2005) utilized four EMG channels to distinguish wrist and four-finger flexions and extensions. Rosenthal et al. (2012) used six bipolar electrodes to identify up to 19 hand postures, and Al-Timemy, Bugmann, Escudero, and Outram (2013) used six bipolar electrodes to identify 15 hand postures. Eight-channel (Cipriani, Controzzi, & Carrozza, 2011) and 10-channel (Bitzer & Van Der Smagt, 2006) sEMG systems with bipolar electrodes were used as well. A four-bipolar-electrode placement on the forearm was carried out by Khokhar, Xiao, and Menon (2010). Wang, Lao, and Zhang (2017) presented a human prosthetic hand utilizing flexure pivots, which was controlled by sEMG signals. They perceived eight hand motions from two-channel EMG signals and progressively controlled the prosthetic turn.

Classification of sEMG signals is not an easy task because sEMG signals are complex and nonlinear. Hence, feature extraction and pattern recognition techniques are very important to differentiate and recognize different sEMG signals. The EMD (Al-Timemy et al., 2013) was used to remove the recurrence area, while decreasing the embedding dimensionality. Following the muscle-contraction discovery, the division of whatever is left of the flag occurs, utilizing overlapping time windows from each section. Various highlights are removed utilizing both the flag as well as its intrinsic mode functions (IMFs) calculated after the utilization of the EMD. Sapsanis, Georgoulas, and Tzes (2013) proposed two strategies for dimensionality reduction before the utilization of the straightforward direct classifier, which endeavors to group each portion to one of the six fundamental developments described. Wojtczak et al. (2009) proposed a technique that examines EMG signals to control prostheses multifunction. Khezri and Jahed (2009) argued the limitations of the current prosthesis hand for closing and opening tasks. They proposed a novel sEMG pattern recognition scheme to enhance performance by including features in time domain, time-frequency domain, and a mixture of these features as a composite representation of sEMG signal features to achieve better classification performance. They used two classifiers to implement this, ANN and FIS. Lisi, Belluco, Cattaneo, and Gini (2011) have set up new plastic prostheses in which the entire prosthetic control depends upon the EMG signal. The results are extremely encouraging with the proposed framework, considering that it can distinguish seven movements with the utilization of only three channels. ACO-based feature selection method was presented by Huang et al. (2012) to classify hand motions by employing sEMG signals. The examinations were carried out on 10 subjects with eight upper limb movements. The proposed framework accomplished impressively high recognition rates in sEMG movement.

1.3 ELECTROCARDIOGRAPHY

ECG is a device that graphically records the tiny electrical activities of the heart's rhythm, heartbeat's rhythm, heartbeat's spreading, and disappearance of the reaction. It records the activity of the heart in a specific period of time via the use of electrodes placed on the body of a patient. It became widely used in clinics for heart disease diagnoses. The ECG signals are recorded by electrodes placed on the surface of the limb and chest to detect the changes in electrical potential difference during the polarization and depolarization of the myocardial fibers. An ECG waveform can be drawn onto graph paper or displayed on a computer screen. The ECG is mostly employed to examine several types of abnormal heart functions including conduction disturbances and arrhythmias (Wasilewski & Poloński, 2012)

ECG signals represent the electrical activities of a heart muscle, containing vital diagnostic information that describes the operation of the heart and other systems, such as the nervous and circulation systems. August Waller, who first realized ECG recording in 1887, employed surface electrodes positioned on the skin and linked to a capillary electrometer. But W. Einthoven recorded the first ECG signal by employing a string galvanometer in 1902. This sort of ECG has been primarily utilized as a helpful technique in the differentiation of atria rhythm irregularities in the last century. ECG has developed as a routine technique in heart diagnostics since the 1940s. An ECG signal is a well-known biomedical signal with high diagnostic abilities. Substantial improvement has been achieved in the last few decades for the processing of ECG signals, with the intent of the extracting valuable diagnostic information (Gacek, 2012).

The ECG can help in defining abnormalities in the human heart, such as cardiovascular disorders (CVDs). Moreover, ECG signal classification has drawn a lot of attention in the last few years. ECG signal classification is a significant diagnosis tool in which feature extraction plays a crucial role. It is the most imperative and common tool for analyzing and describing heart activity. ECG signals can be normal (N), left bundle branch block (LBBB), right bundle branch block (RBBB), atrial premature contraction (APC), premature ventricular complex (PVC), etc. Because ECG is quick and ECG monitoring is capable of being continuous, an accurate analysis of ECG signals is crucial for detecting heart function abnormalities in cardiac patients. The accurate diagnostic includes feature extraction from ECG waveforms and their assessments with known heart diseases to find any variation from normal ECG waveforms. Any heart abnormality characterized by wave shape alterations within an ECG cycle can be detected by these monitoring systems. This ECG monitoring is time-consuming and monotonous for investigating long-term recording waves, even though clinicians should identify any abnormality in the ECG signal's waveform from a normal one. Machine learning methods are able to automate ECG signal analysis by distinguishing between pathological and normal patterns. Research in this area using different machine learning methods such as neural networks (NNs), fuzzy logic, decision trees, k-NNs, GAs, SVMs, and ensemble classifiers are widely used but with different levels of accomplishment. Automatic recognition and classification of heartbeats using different biomedical signal processing techniques are crucial for clinical ECG monitoring. The main objective is to build a framework to accomplish real-time monitoring and to help clinicians make correct decisions (Begg et al., 2008).

A distinctive specification of ECG signals comes from a cyclic occurrence of the P-QRS-T complex. The P-QRS-T part of an ECG signal reflects the single heart cycle, including contraction of atria and ventricles. Because this portion of the ECG signal represents the most vital diagnostic information, it forms a region of interest. A crucial task is to distinguish every segment, including the P-QRS-T part, during the ECG signal analysis. The start and end of positions P and T, Q, R, S, and a position of the center of the ST interval is determined by these points. The most critical one is the realization of point R. The frequency of heartbeat used for diagnosis in cardiology is determined by the time intervals between successive R points. (Gacek, 2012). The P-QRS-T complex functions as the basis for the starting point of ECG data-compression algorithms, and heart rate (HR) detection and classification (Kohler, Hennig, & Orglmeister, 2002). The P-QRS-T morphology describes the mechanical function of the heart, offering an approach to how every chamber is functioning. The waves of depolarization that extend throughout the heart during each cardiac cycle produce electrical signals that travel via the blood up to the body's surface. Acquired ECG signals are amplified by ECG, which records the electrical activity. The ECG is an illustration of the heart's electrical activity recorded over a period of time, and it is the most valuable indicator of cardiac task. Any variation in the transmission of signals throughout the heart can yield irregular electrical currents, which creates a distorted ECG wave shape. This occurrence has been used effectively by clinicians to establish correct diagnosis of CVDs (Begg et al., 2008).

Many researches are being performed with a goal to investigate valuable signal analysis and pattern recognition techniques for computer-aided diagnosis (CAD) based on ECG signals. Previous researches done in this field propose that physiological signals originated from complex self-regulating systems under healthy conditions may have a fractal temporal structure (Bassingthwaighte, Liebovitch, & West, 1994). In other researches (Esgiar & Chakravorty, 2004; Lai & Chan, 1998; Muller, Contento, & Rix, 1992), it was found that ECG signals are appropriate models for self-affined fractal sets, and it is achievable to perform accurate classification by using fractal dimension. In recent researches, a variety of detection algorithms have been suggested. Thakor, Zhu, and Pan (1990) suggested sequential hypothesis testing; whereas threshold-crossing intervals, the auto-correlation function, and the VF-filter were suggested by Clayton, Murray, and Campbell (1993). Algorithms based on NNs were suggested by Yang, Devine, and Macfarlane (1994). Heart rate variability (HRV) are the beat-to-beat HR changes and are thought to be an appropriate indicator of a person's health and heart disorders (Camm et al., 1996). Additional analysis of the HRV incorporates methods from nonlinear mathematics and chaos theory, such as fractal (Malik & Camm, 1995) together with approximate entropy analysis (Tsipouras & Fotiadis, 2004).

ECG-based CAD systems are composed of two functional subparts: feature extraction and pattern recognition. There are various ways for feature extraction. In several studies (de Chazal & Reilly, 2003; Hu, Palreddy, & Tompkins, 1997; Moraes, Seixas, Vilani, & Costa, 2002), feature extraction was done in a time domain. In other studies (Acharya et al., 2004; Minami, Nakajima, & Toyoshima, 1999), feature extraction was done in a frequency domain. Al-Fahoum and Howitt (1999) extracted features by using multiscale decomposition. Osowski and Linh (2001) used statistical measures for feature extraction. A lot of work has also been dedicated to the improvement of classification techniques for these feature sets, such as linear discrimination, decision trees, NNs, and a combination of expert systems (Yu & Chou, 2007).

There are also numerous works done reporting the use of multifractal models for cardiac signal analysis (Ivanov et al., 1999; Wang, Huang, Xie, Wang, & Hu, 2007). Raghav and Mishra (2008) attempted to use local fractal dimension based on the nearest neighbor classification algorithm for ECG signal-based heart disorders, and the achieved results were promising (Mishra & Raghav, 2010). Qin, Chen, Liu, and Lu (2005) used RBFNN to classify ECG signals. Different NN techniques have been used for ECG signal classification (Nadal & de C. Bossan, 1993; Prasad & Sahambi, 2003; Yong, Wenxue, & Yonghong, 2009; Yu & Chou, 2008). Classification has also been performed using SVMs (Acır, 2006; Asl, Setarehdan, & Mohebbi, 2008; Besrour, Lachiri, & Ellouze, 2008; Melgani & Bazi, 2008). Other works were conducted using GAs (Kutlu & Kuntalp, 2011). Özbay, Ceylan, and Karlik (2006) suggested using fuzzy logic as a classification tool. A k-NN classification tool has also been used (Arif & Akram, 2010; Arif, Akram, & Afsar, 2009; Christov, Jekova, & Bortolan, 2005; Karimifard, Ahmadian, Khoshnevisan, &

Nambakhsh, 2006). There have been improvements in the area of signal processing, classification, and interpretation supported by advanced information technologies. Computational intelligence (CI) has appeared as an extremely synergistic, computationally attractive, and practically combined framework supporting intelligent structure design and investigation. The settings of CI achieve a number of computationally and conceptually attractive and exceptional opportunities from the viewpoint of ECG signal processing, interpretation, integration, and classification (Gacek & Pedrycz, 2012).

During ECG recording, a variety of sources have the potential to add noise to the acquired signal and modify it. The occurrence of noise in signal acquisition systems can rigorously change the understanding of the data, resulting in an inaccurate diagnosis. Important noise sources can be electrical power line interference, noise due to instrumentation or electronic devices, impedance changes at the skin/electrode edge, movement artifacts such as electrode movement, baseline drifts caused by respiration, and electrosurgical noise. Hence, accurate preprocessing of the ECG signals is essential to diminish the variety of noise components and significantly improve the (SNR). Therefore, a single band-pass filter having 10 to 25 Hz pass-band is employed for ECG signal filtering (Begg et al., 2008). Because ECG signal denoising and feature extraction provide an indication about several cardiac abnormalities for CVD diagnosis, it received lot of attention from medical societies. Alickovic and Subasi (2015) discussed how ECG beat classification is affected by multiscale PCA (MSPCA) denoising for CVD diagnosis. ECG signals taken from the MIT-BIH Arrhythmia Database were denoised by utilizing MSPCA, and features were extracted utilizing AR Burg method. To get a good classification performance, noise should be removed. The proposed framework demonstrated that MSPCA denoising increases classification accuracy. Moreover, Alickovic and Subasi (2016) proposed another framework where MSPCA was utilized for denoising, DWT for feature extraction, and RF classifier for ECG signal classification, and they achieved a high accuracy rate. Usta and Yildiz (2017) employed RF classifier to classify heart arrhythmia using ECG signals. RF had the most accurate classification, which is the key requirement for detection of all arrhythmia types.

The analysis and interpretation of the ECG signal delivers comprehensive information on the cardiac health status of the patient. Because the analysis of such a lengthy signal by doctors is time-consuming, signal processing techniques present rapid and real-time analysis of ECG signals in an accurate way. Most of the time, arrhythmias do not demonstrate any steady changes in ECG signal; consequently a Holter monitor, where ECG is recorded over a longer period of time, is desirable. Besides, cardiac arrhythmias are one of the most common reasons of death; hence, accurate detection of arrhythmias has been of great interest in biomedical research (Afkhami, Azarnia, & Tinati, 2016). Afkhami et al. (2016) employed the MIT-BIH Arrhythmia Database, including several forms of common arrhythmias, and achieved an overall accuracy of 96.15% in "subject-oriented" schemes and 99.70% in "class-oriented" schemes. DTCWT-based feature extraction technique is proposed by Thomas, Das, and Ari (2015) for automatic classification of cardiac arrhythmias. The results showed that DWT and DTCWT-based feature extraction techniques achieved an overall sensitivity of 91.23% and 94.64%, respectively, for five types of ECG beats of the MIT-BIH Arrhythmia Database.

Li and Zhou (2016a) utilized statistical features extracted from DWT; ICA for dimension reduction; and PNN, k-NN, DT, and SVM for classification. They used tenfold cross-validation and grid-search to optimize parameter γ in the radial basis function (RBF) kernel and penalty factor C in SVM in the training dataset. Li and Zhou (2016b) proposed a method to classify ECG signals utilizing wavelet packet entropy (WPE) and RF. Masetic and Subasi (2016) employed an AR Burg feature extraction algorithm and RF classifier for congestive heart failure (CHF) detection. Two different experiments were designed to classify normal (N) heartbeats and heartbeats with CHF. In the first experiment, they used the ECG signals taken from BeTH Israel Deaconess Medical Center CHF and MIT-BIH Arrhythmia Databases, and in the second experiment, they utilized samples from Physikalisch-Technische Bundesanstalt (PTB) Diagnostic ECG Database, combined with samples from MIT-BIH Arrhythmia Database. Cruz et al. (2016) employed DWT in ECG signal classification, and compared SVM and ANFIS. The experimental result showed that SVM achieved better performance in terms of sensitivity, specificity, accuracy, and training time, whereas ANFIS had the fastest evaluating time. Once the noise variance increased, SVM presented better performance, whereas the performance of ANFIS decreased in terms of noise. Li et al. (2016) proposed a new framework for classification of ECG signals by combining WPD and approximate entropy (ApEn) for feature extraction and the Library for Support Vector Machines (LIBSVM) for classification. The algorithm utilizes the PSO to determine the best parameters. Ai et al. (2015) used generalized N-dimensional ICA (GND-ICA) feature-fusion method based on a multilearning subspace-learning algorithm for ECG heartbeat classification by utilizing MIT-BIH Arrhythmia Database in all experiments. ECG heartbeats include normal beats (N), ventricular ectopic beats (V), ventricular ectopic beats (F), and supraventricular ectopic beats (S). WPD is utilized to extract features and to investigate the relationship between time and frequency. Signals are decomposed into the fourth level, which is composed of approximations and details, and then are employed for further feature fusion.

Kalaiselvi, Sasireka, Senthilkumar, and Maheswari (2016) utilized a DTCWT-based feature set for classifying ECG beats. The performance of the developed technique was compared with DWT feature extraction, and it was determined that the proposed feature set achieved higher recognition performance than the DWT-based feature set. Kiranyaz, Ince, Hamila, and Gabbouj (2015) discussed the classification of ECG heartbeat used CNN to record and detect heart problems. The approaches maintain a robust, fast, and patient-specific scheme with a superior classification performance of heart problems. The classification of the results is more accurate in the ECG analysis by using TF features. Qurraie and Afkhami (2017) developed a novel algorithm based on the TF to get the features and decision tree for arrhythmia classification. Jambukia, Dabhi, and Prajapati (2015) did a comprehensive survey on ECG signal classification, and they observed from the survey that many researchers have used the MIT-BIH Arrhythmia Database as a dataset and NN as a classifier. It is also realized that NNs are good candidates for ECG beat classification in terms of classification accuracy on training and test datasets.

Patidar, Pachori, Upadhyay, and Acharya (2017) implemented different methods for automated detection of alcoholism. The TQWT is utilized to classify ECG signals as normal or alcoholism. A novel Alcoholism Risk Index was proposed to distinguish between normal and alcoholism-affected subjects with a single number. The study was based on an advanced signal processing and machine learning method and achieved 97.02% classification accuracy. The proposed scheme can be used for rehabilitation of alcoholism patients in an accurate way. Moreover, the proposed scheme could reduce the workload of clinicians and could be used in other biomedical applications, such as cardiac, neural systems, eyes, diabetes etc. Also, telehealth care is a novel method for chronic disease management to reduce the workload of cardiologists. Ho et al. (2014) implemented a clinical decision and support system (CDSS) for automated recognition of ECG signals in real-time analysis. In the proposed telehealth care system, ECG signals can automatically be uploaded, and massive data can be added to accurately implement a CDSS. Hence, the proposed telehealth care system can be used as a tool in decision making.

1.4 PHONOCARDIOGRAPHY

PCG is the record of sonic vibrations of heart and blood circulation. The PCG is a sound signal associated with cardiohemic system (the heart and blood together) activity. The PCG signal recording needs a transducer to convert the sound signal into an electrical signal by utilizing microphones, pressure transducers, or accelerometers, which can be positioned on the chest surface. The normal PPG signal indicates the overall state of the heart related to the rhythm and contractility. CVDs show variations or different sounds and murmurs that can be utilized in diagnosis (Rangayyan, 2015).

In the past, different concepts have been proposed to describe the origins of cardiac cycle sounds once it was understood that the strength of the left ventricle's contraction had a substantial effect on the strength of the first heart sound. Heart sounds are formed when a moving column of blood comes to a sudden stop or decelerates significantly. The strength of a heart sound is based on the level of energy that the moving column of blood has accomplished. The sudden deceleration causes a dissipation of energy that results in the creation of vibrations affecting the contiguous cardiohemic mass. The factors included in the creation of the several heart sounds affecting the acceleration and deceleration of columns of blood are different. These should be separately considered for each sound, taking into account the pathophysiology and the physiology of the involved phase of the cardiac cycle (Oweis, Hamad, & Shammout, 2014)

Precise and early diagnosis of CVDs is crucial, which can be realized by heart auscultation. It is a widely used technique to screen for physical anomalies of the cardiovascular system. But perceiving pertinent features and creating a diagnosis based on the heart sounds detected only by a stethoscope is a skill that can take longer to learned and master. The accuracy and effectiveness of a diagnosis based on heart sound auscultation can be enhanced significantly by employing digital signal processing and machine learning techniques to investigate PCG signals (Chen, Wang, Shen, & Choy, 2012; Cherif, Debbal, & Bereksi-Reguig, 2010; Choi, Shin, & Park, 2011; Safara, Doraisamy, Azman, Jantan, & Ranga, 2012).

Several personal health systems have been implemented to effectively distinguish CVDs and to support clinical decisions. Usually, heart sound auscultation and ECG are employed for CVD diagnosis (Carvalho et al., 2011). But some heart diseases are hard to detect using ECG. Because heart sound signals are complex and highly nonstationary in nature, it is not easy to analyze them in an automated way. An auscultation technique employed to support medical doctors with an objective and accurate interpretation of heart sounds can be employed that distinguishes heart sounds (S1, S2, S3, and S4) during the heart cycle. The crucial points for signal analysis are the first heart sound (S1) and the second heart sound (S2), as well as systolic and diastolic periods. Localization (segmentation) of heart sounds must be done before any analysis. Techniques reported in the literature associated with heart sound segmentation include the hidden Markov model. Other methods are based on signal homomorphism, Eigenvalue spectrum, autoregression models, complexity-based segmentation, STFT, and wavelet decomposition. Oweis et al. (2014) proposed a multiband wavelet energy method for the segmentation of S1 and S2 heart sounds. S1 and S2 can be utilized to determine the heart sound type and distinguish abnormal heart sounds (Zhong & Scalzo, 2013). Wavelet decomposition can be successfully used to extract the information hidden in S1 and S2 spectrums by choosing appropriate subbands. The proposed method precisely localizes S1 and S2 to improve the characteristic parameter calculation (Zhong & Scalzo, 2013).

Different methods have been utilized for heart sound segmentation of normal and abnormal hearts, particularly for the S1 and S2 (Babaei & Geranmayeh, 2009; Huiying, Sakari, & Iiro, 1997). In Huiying et al. (1997), DWT was applied to phonocardiographic signals, achieving a 93% accuracy. Zhong and Scalzo (2013) implemented a multiscale wavelet decomposition with a threshold method to improve the accuracy of the detection rate. The suggested algorithm achieved an accuracy of around 92% on abnormal heart sounds. Some approaches have used an unsupervised training process for the task of robust and automatic detection and identification of events of interest in the major heart sounds from unlabeled PCGs. Oweis et al. (2014) employed Teager Energy Operator (TEO) with homomorphic filtering for heart sound segmentation and obtained 98.3% accuracy. The proposed segmentation method divides the signal into periods showing the two dominant heart sounds, S1 and S2, and time durations of the systole and diastole. The proposed method simplifies the presentation of heart sound signals obtained with a digital stethoscope. The resulting signals are more meaningful and easier to judge even for an unskilled doctor, thus eliminating the need for an expert physician. The main advantages of this method are employability in a noisy environment, reduction of the time needed for the examination process, reliability, and ease of use. The proposed method gives a sensitivity level of 99.1%, positive predictivity of 97.7%, specificity of 97.43%, and accuracy of 98.3%.

Medical doctors employ the PCG signals more competently once these signals are presented visually instead of using a conventional stethoscope. A clinician can utilize this significant signal with valuable prognostic and diagnostic information for the diagnosis of CVDs. Even though PCG signal analysis by auscultation is an appropriate clinical tool, the heart sound signals are nonstationary and very complex. Hence the analysis of this signal in a time or frequency domain is not easy. Generally, there are two main procedures required before classification: signal processing and feature extraction. Several signal processing algorithms are utilized in the literature such as discrete Fourier transform, short time Fourier transform, Wigner distribution, Hilbert transform, continuous wavelet transform, DWT, WPT, TQWT, DTCWT, EMD, and ensemble EMD. Actually, the Fourier and wavelet family of transforms are usually employed for PCG signal processing. Because the wavelet achieves a practical decomposition in both the time and frequency domain, it is one of the best transforms to analyze nonstationary and transient signals, such as PCG (Choi et al., 2011). Feature extraction is critical to obtain better classification performance. A different set of features based on wavelet transforms was defined recently to classify various types of heart sounds and murmurs (Safara, Doraisamy, Azman, Jantan, et al., 2012). For example, Ahlstrom, Hult, et al. (2006) extracted energy and entropy from DWT of the PCG signals to classify systolic murmurs. Cherif et al. (2010) proposed a feature extraction method in the TF domain for the investigation of heart sounds. They utilized DWT and WPT in the PCG signal analysis.

Heart murmurs are usually the initial indication of a heart valve abnormality, and they are generally realized at auscultation during the primary health care. To distinguish a normal murmur from a pathological murmur, a smart stethoscope with decision-support capabilities would be of great value. Ahlstrom, Hult, et al. (2006) employed PCG signals acquired from different patients with mitral insufficiency, aortic valve stenosis, or normal murmurs, and the data were analyzed to find an appropriate feature subset for automatic classification of heart murmurs. Furthermore, fractal dimensions, wavelets, Shannon energy, and recurrence quantification analysis were employed for feature extraction. NN classifier achieved 86% correct classifications rate. Gupta, Palaniappan, Swaminathan, and Krishnan (2007) proposed a new technique for the segmentation of PCG signals into single cardiac cycle (S1 and S2) employing K-means clustering and homomorphic filtering. Wavelet transform was used for feature extraction, and NNs were employed for classification of normal, diastolic murmur, and systolic murmur. The S3 and S4 heart sounds are two abnormal heart sound components designated in the heart. Tseng, Ko, and Jaw (2012) proposed an adaptive TF analysis method (Hilbert-Huang Transform) to find the existence of S3 and S4 in PCG signals. By utilizing the proposed algorithm, 90.3% of heart sound cycles with S3 were detected, 9.6% were missed, and 9.6% were false-positive; 94% of S4 were detected, 5.5% were missed, and 16% were false-positive. Yuenyong, Nishihara, Kongprawechnon, and Tungpimolrut (2011) proposed a new framework for heart sound analysis. One of the most crucial steps in PCG signal analysis is segmentation, because of the interference of murmurs. The framework includes envelope detection for the calculation of cardiac cycle lengths by employing autocorrelation of envelope signals. DWT utilized features extraction, principal component analysis was utilized for dimension reduction, and NN bagging predictors were employed for classification. Maximum classification performance employing 10-fold cross-validation was 92%.

Choi (2008) employed WPT and SVM classifier to distinguish normal heart sounds from murmurs;

96% sensitivity and 100% specificity was achieved. Choi and Jiang (2010) proposed a novel PCG spectral analysis technique employing the normalized AR power spectral density curve with the SVM classifier to classify cardiac sound murmurs, normal sound signals, and abnormal sound signals (split sounds, aortic insufficiency, aortic stenosis, atrial fibrillation, mitral stenosis, and mitral regurgitation). As a result, the proposed PCG spectral analysis method achieved 99.5% sensitivity and 99.9% specificity in classifying normal and abnormal heart sounds. Choi et al. (2011) proposed a novel PCG signal analysis technique for mitral and aortic abnormal murmurs using WPD. Normal heart sounds were acquired from subjects without any heart problems by employing a wireless electric stethoscope. Abnormal murmurs were aortic stenosis and mitral stenosis. They utilized the position index and the maximum peak frequency of the WPD coefficient associated with the maximum peak frequency, and the ratios of entropy information and the wavelet energy to accomplish higher accuracy for the heart murmur detection. The proposed abnormal murmur detection technique achieved a classification performance of 99.78% specificity and 99.43% sensitivity. Moreover, many studies classified several categories of murmurs. NN was employed by Ahlstrom, Höglund, et al. (2006) to differentiate normal murmurs from aortic stenosis, and a specificity of 88% and a sensitivity of 90% were achieved. Babaei and Geranmayeh (2009) used a MLPNN for the classification of PCG signals to differentiate murmurs and achieved 94.24% accuracy.

Cardiac auscultatory ability of medical doctors is critical for correct diagnosis of heart diseases. Several abnormal heart sounds with indistinguishable main specifications and diverse details, including ambient noise, are essentially similar, thus it is important to train, evaluate, and improve the skills of medical students in identifying and differentiating the main indications of the cardiac diseases. Babaei and Geranmayeh (2009) employed a useful wavelet-based multiresolution algorithm to extract the key characteristics of three well-known heart valve disorders: pulmonary stenosis, aortic stenosis, and aortic insufficiency sounds. An ANN classifier and Daubechies wavelet filter with five decomposition levels were employed for the most prominent diseases. The proposed ANN achieved 94.42% classification accuracy. Herzig, Bickel, Eitan, and Intrator (2015) proposed new cardiac monitoring based on PCG signal analysis. Specifically, they studied two morphological features and their relations with physical changes. The proposed system was verified using data taken from different patients during laparoscopic surgeries. During laparoscopic surgery, externally induced cardiac stress allowed analysis of each patient with regard to his or her own baseline. Furthermore, they revealed that the proposed features vary during cardiac stress, and the variation is more significant for patients with cardiac complications.

Different features can be extracted based on wavelet transform to distinguish murmurs from normal heart sounds or to differentiate different types of murmurs. Ahlstrom, Hult, et al. (2006) proposed a method including entropy, fractal dimension, and energy to extract features from PCG signals to classify systolic murmurs. Heart murmurs are generally the first sign of pathological variations of the heart valves, and they can be detected during auscultation in primary health care. Separation of a pathological murmur from a physiological murmur is not an easy task, hence, an "intelligent stethoscope" with decision-support capabilities can be of great value. PCG signals were collected from several patients with mitral insufficiency, aortic valve stenosis, or physiological murmurs to find an appropriate feature subset for automatic classification of heart murmurs. Wavelets, Shannon energy, recurrence quantification, and fractal dimensions analysis were employed for feature extraction. The selected multidomain subset achieved the best results, with 86% correct classifications employing an NN classifier. Choi (2008) utilized WPD and the SVM classifier for valvular heart disorder (VHD) detection. The WPE was utilized to include distribution information of energy through the whole frequency range of PCG signals. Moreover, the PCG signal parameters (meanWPE and stdWPE) of the position indices of the terminal nodes were proposed as a feature. The experimental results of the proposed VHD detection method achieved a specificity of 96% and a sensitivity of 100% for both the training and testing data.

Medical doctors can utilize the PCG signal once it is presented visually, rather than by a conventional stethoscope. This signal delivers significant prognostic diagnostic information to the clinician. Although PCG signal analysis by auscultation is suitable as a clinical tool, it is not easy to analyze in the time or frequency domain as they are nonstationary and more complex. Cherif et al. (2010) studied TF feature extraction for the recognition of heart sounds. They highlighted the importance of the choice of wavelet in PCG signal analysis using the DWT and the WPT. The performance of the DWT and the WPT in PCG signal analysis was assessed and discussed. Homaeinezhad, Sabetian, Feizollahi, Ghaffari, and Rahmani (2012) presented a mathematical modeling of the cardiac system with multiple measurement signals: PCG, ECG, and arterial blood pressure (ABP). Moreover, the PCG framework can produce the S4-S1-S2-S3 cycles in normal and cardiac illness conditions such as insufficiency, gallop, regurgitation, and stenosis. The amplitude and frequency content (5–700 Hz) of individual sound and variation patterns can be defined in the PCG model. The three models were realized to produce artificial signals with different cardiac abnormalities for quantitative recognition and performance evaluation of numerous PCG, ECG, and ABP signals based on the DWT, Hilbert transform, geometric features, and the principal components analyzed by geometric index.

Naseri and Homaeinezhad (2012) developed a PCG signal measurement quality assessment technique that consists of three main steps: preprocessing, quality assessment, and advanced measurement algorithms. The preprocessing step consists of normalization, wavelet-based threshold denoising, and baseline wander removal. The quality assessment routine contains the energy and noise level of the PCG signal. The advanced quality measurement step is principally employed by the S1 and S2 sound intervals. The developed algorithm achieved 95.25% classification accuracy. Furthermore, they (Naseri & Homaeinezhad, 2013) proposed a new framework to classify PCG sounds. First, the PCG signal undergoes preprocessing, then two windows were moved into the preprocessed data, and two frequency and amplitude-based features were extracted from these segments in each analysis window. After feature extraction, combining these features created a synthetic decision-making basis. To classify the defined PCG sounds, first S1 and S2 were distinguished. Then, a new decision statistic was utilized to distinguish infrequent S3 and S4 sounds. The proposed framework achieved a positive predictive value of 98.60% and average sensitivity of 99.00%.

Safara, Doraisamy, Azman, and Jantan (2012) combined spectral, temporal, and geometric features for PCG signal classification. The feature set is composed of zero-crossing rates as the temporal feature, and spectral flux, spectral energy entropy, spectral roll-off, and spectral centroid as the spectral features. Moreover, curve length, area under curve, summation of the first order derivatives, summation of the second order derivatives, and centralized mean square values were utilized in the feature set as well. They introduced a new entropy to investigate the heart sounds in classification of five types of heart sounds and murmurs which are normal, mitral stenosis, mitral regurgitation, aortic stenosis, and aortic regurgitation. WPT was employed for extracting features from the heart sound, and the entropy was calculated to derive feature vectors. The proposed scheme achieved 96.94% accuracy with BayesNet. Safara, Doraisamy, Azman, Jantan, and Ramaiah (2013) proposed a multilevel basis selection (MLBS) to preserve the most informative basis of a WPT tree by eliminating the less informative basis by using three exclusion criteria: noise frequency, energy threshold, and frequency range. MLBS achieved an accuracy of 97.56% for classifying normal heart sounds, aortic regurgitation, mitral regurgitation, and aortic stenosis (Safara et al., 2013).

The PCG taken from normal subjects generally includes two separate tones, S1 and S2. Moreover, an auscultation technique utilized to supply medical doctors with objective and precise interpretation of heart sounds can be employed to distinguish four sounds, namely, S1, S2, S3, and S4, during the heart cycle. Ramović, Bandić, Kevrić, Germović, and Subasi (2017) proposed a framework to distinguish four heartbeats efficiently utilizing the combination of multiscale wavelet transform and TEO to improve the accuracy of the recognition process. Wavelet transform was added to the module in the form of creating several combinations for a signal on which TEO is applied. The aim of combining wavelet and TEO was to explore how different details taken from wavelet transform affect the success of TEO in detecting S1, S2, S3, and S4 heart sounds.

1.5 PHOTOPLETHYSMOGRAPHY

PPG is a low-cost optical technique to distinguish blood volume changes in the microvascular bed of tissue. It is frequently employed to take measurements at the skin surface. The PPG waveform contains an AC waveform for synchronous cardiac changes in the blood volume with every heartbeat and is superimposed on a slowly varying DC baseline with several lower frequency components. Even though the source of the components of the PPG signal is not completely realized, it is usually known that they can deliver significant information related to the cardiovascular system. This technique is simple, driven by improvements in computer-based signal analysis techniques and the demand for low-cost, portable technology for community-based clinical settings and primary care. PPG technology has been utilized in a variety of medical devices for measuring blood pressure, oxygen saturation, and cardiac output; evaluating autonomic function; and distinguishing peripheral vascular disease (Allen, 2007).

The PPG waveform, also identified as the pulse oximeter waveform, is a useful clinical waveform. The waveform, which is a highly filtered quantity of light absorbed by the tissue, is an interaction between the cardiovascular, respiratory, and autonomic systems. Pulse oximeters represent the oxygen saturation and HR extracted from PPG measurements with different wavelengths. PPG is an exceptional monitor for cardiac arrhythmia, especially when utilized with ECG. PPG, with minor adjustments, can be employed to assess the ventilator-induced modulations related to hypovolemia. PPG can be analyzed by utilizing digital signal processing algorithms to yield new physical parameters with the extension of potential benefits of PPG. The aim of the clinical researcher is to utilize PPG to determine early therapeutic interventions or to obtain information and therapeutic guidance from the simple PPG (Alian & Shelley, 2014).

PPG signals (Kamal, Harness, Irving, & Mearns, 1989) are attractive for monitoring HR because of their wearable capability compared with straightforward ECG technology. Pulse oximeters are embedded in a small wearable device and can be worn on the wrists, earlobes, or fingertips to collect PPG signals. The pulse oximeters send a light to the skin utilizing light-emitting diodes (LED) and collect the reflected or transmitted light whose intensity are contingent on the amount of blood present in the arteries under the skin. The light intensity changes with the cardiac rhythm as the volume of blood changes with the cardiac cycle, hence it can be utilized to obtain HR information (Khan, Al Hossain, Uddin, Alam, & Hasan, 2016).

PPG monitoring can be achieved at peripheral locations of the body and does not require a conductive gel, unlike ECG in which sticky metal electrodes are located on the chest to monitor electrical activity of the heart. Therefore, PPG sensors are considerably utilized in new applications such as smartwatches and wearable devices, because they can be employed for ordinary heart monitoring by nonspecialist users (Jarchi & Casson, 2016). Different types of wearable sensors, such as smartwatches, can record ECG or PPG signals. Charlton et al. (2016) compared the performances of ECG and PPG to determine how closely the algorithms agreed with a gold standard respiratory rate (RR) measure when operating under ideal conditions. Moreover, they compared algorithm performance with the clinical standard for continuous RR measurement by comparing algorithm performance when using ECG and PPG.

Besides the clinical usage, elderly people and sportsmen need a real-time observation on their HR to avoid any chances of heart seizures and to keep them away from excessive physical action. Although PPG signals can be conveniently monitored in hospital environments, it is difficult to utilize them throughout a fitness program because motion artifacts (MA) will distort the PPG signals. To eliminate MA when PPG signals are employed during physical exercise, three-axis accelerometery data are utilized to model the MA (Lee et al., 2010). Yousefi, Nourani, Ostadabbas, and Panahi (2014) first tried to eliminate the tissue effects by using a two-stage normalized-LMS adaptive noise canceller to eliminate MA. However, the process brings significant enhancements over previously employed algorithms; it necessitates pulse oximeters with diverse wavelengths. López-Silva et al. (2012) on the other hand first proposed a heuristic approach to eliminate this problem. They proposed an algorithm that utilizes previously estimated HR values in HR estimation. Zhang, Pi, and Liu (2015) employed a three-stage framework for substantial MA. In the proposed framework, the subjects must remain unmoving at the beginning to correctly calculate the initial estimates, which the tracker can accomplish accurately. As a result of this work, Zhang (2015) devised an alternative solution to improve the previous results based on the tracking mechanism. Khan, Al Hossain, et al. (2016) tried to eliminate the previously mentioned shortcomings by proposing a new method implemented in two steps. In the first step, a runaway error was eliminated by noise-assisted ensemble EMD as a signal-denoising technique to eliminate dependency on previous estimates. In the second step, recursive least squares (RLS) adaptive filter complemented with a time-domain extraction was utilized.

Elgendi et al. (2015) utilized the PPG signal for heat stress assessment by employing Welch and AR spectral estimation methods. The study designated that using the first and second derivatives of the PPG waveforms were beneficial for defining heat stress level. Both signal-processing techniques presented a distinct discrimination in the frequency domain before and after simulated heat-stress induction once the second derivative was used. Furthermore, the results proved that the Welch method has superior performance than the Yule-Walker method in discriminating before and after the simulated heat-stress inductions.

1.6 OTHER BIOMEDICAL SIGNALS

1.6.1 The Electroneurogram

The electroneurogram (ENG) is electrical activity detected as a stimulus and the related nerve action potential transmitted to the length of a nerve. The ENG, generally collected with needle electrodes, is utilized to check the conduction velocity of the nerve, thus helping in the nerve injury diagnosis. It is likely to calculate the conduction velocity from the distance between two signal waveforms by inspiring a nerve at two different sites, divided by a well-defined distance. The ENG can be measured both noninvasively and invasively (Sörnmo & Laguna, 2005). ENG can be used to measure the conduction velocity of an action potential or stimulus in a nerve. ENGs can be acquired by employing silver-silver-chloride electrodes or concentric needle electrodes on the skin. Conduction velocity can be measured with the related activity in a peripheral nerve by stimulating a motor nerve. The limb is detained in a relaxed position, and a short and strong stimulus is utilized in the form of a pulse to minimize muscle contraction. The variance in the potentials of the ENGs collected over the related muscle achieves the conduction time. Hence, the conduction velocity in the nerve can be determined by knowing the separation distance between the stimulus points (Rangayyan, 2015)

1.6.2 The Electroretinogram

An electroretinogram (ERG) is the electrical potentials produced by the retina of the eye during light stimulation. The ERG can be acquired by employing an electrode, placed in a contact lens, on the cornea. The ERG is beneficial for evaluating the electrical response of the visual cells behind the retina. An abnormal ERG is attained in circumstances such as arteriosclerosis of the retina or detachment of the retina, whereas a normal ERG demonstrates proper responses with increased light intensity (Sörnmo & Laguna, 2005).

1.6.3 The Electrooculogram

The electrooculogram (EOG) is a technique to measure the direction of the gaze from a steady corneal-retinal potential, which is related to the horizontal and vertical actions of the eye. The EOG is particularly important for patients who suffer from sleep disorders, in which the existence of rapid eye movement (REM) is crucial for detecting certain sleep stages. The EOG is recorded when studying a quick and spontaneous oscillation of the eyeballs of patients suffering from vertigo and dizziness. Furthermore, the EOG is beneficial in virtual reality environments in which a device for eye-tracking is desirable (Sörnmo & Laguna, 2005).

1.6.4 The Electrogastrogram

The electrogastrogram (EGG) is a technique to measure the impulses that circulate through the muscles of the stomach that control their contractions. The EGG is utilized when the nerves controlling the muscles or the stomach muscles are not working properly. The EGG is acquired by putting several electrodes over the stomach during fasting and after a meal. A regular "rhythmic" EGG signal is produced by the muscles of the stomach of a normal person with an amplitude that increases after a meal. The normal frequency of the gastric rhythm is around 3 cycles/min. But the rhythm of symptomatic patients is mostly irregular and occasionally without an increase in amplitude that follows a meal (Sörnmo & Laguna, 2005). EGG is a simple and noninvasive technique used as an initial examination to assess the need for endoscopy. The EGG is an electrical signal that propagates through the muscles of the stomach and controls the muscle's contraction. EGG is employed to check if the stomach muscles or the nerves controlling the muscles are working properly or not. EGG is acquired by putting the electrode over the stomach and the electrical signals produced by the stomach muscles are detected by the electrode and stored on a computer for the analysis (Gopu, Neelaveni, Pokumaran, & Shekar, 2010).

1.6.5 The Carotid Pulse

The carotid pulse (CP) goes through the surface of the body at the neck and is recorded over the carotid artery. It is a pulse signal demonstrating the variations in ABP and volume for each heartbeat. As the recording site near to the heart, the CP signal closely looks like the morphology of the pressure signal at the root of the aorta. The CP is a helpful tool to the PCG and can contribute to the identification of S2 and its components. The CP increases abruptly with the ejection of blood from the left ventricle to the aorta, reaching a peak called the percussion wave (P). Then there is a plateau or secondary wave known as the tidal wave (T), produced by an echoed pulse returning from the upper body, following the P wave. Next, closure of the aortic valve causes a notch known as the dicrotic notch (D). The dicrotic wave (DW) follows the dicrotic notch due to an echoed pulse from the lower body. The CP trace is affected by valvular faults such as mitral deficiency and aortic stenosis. CP is not generally utilized in clinical diagnosis and is acquired with ECG and PCG signals. Location of the CP transducer needs to be carefully selected on the neck as near to the carotid artery as possible, where the pulse is the strongest (Rangayyan, 2015)

1.6.6 The Vibromyogram

The vibromyogram (VMG) is a vibration signal and a direct mechanical indicator of contraction of a skeletal muscle. It has been called the acoustic-, phono-, or sound-myogram. Muscle vibrations or sounds are associated with a variation in contraction of the fundamental muscle fibers and can be acquired by utilizing contact microphones or accelerometers located on the muscle surface. The intensity and frequency of the VMG changes directly proportional to the contraction level. The VMG, together with the EMG, can be beneficial in athletic training, muscle contraction, biofeedback, and neuromuscular control (Rangayyan, 2015)

1.7 MACHINE LEARNING METHODS

The extracted features of a medical problem are not enough to completely explain the pathology for many cases. Especially, while describing the problem, utilizing a redundant or suboptimal set of features causes the problem. It is better to assume that there is a nonlinear relation between input and output of the biomedical system, instead of seeking better features. For example, an automatic diagnostic system for detecting neuropathic and myopathic diseases by using EMG waveforms uses processed EMG data as input. The goal is to learn the relationship between the information given to the system and related pathology. After training, when we provide new data to the system, it will recognize the correct pathology. This kind of task can be easily performed by machine learning techniques (Begg et al., 2008).

Fuzzy logic, supervised learning, and evolutionary algorithm are at the core of today's applied machine learning technologies. There are some other technologies that are still being researched such as symbolic learning, reinforcement learning, unsupervised learning, and cognition sciences (Begg et al., 2008; Shavlik & Dietterich, 1990).

The concept of a learning mechanism had attracted philosophers since the time of the Greeks. Philosophers such as Descartes, and psychologists such as Spencer, James, and Thorndike have contributed to the understanding of the learning mechanism (Begg et al., 2008).

Using developments in the engineering and mathematic sciences, fast computing devices were invented based on Von Neumann architecture. Soon, computers were used for learning studies. Hull and Baernstein (1929) said that if muscle reaction systems could be reproduced by some inorganic method, then surely the powers of thinking and inference could be similarly reproduced in a machine. The connectionist theory, which was formed based on Aristotle's theories (400 BC), had been implemented and tested by developing electronic computers during the 1950s. Hence, the "machine learning" field was born and later the term was changed to "computational intelligence" (Begg et al., 2008).

There are numerous CI methods such as supervised learning, unsupervised learning, symbolic learning, and GA. Among these learning paradigms, the one most researched was the supervised learning method, which was based on function estimation. A set of examples is given to the supervised learning formula by an external supervisor with the class label. The system recognizes the hidden relationships between the sample set and desired output. After this training phase, it is possible to predict the output for unknown examples. Reinforcement learning, risk minimization, and stochastic learning are some paradigms in supervised learning. GAs are used for optimization, which is based on evolutionary theory in biological systems. A new method, symbolic learning, is still under development, and the studies are focused on finding a way to teach a machine how to find the relationships between symbols and perceptual learning (Begg et al., 2008).

It seems possible to merge the learning concept of human beings with the development of machine learning techniques after artificial neurons emerged. Today, computers have enough power to process numbers and symbols faster than human beings (Jain, Mao, & Mohiuddin, 1996). Nevertheless, the current computers based on Von Neumann architecture are still insufficient to deal with problems that require complex perception (Begg et al., 2008).

After all, computer implementation shows that it is possible to devise programs that can perform well with today's computer technologies. In biomedicine, mathematical models of living systems are necessary for visualization (Held, Brebbia, & Ciskowski, 1991). Nowadays, machine learning techniques are being used more for modeling biological systems. Today, we know that it is possible to use computers as an electronic and virtual laboratory (Hull & Baernstein, 1929).

REFERENCES

Aarabi, A., & He, B. (2012). A rule-based seizure prediction method for focal neocortical epilepsy. *Clinical Neurophysiology*, *123*(6), 1111–1122.

AbdelMaseeh, M., Chen, T.-W., & Stashuk, D. W. (2016). Extraction and classification of multichannel electromyographic activation trajectories for hand movement recognition. *IEEE Transactions on Neural Systems and Rehabilitation Engineering*, *24*(6), 662–673.

Abdullah, A. A., Subasi, A., & Qaisar, S. M. (2017). Surface EMG signal classification by using WPD and ensemble tree classifiers. In: *Presented at the CMBEBIH 2017: proceedings of the international conference on medical and biological engineering 2017*, Vol. 62. (p. 475). Springer, 475.

Acharya, R., Kumar, A., Bhat, P., Lim, C., Kannathal, N., & Krishnan, S. (2004). Classification of cardiac abnormalities using heart rate signals. *Medical and Biological Engineering and Computing*, *42* (3), 288–293.

Acharya, U. R., Chua, E. C.-P., Chua, K. C., Min, L. C., & Tamura, T. (2010). Analysis and automatic identification of sleep stages using higher order spectra. *International Journal of Neural Systems*, *20*(06), 509–521.

Acharya, U. R., Bhat, S., Faust, O., Adeli, H., Chua, E. C.-P., Lim, W. J. E., & Koh, J. E. W. (2015). Nonlinear dynamics measures for automated EEG-based sleep stage detection. *European Neurology*, *74*(5–6), 268–287.

Acır, N. (2006). A support vector machine classifier algorithm based on a perturbation method and its application to ECG beat recognition systems. *Expert Systems with Applications*, *31*(1), 150–158.

Adeli, H., Ghosh-Dastidar, S., & Dadmehr, N. (2007). A wavelet-chaos methodology for analysis of EEGs and EEG subbands to detect seizure and epilepsy. *IEEE Transactions on Biomedical Engineering*, *54*(2), 205–211.

Afkhami, R. G., Azarnia, G., & Tinati, M. A. (2016). Cardiac arrhythmia classification using statistical and mixture modeling features of ECG signals. *Pattern Recognition Letters*, *70*, 45–51.

Ahlstrom, C., Höglund, K., Hult, P., Häggström, J., Kvart, C., & Ask, P. (2006). Distinguishing innocent murmurs from murmurs caused by aortic stenosis by recurrence quantification analysis. *Proceedings of the Enformatika Transactions on Engineering, Computing and Technology*, 1305–5313.

Ahlstrom, C., Hult, P., Rask, P., Karlsson, J.-E., Nylander, E., Dahlström, U., & Ask, P. (2006). Feature extraction for systolic heart murmur classification. *Annals of Biomedical Engineering*, *34*(11), 1666–1677.

Ai, D., Yang, J., Wang, Z., Fan, J., Ai, C., & Wang, Y. (2015). Fast multi-scale feature fusion for ECG heartbeat classification. *EURASIP Journal on Advances in Signal Processing*, *2015*(1), 46.

Ajiboye, A. B., & Weir, R. F. (2005). A heuristic fuzzy logic approach to EMG pattern recognition for multifunctional prosthesis control. *IEEE Transactions on Neural Systems and Rehabilitation Engineering*, *13* (3), 280–291.

Akram, F., Han, H.-S., & Kim, T.-S. (2014). A P300-based brain computer interface system for words typing. *Computers in Biology and Medicine*, *45*, 118–125.

Akram, F., Han, S. M., & Kim, T.-S. (2015). An efficient word typing P300-BCI system using a modified T9 interface and random forest classifier. *Computers in Biology and Medicine*, *56*, 30–36.

Al-Fahoum, A., & Howitt, I. (1999). Combined wavelet transformation and radial basis neural networks for classifying life-threatening cardiac arrhythmias. *Medical & Biological Engineering & Computing*, *37* (5), 566–573.

Al-Faiz, M. Z., & Miry, A. H. (2012). *Artificial human arm driven by EMG signal.* INTECH Open Access Publisher.

Alian, A. A., & Shelley, K. H. (2014). Photoplethysmography. *Hemodynamic Monitoring Devices*, *28*(4), 395–406. https://doi.org/10.1016/j.bpa.2014.08.006.

Alickovic, E., & Subasi, A. (2015). Effect of multiscale PCA de-noising in ECG beat classification for diagnosis of cardiovascular diseases. *Circuits, Systems, and Signal Processing*, *34*(2), 513–533.

Alickovic, E., & Subasi, A. (2016). Medical decision support system for diagnosis of heart arrhythmia using DWT and random forests classifier. *Journal of Medical Systems*, *40*(4), 108.

Alickovic, E., & Subasi, A. (2018). Ensemble SVM method for automatic sleep stage classification. *IEEE Transactions on Instrumentation and Measurement*, *67*(6), 1258–1265.

Alickovic, E., Kevric, J., & Subasi, A. (2018). Performance evaluation of empirical mode decomposition, discrete wavelet transform, and wavelet packed decomposition for automated epileptic seizure detection and prediction. *Biomedical Signal Processing and Control*, *39*, 94–102.

Allen, J. (2007). Photoplethysmography and its application in clinical physiological measurement. *Physiological Measurement*, *28*(3), R1.

AlOmari, F., & Liu, G. (2014). Analysis of extracted forearm sEMG signal using LDA, QDA, K-NN classification algorithms. *The Open Automation and Control Systems Journal*, *6*, 108–116.

AlOmari, F., & Liu, G. (2015). Novel hybrid soft computing pattern recognition system SVM–GAPSO for classification of eight different hand motions. *Optik - International Journal for Light and Electron Optics*, *126*(23), 4757–4762. https://doi.org/10.1016/j.ijleo.2015.08.170.

Al-Timemy, A. H., Bugmann, G., Escudero, J., & Outram, N. (2013). Classification of finger movements for the dexterous hand prosthesis control with surface electromyography. *IEEE Journal of Biomedical and Health Informatics*, *17*(3), 608–618.

Arif, M., & Akram, M. U. (2010). Pruned fuzzy K-nearest neighbor classifier for beat classification. *Journal of Biomedical Science and Engineering*, *3*(04), 380.

Arif, M., Akram, M. U., & Afsar, F. A. (2009). Arrhythmia beat classification using pruned fuzzy k-nearest neighbor classifier. In: *Presented at the soft computing and pattern recognition, 2009. SOCPAR'09. international conference of,* IEEE, pp. 37–42.

Artameeyanant, P., Sultornsanee, S., & Chamnongthai, K. (2016). An EMG-based feature extraction method using a normalized weight vertical visibility algorithm for myopathy and neuropathy detection. *Springerplus*, *5*(1), 2101.

Asl, B. M., Setarehdan, S. K., & Mohebbi, M. (2008). Support vector machine-based arrhythmia classification using reduced features of heart rate variability signal. *Artificial Intelligence in Medicine*, *44* (1), 51–64.

Baali, H., Khorshidtalab, A., Mesbah, M., & Salami, M. J. (2015). A transform-based feature extraction approach for motor imagery tasks classification. *IEEE Journal of Translational Engineering in Health and Medicine*, *3*, 1–8.

Babaei, S., & Geranmayeh, A. (2009). Heart sound reproduction based on neural network classification of cardiac valve disorders using wavelet transforms of PCG signals. *Computers in Biology and Medicine*, *39*(1), 8–15. https://doi.org/10.1016/j.compbiomed.2008.10.004.

Bajaj, V., & Pachori, R. B. (2013). Automatic classification of sleep stages based on the time-frequency image of EEG signals. *Computer Methods and Programs in Biomedicine*, *112*(3), 320–328.

Bassingthwaighte, J. B., Liebovitch, L. S., & West, B. J. (1994). *Fractal physiology*. New York: Springer-Verlag.

Begg, R., Lai, D. T., & Palaniswami, M. (2008). *Computational intelligence in biomedical engineering*. Boca Raton, FL: CRC Press.

Berry, R. B., Brooks, R., Gamaldo, C. E., Harding, S. M., Marcus, C., & Vaughn, B. (2012). *The AASM manual for the scoring of sleep and associated events. Rules, terminology and technical specifications.* Darien, IL: American Academy of Sleep Medicine.

Besrour, R., Lachiri, Z., & Ellouze, N. (2008). ECG beat classifier using support vector machine. In: *Presented at the information and communication technologies: from theory to applications, 2008. ICTTA 2008. 3rd international conference on* (pp. 1–5): IEEE.

Bigan, C. (1998). A system for neural networks detection and automatic identification of EEG epileptic events. In *IEE Colloquium on Intelligent Methods in Healthcare and Medical Applications (Digest No. 1998/514) (pp. 13/1–13/4)*, York, UK.

Bitzer, S., & Van Der Smagt, P. (2006). Learning EMG control of a robotic hand: towards active prostheses. In: *Presented at the robotics and automation, 2006. ICRA 2006. Proceedings 2006 IEEE international conference on*, IEEE, pp. 2819–2823.

Boyali, A., & Hashimoto, N. (2016). Spectral collaborative representation based classification for hand gestures recognition on electromyography signals. *Biomedical Signal Processing and Control*, *24*, 11–18. https://doi.org/10.1016/j.bspc.2015.09.001.

Bozkurt, M. R., Subasi, A., Koklukaya, E., & Yilmaz, M. (2016). Comparison of AR parametric methods with subspace-based methods for EMG signal classification using stand-alone and merged neural network models. *Turkish Journal of Electrical Engineering and Computer Sciences*, *24*(3), 1547–1559.

Bronzino, J. D. (2000). Principles of electroencephalography. In J. D. Bronzino (Ed.), *The biomedical engineering handbook.* (2nd ed.). Boca Raton, FL: CRC Press.

Cai, L., Wang, Z., & Zhang, H. (1999). An EMG classification method based on wavelet transform [and ANN]. In: *Presented at the [engineering in medicine and biology, 1999. 21st annual conference and the 1999 annual fall meeting of the Biomedical Engineering Society] BMES/EMBS conference, 1999. Proceedings of the first joint*, Vol. 1. (p. 565): IEEE.

Camm, A. J., Malik, M., Bigger, J., Breithardt, G., Cerutti, S., Cohen, R., ... Kleiger, R. (1996). Heart rate variability: standards of measurement, physiological interpretation and clinical use. Task Force of the European Society of Cardiology and the North American Society of Pacing and Electrophysiology. *Circulation*, *93*(5), 1043–1065.

Carvalho, P., Paiva, R. P., Kumar, D., Ramos, J., Santos, S., & Henriques, J. (2011). *A framework for acoustic cardiac signal analysis*: (pp. 151–160). Presented at the BIOSIGNALS.

Charlton, P. H., Bonnici, T., Tarassenko, L., Clifton, D. A., Beale, R., & Watkinson, P. J. (2016). An assessment of algorithms to estimate respiratory rate from the electrocardiogram and photoplethysmogram. *Physiological Measurement*, *37*(4), 610.

Chen, Y., Wang, S., Shen, C.-H., & Choy, F. K. (2012). Intelligent identification of childhood musical murmurs. *Journal of Healthcare Engineering*, *3*(1), 125–139.

Cherif, L. H., Debbal, S. M., & Bereksi-Reguig, F. (2010). Choice of the wavelet analyzing in the phonocardiogram signal analysis using the discrete and the packet wavelet transform. *Expert Systems with Applications*, *37*(2), 913–918. https://doi.org/10.1016/j.eswa.2009.09.036.

Choi, S. (2008). Detection of valvular heart disorders using wavelet packet decomposition and support vector machine. *Expert Systems with Applications*, *35*(4), 1679–1687. https://doi.org/10.1016/j.eswa.2007.08.078.

Choi, S., & Jiang, Z. (2010). Cardiac sound murmurs classification with autoregressive spectral analysis and multi-support vector machine technique. *Computers in Biology and Medicine*, *40*(1), 8–20. https://doi.org/10.1016/j.compbiomed.2009.10.003.

Choi, S., Shin, Y., & Park, H.-K. (2011). Selection of wavelet packet measures for insufficiency murmur identification. *Expert Systems with Applications*, *38*(4), 4264–4271.

Chowdhury, R. H., Reaz, M. B., Ali, M. A. B. M., Bakar, A. A., Chellappan, K., & Chang, T. G. (2013). Surface electromyography signal processing and classification techniques. *Sensors*, *13*(9), 12431–12466.

Christodoulou, C. I., & Pattichis, C. S. (1999). Unsupervised pattern recognition for the classification of EMG signals. *IEEE Transactions on Biomedical Engineering*, *46*(2), 169–178.

Christov, I., Jekova, I., & Bortolan, G. (2005). Premature ventricular contraction classification by the Kth nearest-neighbours rule. *Physiological Measurement*, *26*(1), 123.

Cipriani, C., Controzzi, M., & Carrozza, M. C. (2011). The SmartHand transradial prosthesis. *Journal of Neuroengineering and Rehabilitation*, *8*(1), 29.

Clayton, R. H., Murray, A., & Campbell, R. W. (1993). Comparison of four techniques for recognition of ventricular fibrillation from the surface ECG. *Medical and Biological Engineering and Computing*, *31*(2), 111–117.

Coelho, A. L., & Lima, C. A. (2014). Assessing fractal dimension methods as feature extractors for EMG signal classification. *Engineering Applications of Artificial Intelligence*, *36*, 81–98.

Cruz, C. I. M., Marasigan, J. P., Perez, A. P. G., Pillejera, J. E. V., Veron, N. P., & dela Cruz, A. R. (2016). A comparative study between DWT-ANFIS and DWT-SVM in ECG classification. In: *Region 10 conference (TENCON), 2016 IEEE*, IEEE, pp. 2052–2055.

da Silva-Sauer, L., Valero-Aguayo, L., de la Torre-Luque, A., Ron-Angevin, R., & Varona-Moya, S. (2016). Concentration on performance with P300-based BCI systems: a matter of interface features. *Applied Ergonomics*, *52*, 325–332. https://doi.org/10.1016/j.apergo.2015.08.002.

de Chazal, P., & Reilly, R. B. (2003). Automatic classification of ECG beats using waveform shape and heart beat interval features. In: *Presented at the acoustics, speech, and signal processing, 2003. Proceedings (ICASSP'03). 2003 IEEE international conference on*, Vol. 2. (pp. II–269). IEEE.

De Michele, G., Sello, S., Carboncini, M. C., Rossi, B., & Strambi, S.-K. (2003). Cross-correlation time-frequency analysis for multiple EMG signals in Parkinson's disease: a wavelet approach. *Medical Engineering and Physics*, *25*(5), 361–369.

Dobrowolski, A. P., Wierzbowski, M., & Tomczykiewicz, K. (2012). Multiresolution MUAPs decomposition and SVM-based analysis in the classification of neuromuscular disorders. *Computer Methods and Programs in Biomedicine*, *107*(3), 393–403.

Doroshenkov, L., Konyshev, V., & Selishchev, S. (2007). Classification of human sleep stages based on EEG processing using hidden Markov models. *Biomedical Engineering*, *41*(1), 25–28.

Duckrow, R., & Zaveri, H. (2005). Coherence of the electroencephalogram during the first sleep cycle. *Clinical Neurophysiology*, *116*(5), 1088–1095.

Elgendi, M., Fletcher, R. R., Norton, I., Brearley, M., Abbott, D., Lovell, N. H., & Schuurmans, D. (2015). Frequency analysis of photoplethysmogram and its derivatives. *Computer Methods and Programs in Biomedicine*, *122*(3), 503–512. https://doi.org/10.1016/j.cmpb.2015.09.021.

Enshaeifar, S., Kouchaki, S., Took, C. C., & Sanei, S. (2016). Quaternion singular spectrum analysis of electroencephalogram with application in sleep analysis. *IEEE Transactions on Neural Systems and Rehabilitation Engineering*, *24*(1), 57–67.

Esgiar, A., & Chakravorty, P. (2004). Electrocardiogram signal classification based on fractal features. In: *Presented at the computers in cardiology, 2004*, IEEE, pp. 661–664.

Farwell, L. A., & Donchin, E. (1986). The brain detector-P300 in the detection of deception. In: *Presented at the psychophysiology, SOC psychophysiol RES, Washington, DC 2005* Vol. 23, (p. 434). 434.

Felzer, T., & Freisieben, B. (2003). Analyzing EEG signals using the probability estimating guarded neural classifier. *IEEE Transactions on Neural Systems and Rehabilitation Engineering*, *11*(4), 361–371.

Fraiwan, L., Lweesy, K., Khasawneh, N., Wenz, H., & Dickhaus, H. (2012). Automated sleep stage identification system based on time–frequency analysis of a single EEG channel and random forest classifier. *Computer Methods and Programs in Biomedicine*, *108*(1), 10–19.

Gacek, A. (2012). An introduction to ECG signal processing and analysis. In *ECG signal processing, classification and interpretation* (pp. 21–46). London: Springer.

Gacek, A., & Pedrycz, W. (Eds.). (2012). *ECG signal processing, classification and interpretation: A comprehensive framework of computational intelligence*. London: Springer-Verlag.

Gao, W., Guan, J., Gao, J., & Zhou, D. (2015). Multi-ganglion ANN based feature learning with application to P300-BCI signal classification. *Biomedical Signal Processing and Control*, *18*, 127–137.

Gath, I., & Inbar, G. F. (Eds.). (2013). *Advances in processing and pattern analysis of biological signals*. New York: Springer.

Gokgoz, E., & Subasi, A. (2014). Effect of multiscale PCA de-noising on EMG signal classification for diagnosis of neuromuscular disorders. *Journal of Medical Systems*, *38*(4), 31.

Gokgoz, E., & Subasi, A. (2015). Comparison of decision tree algorithms for EMG signal classification using DWT. *Biomedical Signal Processing and Control*, *18*, 138–144.

Gopu, G., Neelaveni, R., Pokumaran, K., & Shekar, M. G. (2010). An enhanced technique for recording and analysis of electrogastrogram using active electrodes. *Sri Lanka Journal of Bio-Medical Informatics*, *1*(1), 21–27.

Guger, C., Schlogl, A., Neuper, C., Walterspacher, D., Strein, T., & Pfurtscheller, G. (2001). Rapid prototyping of an EEG-based brain-computer interface (BCI). *IEEE Transactions on Neural Systems and Rehabilitation Engineering*, *9*(1), 49–58.

Guger, C., Daban, S., Sellers, E., Holzner, C., Krausz, G., Carabalona, R., … Edlinger, G. (2009). How many people are able to control a P300-based brain–computer interface (BCI)? *Neuroscience Letters*, *462*(1), 94–98.

Gunes, S., Polat, K., Dursun, M., & Yosunkaya, S. (2009). Examining the relevance with sleep stages of time domain features of EEG, EOG, and chin EMG signals. In: *Presented at the biomedical engineering meeting, 2009. BIYOMUT 2009. 14th national* (pp. 1–4): IEEE.

Guo, X., Yang, P., Chen, L., Wang, X.,&Li, L. (2006). Study of the control mechanism of robot-prosthesis based-on the EMG processed. In: *Presented at the intelligent control and automation, 2006. WCICA 2006. The sixth world congress on*, Vol. 2. (pp. 9490–9493). IEEE, pp. 9490–9493.

Gupta, C. N., Palaniappan, R., Swaminathan, S., & Krishnan, S. M. (2007). Neural network classification of homomorphic segmented heart sounds. *Applied Soft Computing*, *7*(1), 286–297. https://doi.org/10.1016/j.asoc.2005.06.006.

Guy, V., Soriani, M. H., Bruno, M., Papadopoulo, T., Desnuelle, C., & Clerc, M. (2018). Brain computer interface with the P300 speller: Usability for disabled people with amyotrophic lateral sclerosis. *Annals of Physical and Rehabilitation Medicine*, *61*(1), 5–11.

Hassan, A. R., Hassan, A. R., & Subasi, A. (2017). A decision support system for automated identification of sleep stages from single-channel EEG signals. *Knowledge-Based Systems, 128*, 115–124.

Hassoun, M. H., Wang, C., & Spitzer, A. (1994a). NNERVE: neural network extraction of repetitive vectors for electromyography. I. Algorithm. *IEEE Transactions on Biomedical Engineering, 41*(11), 1039–1052.

Hassoun, M., Wang, C., & Spitzer, A. (1994b). NNERVE: neural network extraction of repetitive vectors for electromyography. II. Performance analysis. *IEEE Transactions on Biomedical Engineering, 41*(11), 1053–1061.

Hazarika, A., Dutta, L., Boro, M., Barthakur, M., & Bhuyan, M. (2018). An automatic feature extraction and fusion model: application to electromyogram (EMG) signal classification. *International Journal of Multimedia Information Retrieval*, 1–14.

Held, C. M., Heiss, J. E., Estévez, P. A., Perez, C. A., Garrido, M., Algarín, C., & Peirano, P. (2006). Extracting fuzzy rules from polysomnographic recordings for infant sleep classification. *IEEE Transactions on Biomedical Engineering, 53*(10), 1954–1962.

Held, K. D., Brebbia, C. A., & Ciskowski, R. D. (1991). Computers in biomedicine. In: *Presented at the international conference on computers in biomedicine 1991: Southampton, Eng, Computational Mechanics Publications.*

Herzig, J., Bickel, A., Eitan, A., & Intrator, N. (2015). Monitoring cardiac stress using features extracted from S1 heart sounds. *IEEE Transactions on Biomedical Engineering, 62*(4), 1169–1178.

Himanen, S.-L., & Hasan, J. (2000). Limitations of Rechtschaffen and Kales. *Sleep Medicine Reviews, 4*(2), 149–167.

Ho, T.-W., Lai, H.-Y., Wang, Y.-J., Chen, W.-H., Lai, F., Ho, Y.-L., & Hung, C.-S. (2014). A clinical decision and support system with automatically ECG classification in telehealthcare. In: *Presented at the e-health networking, applications and services (Healthcom), 2014 IEEE 16th international conference on* (pp. 293–297): IEEE.

Homaeinezhad, M., Sabetian, P., Feizollahi, A., Ghaffari, A., & Rahmani, R. (2012). Parametric modelling of cardiac system multiple measurement signals: an open-source computer framework for performance evaluation of ECG, PCG and ABP event detectors. *Journal of Medical Engineering & Technology, 36*(2), 117–134.

Hu, Y. H., Palreddy, S., & Tompkins, W. J. (1997). A patient-adaptable ECG beat classifier using a mixture of experts approach. *IEEE Transactions on Biomedical Engineering, 44*(9), 891–900.

Huang, H., Xie, H.-B., Guo, J.-Y., & Chen, H.-J. (2012). Ant colony optimization-based feature selection method for surface electromyography signals classification. *Computers in Biology and Medicine, 42* (1), 30–38.

Huiying, L., Sakari, L., & Iiro, H. (1997). A heart sound segmentation algorithm using wavelet decomposition and reconstruction. In: *Presented at the Engineering in Medicine and Biology Society, 1997. Proceedings of the 19th annual international conference of the IEEE*, Vol. 4. (pp. 1630–1633). IEEE.

Hull, C. L., & Baernstein, H. D. (1929). A mechanical parallel to the conditioned reflex. *Science, 70*(1801), 14–15.

Ikegami, S., Takano, K., Kondo, K., Saeki, N., & Kansaku, K. (2014). A region-based two-step P300-based brain–computer interface for patients with amyotrophic lateral sclerosis. *Clinical Neurophysiology, 125*(11), 2305–2312.

Ivanov, P. C., Amaral, L. A. N., Goldberger, A. L., Havlin, S., Rosenblum, M. G., Struzik, Z. R., & Stanley, H. E. (1999). Multifractality in human heartbeat dynamics. *Nature, 399*(6735), 461.

Jain, A. K., Mao, J., & Mohiuddin, K. M. (1996). Artificial neural networks: a tutorial. *Computer, 29*(3), 31–44.

Jambukia, S. H., Dabhi, V. K., & Prajapati, H. B. (2015). Classification of ECG signals using machine learning techniques: a survey. In: *Computer engineering and applications (ICACEA), 2015 international conference on advances in,* IEEE, pp. 714–721.

Jarchi, D., & Casson, A. J. (2016). Description of a database containing wrist PPG signals recorded during physical exercise with both accelerometer and gyroscope measures of motion. *Data, 2*(1), 1.

Jiang, M., Wang, R., Wang, J., & Jin, D. (2006). A method of recognizing finger motion using wavelet transform of surface EMG signal. In: *Presented at the Engineering in Medicine and Biology Society, 2005. IEEE-EMBS 2005. 27th annual international conference of the* (pp. 2672–2674): IEEE.

Jo, H. G., Park, J. Y., Lee, C. K., An, S. K., & Yoo, S. K. (2010). Genetic fuzzy classifier for sleep stage identification. *Computers in Biology and Medicine, 40*(7), 629–634.

Ju, Z., Ouyang, G., Wilamowska-Korsak, M., & Liu, H. (2013). Surface EMG based hand manipulation identification via nonlinear feature extraction and classification. *IEEE Sensors Journal, 13*(9), 3302–3311.

Kalaiselvi, A., Sasireka, M., Senthilkumar, A., & Maheswari, S. (2016). Complex wavelet transform based cardiac arrhythmia classification. *International Journal of Emerging Technology in Computer Science & Electronic, 20*(2).

Kamal, A., Harness, J., Irving, G., & Mearns, A. (1989). Skin photoplethysmography—a review. *Computer Methods and Programs in Biomedicine, 28*(4), 257–269.

Kamali, T., Boostani, R., & Parsaei, H. (2013). A hybrid classifier for characterizing motor unit action potentials in diagnosing neuromuscular disorders. *Journal of Biomedical Physics & Engineering, 3*(4), 145.

Kamali, T., Boostani, R., & Parsaei, H. (2014). A multi-classifier approach to MUAP classification for diagnosis of neuromuscular disorders. *IEEE Transactions on Neural Systems and Rehabilitation Engineering, 22*(1), 191–200.

Karimi, M., Pourghassem, H., & Shahgholian, G. (2011). A novel prosthetic hand control approach based on genetic algorithm and wavelet transform features. In: *Presented at the signal processing and its applications (CSPA), 2011 IEEE 7th international colloquium on* (pp. 287–292): IEEE.

Karimifard, S., Ahmadian, A., Khoshnevisan, M., & Nambakhsh, M. (2006). Morphological heart arrhythmia detection using Hermitian basis functions and kNN classifier. In: *Presented at the Engineering in Medicine and Biology Society, 2006. EMBS'06. 28th annual international conference of the IEEE* (pp. 1367–1370): IEEE.

Katsis, C. D., Goletsis, Y., Likas, A., Fotiadis, D. I., & Sarmas, I. (2006). A novel method for automated EMG decomposition and MUAP classification. *Artificial Intelligence in Medicine, 37*(1), 55–64.

Kaur, G., Arora, A., & Jain, V. (2010). EMG diagnosis via AR modeling and binary support vector machine classification. *International Journal of Engineering, Science and Technology, 2*(6), 1767–1772.

Kayikcioglu, T., & Aydemir, O. (2010). A polynomial fitting and k-NN based approach for improving classification of motor imagery BCI data. *Pattern Recognition Letters, 31*(11), 1207–1215.

Kerem, D. H., & Geva, A. B. (2017). state identification and forecasting of acute pathology using unsupervised fuzzy clustering of EEG temporal patterns. In H. N. Teodorescu, A. Kandel, & L. Jain (Eds.), *Fuzzy and neuro-fuzzy systems in medicine (pp. 19–68).* Boca Raton, IL: CRC Press.

Kevric, J., & Subasi, A. (2017). Comparison of signal decomposition methods in classification of EEG signals for motor-imagery BCI system. *Biomedical Signal Processing and Control*, *31*, 398–406.

Khan, E., Al Hossain, F., Uddin, S. Z., Alam, S. K., & Hasan, M. K. (2016). A robust heart rate monitoring scheme using photoplethysmographic signals corrupted by intense motion artifacts. *IEEE Transactions on Biomedical Engineering*, *63*(3), 550–562.

Khan, M., Singh, J., & Tiwari, M. (2016). A multi-classifier approach of EMG signal classification for diagnosis of neuromuscular disorders. *International Journal of Computer Applications*, *133*(4), 13–18.

Khezri, M., & Jahed, M. (2009). An exploratory study to design a novel hand movement identification system. *Computers in Biology and Medicine*, *39*(5), 433–442.

Khokhar, Z. O., Xiao, Z. G., & Menon, C. (2010). Surface EMG pattern recognition for real-time control of a wrist exoskeleton. *Biomedical Engineering Online*, *9*(1), 41.

Kiranyaz, S., Ince, T., Hamila, R., & Gabbouj, M. (2015). Convolutional neural networks for patient-specific ECG classification. In: *Engineering in Medicine and Biology Society (EMBC), 2015 37th annual international conference of the IEEE* (pp. 2608–2611): IEEE.

Klem, G. H., Lüders, H. O., Jasper, H., & Elger, C. (1999). The ten-twenty electrode system of the International Federation. *Electroencephalography and Clinical Neurophysiology*, *52*(3), 3–6.

Kohler, B.-U., Hennig, C., & Orglmeister, R. (2002). The principles of software QRS detection. *IEEE Engineering in Medicine and Biology Magazine*, *21*(1), 42–57.

Kurzynski, M., Krysmann, M., Trajdos, P., & Wolczowski, A. (2016). Multiclassifier system with hybrid learning applied to the control of bioprosthetic hand. *Computers in Biology and Medicine*, *69*, 286–297. https://doi.org/10.1016/j.compbiomed.2015.04.023.

Kutlu, Y., & Kuntalp, D. (2011). A multi-stage automatic arrhythmia recognition and classification system. *Computers in Biology and Medicine*, *41*(1), 37–45.

Lafuente, V., Gorriz, J. M., Ramirez, J., & Gonzalez, E. (2017). P300 brainwave extraction from EEG signals: an unsupervised approach. *Expert Systems with Applications*, *74*, 1–10.

Lai, K., & Chan, K. (1998). Real-time classification of electrocardiogram based on fractal and correlation analyses. In: *Presented at the Engineering in Medicine and Biology Society, 1998. Proceedings of the 20th annual international conference of the IEEE* (pp. 119–122): IEEE.

Leamy, D. J., Kocijan, J., Domijan, K., Duffin, J., Roche, R. A., Commins, S., ... Ward, T. E. (2014). An exploration of EEG features during recovery following stroke–implications for BCI-mediated neurorehabilitation therapy. *Journal of Neuroengineering and Rehabilitation*, *11*(1), 9.

Lee, B., Han, J., Baek, H. J., Shin, J. H., Park, K. S., & Yi, W. J. (2010). Improved elimination of motion artifacts from a photoplethysmographic signal using a Kalman smoother with simultaneous accelerometry. *Physiological Measurement*, *31*(12), 1585.

Lee, Y.-R., & Kim, H.-N. (2018). A data partitioning method for increasing ensemble diversity of an eSVM-based P300 speller. *Biomedical Signal Processing and Control*, *39*, 53–63.

Li, D., Zhang, H., Khan, M. S., & Mi, F. (2018). A self-adaptive frequency selection common spatial pattern and least squares twin support vector machine for motor imagery electroencephalography recognition. *Biomedical Signal Processing and Control*, *41*, 222–232.

Li, H., Feng, X., Cao, L., Li, E., Liang, H., & Chen, X. (2016). A new ECG signal classification based on WPD and ApEn feature extraction. *Circuits, Systems, and Signal Processing*, *35*(1), 339–352.

Li, T., & Zhou, M. (2016). ECG classification using wavelet packet entropy and random forests. *Entropy*, *18*(8), 285.

Li, Y., Wong, K. M., & de Bruin, H. (2012). Electroencephalogram signals classification for sleep-state decision—a Riemannian geometry approach. *IET Signal Processing*, *6*(4), 288–299.

Liang, S.-F., Kuo, C.-E., Hu, Y.-H., Pan, Y.-H., & Wang, Y.-H. (2012). Automatic stage scoring of single-channel sleep EEG by using multiscale entropy and autoregressive models. *IEEE Transactions on Instrumentation and Measurement*, *61*(6), 1649–1657.

Lisi, G., Belluco, P., Cattaneo, D., & Gini, G. (2011). From the classification of EMG signals to the development of a new lower arm prosthesis. *IFAC Proceedings Volumes*, *44*(1), 6493–6498.

Liu, M., Wu, W., Gu, Z., Yu, Z., Qi, F., & Li, Y. (2018). Deep learning based on batch normalization for P300 signal detection. *Neurocomputing*, *275*, 288–297.

Liu, Y., Zhou, W., Yuan, Q., & Chen, S. (2012). Automatic seizure detection using wavelet transform and SVM in long-term intracranial EEG. *IEEE Transactions on Neural Systems and Rehabilitation Engineering*, *20*(6), 749–755.

Long, X., Fonseca, P., Foussier, J., Haakma, R., & Aarts, R. M. (2014). Sleep and wake classification with actigraphy and respiratory effort using dynamic warping. *IEEE Journal of Biomedical and Health Informatics*, *18*(4), 1272–1284.

López-Silva, S. M., Giannetti, R., Dotor, M. L., Silveira, J. P., Golmayo, D., Miguel-Tobal, F., ... Martín-Escudero, P. (2012). Heuristic algorithm for photoplethysmography heart rate tracking during athletes maximal exercise test. *Journal of Medical and Biological Engineering*, *32*, 181–188.

Loudon, G., Jones, N., & Sehmi, A. (1992). New signal processing techniques for the decomposition of EMG signals. *Medical and Biological Engineering and Computing*, *30*(6), 591–599.

Lum, P. S., Burgar, C. G., Shor, P. C., Majmundar, M., & Van der Loos, M. (2002). Robot-assisted movement training compared with conventional therapy techniques for the rehabilitation of upper-limb motor function after stroke. *Archives of Physical Medicine and Rehabilitation*, *83*(7), 952–959.

Malik, M., & Camm, A. J. (1995). *Heart rate variability.* Futura Publishing Company.

Martín-Smith, P., Ortega, J., Asensio-Cubero, J., Gan, J. Q., & Ortiz, A. (2017). A supervised filter method for multi-objective feature selection in EEG classification based on multi-resolution analysis for BCI. *Neurocomputing*, *250*, 45–56.

Masetic, Z., & Subasi, A. (2016). Congestive heart failure detection using random forest classifier. *Computer Methods and Programs in Biomedicine*, *130*, 54–64.

McFarland, D. J., & Wolpaw, J. R. (2017). EEG-based brain–computer interfaces. *Synthetic Biology and Biomedical Engineering/Neural Engineering*, *4*, 194–200. https://doi.org/10.1016/j.cobme.2017.11.004.

Melgani, F., & Bazi, Y. (2008). Classification of electrocardiogram signals with support vector machines and particle swarm optimization. *IEEE Transactions on Information Technology in Biomedicine*, *12*(5), 667–677.

Merletti, R., & Di Torino, P. (1999). Standards for reporting EMG data. *Journal of Electromyography and Kinesiology*, *9*(1), 3–4.

Miao, M., Wang, A., & Liu, F. (2017). Application of artificial bee colony algorithm in feature optimization for motor imagery EEG classification. *Neural Computing and Applications*, 1–15.

Minami, K., Nakajima, H., & Toyoshima, T. (1999). Real-time discrimination of ventricular tachyarrhythmia with Fourier-transform neural network. *IEEE Transactions on Biomedical Engineering*, *46*(2), 179–185.

Mishra, A. K., & Raghav, S. (2010). Local fractal dimension based ECG arrhythmia classification. *Biomedical Signal Processing and Control*, *5*(2), 114–123.

Mishra, V. K., Bajaj, V., Kumar, A., Sharma, D., & Singh, G. (2017). An efficient method for analysis of EMG signals using improved empirical mode decomposition. *AEU International Journal of Electronics and Communications*, *72*, 200–209.

Moraes, J., Seixas, M., Vilani, F., & Costa, E. (2002). A real time QRS complex classification method using Mahalanobis distance. In: *Presented at the computers in cardiology, 2002*, IEEE, pp. 201–204.

Muller, P., Contento, N., & Rix, H. (1992). Fractal dimension on ECG. In: *Presented at the engineering in Medicine and Biology Society, 1992 14th annual international conference of the IEEE*, Vol. 3. (pp. 977–978). IEEE.

Nadal, J., & de C. Bossan, M. (1993). Classification of cardiac arrhythmias based on principal component analysis and feedforward neural networks. In: *Presented at the computers in cardiology 1993, Proceedings of the IEEE*, pp. 341–344.

Naik, G. R., & Kumar, D. K. (2010). Twin SVM for gesture classification using the surface electromyogram. *IEEE Transactions on Information Technology in Biomedicine*, *14*(2), 301–308.

Naik, G. R., Selvan, S. E., & Nguyen, H. T. (2016). Single-channel EMG classification with ensemble-empirical-mode-decomposition-based ICA for diagnosing neuromuscular disorders. *IEEE Transactions on Neural Systems and Rehabilitation Engineering*, *24*(7), 734–743.

Naseri, H., & Homaeinezhad, M. (2012). Computerized quality assessment of phonocardiogram signal measurement-acquisition parameters. *Journal of Medical Engineering & Technology*, *36*(6), 308–318.

Naseri, H., & Homaeinezhad, M. (2013). Detection and boundary identification of phonocardiogram sounds using an expert frequency-energy based metric. *Annals of Biomedical Engineering*, *41*(2), 279–292.

Nijboer, F., Sellers, E., Mellinger, J., Jordan, M. A., Matuz, T., Furdea, A., … Vaughan, T. (2008). A P300-based brain–computer interface for people with amyotrophic lateral sclerosis. *Clinical Neurophysiology*, *119*(8), 1909–1916.

Oskoei, M. A., & Hu, H. (2007). Myoelectric control systems—a survey. *Biomedical Signal Processing and Control*, *2*(4), 275–294.

Osowski, S., & Linh, T. H. (2001). ECG beat recognition using fuzzy hybrid neural network. *IEEE Transactions on Biomedical Engineering*, *48*(11), 1265–1271.

Oweis, R. J., Hamad, H., & Shammout, M. (2014). Heart sounds segmentation utilizing teager energy operator. *Journal of Medical Imaging and Health Informatics*, *4*(4), 488–499.

Özbay, Y., Ceylan, R., & Karlik, B. (2006). A fuzzy clustering neural network architecture for classification of ECG arrhythmias. *Computers in Biology and Medicine*, *36*(4), 376–388.

Pang, C. C., Upton, A. R., Shine, G., & Kamath, M. V. (2003). A comparison of algorithms for detection of spikes in the electroencephalogram. *IEEE Transactions on Biomedical Engineering*, *50*(4), 521–526.

Parsaei, H., & Stashuk, D. W. (2013). EMG signal decomposition using motor unit potential train validity. *IEEE Transactions on Neural Systems and Rehabilitation Engineering*, *21*(2), 265–274.

Patidar, S., Pachori, R. B., Upadhyay, A., & Acharya, U. R. (2017). An integrated alcoholic index using tunable-Q wavelet transform based features extracted from EEG signals for diagnosis of alcoholism. *Applied Soft Computing*, *50*, 71–78.

Pattichis, C. S., & Elia, A. G. (1999). Autoregressive and cepstral analyses of motor unit action potentials. *Medical Engineering and Physics*, *21* (6), 405–419.

Pattichis, C. S., Schizas, C. N., & Middleton, L. T. (1995). Neural network models in EMG diagnosis. *IEEE Transactions on Biomedical Engineering*, *42*(5), 486–496.

Pfurtscheller, G., Leeb, R., Keinrath, C., Friedman, D., Neuper, C., Guger, C., & Slater, M. (2006). Walking from thought. *Brain Research*, *1071*(1), 145–152.

Phinyomark, A., Quaine, F., Charbonnier, S., Serviere, C., Tarpin-Bernard, F., & Laurillau, Y. (2013). EMG feature evaluation for improving myoelectric pattern recognition robustness. *Expert Systems with Applications*, *40*(12), 4832–4840.

Podrug, E., & Subasi, A. (2015). Surface EMG pattern recognition by using DWT feature extraction and SVM classifier. In: *The 1st conference of Medical and Biological Engineering in Bosnia and Herzegovina (CMBEBIH 2015). Sarajevo, Bosnia and Herzegovina.*

Prasad, G. K., & Sahambi, J. (2003). Classification of ECG arrhythmias using multi-resolution analysis and neural networks. In: *Presented at the TENCON 2003. Conference on convergent technologies for the Asia-Pacific Region*, Vol. 1. (pp. 227–231). IEEE.

Preedy, V. R., & Peters, T. J. (Eds.). (2002). *Skeletal muscle: Pathology, diagnosis and management of disease*. Cambridge: Cambridge University Press.

Preston, D. C., & Shapiro, B. E. (2012). *Electromyography and neuromuscular disorders: Clinical-electrophysiologic correlations*. London: Elsevier Health Sciences.

Qin, Z., Chen, J., Liu, Y., & Lu, J. (2005). Evolving RBF neural networks for pattern classification. In: *Presented at the international conference on computational and information science*, Springer, pp. 957–964.

Qurraie, S. S., & Afkhami, R. G. (2017). ECG arrhythmia classification using time frequency distribution techniques. *Biomedical Engineering Letters*, *7*(4), 325–332.

Rafiee, J., Rafiee, M., Yavari, F., & Schoen, M. (2011). Feature extraction of forearm EMG signals for prosthetics. *Expert Systems with Applications*, *38*(4), 4058–4067.

Raghav, S., & Mishra, A. K. (2008). Fractal feature based ECG arrhythmia classification. In: *Presented at the TENCON 2008-2008 IEEE region 10 conference* (pp. 1–5): IEEE.

Raghunathan, S., Jaitli, A., & Irazoqui, P. P. (2011). Multistage seizure detection techniques optimized for low-power hardware platforms. *Epilepsy & Behavior*, *22*, S61–S68.

Ramadan, R. A., & Vasilakos, A. V. (2017). Brain computer interface: control signals review. *Neurocomputing*, *223*, 26–44.

Ramović, A., Bandić, L., Kevrić, J., Germović, E., & Subasi, A. (2017). Wavelet and Teager energy operator (TEO) for heart sound processing and identification. In: *CMBEBIH 2017*, Springer pp. 495–502.

Rangayyan, R. M. (2015). *Biomedical signal analysis* (Vol. 33). New York: John Wiley & Sons.

Rao, R. P., & Scherer, R. (2010). Statistical pattern recognition and machine learning in brain–computer interfaces. In *Statistical signal processing for neuroscience and neurotechnology* (pp. 335–367). Elsevier.

Rasheed, S., Stashuk, D., & Kamel, M. (2008). A software package for interactive motor unit potential classification using fuzzy k-NN classifier. *Computer Methods and Programs in Biomedicine*, *89*(1), 56–71.

Rasool, G., Bouaynaya, N., Iqbal, K., & White, G. (2014). Surface myoelectric signal classification using the AR-GARCH model. *Biomedical Signal Processing and Control*, *13*, 327–336.

Rechtschaffen, A., & Kales, A. (1968). *A manual of standardized terminology, techniques and scoring system for sleep stages of human subjects*. Washington, DC: Public Health Service, US Government Printing Office.

Rechy-Ramirez, E. J., & Hu, H. (2015). Bio-signal based control in assistive robots: a survey. *Digital Communications and Networks*, *1* (2), 85–101. https://doi.org/10.1016/j.dcan.2015.02.004.

Richfield, E. K., Cohen, B. A., & Albers, J. W. (1981). Review of quantitative and automated needle electromyographic analyses. *IEEE Transactions on Biomedical Engineering*, (7), 506–514.

Riener, R., Nef, T., & Colombo, G. (2005). Robot-aided neurorehabilitation of the upper extremities. *Medical and Biological Engineering and Computing*, *43*(1), 2–10.

Robertson, G., Caldwell, G., Hamill, J., Kamen, G., & Whittlesey, S. (2014). *Research methods in biomechanics (2nd ed.)*. Windsor, ON: Human Kinetics.

Rodríguez-Sotelo, J. L., Osorio-Forero, A., Jiménez-Rodríguez, A., Cuesta-Frau, D., Cirugeda-Roldán, E., & Peluffo, D. (2014). Automatic sleep stages classification using EEG entropy features and unsupervised pattern analysis techniques. *Entropy*, *16*(12), 6573–6589.

Rosenthal, V. D., Bijie, H., Maki, D. G., Mehta, Y., Apisarnthanarak, A., Medeiros, E. A., … Khader, I. A. (2012). International Nosocomial Infection Control Consortium (INICC) report, data summary of 36 countries, for 2004-2009. *American Journal of Infection Control*, *40* (5), 396–407.

Ryan, D., Townsend, G., Gates, N., Colwell, K., & Sellers, E. (2017). Evaluating brain-computer interface performance using color in the P300 checkerboard speller. *Clinical Neurophysiology*, *128*(10), 2050–2057.

Safara, F., Doraisamy, S., Azman, A., & Jantan, A. (2012). Heart sounds clustering using a combination of temporal, spectral and geometric features. In: *Presented at the Computing in Cardiology (CinC), 2012* (pp. 217–220): IEEE.

Safara, F., Doraisamy, S., Azman, A., Jantan, A., & Ranga, S. (2012). Wavelet packet entropy for heart murmurs classification. *Advances in Bioinformatics*, 6 Article ID 327269.

Safara, F., Doraisamy, S., Azman, A., Jantan, A., & Ramaiah, A. R. A. (2013). Multi-level basis selection of wavelet packet decomposition tree for heart sound classification. *Computers in Biology and Medicine*, *43*(10), 1407–1414.

Sanei, S. (2013). *Adaptive processing of brain signals*. Chichester: John Wiley & Sons.

Sapsanis, C., Georgoulas, G., & Tzes, A. (2013). EMG based classification of basic hand movements based on time-frequency feature. In: *Presented at the control & automation (MED), 2013 21st Mediterranean conference on* (pp. 716–722): IEEE.

Schizas, C. N., & Pattichis, C. S. (1997). Learning systems in biosignal analysis. *Biosystems*, *41*(2), 105–125.

Sejnowski, T. J., & Destexhe, A. (2000). Why do we sleep? *Brain Research*, *886*(1–2), 208–223.

Sellers, E. W., Krusienski, D. J., McFarland, D. J., Vaughan, T. M., & Wolpaw, J. R. (2006). A P300 event-related potential brain–computer interface (BCI): the effects of matrix size and inter stimulus interval on performance. *Biological Psychology*, *73*(3), 242–252.

Sengur, A., Akbulut, Y., Guo, Y., & Bajaj, V. (2017). Classification of amyotrophic lateral sclerosis disease based on convolutional neural network and reinforcement sample learning algorithm. *Health Information Science and Systems*, *5*(1), 9.

Shahriari, Y., & Erfanian, A. (2013). Improving the performance of P300-based brain–computer interface through subspace-based filtering. *Neurocomputing*, *121*, 434–441.

Shavlik, J. W., & Dietterich, T. G. (1990). *Readings in machine learning*. San Mateo, CA: Morgan Kaufmann.

Shimada, T., Shiina, T., & Saito, Y. (2000). Detection of characteristic waves of sleep EEG by neural network analysis. *IEEE Transactions on Biomedical Engineering*, *47*(3), 369–379.

Soleimani-B, H., Lucas, C., Araabi, B. N., & Schwabe, L. (2012). Adaptive prediction of epileptic seizures from intracranial recordings. *Biomedical Signal Processing and Control*, *7*(5), 456–464.

Sörnmo, L., & Laguna, P. (2005). *Bioelectrical signal processing in cardiac and neurological applications*. (Vol. 8). San Diego, CA: Academic Press.

Stickgold, R. (2005). Sleep-dependent memory consolidation. *Nature*, *437* (7063), 1272.

Subasi, A. (2005). Automatic recognition of alertness level from EEG by using neural network and wavelet coefficients. *Expert Systems with Applications*, *28*(4), 701–711.

Subasi, A. (2007). EEG signal classification using wavelet feature extraction and a mixture of expert model. *Expert Systems with Applications*, *32*(4), 1084–1093.

Subasi, A. (2012a). Classification of EMG signals using combined features and soft computing techniques. *Applied Soft Computing*, *12*(8), 2188–2198.

Subasi, A. (2012b). Medical decision support system for diagnosis of neuromuscular disorders using DWT and fuzzy support vector machines. *Computers in Biology and Medicine*, *42*(8), 806–815.

Subasi, A. (2013). Classification of EMG signals using PSO optimized SVM for diagnosis of neuromuscular disorders. *Computers in Biology and Medicine*, *43*(5), 576–586.

Subasi, A. (2015). A decision support system for diagnosis of neuromuscular disorders using DWT and evolutionary support vector machines. *Signal, Image and Video Processing*, *9*(2), 399–408.

Subasi, A., & Gursoy, M. I. (2010). EEG signal classification using PCA, ICA, LDA and support vector machines. *Expert Systems with Applications*, *37*(12), 8659–8666.

Subasi, A., Yilmaz, M., & Ozcalik, H. R. (2006). Classification of EMG signals using wavelet neural network. *Journal of Neuroscience Methods*, *156*(1), 360–367.

Subasi, A., Kevric, J., & Canbaz, M. A. (2017). Epileptic seizure detection using hybrid machine learning methods. *Neural Computing and Applications*, 1–9.

Tavakoli, M., Benussi, C., & Lourenco, J. L. (2017). Single channel surface EMG control of advanced prosthetic hands: a simple, low cost and efficient approach. *Expert Systems with Applications*, *79*, 322–332.

Thakor, N. V., Zhu, Y.-S., & Pan, K.-Y. (1990). Ventricular tachycardia and fibrillation detection by a sequential hypothesis testing algorithm. *IEEE Transactions on Biomedical Engineering*, *37*(9), 837–843.

Thomas, M., Das, M. K., & Ari, S. (2015). Automatic ECG arrhythmia classification using dual tree complex wavelet based features. *AEU International Journal of Electronics and Communications*, *69*(4), 715–721.

Togawa, T., Tamura, T., & Oberg, P. A. (1997). *Biomedical transducers and instruments*. Boca Raton, FL: CRC Press.

Tsai, A.-C., Hsieh, T.-H., Luh, J.-J., & Lin, T.-T. (2014). A comparison of upper-limb motion pattern recognition using EMG signals during dynamic and isometric muscle contractions. *Biomedical Signal Processing and Control*, *11*, 17–26.

Tseng, Y.-L., Ko, P.-Y., & Jaw, F.-S. (2012). Detection of the third and fourth heart sounds using Hilbert-Huang transform. *Biomedical Engineering Online*, *11*(1), 8.

Tsipouras, M. G., & Fotiadis, D. I. (2004). Automatic arrhythmia detection based on time and time–frequency analysis of heart rate variability. *Computer Methods and Programs in Biomedicine*, *74*(2), 95–108.

Usta, N., & Yildiz, M. (2017). Classification of ECG arrhythmia with machine learning techniques. In: *Machine learning and applications (ICMLA), 2017 16th IEEE international conference on* (pp. 546–549): IEEE.

Vasilyev, A., Liburkina, S., Yakovlev, L., Perepelkina, O., & Kaplan, A. (2017). Assessing motor imagery in brain-computer interface training: psychological and neurophysiological correlates. *Neuropsychologia*, *97*, 56–65.

Vidovic, M. M.-C., Hwang, H.-J., Amsüss, S., Hahne, J. M., Farina, D., & Müller, K.-R. (2016). Improving the robustness of myoelectric pattern recognition for upper limb prostheses by covariate shift adaptation. *IEEE Transactions on Neural Systems and Rehabilitation Engineering*, *24*(9), 961–970.

Wang, G., Huang, H., Xie, H., Wang, Z., & Hu, X. (2007). Multifractal analysis of ventricular fibrillation and ventricular tachycardia. *Medical Engineering and Physics*, *29*(3), 375–379.

Wang, J., Gu, Z., Yu, Z., & Li, Y. (2017). An online semi-supervised P300 speller based on extreme learning machine. *Neurocomputing*, *269*, 148–151.

Wang, N., Lao, K., & Zhang, X. (2017). Design and myoelectric control of an anthropomorphic prosthetic hand. *Journal of Bionic Engineering*, *14*(1), 47–59.

Wasilewski, J., & Polonski, L. (2012). An introduction to ECG interpretation. In A. Gacek, & W. Pedrycz (Eds.), *ECG signal processing, classification and interpretation*. London: Springer.

Williamson, J. R., Bliss, D. W., Browne, D. W., & Narayanan, J. T. (2012). Seizure prediction using EEG spatiotemporal correlation structure. *Epilepsy & Behavior*, *25*(2), 230–238.

Wojtczak, P., Amaral, T. G., Dias, O. P., Wolczowski, A., & Kurzynski, M. (2009). Hand movement recognition based on biosignal analysis. *Engineering Applications of Artificial Intelligence*, *22*(4), 608–615.

Wołczowski, A., & Zdunek, R. (2017). Electromyography and mechanomyography signal recognition: experimental analysis using multi-way array decomposition methods. *Biocybernetics and Biomedical Engineering*, *37*(1), 103–113. https://doi.org/10.1016/j.bbe.2016.09.004.

Wolpaw, J. R., Birbaumer, N., McFarland, D. J., Pfurtscheller, G., & Vaughan, T. M. (2002). Brain–computer interfaces for communication and control. *Clinical Neurophysiology*, *113*(6), 767–791.

Wu, H., Talmon, R., & Lo, Y.-L. (2015). Assess sleep stage by modern signal processing techniques. *IEEE Transactions on Biomedical Engineering*, *62*(4), 1159–1168.

Xing, K., Yang, P., Huang, J., Wang, Y., & Zhu, Q. (2014). A real-time EMG pattern recognition method for virtual myoelectric hand control. *Neurocomputing*, *136*, 345–355.

Yan, T., Cempini, M., Oddo, C. M., & Vitiello, N. (2015). Review of assistive strategies in powered lower-limb orthoses and exoskeletons. *Robotics and Autonomous Systems*, *64*, 120–136.

Yan, Z., Wang, Z., & Xie, H. (2008). The application of mutual information-based feature selection and fuzzy LS-SVM-based classifier in motion classification. *Computer Methods and Programs in Biomedicine*, *90*(3), 275–284.

Yang, T.-F., Devine, B., & Macfarlane, P. (1994). Artificial neural networks for the diagnosis of atrial fibrillation. *Medical and Biological Engineering and Computing*, *32*(6), 615–619.

Yong, Z., Wenxue, H., & Yonghong, X. (2009). ECG beats feature extraction based on geometric algebra computational intelligence and software engineering. In: *Presented at the international conference on computational intelligence and software engineering*Vol. 2009. (pp. 1–3).

Young, A. J., Smith, L. H., Rouse, E. J., & Hargrove, L. J. (2013). Classification of simultaneous movements using surface EMG pattern recognition. *IEEE Transactions on Biomedical Engineering*, *60*(5), 1250–1258.

Yousefi, J., & Hamilton-Wright, A. (2014). Characterizing EMG data using machine-learning tools. *Computers in Biology and Medicine*, *51*, 1–13.

Yousefi, R., Nourani, M., Ostadabbas, S., & Panahi, I. (2014). A motion-tolerant adaptive algorithm for wearable photoplethysmographic biosensors. *IEEE Journal of Biomedical and Health Informatics*, *18*(2), 670–681.

Yu, S.-N., & Chou, K.-T. (2007). A switchable scheme for ECG beat classification based on independent component analysis. *Expert Systems with Applications*, *33*(4), 824–829.

Yu, S.-N., & Chou, K.-T. (2008). Integration of independent component analysis and neural networks for ECG beat classification. *Expert Systems with Applications*, *34*(4), 2841–2846.

Yuan, Q., Zhou, W., Liu, Y., & Wang, J. (2012). Epileptic seizure detection with linear and nonlinear features. *Epilepsy & Behavior*, *24*(4), 415–421.

Yuenyong, S., Nishihara, A., Kongprawechnon, W., & Tungpimolrut, K. (2011). A framework for automatic heart sound analysis without segmentation. *Biomedical Engineering Online*, *10*(1), 13.

Zarei, R., He, J., Siuly, S., & Zhang, Y. (2017). A PCA aided cross-covariance scheme for discriminative feature extraction from EEG signals. *Computer Methods and Programs in Biomedicine*, *146*, 47–57.

Zhang, H., Guan, C., & Wang, C. (2008). Asynchronous P300-based brain-computer interfaces: a computational approach with statistical models. *IEEE Transactions on Biomedical Engineering*, *55*(6), 1754–1763.

Zhang, Z., & Sup, F. (2014). Activity recognition of the torso based on surface electromyography for exoskeleton control. *Biomedical Signal Processing and Control*, *10*, 281–288. https://doi.org/10.1016/j.bspc.2013.10.002.

Zhang, Z. (2015). Photoplethysmography-based heart rate monitoring in physical activities via joint sparse spectrum reconstruction. *IEEE Transactions on Biomedical Engineering*, *62*(8), 1902–1910.

Zhang, Z., Pi, Z., & Liu, B. (2015). TROIKA: a general framework for heart rate monitoring using wrist-type photoplethysmographic signals during intensive physical exercise. *IEEE Transactions on Biomedical Engineering*, *62*(2), 522–531.

Zhong, J., & Scalzo, F. (2013). Automatic heart sound signal analysis with reused multi-scale wavelet transform. *International Journal of Engineering Science*, *2*(50), 50–57.

Zhu, G., Li, Y., & Wen, P. P. (2014). Analysis and classification of sleep stages based on difference visibility graphs from a single-channel EEG signal. *IEEE Journal of Biomedical and Health Informatics*, *18*(6), 1813–1821.

Chapter 2

Biomedical Signals

2.1 THE ELECTROENCEPHALOGRAM

2.1.1 Introduction

Electrical signals produced by the brain characterize brain function and the status of the entire body. This provides the motivation to utilize digital signal-processing techniques for electroencephalogram (EEG) signals acquired from the brain of a human subject. The physiological characteristics of brain activities have numerous characteristics related to the original sources, the medium, and their actual patterns. The medium describes the path from the neurons, which are signal sources, to the electrodes, which are the sensors in which some form of source mixture is collected. An understanding of neurophysiological properties and neuronal functions of the brain, together with the working principle underlying the signal generation and acquisition, is useful when dealing with these signals for recognition, analysis, and treatment of brain disorders. EEG presents a way for diagnosing several neurological disorders and abnormalities in the human body. The recorded EEG signals from a human can be utilized for several clinical problems such as:

(a) monitoring alertness, coma, and brain death;
(b) locating areas of damage following head injury, stroke, and tumor;
(c) testing afferent pathways (by evoked potentials);
(d) monitoring cognitive engagement (alpha rhythm);
(e) producing biofeedback situations;
(f) controlling anaesthesia depth (servo anaesthesia);
(g) investigating epilepsy and locating seizure origin;
(h) testing epilepsy drug effects;
(i) assisting in experimental cortical excision of epileptic focus;
(j) monitoring the brain development;
(k) testing drugs for convulsive effects;
(l) investigating sleep disorders and physiology;
(m) investigating mental disorders;
(n) providing a hybrid data recording system together with other imaging modalities.

This list reveals that EEG has rich potential for analysis, fulfilling the need for advanced signal-processing techniques to support the clinician's interpretation (Sanei & Chambers, 2013).

2.1.2 The Nervous System

The nervous system collects and processes information from many parts of the body. It also ensures that both internal and external changes are handled correctly and quickly. The nervous system is commonly divided into the central nervous system (CNS) and the peripheral nervous system (PNS). The CNS consists of the spinal cord and the brain, whereas the PNS connects those two to the body's organs and sensory systems. Therefore, the CNS and the PNS are closely integrated. The CNS processes sensory inputs from the PNS and sends responses to the body's organs via the PNS. The nerves that carry signals to the CNS are called sensory nerves, whereas the nerves that carry signals from the CNS are called motor nerves because these signals can provoke muscle contraction (Sörnmo & Laguna, 2005).

2.1.3 The Brain

The brain, which is an extremely complex structure containing billions of interrelated neurons, represents the control center of the CNS. It controls a vast majority of our movements in conjunction with the spinal cord. Structurally, a human brain includes up to 100 functional area sections and is around 17 cm anterior to posterior. The brain's compartments have specific functions. The back region controls vision, whereas the frontal region controls movement. The number of neurons within a region determines information content and functional capability. More neurons mean more information-processing capability. The brain frequently communicates with different body systems. It also handles processing to control the body's actions. The brain transmits and receives signals through nerves in combination with the spinal cord. The cerebellum, being close to the spinal cord, controls movements by muscle activation. Thus, the cerebellum is crucial for human EEG signal analysis. The motor cortex, where action potentials are initiated, covers a large area of the brain (Deutsch & Deutsch, 1993).

2.1.4 Electroencephalography

The human brain has been the focus of extended research because it represents the most composite organic matter known to mankind. Brain functionality has been examined

Practical Guide for Biomedical Signals Analysis Using Machine Learning Techniques. https://doi.org/10.1016/B978-0-12-817444-9.00002-7

from low-level molecular and chemical properties to high-level characteristics such as learning and memory. It is well known that the brain is related to the generation of electrical activity. In 1875, Richard Caton was able to record microvolt electrical signals on the cerebral cortex of rabbits and dogs. A half century later, Hans Berger recorded electrical "brainwaves" by attaching electrodes to the human scalp for the first time in history. These experiments turned out to be the basis of EEG. EEG today represents an important noninvasive medical tool for analyzing numerous functional brain disorders and for better understanding of the human brain in general. The person interpreting the EEG must draw conclusions based on the frequency, amplitude, morphology, and spatial distribution of the brainwaves. However, the diversity of EEG patterns cannot be completely explained by any single mathematical or biological model available today. Therefore, EEG interpretation basically remains a phenomenological medical discipline (Barlow, 1993) (Fig. 2.1).

EEGs are recordings of the electrical potentials created by the brain, usually less than 300 μV. Since 1924 and Hans Berger's experiment, the study of EEG has been conducted mostly in research facilities and medical settings with the aim of detecting pathologies and epilepsies. For many years, an electroencephalographer, an individual trained to qualitatively differentiate normal and abnormal EEG activity within pretty long EEG records, was the only person qualified for EEG visual interpretation. Researchers and clinicians were left out, covered in a bunch of EEG paper records. However, the arrival of modern powerful computers and related technologies opened a whole new door of possibilities for applying various methods to quantify EEG signals (Bronzino, 1999).

The interpretation is considerably facilitated by a selection of digital signal-processing methods with different purposes such as noise reduction, extraction of information not visually obtainable, or quantification of different signal characteristics. However, this interpretation is far from completely automated. Signal-processing methods belong to one of the two general groups: analysis of spontaneous brain activity and analysis of brain potentials evoked by many cognitive and sensory stimuli. The former certainly provides a better understanding of EEG. Recently, the study of brain function has been improved with the usage of several imaging modalities; positron emission tomography (PET), single photon emission computed tomography (SPECT), and magnetic resonance imaging (MRI) are able to create good resolution for two- or three-dimensional images. These modalities provide comprehensive information on anatomy and blood flow in various brain regions. Even though EEG has lost a portion of its supremacy in the medical routine due to these modalities, it still remains a very powerful tool for the analysis of many diseases such as epilepsy, sleep disorders, and dementia. Moreover, the

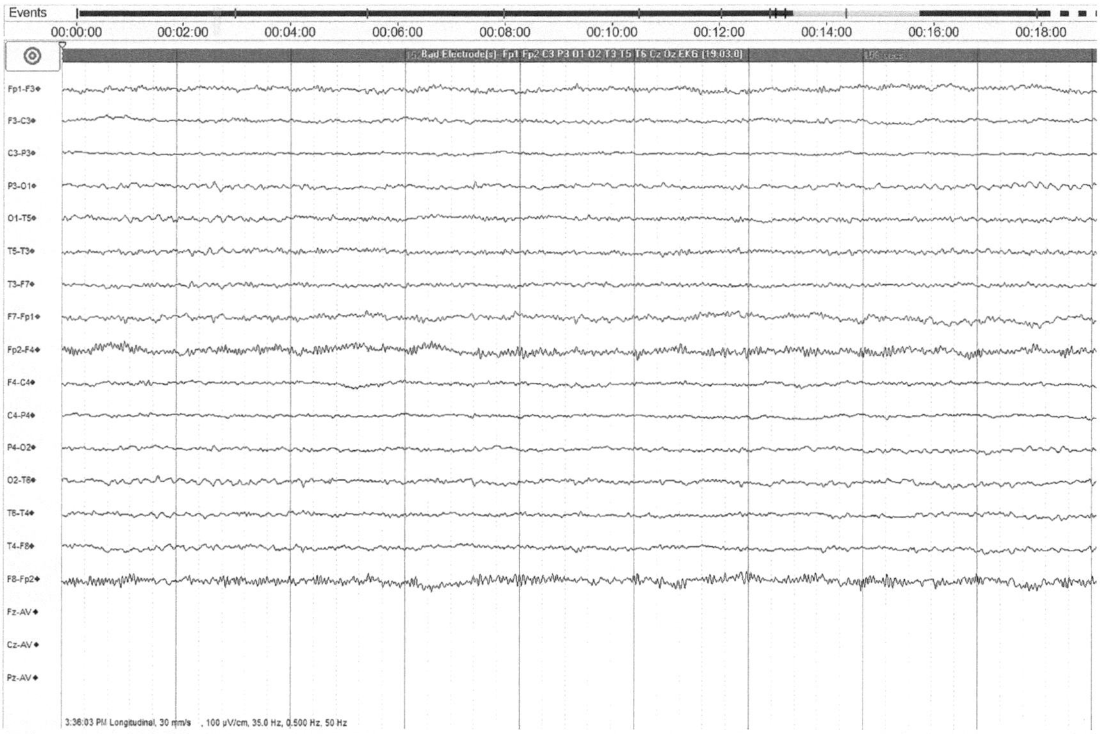

FIG. 2.1 A typical set of EEG signals of normal adult brain activity.

EEG signal is essential for real-time monitoring of progress of patients with encephalopathies or those in a coma. In these applications, the temporal resolution of the EEG is unmatched by the previously mentioned imaging modalities. Furthermore, the overall cost related to recording equipment and skilled workers required for managing the instrumentation is significantly lower than the cost related to neuroimaging. For a simple recording system, the technical demands on instrumentation are quite modest. They are limited to a set of electrodes, a signal amplifier, and a PC for data storage, signal analysis, and graphical demonstration (Sörnmo & Laguna, 2005).

The magnetoencephalogram (MEG) is another noninvasive method that measures the weak magnetic field of mass neural activity by using the SQUID, an extremely sensitive magnetic field sensor. The magnetic field is less distorted by the skull than the electrical potential, which is the main advantage of the MEG over EEG. The MEG was initially supposed to deliver information independent of the EEG (Plonsey, 1972). However, Malmivuo, Suikho, and Eskola (1997) showed that EEG and MEG signals have strong interdependence. These two recording technologies, plus three imaging techniques stated earlier, should eventually be used to complement each other due to their strengths and weaknesses (Dale & Sereno, 1993).

2.1.5 Historical Perspective

In 1875, Richard Caton employed a galvanometer and positioned two electrodes over the scalp of a human subject to record brain activity in the form of electrical signals. The concepts of EEG were then was utilized to represent the electrical neural activity of the brain. In 1920, Hans Berger discovered the existence of human EEG signals and began to work with a string galvanometer. Berger recognized the alpha rhythm as the main element of the EEG signals and the alpha blocking response. During the 1930s, he realized the first EEG recording of sleep spindles and then described the effect of hypoxia on the human brain, the nature of several diffuse and localized brain disorders, and gave an inkling of epileptic discharges. Moreover, he also dealt with cerebral localization, especially the localization of brain tumors. He also discovered a relationship between mental activities and changes in the EEG signals. The history of EEG has been a continuous process that began in the early 1930s and has brought daily improvement to experimental, clinical, and computational efforts for the analysis, recognition, detection, and treatment of a huge number of physiological and neurological anomalies of the brain and CNS. Currently, EEG signals are collected invasively and noninvasively by employing completely computerized systems. The EEG machines involve many signal-processing tools, precise and delicate measurement electrodes, and sufficient memory for long-term recordings of many hours (Sanei & Chambers, 2013).

Berger also noted that these brainwaves exhibited certain regularities rather than being completely random. For instance, he observed that these brainwaves were slower during sleep (<3 Hz) and faster during the waking period (15–25 Hz). He concluded that the brain's activity changes in a familiar fashion if the subject changes status such as from relaxation to alertness. Berger noted a noticeable increase in the amplitude of these brainwaves during sudden seizures and concluded that certain pathologic conditions may significantly affect these brainwaves. Despite all these insights, Berger's paper published in 1929 did not produce much attention. Until British investigators carried out and confirmed similar findings, the efforts of this outstanding researcher were basically overlooked (Bronzino, 1999).

In 1934, the idea of "human brainwaves" was actually accepted and the study of EEG was put on a firm foundation when Adrian and Matthews published their paper confirming Berger's findings. One of their main contributions was the identification of an "alpha rhythm" in the EEG, a regular oscillation at around 10–12 Hz recorded from the occipital lobes. They also found that, if a subject exhibited any type of attention or alertness, the alpha rhythm would disappear. An outstanding historical review of the early era of brain research can be found in an interesting text by Brazier (1968).

2.1.6 EEG Recording Techniques

Scalp recordings of the EEG allow measurement of potential variations between a signal and a reference electrode over time (Kondraske, 1986). The EEG is quite challenging for an untrained eye to interpret when compared to other biopotentials such as the electrocardiogram (ECG). This is partially due to electrode placement and the spatial mapping of functions onto different regions. The International Federation in Electroencephalography and Clinical Neurophysiology, aware that some standardization was needed, accepted the 10/20 electrode placement system (Jasper, 1958). This specific electrode montage employs 21 electrodes attached to the scalp surface at positions determined by certain anatomical reference points. Relative distances between different electrode locations on the skull perimeter are indicated by the numbers 10 and 20, which represent percentages (Fig. 2.2). Both bipolar and unipolar electrodes are used in the medical routine. The latter type needs a reference electrode located distantly or taken as the average of all electrodes (Sörnmo & Laguna, 2005).

The interelectrode distance with the 10/20 system is around 4.5 cm for a typical adult skull. If brain mapping is of importance, then better spatial resolution may be necessary. Brain mapping constitutes a technique where the EEG is represented as a topographic map projected onto

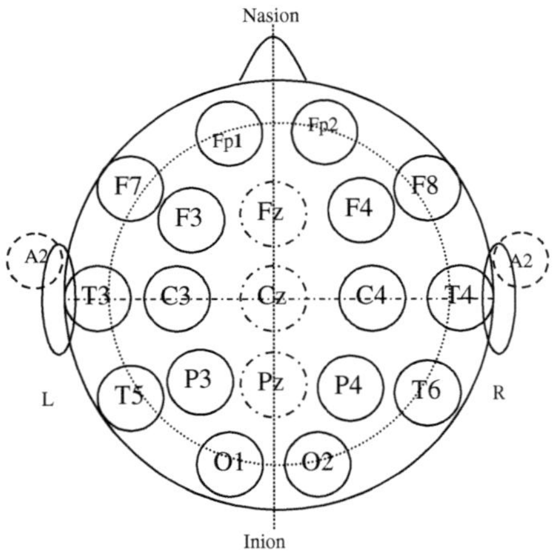

FIG. 2.2 The international 10/20 system for recording clinical EEG. The top of the nose (nasion) and the back of the skull (inion) define the anatomical reference points. The letters F, P, C, T, O, and A represent frontal, parietal, central, temporal, occipital, and auricle, respectively. Even-numbered electrodes are on the right side, whereas odd-numbered electrodes are on the left side. Z (zero) is along the midline.

the scalp. Using a small number of electrodes would incorrectly represent the brain's electrical activity due to aliasing in the spatial domain. To provide a satisfactory amount of detail, it has been specified that more than 64 electrodes should be used in brain mapping applications (Gevins & Bressler, 1988). Besides the standard 10/20 system, electrodes for eye movement monitoring, ECG, and muscle activity are crucial for discrimination of various states of vigilance or behavior. Recording equipment for EEG can be simple or elaborate. Scale electrodes consist of Ag-AgCl disks 1–3 mm in diameter, with lengthy flexible leads plugged into an amplifier. Hair and the trouble of mechanically stabilizing the electrode make it difficult to obtain low-impedance contact at the electrode-skin interface (less than 10 kΩ). Conductive electrode paste helps overcome these difficulties. For mechanical stability, contact cement is usually used to fix gauze over the electrodes. To relieve some stress and pressure, leads are typically taped to the subject. To obtain lower electrode impedance, small abrasions of the skin are sometimes performed. However, sensitive subjects complain about the irritation, which is sometimes followed by pain and infection (Bronzino, 1999).

Electrodes are the main problem for obtaining long-term recordings required for epileptic seizure monitoring. Needle electrodes are sometimes useful. They must be implanted between the skull and scalp surface, significantly increasing the risk of infection. Sometimes a more acceptable method is the use of electrodes with tiny self-contained amplifiers. Although they provide a low-impedance source to interconnecting leads, they are not a cheap option to choose. Regardless of various efforts to simplify electrode application and assure long-term stability, no single technique has been generally accepted. The procedure of measuring impedance between electrode pairs is strongly recommended as good practice, because high impedance causes distortions hardly separated from real EEG signals. Actually, some commercially available EEG devices contain built-in electrode impedance monitors. Typical DC ohmmeters should not be used because they cause a polarizing current resulting in accumulation of noise at the skin-electrode interface. Signal amplitudes of up to 10 μV can be acquired from carefully applied electrodes. With the intention of increasing signal strength to a suitable level usable as input to a recording device, significant amplification with gains approaching 106 is necessary. Because of the length of the electrodes, as well as the noisy surroundings in which recordings are usually carried out, very powerful amplifiers must be used to ensure high-quality EEG recordings. Therefore, differential amplifiers with high input impedance and high common-mode rejection ratios are crucial (Bronzino, 1999).

Special electrically isolated rooms in some facilities reduce environmental electrical noise, especially 50 or 60 Hz alternating current (AC) line noise. However, low-pass filters in the amplifier can help minimize line noise because the information of importance in the EEG lies in frequency bands below 40 Hz. If the low-pass cutoff is beyond the line noise frequency, numerous EEG amplifiers incorporate a notch filter. A notch filter is specific only for frequencies in a narrow band around line noise frequency. Using a dummy source such as a fixed 100 kΩ resistor attached to the electrodes might help remove or reduce the effect of 50 or 60 Hz sources. The output of the differential amplifier, having a dummy source as one of the input signals, represents only interfering sources' influence. If we can minimize the noise at a minimum of a factor 10 less than the EEG signals, we can probably obtain uncontaminated EEG records. Different types of recording equipment can obtain a temporary or permanent record of the EEG. Pen or multichannel chart recorders are the most common recording instruments, as they are an essential part of many commercially available EEG instruments. Recordings are taken with the moving pen on a long sheet of continuous paper that can be chosen to move at one of several constant speeds, which is carefully chosen according to the monitoring situation. Slow speeds of around 10 mm/s are reserved for spotting the spiking related with seizure activity, whereas faster speeds, which can go up to 120 mm/s, are used to detect the manifestation of individual frequency bands in the EEG (Bronzino, 1999).

Another type of recording device is a computer. One or multiple channels of analog EEG signals are constantly sampled at a fixed time interval, after which an analog-to-digital (A/D) converter translates each sample into a digital representation. Each sample is then saved in the computer's memory. The minimum amplitude that can be sampled determines the A/D converter's resolution. This amplitude can be calculated when the range of the A/D converter is divided by 2b, where b equals the number of bits of the A/D converter. For instance, an A/D converter with a voltage range of ±10 V and 16-bit resolution can resolve sample amplitudes as small as 0.3 mV. If a proper combination of amplification and A/D converter sensitivity is selected, the smallest signal required can be resolved together with the prevention of clipping of the largest amplitudes. These samples carry all the information in the waveform if they are picked up at a sampling rate that is twice the highest frequency component of interest in the sampled signal or more. This rule is known as the Nyquist rule. We can use a low-pass filter to make sure that the signal is band-limited and choose a cutoff frequency as the highest frequency of interest. However, the sampling rate that is at least twice the filter's cutoff frequency is typically selected because physical filters do not have ideal characteristics. Moreover, we can use digital filters as soon as the signal is converted to a digital form (Bronzino, 1999).

Due to storage requirements, real-time computer recordings are useful only when the EEG is instantaneously processed or for short-term recordings. A sampling rate of 256 Hz of the 16-bit A/D converter produces 256 new data points every second. An EEG sample with a duration of 16 s contains 4096 data points, whereas a recording period of 1 h produces 921,600 points. Therefore, computer memory turns out to be an important aspect for deciding on the number of channels of EEG activity and the time length in online computer recording applications. However, additional data processing may include (Bronzino, 1999):

- compression for more effective storage;
- producing reduced features set containing only the data of interest;
- feature extraction followed by pattern recognition.

2.1.7 The EEG Measured on the Scalp

The cerebral cortex's group electrical activity is generally denoted as a rhythm due to its oscillatory, repetitive behavior. We cannot quantify the electrical activity of a particular neuron using electrodes on the scalp. The electrical signal is weakened as it propagates toward the electrodes through dense tissue layers. However, we can measure the combined activity of billions of neurons, resulting in a suitably strong electrical field at a depth of several millimeters. Currents flowing during synaptic excitation of the dendrites primarily generate this electrical field. Numerous different EEG rhythms exist. Degrees of concentration, waking, and sleeping are some of the states of a subject affecting these rhythms. The rhythms are conventionally characterized by their relative amplitude and frequency range. The degree of synchronization between the neurons influences the amplitude of the EEG signal. During synchronous excitation, the signals from individual neurons will be summed producing a large amplitude signal on the scalp. A rhythmic EEG signal, having large amplitudes at a certain rate, is produced as a result of repetitive synchronous excitation. In contrast, an arrhythmic look of the EEG with low amplitudes is a result of asynchronous excitation of the neurons and not the number of neurons. Therefore, different EEG amplitudes depend on the time distribution of the neuronal input (Sörnmo & Laguna, 2005).

The thalamus, a part of the brain with neurons possessing pacemaker properties, is partially responsible for the oscillatory rate of an EEG rhythm. These neurons possess the inherent capability to produce a rhythmic firing pattern. Organized interactions between neurons in a specific area of the cortex, such as a feedback mechanism that may happen in a neuronal network, are another reason for the rhythmic behavior. Low frequency/large amplitude rhythms are related to sleepiness and sleep states with no dreaming, whereas high frequency/low amplitude rhythms are related to awareness and dream sleep. This relationship is logical because, when the cortex is involved in information processing whether produced by sensory input or by some internal procedure, the activity level of cortical neurons is comparatively high but comparatively unsynchronized. In other words, each neuron, or very small group of neurons, is dynamically included in a slightly dissimilar feature of a complicated cognitive task; it fires rapidly but not quite at the same time as most of its neighbors. This achieves low synchrony; hence the EEG amplitude is small. On the other hand, during deep sleep, cortical neurons are not involved in information processing, and large numbers of them are phasically excited by a common, rhythmic input. In this circumstance, synchrony is higher; hence the EEG amplitude is large (Bear, Connors, & Paradiso, 1996). Despite several hypotheses, the meaning of various brain rhythms basically remains unclear. However, a really convenient clinical methodology in exploring functional states of the brain lies in the quantification of EEG rhythm characteristics (Sörnmo & Laguna, 2005).

2.1.8 EEG Rhythms and Waveforms

Signals recorded from the scalp have a frequency content not exceeding 40 Hz and amplitudes in a microvolt range up to nearly 100 μV. EEC rhythms are conventionally categorized into five frequency bands. The understanding of these bands depends on the mental state of the subject and his or her age. For instance, the EEG of an adult is significantly different from that of a newborn, containing noticeably lower frequencies. The frequency bands listed in the following section offer a clinically useful classification of different rhythms (Sörnmo & Laguna, 2005), although they are somewhat imprecise (the definitions of bands are taken from Niedermayer (1999c).

Delta rhythm (up to 4 Hz). The delta rhythm has a large amplitude, is usually perceived during deep sleep, and unobserved in the awake normal adult. However, it can be observed in an unhealthy adult, indicating brain disease (encephalopathy) or cerebral damage.

Theta rhythm (4–7 Hz). The theta rhythm arises in certain sleep stages and during sleepiness.

Alpha rhythm (8–13 Hz). The alpha rhythm is most noticeable in relaxed and awake normal subjects with closed eyes. If the eyes are open, this activity is inhibited. Therefore, the occipital regions are areas of the brain where this rhythm has the largest amplitude.

Beta rhythm (14–30 Hz). The beta rhythm, a fast rhythm with low amplitude, is mostly observed in the central and frontal regions of the scalp during some sleep stages.

Gamma rhythm (greater than 30 Hz). The gamma rhythm is associated with active information processing state of the cortex. This rhythm can be observed during finger movements with an electrode at the sensorimotor area connected to a high-sensitivity recording device.

One rhythm does not exist all the time. During long time intervals, an arrhythmic look of the EEG may prevail. Some of these rhythms (primarily gamma rhythms) last only for a few seconds, whereas most of them can last up to a number of minutes (Sörnmo & Laguna, 2005).

Spikes and sharp waves (SSWs) represent transient waveforms with an unpredictable temporal pattern (uncontrollable activity). They point to abnormal neuronal activities that can be usually found in patients with epileptic seizures (Niedermayer, 1999a). Therefore, SSWs are usually denoted as interictal (normal) because they arise between ictal (seizure) events. Both types of waveforms are normally described as having a sudden initial upstroke, with its duration as the only difference. A sharp wave has a duration in the range 70–200 ms, whereas a spike is no longer than 70 ms. Specific artifacts such as cardiac activity can affect a normal EEG so a heartbeat can sometimes be incorrectly classified as SSWs (Sörnmo & Laguna, 2005). Spikes can arise in numerous types of runs or as isolated events. Those runs are denoted as spike-wave complexes consisting of a spike accompanied with a slow wave (Niedermayer, 1999a). Spike-wave complexes take place at rates less than 6 Hz.

Sleep rhythms: The brain has three vital functional states: conscious state (awake or alert state), sleep without rapid eye movement (REM), and sleep with REM. These two sleep states occur numerous times throughout the night. Non-REM sleep represents an "idle" state during which the brain rests and relaxes. A great synchrony between neurons results in slow, large amplitude EEG rhythms during non-REM sleep (Sörnmo & Laguna, 2005). According to the sleep depth, we can divide non-REM sleep state into four different phases (Table 2.1).

TABLE 2.1 Characteristics of Four Non-REM Sleep Phases and REM Sleep

Phase	Depth	Waveforms
1	Sleepiness	Alpha dropout, vertex waves
2	Light sleep	Vertex waves, spindles, K-complexes
3	Deep sleep	Slowing, K-complexes, some spindles
4	Very deep sleep	Slowing, some K-complexes
REM	REM sleep	Desynchronization with faster frequencies

Several transient waveforms characterizing different sleep phases may occur: vertex waves, sleep spindles, and K-complexes. Vertex waves arise during the first two non-REM sleep phases (drowsiness and light sleep) and represent reactions to outside stimuli such as sounds. Sleep spindles represent eruptions of alpha-like activity shorter than a second. The K-complexes are a combination of sleep spindles and vertex waves. REM sleep represents an active state during which the brain is most likely occupied with dreaming. Thus, the EEG during REM sleep looks a lot like the EEG of the waking brain with present beta rhythms. The unbalanced rapid movement of the eyes with closed lids generates a sawtooth pattern in the EEG that can be acquired if the electrodes are located near the eyes. Ictal EEG, characterized by an irregular rhythm with a rapid increase in amplitude, represents the EEG during an epileptic seizure. Additionally, a rapid change in frequency evolving into a spiky wave rhythm indicates the onset of a seizure. The ictal EEG may show significant changeability from seizure to seizure. Therefore, the manual or automatic detection of epileptic seizures is quite a challenging task (Sörnmo & Laguna, 2005).

The EEG has demonstrated its power in monitoring and diagnosing several diseases or disorders such as:

- Epilepsy
- Alzheimer disease
- Huntington's disease
- Sleep disorders
- Brain-computer interface (BCI)

2.1.9 Uses of EEG Signals in Epileptic Seizure Detection and Prediction

EEG analysis has improved considerably with the extensive use of mathematical modeling and machine learning tools. Machine learning tools have also enabled the classification of patterns within the EEG to enhance recognition, making EEG signals valuable for recognition of brain disorders and primary pathologies. Therefore, several studies on the characteristics of EEG signals related to neurological diseases have been carried out (Begg, Lai, & Palaniswami, 2007).

Epilepsy is a neurological disorder that affects more than 50 million people worldwide. It is the second most common neurological disorder (D'Alessandro et al., 2003) after stroke. Epilepsy is a severe disease characterized by temporary changes in the bioelectrical functioning of the brain. These fluctuations cause irregular neuronal synchronization and seizures that affect awareness, sensation, or movement. Epileptic seizures are triggered by extremely synchronized activity of large groups of neurons. Among the wide spectrum of mechanisms that result in such a pathologic activation are brain injuries of various types, physical abnormalities of the cerebral cortex, and physiological conditions resulting in network excitability changes. Depending on the localization and extent of ictal epileptic activity, epileptic seizures can cause a variety of temporary changes in perception and behavior (Meier, Dittrich, Schulze-Bonhage, & Aertsen, 2008). Epileptic seizures represent unexpected bursts of wild electrical activity in a group of cerebral cortex neurons. Due to the location of the focus (origin) of the electrical activity and sequential enrollment of various brain regions, epileptic seizures may manifest in numerous ways. For instance, some sort of auditive or visual sensation follows seizures that has its focus located in the sensory regions of the cortex. A group of neurons with reduced functionality is referred to as the epileptic focus (Sörnmo & Laguna, 2005).

Seizure EEG signals contain characteristic patterns that health professionals use to distinguish them from normal (nonseizure) EEG signals. Therefore, their detection may be used to respond to a forthcoming or ongoing seizure. Also, automated recognition techniques have been tested to decrease the amount of data and enable quicker and more accurate detection of the pathological EEG waveforms that characterize epileptic seizures. In addition, a few techniques have been suggested to identify spikes in the EEG to predict epileptic events (Begg et al., 2007).

During normal conditions, we have a stable relationship between inhibitory and excitatory signals. The former signals prevent neurons from firing and thus decrease the brain's electrical activity, whereas the latter signals force neurons to fire. Nonetheless, an important cause of epilepsy lies in an impaired balance between these two actions. The reason for producing this imbalance or unstable condition hides inside the neurotransmitters responsible for chemical transfer of the signals in the synapse. Consequently, bursts of wild electrical activity will arise when the inhibitory neurotransmitters are inactive, or the excitatory ones are excessively active. This neurotransmitter imbalance can be improved by increasing inhibitory activity or decreasing excitatory activity, which is the main challenge of antiepileptic drugs (Sörnmo & Laguna, 2005).

Epileptic seizures, although rarely causing long-lasting injuries or death, may result in loss of consciousness or cause slight mental confusion. Seizures possess an extremely variable time duration and rate of occurrence. Their duration may range from several seconds to several minutes. There are epileptic patients who experience just a few seizures during their entire life, whereas some of them have several seizures during a single day. Therefore, Niedermayer (1999b) created a scheme to classify seizures into groups based on EEG characteristics. Groups are formed based on the epileptic seizure focus (origin): primary generalized seizures include the whole brain, yet partial seizures start in a limited brain area. The latter group is associated with a single epileptic focus, which cannot be said for the former group. Therefore, we can treat some partial seizures by eliminating a small part of the cortex during surgery. To make sure we correctly bordered the location of the epileptic focus, a series of very systematic and in-depth studies and examinations must be carried out before surgery. A partial seizure may sometimes progress to other brain areas. Such seizure is denoted as a partial seizure with secondary generalization (Sörnmo & Laguna, 2005).

Pathological conditions such as brain injury and tumors, stroke, infections, and genetic factors may all cause epilepsy. But the exact cause of epilepsy cannot be determined for most epileptic patients. Although imaging techniques such as MRI can identify physical defects of the brain, EEG grew into the routine procedure for epilepsy diagnosis. In a fairly silent and dark room, a subject's EEG is recorded for 30 minutes in clinical investigation. To examine the changes in the EEG associated with light, the subject is requested to open and close the eyes. Finally, two activation methods are used to provoke waveforms related to epilepsy. The subject is faced with a flashing strobe light (photic stimulation) or told to breathe quickly and deeply (hyperventilation). Sleep deficiency may also be taken into account. The EEG acquired in clinical investigation may point toward seizures, even though that EEG is often recorded between seizures (the interictal EEG). The manifestation of SSWs in one of the brain's lobes may indicate that partial seizures started in that specific lobe. Likewise, the existence of primary generalized seizures may be spotted if spike-wave complexes are present in both brain hemispheres (Sörnmo & Laguna, 2005).

It is usually required to record the EEG during long periods to catch an ictal EEG. In this situation, the subject is often captured on film in the hospital for a few days. The neurologist can therefore draw a parallel between EEG and visual recordings to improve judgment. This form of recording is denoted as a video EEG. Another form of recording can be carried out at home during at least one day and is denoted as an ambulatory EEG. The EEG is recorded during daily situations using a small digital recording device positioned on a belt around the subject's stomach (Waterhouse, 2003). This is a cheaper method than video EEG and contains both sleeping and waking phases. Itching is occasionally caused because several electrodes need to be attached to the scalp for a long time. If the subject scratches the head, a noise is introduced into the EEG recording. Noise can also be caused by blinking or grimacing, which may sometimes look like waveforms of a physiological origin making the analysis of an ambulatory EEG quite difficult (Sörnmo & Laguna, 2005).

Health specialists use distinctive patterns within ictal (seizure) EEG waveforms to differentiate them from interictal (nonseizure) EEG waveforms. Any form of long-term EEG monitoring creates huge amounts of data as a result. This data requires a lot of time to be properly analyzed. To reduce the amount of data and allow quicker and better detection of abnormal EEG signals related to epileptic seizures, automatic detection systems have been tested. Moreover, a few methods have been suggested to predict epileptic seizures by discovering spikes in the EEG. Several signal-processing techniques that consider the mathematical representation of interictal and ictal data are essential for the design of these detection algorithms (Khan & Gotman, 2003). Noise and artifact elimination is an additional factor that should be taken into consideration. An effective seizure prediction algorithm may warn a patient wearing an ambulatory recording device to consider proper security actions before the seizure occurs (Sörnmo & Laguna, 2005).

A medical application designed to control seizures is comprised of two systems (Winterhalder, Maiwald, Voss, & Aschenbrenner-Scheibe, 2003):

- A seizure prediction algorithm that raises an alarm when it senses an upcoming seizure
- A system that acts to control a seizure

Systems that control seizures include an effective anticonvulsive agent in the epileptic focus or electrical stimulation of the vagus nerve (Elger, 2001). Moreover, one simple alert of an upcoming seizure may be enough for a patient to leave dangerous situations or actions, such as climbing stairs or playing sports. For further development of seizure prediction approaches, three limitations must be addressed (Maiwald et al., 2004).

Valuation measures of seizure prediction approaches differ; also the term "seizure prediction" is not explicitly defined. Researchers haven't paid too much attention to the dependency between sensitivity and the false prediction rate or specificity because the performance of many prediction approaches is characterized only by the sensitivity. High sensitivity is one of the goals when developing seizure prediction methods. A missed alarm for a seizure may cause dangerous situations. However, a lot of false predictions may cause side effects as a result, for example, delivering too much medication or electrical stimulation. A patient may also be unaccepting of a seizure-warning device and stop using it. Therefore, satisfactory sensitivity and specificity must be achieved by seizure-prediction methods. A valuation of several algorithms has shown that high prediction sensitivity cannot be achieved without accepting low specificity (Aschenbrenner-Scheibe et al., 2003; Maiwald et al., 2004; Winterhalder et al., 2003). Comparing the performance of prediction methods is quite difficult because they have been developed and tested on different EEG databases.

Seizure prediction is still qualified as a "long and winding road" regardless of the decades of research (Mormann, Andrzejak, Elger, & Lehnertz, 2007). Generally, seizure prediction methods are based on the extraction of EEG features calculated over a short time window of a few seconds to a few minutes. There exist univariate measures, calculated separately for each EEG channel, and bivariate (or multivariate) measures, which quantify some relationship between two or more EEG channels. Many univariate features have been investigated for seizure prediction (Esteller et al., 2005; Harrison, Frei, & Osorio, 2005; Jerger, Weinstein, Sauer, & Schiff, 2005; Jouny, Franaszczuk, & Bergey, 2005). However, none have succeeded in that task, proving the superiority of bivariate measurements for seizure prediction (Mormann, Kreuz, Rieke, Andrzejak, & Kraskov, 2005).

EXAMPLE 2.1 The following MATLAB code was used to plot normal and epileptic EEG signals. For this example, Bonn EEG dataset will be used. You can download the data from the following website: http://epileptologie-bonn.de/cms/front_content.php?idcat=193&lang=3&changelang=3

Dataset Information: This epilepsy dataset is the widely used EEG dataset that has five sets denoted A–E, each containing 100 single-channel EEG segments of 23.6-s duration. Sets A and B are composed of segments taken from EEG recordings of five healthy volunteers employing a standardized electrode placement scheme (cf. Fig. 2.2). Subjects were relaxed in an awakened state with eyes open (A-Z.zip) and eyes closed (B-O.zip), respectively. Sets C-N.zip, D-F.zip, and E-S.zip were created from an EEG archive of presurgical diagnosis. EEG data collected from subjects who had attained complete seizure control after resection of one of the hippocampal formations were properly identified to be the epileptogenic zone. Set D was taken from the epileptogenic zone and those in set C from the hippocampal formation of the opposite hemisphere of the brain. Although sets C and D include only activity recorded during seizure-free intervals, set E only includes seizure activity. Here, segments were chosen from all recording revealing ictal activity. After 12-bit analog-to-digital conversion, the data were sampled at 173.61 Hz. Band-pass filter settings were 0.53–40 Hz (12 dB/oct.). Exemplary EEGs are depicted in Fig. 2.3. The following MATLAB code is used to plot each signal:

```
%% Ch2_EX_1.m
clc
clear
%Load Sample EEG Data downloaded from the web site
%http://epileptologie-bonn.de/cms/front_content.php?idcat=193&lang=3&changelang=3
load AS_BONN_ALL_EEG_DATA_4096.mat

%Plot Normal EEG Signal
subplot(2,1,1);
plot(A(:,1));
title('Normal EEG Signal')
ylabel('Amplitude')
axis ([0 4096 -inf inf])

%Plot Epileptic Seizure
subplot(2,1,2);
plot(E(:,1))
title('Epileptic Seizure')
xlabel('Samples')
ylabel('Amplitude')
axis ([0 4096 -inf inf])
```

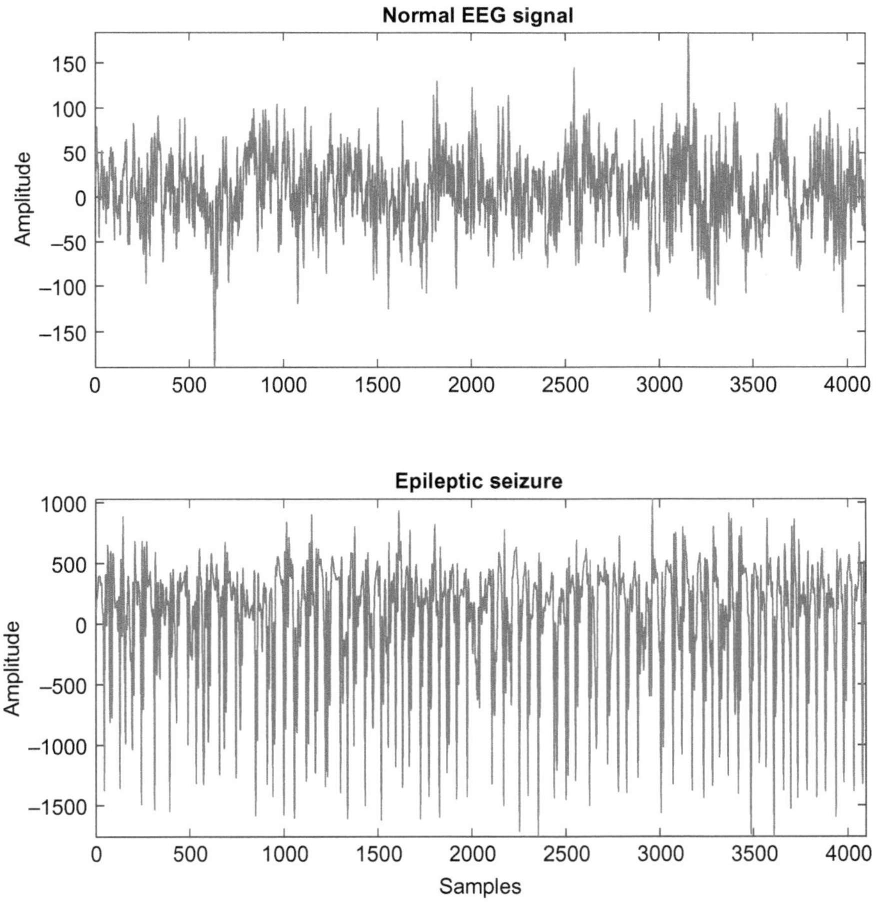

FIG. 2.3 Drawing of normal EEG signal and seizure activity.

EXAMPLE 2.2 The following MATLAB code was used to plot the interictal, preictal, and ictal EEG signals. For this example, EEG dataset collected at the Children's Hospital Boston was used. You can download the data from the following website: https://www.physionet.org/pn6/chbmit/

Dataset Information: This database, collected at the Children's Hospital Boston, consists of EEG recordings taken from pediatric subjects with intractable seizures. Recordings, grouped into 23 cases, were recorded from 22 subjects (5 males, ages 3–22; and 17 females, ages 1.5–19). All signals were sampled at 256 Hz with 16-bit resolution. The international 10–20 system of EEG electrode positions and nomenclature was employed for these recordings. In a few records, other signals are also recorded, such as an ECG signal and a vagal nerve stimulus (VNS) signal. The file RECORDS includes a list of all 664 .edf files involved in this collection, and the file RECORDS-WITH-SEIZURES lists the 129 of those files that contain one or more seizures. Exemplary EEGs are depicted in Fig. 2.4. The following MATLAB code is used to plot each signal:

```
%% Ch2_EX_2.m
clc
clear
%Load Sample EEG Data downloaded from the web site
%https://www.physionet.org/pn6/chbmit/
load AS_PHYSIONET_EEG_DATA.mat

%Plot INTERICTAL EEG Signal
subplot(3,1,1);
plot(EEGINTERICTAL(:,2));
title('Interictal EEG Signal')
ylabel('Amplitude')
axis ([0 2048 -inf inf])
```

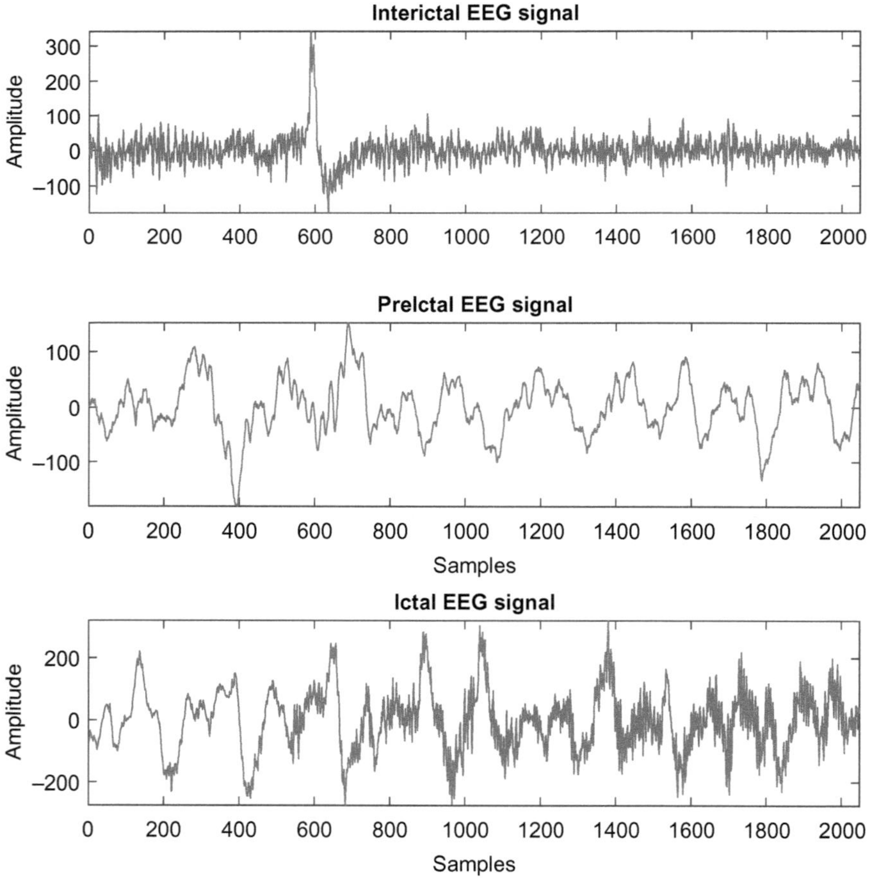

FIG. 2.4 Drawing of interictal, preictal, and ictal EEG signals.

```
%Plot PreIctal EEG Signal
subplot(3,1,2);plot(EEGPREICTAL(:,2))
title('PreIctal EEG Signal')
xlabel('Samples')
ylabel('Amplitude')
axis ([0 2048 -inf inf])

%Plot Ictal EEG Signal
subplot(3,1,3);plot(EEGICTAL(:,2))
title('Ictal EEG Signal')
xlabel('Samples')
ylabel('Amplitude')
axis ([0 2048 -inf inf])
```

2.1.10 Uses of EEG Signals in Brain-Computer Interfacing

A motor function injury can be one of the most disturbing results to the nervous system. Growth of the BCI during the last decades has enabled interactions or control of external devices such as computers and artificial prostheses with EEG. In a straightforward BCI framework, the user is trained to imagine movement of different body parts, and the computer recognizes different patterns of the simultaneously recorded EEG activity. BCI provides a link between the brain and the physical world without any physical contact. The main goal of BCI is to employ the electrical signals produced by the neurons of the brain and generate the signals necessary to control some external devices. The most imperative application is to energize paralyzed organs or bypass the disabled parts of the human body. BCI systems may appear as a unique interaction mode for people with different neuromuscular disorders, such as stroke, amyotrophic lateral sclerosis (ALS), spinal cord injury, and cerebral palsy (Sanei, 2013).

EEG signals are employed in BCI to interpret them into commands for external device control, such as computers, tablets, mobile phones, wheelchairs, and prostheses. BCI framework offers a new connection between the brain and the world that circumvents the human body. Hence, this framework can be employed to improve the life-quality of people with disabilities (Gao, Guan, Gao, & Zhou, 2015). BCI systems are conventionally implemented to support people who have different anomalies such as a locked-in state. Because locked-in patients are reliant on others, BCI can offer them better quality of life with greater autonomy. Recently, researchers were interested in different applications of BCI technology to enhance neurorehabilitation for people with strokes and other chronic disorders (McFarland & Wolpaw, 2017).

The main goal of EEG signal processing in BCI is to implement a framework to create an interaction between a person and an external device such as a robotic arm, a wheelchair, or a speller without physical contact. Such a framework can help handicapped people and enable them to perform their regular activities. Actually, the disease does not affect cognitive functions such as seeing, hearing, and understanding what is occurring, but it affects the control of muscles. Any normal way of interaction with the environment is lost when the motor pathway is lost. BCI offers an option for communication in such situations. Hence, a BCI is a direct interaction between the brain and a machine/computer receiving instructions directly from the brain without physical involvement (Graimann, Allison, & Pfurtscheller, 2009). Fig. 2.5 represents EEG-controlled assistive devices.

BCI systems are categorized as invasive (intracranial) and noninvasive in terms of signal acquisition. Noninvasive BCI systems employ EEG signals to interpret a subject's thoughts or intentions into a device-control signal that enables a disabled person to interact with a computer, wheelchair, or prosthesis. The mental practice of hand or foot movement can be achieved by motor imagery (MI) without executing it (Zarei, He, Siuly, & Zhang, 2017). P300 event-related potential (ERP) (da Silva-Sauer, Valero-Aguayo, de la Torre-Luque, Ron-Angevin, & Varona-Moya, 2016; Gao et al., 2015; Lafuente, Gorriz, Ramirez, & Gonzalez, 2017; Lee & Kim, 2018; Sellers, Krusienski, McFarland, Vaughan, & Wolpaw, 2006; Shahriari & Erfanian, 2013), slow cortical potentials, oscillations in alpha and beta range (Guger et al., 2001; Pfurtscheller et al., 2006), and steady-state visual evoked potentials (SSVEP) (Zhang, Guan, & Wang, 2008) have been employed in BCI systems. The common BCI technique is the P300-based ERP, which happens about 300 ms after an occasional or substantial stimulus is received (da Silva-Sauer et al., 2016; Wolpaw, Birbaumer, McFarland, Pfurtscheller, & Vaughan, 2002). Most of the time, MI approaches are used for BCI systems based on oscillations to produce event-related synchronization (ERS) and event-related desynchronization (ERD) in the α and β frequency bands of the EEG. These kinds of BCI

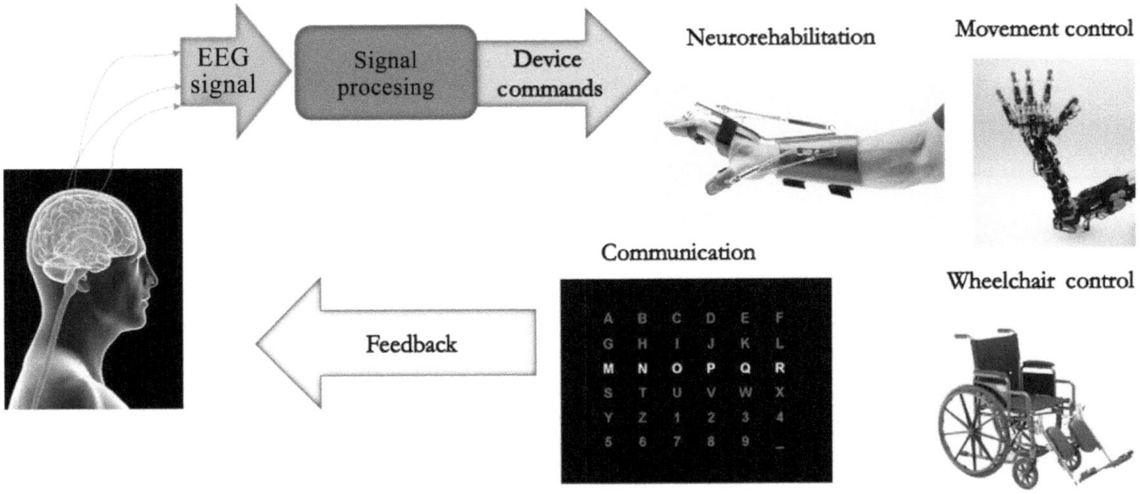

FIG. 2.5 Representation of EEG-controlled assistive devices.

systems are employed for navigation of wheelchairs or cursor control on computer screens. Consequently, many MI-based BCI methods such as foot movement, left/right hand movement, tongue movement, and/or mental counting are implemented. Flickering lights, which carry EEG oscillations with the same frequency as the stimulation source, are employed in SSVEP-based BCI systems. SSVEP schemes employ different stimulation frequencies to control robots or mobile phones. On the other hand, P300-based BCI systems employ the P300 component in the EEG to generally enhance the communication speed of spelling devices (Guger et al., 2009). The computer's actions can be controlled by the subject's thoughts after the training phase. Hence, BCI systems mainly employ EEG signals to control a wheelchair, a robotic arm, or a computer cursor (Sanei, 2013). Farwell and Donchin (1986) initially implemented a BCI system that employed the oddball paradigm to allow the subject to spell words on a computer by successively selecting alphabet letters. In this model, all the alphabet letters and other useful symbols are displayed on the computer screen allowing the subject to select a required character by visually concentrating on the relevant cell for a short time; in the meantime, the column or row representing the focused cell flashes producing a P300 component in the EEG (Shahriari & Erfanian, 2013).

BCI constructed on selective attention needs peripheral stimuli, which can be diverse tangible stimulations, different tones, or flashing lights with different frequencies. In an archetypal BCI system, each stimulus is used to control the BCI system. The subject should pay attention to the related stimulus to choose a command. Visual attention can be realized with P300 potentials or SSVEPs. A P300-based BCI system employs stimuli of flash sequence, which are generally letters or symbols used to control a robot arm, mobile robot, or cursor. Discriminatory attention to a definite flashing letter or symbol produces a brain pattern termed P300 seen about 300 ms after the stimulus. This P300 is detected by the BCI system to define the letter or symbol the user selected (Graimann et al., 2009).

Limb movement or a single muscle contraction alters brain activity, and imagining the movement also alters the so-called sensorimotor rhythms. The decrease of oscillatory activity is ERD and the increase of oscillatory activity is ERS. MI is the movement imagination without performing the actual movement, forming ERD/ERS patterns. Hence, BCI can control by imagining right hand, left hand, feet, and tongue movements. MI BCI is an ability that must be learned contrary to the BCI based on selective attention, and it does not depend on external stimuli (Graimann et al., 2009). Hence, it is crucial to correctly recognize distinct MI classes associated with a realized task using classification methods, because an MI-based BCI system converts the subject's motor intention into a command signal by means of MI states (Siuly, Li, & Zhang, 2016).

EXAMPLE 2.3 The following MATLAB code is used to plot right hand and foot imagery EEG signals. For this example, BCI Motor Imagery Dataset IVa from the BCI competition will be used. You can download the data from the following website: http://www.bbci.de/competition/iii/

Dataset Information: This dataset was taken from five healthy subjects. Subjects were seated in a relaxed chair with arms resting on armrests. This dataset includes only data from the four initial sessions without feedback. Visual cues indicated which of the following three motor imageries the subject should perform for 3.5 s: (L) left hand, (R) right hand, (F) right foot. The presentation of target cues was intermitted by periods of random length, 1.75–2.25 s, where the subject could relax. There were two types of visual stimulation: (1) targets were designated by letters seeming behind a fixation cross and (2) a randomly moving object designated targets. From subjects al and aw, two sessions of both types were acquired, whereas from the other subjects, three sessions of type (2) and one session of type (1) were acquired. Continuous EEG signals are collected from 118 channels, and markers designated the time points of 280 cues for each of the five subjects (aa, al, av, aw, ay). Only cues for the classes "right" and "foot" are provided for the competition. The EEG signals were recorded by employing BrainAmp amplifiers and a 128-channel Ag/AgCl electrode cap from ECI. 118 EEG channels were measured at positions of the extended international 10/20 system. Signals were band-pass filtered between 0.05 and 200 Hz and then digitized at 1000 Hz with 16-bit (0.1 μV) accuracy. The downsampled EEG signals at 100 Hz also exist. The data was collected and prepared by Dornhege, Blankertz, Curio, and Muller (2004). Exemplary EEGs are depicted in Fig. 2.6. The following MATLAB code was used to plot each signal:

```
%% Ch2_EX_3.m
clc
clear
%BCI Competition III Data Set IVa
%Load Sample EEG Data downloaded from the web site
%http://www.bbci.de/competition/iii/
load BCI_III_IVa_aa_train.mat

%Plot Right Hand Imagery Movement EEG Signal
subplot(2,1,1);
plot(w(111,:));
title('Right Hand Imagery Movement EEG Signal')
ylabel('Amplitude')
axis ([0 350 -inf inf])

%Plot Foot Imagery Movement EEG Signal
subplot(2,1,2);
plot(w(112,:))
title('Foot Imagery Movement EEG Signal')
xlabel('Samples')
ylabel('Amplitude')
axis ([0 350 -inf inf])
```

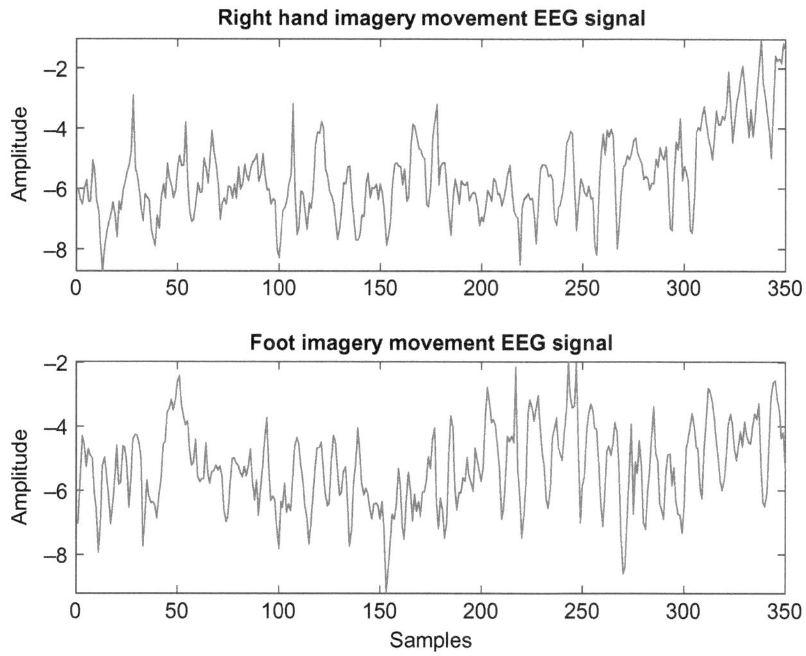

FIG. 2.6 Drawing of right hand and foot imagery EEG signals.

2.1.11 Uses of EEG Signals in Migraine Detection

Migraine is a persistent neurological disorder with highly severe symptoms such as throbbing pain in one or both sides of the brain and sensitivity to the light. Importantly, it is the third most prevalent disease, affecting one in every seven persons. It is worth mentioning that the migraine is globally ranked as the seventh most disabling disease and the first among the neurological disorders (Trust, 2014). However, migraine would often be misdiagnosed due to its overlapping symptoms with other diseases such as tension headache, epilepsy, and strokes. In the past decades, a plethora of studies have been conducted aimed at highly accurate migraine identification. One particular method from these studies has already shown promising results, which is flash stimulation. The basis for this approach is the analysis of the subject's neural responses, recorded with multichannel EEG under flash stimulation at different frequencies for variable amounts of time in seconds (Akben, Subasi, & Tuncel, 2010).

EXAMPLE 2.4 The following MATLAB code is used to plot each EEG signal. For this example, Healthy and Migraineur EEG data collected from Neurology Department at Kahramanmaras Sutcu Imam University was used.

Dataset Information: An 18-channel (10–20 system) Nicolet One Machine was used to record the data. EEG data was collected from Neurology Department at Kahramanmaras Sutcu Imam University from migraineurs and healthy subjects. Fifteen migraineurs (two males, thirteen females) with ages ranging from 20 to 34 years (mean age ± standard deviation: 27 ± 4.4 years), diagnosed according to the diagnostic criteria proposed by the International Headache Society (IHS), and fifteen control subjects (five males, ten females) with ages ranging from 19 to 35 years (mean age ± standard deviation: 26 ± 5.3 years) participated in the experiment. The experiment was conducted in a dimly lit room. None of the participants had taken any drugs before the recordings. All participants were in the interictal (pain-free) state while in a couchant position. 10–20 EEG system used to collect the EEG data. EEG signals were collected at 256 Hz sampling frequency. Each stimulation was 30 s long at a frequency of 4 Hz. This dataset was used in (Akben et al., 2010; Akben, Subasi, & Tuncel, 2011), where a different analysis strategy was considered. Exemplary EEGs are depicted in Fig. 2.7. The following MATLAB code was used to plot each signal:

```
%% Ch2_EX_4.m
clc
clear
%Migraineur EEG Data Set
%Load Sample EEG Data
load Migraine.mat
```

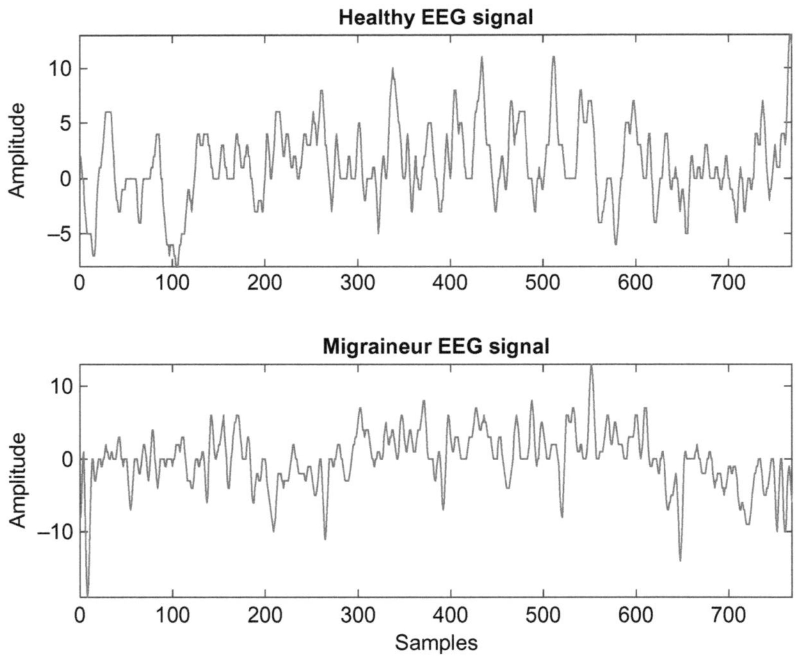

FIG. 2.7 Drawing of healthy and migraineur EEG signals.

```
%Plot Healthy EEG Signal
subplot(2,1,1);
plot(Healthy(:,1));
title('Healthy EEG Signal')
ylabel('Amplitude')
axis ([0 768 -inf inf])

%Plot Migraineur EEG Signal
subplot(2,1,2);
plot(Migraineur(:,1))
title('Migraineur EEG Signal')
xlabel('Samples')
ylabel('Amplitude')
axis ([0 768 -inf inf])
```

2.1.12 Uses of EEG Signals in Source Localization

The brain is separated into a large number of regions that produce a synaptic electric current or local magnetic field. Brain activities can be thought to create signal sources that are either spontaneous or related to the normal rhythms of the brain as a result of brain stimulation or corresponding to physical movements. Localization of brain signal sources using only EEGs has been an active area of research during the last few years. Such source localization is needed to study the brain's functional, mental, physiological, and pathological abnormalities, and even problems corresponding to various body disabilities, and finally to determine the sources of abnormalities such as tumors and epilepsy. Although general localization of brain sources is a problematic task, there are some simple circumstances in which the localization problem can be simplified and accomplished (Sanei & Chambers, 2013).

Based on the corresponding channel position that created the abnormal signals, the focus of epilepsy is defined by clinical diagnosis and surgical results. Hence, EEG is a decisive method for epilepsy evaluation (Xiaoying et al., 2012). The main issue for neurosurgery is the noninvasive initial seizure discharge localization, defining the region of the brain that is the abnormal activity source. Even though epilepsy diagnosis depends on an individual's medical history, EEG is crucial for detection and diagnosis (Xiaoying et al., 2012).

EEG signals recorded from the brain help us understand brain functions and malfunctions (Pereda, Quian Quiroga, & Bhattacharya, 2005; Stam, 2005). The goal of these records is to localize the areas of the brain in which seizures begin and to evaluate if the patient may benefit from neurosurgical resection of these brain areas. Hence, these kinds of intracranial epilepsy patient recordings are a challenging area of applications for the analysis of the signal (Lehnertz, 2008). There is an increasing indication that this interdisciplinary analysis may give helpful information about epileptic focus localization even from the seizure-free interval (Andrzejak, Chicharro, Lehnertz, & Mormann, 2011; Andrzejak, Widman, Lehnertz, David, & Elger, 2006; Varotto, Tassi, Franceschetti, Spreafico, & Panzica, 2012). This is very important, because every seizure can be a health-threatening event, and neurologists are trying to decrease the occurrence of seizures. There is increasing indication of surrogates that may be crucial for an effective evaluation of epileptic seizures from EEG recordings.

Seizure detection in EEG recordings can be helpful in epileptogenic foci localization (Robyn, Mingui, Mark, & Robert, 2000). Different methods are used to analyze the ictal EEG (Corsini, Shoker, Sanei, & Alarcon, 2006; Feldt, Osterhage, Mormann, Lehnertz, & Zochowski, 2007; Swiderski, Osowski, Cichocki, & Rysz, 2007) for seizure detection and prediction. Invasive seizures observed with depth electrode arrays may help localization of seizures. However, even with these techniques, the sources of the seizure cannot be fully defined every time by visual analysis of EEG recordings (Schlkopf & Smola, 2002). For the epileptic focus localization, analysis of EEG signals is very important (Rutkowski, Zdunek, & Cichocki, 2006; Swiderski, Osowski, Cichocki, & Rysz, 2008).

The intracranial EEG recordings taken from subjects suffering from pharmacoresistant focal-onset epilepsy were used. The aim of these signals was to determine the part of the brain to be surgically removed in an epilepsy patient. This results in two different types of signals: the first type recorded from areas of the brain where the changes of ictal EEG signal were noticed ("focal signals") and the second type recorded from areas of the brain that were not included at the seizure's onset ("nonfocal signals") (Ralph, Kaspar, & Rummel, 2012). The ictal areas might create an interictal spike at the time of its onset and, therefore, the clusters covering most spikes start at the most active zone. Localization of this zone and focus of activity in the brain determined by these spike clusters could provide evidence for exact locations of epileptogenic tissue in the surgical evaluation (Pedro et al., 2012).

EXAMPLE 2.5 The following MATLAB code is used to plot focal and nonfocal EEG signals. For this example, epileptic source localization data was used. You can download the data from the following website: http://ntsa.upf.edu/downloads/andrzejak-rg-schindler-k-rummel-c-2012-nonrandomness-nonlinear-dependence-and

Dataset Information: There are two types of files with F and N involving focal and nonfocal signal pairs, respectively. Each Zip file involves 750 separate text files. The number in the file name is related to the index of the signal pair included in this file. Every text file includes one distinct signal pair. The *x*-signal is included in the first column, and the *y*-signal is involved in the second column. Commas divide the two columns. All files have 10,240 rows. Succeeding rows are related to the subsequent samples. The files have no headers (Andrzejak, Schindler, & Rummel, 2012). Exemplary EEGs are depicted in Fig. 2.8. The following MATLAB code was used to plot each signal:

```
%% Ch2_EX_5.m
clc
clear
%Load Sample EEG Data downloaded from the web site
%http://ntsa.upf.edu/downloads/andrzejak-rg-schindler-k-rummel-c-2012-
nonrandomness-nonlinear-dependence-and
load FOCAL_NFOCAL.mat

%Plot Focal EEG Signal
subplot(2,1,1);
plot(focal(1:4096,1));
title('Focal EEG Signal')
ylabel('Amplitude')
axis ([0 4096 -inf inf])

%Plot Non-Focal EEG Signal
subplot(2,1,2);
plot(nfocal(1:4096,1))
title('Non Focal EEG Signal')
xlabel('Samples')
ylabel('Amplitude')
axis ([0 4096 -inf inf])
```

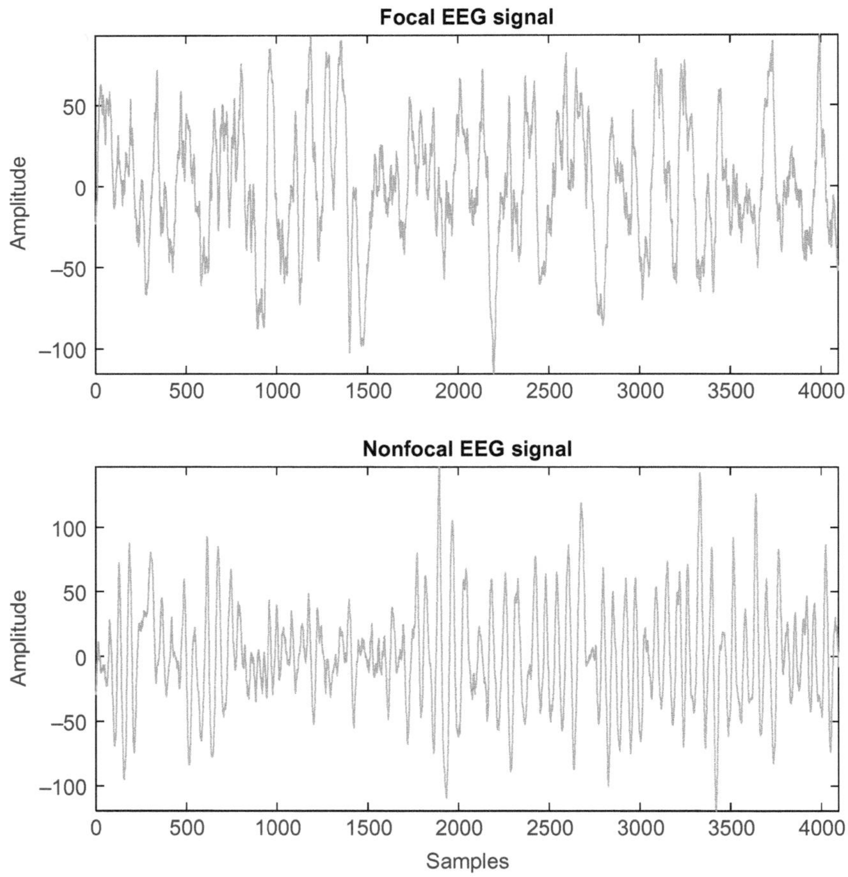

FIG. 2.8 Drawing of focal and nonfocal EEG signals.

2.1.13 Uses of EEG Signals in Sleep

Sleep is a passive condition in which the brain is separated from the rest of the body. EEG allows a researcher to collect electrical activity of the brain during sleep. Lack of sleep affects the function of the brain. With continuous lack of adequate sleep, the part of the brain that controls memory, language, planning, and sense of time can be affected, and judgment weakens. The sleep state is an imperative indication of diagnosing psychological abnormality and mental disease. Although sleep was thought to be a passive state, it is a dynamic procedure, and our brains are active during the sleep. Our mental and physical health, and immune system are affected by sleep. Brain activity during sleep and wakefulness achieves several stimulating and preventative forces produced by the brain. Sleep is essential for the brain to remain healthy. Sleep deprivation results in a drowsy individual unable to concentrate. Moreover, it causes weakening of memory and physical performance, and a reduced ability to implement mathematical calculations and other mental tasks. If sleep deprivation continues, hallucinations and mood swings might occur. Sleep deprivation has a main effect on cognitive functioning as well as on emotional and physical health. Disorders such as sleep apnea that creates extreme daytime sleepiness have been linked to high blood pressure and stress. Sleep helps individuals keep optimal emotional and social functioning while awake; during sleep, rest is given to the parts of the brain that control emotions and social interactions (Sanei, 2013).

There are two different sleep states that alternate in cycles, and each state is considered a different type of EEG activity. Sleep is composed of nonrapid eye movement (NREM) and REM sleep. NREM is further subdivided into four stages: I (drowsiness), II (light sleep), III (deep sleep), and IV (very deep sleep). During the night, the NREM and REM stages of sleep alternate. Stages I, II, III, and IV are followed by REM sleep. A complete sleep cycle from the beginning of stage I to the end of REM sleep generally takes about 1½ h. The resultant sleep is comparatively short and, for most cases, a duration of 10–30 min is enough (Sanei & Chambers, 2013).

NREM Sleep: Stage I (drowsiness) is a very light sleep and is thought as a transition between wakefulness and sleep. Drowsiness demonstrates age-dependent changes. The theta activity of late infancy and early childhood is a noteworthy representative of such ages. Later in childhood and in the decreasing years of life, the drowsiness contains a large amount of slowed activity mixed with the posterior alpha rhythm. However, gradual or abrupt alpha dropout in adults represents the onset of drowsiness. Deep drowsiness contains the appearance of vertex waves. Before the appearance of the first spindle trains, vertex waves occur in the transition from stage I to II. A vertex wave is a compound potential, a small spike discharge of positive polarity followed by a large negative wave that is a typical discharge wave. It might happen as an isolated event with larger amplitude than a normal EEG. In old persons, they become small, inconspicuous, and hardly visible. Another indication of deep drowsiness is the positive occipital sharp transients (POST) of sleep. Spindles (sigma activity) happen mainly in the frontal lobe of the brain. They are recognized as a group of rhythmic waves represented by gradually increasing, then progressively decreasing amplitude. Stage II of sleep arises throughout the sleep period and signifies 40%–50% of the total sleep time. During stage II, brainwaves slow down with infrequent bursts of rapid waves. There is no eye movement during this stage. Slow frequencies ranging from 0.7 to 4 Hz are generally predominant; their voltage is high, with a very prominent occipital peak in small children that gradually falls as age increases. K-complexes occur in stage II and create a substantial response to arousing stimuli. For wave morphology, the K-complex contains an initial sharp component, followed by a slow component that fuses with a superimposed fast component. In stage III, delta waves begin to appear. They are combined with smaller and faster waves. Sleep spindles still exist around 12–14 Hz but progressively vanish as the sleep becomes deeper. In stage IV, delta waves are the main waves seen in the brain. Delta or slow wave sleep (SWS) cannot be recognized easily during routine EEG. Stages III and IV are generally differentiated from each other by the percentage of delta activity. Both of them denote up to 20% of total sleep time. During stages III and IV, all eye and muscle movement stops. It is not easy to wake up somebody during these two stages. If somebody is awakened during deep sleep, s/he does not regulate instantly and frequently feels sleepy and confused for several minutes after waking up (Sanei & Chambers, 2013).

REM Sleep: REM sleep contains 20%–25% of the total sleep, follows NREM sleep, and occurs four or five times during a normal 8- to 9-h sleep interval. The first REM sleep interval of the night might be less than 10 min in duration, whereas the last interval can exceed 60 min. REM sleep is generally related to dreaming. During REM sleep, the eyeballs move quickly, the breathing and heart rates are quick and irregular, blood pressure increases, and there is a loss of muscle tone (paralysis). The brain is very active during REM sleep, and the whole brain metabolism might be increased by as much as 20%. EEG activity acquired in the brain during REM sleep is similar to wakefulness. As the first phase of REM cannot be seen before 60–90 min after the start of sleep, the assessment of REM sleep takes a longer waiting period. The EEG in the REM stage presents low voltage activity with a slower rate of alpha rhythms (Sanei & Chambers, 2013).

Even though physical activities of body are covert to a great extent, internal brain activities during sleep have an influence that cannot be easily understood or explained. A set of highly complex patterns occurs in the human brain during sleep. Recently, a large amount of effort has been devoted to sleep analysis studies and their relationship with other psychological states. So far, very little is known about sleep. Unfortunately, the number of subjects suffering from sleep disorders is not insignificant, and it affects subjects' everyday activities and health conditions in different ways. Objective sleep monitoring and analysis is usually done by an expert examining several sleep stages for an entire night. Currently, polysomnography (PSG) is the "paragon of excellence" in sleep analysis. PSG can be thought as a multivariate system that records several biomedical signals, such as electrooculogram (EOG), EEG, ECG, and electromyogram (EMG), simultaneously. As soon as the biomedical signals are recorded, the next common step could be to agree on how the recorded signal should be scored. Rechtschaffen and Kales (R & K) (Rechtschaffen & Kales, 1968) proposed a guide for classification of sleep that later became a gold standard and is now employed as a tool for classification of sleep stages despite its weaknesses (Himanena & Hasan, 2004; Hirshkowitz, 2000). The R & K standard was further developed by the American Academy of Sleep Medicine (AASM) (Iber, Ancoli-Israel, Chesson, & Quan, 2007). Consequently, sleep scoring differentiates wakefulness (W) and one of two sleep stages: REM and NREM. NREM can be further separated into four different stages enumerated as 1, 2, 3, and 4. Often NREM 3 and NREM 4 are combined into one sleep stage called slow-wave sleep (SWS). However, there might still be certain disagreement in manual scoring of sleep stages as a result of experts' biased decisions and training education. Thus, there is a requirement to create an objective, nonbiased, automated sleep scoring system (Alickovic & Subasi, 2018).

EXAMPLE 2.6 The following MATLAB code was used to plot different sleep stages of EEG signals. For this example, the Sleep-EDF Database (Expanded) was used. You can download the data from the following website: https://physionet.org/pn4/sleep-edfx/

Dataset Information: The PSG dataset includes hypnograms (expert annotations of sleep stages), which come from two studies. The *PSG.edf files are whole-night polysomnographic sleep recordings including EEG (Fpz-Cz and Pz-Oz channels), EOG, submental chin EMG, and an event marker; the SC*PSG.edf files involve rectal body temperature and oronasal respiration. The *Hypnogram.edf files involve annotations of the sleep patterns relevant to the PSGs. These hypnograms includes different sleep stages: W, R, 1, 2, 3, 4, M (Movement time), and ? (not scored). All hypnograms were manually scored by well-trained technicians according to the Rechtschaffen and Kales manual (Rechtschaffen & Kales, 1968). The SC* files (SC = Sleep Cassette) were taken from 79 healthy Caucasians aged 25–101 without any sleep-related medication. Two PSGs of about 20 h each were recorded during two subsequent day-night periods at the subjects' homes. The EOG and EEG sampling frequency was 100 Hz. The submental-EMG signal, oronasal airflow, rectal body temperature, and event marker were sampled at 1 Hz. The ST* files (ST = Sleep Telemetry) were obtained from 22 Caucasian males and females without other medication. Subjects had mild difficulty falling asleep but were otherwise healthy. The PSGs of about 9 h were collected in the hospital for two nights, one of which was after temazepam intake. EOG, EMG, and EEG signals were sampled at 100 Hz, and the event marker at 1 Hz. Exemplary EEGs are depicted in Fig. 2.9. The following MATLAB code was used to plot each signal:

```
%% Ch2_EX_6.m
clc
clear
%Load Sample EEG Data downloaded from the web site
%https://physionet.org/pn4/sleep-edfx/
load SC4001E0-PSG.mat

%Plot AWAKE EEG Signal
subplot(3,2,1);
plot(sleepW(:));
title('AWAKE EEG Signal')
ylabel('Amplitude')
axis ([0 256 -inf inf])

%Plot SLEEP STAGE I EEG Signal
subplot(3,2,2);
plot(sleepI(:))
title('SLEEP STAGE I EEG Signal')
xlabel('Samples')
ylabel('Amplitude')
axis ([0 256 -inf inf])
```

```
%Plot SLEEP STAGE II EEG Signal
subplot(3,2,3);
plot(sleepII(:))
title('SLEEP STAGE II EEG Signal')
xlabel('Samples')
ylabel('Amplitude')
axis ([0 256 -inf inf])

%Plot SLEEP STAGE III EEG Signal
subplot(3,2,4);
plot(sleepIII(:))
title('SLEEP STAGE III EEG Signal')
xlabel('Samples')
ylabel('Amplitude')
axis ([0 256 -inf inf])

%Plot SLEEP STAGE IV EEG Signal
subplot(3,2,5);
plot(sleepIV(:))
title('SLEEP STAGE IV EEG Signal')
xlabel('Samples')
ylabel('Amplitude')
axis ([0 256 -inf inf])

%Plot REM SLEEP EEG Signal
subplot(3,2,6);
plot(sleepR(:))
title('REM SLEEP EEG Signal')
xlabel('Samples')
ylabel('Amplitude')
axis ([0 256 -inf inf])
```

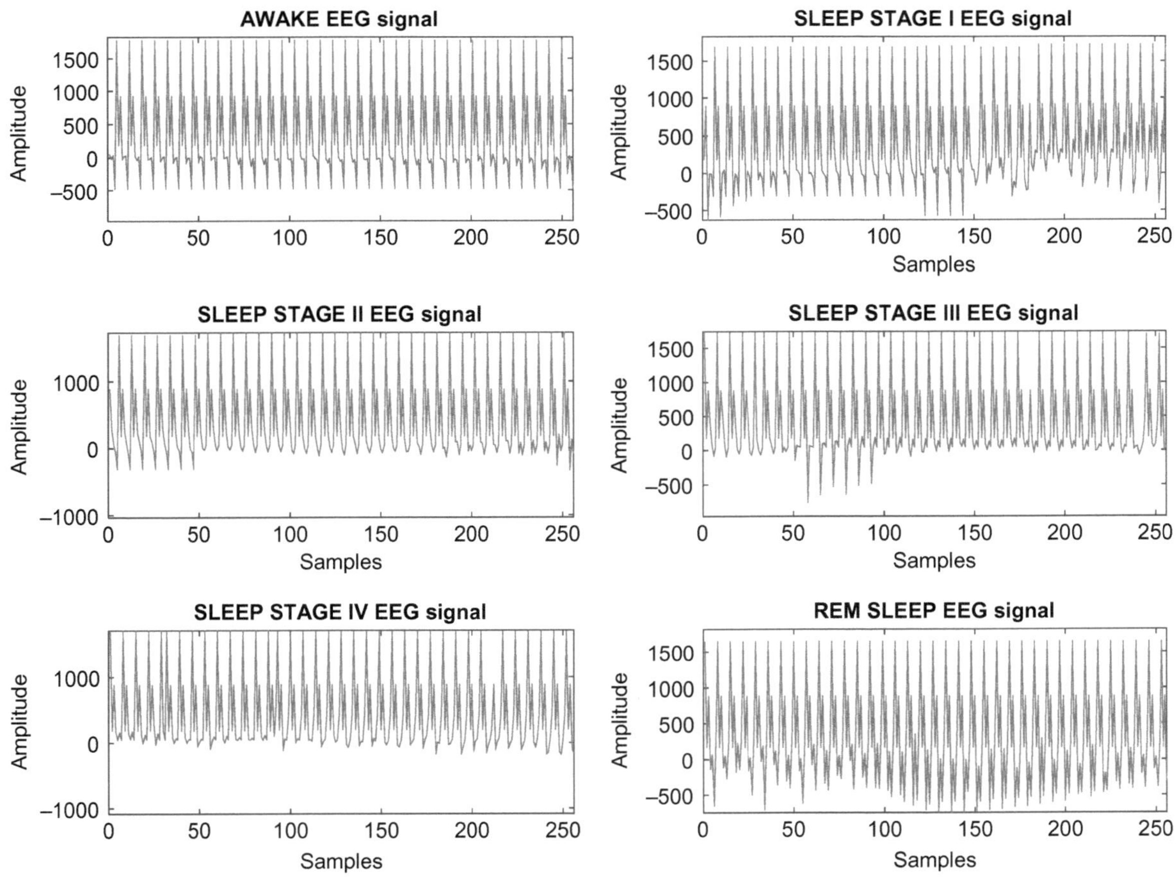

FIG. 2.9 Drawing of EEG signals of different sleep stages.

2.1.14 Uses of EEG Signal for Emotion Recognition

Emotion is a psychophysiological process activated by conscious and/or unconscious perception of a human being and is frequently related to disposition, mood, personality, temperament, and motivation. Emotions play an important role in human communication and can be expressed either verbally through emotional vocabulary or by expressing nonverbal cues such as intonation of voice, facial expressions, and gestures (Koelstra et al., 2012).

More common emotions, such as anger, happiness, surprise, fear, shame, disgust, sadness, interest, contempt, suffering, love, tension, and mirth, have been investigated by researchers in emotion understanding, control, and regulation mainly by analyzing the brain's signals and images. EEG signal processing and analysis techniques have been used for detection and recognition of emotions. Studies in engineering, computer science, neuroscience, and psychology aim to develop devices that identify human affect, and monitor and model emotions. In computer science, affective assessment is a division of the study developing machine learning that deals with the design of systems and devices for processing, interpretation, and recognition of human emotions. Therefore, new techniques can be produced to indirectly assess stress, frustration, and mood through natural interaction and conversation, or by making the computers more emotionally intelligent, especially when responding to a person's frustration in a way that reduces negative feelings, which can be a required research direction in emotion recognition and regulation. EEG helps connectivity, localization, and synchronization of the brain for different emotions to extract and process highly valuable emotion-related information. Recognition and classification of emotional EEG signals are generally complex, and advanced signal-processing methods are required (Sanei, 2013). The goal of affective computing is to fill this gap by detecting emotional cues occurring during human-computer interaction and synthesizing emotional responses. Characterizing multimedia content with relevant, reliable, and discriminating tags is vital for multimedia information retrieval (Koelstra et al., 2012)

Emotion research is an interdisciplinary field that incorporates research in computer science, psychology, neuroscience, and cognitive science. For neuroscience, researchers aim to find out the neural circuits and brain mechanisms of emotion processing. For psychology, there are numerous basic theories of emotion from different scientists, and it is imperative to create computational models of emotion. The modeling and detection of human emotions are the main studies of sentimental computing employing machine learning and pattern recognition techniques. Among numerous methods for emotion recognition, the technique based on EEG signals is more consistent because of its objective evaluation and high accuracy in comparison with other external appearance clues such as facial expression. Understanding the brain's response under dissimilar emotional states can primarily improve the computational models for emotion recognition. Several psychophysiology researches have confirmed the correlations between human emotions and EEG signals (Sammler, Grigutsch, Fritz, & Koelsch, 2007). Furthermore, with the rapid progress of wearable devices and dry electrode techniques (Chi et al., 2012; Grozea, Voinescu, & Fazli, 2011; Huang, Wu, Wong, & Lin, 2015), it is now possible to realize EEG-based emotion recognition from laboratories to real-world applications, such as driving fatigue detection and mental state monitoring (Liu, Chiang, & Hsu, 2013; van Erp, Lotte, & Tangermann, 2012; Zheng & Lu, 2015).

EXAMPLE 2.7 The following MATLAB code was used to plot EEG signals with negative, positive, and neutral emotions. For this example, SEED Emotion Recognition data was used. You can download the data from the following website: http://bcmi.sjtu.edu.cn/~seed/index.html

Dataset Information: The SEED database contains EEG signals of 15 subjects who were recorded while they were watching emotional film clips. Every subject was asked to carry out the experiments in three sessions. There were 45 total experiments in this dataset. These used emotional film clips as stimuli in the experiments. The choice of criteria for film clips were: (a) the length of the whole experiment must not be too long in case it would make subjects fatigue; (b) the videos must be understood without explanation; and (c) the videos must elicit a single desired target emotion. Different film clips (positive, neutral, and negative emotions) were selected to receive the highest match across participates. The length of each film clip was about 4 min. Each film clip was well designed to achieve coherent emotion and maximize emotional meanings. There were 15 total trials for each experiment. There was a 15 s hint before every clip and 10 s feedback after every clip. The order of demonstration was organized in a way that two film clips targeting the same emotion were not presented successively. For the feedback, participants are asked to describe their emotional reactions to every film clip by carrying out the questionnaire instantly after watching every clip. The EEG signals of each subject were saved as separate files with the name of the subjects and the date. Fifteen subjects (7 males and 8 females; Mean: 23.27, SD: 2.37) participated in the experiments. These files included a preprocessed, downsampled, and segmented version of the EEG data in MATLAB (.mat file). The data was downsampled to 200 Hz. A band-pass frequency filter from 0 to 75 Hz was applied. The EEG segments were extracted related to every movie. There are totally 45 .mat files, one for per experiment. Every person carried out the experiment three times within a week. Every subject file includes 16 arrays; 15 arrays included preprocessed and segmented EEG data of 15 trials in one experiment. An array name, LABELS, contains the label of the corresponding emotional labels (−1 for negative, 0 for neutral, and +1 for positive). The EEG was acquired according to the international 10-20 system for 62 channels. Exemplary EEGs are depicted in Fig. 2.10. The following MATLAB code was used to plot each signal:

```
%% Ch2_EX_7.m
clc
clear
%Load Sample EEG Data downloaded from the web site
%http://bcmi.sjtu.edu.cn/~seed/index.html
load EMOTIONSDAT.mat

%Plot NEGATIVE EEG Signal
subplot(3,1,1);
plot(NEGATIVE(1,:));
title('NEGATIVE EMOTION EEG Signal')
ylabel('Amplitude')
axis ([0 4096 -inf inf])

%Plot POSITIVE EEG Signal
subplot(3,1,2);
plot(POSITIVE(1,:))
title('POSITIVE EMOTION EEG Signal')
xlabel('Samples')
ylabel('Amplitude')
axis ([0 4096 -inf inf])

%Plot NEUTRAL EEG Signal
subplot(3,1,3);
plot(NEUTRAL(1,:))
title('NEUTRAL EMOTION EEG Signal')
xlabel('Samples')
ylabel('Amplitude')
axis ([0 4096 -inf inf])
```

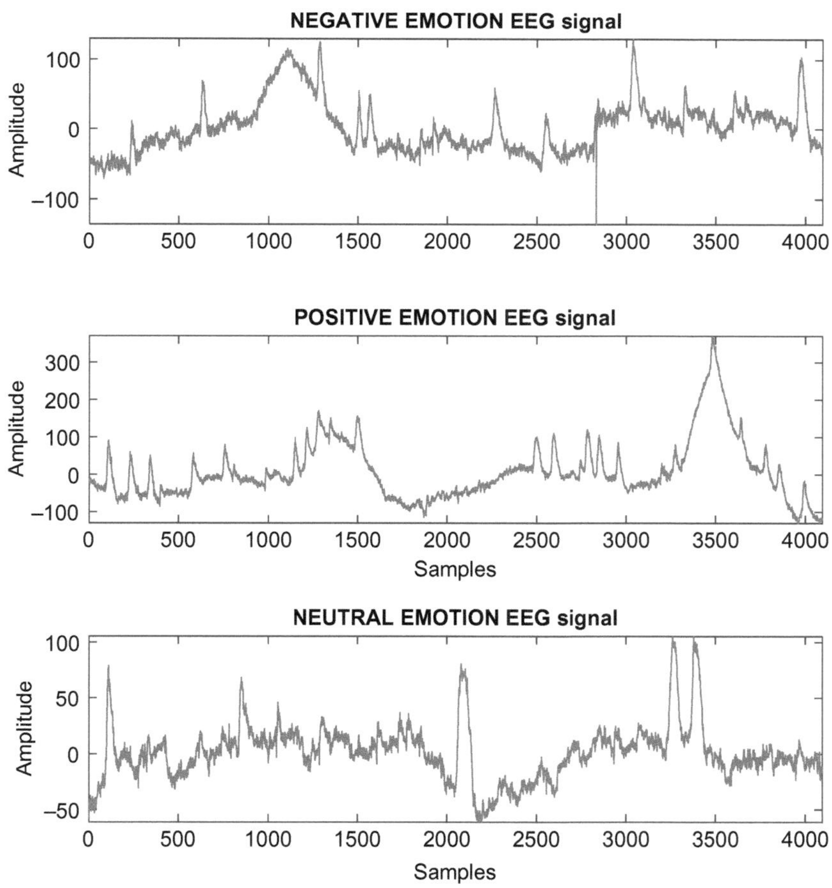

FIG. 2.10 Drawing of EEG signals with positive, negative, and neutral emotions.

2.1.15 Freiburg EEG Database for Epileptic Seizure Prediction and Detection

The Freiburg EEG database was recorded at the Epilepsy Center of the University Hospital of Freiburg, and it consists of invasive EEG recordings of 21 patients suffering from pharmacorefractory (or medically intractable) focal epilepsy. Every subject had between two and five seizures, with an average value of approximately 4.2 seizures per patient. For 21 patients, around 17 Gigabytes of EEG data is available. Totally, 88 seizures including preictal data and at least 24 h of seizure-free (interictal) data are available. More precisely, 509 h of interictal and 73 h of preictal or ictal EEG data is available. Epilepsy doctors visually inspected the EEG data to find the epileptic focus required for surgery. Recordings over several days were sampled using 60 channels with at least 256 Hz (Uni-Freiburg, 2011).

The epileptic focus was located in the hippocampus in eight patients. Eleven patients had the epileptic focus located in neocortical brain structures, whereas two patients had the epileptic focus in both. Intracranial grid, strip-, and depth-electrodes were used to record directly from focal areas. These electrodes were also responsible for obtaining fewer artifacts and high signal-to-noise ratio. Subdural electrodes were implanted via burr holes or open craniotomy, and depth electrodes were implanted stereotactically. Neurofile NT digital video EEG system with 128 channels, 256 Hz sampling rate, and a 16 bit analogue-to-digital converter was used to sample the EEG data. A notch filter removed the line noise (50 Hz). There were gaps of 1–3 s between recording blocks because recording had to be momentarily stopped for technical reasons. Every patient contains "interictal" and "ictal" datasets. The former dataset covers approximately 1 day of seizure-free EEG recordings, whereas the latter is comprised of files with epileptic activity and at least 50 min preictal data. Only 13 patients contained at least 1 day of continuous interictal recordings. Remaining patients did not have a seizure-free day during monitoring. In those cases, interictal EEG data segments with duration of less than 24 h were linked together, producing at least 1 day of seizure-free EEG recordings per patient. The recordings of six electrode contacts were available for each patient: three focal and three extrafocal (Uni-Freiburg, 2011).

All times of day were included when recording EEG data. There were 243 h of total interictal recording that was recorded between 9 a.m. and 9 p.m., whereas the remaining 265 h belonged to the time period between 9 p.m. and 9 a.m. The average time between the end of the interictal period and the following seizure was 9 h 36 min, whereas the minimum time was 1 h 18 min for patient 14. This time interval was greater than 3 h for all the remaining patients. Similarly, the average time between the last seizure and the beginning of the following interictal period was 5 h 18 min (Schelter et al., 2006). For every subject, a qualified epileptologist chose six contacts of all implanted electrodes after visual inspection of the raw data. Three contacts were chosen from the areas included early in ictal/seizure activity, and the remaining three were selected from areas not included or included later during seizure. Moreover, certified epileptologists determined the onset and end of all 88 seizures after visual examination based on detection of distinctive seizure patterns in intracranial EEG prior to clinically manifest seizures (Uni-Freiburg, 2011).

2.2 THE ELECTROMYOGRAM

2.2.1 Introduction

The human skeletal-muscular system first and foremost provides the forces needed to perform a variety of activities. Such a system consists of two subsystems, that is, the nervous system and the muscular system, together creating the neuromuscular system (Begg et al., 2007). Motion and arrangement of limbs are administered by electrical signals propagating back and forth among the muscles and the PNS and CNS (Bronzino, 1999). The CNS administers via nerve signals and muscle excitation. Nerves might be understood as wires conducting electrical currents; the nerve heads (nuclei) begin in the spinal column, and their long axonal bodies enlarge distant and deep, exciting single motor entities in different muscles. The skeletal-muscular system is composed of muscle sets connected to bones via tendons, and a motion is performed once nerve signals produce muscle contractions or relaxations that either attract or repel the bone (Begg et al., 2007).

After pathologic circumstances occur in the motor system, either in the spinal cord, the motor neurons, the muscle, or the neuromuscular junctions, the properties of the electrical signals in the muscle are altered. Cautious recording and analysis of electrical signals in the muscle (via EMGs) may therefore offer important help in finding and diagnosing irregularities not just in the muscles but moreover in the entire motor system. EMG is the recording and understanding of these muscle activity's potentialities. Initially, EMGs were recorded mainly for investigative or diagnostic aims, but with the progression of bioelectric technology, EMGs also turned out to be a basic instrument in accomplishing artificial control of limb motion, such as functional electrical stimulation (FES) and rehabilitation (Bronzino, 1999).

Neuromuscular illnesses consist of abnormalities initially appearing in the nervous system, in the neuromuscular junctions, and in the muscle fibers. These abnormalities have distinct degrees of harshness, from negligible damage to muscle, to amputation caused by neuron or muscle death. In more harsh illnesses such as ALS, death is generally assured. Early and correct diagnosis is essential for enhanced patient prediction and increased chance for rehabilitation. In a majority of cases, clinical testing is insufficient to detect and prevent disorders from spreading (Preston & Shapiro, 2012) because many dissimilar abnormalities could be the result of a specific symptom. The correct diagnosis of a disorder is, for that reason, of supreme significance so more purposeful healing can be achieved. At present, electrodiagnostic (EDX) studies composed of nerve conduction studies (NCS) and EMG are employed for assessment and identification of patients with neuromuscular diseases. EMG was first used as a method of analyzing neuromuscular states established on cell action capabilities throughout muscle activity. Characteristics of the EMG waveform (0.01–10 mV and 10–2000 Hz on average) can indicate the position, etiology, and kind of anatomy. As an example, an EMG pulse interval shows the position and network of the metabolic status of the muscle fiber (Emly, Gilmore, & Roy, 1992), yet irregular spikes can show myopathy. Notice, on the other hand, that EDX methods help doctors in illness discovery, however they are hardly ever helpful for confirming the diagnosis and in difficult situations, more invasive tools like muscle biopsies or more enhanced imaging methods like ultrasound or MRI are necessary. The understanding of EMG is, as a rule, performed by trained expert neurologists who, besides examining EMG waveforms, employ other methods such as needle conduction researches and muscle acoustics. Trouble appears once there are hardly any specialists to meet the demand of patients and, consequently, it is imperative to build up automated diagnostic systems established on EMG readings. This imperative offers a range for the usage of machine learning methods for the discovery and classification of neuromuscular irregularities based on EMG processing. These smart systems will help doctors in the discovery of abnormalities of the neuromuscular system. Other usages of EMG signal processing such as motion control of prostheses, ergonomics, and intention detection will also make use of enhancements in machine learning methods. The goal of smart diagnostic and artificially administered neuromuscular systems is to primarily preprocess the raw EMG signal and, after that, extract the characteristic data or features. Features that may be extracted include time and frequency domain data, Fourier coefficients, autoregressive coefficients, wavelet coefficients, and a wide range of quantities derived from other signal-processing methods. These data may then be employed as input data for classifiers such as decision trees and support vector machines (SVMs), which may classify neuromuscular diseases or find the intention to propagate. The challenge in this case is to invent signal-processing methods that protect or confine significant discriminatory data so as to give a good quality set of features for classification (Begg et al., 2007).

The EMG signal points out the electrical state of skeletal muscles and holds the data related to the construction and purpose of the muscles that initiate various body parts to activate. The EMG signal carries data related to the controller effect of the CNS and PNS on the muscles. Intrinsically, the EMG signal gives an extremely helpful description of the neuromuscular system because numerous pathological procedures, regardless of how they appear in the nervous system or the muscle, are apparent by the changes in the signal's features. In the last few decades, the quality of EMG signal understanding has been significantly enhanced, and with the improvement of new signal-processing methods, we now better understand the muscles' electrophysiology (Sörnmo & Laguna, 2005). EMG analysis is more precise than clinical tests in determining the kinds of muscle fibers included in an abnormality. EMG analysis can show irregular physical sensory nerve phenomenon even though the patient still has normal motor-nerve conduction. Moreover, EMG analysis may help a medical worker in diagnosing without a need for a muscle biopsy (Begg et al., 2007). Progress in engineering and technology have taken EMG beyond the conventional diagnostics to consider its use in various areas such as ergonomics, exercise physiology, rehabilitation, movement analysis, biofeedback, and myoelectric control of prostheses as well (Sörnmo & Laguna, 2005).

2.2.2 The Electromyograph and Instrumentation

The basic recorder of motor unit action potential (MUAP) is the EMG, a tool employed for recording the action potential of the motor unit. There exist two fundamental kinds of EMGs: needle or fine wire EMG and surface EMG (sEMG). EMG tools are comprised of electrodes, a signal acquisition system with audio and visual screens, biosignal amplifiers, and low-pass and band-pass filters. A satisfactory EMG system also results in minimal noise and artifacts. Almost all EMG tools have standardized settings for signal features, for example, gain, common mode rejection ratio, input impedance, and filter bandwidth (Begg et al., 2007; Preedy & Peters, 2002).

2.2.3 EMG Electrodes

Needle electrodes or intramuscular wire electrodes are favored over surface electrodes because they can enter the individual motor units straightaway and evaluate the coming action potentials more precisely. The most widely employed needle electrode is the concentric ring needle with a monopolar single electrode. Other needle electrodes are used by single-fiber EMG (sFEMG) and macro-EMG (Togawa, Tamura, & Oberg, 1997). Different from needle electrodes, wire electrodes are more simply bent and used for kinesiology works. The wires are physically removed after the recording. The deepness and position of injection is significant to acquire correct interpretations. Usually, a ground electrode is employed, and evaluations are made concerning ground voltage. The ground electrode is usually positioned in an area having no muscle motion. The disadvantage of employing these electrodes is the discomfort to patients or test subjects. sEMG electrodes are a more efficient option than needle and wire techniques because they are confined. These electrodes are composed of Ag/AgCl and found in the shapes of bars or discs. It is recommended to apply gel or paste to acquire a desirable interface for signal acquiring contrary to use of dry or separated electrodes that have built-in higher noise extortion (Begg et al., 2007).

2.2.4 Signal Acquisition

EMG data are recorded in the hospital EMG lab by an electromyographer employing either noninvasive sEMG or needle EMG relying on the sensitiveness of the equipment used. The doubted area of irregularities is determined for EMG recording. After that, the EMG is recorded for a planned amount of time, and then the acquired signal is differentially amplified, band-pass filtered, and digitized. All EMG recordings either from sEMG or needle EMG display alterations in voltage potentials because of intracellular activity, either in the nerve or muscle (Preston & Shapiro, 2012). The procedure where an intracellular electric potential is sent via the extracellular fluid and tissue is called volume conduction. Volume conduction can be presented as near-field or far-field potentials via layers of skin or fat. EMG investigation of skeletal muscle contains four elements:

a. *Insertional activity*: The injection and motion of the needle electrode in the muscle appears as explosions of short potentials called insertional activity.
b. *Spontaneous activity*: Spontaneous activity recorded with the unmoving muscle, in other words, with the needle not moving in a relaxed muscle.
c. *MUAP morphology and stability*: When the beginning action has been determined, the MUAP waveform and shape may be processed. The shape of the individual MUAPs may offer ideas as to the etiology of the abnormality, if their physiological generation is correctly interpreted.
d. *MUAP recruitment*: The amount of motor units recruited by the CNS may also be deducted from the recorded EMG waveforms. These can evaluate physical strength and are helpful for the discovery of neuromuscular diseases (Begg et al., 2007).

Routine oscilloscope settings consist of a sweep speed going from 2 to 20 ms/cm and an optimum gain to maximize the recorded potentials without shortening the extremum. The sensitivity varies from 50 to 500 μV/cm for insertional and spontaneous activities and from 100 μV to 1 mV/cm for motor unit potentials. Apparently, a smaller amplification is sufficient for the examination of larger potentials. Most researchers apply the low-frequency filter of 10–20 Hz and high-frequency filter of 10 kHz, and also some choose to decrease the lower limit to 2 Hz or less while finding the waveform of motor unit potentials (Bronzino, 1999).

2.2.5 Signal Amplification and Filtering

Subsequent to amplification, the signal is primary band-pass filtered, sampled, and then low-pass filtered again. Butterworth or Chebyshev filters are commonly used in the equipment, and there is rarely the requirement to define the cutoff or band-pass frequencies. This is principally due to a larger allocation of the EMG signal power known to be part of the 5- to 500-Hz frequency scope of the power density spectra. sEMG recordings are usually in the range of the 10- to 350-Hz band in as much as filtering within 10–150 Hz or 50–350 Hz, for example, it is not desired because the parts of the signal's power higher than 150 Hz and lower than 50 Hz are removed. Intramuscular recording ought to be formed with the suitable rising of the high-frequency cut-off to a low point of 450 Hz or by using a band-pass filter of 10–450 Hz. The previously described levels are perfect filter settings that can be attuned if full condition is provided for working so (Begg et al., 2007).

2.2.6 Signal Digitization

After the analog signal is gathered, amplified, and band-pass filtered, it is ready to be digitized and stored on a computer for further processing. During digitization, sampling is accomplished at the Nyquist rate that is at least double the highest frequency cut-off of the band-pass filter. For instance, if a band-pass filter of 10–350 Hz is employed, the lowest sampling rate ought to be 700 Hz (350×2) or possibly higher to enhance accuracy and resolution. In certain cases, a low-pass filter is employed to primary smooth the EMG signal so that a lower sampling rate (50–100 Hz) may be applied for digitization. The raw unfiltered EMG signal may be stored in the computer applying a sampling rate larger than 2.5 kHz as well. It should be noticed that the signal should be influenced by high-frequency noise and band-pass filtered as explained earlier. The changeover of the analog EMG signal to digital signal may be achieved by using an A/D converter having 8- to 12-bits resolution. Digitized signals may be additionally normalized, for example in force/torque experiments where normalization is performed with respect to the maximum voluntary contraction (MVC). This process asks the subjects to be in good order, and trained to bypass variability and experimental incompatibilities. Normalization can also be accomplished comparative to joint angle or muscle length, velocity of reduction or extension, or applied load (Begg et al., 2007).

2.2.7 The Motor Unit Action Potential

The contraction of muscle tissue activates various body organs such as the eyes and limbs, and propagates liquids inside the body. Being dependent on its aims, the muscle can be classified as skeletal, smooth, or cardiac. Skeletal muscle is linked to the skeleton and creates motion and location of the body, although smooth muscle is located inside the intestines and blood vessels. In skeletal muscle, contraction is regulated by electrical impulses, such as action potentials that travel along the CNS and PNS, and the muscles. The action potentials are sent down the axons of the motor neurons, develop in the brain or the spinal cord, and sent to the muscle fibers. Every motor neuron is linked to muscle fibers via a specialized synapse called the neuromuscular junction, which permits action potentials to excite muscle contraction. Considered jointly, a motor neuron and the fibers to which it is linked (excites) consist of a motor unit and stands for a functional contraction unit. Regarding the aims of the muscle, a single motor unit consists of several muscle fibers or more than a thousand muscle fibers [1, 2]. Muscles controlling precise motions, for instance, those of an eye or a finger, have a fewer number of muscle fibers per motor unit than muscles regulating gross motions, for instance, those while running and exercising. The contraction of a muscle fiber is started at the moment when neuronal action potentials arrive at the neuromuscular junction and fire action potentials that disperse on the volatile membranes of the muscle fiber. MUAP outcomes from spatial and temporal accomplishments are the sum of the single action potentials as they disperse via the various muscle fibers of an individual motor unit. The EMG signal outcomes, on the other hand, are from the addition of the various MUAPs close enough to the recording electrode. The amount of MUAPs inside the pick-up (discovery) area of the electrode is dependent on the chosen type of electrode and is generally larger than one because fibers of various motor units are positioned in the whole muscle. Motor unit achievement is a basic muscular procedure where the force applied by muscle contraction is regulated by the CNS using spatial and temporal accomplishment of motor units. Spatial accomplishment represents the force being increased by using more motor units, whereas temporal accomplishment means that force is raised by the firing of action potentials at quicker rates. Even though both kinds of accomplishment can happen at the same moment, spatial accomplishment prevails from lower levels of muscle contraction until a majority of motor units have been achieved. At high levels of muscle contraction, temporal accomplishment prevails and causes the motor units to have firing rates around 50 Hz or quicker. A high firing rate means that each MUAP waveform cannot be distinguished any further because temporal superimposition, and the outcoming EMG signal, displays a noise-like, random occurrence, thought as an interference pattern. Motor units are commonly achieved considering their magnitude, beginning with the most diminutive motor units, with bigger units increasingly achieved for raising the capability of muscle contraction. As a consequence, primary activation is not strong and is preceded by a smooth incline in strength via the consecutive summation of larger, more powerful motor units. Investigation of motor unit accomplishment at low levels of muscular contraction is significant for identification of abnormalities connected to the nervous system (neuropathy) and muscular tissue (myopathy). The accomplishment pattern is generally determined in terms of the firing rate of the initial motor unit when the second motor unit is reached. Without any problems, the accomplishment rate is distinguished because of the occupancy of a fewer number of active motor units. A neuropathic accomplishment pattern is defined by a firing rate of the first motor unit that passes 20 Hz prior to the second motor unit appearance. However, a myopathic muscle is defined by an accomplishment pattern where fewer motor units are already activated at minimum self-imposed contraction to generate the required force; therefore, a myopathic accomplishment pattern is linked to various MUAPs of various shapes (Sörnmo & Laguna, 2005).

The activities to determine a motor unit potential consist of amplitude, rise time, duration, phases, stability, and territory. A broad range of neuromuscular abnormalities changes the waveform in various but distinguishing combinations. Therefore, these kinds of abnormalities assist to differentiate primary muscle disorders from disorders of neuromuscular junction and lower motor neurons. A lessening in spike duration and amplitude describes motor unit potentials in myopathies connected to a random loss of single fibers (Kimura, 1989).

The data in an intramuscular EMG signal can be divided into two simple types, that is to say, morphology and firing pattern, a most helpful method to distinguish the two when models are developed and processing algorithms for the EMG signal. Some essential characteristics of each type can be summarized as follows:

Morphology: A morphologic characterization of the MUAP waveform is necessary to perform typical clinical investigation and consists of the next parameters taken from the intramuscular EMG. Amplitude is the peak-to-peak amplitude of the MUAP and is in the range of 0.25–5 mV for a healthy case. The amplitude is evaluated by the amount of active muscle fibers inside the nearest direct electrode. Generally, large amplitudes are connected to neuropathies and small amplitudes to myopathies (Sörnmo & Laguna, 2005). Clinical education and computer simulation show that surface measuring can assist in distinguishing neuropathy from myopathy. Considering the amplitude, a larger amount of muscle fibers inside a 2-mm radius of the electrode tip provides this evaluation. The value, on the other hand, alters noticeably, with a tenuous activity of the recording electrode chiefly representing an amplitude alteration. The ratio between the surface and amplitude evaluates the "thickness" of the potential, which changes much less with alterations in electrode location. The mix of amplitude and surface/amplitude ratio significantly enhances discrimination, finding around 70% of neurogenic alterations, compared to just 15%–30% by duration criteria alone (Kimura, 1989).

The number of phases of a MUAP waveform shows the degree of disarrangement of action potentials traveling in various individual fibers. Although the action potential of a single fiber consists of just one or two phases, the recording electrode discovers the addition of action potentials of numerous fibers that, because of changing degrees of disarrangement, may produce an MUAP with many phases. Polyphasic MUAPs consist of more than four phases and are largely seen in neuropathic and myopathic circumstances (Sörnmo & Laguna, 2005). In neurogenic disorders, polyphasic MUAPs appear because of the slow conduction velocity in infantile nerve sprouts or slow conduction velocity in reinnervated but still atrophied muscle fibers. Alterations in muscle fiber size may also result in polyphasic MUAPs in myopathic disorders (Bronzino, 1999).

The number of phases, evaluated by determining and counting positive and negative peaks to and from the baseline, is equal to the amount of baseline crossings plus one. Usually, motor unit potentials consist of four or less phases. Polyphasic motor unit potentials, having more than four phases, outcomes from desynchronized discharges of individual muscle fibers, likely showing fiber size changeability more than random fibers loss. These potentials do not pass 5%–15% of the overall population in a healthy muscle, if they are recorded using a concentric needle electrode. Polyphasic activities usually occur when a monopolar needle is used, although no research has found the precise occurrence. Some action potentials demonstrate a couple of "turns" or directional alterations with no baseline crossing. These rough action potentials or, less suitably, complex or pseudopolyphasic potentials, show desynchronization between discharging muscle fibers as well (Kimura, 1989).

Duration is termed as the onset and end of the MUAP waveform, that is, the time interval among the initial and last appearance of the waveform already passing the defined amplitude threshold, for example, 5 μV. The duration is dependent on the amount of muscle fibers inside the motor unit and rises as the amount of fibers rises. The average MUAP duration is 2–10 ms. Yet because the motor unit region might be larger than the pick-up range of the electrode, MUAP duration does not give data related to the overall magnitude of a large motor unit. MUAP duration will rise if a motor unit has a increased amount of fibers because of the reinnervation (Bronzino, 1999).

Duration is measured between the first take-off and arrival to the baseline. Different from the spike amplitude, and completely determined by a very minor amount of muscle fibers close to the electrode, the duration of a motor unit potential reveals the movement from a larger amount of muscle fibers inside the interest zone of the recording are in a concentric needle, spreading 2.0–2.5 mm from the core. For that reason, a minor needle change or rotation does not affect the duration as much as the amplitude. The duration usually is the range between 5 and 15 ms, with regard to the subject age. In one study, the values obtained at the ages of 3 and 75 years were 7.3 and 12.8 ms in biceps brachii, 9.2 and 15.9 ms in tibialis anterior, and 4.3 and 7.5 ms in the facial muscles, respectively. Another study working with four proximal and distal muscles of the upper and lower limbs in 111 healthy subjects between 20 and 80 years of age discovered no noticeable rise of the average duration before the age of 55. Cases aged more than 55 revealed a minor affinity to larger duration. The usage of a wide-open amplifier band-pass, joined with improved signal-to-noise ratio (SNR), outcomes in a much lengthier duration, up to 30 ms recorded either with a single-fiber electrode or a macroelectrode. Considering this condition, the net time of single action potential from end-plate zone to musculotendinous junction can show complete duration of MUAP (Kimura, 1989).

Rise time is a rising function of the space between the electrode and the nearest active muscle fiber. A petite rise time combined with minor MUAP amplitude can, for that reason, show that the amplitude is decreased because of fiber atrophy more than because of a huge space between the electrode and the nearest fiber. The rise time is measured as a time lag from the first positive peak to the following negative peak, and assists to find an approximate length among the recording tip of the electrode and the discharging motor unit. A far-away unit takes a higher rise time due to the resistance and capacitance of the overriding tissue performing as a high-frequency filter. This type of discharge is accompanied by a dull sound, showing the requirement to relocate the electrode nearer to the source. Overall, a rise time smaller than 500 μs confirms recording from inside the motor unit region, however there are some arguments for reduced border criteria. This kind of a motor unit results in a sharp, crisp sound at the loudspeaker that gives a significant idea to the nearness of the unit to electrode. The evaluation of the rise time approves the appropriateness of the recorded potential for numerical amplitude evaluation (Kimura, 1989).

Area indicates the amount of fibers neighboring the electrode; on the other hand, different from MUAP amplitude, MUAP area is determined by MUAP duration and is, for that reason, affected by fibers in a larger area matched to that of MUAP amplitude (Bronzino, 1999).

Turns stands for the complication of the MUAP, such as the number of phases; on the other hand, because a usable turn does not need a baseline crossing such as valid phase, the number of turns is more elusive to alterations in the MUAP waveshape. To differentiate accurate turns from signal noise, consecutive turns have to be offset by a minimal amplitude difference (Bronzino, 1999).

In a healthy subject, MUAP morphology generally stays stabilized from discharge to discharge. Destabilized morphology can still appear when the neuromuscular junction is broken, outcoming in known nerve action potentials obstructed from transmitting. Therefore, some of muscle fibers of the motor unit might not be actuated. In these kinds of cases, MUAP may in reality increase twice from one discharge to following one and, from time to time, can be followed by alteration in the number of phases. Gradual alterations in MUAP morphology can be seen when the biochemical characteristics of a motor unit alter, for instance, happening during muscle tiredness, or when the relative position of the electrode versus the motor unit alters (Sörnmo & Laguna, 2005).

Firing pattern: A muscle contraction is maintained via continual stimulation of the motor units, with every motor unit connected to a specific train of MUAPs. As the number of active motor units rises, the recording electrode finds a rising amount of concurrent MUAP trains, and single MUAP trains turn to be harder to distinguish from each other. The MUAP's repeating timing determines the firing pattern and carries data on how the CNS regulates the motor unit. The firing pattern is generally regular (rhythmic) having quite little changeability of consecutive interfering intervals, for example, the times among consecutive MUAPs have the same source while contracting. In the case of regular firing patterns, it is frequently satisfactory to describe the pattern by mean firing rate, characterized as the inverse of the average duration of the interfering intervals. Under voluntary muscle control, a motor unit ordinarily starts firing at a rate of 4–5 Hz, even though lower rates can be seen because of the induced activity behind intentional control, and its rate raises as the force increases. The biceps brachii might illustrate the average firing rate for normal muscle where the frequency of the motor units is between 7 and 12 Hz at the lowest level of muscle contraction. This level is defined as a percentage of MVC that stands for a subject-specific contraction measure. While rising isometric I force, the average firing rate of the biceps brachii rises to obtain an upper limit of around 20 Hz at 100% MVC. When evaluating motor unit irregularities, it is generally essential to consider that various muscles have their own particular firing rates, and also the rates when additional motor units are to be recruited. Although the average firing rate is a basic parameter for investigation of intramuscular EMGs, it is not desirable when describing abnormal firing patterns seen while there are dynamic contractions with fast alterations in the force level. The identical restriction also refers to the description of firing patterns related to impaired neuromuscular transmitting, where the interfering intervals expose augmented changeability. Therefore, alternative rhythm parameters are guaranteed that more suitably show the transient characteristics of a firing pattern (Sörnmo & Laguna, 2005).

2.2.8 Myoelectric Signal Recording

Myoelectric activity is evaluated invasively by employing a needle electrode or noninvasively by positioning an area electrode on the skin covering the muscle. The needle EMG is a regular clinical way applied for diagnostic goals, whereas the sEMG is studied and computerized in many applications such as prosthesis control, ergonomics, movement and gait analysis, and sports medicine. The needle EMG is evaluated by placing a needle electrode through the skin straight into the muscle. This method gives a high-resolution, localized characterization of the muscle's electrical activity, although it is relatively unpleasant for the patient. The needle electrode's pick-up area depends on its particular system and can comprise of only one or two muscle fibers. The monopolar needle electrode asks for a reference electrode positioned further away from the needle insertion point at an electrically neutral site such as over a bone. The concentric needle electrode bypasses the reference electrode by citing the electrode active surface to the needle cannula. Both types of needle electrodes are employed in clinical EMG signal recordings. By inserting several needle electrodes at various positions, a more careful description of a large muscle's electrical activity may be received. The sEMG shows the gross activity generated by a huge number of motor units. Its spatial resolution is more controlled than that of the needle EMG, and the MUAP high-frequency content becomes smoothed. The positioning of surface electrodes is dependent on the muscle under consideration and includes factors such as muscle fiber placement, anatomical landmarks, and minimization of electrical cross-talk from other muscles. The sEMG is principally employed when the activation time and the signal amplitude consist of the desirable data, for instance, regarding researches of motor behavior or myoelectric prosthesis regulation. The sEMG does not, in general, permit the discovery of single MUAPs, even though MUAP trains can be found at low levels of muscle contraction. An additional improvement of the surface electrode method is the linear electrode array, created to give spatial characterization of myoelectric activity. It is possible to record the sEMG at lower sampling rates than the needle EMG because the middle tissue among the motor units and the surface electrode works as a low-pass filter of the electrical signal. Most of the spectral power of the sEMG is less than 400–500 Hz, meaning that a sampling rate of 1 kHz or higher is needed. In the case of the needle EMG, the sampling rate should be selected so that various MUAP waveforms that contain frequencies ranging up to 10 kHz are correctly regenerated; thus, a frequently used sampling rate is 50 kHz. The recording of an EMG is connected to various forms of noise and artifacts that, to different degrees, restrain the quality of the signal. In the sEMG, electrode motion artifact outcomes from relative propagation among the skin and the electrode and by distortion of the skin under the electrode. Because this form of artifact is low frequency in nature, having majority of its spectral components less than 20 Hz, it can be decreased by using high-pass filtering with no important changes in the EMG signal's spectral content. Like in the conditions when an EEG signal is recorded, inadequate protection of the EMG electrode cable makes it sensitive to electromagnetic fields resulting from currents propagating in close electrical devices or power lines. As a consequence, the recorded EMG signal will be comprised of power line interference at a frequency of 50–60 Hz. Filtering methods are used for removal of a narrowband noise component. Nevertheless, these filtering methods can be applied for the processing of EMG signals as well (Sörnmo & Laguna, 2005).

2.2.9 Neuromuscular Disorders

Neuromuscular disorders are commonly anomalies found around the PNS. They can be categorized depending on the location and reason of the disorders. Two main disorders are neuropathy and myopathy, which are described in this section (Begg et al., 2007).

Neuropathy is a term to define a disease of nerves that cause pain and some disability. The cause of neuropathic disorders differs, which includes injury, infection, diabetes, alcohol abuse, and cancer chemotherapy. It can be categorized as mononeuropathy and polyneuropathy. When the neuropathic disorder is caused by the malfunction of a single nerve, it is called mononeuropathy. Some types of neuropathies are median nerve entrapment, ulnar neuropathy, radial neuropathy, peroneal neuropathy, femoral neuropathy, facial mononeuropathy, and trigeminal neuropathy. Median nerve entrapment is the most common mononeuropathy found in the arms and hand. The nerve is compressed or placed in an unusual position, which disturbs the function and causes pain. EMG is generally used to distinguish median nerve entrapment from lesions of the proximal median nerve, brachial plexus, and cervical nerve roots. Ulnar neuropathy is caused by compression of the ulnar nerve at the elbow, which is difficult to find a real location using either EMG or the other EDX methods. Radial neuropathy disorders are at the radial nerve, which extends from the brachial plexus to the wrist. Radial neuropathy is less studied than ulnar neuropathy. The three general lesions that cause radial neuropathy are the spinal grove, axilla lesion, and lesions of the posterior interosseous and superficial radial sensory nerves. Peroneal neuropathy is the most common neuropathy in the lower extremities of the body. Peroneal neuropathy is a problem with the peroneal nerve where it occurs at the fibular neck of the calf. It causes sensory disturbances of the calf and foot region, usually causing a lack of feeling and weakness. Femoral neuropathy is disorders at the femoral nerve, which runs parallel to the femoral artery and innervates the quadriceps muscle. Generally, it is not observed by EMG diagnosis, because it has similar signs as the lumbar plexus lesion in peroneal neuropathy. Usually, it causes buckling knees, difficulty with lifting objects, drifting of feet, and some deadness and weakness at the thigh and calf region. Facial mononeuropathy is facial palsy, which generally known as the idiopathic Bell's palsy. This problem is related to some other disorders such as diabetes, leprosy, stroke, and herpes zoster. Weakness in the lower facial region and facial spasm are some symptoms of facial neuropathy. Trigeminal neuropathy is disorder at the trigeminal nerve, which carries the nerve impulse to the face and motor fibers. The patient with trigeminal neuropathy has difficulty in chewing and numbness over the ipsilateral face.

When a group of nerves or all nerves of a peripheral nerve are affected, it is termed polyneuropathy. Polyneuropathies are similar because of the insufficient manner where sensory nerves react to failure. EMG diagnosis is not beneficial for polyneuropathy, because patients with polyneuropathy have normal electrophysiological characteristics. Myopathy is a disorder generally related to the skeletal muscle caused by injury of a muscle group or some genetic mutation. Myopathy avoids the normal task of affected muscles. Therefore, the patient suffering with myopathic diseases has weak muscles and, depending on severity of disorder, has a problem with performing regular tasks or finds it impossible to make any movement by using the affected muscles. Some common types of myopathies are muscular dystrophy (MD), Becker muscular dystrophy (BMD), Duchenne muscular dystrophy (DMD), congenital muscular dystrophy, Emery-Dreifuss muscular dystrophy (EDMD), myotonic muscular dystrophy (MMD), distal muscular dystrophy (DD), limb-girdle muscular dystrophy (LGMD), facioscapulohumeral muscular dystrophy, and oculopharyngeal muscular dystrophy (OPMD). MD is the most common form of myopathy, which causes muscle weakness and degeneration. It can be hereditary or acquired. While reviewing EMG readings, it is characterized by short MUAP duration and reduced amplitude. Congenital muscular dystrophy (CMD) occurs before birth because of genetic mutation. The necessary proteins for muscles are not produced. It can cause mental slowness and disability for life. DMD influences involuntary muscle such as the heart muscle. It occurs in children of all ages and is genetic in origin. This disorder causes muscles to waste away because of a lack of dystrophin protein. It is severe, and survival after the age 30 is grim. BMD is similar to DMD, but it starts in adolescence and may cause heart failure. EDMD occurs in children of all ages and is caused by gene mutation. It causes weakness and wasting upper body. MMD arises at adulthood and is caused by gene defect. It grows slowly with starting weakness at the facial muscles and moves to other parts of the body. DD affects young to middle-aged adults and is caused by gene mutation. It occurs in the leg but is not a fatal disease. LGMD occurs at the waist and lower limbs. It grows slowly and causes some cardiopulmonary problems and wasting of the limbs. OPMD is caused by gene defection or mutation and affects the face and eyes (Begg et al., 2007).

2.2.10 Uses of EMG Signals in Diagnosis of Neuromuscular Disorders

The needle EMG is the usual clinical recording method employed for diagnosis of neuromuscular pathology. When, for example, a patient goes to a doctor for muscle weakness, recording of the needle EMG during contraction of specific muscles will be performed. The morphology of single MUAP waveforms gives the necessary clinical data regarding the muscle's skill in response to the CNS. This data may assist to identify irregular activity in situations such as muscle irritation, injury to nerves in the arms and legs, pinched nerves, and MD. The needle EMG is also used for nerve wounds and can be employed to find out if the wound heals and goes back to normal with complete muscle reactivity, for instance, by analyzing alterations in motor unit accomplishment over a definite time period. The diagnostic EMG is comprised of the investigation of unplanned motor action ocurring throughout muscle relaxation. In ordinary situations, the muscle is electrically quiet during a relaxation period; on the other hand, irregular unplanned waveforms and waveform patterns can be produced that are connected with spontaneous muscular activities and seizures (Sörnmo & Laguna, 2005).

EXAMPLE 2.8 The following MATLAB code was used to plot healthy, ALS, and myopathic EMG signals. For this example, EMGLAB data was used. You can download the data from the following website: http://www.emglab.net/emglab/Signals/signals.php

Dataset Information: EMG signals were taken from EMGLAB website (http://www.emglab.net/). The clinical EMG signals were acquired under normal conditions for MUAP analysis. The EMG signals were recorded at low voluntary and constant level of con traction with a standard concentric needle electrode. The EMG signals were filtered between 2 Hz and 10 kHz, and consist of a control group, a group of patients with ALS, and a group with myopathy. There were 10 normal subjects (four females and six males) aged 21–37 years in the control group. There were eight patients (four females and four males) between 35 and 67 years old in the ALS group. There were seven patients (two females and five males) between 19 and 63 years old in the myopathy group (Nikolic, 2001). Exemplary EMGs are depicted in Fig. 2.11. The following MATLAB code was used to plot each signal:

```
%% Ch2_EX_8.m
clc
clear
%Load Sample EMG Data downloaded from the web site
%http://www.emglab.net/emglab/Signals/signals.php
load EMGDAT.mat

%Plot HEALTHY EMG Signal
subplot(3,1,1);
plot(CON(:,2));
title('HEALTHY EMG Signal')
ylabel('Amplitude')
axis ([0 8192 -inf inf])

%Plot ALS EMG Signal
subplot(3,1,2);
plot(ALS(:,2))
title('ALS EMG Signal')
xlabel('Samples')
ylabel('Amplitude')
axis ([0 8192 -inf inf])

%Plot MYOPATHY EMG Signal
subplot(3,1,3);
plot(MYO(:,2))
```

```
title('MYOPATHY EMG Signal')
xlabel('Samples')
ylabel('Amplitude')
axis ([0 8192 -inf inf])
```

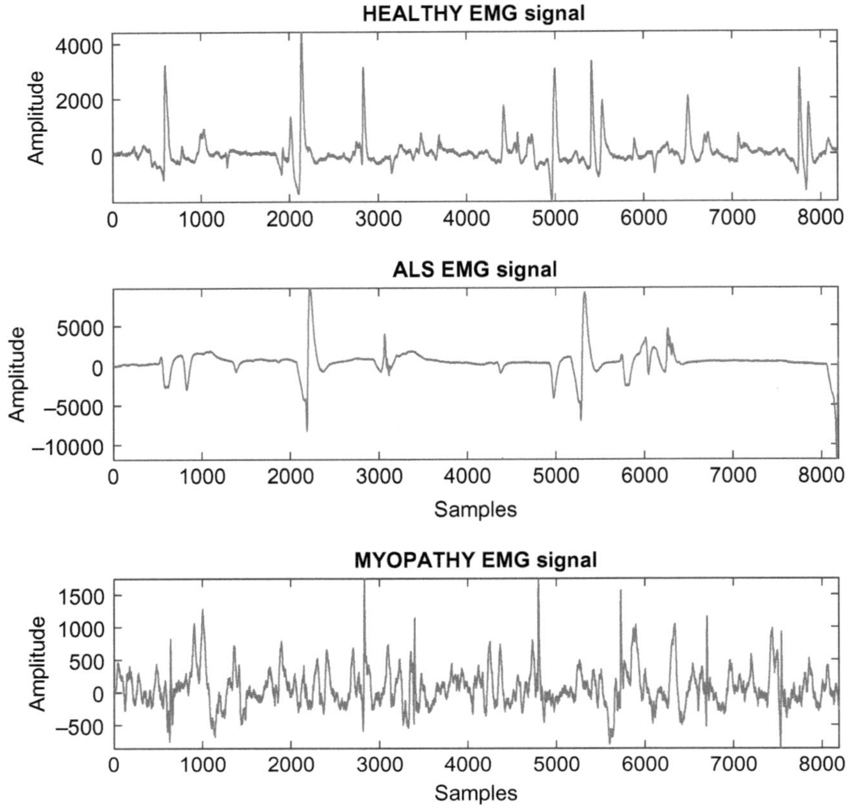

FIG. 2.11 Drawing of three different EMG signals (healthy, ALS, and myopathy).

2.2.11 Uses of EMG Signals in Prosthesis Control

Myoelectric-controlled prostheses are employed by persons lacking upper limbs or with amputations. The EMG signal is received from surface electrodes placed on muscles, which are brought to the prosthesis; the signal's properties are analyzed and interpreted to activate the required function. Considering the category of prosthesis, the regulation data goes from simple on/off commands produced by only one muscle to complex multifunction commands produced by an assembly of muscles. The single-muscle regulator is commonly found on the EMG amplitude in such a way that muscle contractions of various strengths, as shown by various amplitudes, may distinguish among hand closing and opening, or elbow flexion and extension. A multifunction prosthesis mixes the usage of several electrodes over diverse muscle assemblies using improved signal-processing algorithms to increase the amount of data capable of being extracted with regard to the active muscle state. The multifunction prosthesis accomplishes finer accuracy of the user's intention by studying transient signal patterns based on contractions, applying time, frequency, or time-frequency methods. It is essential to understand that the improvement of algorithms for prosthetic control is connected to a real-time restraint (Sörnmo & Laguna, 2005). Human hands are very important because they play an important role for grasping and manipulating different objects. Even the loss of a single hand affects human activity, and a prosthetic hand is a solution for equipping the armless subject. Currently, muscle signals control hand prostheses. Control is possible after hand amputation because a substantial amount of the muscles exist in the arm stump to control a prosthesis (Kurzynski, Krysmann, Trajdos, & Wolczowski, 2016).

Biomedical signals are the set of body signals that characterize a physical variable of interest. A biomedical signal such as EMG is used to control the prostheses' movement. Upper limb prosthesis structures are primarily based on myoelectric control, identifying EMG signals that arise during muscle contraction on the skin surface. Because most of the muscles that produce finger motion is left in the stump after a hand amputation, these muscles are employed for control of prosthesis motion (Wojtczak, Amaral, Dias, Wolczowski, & Kurzynski, 2009). Because the analysis of EMG signals is used widely in medical diagnostics, sports, rehabilitation, and prosthesis control, the recognition of EMG signals offers substantial support for the automation of human tasks. A novel application of EMG signal classification is the individual skeletal muscle contraction in which the relevant EMG signals are employed to control a machine action after classification. But the challenge is the prosthetic hand control where the hand is multitalented to accomplish different movements, such as gesticulation or the playing of musical instruments by grasping and manipulating several objects. The loss of a hand intensely decreases opportunities to work, and the loss of both hands essentially eliminates independent work. The goal of a prosthetic hand is to moderately re-establish the function of the lost limb, particularly its working functions and ability to perform different movements and to achieve different finger configurations. The muscle signals, which usually control contemporary prostheses, are related to healthy hand and finger movements and should be acquired noninvasively by appropriate sensors positioned above the muscles on the skin. Hence, EMG-controlled prosthetics use surface signals taken from the hand stump muscles and then distinguishing the category of intended prosthesis action by classifying them (Wołczowski & Zdunek, 2017).

People with disability conditions might experience different modern life technologies than people without disabilities. Doorknobs, kitchen tools, or shirt buttons that do not have any effect on a large portion of individuals might end up as an obstacle for somebody who has a disability. Replacing a doorknob with a lever door handle might make a huge difference to that individual, and would be welcomed and valued toward others. A straightforward buttonhook device, although it is not handy for most people, might help somebody who faces difficulties in manipulating buttons. Certain innovations have overcome independence obstacles for disabled persons, whereas other technologies provide the ability to overcome environmental barriers (Kim, Choi, Moon, & Mun, 2011). These supportive technologies might work by enhancing a person's abilities. The hand is one of the most important elements of a human being, used as a basic element in the sense of feeling. The hand is used in real-life experience to feel appearances and textures, and perform basic lifting functions. With a controlled muscle movement, for example, the basic function of the arm is to grip, lift, wave, and perform other rotational motions of the arm. For amputees, a prosthesis hand is an artificial device used to replace a missing part. The hand prosthesis is an improved though functional hand which can be equipped and manipulated to perform different functions. For instance, the myoelectric arm uses a controlled muscle contraction generated from an electrical charge to transfer and strengthen the control center. In this manner, with a controlled movement, an amputee can perform normal functions of the arm such as gripping, feeling, and waving among other hand-related movements.

EXAMPLE 2.9 The following MATLAB code was used to plot the sEMG signals related to basic hand movements. For this example, sEMG for Basic Hand Movements Dataset was used. You can download the data from the following website: https://archive.ics.uci.edu/ml/datasets/sEMG+for+Basic+Hand+movements#

Dataset Information: The data were sampled at 500 Hz, band-pass filtered using a Butterworth Band-Pass filter with low and high cut-off at 15 and 500 Hz, respectively, and a notch filter at 50 Hz to eliminate line interference artifacts. The signal was recorded from two differential EMG sensors, and the signals were conducted to a 2-channel EMG system by Delsys Bagnoli Handheld EMG Systems. The experiments include freely and repeatedly grasping of different items that were crucial to conduct the hand movements. The force and speed were purposely left to the subject's will. There were two forearm sEMG electrodes, Flexor Capri Ulnaris and Extensor Capri Radialis, Longus and Brevis, detained in place by elastic bands with the reference electrode in the middle to collect information about muscle activation. Five healthy subjects (two males and three females) of the same age approximately (20- to 22-year-old) conducted six grasps for 30 times each. The measured time was 6 sec. There exists a mat file for each subject. The subjects were required to repeatedly carry out the following six movements that can be thought as daily hand grasps:

(a) Spherical: for holding spherical tools
(b) Tip: for holding small tools
(c) Palmar: for grasping with palm facing the object
(d) Lateral: for holding thin, flat objects
(e) Cylindrical: for holding cylindrical tools
(f) Hook: for supporting a heavy load

More information can be found in Sapsanis, Georgoulas, and Tzes (2013). Exemplary EMGs are depicted in Fig. 2.12. The following MATLAB code was used to plot each signal:

```
%% Ch2_EX_9.m
clc
clear
%sEMG for Basic Hand movements Data Set
%Load Sample sEMG Data downloaded from the web site
%https://archive.ics.uci.edu/ml/datasets/sEMG+for+Basic+Hand+movements#
load sEMG_UCI_BHM.mat

%Plot Spherical sEMG Signal
subplot(3,2,1);
plot(F1_spher_ch1(1,:));
title('Spherical sEMG Signal')
ylabel('Amplitude')
axis ([0 3000 -inf inf])

%Plot Tip sEMG Signal
subplot(3,2,2);
plot(F1_tip_ch1(1,:))
title('Tip sEMG Signal')
xlabel('Samples')
ylabel('Amplitude')
axis ([0 3000 -inf inf])

%Plot Palmar sEMG Signal
subplot(3,2,3);
plot(F1_palm_ch1(1,:))
title('Palmar sEMG Signal')
xlabel('Samples')
ylabel('Amplitude')
axis ([0 3000 -inf inf])

%Plot Lateral sEMG Signal
subplot(3,2,4);
plot(F1_lat_ch1(1,:))
title('Lateral sEMG Signal')
xlabel('Samples')
ylabel('Amplitude')
axis ([0 3000 -inf inf])

%Plot Cylindrical sEMG Signal
subplot(3,2,5);
plot(F1_cyl_ch1(1,:))
title('Cylindrical sEMG Signal')
xlabel('Samples')
ylabel('Amplitude')
axis ([0 3000 -inf inf])

%Plot Hook sEMG Signal
subplot(3,2,6);
plot(F1_hook_ch1(1,:))
title('Hook sEMG Signal')
xlabel('Samples')
ylabel('Amplitude')
axis ([0 3000 -inf inf])
```

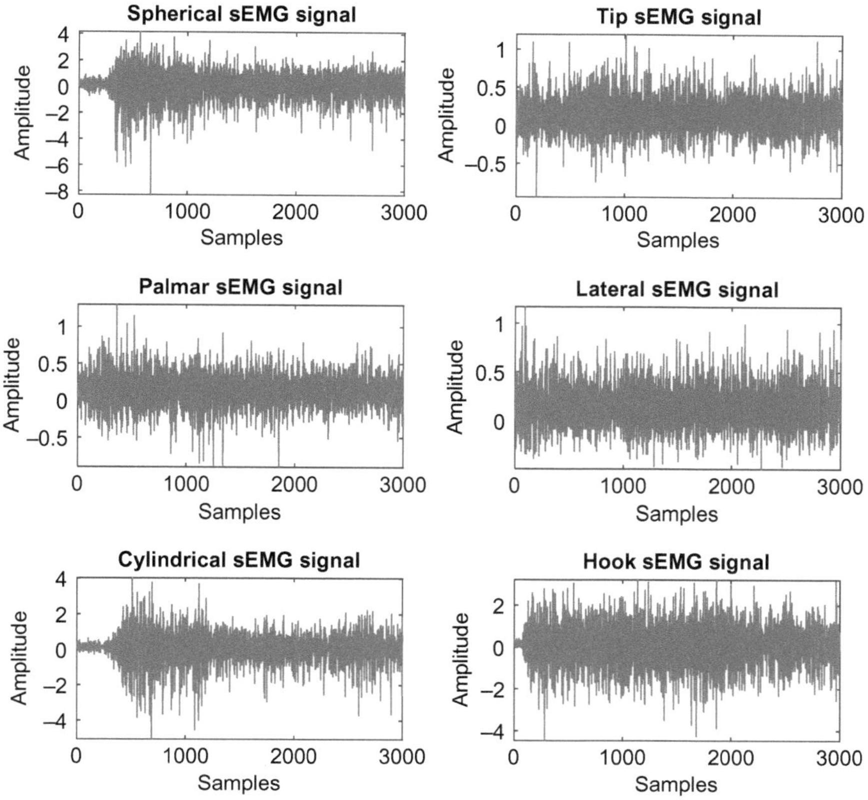

FIG. 2.12 Drawing of sEMG signals taken from six different hand movements.

2.2.12 Uses of EMG Signals in Rehabilitation Robotics

EMG-controlled assistive devices are also used for stroke treatment in intensive therapy to help with rehabilitation. In such cases, intentions of a patient's activity can be achieved using the sEMG (Lum, Burgar, Shor, Majmundar, & Van der Loos, 2002; Riener, Nef, & Colombo, 2005). With recent technological improvements, active exoskeleton robots assist with rehabilitation applications, human power augmentation, assistive robotics, impairment evaluation, and haptic communication in virtual and teleoperated environments. The human body commands must be understood by these robots to help humans. Therefore, EMG signals should be acquired and analyzed for control of the exoskeleton robot (Sasaki, Noritsugu, & Takaiwa, 2005). Exoskeleton robots have two controllers working together simultaneously: the robot controller and the human muscle. Upper-limb exoskeleton robots need to be controlled differently than conventional industrial and field robots, because humans operate both commands, and the control system employs these commands as part of its decision-making components. The exoskeleton maturely implements the real decisions of the human operator. However, there are still challenges in making decisions based on the motion intention identifications of the robot user (Abdullah, Subasi, & Qaisar, 2017).

The best approach to design an upper limb exoskeleton robot controller is to focus on the controller input information. In modern techniques, the input consists of human biomedical signals and platform-independent control signals. Various strategies are applied in different fields of applications. The EMG signals have been successfully used as human biomedical signal inputs to some exoskeleton developments such as upper-limb exoskeleton robots (Lo & Xie, 2012). For example, in Gopura, Kiguchi, and Li (2009), a muscle model based on EMG control was proposed to control a 7DOF (Degrees of Freedom) upper-limb exoskeleton robot. The user can adapt the method. Most upper-limb disabled people can use it. Control methods for upper-limb exoskeleton robots using EMG are mostly of a binary (on/off) nature (Lenzi et al., 2009). A good design would allow a high accuracy of motion intention classification for even a physically weak person who cannot properly generate daily motions (Abdullah et al., 2017).

EXAMPLE 2.10 The following MATLAB code was used to plot the sEMG signals related to different physical actions. For this example, sEMG Physical Action Dataset was used. You can download the data from the following website: https://archive.ics.uci.edu/ml/datasets/EMG + Physical + Action + Data + Set

Dataset Information: One female and three male subjects (aged 25–30 years) who have experienced aggression in circumstances such as physical fighting were employed in the experiment. Every subject had to implement 10 normal and 10 aggressive activities through 20 distinct experiments. The Essex robotic arena was the main experimental hall in which the data collection took place. The subjects' performance was collected by the Delsys EMG apparatus, interfacing human activity with myoelectrical contractions. Based on this context, the data recording procedure included eight skin-surface electrodes located on the upper arms (biceps and triceps) and upper legs (thighs and hamstrings). The overall number of electrodes was eight that corresponded to eight input time series, one for a muscle channel. Each time series contains ~10,000 samples (~15 actions per experimental session for each subject). Exemplary EMGs are depicted in Fig. 2.13. The following MATLAB code was used to plot each signal:

```
%% Ch2_EX_10.m
clc
clear
%EMG Physical Action Data Set
%Load Sample sEMG Data downloaded from the web site
%https://archive.ics.uci.edu/ml/datasets/EMG+Physical+Action+Data+Set
load sEMG_UCI_PA_NOR.mat

%Plot Bowing sEMG Signal
subplot(5,2,1);
plot(Bowing(1:4096,1));
title('Bowing sEMG Signal')
ylabel('Amplitude')
axis ([0 4096 -inf inf])

%Plot Clapping sEMG Signal
subplot(5,2,2);
plot(Clapping(1:4096,1))
title('Clapping sEMG Signal')
xlabel('Samples')
ylabel('Amplitude')
axis ([0 4096 -inf inf])

%Plot Handshaking sEMG Signal
subplot(5,2,3);
plot(Handshaking(1:4096,1))
title('Handshaking sEMG Signal')
xlabel('Samples')
ylabel('Amplitude')
axis ([0 4096 -inf inf])

%Plot Hugging sEMG Signal
subplot(5,2,4);
plot(Hugging(1:4096,1))
title('Hugging sEMG Signal')
xlabel('Samples')
ylabel('Amplitude')
axis ([0 4096 -inf inf])

%Plot Jumping sEMG Signal
subplot(5,2,5);
plot(Jumping(1:4096,1))
title('Jumping sEMG Signal')
xlabel('Samples')
ylabel('Amplitude')
axis ([0 4096 -inf inf])

%Plot Running sEMG Signal
subplot(5,2,6);
plot(Running(1:4096,1))
title('Running sEMG Signal')
```

```
xlabel('Samples')
ylabel('Amplitude')
axis ([0 4096 -inf inf])

%Plot Seating sEMG Signal
subplot(5,2,7);
plot(Seating(1:4096,1))
title('Seating sEMG Signal')
xlabel('Samples')
ylabel('Amplitude')
axis ([0 4096 -inf inf])

%Plot Standing sEMG Signal
subplot(5,2,8);
plot(Standing(1:4096,1))
title('Standing sEMG Signal')
xlabel('Samples')
ylabel('Amplitude')
axis ([0 4096 -inf inf])

%Plot Walking sEMG Signal
subplot(5,2,9);
plot(Walking(1:4096,1))
title('Walking sEMG Signal')
xlabel('Samples')
ylabel('Amplitude')
axis ([0 4096 -inf inf])

%Plot Waving sEMG Signal
subplot(5,2,10);
plot(Waving(1:4096,1))
title('Waving sEMG Signal')
xlabel('Samples')
ylabel('Amplitude')
axis ([0 4096 -inf inf])
```

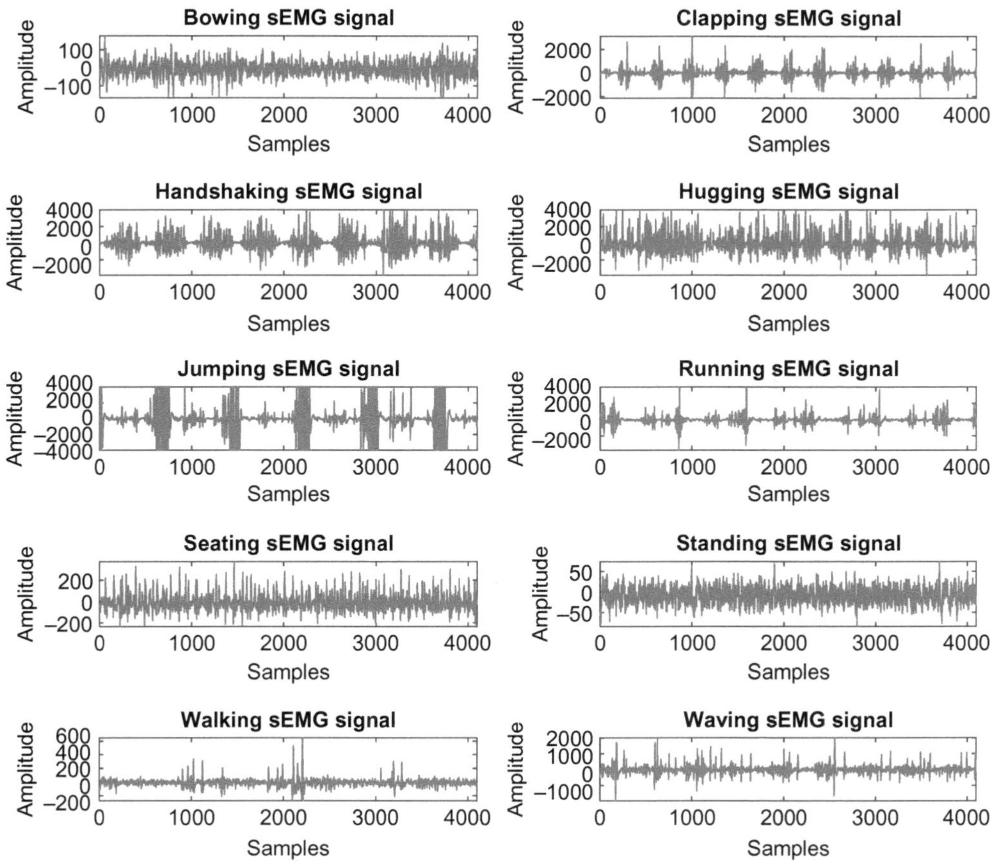

FIG. 2.13 Drawing of sEMG signals for 10 different physical actions.

2.2.13 Other EMG Applications

Kinesiology: Kinesiology refers to the investigation of body movement and has the goal of understanding the processes regulating movement. The sEMG is vital for a number of characteristics of kinesiology, comprising the learning of motor regulation approaches, mechanics of muscle contraction, and gait. To illustrate, the EMG pattern recorded during gait is described by consecutive spurts reflecting periods of muscle activation. Definition of the times for each spurt's beginning and termination, followed by an investigation of the outcoming beginning/termination timing pattern, is a vital duty for clinical evaluation of numerous movement abnormalities. The EMG pattern analysis is eased by signal-processing methods that automate the definition of spurts and describe the beginning/termination timing pattern (Sörnmo & Laguna, 2005).

Ergonomics: The amplitude of the sEMG signal offers an appreciated, numerical evaluation of muscle load, frequently employed to evaluate a physical load while working. In ergonomic study, the EMG signal is frequently recorded for the duration of light, repetitious effort with special interest to evaluate the movement of explicit muscles at known work locations. After being mixed with other ergonomic data, the consequence of the myoelectric signal study can eventually assist to bypass work-related irregularities, construct enhanced workplaces, and enhance efficiency. The investigation of muscle weakness is fundamental in ergonomics because of the natural desire to bypass weakness in work conditions. A valuable explanation of muscle weakness is the situation when a subject no longer can hold a requisite force. Regarding signal-processing methods, a weakened muscle contraction is demonstrated by a gradual rise in sEMG signal amplitude and noteworthy fluctuations in its power spectrum (Sörnmo & Laguna, 2005).

2.3 THE ELECTROCARDIOGRAM

2.3.1 Introduction

The ECG is the recording of electrical activities originating from the heart on the body surface. Waller first noticed it in 1889 (Waller, 1889). He used his pet bulldog as the signal source and a capillary electrometer as the recording device. Later, in 1903, Einthoven improved this technology by using the string galvanometer as the recording device and using human subjects with a variety of cardiac irregularities. He is primarily meritorious for introducing a number of concepts still in use today: counting the classification of a variety of waves, defining a number of standard recording sites using the arms and legs, and creating the initial theoretical assembly through which the heart is modeled as a single-time varying dipole (Einthoven, 1903). The acronym "EKG" came from Einthoven's native Dutch language, because the root word "cardio" is spelled with a k. With the aim of tracing an ECG waveform, a differential recording between two points on the body surface is performed. Conventionally, every differential recording is called a lead. Einthoven specified three leads named with the Roman numerals I, II, and III. Because the body is understood to be merely resistive at ECG frequencies, the four limbs can be considered as wires connected to the torso. For this reason, lead I may possibly be recorded from the person's shoulders without ruining any of the cardiac data (Berbari, 2000).

The progress of the ECG continued for 30 years when F. N. Wilson appended concepts of a "unipolar" recording (WIlson, Johnson, & Hill, 1934). He designed a reference point by connecting three limbs together and finding the average of their potentials, discovering that personal recording site on the limbs or chest surface can be differentially recorded having an identical reference point. He broadened the biophysical models to design the concept of the cardiac origin placed inside the volume conductor of the body. He incorrectly considered that the central terminal was a true ground potential. On the other hand, from the mid-1930s until today, the 12 leads are made up of 3 limb leads; 3 leads in which the limb potentials are referenced to a modified Wilson terminal (the amplified leads (Goldberger, 1942)); and 6 leads located across the chest and referenced to the Wilson terminal, which form the basis of the standard 12-lead ECG. These sites are historically based, have a built-in redundancy, and are not best suited for every kind of cardiac event. An ECG will record the voltage difference from any two sites. The ECG signals are usually in the range of ± 2 mV and need a recording bandwidth of 0.05–150 Hz. Complete technical specification for ECG equipment is suggested by both the American Heart Association (EC11a, 1984) and the Association for the Advancement of Medical Instrumentation (Bailey, Berson, & Garson, 1990).

Body surface mapping indicates the usage of numerous recording sites (>64) set on the body so that isopotential surfaces could be calculated and analyzed over time. This concept still has a role in many researches. Other subgroups of the 12-lead ECG are used in limited-mode recording events such as tape-recorded ambulatory ECG (typically 2 leads), in intensive care monitoring at the bedside (typically 1 or 2 leads), or telemetered within regions of the hospital from patients who are not restrained to bed (1 lead). The recording electronics of these ECG systems have come from the typical progress of contemporary instrumentation, for example, vacuum tubes, transistors, integrated chips, and microprocessors. Usage of computers with the ECG for machine reading was one of the first usages of computers in medicine (Jenkins, 1981). Of importance in computer-based systems was substitution of a human reader and explanation of the standard waves and intervals. Initially this was performed by connecting the ECG machine to a centralized computer using phone lines. The contemporary ECG machine is entirely integrated with an analog front end, a 12- to 16-bit analog-to-digital (A/D) converter, a computational microprocessor, and dedicated input-output (I/O) processors. These systems find a dimensional matrix derived from the 12 lead signals and examine this matrix using a set of rules to attain the final set of interpretive statements (Pryor, Drazen, & Laks, 1980). The better hospital-based system will record these changes and keep a large database of all ECGs accessible by all permutations of parameters, for example, all females, older than age 30, or with an inferior congenital heart disease (Berbari, 2000).

There are hundreds of demonstrative approaches where a particular diagnosis is made for every ECG, but there are only about five or six major classification sets for which ECG is used. The initial step in ECG analysis requires computation of the rate and rhythm for the atria and ventricles, including any conduction instability, either in the connection among the different chambers or within the chambers themselves. Then feature identification connected to the presence or absence of scarring due to a myocardial infarction would be performed. The ECG has been a principal method for evaluating chamber size or growth, but one might argue that more precise data in this area would be received by noninvasive imaging technologies (Berbari, 2000).

In recent times, the high-resolution ECG has been expanded, by which the digitized ECG is signal-averaged to lessen random noise (Berbari, Lazzara, Samet, & Scherlag, 1973; Berbari, Lazzara, & Scherlag). This concept, tied with postaveraging high-pass filtering, is used to identify and compute low-level signals ($\sim$1.0 μV) not visible with standard procedures. This computer-based concept has allowed the recording of actions prognostic of prospective life-threatening cardiac actions (Berbari, 2000; Simson, 1981).

2.3.2 Electrocardiogram Signals

Every ECG instrument has horizontal and vertical calibration lines. Horizontal lines are used for voltage calculation and are scaled such that 1 mV is equal to a deflection of 10 tiny segments in either the upward (positive) or downward (negative) direction. Vertical lines on the ECG refer to time calibration for the period of the signal and are usually calibrated so that 2.5 cm in the horizontal direction represents 1 s. Each horizontal section is divided into five segments representing 0.2 s with smaller units further subdivided into 0.04 s segments. ECG traces are generally recorded on specialized ECG paper printed with thin (Julian, Campbell-Cowan, & McLenachan, 2005) vertical lines 1 mm distant interspaced with thick vertical lines at 5 mm distance. This separation is comparable to that of the instrumentation itself and can be employed to compute the heart rate. Assuming a regular heartbeat, the heart rate can be calculated by counting the number of little squares between two consecutive R waves and dividing by 1500; on the other hand, one can add up the large squares and divide by 300 (Begg et al., 2007).

2.3.3 Physiology

The heart consists of four chambers; the upper two chambers are called atria, and the lower two chambers are called ventricles. The atria are thin-walled, low-pressure pumps that collect blood from the venous circulation. Positioned in the upper right atrium are a group of cells that operate as the main pacemaker of the heart. Throughout a composite alteration of ionic concentration crossways, that is, the cell membranes (the present supply), an extracellular potential field is created, which excites bordering cells, and a cell-to-cell transmission of electrical actions occurs. Because the body operates as a purely resistive medium, these potential fields expand to the body surface (Geselowitz, 1989). The nature of the body surface waves is dependent on the total of tissue activating at one time and the relative speed and direction of the activation wave front. Thus the pacemaker potentials produced by a tiny tissue mass are not visible on the ECG. As the activation wave front faces the amplified mass of atrial muscle, the beginning of electrical activity is noticed on the body surface, and the initial ECG wave of the cardiac cycle is visible. This is the P wave, and it symbolizes the start of the atria. Transmission of the cardiac impulse continues from the atria through a sequence of dedicated cardiac cells (the A-V node and the His-Purkinje system), which once more are too undersized in total mass to produce a signal large enough to be visible on the typical ECG. There is a small, fairly isoelectric section following the P wave. Once the large muscle mass of the ventricles is wound up, a fast and big deflection is visible on the body surface. The excitement of the ventricles initiates their contraction and gives the most important force for circulating blood to the organs of the body. This great wave seems to include several components. The original downward deflection is the Q wave, the original upward deflection is the R wave, and the terminal downward deflection is the S wave. The polarity and real existence of these three components rely on the site of the leads on the body as well as a large number of irregularities that might exist. Generally, the big ventricular waveform is commonly called the QRS complex despite its structure. Following the QRS complex is one smaller fairly isoelectric section. Following this smaller section, the ventricles go back to their electrical resting condition, and a wave of repolarization is visible as a low-frequency signal identified as the T wave. In a few persons, a little peak happens at the end or after the T wave, which is the U wave. Its source has never been fully recognized, but it is thought to be a repolarization potential (Berbari, 2000). The heart is one of the main organs of the human body, crucial to our existence. It is basically a huge pump, its only function is to keep blood circulating and keep the organs alive (Begg et al., 2007).

2.3.4 The ECG Waveform

The QRS complex is an electrical ventricular system and is the most well-known waveform showing electrical activity inside the heart. It is the basis for automatic recognition of heart rate and also as an access point for classification schemes and ECG data-compression algorithms (Kohler, Hennig, & Orglmeister, 2002). The QRS complex morphology describes the mechanical action of the heart, offering a view into how each chamber functions. The waves of depolarization extending all the way through the heart via each cardiac cycle produce electrical impulses. These impulses travel via a variety of body fluids, that is, blood, up to the body's surface where they can be recorded using surface electrodes. These signals are then sent to an ECG. The main characteristics of the QRS wave that describe significant data related to cardiac health are as follows (Begg et al., 2007):

(a) P wave
(b) QRS complex
(c) T wave
(d) QRS intervals

P Wave: The original P wave is due to electrical action coming from atrial contraction (systole). In cardiovascular diseases (CVDs), the P wave can turn out to be unclear and shown as irregular. An inverted P wave resulting from positive to negative altered voltage symbolizes that polarization of the atria is irregular (Julian et al., 2005). This indicates that the source of the pacemaker signal is not in the SA node but could arrive from somewhere else, for example, the atrium or the AV node. If the P wave has a seemingly broadened or notched characteristic, this shows a stoppage in the depolarization of the left atrium, which could indicate trouble in the conduction system. Right atrial growth (P pulmonale) can direct high P waves, usually growing more than 3 mm on the ECG trace (Begg et al., 2007).

QRS Complex: The QRS complex is the ventricular contraction (systole) consisting of the Q wave, which is the first negative deviation, followed by the R wave, a positive (upward) deviation. Any negative deflection following immediately after the R portion is termed the S wave. Models of abnormal QRS complexes could reveal a couple of R waves present or cases where the R part is not present. This wave's analysis is dependent on the ECG electrode/lead recording. Unusually large Q waves could indicate MI, opposite to a healthy Q wave, which is not normally higher than 2 mm in amplitude or 0.03 s in width. The QRS complex is generally not larger than 0.1 s and on average is of 0.06 to 0.08 s duration (Begg et al., 2007).

T Wave: Subsequent to ventricular contraction, the ventricles relax (diastole) producing the T wave. Contrasted with the earlier two wave segments, the T wave is a repolarization wave, usually lasting for 0.25–0.35 s subsequent to ventricular depolarization. In this phase, the lower heart chambers are relaxing electrically and getting ready for their subsequent muscle contraction. Atria repolarization is also present, but the signal amplitude is generally covered by the larger QRS complex because of ventricular contraction and is consequently hard to notice. The T wave goes in the same direction as the QRS component because repolarization happens from epicardium to endocardium, in a direction reverse to depolarization, which continues from endocardium to epicardium. T waves are generally not higher than 5 mm in the typical leads. Unusually high T waves may indicate MI, whereas smoothened T waves can be sign of myxedema or hypokalemia (Begg et al., 2007).

QRS Segment Intervals: Besides the wave shapes, the time intervals are essential in the assessment of cardiac health. Among the start of the P wave and the start of the QRS complex is the PQ interval; because the Q wave is frequently not present, this interval is also called the PR interval and stays for the time among the start of atrial contraction and the start of ventricular contraction (normally about 0.16 s). In cases where heart disease is present, particularly with scarred or reddened heart tissue, a longer PR interval may be noticed as more time is necessary for the depolarization wave to extend through the atrial myocardium and the AV node. Reduced PR intervals could represent that the impulse is coming in the junction tissue or could be because the Wolff-Parkinson-White syndrome (Begg et al., 2007; Julian et al., 2005).

The ST segment is an additional interval significant for recognition of many CVDs. As a rule, it comes as a leveled straight line among the QRS complex and the T wave. An eminent or dejected ST segment (relying on the ECG lead being noticed) is an indication of MI because the heart muscle is injured or does not collect enough blood, resulting in a disorder in ventricular repolarization. Pericarditis can be discovered by noticing ST segments concaving upward over a many cardiac cycles. In digitalis therapy, it has been found that ST segments are depressed with a tender drooping, whereas the T wave stays unchanged or flattened (Begg et al., 2007).

The QT interval lasts roughly 0.35 s and gives relevant data concerning the state of ventricular contractions. Its duration decreases as the heart rate goes up, and typically the QT interval does not last more than half the time between the earlier RR interval for rates among 60 and 90 beats per min. From time to time, this is hard to compute because it cannot be instantly noticed. But extended QT intervals can point to the danger of ventricular tachycardia (VT) or the occurrence of certain drugs, for example, antidepressants (Begg et al., 2007).

2.3.5 Heart Diseases

There are different heart arrhythmias. In this chapter, four major heart arrhythmias are selected because these are the most frequent arrhythmias. These are premature ventricular contraction (PVC), atrial premature contraction (APC), right bundle branch block (RBBB), and left bundle branch block (LBBB). In the following text, a short description of these four arrhythmias will be given.

Premature Ventricular Contraction: PVCs are the result of an irritable ventricular focus. They may be uniform (identical form) or multiform (different forms). Typically, PVC is followed by a full compensating break because the sinus node timing is not interrupted. In general, the sinus rate gives the subsequent sinus impulse on time. Quite the opposite, PVC may be followed by a noncompensatory pause if the PVC comes into the sinus node and restarts its timing; this causes the subsequent sinus P wave to show up earlier than expected. Its rate is dependent on the rate of the original rhythm. Rhythm is irregular whenever PVC occurs. PVC is not related to P wave and PR interval. QRS lasts longer than 0.1 s.

Atrial Premature Contraction: APC, also called PAC in the literature, is another important heart arrhythmia. A single contraction happens before the subsequent estimated sinus contraction. After the APC, sinus rhythm typically resumes. Its rate is dependent on the rate of underlying rhythm, and rhythm is abnormal whenever APC occurs. P wave is present and may have various shapes. PR interval is also present and varies in PAC. Duration of QRS is normal, ranging from 0.06 to 0.1 s.

Right Bundle Branch Block: Third most frequent heart arrhythmia is RBBB. QRS complex lasts more than 0.10 s. QRS axis may be normal but may also deviate to the right. It broads S wave in leads I, aVL, V5, and V6. Its RSR pattern in lead V1 with R is higher than R QRS pattern in V5 and V6. ST-T is fuzzy and in reverse direction to terminal segment of QRS (this is not ST elevation or ST depression).

Left Bundle Branch Block: Fourth arrhythmia that will be discussed is LBBB. It is similar to RBBB. QRS complex lasts more than 0.1 s. It is mainly negative in leads V1 and V2 and positive in V5 and V6. Absence of small, normal Q waves is noticed in I, aVL, V5, and V6. Its wide monophasic R waves are found in I, aVL, V1, V5, and V6 (Jones, 2008).

2.3.6 Uses of ECG Signals in Diagnosis of Heart Arrhythmia

CVDs are one of the main causes of mortality worldwide. The creation of accurate and fast techniques for automated ECG heartbeat signal classification is vital for clinical diagnosis of different CVDs (Thaler, 1999), for example, an arrhythmia. Notion arrhythmia is employed to represent a group of circumstances in which irregular electrical activities coming from heart are characterized by ECG beats or patterns (Chazal, O'Dwyer, & Reilly, 2004; Pan & Tompkins, 1985). ECG is an effective, simple, noninvasive tool for heart disease recognition. Medical doctors investigate several waveforms based on their characteristics (amplitude, polarity, etc.) and diagnose and treat based on this investigation.

EXAMPLE 2.11 The following MATLAB code was used to plot the normal, APC, PCV, LBBB, and RBBB ECG signals. For this example, MIT-BIH Arrhythmia Database was used. You can download the data from the following website: https://www.physionet.org/physiobank/database/mitdb/

Dataset Information: Boston's Beth Israel Hospital and MIT have maintained a study on arrhythmia analysis and related subjects. One of the first chief achievements of this effort was the MIT-BIH Arrhythmia Databas, which started distributing in 1980. The database was the first publicly accessible dataset of standard test material for the assessment of arrhythmia detectors and has been employed for this aim as well as for basic research into cardiac dynamics at more than 500 sites worldwide. The MIT-BIH Arrhythmia Database includes 48 half-hour extracts of 2-channel ambulatory ECG recordings, taken from 47 subjects studied by the BIH Arrhythmia Laboratory between 1975 and 1979. Twenty-three signals were selected randomly from a set of 4000 24-h ambulatory ECG signals recorded from a mixed population of inpatients (about 60%) and outpatients (about 40%) at Boston's Beth Israel Hospital; the remaining 25 recordings were chosen from the same set to contain less common but clinically important arrhythmias that would not be well represented in a small random sample. The ECG signals were sampled at 360 Hz per channel with 11-bit resolution over a 10-mV range. Two or more cardiologists individually interpreted each record; disagreements were resolved to get the computer-readable reference annotations for each beat (approximately 110,000 annotations in all) included with the database. The entire MIT-BIH Arrhythmia Database has been freely accessible since PhysioNet's inception in September 1999. Exemplary ECGs are depicted in Fig. 2.14. The following MATLAB code was used to plot each signal:

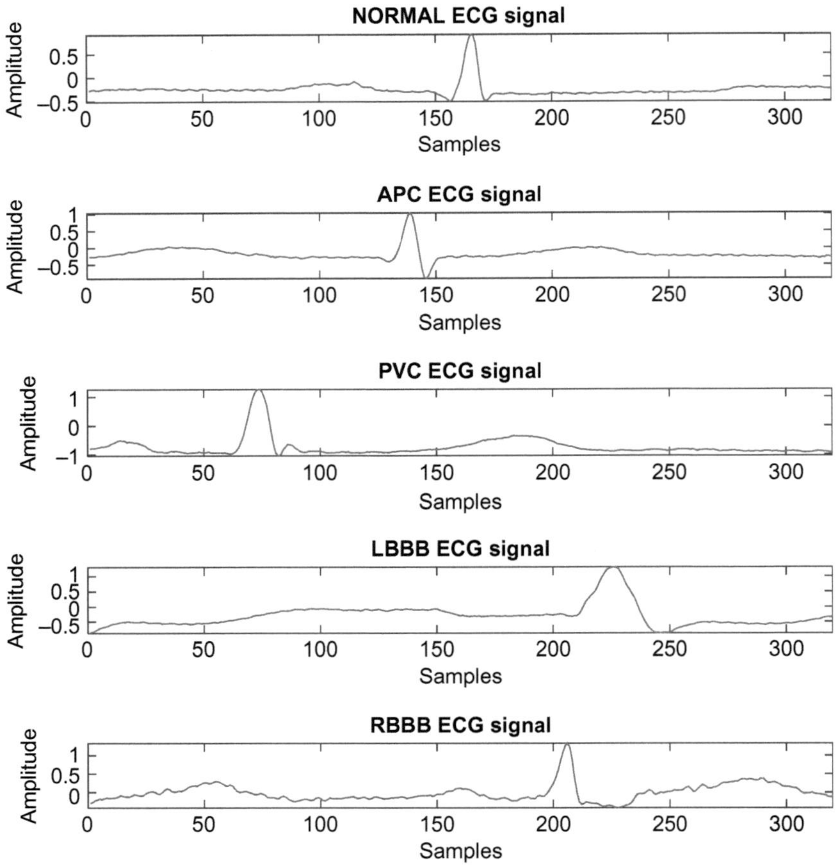

FIG. 2.14 Drawing of different ECG signals (normal, APC, PVC, LBBB, and RBBB).

```
%% Ch2_EX_11.m
clc
clear
%Load Sample ECG Data downloaded from the web site
%https://www.physionet.org/physiobank/database/mitdb/
load MITBIH_ECG.mat

%Plot NORMAL ECG Signal
subplot(5,1,1);
plot(ECGN(:,1));
title('NORMAL ECG Signal')
ylabel('Amplitude')
axis ([0 320 -inf inf])

%Plot APC ECG Signal
subplot(5,1,2);
plot(ECGAPC(:,1))
title('APC ECG Signal')
xlabel('Samples')
ylabel('Amplitude')
axis ([0 320 -inf inf])

%Plot PVC ECG Signal
subplot(5,1,3);
plot(ECGPVC(:,1))
title('PVC ECG Signal')
```

```
xlabel('Samples')
ylabel('Amplitude')
axis ([0 320 -inf inf])

%Plot LBBB ECG Signal
subplot(5,1,4);
plot(ECGLBBB(:,1))
title('LBBB ECG Signal')
xlabel('Samples')
ylabel('Amplitude')
axis ([0 320 -inf inf])

%Plot RBBB ECG Signal
subplot(5,1,5);
plot(ECGRBBB(:,1))
title('RBBB ECG Signal')
xlabel('Samples')
ylabel('Amplitude')
axis ([0 320 -inf inf])
```

EXAMPLE 2.12 The following MATLAB code was used to plot the normal, APC, PCV, and RBBB ECG signals. For this example, St. Petersburg Institute of Cardiological Technics 12-lead Arrhythmia Database data was used. You can download the data from the following website: https://www.physionet.org/pn3/incartdb/

Dataset Information: This database contains of 75 annotated ECG signals taken from 32 Holter records. Every signal is 30 min long and contains 12 standard leads, sampled at 257 Hz, with gains varying from 250 to 1100 analog-to-digital converter units per millivolt. Gains for each signal is specified in its .hea file. The reference annotation files contain more than 175,000 beat annotations in all. The original ECG signals were recorded from patients undergoing tests for coronary artery disease (17 men and 15 women, aged 18–80; mean age: 58). None of the patients had pacemakers; most had ventricular ectopic beats. In choosing the ECG signals to be contained in the database, preference was given to subjects with ECGs composed of arrhythmias, coronary artery disease, ischemia, and conduction abnormalities. The annotations were formed by an automated algorithm and then revised manually, following the standard PhysioBank beat annotation definitions. The algorithm usually places beat annotations in the middle of the QRS complex; the locations have not been manually corrected, and there might be rare misaligned annotations as a result. This database was donated by the St. Petersburg Institute of Cardiological Technics (Incart), St. Petersburg, Russia. Exemplary ECGs are depicted in Fig. 2.15. The following MATLAB code was used to plot each signal:

```
%% Ch2_EX_12.m
clc
clear
%Load Sample ECG Data downloaded from the web site
%https://www.physionet.org/pn3/incartdb/
load PetersburgECG.mat

%Plot NORMAL ECG Signal
subplot(4,1,1);
plot(ECGN(:,1));
title('NORMAL ECG Signal')
ylabel('Amplitude')
axis ([0 1000 -inf inf])

%Plot APC ECG Signal
subplot(4,1,2);
plot(ECGAPC(:,1))
```

```
title('APC ECG Signal')
xlabel('Samples')
ylabel('Amplitude')
axis ([0 1000 -inf inf])

%Plot PVC ECG Signal
subplot(4,1,3);
plot(ECGPVC(:,1))
title('PVC ECG Signal')
xlabel('Samples')
ylabel('Amplitude')
axis ([0 1000 -inf inf])

%Plot RBBB ECG Signal
subplot(4,1,4);
plot(ECGRBBB(:,1))
title('RBBB ECG Signal')
xlabel('Samples')
ylabel('Amplitude')
axis ([0 1000 -inf inf])
```

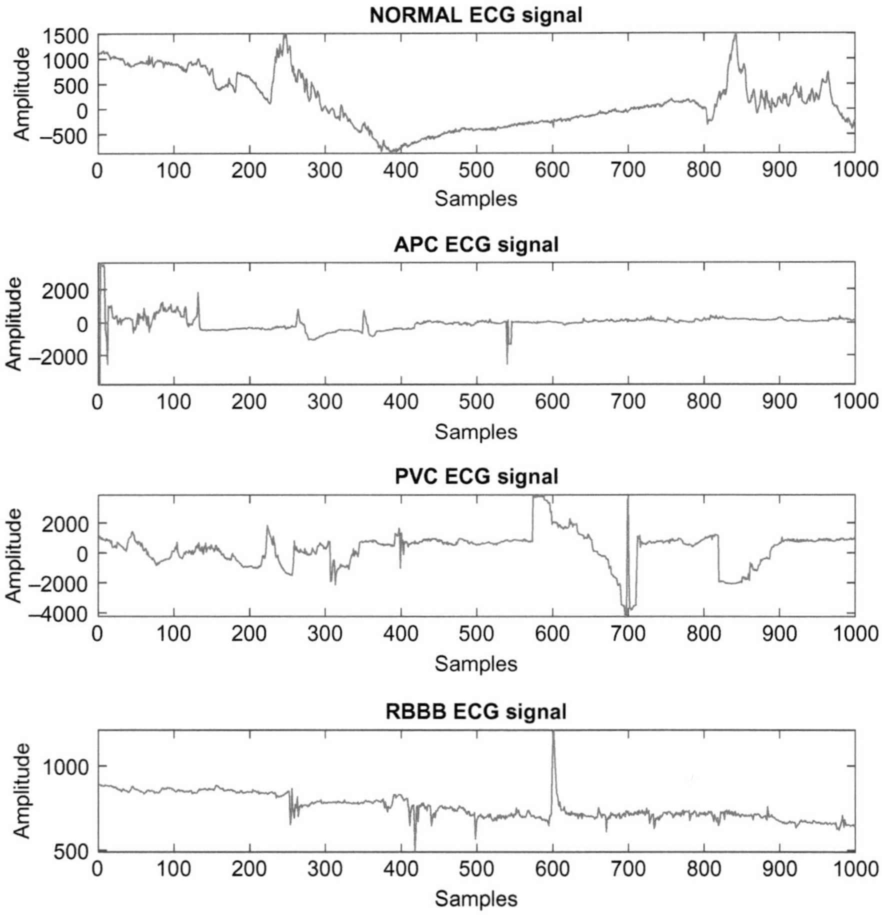

FIG. 2.15 Drawing of different ECG signals (normal, APC, PVC, and RBBB).

2.3.7 Uses of ECG Signals in Congestive Heart Failure Detection

The human heart is the most crucial muscle in human body, which together with blood vessels forms the cardiovascular system. It pumps blood into every cell of the human body. Moreover, heart muscle is the engine of the human body (Horobin, 2008). Heart failure is a common syndrome that progresses gradually but causes cardiac dysfunction due to the failure of the heart to effectively pump blood to all the cells of the human body. The heart deteriorates through heart attacks, long-term high blood pressure, or an anomaly of one of the heart valves. Hence, heart failure is usually not realized until it comes to the more progressive phase, called congestive heart failure (CHF), which causes fluid to flow to the lungs, feet, and abdominal cavity. CHF is a disorder that can result from heart diseases, such as coronary artery disease, damage to the heart after a heart attack, high blood pressure, valvular heart disease, diabetes, and even alcoholism (Masetic & Subasi, 2016; Quinn, 2006).

Considering that there is no exact diagnosis of heart failure, medical diagnosis such as history or physical examinations, ECG, echocardiography, or chest radiography is critical for recognizing CHF. ECG is a noninvasive tool that records electrical activity of the heart and reveals the heartbeat irregularities. It safely studies and records the electrical impulses of the heart, representing possible damage of the heart or irregularities of its heartbeats (Passanisi, 2004). Therefore, ECG is an imperative tool for defining the function and health of the cardiovascular system. Furthermore, it is substantial to describe precise and appropriate diagnosis of physicians to circumvent more damage and to define appropriate approaches and techniques (Son, Kim, Kim, Park, & Kim, 2012). However, a problem arises when there is an inadequate number of physicians to meet the needs of patients. Hence, it is essential to implement an efficient and automated diagnostic system based on ECG signals, together with the application of machine learning techniques for classification of heart diseases. These diagnostic systems support medical experts in distinguishing the anomalies in the cardiovascular system. The diagnostic system first processes the ECG signals taken from several subjects, which are then decomposed into a few applications that perform feature extraction. Extracted ECG signals are employed to distinguish different kinds of heart failure by utilizing several machine learning techniques (Masetic & Subasi, 2016; Subasi, 2013).

According a number of studies implemented, the field of heartbeat classification employing several methods is widespread. Beth Israel Deaconess Medical Center (BIDMC) CHF Database was employed in many researches. Baim et al. (1986)) employed this database to represent the effect of treatment with oral milrinone. Thuraisingham (2009) employed BIDMC CHF Database in distinguishing CHF from normal heartbeats, utilizing k-nearest neighbor algorithm and features from the second-order difference plot taken from Holter monitor cardiac RR intervals. The same database was employed by Kuntamalla and Reddy (2010), and they used sequential trend analysis to discriminate CHF patients from subjects with normal heartbeats. Moreover, Kuntamalla and Reddy (2014) employed multiscale entropy (MSE) to heart rate variability (HRV) signals, to distinguish healthy, young, and elderly subjects from CHF patients. They used Reduced Data Dualscale Entropy Analysis to decrease the data size for clear differentiation of subjects. Hossen and Al-Ghunaimi (2008) employed a technique based on recognition of power spectral density patterns of decomposed subbands of RR interval to recognize patients with CHF. They used 12 subjects from the BIDMC CHF Database. The same authors, Hossen and Al-Ghunaimi (2007) employed different wavelet-decomposition filters with soft-decision algorithm to evaluate the power spectral density of RR-interval data for screening patients with CHF. Yu and Lee (2012) focused on choosing the best feature selector to achieve high accuracy in identifying CHF. The authors employed conditional mutual information feature selector with SVM classifier. Isler and Kuntalp (2007) combined classical HRV and wavelet entropy measures to differentiate healthy patients from subjects with CHF. They utilized a genetic algorithm to choose the best ones among all possible combinations of measures. Moreover, they employed k-nearest neighbor classifier with several k values. Similar study was implemented by Asyali (2003), and the author employed linear discriminant analysis to observe the differentiation power of nine long-term HRV measures, validating the result by employing Bayesian classifier. Pecchia, Melillo, Sansone, and Bracale (2011) studied the differentiation power of HRV in distinguishing normal subjects and subjects with CHF. They employed time and frequency analysis to measure HRV features. The result was assessed by using classification and regression tree (CART) classifier. Also, the authors presented two nonstandard features: average of normal intervals and low/high frequencies recorded over 24 h (Masetic & Subasi, 2016).

EXAMPLE 2.13 The following MATLAB code was used to plot CHF ECG signals. For this example, the BIDMC CHF Database was used. You can download the data from the following website: https://www.physionet.org/physiobank/database/chfdb/

Dataset Information: This database includes long-term ECG signals taken from 15 subjects (11 men, aged 22–71, and 4 women, aged 54–63) with severe CHF (NYHA class 3–4). This group of subjects was part of a larger study group taking conventional medical therapy before taking the oral inotropic agent, milrinone. The different signals are each about 20 h in duration and include two ECG signals each sampled at 250 Hz with 12-bit resolution over a range of ± 10 mV. The original analog signals were recorded at

Boston's Beth Israel Hospital (now the Beth Israel Deaconess Medical Center) employing ambulatory ECG recorders with a typical recording bandwidth of approximately 0.1–Hz. Annotation files (with the suffix.ecg) were prepared employing an automated detector and have not been corrected manually. Exemplary ECGs are depicted in Fig. 2.16. The following MATLAB code was used to plot each signal:

```
%% Ch2_EX_13.m
clc
clear
%Load Sample Congestive Heart Failure (CHF) ECG Data downloaded from the web site
%https://www.physionet.org/physiobank/database/chfdb/
load chf01m.mat

%Plot Congestive Heart Failure (CHF) ECG signal
subplot(2,1,1);
plot(val(1,1:2048));
title('Congestive Heart Failure (CHF) ECG signals')
xlabel('Number of Samples')
ylabel('Amplitude')
legend('Channel I')
axis ([0 2048 -inf inf])

%Plot Congestive Heart Failure (CHF) ECG signal
subplot(2,1,2);
plot(val(2,1:2048));
xlabel('Number of Samples')
ylabel('Amplitude')
legend('Channel II')
axis ([0 2048 -inf inf])
```

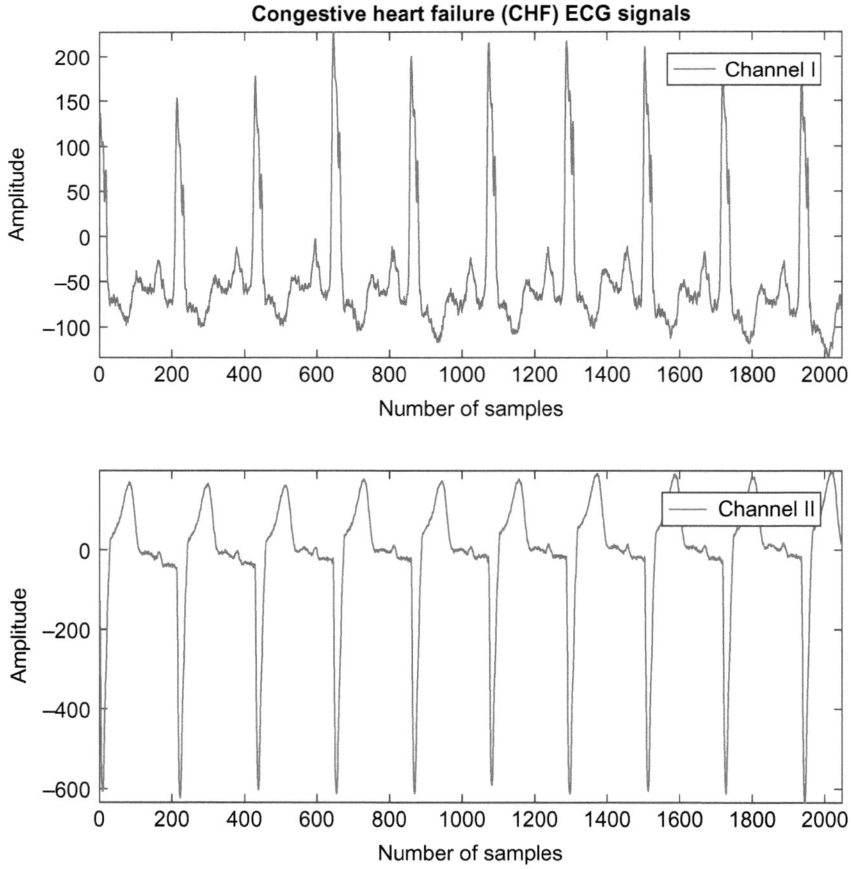

FIG. 2.16 Drawing of CHF ECG signals.

2.3.8 Uses of ECG Signals in Sleep Apnea Detection

Obstructive sleep apnea (OSA) is a type of sleep-disordered breathing represented by repeated cycles of partial (hypopnea) or complete (apnea) failure of the pharyngeal airway (Baguet, Barone-Rochette, Tamisier, Levy, & Pépin, 2012). It disturbs about 10% of middle-aged adults, mainly overweight or obese men. If untreated, OSA can cause cognitive impairment, somnolence, ischemic heart disease, cardiovascular morbidity, and mortality. Nevertheless, almost 80% of people with such conditions persist undiagnosed and consequently untreated because of the various problems related to present diagnostic techniques (Jin & Sánchez-Sinencio, 2015). Conventionally, OSA is diagnosed by expert physicians by manual observation of PSG signals. OSA screening based on visual inspection is challenging. First, PSG is expensive and requires the subject to sleep in a special sleep center. Due to the unfamiliar environment of the sleep center, the subject's quality of sleep is degraded. Second, manual screening of OSA is time-consuming, burdensome, reliant on expensive human resources and physicians' level of expertise and experience, and error-prone due to fatigue. Hence, there is a need for a computer-aided OSA detection scheme. ECG signals are extensively employed for computerized OSA detection (Hassan & Haque, 2016)

EXAMPLE 2.14 The following MATLAB code was used to plot the ECG signals related to the sleep apnea. For this example, data for development and evaluation of ECG-based apnea detectors database was used. The data can be downloaded from the following website:https://www.physionet.org/physiobank/database/apnea-ecg/

Dataset Information: This database is described by Penzel, Moody, Mark, Goldberger, and Peter (2000). The data consist of 70 records, divided into a LEARNING SET of 35 records, and a TEST SET of 35 records. Signals vary in length from slightly less than 7 h to nearly 10 h each. Each signal contains a continuous digitized ECG signal, a set of apnea annotations, and a set of machine-generated QRS annotations. The digitized ECG signals were sampled at 100 Hz with 16 bits per sample. Exemplary ECGs are depicted in Fig. 2.17. The following MATLAB codewas used to plot each signal:

```
%% Ch2_EX_14.m
clc
clear
%Load Sample Sleep Apnea ECG Data downloaded from the web site
%https://www.physionet.org/physiobank/database/apnea-ecg/
load ucddb002_recm.mat

%Plot Sleep Apnea ECG signal
subplot(2,1,1);
plot(val(1,10000:15000));
title('Sleep Apnea ECG signal')
xlabel('Number of Samples')
ylabel('Amplitude')
legend('Channel I')
axis ([0 5000 -inf inf])

%Plot Sleep Apnea ECG signal
subplot(2,1,2);
plot(val(2,10000:15000));
title('Sleep Apnea ECG signal')
xlabel('Number of Samples')
ylabel('Amplitude')
legend('Channel II')
axis ([0 5000 -inf inf])
```

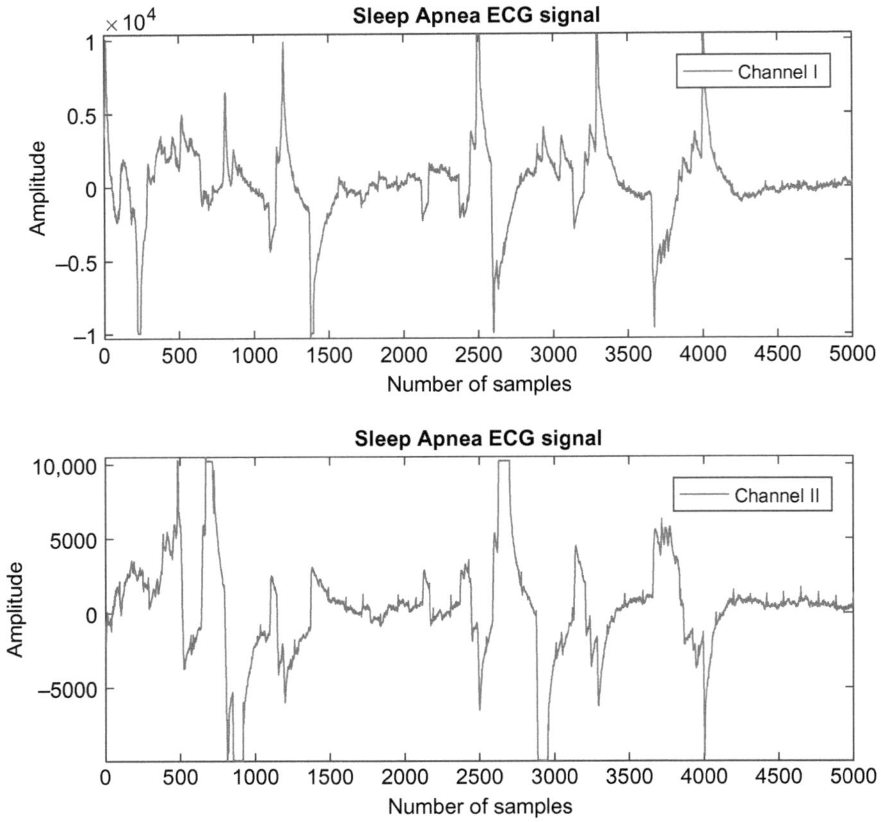

FIG. 2.17 Drawing of sleep apnea ECG signals.

2.3.9 Uses of ECG Signals in Fetal Analysis

Fetal health conditions can be evaluated utilizing ECG signals, which allows discovery of possible distress or congenital heart defects in early stages of pregnancy and during delivery (Kennedy, 1998). Fetal heart rate (HR) monitoring and recognition of the fetal ECG (f-ECG) provides early diagnosis of fetal arrhythmias, making it possible to treat them with drug administration or to preschedule the delivery. In abdominal f-ECG recording, the electrical signal generated by the fetal heart is measured by noninvasive electrodes placed on the mother's abdomen surface. This type of measurement is obviously appropriate for long-term observation as well as for at-home monitoring during all pregnancy (Da Poian, Bernardini, & Rinaldo, 2016).

EXAMPLE 2.15 The following MATLAB code was used to plot the f-ECG signals. For this example, Abdominal and Direct Fetal Electrocardiogram Database was used, and the data can be downloaded from the following website: https://physionet.org/physiobank/database/adfecgdb/

Dataset Information: The Abdominal and Direct Fetal Electrocardiogram Database includes multichannel f-ECG signals taken from five different women in labor, between 38 and 41 weeks of gestation. The signals were recorded in the Department of Obstetrics at the Medical University of Silesia by means of the KOMPOREL system for acquisition and analysis of f-ECG. Every signal contains four differential signals recorded from maternal the abdomen and direct f-ECG registered from the fetal head. The acquisition of direct f-ECG was implemented with a typical spiral electrode, usually employed in a direct f-ECG channel of popular fetal monitors. The signal bandwidth was 1–150 Hz and sampled at 1 kHz with 16-bits resolution (Kotas, Jezewski, Horoba, & Matonia, 2011). Exemplary ECGs are depicted in Fig. 2.18. The following MATLAB code was used to plot each signal:

```
%% Ch2_EX_15.m
clc
clear
%Abdominal and Direct Fetal Electrocardiogram Database
%Load Sample Abdominal and Direct Fetal ECG Database downloaded from the web site
%https://physionet.org/physiobank/database/adfecgdb/
load r01_edfm.mat

%Plot Fetal Electrocardiogram Signal
subplot(2,1,1);
plot(val(1,1:2048));
title('Fetal Electrocardiogram Signals')
xlabel('Samples')
ylabel('Amplitude')
legend('Channel I')
axis ([0 2048 -inf inf])

%Plot Fetal Electrocardiogram Signal
subplot(2,1,2);
plot(val(2,1:2048));
xlabel('Samples')
ylabel('Amplitude')
legend('Channel II')
axis ([0 2048 -inf inf])
```

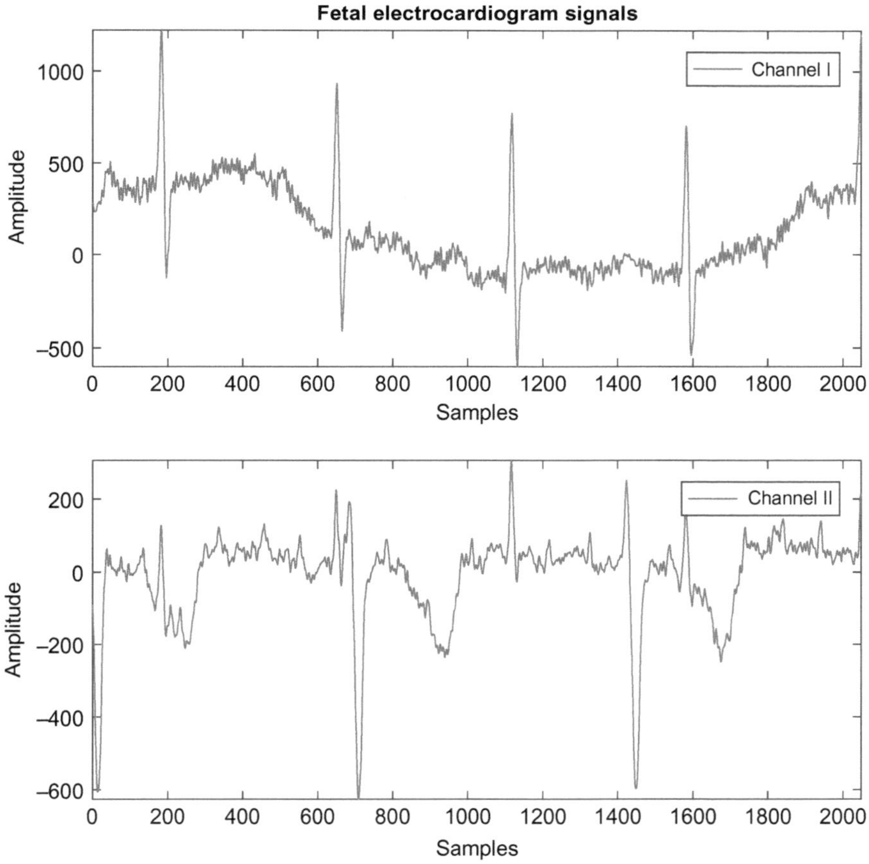

FIG. 2.18 Drawing of fetal ECG signals.

2.4 PHONOCARDIOGRAM

The heart sound signal is a traditional biomedical signal, because the stethoscope is the primary instrument utilized by doctors. Phonocardiogram (PCG) signals produce valuable information related to the heart valves, and consequently several heart diseases can be detected by utilizing this signal (Jia, Song, Tao, & Lu, 2012). Two of the loudest heart sounds are the first (S1) and the second (S2) sounds; the time interval between S2 and S1 is termed diastole and the time interval between S1 and S2 is termed systole. Normal heart sounds produced by the heart valves are low-frequency transient signals, whereas abnormal heart sounds such as heart murmurs are high-frequency, noise-like sounds (Ergen, Tatar, & Gulcur, 2012). Heart murmurs are generated by turbulence in blood flow through narrow cardiac valves or reflow through the atrioventricular valves. Congenital heart faults or detected heart valve diseases are generally the source of abnormal heart murmurs. Aortic stenosis, aortic regurgitation, mitral regurgitation, and mitral stenosis are among the most frequently seen pathological types of murmurs. Research on PCG signals using biomedical signal processing methods is on the rise due to the ability of PCG recordings to characterize significant features of the heart sounds. Many studies have been focused on examining the possibility of the PCG signal classification toward the diagnosis of heart valve disorders (Kao & Wei, 2011). Heart sound is the result of turbulent blood flow and vibrating cardiovascular structures, which propagate to the chest. These vibrations are naturally a consequence of myocardial and valvular events, which are affected by the hemodynamics, function, and electrical activity of the cardiac muscle. These have a direct influence on the spectral, morphological, and timing features of the main heart sounds (S1, S2, and S3) (Carvalho et al., 2011).

In the past, several theories have been posited to describe the source of sounds during the cardiac cycle. Once it was understood that the strength of contraction of the left ventricle had a substantial effect on the strength of the first heart sound, the myocardial theory of the source of the sound was hypothesized. All heart sounds are achieved when a moving column of blood comes to an abrupt stop or slows down significantly. The strength of a heart sound will depend on the level of energy the moving column of blood has accomplished. The abrupt slowing down causes dissipation of energy that produces vibrations affecting the contiguous cardiohemic mass. The factors affecting the acceleration and deceleration of columns of blood included in the creation of the several heart sounds are diverse. These must be taken into consideration for each sound separately, considering the pathophysiology and the physiology of the phase of the cardiac cycle involved (Oweis, Hamad, & Shammout, 2014)

Sound is produced and transmitted by vibrations. These vibrations consist of a series of waves, which are made up of compressions (areas of increased pressure) and subsequent rarefactions (areas of decreased pressure). These waves travel through solid, liquid, or gaseous media. In general, the ease and speed of transmission of sound waves are inversely proportional to the density of the media in which they are traveling. We generally use a stethoscope to transmit the cardiovascular sounds from the chest wall to our ears. Cardiovascular sound can be divided into brief, circumscribed sounds or transients (heart sounds and clicks) and longer combinations of vibrations (heart murmurs) (Ranganathan, Sivaciyan, & Saksena, 2007)

In some situations, a third heart sound (S3) might be heard, related to abrupt termination of the ventricular rapid-filling phase. Because the ventricles are full of blood and their walls are relaxed during this part of diastole, the vibrations of S3 are of very low frequency. In late diastole, a fourth heart sound (S4) may be heard occasionally, produced by atrial contractions transferring blood into the enlarged ventricles. In addition to these sounds, valvular clicks and snaps are rarely heard (Rangayyan, 2015).

2.4.1 Heart Murmurs

The intervals between S1 and S2 (ventricular systole), and S2 and S1 (ventricular diastole) of the next cycle are normally silent. Murmurs, which may happen in these intervals, are produced by certain cardiovascular imperfections and disorders. Murmurs are high-frequency, noise-like sounds that occur when the velocity of blood becomes high as it flows through an irregularity. Typical conditions in the cardiovascular system that produce turbulence in blood flow are valvular stenosis and insufficiency. A valve can be insufficient when it cannot close well and achieves reverse leakage or regurgitation of blood through a narrow opening. Systolic murmurs (SM) are generated by conditions such as ventricular septal defect, aortic stenosis (AS), pulmonary stenosis (PS), mitral insufficiency (MI), and tricuspid insufficiency (TI). Semilunar valvular stenosis (aortic stenosis, pulmonary stenosis) generates an obstruction in the path of blood ejected during systole. AV valvular insufficiency generates regurgitation of blood to the atria during ventricular contraction. Diastolic murmurs (DM) are generated by conditions such as aortic or pulmonary insufficiency, and mitral or tricuspid stenosis. Other conditions generating murmurs are atrial septal defect (ASD), patent ductus arteriosus (PDA), as well as certain physiological or functional conditions that cause increased cardiac output or blood velocity (Rangayyan, 2015).

2.4.2 First Heart Sound (S1)

The first heart sound happens at the beginning of ventricular contraction. To better understand the physiology of the first heart sound, the cardiac events that happen around the time of the first heart sound need to be distinguished. At the end of diastole, the atrium contracts and achieves an extra stretch and filling of the ventricle. This is instantly followed by the ventricular contraction. Once the ventricular pressure increases and surpasses the atrial pressure, the mitral and tricuspid valve leaflets turn out to be apposed and close. As the ventricular pressures continue to increase and surpass that of the aorta and the pulmonary artery, the semilunar valves open and the ejection phase begins. All these actions happen in quick sequence over a short period of time and subsidize the production of the first heart sound. S1 is comparatively wide and is composed of many components that overlap each other. These components are "atrial," "mitral," "tricuspid," and "aortic." The strength of S1 is apparently associated with the strength or intensity of the distinct components (Ranganathan et al., 2007).

2.4.2.1 *Atrial Component*

The energy of the column of blood pushed by the atrial contraction dissipates as the column slows down against the ventricular walls. This slowing down is steady in most normal subjects because of the good compliance and distensibility of the ventricles. Hence, the sound produced has a very low frequency and is not audible. Nevertheless, when the PR interval on the ECG is short, this component can happen very close to the beginning of the ventricular contraction and essentially becomes a part of the first heart sound and contributes to its duration. Aside from this, it has no clinical implication (Ranganathan et al., 2007).

2.4.2.2 *Mitral Component*

This is the most imperative component of S1. It corresponds in timing to the closure of the mitral valve leaflets. But the apposition of the valve leaflets cannot yield the sound itself. Once the ventricle begins contracting, the pressure increases and imparts energy into the mass of blood within its cavity. Once the pressure in the ventricle surpasses that of the atrium, the column of blood is put into motion. Meanwhile the aortic pressure is much higher than the atrial pressure, and the aortic valve remains closed at this time; the blood contained in the ventricle can only rush toward the atrium. Because the anatomical structure of the mitral valve is like a parachute, the leaflets are raised by the moving blood into a closed position. The papillary muscles contracting and pulling on the chordae tendineae stop leaflet eversion into the atrium. The closed leaflets are detained back by the papillary muscles, attaining the limits of their stretch, stopping the column of blood from moving into the atrium. This abrupt slowing down of the column of blood produces the mitral component, or M1. The energy dissipation produces vibrations of the column of blood as well as the surrounding structures such as the mitral valve structures and the ventricular wall. The mechanism of sound creation of this M1 is identical to the sound created by a parachute filling with wind as it stretches and produces the deceleration of the moving mass of air or by the sail of a sailboat that snaps when filled with a gust of wind (Ranganathan et al., 2007).

2.4.2.3 *Tricuspid Component*

This component is apparently analogous to the M1 for similar cardiac events that happen including the right-sided structures, such as the tricuspid valve leaflets and the right ventricular wall. But these events happen at much lower pressures and are slightly delayed. The effects of the mechanical events of the right ventricle start slightly later than that of the left ventricle. Consequently, the tricuspid component (T1) follows the M1. It must be distinguished that this component arises because the lower pressure is generally low in frequency. The T1 in the normal adult subjects, although it can be recordable, might subsidize to the duration of S1 but not be audible as a distinct component (Ranganathan et al., 2007).

2.4.2.4 *Aortic Component*

The aortic component (A1) is generally the second component of audibly split S1 in adults. After mitral and tricuspid valve closures, the ventricular pressure continues to increase during the phase of isovolumic contraction. Once the pressure surpasses the aortic and pulmonary diastolic pressures, the ejection phase starts as the semilunar valves open. The column of blood ejected into the aorta as it hits the aortic walls slows down, and once slowed will result in an audible sound (Ranganathan et al., 2007).

2.4.3 Second Heart Sound (S2)

The second heart sound happens at the end of the ejection phase of systole. It is associated with the closure of the semilunar valves. As the blood is ejected into the aorta and the pulmonary artery during systole (stroke volume), the aortic and the pulmonary pressures increase, and these two vessels become swollen. At the end of systole, as the ventricular pressures start to decrease, the elastic components of the great vessels preserving a higher pressure results in a pressure gradient that drives the columns of blood back into the ventricles. Because of the lower resistance with the dropping ventricular pressures compared to the periphery, the columns of blood in the great vessels flow toward the ventricles at this time. The opposite flow of the columns of blood in the aorta and the pulmonary artery parachutes the cusps of the aortic and the pulmonary valves, closing them. The abrupt slowing down of the columns of blood against the closed semilunar valves results in dissipation of energy, causing the A2 and the P2 components of S2 (Ranganathan et al., 2007).

The S2 is typically louder, crisper, and shorter in duration compared to S1, because the semilunar valve closures happen at much higher pressures than the AV valves, and the dissipated energy in the columns of blood is much larger. In normal young subjects, one can repeatedly hear both components of S2 (A2 and P2). Hence, the S2 can be heard as a split sound. A2 is the first of the two components. The pulmonary arterial bed is bigger and produces noticeably less resistance to forward flow. This makes the inclination to reverse flow happen later and slower compared to the left side. Moreover, the lower pressures accomplished by the right ventricle during systole may cause a slower rate of relaxation of the right ventricle compared to the left ventricle. Consequently, the P2 component happens later. The A2 component is usually heard over the true aortic area, but P2 is heard over the second and the third left intercostal spaces near the sternal border. If the loudness of the two components is compared, the A2 is always louder than the P2 because the systemic arterial resistance is normally 10 times higher than the pulmonary arterial resistance (Ranganathan et al., 2007).

2.4.4 Third Heart Sound (S3)

After the opening of the mitral and the tricuspid valves, blood flows into the ventricles from the atria during diastole. Diastolic filling of the ventricle is separable into three phases, an early rapid filling phase, followed by a slow filling phase or diastasis, and at the end by the atrial contraction phase. The peak of this filling period is characterized by a sound known as the third heart sound. S3 happens at the end of the rapid filling phase of diastole. This phase of diastolic expansion is normally quick and dynamic. However, there is an inclination for the moving column of blood entering the ventricle during the rapid-filling phase to slow down toward the end of this period (Ranganathan et al., 2007).

2.4.5 Fourth Heart Sound (S4)

The atrium usually contracts at the end of diastole and achieves an extra stretch to the ventricles. Once the ventricular compliance is decreased because of factors such as ischemia, infiltrates, fibrosis, infarction, or hypertrophy, this arouses a more dynamic contraction from the atrium. The improved force of atrial contraction also increases the atrial's wave pressure peak. This increased pressure head inclines to accelerate the diastolic inflow at the end of diastole. Because the ventricular compliance is decreased, the accelerated inflow at this phase of diastole is quickly slowed down. This sudden slowing down of the column of blood entering the ventricle at end-diastole leads to dissipation of energy that produces a sound. The sound is referred to as atrial gallop and is called S4. The sound, produced at low pressures of diastole, has a low frequency like S3. But the timing is not similar; it is closer to S1 (Ranganathan et al., 2007).

2.4.6 Uses of PCG Signals in Diagnosis of Heart Diseases

CVDs are the main reason of death worldwide. To successfully detect CVD, different personal health systems have been developed to improve detection and collection of data for clinical decision support. Usually, ECG and heart sound (HS) auscultation are employed for CVD diagnosis (Carvalho et al., 2011). But some of the heart diseases are hard to detect using ECG. Because heart sound signals are complex and highly nonstationary in nature, it is difficult to examine them in an automated way. An auscultation method employed to support clinicians with precise and objective interpretation of heart sounds can be utilized to detect four sounds, namely, S1, S2, S3, and S4, during the heart cycle. The crucial components for signal analysis are S1 and S2, as well as systolic and diastolic periods. Localization (segmentation) of heart sounds must be done before any analysis (Oweis et al., 2014). Duration of S1 and S2 may help determine the heart sound type and detect abnormal heart sounds, and the amplitude of S1 reveals valuable information about myocardial contractility ability (Zhong & Scalzo, 2013).

The heart sound signals (heartbeats) taken from normal subjects generally comprise two separate tones, S1 and S2. Moreover, an auscultation method employed to deliver clinicians with precise and objective interpretation of heart sounds can be utilized to distinguish four sounds, namely, S1, S2, S3, and S4, during the heart cycle (Ramović, Bandić, Kevrić, Germović, & Subasi, 2017). Precise and early diagnosis of cardiac disorders is of great importance that can be achieved through heart auscultation. It is the most common and widely used technique to screen for structural abnormalities of the cardiovascular system. Distinguishing related features and creating an analysis based on the sounds heard through a stethoscope is an ability that can take years to be learned and improve. The effectiveness and precision of analysis based on heart sound auscultation can be enhanced significantly by employing biomedical signal-processing techniques to investigate the PCG signals (Chen, Wang, Shen, & Choy, 2012; Cherif, Debbal, & Bereksi-Reguig, 2010; Choi, Shin, & Park, 2011; Safara, Doraisamy, Azman, Jantan, & Ranga, 2012).

PCG signals can deliver significant information relevant to the performance of heart valves, thus it has potential for distinguishing several heart diseases (Jia et al., 2012). Normal heart sounds are low-frequency transient signals generated by the heart valves, whereas pathological heart sounds, such as heart murmurs, are high-frequency, noise-like sounds (Ergen et al., 2012). Heart murmurs are generated as a result of turbulence in blood flow through narrow cardiac valves or reflow through the atrioventricular valves.

Congenital heart defects or acquired heart valve diseases are frequently the source of abnormal heart murmurs. Aortic regurgitation, aortic stenosis, mitral regurgitation, and mitral stenosis are among the most common pathological types of murmurs. Studies on PCG signals employing biomedical signal-processing techniques are on the rise due to the ability of PCG signals to characterize significant features of the heart sounds (Kao & Wei, 2011).

EXAMPLE 2.16 The following MATLAB code was used to plot the PCG signals. For this example, PCG Phonogram Dataset was used. You can download the data from the following website: https://www.prweb.com/releases/2004/09/prweb162074.htm

Exemplary PCG signals are depicted in Fig. 2.19. The following MATLAB code was used to plot each signal:

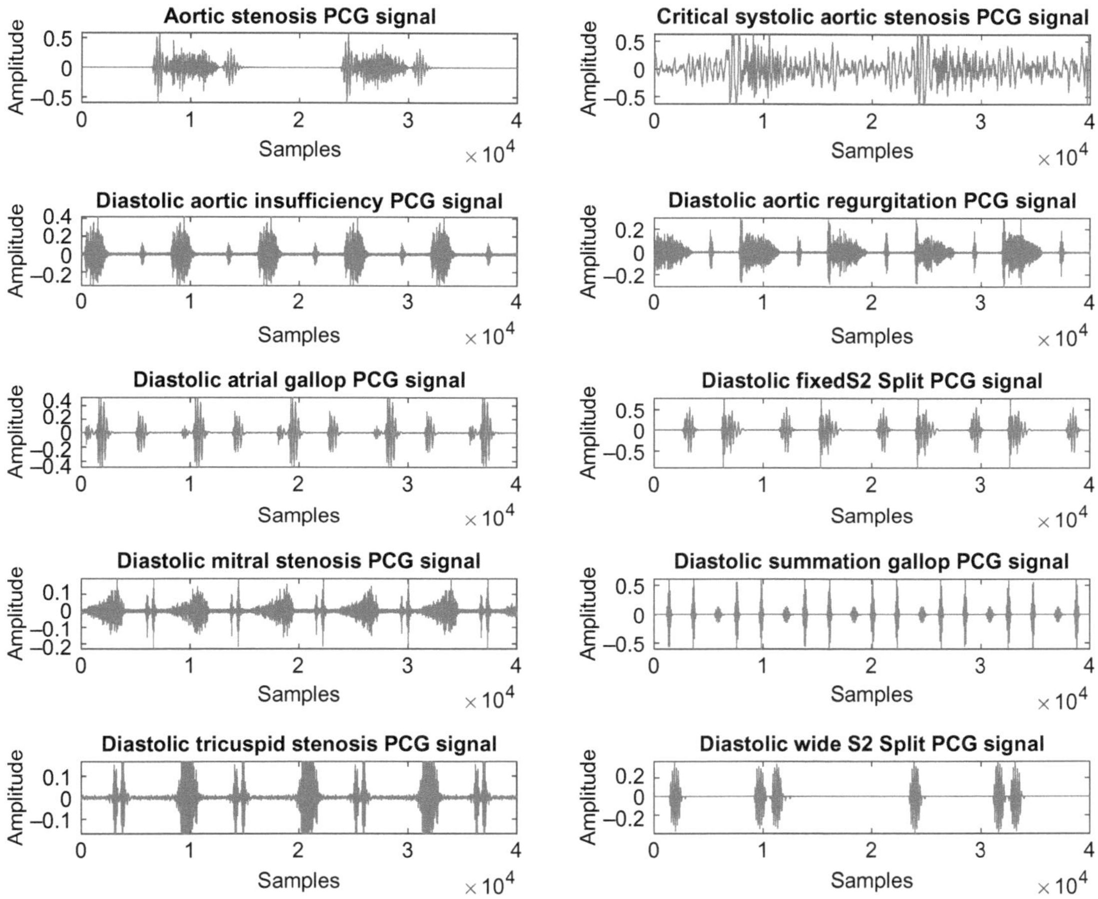

FIG. 2.19 Drawing of different PCG signals.

```
%% Ch2_EX_16.m
clc
clear
%PCG Phonogram Data Set from BioSignetics Corporation
%Load Sample PCG Data downloaded from the web site
%https://www.prweb.com/releases/2004/09/prweb162074.htm
load PCG.mat

%Plot Aortic_Stenosis PCG Signal
subplot(5,2,1);
plot(Aortic_Stenosis_2(1:40000,1));
title('Aortic Stenosis PCG Signal')
ylabel('Amplitude')
axis ([0 40000 -inf inf])
```

```
%Plot Critical_Systolic_Aortic_Stenosis PCG Signal
subplot(5,2,2);
plot(Critical_Systolic_Aortic_Stenosis(1:40000,1))
title('Critical Systolic Aortic Stenosis PCG Signal')
xlabel('Samples')
ylabel('Amplitude')
axis ([0 40000 -inf inf])

%Plot Diastolic_Aortic_Insufficiency PCG Signal
subplot(5,2,3);
plot(Diastolic_Aortic_Insufficiency(1:40000,1))
title('Diastolic Aortic Insufficiency PCG Signal')
xlabel('Samples')
ylabel('Amplitude')
axis ([0 40000 -inf inf])

%Plot Diastolic_Aortic_Regurgitation PCG Signal
subplot(5,2,4);
plot(Diastolic_Aortic_Regurgitation(1:40000,1))
title('Diastolic Aortic Regurgitation PCG Signal')
xlabel('Samples')
ylabel('Amplitude')
axis ([0 40000 -inf inf])

%Plot Diastolic_Atrial_Gallop PCG Signal
subplot(5,2,5);
plot(Diastolic_Atrial_Gallop(1:40000,1))
title('Diastolic Atrial Gallop PCG Signal')
xlabel('Samples')
ylabel('Amplitude')
axis ([0 40000 -inf inf])

%Plot Diastolic_FixedS2_Split PCG Signal
subplot(5,2,6);
plot(Diastolic_FixedS2_Split_II(1:40000,1))
title('Diastolic FixedS2 Split PCG Signal')
xlabel('Samples')
ylabel('Amplitude')
axis ([0 40000 -inf inf])

%Plot Diastolic_Mitral_Stenosis PCG Signal
subplot(5,2,7);
plot(Diastolic_Mitral_Stenosis(1:40000,1))
title('Diastolic Mitral Stenosis PCG Signal')
xlabel('Samples')
ylabel('Amplitude')
axis ([0 40000 -inf inf])

%Plot Diastolic_Summation_Gallop PCG Signal
subplot(5,2,8);
plot(Diastolic_Summation_Gallop(1:40000,1))
title('Diastolic Summation Gallop PCG Signal')
xlabel('Samples')
ylabel('Amplitude')
axis ([0 40000 -inf inf])
```

```
%Plot Diastolic_Tricuspid_Stenosis PCG Signal
subplot(5,2,9);
plot(Diastolic_Tricuspid_Stenosis(1:40000,1))
title('Diastolic Tricuspid Stenosis PCG Signal')
xlabel('Samples')
ylabel('Amplitude')
axis ([0 40000 -inf inf])

%Plot Diastolic_Wide_S2_Split PCG Signal
subplot(5,2,10);
plot(Diastolic_Wide_S2_Split(1:40000,1))
title('Diastolic Wide S2 Split PCG Signal')
xlabel('Samples')
ylabel('Amplitude')
axis ([0 40000 -inf inf])
```

2.5 PHOTOPLETHYSMOGRAPHY

Photoplethysmography (PPG), also known as the pulse oximeter waveform, is a simple and low-cost optical tech nique that can be utilized to detect blood volume changes in the microvascular bed of tissue. It is frequently employed noninvasively to take measurements at the skin surface. PPG signals (Allen, 2007; Kamal, Harness, Irving, & Mearns, 1989) are becoming popular for monitoring heart rate (HR) because of their wearable implementation com pared to conventional ECG technology. These signals are obtained through pulse oximeters embedded in a small wearable device worn on the earlobes, fingertips, or at the wrists. The pulse oximeters cast light on the wearer's skin through light emitting diodes (LED) and acquire the reflected or transmitted light whose intensity depends on the volume of blood existing in the arteries under the skin. As the volume of blood changes with the cardiac cycle, the light intensity also changes with the cardiac rhythm, which can be employed to extract HR information. Originally PPG signals were employed to monitor patients' HRs 24 hours in hospital beds, mostly utilizing the fingertip-type devices (Khan, Al Hossain, Uddin, Alam, & Hasan, 2016).

The PPG waveform contains a pulsatile (AC) physiological waveform for synchronous cardiac changes in the blood volume with each heartbeat and is superimposed on a slowly varying (DC) baseline with several lower frequency components attributed to respiration, sympathetic nervous system activity, and thermoregulation. Even though the source of the components of the PPG signal are not fully understood, it is usually known that they can provide valuable information about the cardiovascular system. Recently, there has been great interest in this method, driven by the demand for simple, portable, and low-cost technology for the primary care and community-based clinical environments, wide accessibility of small and low-cost semiconductor components, and the development of computer-based pulse wave analysis techniques. PPG technology has been utilized in a wide range of medical devices for measuring blood pressure, oxygen saturation, and cardiac output; evaluating autonomic function; and detecting peripheral vascular disease (Allen, 2007).

A PPG signal from any location on the skin can be divided into an oscillating (AC) and steady-state (DC) component, and their amplitudes are reliant on the structure and flow in the vascular bed. A good emitter for use in a PPG of skin blood flow is one in the frequency range 600–700 nm, and the best signal for AC analysis can be taken from the finger pulp. The frequency range of the data acquisition should be from 0.01 to 15 Hz, then all the information in the signal can be extracted about the autonomic nervous system control of the cardiovascular system, particularly between 0.01 and 2 Hz (Kamal et al., 1989).

The waveform, as presented on the modern pulse oximeter, is an amplified and highly filtered measurement of light absorption by the local tissue over time. It is adjusted by medical device manufacturers to emphasize its pulsatile components. All modern pulse oximeters extract and present the heart rate and oxygen saturation taken from the PPG measurements at multiple wavelengths. Hence, the PPG is an exceptional monitor for cardiac arrhythmia, mostly when employed in conjunction with ECG. The PPG can be employed to measure the ventilator-induced modulations with slight modifications in the display of the PPG. Exploration of PPG is underway to analyze its use in advanced digital signal-processing techniques to achieve a new physiologic parameter (Alian & Shelley, 2014)

Apart from clinical scenarios, many sportsman, exercisers, and elderly people need real-time mobile observation of their HR, to confine them from extreme physical action and thereby to prevent any chances of heart attack. When PPG signals are employed during physical exercise, frequently three-axis accelerometry data are acquired (Lee et al., 2010). Although PPG signals can monitor HR rather suitably in hospital environments, trying to include them during fitness programs yields a great challenge, because the signals are heavily corrupted by motion artifacts (MA) (Khan et al., 2016).

PPG is a well-known, noninvasive, heart monitoring technique. Unlike ECG, PPG monitoring can be realized at peripheral sites on the body and does not require a conductive gel to have good body contact. Hence, PPG sensors are finding considerable new applications in wearable devices and smartwatches (Jarchi & Casson, 2016)

EXAMPLE 2.17 The following MATLAB code was used to plot the PPG and related ECG signal. For this example, Wrist PPG During Exercise Dataset was used. You can download the data from the following website: https://physionet.org/physiobank/database/wrist/

Dataset Information: This database includes wrist PPGs collected throughout running, walking, and bike riding. Synchronized motion evaluations are collected utilizing gyroscopes and accelerometers to produce multiple options for the removal of motion interference from the PPG traces. Measurements were collected employing an ECG unit placed on the chest together with a PPG and Inertial Measurement Unit placed on the left wrist while participants employed an indoor treadmill and exercise bike. Single-channel, two -electrode, ECG signals were acquired utilizing an Actiwave (CamNtech, Cambridge, UK) recorder and pregelled, self-adhesive, Silver-Silver Chloride (Ag/AgCl) electrodes, which are standard for ECG monitoring. PPG and motion data were collected employing a Shimmer 3 GSR+ unit (Shimmer Sensing, Dublin, Ireland). This includes a gyroscope, a low-noise accelerometer, a wide-range accelerometer, and a magnetometer combined into a single package. The PPG sensor was glued to the main Shimmer unit to achieve a rigid connection and permit the movement sensors inside the main Shimmer unit to precisely acquire the movement of the PPG sensor. The combined unit was then located on the left wrist in almost the position of a standard watch. All signals were sampled at 256 Hz. Records from eight participants are present (three male, five female), aged 22–32 (mean 26.5). For the walking and running records, the database contains the raw PPG and motion signals existing after segmentation into the suitable activity. The cycling PPG signals were low-pass filtered employing a second-order IIR Butterworth digital filter with ~15 Hz cutoff frequency. Further details and illustrative example signals can be found in the reference paper (Jarchi & Casson, 2016). Exemplary PPGs are depicted in Fig. 2.20. The following MATLAB code was used to plot each signal:

```
%% Ch2_EX_17.m
clc
clear
%Wrist PPG During Exercise
%Load Sample PPG Data downloaded from the web site
%https://physionet.org/physiobank/database/wrist/
load s1_high_resistance_bikem.mat

%Plot ECG Signal
subplot(2,1,1);
plot(val(1,4097:8192));
title('ECG Signal')
xlabel('Samples')
ylabel('Amplitude')
axis ([0 4096 -inf inf])

%Plot PPG Signal
subplot(2,1,2);
plot(val(2,4097:8192))
title('PPG Signal')
xlabel('Samples')
ylabel('Amplitude')
axis ([0 4096 -inf inf])
```

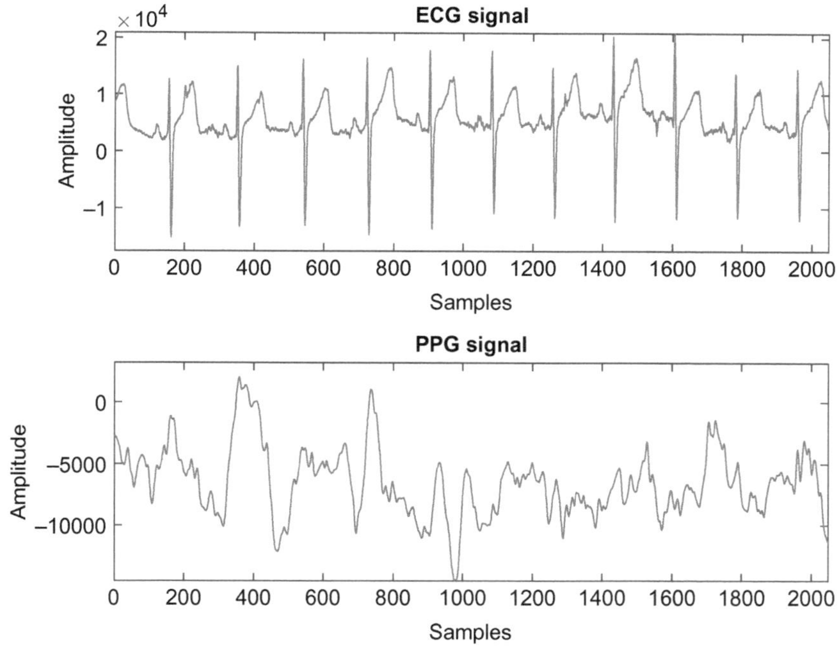

FIG. 2.20 Drawing of ECG and PPG signals.

2.6 OTHER BIOMEDICAL SIGNALS

2.6.1 Electroneurogram

The electroneurogram (ENG) measures consequences of the stimulation of a peripheral nerve with an electric shock so the response along the nerve can be measured. The ENG, generally collected with needle electrodes, is employed to measure conduction velocity of the nerve to help with nerve injury diagnosis. The conduction velocity can be estimated from the distance the resulting two signal waveforms are separated (Sörnmo & Laguna, 2005).

2.6.2 Electroretinogram

An electroretinogram (ERG) is utilized for studying the electrical potentials produced by the retina of the eye during light stimulation. The ERG is acquired by inserting the exploring electrode, encapsulated in a contact lens, on the cornea. The ERG is beneficial for evaluating the electrical response of the rods and cones. A normal ERG presents suitable responses with increased light intensity, whereas an abnormal ERG is attained in conditions such as arteriosclerosis of the retina or detachment of the retina (Sörnmo & Laguna, 2005).

2.6.3 Electrooculogram

The EOG is the recording of the steady corneal-retinal potential that is related to vertical and horizontal movements of the eye to quantify the direction of the gaze. The EOG is of particular interest in people with sleep disorders in which the occurrence of REM is crucial to determine the certain sleep stages. The EOG is acquired when studying a rapid, involuntary oscillation of the eyeballs of patients suffering from vertigo and dizziness. The EOG is also beneficial in virtual reality applications in which a device for eye-tracking may be required (Sörnmo & Laguna, 2005).

2.6.4 Electrogastrogram

The electrogastrogram (EGG) is a recording of the impulses that propagate through the muscles of the stomach to evaluate their contractions. The EGG is studied when the muscles of the stomach or the nerves directing the muscles are not working properly. The EGG is acquired by putting a number of electrodes over the stomach during fasting and after the eating. In normal subjects, a steady "rhythmic" signal is produced by the muscles of the stomach, having an amplitude that increases after eating; the normal frequency of the gastric rhythm is roughly 3 cycles/minute. But in symptomatic patients, the rhythm is usually irregular and occasionally without an increase in the amplitude after the eating (Sörnmo & Laguna, 2005).

The surface EGG represents the overall electrical activity of the stomach, including the electrical control activity and the electrical response activity. Gastric dysrhythmia or arrhythmia can be detected via analysis of the EGG. Precise and consistent measurement of the electrical activity of the stomach necessitates implantation of electrodes within the stomach that limits practical usage (Rangayyan, 2015).

2.6.5 Carotid Pulse

The carotid pulse (CP) is a pressure signal acquired over the carotid artery as it passes near the surface of the body at the neck. It delivers a pulse signal signifying the variations in arterial blood pressure and volume with each heartbeat. Because of the proximity of the recording site to the heart, the CP signal closely looks like the morphology of the pressure signal at the root of the aorta. The CP is a beneficial assistant to the PCG and can help in the recognition of S2 and its components. The CP increases sharply with the ejection of blood from the left ventricle to the aorta, reaching a peak called the percussion (P) wave. This is followed by a plateau or a secondary wave known as the tidal (T) wave, resulting from a reflected pulse returning from the upper body. Next, closure of the aortic valve results in a notch known as the dicrotic (D) notch. The dicrotic notch may be followed by the dicrotic wave (DW) because of the reflected pulse from the lower body. The carotid pulse trace is affected by valvular imperfections such as mitral insufficiency and aortic stenosis; however, it is not commonly employed in clinical diagnosis. The CP signals have a nominal bandwidth of 0–100 Hz and usually recorded with the PCG and ECG signals. Location of the CP transducer needs careful choice of a location on the neck as close to the carotid artery as possible, in which the pulse is felt the strongest, usually by a trained technician (Rangayyan, 2015).

2.6.6 Vibromyogram

The vibromyogram (VMG) is the direct mechanical manifestation of contraction of a skeletal muscle and is a vibration signal that escorts the EMG. The signal has also been called the sound-, acoustic-, or phonomyogram. Muscle sounds or vibrations are associated with the change in contraction of the fundamental muscle fibers, and can be acquired utilizing a contact microphone or accelerometer located on the muscle surface. The intensity and frequency of the VMG can change directly proportional to the contraction level. The VMG, along with the EMG, can be beneficial in studies associated with neuromuscular control, athletic training, muscle contraction, and biofeedback. Nevertheless, VMG signal analysis is not as well established or as popular as EMG analysis (Rangayyan, 2015).

2.6.7 Vibroarthrogram

The knee joint is formed between the femur, the patella, and the tibia. The knee joint is the main articulation in the human body which can efficiently change from 0 degrees extension to 135 degrees flexion, together with 20–30 degrees rotation of the flexed leg on the femoral condyles. The knee joint is known as a synovial joint, as it encompasses a lubricating substance known as the synovial fluid. The patella (knee cap), a sesamoid bone, protects the joint, and is exactly associated with the slide in the groove (trochlea) of the femur during leg movement. The knee derives its physiological movement and typical rolling-gliding mechanism of flexion and extension from its six degrees of freedom. Considerable noise is frequently related to degeneration of knee-joint surfaces. The vibroarthrogram (VAG) is the vibration signal taken from a joint during movement. Normal joint surfaces are smooth and yield little or no sound, whereas joints affected by osteoarthritis and other degenerative disorders suffer from tendon loss and yield grinding sounds. Detection of knee-joint disorders via the analysis of VAG signals may support physicians to circumvent needless exploratory surgery, and also aid better selection of patients who would benefit from surgery. The VAG signal is not yet well understood and is a difficult signal to investigate due to its complex nonstationary characteristics (Rangayyan, 2015).

REFERENCES

Abdullah, A. A., Subasi, A., & Qaisar, S. M. (2017). Surface EMG signal classification by using WPD and ensemble tree classifiers. In: *Presented at the CMBEBIH 2017: Proceedings of the international conference on medical and biological engineering 2017*, Vol. 62. (p. 475). Springer.

Akben, S. B., Subasi, A., & Tuncel, D. (2010). *Analysis of repetitive flash stimulation frequencies and record periods to detect migraine using artificial neural network: Vol. 36.* (pp. 925–931). Springer.

Akben, S. B., Subasi, A., & Tuncel, D. (2011). Analysis of EEG signals under flash stimulation for migraine and epileptic patients. *Journal of Medical Systems*, *35*(3), 437–443.

Alian, A. A., & Shelley, K. H. (2014). Photoplethysmography. *Hemodynamic Monitoring Devices*, *28*(4), 395–406. https://doi.org/10.1016/j.bpa.2014.08.006.

Alickovic, E., & Subasi, A. (2018). Ensemble SVM method for automatic sleep stage classification. *IEEE Transactions on Instrumentation and Measurement*.

Allen, J. (2007). Photoplethysmography and its application in clinical physiological measurement. *Physiological Measurement*, *28*(3), R1.

Andrzejak, R. G., Schindler, K., & Rummel, C. (2012). Nonrandomness, nonlinear dependence, and nonstationarity of electroencephalographic recordings from epilepsy patients. *Physical Review E*, *86*(4), 046206.

Andrzejak, R. G., Chicharro, D., Lehnertz, K., & Mormann, F. (2011). Using bivariate signal analysis to characterize the epileptic focus: the benefit of surrogates. *Physical Review E*, (046203), 83.

Andrzejak, R. G., Widman, G., Lehnertz, K., David, P., & Elger, C. E. (2006). Improved spatial characterization of the epileptic brain by focusing on nonlinearity. *Epilepsy Research*, *69*, 30.

Aschenbrenner-Scheibe, R., Maiwald, T., Winterhalder, M., Voss, H., Timmer, J., & Schulze-Bonhage, A. (2003). How well can epileptic seizures be predicted? An evaluation of a nonlinear method. *Brain*, 2616–2626.

Asyali, M. H. (2003). Discrimination power of long - term heart rate variability measures. In: *Engineering in medicine and biology society:*

Proceeding of the 25th annual international conference of the IEEE EMBS, Cancum, pp. 200–203.

Baguet, J.-P., Barone-Rochette, G., Tamisier, R., Levy, P., & Pépin, J.-L. (2012). Mechanisms of cardiac dysfunction in obstructive sleep apnea. *Nature Reviews Cardiology*, *9*(12), 679.

Bailey, J. J., Berson, A. S., & Garson, A. (1990). Recommendations for standardization and specifications in automated electrocardiography: bandwidth and digital signal processing: a report for health professionals by an ad hoc writing group of the Committee on Electrocardiography and Cardiac electrophysiology of the Council on Clinical Cardiology, American Heart Association. *Circulation*, *81*, 730.

Baim, D. S., Colucci, W. S., Monrad, E. S., Smith, H. S., Wright, R. F., Lanoue, A., … Braunwald, E. (1986). Survival of patients with severe congestive heart failure treated with oral milrinone. *Journal of American College of Cardiology*, *7*(3), 661–670.

Barlow, J. S. (1993). *The electroencephalogram. Its patterns and origins*. Cambridge, MA: The MIT Press.

Bear, M. F., Connors, B. W., & Paradiso, M. A. (1996). *Neuroscience. Exploring the brain*. Baltimore: Williams & Wilkins.

Begg, R., Lai, D. T., & Palaniswami, M. (2007). *Computational intelligence in biomedical engineering*. CRC Press.

Berbari, E. (2000). Principles of electrocardiography. In J. D. Bronzino (Ed.), *The biomedical engineering handbook*. 2nd ed.: CRC Press LLC.

Berbari, E. J., Lazzara, R., Samet, P., & Scherlag, B. J. (1973). Noninvasive technique for detection of electrical activity during the PR segment. *Circulation*, *48*, 1006.

Brazier, M. (1968). *Electrical activity of the nervous system*. Baltimore: Williams & Wilkins.

Bronzino, J. D. (1999). *Biomedical engineering handbook*. Vol. 2. CRC Press.

Carvalho, P., Paiva, R. P., Kumar, D., Ramos, J., Santos, S., & Henriques, J. (2011). *A framework for acoustic cardiac signal analysis*: (pp. 151–160). Presented at the BIOSIGNALS.

Chen, Y., Wang, S., Shen, C.-H., & Choy, F. K. (2012). Intelligent identification of childhood musical murmurs. *Journal of Healthcare Engineering*, *3*(1), 125–139.

Chazal, P., O'Dwyer, M., & Reilly, R. B. (2004). Automatic classification of heartbeats using ECG morphology and heartbeat interval features. *IEEE Transactions on Biomedical Engineering*, *51*(7), 1196–1206.

Cherif, L. H., Debbal, S. M., & Bereksi-Reguig, F. (2010). Choice of the wavelet analyzing in the phonocardiogram signal analysis using the discrete and the packet wavelet transform. *Expert Systems With Applications*, *37*(2), 913–918. https://doi.org/10.1016/j.eswa.2009.09.036.

Chi, Y. M., Wang, Y.-T., Wang, Y., Maier, C., Jung, T.-P., & Cauwenberghs, G. (2012). Dry and noncontact EEG sensors for mobile brain–computer interfaces. *IEEE Transactions on Neural Systems and Rehabilitation Engineering*, *20*(2), 228–235.

Choi, S., Shin, Y., & Park, H.-K. (2011). Selection of wavelet packet measures for insufficiency murmur identification. *Expert Systems With Applications*, *38*(4), 4264–4271.

Corsini, J., Shoker, L., Sanei, S., & Alarcon, G. (2006). Epileptic seizure predictability from scalp EEG incorporating constrained blind source separation. *IEEE Transactions on Biomedical Engineering*, *56*, 790–799.

Da Poian, G., Bernardini, R., & Rinaldo, R. (2016). Separation and analysis of fetal-ECG signals from compressed sensed abdominal ECG recordings. *IEEE Transactions on Biomedical Engineering*, *63*(6), 1269–1279.

Dale, A. M., & Sereno, M. J. (1993). Improved localization of cortical activity by combining EEG and MEG with MRI cortical surface reconstruction. *Journal of Cognitive Neuroscience*, 162–176.

D'Alessandro, M., Esteller, R., Vachtsevanos, G., Hinson, A., Echauz, J., & Litt, B. (2003). Epileptic seizure prediction using hybrid feature selection over multiple intracranial EEG electrode contacts: a report of four patients. *IEEE Transactions on Biomedical Engineering*, *50* (5), 603–615.

da Silva-Sauer, L., Valero-Aguayo, L., de la Torre-Luque, A., Ron-Angevin, R., & Varona-Moya, S. (2016). Concentration on performance with P300-based BCI systems: a matter of interface features. *Applied Ergonomics*, *52*, 325–332. https://doi.org/10.1016/j.apergo.2015.08.002.

Deutsch, S., & Deutsch, A. (1993). *Understanding the nervous system: An engineering perspective*. New York: IEEE Press.

Dornhege, G., Blankertz, B., Curio, G., & Muller, K.-R. (2004). Boosting bit rates in noninvasive EEG single-trial classifications by feature combination and multiclass paradigms. *IEEE Transactions on Biomedical Engineering*, *51*(6), 993–1002.

EC11a, A. (1984). *Voluntary standard for diagnostic electrocardiographic devices*. Arlington: Association for the Advancement of Medical Instrumentation.

Einthoven, W. (1903). Die galvanometrische Registrirung des menschlichen Elektrokardiogramms, zugleich eine Beurtheilung der Anwendung des Capillar-Elecktrometers in der Physiologie. *Pflugers Arch Ges Physiol*, *99*, 472.

Elger, C. E. (2001). Future trends in epileptology. *Current Opinion in Neurology*, *14*, 185–186.

Emly, M., Gilmore, L. D., & Roy, S. H. (1992). Electromyography. *IEEE Potentials*, *11*(2), 25–28.

Ergen, B., Tatar, Y., & Gulcur, H. O. (2012). Time-frequency analysis of phonocardiogram signals using wavelet transform: a comparative study. *Computer Methods in Biomechanics and Biomedical Engineering*, *15*(4), 371–381.

Esteller, R., Echauz, J., D'Alessandro, M., Worell, G., Cranstoun, S., Vachtsevanos, G., & Litt, B. (2005). Continuous energy variation during the seizure cycle: towards an online accumulated energy. *Clin Neurophy*, 517–526.

Feldt, S., Osterhage, H., Mormann, F., Lehnertz, K., & Zochowski, M. (2007). Internetwork and intra-network communications during bursting dynamics: applications to seizure prediction. *Phys Rev E*, *76*.

Farwell, L. A., & Donchin, E. (1986). The brain detector-P300 in the detection of deception. In: *Presented at the psychophysiology, Vol. 23, Soc Psychophysiol Res 1010 Vermont Ave NW Suite 1100, Washington, DC2005*, (p. 434). 434.

Gao, W., Guan, J., Gao, J., & Zhou, D. (2015). Multi-ganglion ANN based feature learning with application to P300-BCI signal classification. *Biomedical Signal Processing and Control*, *18*, 127–137.

Geselowitz, D. B. (1989). On the theory of the electrocardiogram. *Proceedings of IEEE*, *77*, 857–876.

Gevins, A. S., & Bressler, S. L. (1988). Functional topography of the human brain. In G. Pfurtscheller, & F. L. Silva (Eds.), *Functional brain mapping* (pp. 99–116). Toronto: Hans Huber Publishers.

Goldberger, E. (1942). A simple, indifferent, electrocardiographic electrode of zero potential and a technique of obtaining augmented, unipolar, extremity leads. *American Heart Journal*, *23*, 483.

Gopura, R. A. R. C., Kiguchi, K., & Li, Y. (2009). SUEFUL-7: A 7DOF upper-limb exoskeleton robot with muscle-model-oriented EMG-based control. In: *Presented at the intelligent robots and systems, 2009. IROS 2009. IEEE/RSJ international conference on IEEE*, pp. 1126–1131.

Graimann, B., Allison, B., & Pfurtscheller, G. (2009). Brain-computer interfaces: a gentle introduction. In *Brain-computer interfaces* (pp. 1–27). Springer.

Grozea, C., Voinescu, C. D., & Fazli, S. (2011). Bristle-sensors—low-cost flexible passive dry EEG electrodes for neurofeedback and BCI applications. *Journal of Neural Engineering*, *8*(2).

Guger, C., Daban, S., Sellers, E., Holzner, C., Krausz, G., Carabalona, R., ... Edlinger, G. (2009). How many people are able to control a P300-based brain-computer interface (BCI)? *Neuroscience Letters*, *462*(1), 94–98.

Guger, C., Schlogl, A., Neuper, C., Walterspacher, D., Strein, T., & Pfurtscheller, G. (2001). Rapid prototyping of an EEG-based brain-computer interface (BCI). *IEEE Transactions on Neural Systems and Rehabilitation Engineering*, *9*(1), 49–58.

Hassan, A. R., & Haque, M. A. (2016). Computer-aided obstructive sleep apnea screening from single-lead electrocardiogram using statistical and spectral features and bootstrap aggregating. *Biocybernetics and Biomedical Engineering*, *36*(1), 256–266.

Harrison, M., Frei, M. G., & Osorio, I. (2005). Accumulated energy revisited. *Clinical Neurophysiology*, 527–531.

Himanena, S.-L., & Hasan, J. (2004). Limitations of Rechtschaffen and Kales. *Sleep Medicine Reviews*, *4*(2), 149–167.

Hirshkowitz, M. (2000). Standing on the shoulders of giants: the Standardized Sleep Manual after 30 years. *Sleep Medicine Reviews*, *4*(2), 169–170.

Horobin, W. (Ed.). (2008). *Diseases and disorders*. Marshall Cavendish Corporation.

Hossen, A., & Al-Ghunaimi, B. (2007). A wavelet-based soft decision technique for screening of patients with congestive heart failure. *Biomedical Signal Processing and Control*, *2*, 135–143.

Hossen, A., & Al-Ghunaimi, B. (2008). *Identification of patients with congestive heart failure by recognition of sub-bands spectral patterns*: (pp. 21–24). World Academy of Science, Engineering and Technology.

Huang, Y.-J., Wu, C.-Y., Wong, A. M.-K., & Lin, B.-S. (2015). Novel active comb-shaped dry electrode for EEG measurement in hairy site. *IEEE Transactions on Biomedical Engineering*, *62*(1), 256–263.

Iber, C., Ancoli-Israel, S., Chesson, A. L., & Quan, S. F. (2007). *The AASM manual for scoring of sleep and associated events-rules: Terminology and technical specification*. Darien, IL, USA: American Academy of Sleep Medicine.

Isler, Y., & Kuntalp, M. (2007). Combining classical HRV indices with wavelet entropy measures improves to performance in diagnosing congestive heart failure. *Computers in Biology and Medicine*, *37*, 1502–1510.

Jasper, H. (1958). The ten twenty electrode system of the international federation. *Electroencephalography and Clinical Neurophysiology*, 371–375.

Jarchi, D., & Casson, A. J. (2016). Description of a database containing wrist PPG signals recorded during physical exercise with both accelerometer and gyroscope measures of motion. *Data*, *2*(1), 1.

Jenkins, J. M. (1981). Computerized electrocardiography. *CRC Critical Reviews in Bioengineering*, *6*.

Jerger, K., Weinstein, S., Sauer, T., & Schiff, S. (2005). Multivariate linear discrimination of seizures. *Clinical Neurophysiology*, 545–551.

Jia, L., Song, D., Tao, L., & Lu, Y. (2012). Heart sounds classification with a fuzzy neural network method with structure learning. In *Presented at the international symposium on neural networks* (pp. 130–140). Springer.

Jin, J., & Sánchez-Sinencio, E. (2015). A home sleep apnea screening device with time-domain signal processing and autonomous scoring capability. *IEEE Transactions on Biomedical Circuits and Systems*, *9*(1), 96–104.

Jones, S. A. (2008). *ECG success exercises in ECG interpretation*. Philadelphia: F.A. Davis Company.

Jouny, C., Franaszczuk, P., & Bergey, G. (2005). Signal complexity and synchrony of epileptic seizures: is there an identifiable preictal period. *Clinical Neurophysiology*, 552–558.

Julian, D. G., Campbell-Cowan, J., & McLenachan, M. J. (2005). *Cardiology* 8th ed.Edinburgh, NY: Saunders.

Kamal, A., Harness, J., Irving, G., & Mearns, A. (1989). Skin photoplethysmography—a review. *Computer Methods and Programs in Biomedicine*, *28*(4), 257–269.

Kao, W.-C., & Wei, C.-C. (2011). Automatic phonocardiograph signal analysis for detecting heart valve disorders. *Expert Systems With Applications*, *38*(6), 6458–6468.

Kennedy, R. G. (1998). Electronic fetal heart rate monitoring: retrospective reflections on a twentieth-century technology. *Journal of the Royal Society of Medicine*, *91*(5), 244–250.

Khan, E., Al Hossain, F., Uddin, S. Z., Alam, S. K., & Hasan, M. K. (2016). A robust heart rate monitoring scheme using photoplethysmographic signals corrupted by intense motion artifacts. *IEEE Transactions on Biomedical Engineering*, *63*(3), 550–562.

Khan, Y. U., & Gotman, J. (2003). Wavelet based automatic seizure detection in intra-cerebral electroencephalogram. *Clinical Neurophysiology*, 898–908.

Kim, K. S., Choi, H. H., Moon, C. S., & Mun, C. W. (2011). Comparison of *k*-nearest neighbor, quadratic discriminant and linear discriminant analysis in classification of electromyogram signals based on the wrist-motion directions. *Current Applied Physics*, *11*(3), 740–745.

Kimura, J. (1989). *Electrodiagnosis in diseases of nerve and muscle: Principles and practice* (2nd ed.). Quest-Meridien Ltd. F.A. Davis.

Koelstra, S., Muhl, C., Soleymani, M., Lee, J.-S., Yazdani, A., Ebrahimi, T., ... Patras, I. (2012). Deap: a database for emotion analysis; using physiological signals. *IEEE Transactions on Affective Computing*, *3*(1), 18–31.

Kohler, B., Hennig, C., & Orglmeister, R. (2002). The principles of software QRS detection. *IEEE Engineering in Medicine and Biology Magazine*, *21*(1), 42–57.

Kotas, M., Jezewski, J., Horoba, K., & Matonia, A. (2011). Application of spatio-temporal filtering to fetal electrocardiogram enhancement. *Computer Methods and Programs in Biomedicine*, *104*(1), 1–9.

Kondraske, G. V. (1986). Neurophysiological measurements. In J. D. Bronzino (Ed.), *Biomedical engineering and instrumentation* (pp. 138–179). Boston: PWS Publishing.

Kuntamalla, S., & Reddy, L. R. (2010). Detecting congestive heart failure using heart rate sequential trend analysis plot. *International Journal of Engineering Science and Technology*, *2* (12), 7329–7334.

Kuntamalla, S., & Reddy, L. R. (2014). Reduced data dualscale entropy analysis of HRV signals for improved congestive heart failure detection. *Measurement Science Review*, *14*(5), 294–301.

Kurzynski, M., Krysmann, M., Trajdos, P., & Wolczowski, A. (2016). Multiclassifier system with hybrid learning applied to the control of bioprosthetic hand. *Computers in Biology and Medicine*, *69*, 286–297. https://doi.org/10.1016/j.compbiomed.2015.04.023.

Lafuente, V., Gorriz, J. M., Ramirez, J., & Gonzalez, E. (2017). P300 brainwave extraction from EEG signals: an unsupervised approach. *Expert Systems With Applications*, *74*, 1–10.

Lee, B., Han, J., Baek, H. J., Shin, J. H., Park, K. S., & Yi, W. J. (2010). Improved elimination of motion artifacts from a photoplethysmographic signal using a Kalman smoother with simultaneous accelerometry. *Physiological Measurement*, *31*(12), 1585.

Lee, Y.-R., & Kim, H.-N. (2018). A data partitioning method for increasing ensemble diversity of an eSVM-based P300 speller. *Biomedical Signal Processing and Control*, *39*, 53–63.

Lehnertz, K. (2008). Epilepsy and nonlinear dynamics. *Journal of Biological Physics*, (253), 34.

Lenzi, T., De Rossi, S., Vitiello, N., Chiri, A., Roccella, S., Giovacchini, F., … Carrozza, M. C. (2009). The neuro-robotics paradigm: NEURARM, NEUROExos, HANDEXOS. In: *Presented at the Engineering in Medicine and Biology Society, 2009. EMBC 2009. Annual International Conference of the IEEE*, IEEE, pp. 2430–2433.

Liu, N.-H., Chiang, C.-Y., & Hsu, H.-M. (2013). Improving driver alertness through music selection using a mobile EEG to detect brainwaves. *Sensors*, *13*(7), 8199–8221.

Lo, H. S., & Xie, S. Q. (2012). Exoskeleton robots for upper-limb rehabilitation: state of the art and future prospects. *Medical Engineering and Physics*, *34*(3), 261–268.

Lum, P. S., Burgar, C. G., Shor, P. C., Majmundar, M., & Van der Loos, M. (2002). Robot-assisted movement training compared with conventional therapy techniques for the rehabilitation of upper-limb motor function after stroke. *Archives of Physical Medicine and Rehabilitation*, *83*(7), 952–959.

Maiwald, T., Winterhalder, M., Aschenbrenner-Scheibe, R., Voss, H. U., Schulze-Bonhage, A., & Timmer, J. (2004). Comparison of three nonlinear seizure prediction methods by means of the seizure prediction characteristic. *Physica D*, *194*, 357–368.

Malmivuo, J., Suikho, V., & Eskola, H. (1997). Sensitivity distributions of EEG and MEG measurements. *IEEE Transactions on Biomedical Engineerin*, 196–208.

Masetic, Z., & Subasi, A. (2016). Congestive heart failure detection using random forest classifier. *Computer Methods and Programs in Biomedicine*, *130*, 54–64.

McFarland, D. J., & Wolpaw, J. R. (2017). EEG-based brain-computer interfaces. *Synthetic Biology and Biomedical Engineering/Neural Engineering*, *4*, 194–200. https://doi.org/10.1016/j.cobme.2017.11.004.

Meier, R., Dittrich, H., Schulze-Bonhage, A., & Aertsen, A. (2008). Detecting epileptic seizures in long-term human EEG: a new approach to automatic online and real-time detection and classification of polymorphic seizure patterns. *Journal of Clinical Neurophysiology*, 119–131.

Mormann, F., Andrzejak, R. G., Elger, C. E., & Lehnertz, K. (2007). Seizure prediction: the long and winding road. *Brain*, *130*, 314–333.

Mormann, F., Kreuz, T., Rieke, C., Andrzejak, R., & Kraskov, A. (2005). On the predictability of epileptic seizures. *Clinical Neurophysiology*, 569–587.

Muthuswamy, J. (2004). Biomedical signal analysis. *Standard handbook of biomedical engineering and design*Vol. 14, 18–1.

Nikolic, M. (2001). *Findings and firing pattern analysis in controls and patients with myopathy and amytrophic lateral sclerosis.* Copenhagen: Faculty of Health Science, University of Copenhagen.

Niedermayer, E. (1999a). Abnormal EEG patterns: epileptic and paroxysmal. In E. Niedermayer, & F. L. Silva (Eds.), *Electroencephalography. Basic principles, clinical applications and related fields* (pp. 235–260). Baltimore: Williams & Wilkins.

Niedermayer, E. (1999b). Epileptic seizure disorders. In E. Niedermayer, & F. L. Silva (Eds.), *Electroencephalography. Basic principles, clinical applications and related fields* (pp. 476–585). Baltimore: Williams & Wilkins.

Niedermayer, E. (1999c). The normal EEG of the waking adult. In E. Niedermayer, & F. L. Silva (Eds.), *Electroencephalography. Basic principles, clinical applications and related fields* (pp. 149–173). Baltimore: Williams & Wilkins.

Oweis, R. J., Hamad, H., & Shammout, M. (2014). Heart sounds segmentation utilizing Teager Energy Operator. *Journal of Medical Imaging and Health Informatics*, *4*(4), 488–499.

Pan, J., & Tompkins, W. J. (1985, May). A real-time QRS detection algorithm. *IEEE Transactions on Biomedical Engineering*, *BME-32*(3), 230–236.

Passanisi, C. (2004). *Electrocardiography.* New York: Delmar Learning.

Pecchia, L., Melillo, P., Sansone, M., & Bracale, M. (2011). Discrimination power of short-term heart rate variability measures for CHF assessment. *IEEE Transactions on Information Technology in Biomedicine*, *15*(1), 40–46.

Pedro, C.-C. E., Joyce, W. Y., Leo, C. L., Kirk, S., Sandy, D., & Marc, N. R. (2012). Quantification and localization of EEG interictal spike activity in patients with surgically removed epileptogenic foci. *Clinical Neurophysiology*, *123*, 471–485.

Penzel, T., Moody, G. B., Mark, R. G., Goldberger, A. L., & Peter, J. H. (2000). The apnea-ECG database. In: *Presented at the Computers in cardiology 2000*, IEEE, pp. 255–258.

Pereda, E., Quian Quiroga, R., & Bhattacharya, J. (2005). Nonlinear multivariate analysis of neurophysiological signals. *Progress in Neurobiology*, *77*, 1), 1–37.

Pfurtscheller, G., Leeb, R., Keinrath, C., Friedman, D., Neuper, C., Guger, C., & Slater, M. (2006). Walking from thought. *Brain Research*, *1071*(1), 145–152.

Plonsey, R. (1972). Capabilities and limitations of electrocardiography and magnetocardiography. *IEEE Transactions on Information Technology in Biomedicine*, 239–244.

Preedy, V., & Peters, T. (2002). *Skeletal muscle: Pathology, diagnosis and management of disease.* Cambridge: Cambridge University Press.

Preston, D. C., & Shapiro, B. E. (2012). *Electromyography and neuromuscular disorders: Clinical-electrophysiologic correlations.* London: Elsevier Health Sciences.

Pryor, T. A., Drazen, E., & Laks, M. (1980). *Computer systems for the processing of diagnostic electrocardiograms.* Los Alamitos, CA: IEEE Computer Society Press.

Quinn, C. E. (2006). *100 questions and answers about congestive heart failure.* Jones and Bartlett Publishers, Inc.

Ralph, A. G., Kaspar, S., & Rummel, C. (2012). Nonrandomness, nonlinear dependence, and nonstationarity of electroencephalographic. *Physical Review E*, *86*.

Rechtschaffen, A., & Kales, A. (1968). *A manual of standardized terminology, techniques, and scoring system for sleep stages of human subjects.* Los Angeles, CA: UCLA, Brain Research Institute/Brain Information Service.

Ramović, A., Bandić, L., Kevrić, J., Germović, E., & Subasi, A. (2017). Wavelet and Teager Energy Operator (TEO) for heart sound processing and identification. In *CMBEBIH 2017* (pp. 495–502). Springer.

Ranganathan, N., Sivaciyan, V., & Saksena, F. B. (2007). *The art and science of cardiac physical examination: With heart sounds and pulse wave forms on CD*. Springer Science & Business Media.

Rangayyan, R. M. (2015). *Biomedical signal analysis*. (Vol. 33). John Wiley & Sons.

Riener, R., Nef, T., & Colombo, G. (2005). Robot-aided neurorehabilitation of the upper extremities. *Medical and Biological Engineering and Computing, 43*(1), 2–10.

Robyn, B. R., Mingui, S., Mark, S. L., & Robert, S. J. (2000). Detection of seizure foci by recurrent neural networks. In: *22nd annual EMBS international conference, Chicago, IL*.

Rutkowski, T. M., Zdunek, R., & Cichocki, A. (2006). Multichanned EEG brain activity patterns analysis in time frequency domain with nonnegative matric factorization support. In: *Third international conference on brain-inspired information technology. Hibikino, Kitakyushu, Japan.*

Safara, F., Doraisamy, S., Azman, A., Jantan, A., & Ranga, S. (2012). Wavelet packet entropy for heart murmurs classification. *Advances in Bioinformatics, 2012*.

Sammler, D., Grigutsch, M., Fritz, T., & Koelsch, S. (2007). Music and emotion: electrophysiological correlates of the processing of pleasant and unpleasant music. *Psychophysiology, 44*(2), 293–304.

Sanei, S. (2013). *Adaptive processing of brain signals*. John Wiley & Sons.

Sanei, S., & Chambers, J. A. (2013). *EEG signal processing*. John Wiley & Sons.

Sapsanis, C., Georgoulas, G., & Tzes, A. (2013). EMG based classification of basic hand movements based on time-frequency features. In: *Presented at the Control & Automation (MED), 2013 21st Mediterranean Conference on IEEE*, pp. 716–722.

Sasaki, D., Noritsugu, T., & Takaiwa, M. (2005). Development of active support splint driven by pneumatic soft actuator (ASSIST). Presented at the Robotics and Automation, 2005. ICRA 2005. Proceedings of the 2005 IEEE International Conference on IEEE, pp. 520–525.

Schelter, B., Winterhalder, M., Maiwald, T., Brandt, A., Schad, A., Timmer, J., & Schulze-Bonhage, A. (2006). Do false predictions of seizures depend on the state of vigilance? A report from two seizure-prediction methods and proposed remedies. *Epilepsia, 47*(12), 2058–2070.

Schlkopf, B., & Smola, A. (2002). Learning with kernels: support vector machines, regularization, optimization and beyond. In *Adaptive computation and machine learning*. Cambridge, MA: MIT Press.

Sellers, E. W., Krusienski, D. J., McFarland, D. J., Vaughan, T. M., & Wolpaw, J. R. (2006). A P300 event-related potential brain-computer interface (BCI): the effects of matrix size and inter stimulus interval on performance. *Biological Psychology, 73*(3), 242–252.

Shahriari, Y., & Erfanian, A. (2013). Improving the performance of P300-based brain-computer interface through subspace-based filtering. *Neurocomputing, 121*, 434–441.

Simson, M. B. (1981). Use of signals in the terminal QRS complex to identify patients with ventricular tachycardia after myocardial infarction. *Circulation, 64*, 235–242.

Siuly, S., Li, Y., & Zhang, Y. (2016). *EEG signal analysis and classification*. Springer.

Son, C.-S., Kim, Y.-N., Kim, H.-S., Park, H.-S., & Kim, M.-S. (2012). Decision-making model for early diagnosis of congestive heart failure using rough set and decision tree approaches. *Journal of Biomedical Informatics*, 999–1008.

Sörnmo, L., & Laguna, P. (2005). *Bioelectrical signal processing in cardiac and neurological applications*. (Vol. 8). Academic Press.

Stam, C. J. (2005). Nonlinear dynamical analysis of EEG and MEG: review of an emerging field. *Clinical Neurophysiology, 16*(2266).

Subasi, A. (2013). Classification of EMG signals using PSO optimized SVM for diagnosis of neuromuscular disorders. *Computers in Biology and Medicine, 43*, 576–586.

Swiderski, B., Osowski, S., Cichocki, A., & Rysz, A. (2007). Epileptic seizure characterization by Lyapunov exponent of EEG signal. *Compel, 26*, 1276–1287.

Swiderski, B., Osowski, S., Cichocki, A., & Rysz, A. (2008). Single-Class SVM Classifier for localization of epileptic focus on the basis of EEG. In: *International joint conference on neural networks*.

Thaler, M. S. (1999). *The only EKG book you'll ever need*. (Vol. 3). Philadelphia, PA: Lippincott Williams & Wilkins.

Thuraisingham, R. A. (2009). A classification system to detect congestive heart failure using second-order difference plot of RR intervals. *Cardiology Research and Practice*.

Togawa, T., Tamura, T., & Oberg, P. A. (1997). *Biomedical transducers and instruments*. Boca Raton, FL: CRC Press.

Trust, T. M. (2014). *Facts and figures*. Retrieved April 7, 2016 from: (2014). https://www.migrainetrust.org/about-migraine/migraine-what-is-it/facts-figures/.

Uni-Freiburg (2011). *Seizure prediction project Freiburg*. Retrieved October 2, 2011 from:(2011). https://epilepsy.uni-freiburg.de/freiburg-seizure-prediction-project/eeg-database.

Varotto, G., Tassi, L., Franceschetti, S., Spreafico, R., & Panzica, F. (2012). Epileptogenic networks of type II focal cortical dysplasia: a stereo-EEG study. *Neuroimage, 61*, 591.

Waller, A. D. (1889). One of the electromotive changes connected with the beat of the mammalian heart, and the human heart in particular. *Philosophical Transactions of the Royal Society B, 180*, 169.

van Erp, J., Lotte, F., & Tangermann, M. (2012). Brain-computer interfaces: beyond medical applications. *Computer, 45*(4), 26–34.

Waterhouse, E. (2003). New horizons in ambulatory electroencephalography. *IEEE Engineering in Medicine and Biology Magazine*, 74–80.

WIlson, F. N., Johnson, F. S., & Hill, I. G. (1934). The interpretation of the falvanometric curves obtained when one electrode is distant from the heart and the other near or in contact with the ventricular surface. *American Heart Journal, 10*, 176.

Winterhalder, M., Maiwald, T., Voss, H. U., & Aschenbrenner-Scheibe, R. (2003). The seizure prediction characteristic: a general framework to assess and compare seizure prediction methods. *Epilepsy and Behavior*, 318–325.

Wojtczak, P., Amaral, T. G., Dias, O. P., Wolczowski, A., & Kurzynski, M. (2009). Hand movement recognition based on biosignal analysis. *Engineering Applications of Artificial Intelligence, 22*(4), 608–615.

Wołczowski, A., & Zdunek, R. (2017). Electromyography and mechanomyography signal recognition: experimental analysis using multi-way array decomposition methods. *Biocybernetics and Biomedical Engineering, 37*(1), 103–113. https://doi.org/10.1016/j.bbe.2016.09.004.

Wolpaw, J. R., Birbaumer, N., McFarland, D. J., Pfurtscheller, G., & Vaughan, T. M. (2002). Brain-computer interfaces for communication and control. *Clinical Neurophysiology, 113*(6), 767–791.

Xiaoying, T., Li, X., Weifeng, L., Yuhua, P., Duanduan, C., Tianxin, G., & Yanjun, Z. (2012). Analysis of frequency domain of EEG signals in clinical location of epileptic focus. *Clinical EEG Neuroscience*.

Yu, S.-N., & Lee, M.-Y. (2012). Conditional mutual information-based feature selection for congestive heart failure recognition using heart rate variability. *Computer Methods and Programs in Biomedicine, 1*(08), 299–309.

Zarei, R., He, J., Siuly, S., & Zhang, Y. (2017). A PCA aided cross-covariance scheme for discriminative feature extraction from EEG signals. *Computer Methods and Programs in Biomedicine*, *146*, 47–57.

Zhang, H., Guan, C., & Wang, C. (2008). Asynchronous P300-based brain-computer interfaces: a computational approach with statistical models. *IEEE Transactions on Biomedical Engineering*, *55*(6), 1754–1763.

Zheng, W.-L., & Lu, B.-L. (2015). Investigating critical frequency bands and channels for EEG-based emotion recognition with deep neural networks. *IEEE Transactions on Autonomous Mental Development*, *7*(3), 162–175.

Zhong, J., & Scalzo, F. (2013). Automatic heart sound signal analysis with reused multi-scale wavelet transform. *International Journal of Engineering and Science*, *2*(50), 50–57.

FURTHER READING

Schelter, B., Winterhalder, M., Drentrup, H. F., Wohlmuth, J., Nawrath, J., Brandt, A., … Timmer, J. (2007). Seizure Prediction: the impact of long prediction horizons. *Epilepsy Research*, 213–217.

Townsend, N., Luengo-Fernandez, R., Leal, J., Gray, A., & Nichols, M. (2012). *European cardiovascular disease statistics 2012*. European Heart Health Strategy II Project, European Heart Network (Brussels), European Society of Cardiology (Sophia Antipolis).

Chapter 3

Biomedical Signal Processing Techniques

3.1 INTRODUCTION TO SPECTRAL ANALYSIS

Sometimes, the frequency domain of a waveform provides more useful information than its representation in a time domain. Therefore, interesting and useful diagnosis properties can be obtained if we examine many of the biological signals in the frequency domain. Heart rate, electromyogram (EMG), electroencephalogram (EEG), electrocardiogram (ECG), eye movements, acoustic heart sounds, and stomach sounds are several of such biomedical signals (Semmlow, 2004). Spectral analysis is defined as determining the frequency content of a waveform. It proves to be a powerful method for characterization of many biomedical signals. This technique was introduced to provide a more in-depth representation of EEG activity than what could be accomplished by simple analysis techniques such as one-dimensional histograms of the samples. Signal decomposition, in terms of sine and cosine functions, was found suitable bearing in mind the oscillatory behavior of many EEG rhythms. From a computational perspective, it was a feasible solution as well. In essence, spectral analysis based on the Fourier transform (FT) produces a set of coefficients that characterize the power spectrum and correlates the signal with sines/cosines of different frequencies. The power spectrum can be used to easily obtain the power of a certain frequency band. This property can help us conclude if an alpha rhythm or other rhythms are present in EEG signal (Sörnmo & Laguna, 2005). Numerous useful methodologies for frequency decomposition have been developed over the past years (Marple & Marple, 1987). In the same way a prism decomposes light into a color spectrum, we can consider a spectral analysis as a mathematical prism decomposing a waveform into frequencies (Hubbard, 1998).

There is a good selection of spectral analysis methods, each having different strengths and weaknesses. These techniques can be separated into two categories: classical methods (FT) and modern methods such as the ones based on the model parameter estimation (Semmlow, 2004). Therefore, spectral analysis based on FT is usually denoted as nonparametric, because no parametric modeling assumptions are made (Dumermuth & Flühler, 1967).

Generally, spectral analysis can be applied to signals gathered during natural waking activity at rest, which are rather short and free of artifacts (da Silva, Niedermeyer, & da Silva, 1987). For an accurate determination of the spectrum, the signal needs to be periodic (or of limited length) and noise-free. In many biomedical applications, however, the signal is infinite or simply a portion of it is accessible for analysis. Also, a waveform often contains significant amounts of artifact and noise along with the signal. Taking these two constraints into consideration, it is evident that all spectral analysis methods are approximate, producing estimates of the true spectrum. Therefore, various spectral analysis methods try to improve the estimation accuracy of particular spectral features (Semmlow, 2004).

There are two important questions that need to be answered by intelligent applications of spectral analysis methods:

- Which spectral features are of potential interest?
- What techniques deliver the most accurate determination of those features?

Global features such as the spectral estimate are likely to be of interest, as well as the local features of the spectrum denoted as parametric estimates. However, good local estimators are poor spectral estimators and vice versa (Semmlow, 2004). In the following pages, the spectral estimation techniques are described briefly and implemented in MATLAB.

3.2 POWER SPECTRAL DENSITY

3.2.1 Continuous-Time Fourier Series Analysis

The classical FT technique is one of the most popular among numerous spectral estimation methods currently available. The FT is based on the fact that sine and cosine functions have energy at only one frequency. The main idea is to break down a waveform into a series of sinusoids of different frequencies. Frequency component of the waveform at these frequencies is then directly proportional to the amplitude of these sinusoids (Semmlow, 2004).

Practical Guide for Biomedical Signals Analysis Using Machine Learning Techniques. https://doi.org/10.1016/B978-0-12-817444-9.00003-9

Any periodic waveform can be represented by a series of sines and cosines of the same frequency as the waveform frequency and its multiples. It can also be represented in a form of a single sine wave of appropriate amplitude and phase angle. In the former case, to find the appropriate amplitude of frequency components, we need to multiply the waveform with the sine and cosine family, and integrate over time. The following two equations explain this process (Semmlow, 2004):

$$a(m) = \frac{2}{T}\int_{-\frac{T}{2}}^{\frac{T}{2}} x(t)\cos(2\pi m f_T t)dt \quad (3.1)$$

$$b(m) = \frac{2}{T}\int_{-\frac{T}{2}}^{\frac{T}{2}} x(t)\sin(2\pi m f_T t)dt \quad (3.2)$$

where T is the period or complete time length of the waveform, $f_T = 1/T$, and m is set of integers, $m = 1, 2, 3, \ldots$, denoting the family member and producing harmonically related frequencies, mf_T.

Because waveforms are represented by harmonically related sinusoids, the Fourier series analysis is sometimes denoted as harmonic decomposition. The sinusoids have valid frequencies only at values of m/T, having the same frequency as the waveform for $m = 1$. Higher multiples, $m > 1$, are called harmonics. The FT Fourier series constitute a bilateral transform for periodic functions. It means that, once the FT is applied to a waveform to get the sinusoidal components, those components can be used to reconstruct the original waveform (Semmlow, 2004):

$$x(t) = a(0)/2 + \sum_{m=1}^{\infty} a(m)\cos(2\pi m f_T t) + b(m)\sin(2\pi m f_T t) \quad (3.3)$$

$$x(t) = a(0)/2 + \sum_{m=1}^{\infty} c(m)\sin(2\pi m f_T t + \Theta_m) \quad (3.4)$$

where $c(m) = \sqrt{a^2(m) + b^2(m)}$ and $\Theta_m = \tan^{-1}(b(m)/a(m))$.

However, a finite summation in Eq. (3.4) is often very accurate because the number of sinusoidal components with significant amplitudes is limited for most real waveforms.

There are waveforms that are symmetrical or antisymmetrical about $t = 0$ or y-axis. If the waveform has mirror symmetry about $t = 0$, it is called an even function, $x(t) = x(-t)$. Multiplication of an even function with an odd function such as sine produces an odd function, making all $b(m)$ terms equal to zero (see Eq. 3.2). When the waveform is antisymmetrical about $t = 0$, $x(t) = -x(t)$, which is termed as an odd function, multiplication with cosines will produce an odd function, thus zeroing all $a(m)$ coefficients (see Eq. 3.1). Functions with half-wave symmetry will have both $a(m)$ and $b(m)$ terms equal to zero for even m. Those functions have the following property, $x(t) = x(T - t)$, and can be either odd or even (Semmlow, 2004). When computing the coefficients by hand, these properties of symmetry may come in handy for reducing complexity and calculation time.

3.2.2 Discrete-Time Fourier Series Analysis

The discrete-time Fourier series analysis comes as an extension of the continuous analysis we have just explained. However, it has been modified by two very important processes: sampling and windowing. The sampling (Nyquist) theorem says that the analogue waveform must be bandlimited in a way that its highest frequency fMAX is less than *fs*/2, where fs represents the Nyquist frequency. In other words, the sampling frequency must be greater or equal to the Nyquist frequency, *fs*, if we want the digitized data to carry all relevant information of the original analogue waveform. The discrete Fourier series of any analogue waveform has three properties: it is theoretically infinite, periodic, and symmetric about *fs*/2. Hence, the frequency range of 0 to *fs*/2 holds all relevant information (Semmlow, 2004).

Windowing is a process in which the digitized waveform must necessarily be truncated due to storage and other constraints. This process can also be considered as multiplying the data using some window shape. The default window shape is rectangular, where the resulting digitized waveform does not undergo any additional shaping (Semmlow, 2004). The equations for discrete-time Fourier series analysis are very similar to their continuous counterparts. The integration is replaced by summation and sinusoidal terms using complex variables notation and are expressed by Euler's identity:

$$e^{jx} = \cos x + j\sin x \quad (3.5)$$

where j represents $\sqrt{-1}$ because i is reserved for current in engineering.

The expression for the discrete FT now becomes:

$$X(m) = \sum_{n=0}^{N-1} x(n)e^{(-j2\pi mn/N)}$$

$$X(m) = \sum_{n=0}^{N-1} x(n)\cos(2\pi mn/N) - j\sum_{n=0}^{N-1} x(n)\sin(2\pi mn/N) \quad (3.6)$$

where m is the family member or the harmonic number and N represents the total number of points (the length of the digital data).

From a computational point of view, $X(m)$ must be allowed to have both positive and negative values for m, $m = -N/2, \ldots, N/2 - 1$. Negative values correspond to negative frequencies that have no physical meaning. The inverse FT can be calculated as:

$$x(n) = \frac{1}{N}\sum_{n=0}^{N-1} X(m)e^{j2\pi mn/N} \tag{3.7}$$

The sampling interval T_s is explicitly incorporated in some versions of Eqs. (3.6), (3.7). Other approaches of scaling these equations can be easily found in the available literature.

Because $X(m)$ is a function of m, we need to convert it to frequency:

$$f = f_m = mf_T = m/T = m/NT \tag{3.8}$$

Substituting $m/N = fT_s$ into Eq. (3.6), the expression for the discrete FT becomes:

$$X(f) = \sum_{n=0}^{N-1} x(n)e^{(-j2\pi nfT_s)} \tag{3.9}$$

For a periodic waveform, the previously mentioned method yields a true spectrum of that waveform, and such analysis is usually called Fourier series analysis. If the waveform of interest is aperiodic or truncated, FT analysis is the more proper term to be used to describe such analysis. Because we talk about the same algorithms in both cases, all spectral analyzes built on decomposition of a waveform into sine and cosine functions are generally called FT (Semmlow, 2004).

At first, the FT was implemented by direct application of Eq. (3.6) or (3.9). Now, a more computationally efficient algorithm called the fast Fourier transform (FFT) is used. Its main advantage is the reduction of the number of required calculations from N^2 to $2 \log N$ (Fig. 3.1).

EXAMPLE 3.1. The following MATLAB code was used to plot the EEG signal and its FFT spectrum. You can download data from the following website:
http://epileptologie-bonn.de/cms/front_content.php?idcat=193&lang=3&changelang=3

```
%Ch3_Ex1_EEG_FFT.m
%The following MATLAB code can be used to plot
%the EEG signal and its FFT spectrum.
clc
clear
%Load Sample EEG Data downloaded from the web site
%http://epileptologie-bonn.de/cms/front_content.php?idcat=193&lang=3&changelang=3
load AS_BONN_EEG_DATA_4096.mat

%%
Fs = 173.6;              % Sampling frequency
T = 1/Fs;                % Sampling period
L = length(A(:,1));      % Length of signal
t = (0:L-1)*T;           % Time vector
%%
%Plot the Original EEG Signal
subplot(2,1,1);
plot(Fs*t(1:L),A(1:L,1))
title('Original EEG Signal')
xlabel('Number of Samples')
ylabel('X(t)')
axis ([0 L -inf inf])

%%
% Compute the Fourier transform of the signal.

Y = fft(A(1:L,1));
%%
% Compute the two-sided spectrum |P2|. Then compute the single-sided
% spectrum |P1| based on |P2| and the even-valued signal length |L|.
```

Continued

EXAMPLE 3.1.—cont'd

```
P2 = abs(Y/L);
P1 = P2(1:L/2+1);
P1(2:end-1) = 2*P1(2:end-1);
%%
% Define the frequency domain |f| and plot the spectrum |P|.
subplot(2,1,2);
f = Fs*(0:(L/2))/L;
plot(f,P1)
title('Spectrum of EEG Signal')
xlabel('Frequency (Hz)');
ylabel('|P(f)|')
```

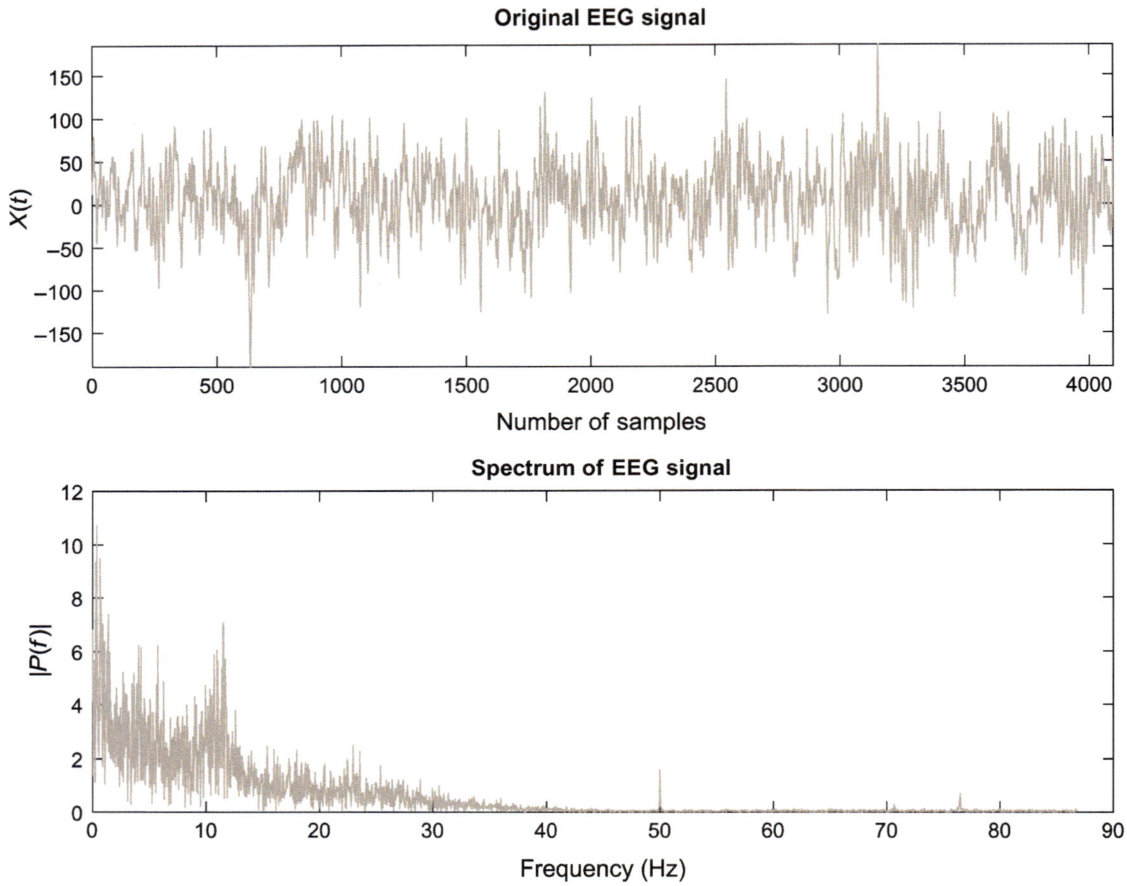

FIG. 3.1 Drawing of healthy EEG signal and its FFT spectrum.

EXAMPLE 3.2. The following MATLAB code was used to plot the EMG signal and its FFT spectrum. For this example, EMGLAB data was used. You can download the data from the following website: http://www.emglab.net/emglab/Signals/signals.php

```
%Ch3_Ex2_EMG_FFT.m
%The following MATLAB code can be used to plot
%the EMG signal and its FFT spectrum.
clc
clear
%Load Sample EMG Data downloaded from the web site
%http://www.emglab.net/emglab/Signals/signals.php
load EMGDAT.mat
%%
```

```
Fs = 10000;              % Sampling frequency
T = 1/Fs;                % Sampling period
L = length(CON(:,2));    % Length of signal
t = (0:L-1)*T;           % Time vector
%%
%Plot the Original EMG Signal
subplot(2,1,1);
plot(Fs*t(1:L),CON(:,2))
%plot(t,CON(:,2))
title('Original EMG Signal')
xlabel('Number of Samples')
ylabel('X(t)')
axis ([0 L -inf inf])

%%
% Compute the Fourier transform of the signal.
Y = fft(CON(:,1));

%%
% Compute the two-sided spectrum |P2|. Then compute the single-sided
% spectrum |P1| based on |P2| and the even-valued signal length |L|.

P2 = abs(Y/L);
P1 = P2(1:L/2+1);
P1(2:end-1) = 2*P1(2:end-1);
%%
% Define the frequency domain |f| and plot the spectrum |P|.
subplot(2,1,2);
f = Fs*(0:(L/2))/L;
plot(f,P1)
title('Spectrum of EMG Signal')
xlabel('Frequency (Hz)');
ylabel('|P(f)|')
```

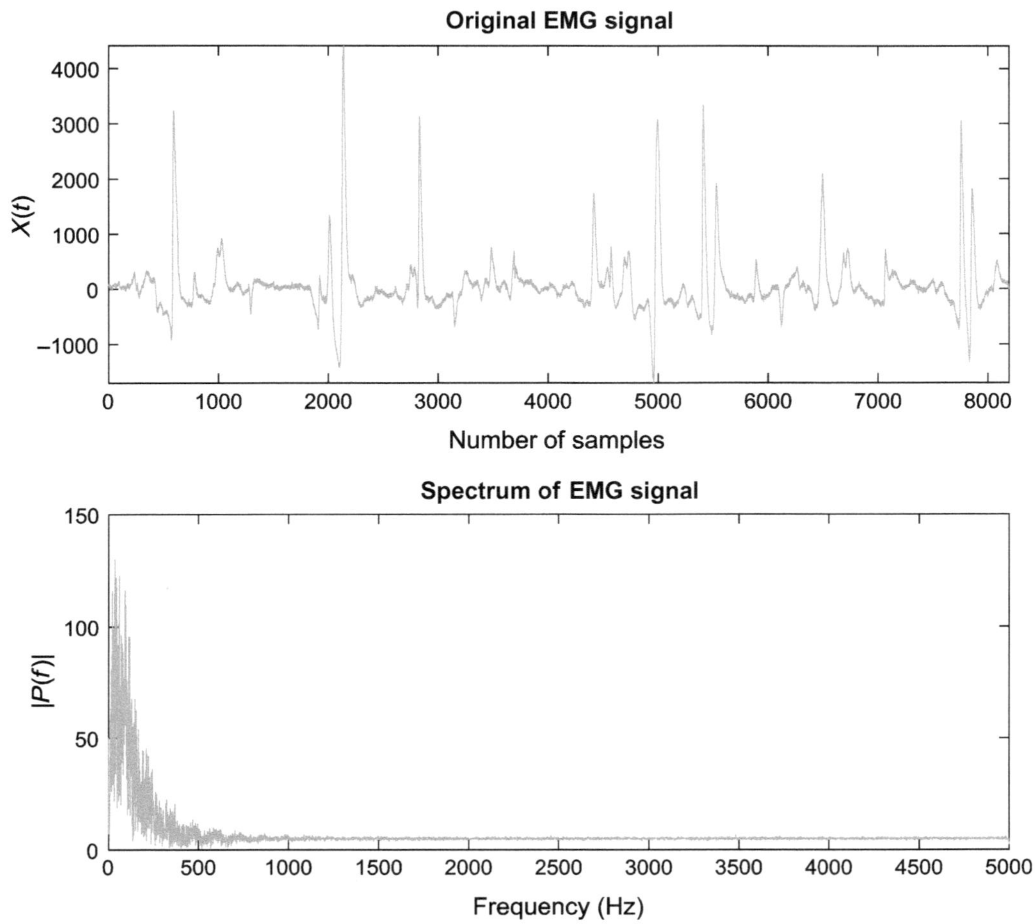

FIG. 3.2 Drawing of healthy EMG signal and its FFT spectrum.

EXAMPLE 3.3. The following MATLAB code was used to plot the ECG signal and its FFT spectrum. For this example, MIT-BIH Arrhythmia Database was used. You can download the data from the following website: https://www.physionet.org/physiobank/database/mitdb/

```
%Ch3_Ex3_ECG_FFT.m
%The following MATLAB code can be used to plot
%the ECG signal and its FFT spectrum.
clc
clear
%Load Sample ECG Data downloaded from the web site
%https://www.physionet.org/physiobank/database/mitdb/
load MITBIH_ECG.mat
%%
Fs = 320;                 % Sampling frequency
T = 1/Fs;                 % Sampling period
L = length(ECGN(:,1));    % Length of signal
t = (0:L-1)*T;            % Time vector
%%
%Plot NORMAL ECG Signal
subplot(2,1,1);
plot(ECGN(:,1));
title('Original ECG Signal')
xlabel('Number of Samples')
ylabel('Amplitude')
ylabel('X(t)')
axis ([0 L -inf inf])
%%
% Compute the Fourier transform of the signal.
Y = fft(ECGN(:,1));
%%
% Compute the two-sided spectrum |P2|. Then compute the single-sided
% spectrum |P1| based on |P2| and the even-valued signal length |L|.
P2 = abs(Y/L);
P1 = P2(1:L/2+1);
P1(2:end-1) = 2*P1(2:end-1);
%%
% Define the frequency domain |f| and plot the spectrum |P|.
subplot(2,1,2);
f = Fs*(0:(L/2))/L;
plot(f,P1)
title('Spectrum of ECG Signal')
xlabel('Frequency (Hz)');
ylabel('|P(f)|')
```

See Fig. 3.3.

EXAMPLE 3.4. The following MATLAB code was used to plot the PCG signal and its FFT spectrum. For this example, PCG Phonogram Dataset was used. You can download the data from the following website: https://www.prweb.com/releases/2004/09/prweb162074.htm

```
%Ch3_Ex4_PCG_FFT.m
%The following MATLAB code can be used to plot
%the PCG signal and its FFT spectrum.
clc
clear
%PCG Phonogram Data Set from BioSignetics Corporation
%Load Sample PCG Data downloaded from the web site
%https://www.prweb.com/releases/2004/09/prweb162074.htm
load PCG.mat
%%
```

```
Fs = 22000;             % Sampling frequency
T = 1/Fs;               % Sampling period
L = length(Aortic_Stenosis_2(1:40000,1));    % Length of signal
t = (0:L-1)*T;          % Time vector
%%
%Plot Aortic_Stenosis PCG Signal
subplot(2,1,1);
plot(Aortic_Stenosis_2(1:40000,1));
title('Original Aortic Stenosis PCG Signal')
xlabel('Number of Samples')
ylabel('Amplitude')
ylabel('X(t)')
axis ([0 L -inf inf])
%%
% Compute the Fourier transform of the signal.
Y = fft(Aortic_Stenosis_2(1:40000,1));
%%
% Compute the two-sided spectrum |P2|. Then compute the single-sided
% spectrum |P1| based on |P2| and the even-valued signal length |L|.
P2 = abs(Y/L);
P1 = P2(1:L/2+1);
P1(2:end-1) = 2*P1(2:end-1);
%%
% Define the frequency domain |f| and plot the spectrum |P|.
subplot(2,1,2);
f = Fs*(0:(L/2))/L;
plot(f,P1)
title('Spectrum of PCG Signal')
xlabel('Frequency (Hz)');
ylabel('|P(f)|')
```

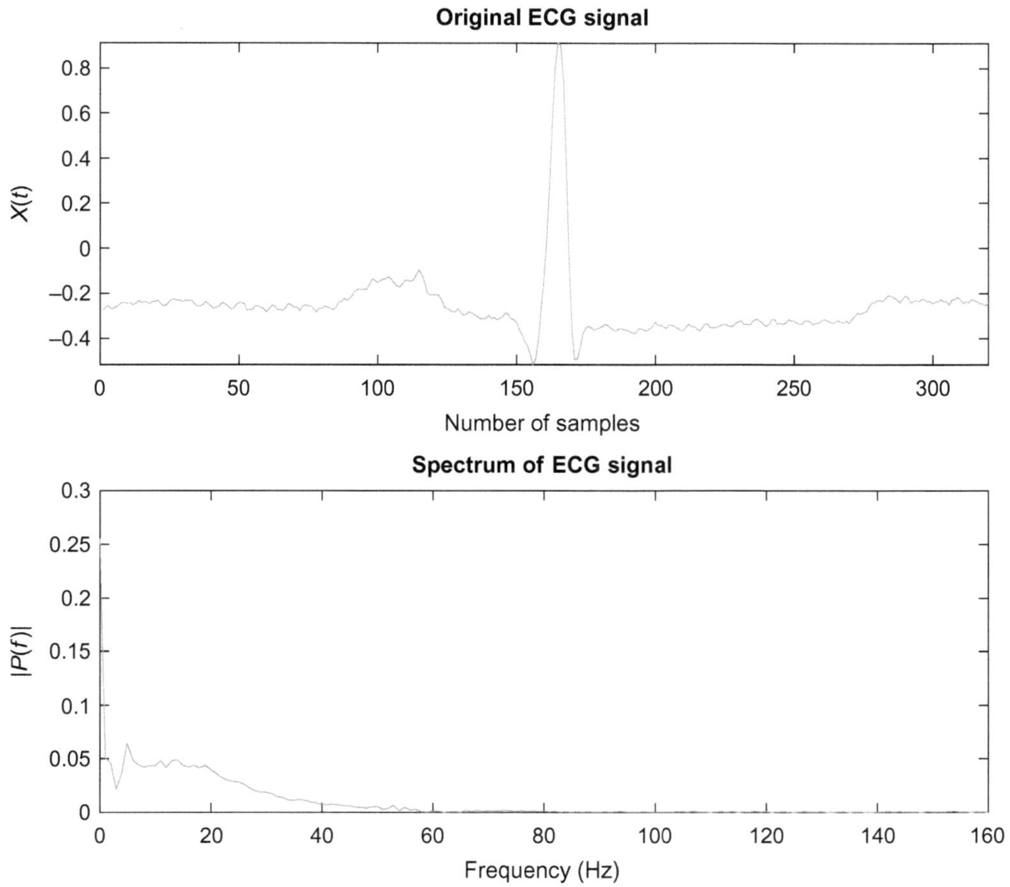

FIG. 3.3 Drawing of healthy ECG signal and its FFT spectrum.

See Fig. 3.4.

EXAMPLE 3.5. The following MATLAB code was used to plot the PPG signal and its FFT spectrum. For this example, Wrist PPG During Exercise dataset was used. You can download the data from the following website: https://physionet.org/physiobank/database/wrist/

```
%Ch3_Ex5_PPG_FFT.m
%The following MATLAB code can be used to plot
%the PPG signal and its FFT spectrum
clc
clear
%Wrist PPG During Exercise
%Load Sample PPG Data downloaded from the web site
%https://physionet.org/physiobank/database/wrist/
load s1_high_resistance_bikem.mat
%%
Fs = 256;            % Sampling frequency
T = 1/Fs;             % Sampling period
L = length(val(2,4097:8192));     % Length of signal
t = (0:L-1)*T;        % Time vector
%%
%Plot PPG Signal
subplot(2,1,1);
plot(val(2,4097:8192))
title('Original PPG Signal')
xlabel('Number of Samples')
ylabel('Amplitude')
ylabel('X(t)')
axis ([0 L -inf inf])
%%
% Compute the Fourier transform of the signal.
Y = fft(val(2,4097:8192));
%%
% Compute the two-sided spectrum |P2|. Then compute the single-sided
% spectrum |P1| based on |P2| and the even-valued signal length |L|.
P2 = abs(Y/L);
P1 = P2(1:L/2+1);
P1(2:end-1) = 2*P1(2:end-1);
%%
% Define the frequency domain |f| and plot the spectrum |P|.
subplot(2,1,2);
f = Fs*(0:(L/2))/L;
plot(f,P1)
title('Spectrum of PPG Signal')
xlabel('Frequency (Hz)');
ylabel('|P(f)|')
```

See Fig. 3.5.

3.2.3 Frequency Resolution

Looking at Eq. (3.6), the number of points produced by the transform is equal to N. Let us recall that the same number of points enters the transform because N represents the length of digitized data. As a result of having half spectra with negative frequencies, and because the sampling frequency f_s is twice the highest frequency (highest harmonic) in the waveform, every point in the spectra represents a frequency increment of:

$$\frac{f_s}{N}=\frac{1}{NT_s}=\frac{1}{T}=f_T \tag{3.10}$$

Frequency resolution (spacing) of the spectrum produced by the FT cannot be altered by changing the sampling frequency/interval because the number of points N is directly/inversely proportional to f_s/T_s. In the case of aperiodic

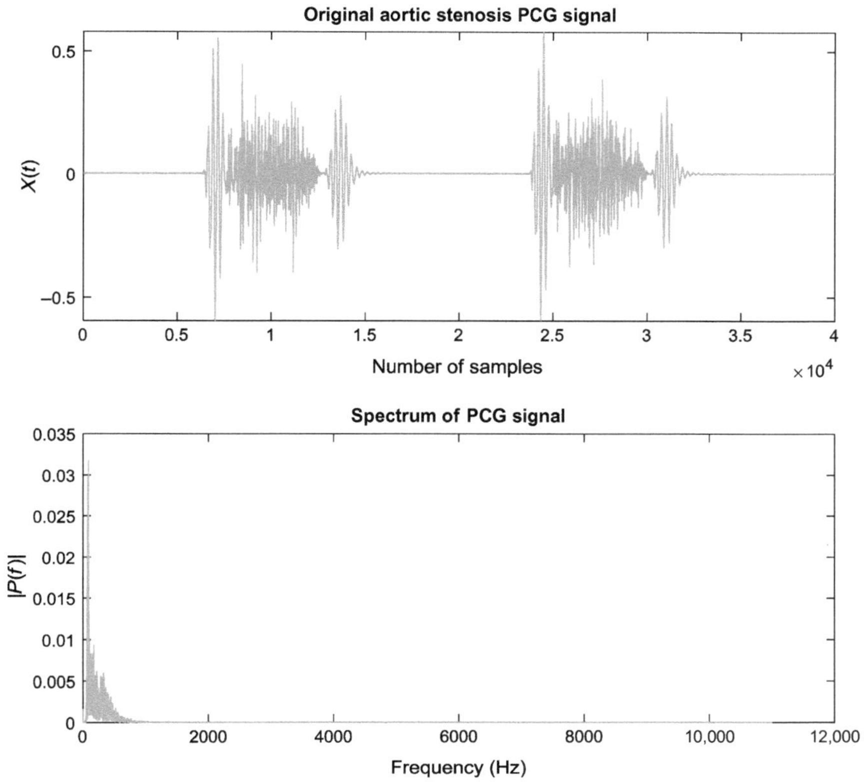

FIG. 3.4 Drawing of stenosis PCG signal and its FFT spectrum.

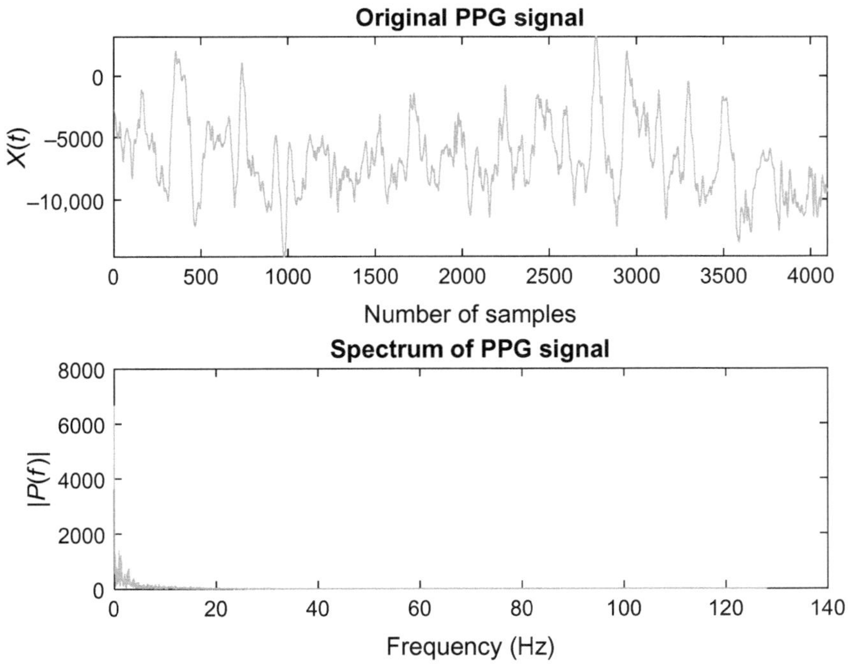

FIG. 3.5 Drawing of PPG signal and its FFT spectrum.

functions, expanding the time segment T_{ap} automatically increases the period T, improving the frequency resolution. Unfortunately, expanded T_{ap} might not be an appropriate solution because additional time segments may include unwanted frequency components and pollute the output of FT. However, there is a way of enhancing frequency resolution while keeping the period and sampling frequency fixed. This procedure is termed zero padding. The idea is to increase N even after the waveform has been sampled. Zero padding assumes that the portion of the waveform outside interval T is always zero and basically includes some part of it. This procedure can also be thought of as increasing period T while the number of points that essentially provide information remains the same, which could not be achieved by expanding T_{ap}. Actually, zero padding does not improve the original resolution of the transform but does offer an interpolated transform with a smoother appearance. Furthermore, it may eliminate confusion in situations when a center frequency of a narrowband signal lies between the frequency evaluation points. Consequently, zero padding facilitates the estimation of peak frequencies in the spectra (Semmlow, 2004).

3.2.4 Windowing Techniques

A waveform often represents a portion of a considerably longer time series. Biomedical waveforms such as EEG or ECG last over the lifetime of the subject. Apparently, only a segment of these waveforms can be stored in computer memory, paying attention to how the waveform is truncated. Generally, a portion is basically cut out from the overall waveform and stored in the memory without any further shaping, which is equivalent to a rectangular window. The analysis is now limited to the windowed segment of the waveform (Semmlow, 2004).

Other window shapes, sometimes denoted as tapering functions, are recommended in some situations. Their application involves multiplying the waveform by the desired shape. By taking the FT of the window itself, we can conclude that all windows produce artifact, as the window's frequency characteristics become part of the resultant frequency spectrum. Hanning, Blackman, Kaiser, Hamming, and the triangular windows are perhaps the most popular. Their equations are given in the following sections (Semmlow, 2004):

Triangular window:
for odd n:

$$w(k)=\begin{cases}2k/(n-1) & 1\le k\le (n+1)/2\\ 2(n-k-1)/(n+1) & (n+1)/2\le k\le n\end{cases}\tag{3.11}$$

for even n:

$$w(k)=\begin{cases}(2k-1)/n & 1\le k\le n/2\\ 2(n-k+1)/n & n/2+1\le k\le n\end{cases}\tag{3.12}$$

Hamming window:

$$w(k+1)=0.54-0.46(2\pi k/(n-1))\quad k=0,1,\ldots,n-1\tag{3.13}$$

All these window functions are easily implemented in MATLAB within the Signal Processing Toolbox. Selecting the most suitable window for a specific application represents one of many other phases of signal analysis. The most appropriate window is often chosen based on experimentation with the actual data, or we can look at what spectral features are important to us. If we have two closely spaced narrowband signals, then the rectangular window is desired because it has the narrowest mainlobe. If we want to resolve moderately spaced strong and weak signals, the sidelobes of the strong signal should be prevented from overriding the weak signal. For this situation, a window with fast decay in the sidelobes is desired. If there are two signals of reasonable strength, one close and the other more distant from a weak signal, the best choice may be a window with reasonably decaying sidelobes and a reasonably narrow mainlobe (Semmlow, 2004).

3.2.5 Periodogram Power Spectral Density

The power spectrum of the stationary signal $x(n)$ is generally defined as the FT of the autocorrelation function (the definition of autocorrelation function will be introduced later). The equation for the power spectrum, in continuous and discrete notation, is listed here:

$$\begin{aligned}PS(f)&=\int_0^T r_{xx}(\tau)e^{-j2\pi f\tau}d\tau\\ PS(f)&=\sum_{k=0}^{N-1} r_{xx}(k)e^{-j2\pi kfT_S}\end{aligned}\tag{3.14}$$

where $r_{xx}(\tau)$ or $r_{xx}(k)$ is the autocorrelation function. Because the autocorrelation function has mirror symmetry, this equation can be simplified to contain real cosine terms, only the sine terms $b(m)$ will be zero:

$$\begin{aligned}PS(f)&=\int_0^T r_{xx}(\tau)\cos(2\pi f\tau)d\tau\\ PS(f)&=\sum_{k=0}^{N-1} r_{xx}(k)\cos(2\pi kfT_S)\end{aligned}\tag{3.15}$$

The direct method is inspired by the fact that the energy of an analogue signal $x(t)$ is equal to the magnitude of the signal squared, integrated over time:

$$E=\int_{-\infty}^{+\infty}|x(t)|^2dt\tag{3.16}$$

Using an extension of Parseval's theorem, we can easily show that:

$$\int_{-\infty}^{+\infty}|x(t)|^2dt=\int_{-\infty}^{+\infty}|X(f)|^2df\tag{3.17}$$

The term $|X(f)|^2$ represents the energy density function over frequency, also called the power spectrum (energy spectral density or power spectral density (PSD)). Therefore, the power spectrum is computed as the magnitude squared of the FT of the waveform of interest $x(t)$:

$$PS(f) = |X(f)|^2 \tag{3.18}$$

Instead of applying the FFT to the entire waveform to calculate the power spectrum, we often use a technique called averaging, which enhances the statistical properties of the result. In such cases, calculation of the power spectrum actually represents an estimation process. An average periodogram is the common term used to designate that the power spectrum is calculated using the FT followed by averaging. To produce statistically reliable spectral estimates with high resolution, we need to take into account the necessary trade-offs that result from selecting the window shape and the averaging approach. We can use these two techniques combined or separately (Stoica & Moses, 1997). In reality, the selection of window shape and averaging approach is usually done by trial and error (Semmlow, 2004).

Having already described windowing technique, averaging is generally implemented by dividing the waveform into sections (possibly overlapping) and calculating the FT for each of the sections. An average of the FTs acquired from the sections represents the final spectrum. Selecting a short section length (a small N) delivers more sections for averaging. This enhances the reliability of the spectral estimate but also reduces frequency resolution as a result of smaller N chosen (Semmlow, 2004).

EXAMPLE 3.6. The following MATLAB code was used to plot the EEG signal and its periodogram spectrum. You can download data from the following website:
http://epileptologie-bonn.de/cms/front_content.php?idcat=193&lang=3&changelang=3

```
%Ch3_Ex6_EEG_Per.m
%The following MATLAB code is used to plot
%the EEG signal and its Periodogram spectrum.
clc
clear
%Load Sample EEG Data downloaded from the web site
%http://epileptologie-bonn.de/cms/front_content.php?idcat=193&lang=3&changelang=3
load AS_BONN_ALL_EEG_DATA_4096.mat

%%
Fs = 173.6;            % Sampling frequency
T = 1/Fs;              % Sampling period
L = length(Normal_Eyes_Open(:,1));    % Length of signal
t = (0:L-1)*T;         % Time vector
%%
%Plot the Original EEG Signal
subplot(2,1,1);
plot(Fs*t(1:L),Normal_Eyes_Open(:,1))
title('Original EEG Signal')
xlabel('Numeber of Samples')
ylabel('X(t)')
axis ([0 L -inf inf])

%%
% Obtain the periodogram of the signal using periodogram. Plot the periodogram.
[Pxx,F] = periodogram(Normal_Eyes_Open(:,1),rectwin(length(Normal_Eyes_Open(:,1))),
length(Normal_Eyes_Open(:,1)),Fs);
subplot(2,1,2);
f = Fs*(0:(L/2))/L;
plot(f,Pxx)
title('Periodogram of EEG Signal')
xlabel('Frequency (Hz)');
ylabel('|P(f)|')
```

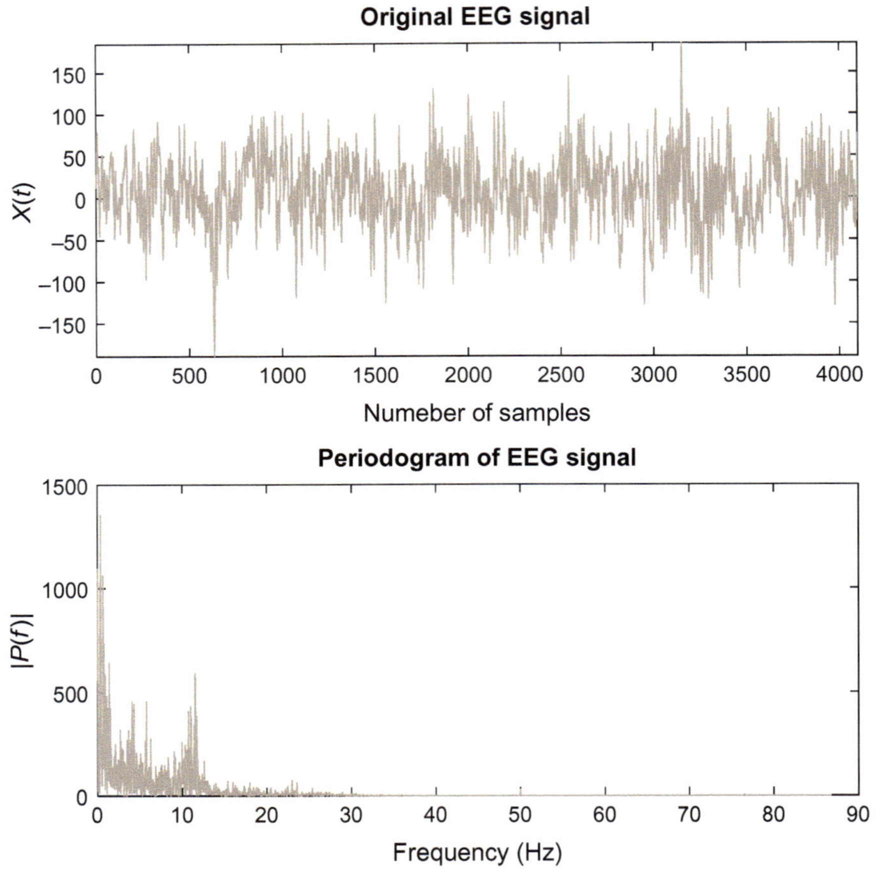

FIG. 3.6 Drawing of healthy EEG signal and its periodogram spectrum.

EXAMPLE 3.7. The following MATLAB code was used to plot the EMG signal and its periodogram spectrum. For this example, EMGLAB data was used. You can download the data from the following website: http://www.emglab.net/emglab/Signals/signals.php

```
%Ch3_Ex7_EMG_Per.m
%The following MATLAB code is used to plot
%the EMG signal and its Periodogram spectrum.
clc
clear
%Load Sample EMG Data downloaded from the web site
%http://www.emglab.net/emglab/Signals/signals.php
load EMGDAT.mat
%%
Fs = 10000;              % Sampling frequency
T = 1/Fs;                % Sampling period
L = length(CON(:,2));      % Length of signal
t = (0:L-1)*T;           % Time vector
%%
%Plot the Original EMG Signal
subplot(2,1,1);
plot(Fs*t(1:L),CON(:,2))
%plot(t,CON(:,2))
title('Original EMG Signal')
xlabel('Number of Samples')
ylabel('X(t)')
axis ([0 L -inf inf])
%%
% Obtain the periodogram of the signal using periodogram. Plot the
periodogram.
[Pxx,F] = periodogram(CON(:,2),rectwin(L),length(CON(:,2)),Fs);
subplot(2,1,2);
f = Fs*(0:(L/2))/L;
plot(f,Pxx)
title('Periodogram of EMG Signal')
xlabel('Frequency (Hz)');
ylabel('|P(f)|')
```

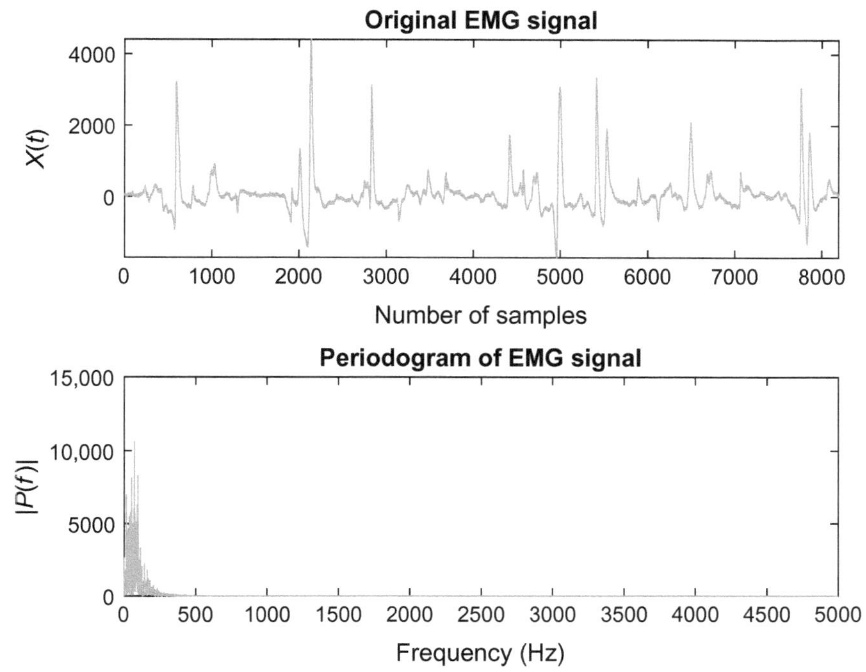

FIG. 3.7 Drawing of healthy EMG signal and its periodogram spectrum.

EXAMPLE 3.8. The following MATLAB code was used to plot the ECG signal and its periodogram spectrum. For this example, MIT-BIH Arrhythmia Database was used. You can download the data from the following website: https://www.physionet.org/physiobank/database/mitdb/

```
%Ch3_Ex8_ECG_Per.m
%The following MATLAB code is used to plot
%the ECG signal and its Periodogram spectrum.
clc
clear
%Load Sample ECG Data downloaded from the web site
%https://www.physionet.org/physiobank/database/mitdb/
load MITBIH_ECG.mat
%%
Fs = 320;                % Sampling frequency
T = 1/Fs;                  % Sampling period
L = length(ECGN(:,1));       % Length of signal
t = (0:L-1)*T;             % Time vector
%%
%Plot NORMAL ECG Signal
subplot(2,1,1);
plot(ECGN(:,1));
title('Original ECG Signal')
xlabel('Number of Samples')
ylabel('Amplitude')
ylabel('X(t)')
axis ([0 L -inf inf])
%%
% Obtain the periodogram of the signal using periodogram. Plot the periodogram.
[Pxx,F] = periodogram(ECGN(:,1),hamming(L),L,Fs);
subplot(2,1,2);
f = Fs*(0:(L/2))/L;
plot(f,Pxx)
title('Periodogram of ECG Signal')
xlabel('Frequency (Hz)');
ylabel('|P(f)|')
axis ([0 50 -inf inf])
```

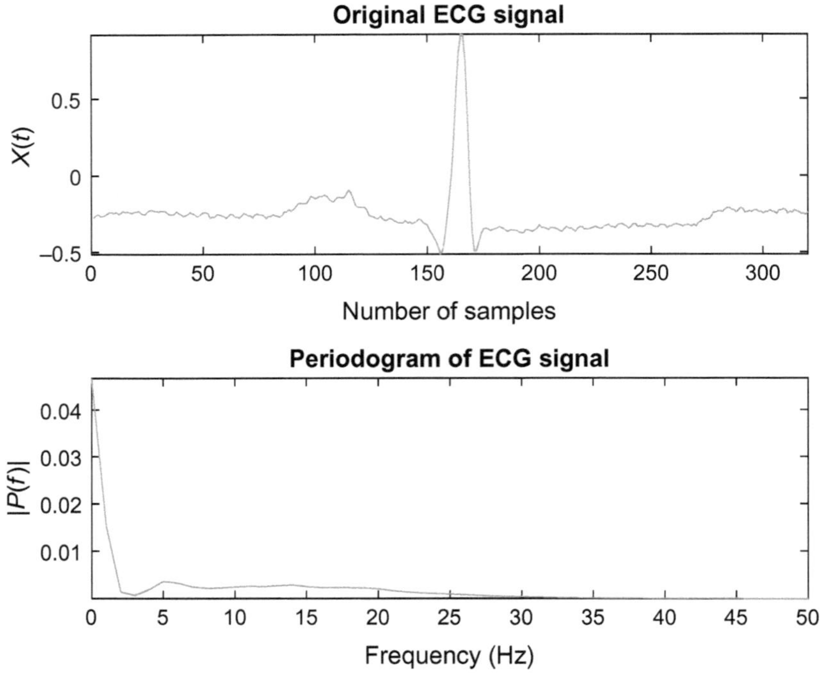

FIG. 3.8 Drawing of healthy ECG signal and its periodogram spectrum.

EXAMPLE 3.9. The following MATLAB code was used to plot the PCG signal and its periodogram spectrum. For this example, PCG Phonogram Dataset was used. You can download the data from the following website: https://www.prweb.com/releases/2004/09/prweb162074.htm

```
%Ch3_Ex9_PCG_Per.m
%The following MATLAB code is used to plot
%the PCG signal and its Periodogram spectrum.
clc
clear
%PCG Phonogram Data Set from BioSignetics Corporation
%Load Sample PCG Data downloaded from the web site
%https://www.prweb.com/releases/2004/09/prweb162074.htm
load PCG.mat
%%
Fs = 22000;            % Sampling frequency
T = 1/Fs;              % Sampling period
L = length(Aortic_Stenosis_2(1:40000,1));    % Length of signal
t = (0:L-1)*T;         % Time vector
%%
%Plot Aortic_Stenosis PCG Signal
subplot(2,1,1);
plot(Aortic_Stenosis_2(1:40000,1));
title('Original Aortic Stenosis PCG Signal')
xlabel('Number of Samples')
ylabel('Amplitude')
ylabel('X(t)')
axis ([0 L -inf inf])
%%
% Obtain the periodogram of the signal using periodogram. Plot the periodogram.
[Pxx,F] = periodogram(Aortic_Stenosis_2(1:40000,1),hamming(L),L,Fs);
subplot(2,1,2);
f = Fs*(0:(L/2))/L;
plot(f,Pxx)
title('Periodogram of PCG Signal')
xlabel('Frequency (Hz)');
ylabel('|P(f)|')
axis ([0 1000 -inf inf])
```

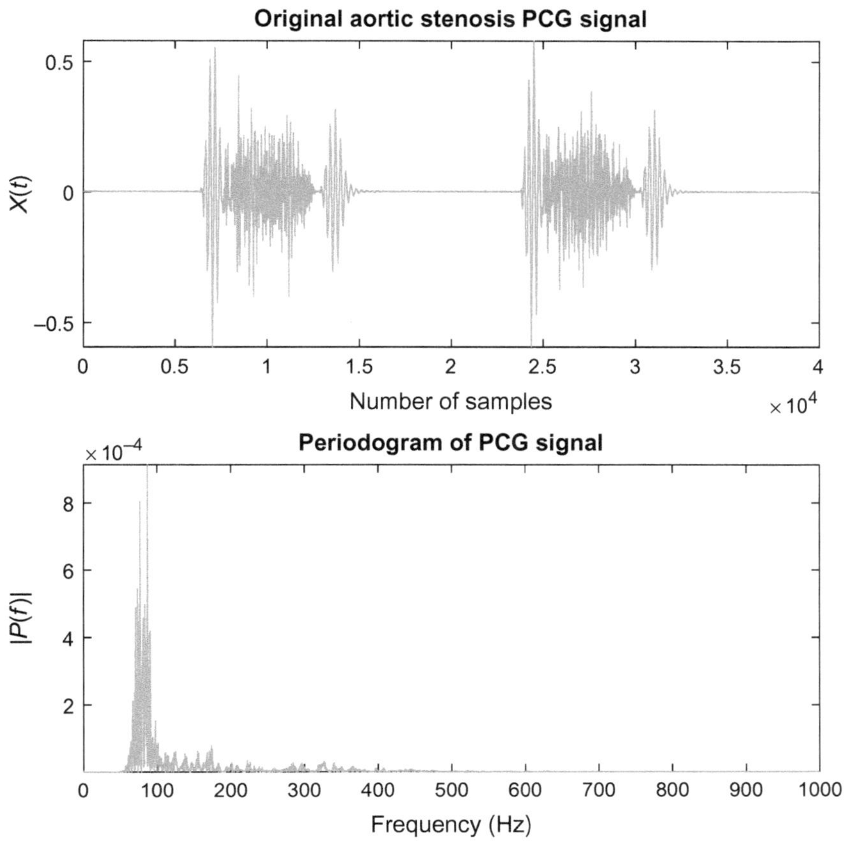

FIG. 3.9 Drawing of aortic stenosis PCG signal and its periodogram spectrum.

EXAMPLE 3.10. The following MATLAB code is used to plot the PPG signal and its periodogram spectrum. For this example, Wrist PPG During Exercise dataset was used. You can download the data from the following website: https://physionet.org/physiobank/database/wrist/

```
%Ch3_Ex10_PPG_Per.m
%The following MATLAB code is used to plot
%the PPG signal and its Periodogram spectrum.
clc
clear
%Wrist PPG During Exercise
%Load Sample PPG Data downloaded from the web site
%https://physionet.org/physiobank/database/wrist/
load s1_high_resistance_bikem.mat
%%
Fs = 256;          % Sampling frequency
T = 1/Fs;           % Sampling period
L=length(val(2,4097:8192));    % Length of signal
t = (0:L-1)*T;       % Time vector
%%
%Plot PPG Signal
subplot(2,1,1);
plot(val(2,4097:8192))
title('Original PPG Signal')
xlabel('Number of Samples')
ylabel('Amplitude')
ylabel('X(t)')
axis ([0 L -inf inf])
```

Continued

EXAMPLE 3.10.—cont'd

```
%%
% Obtain the periodogram of the signal using periodogram. Plot the periodogram.
[Pxx,F] = periodogram(val(2,4097:8192),hamming(L),L,Fs);
subplot(2,1,2);
f = Fs*(0:(L/2))/L;
plot(f,Pxx)
title('Periodogram of PPG Signal')
xlabel('Frequency (Hz)');
ylabel('|P(f)|')
axis ([0 10 -inf inf])
```

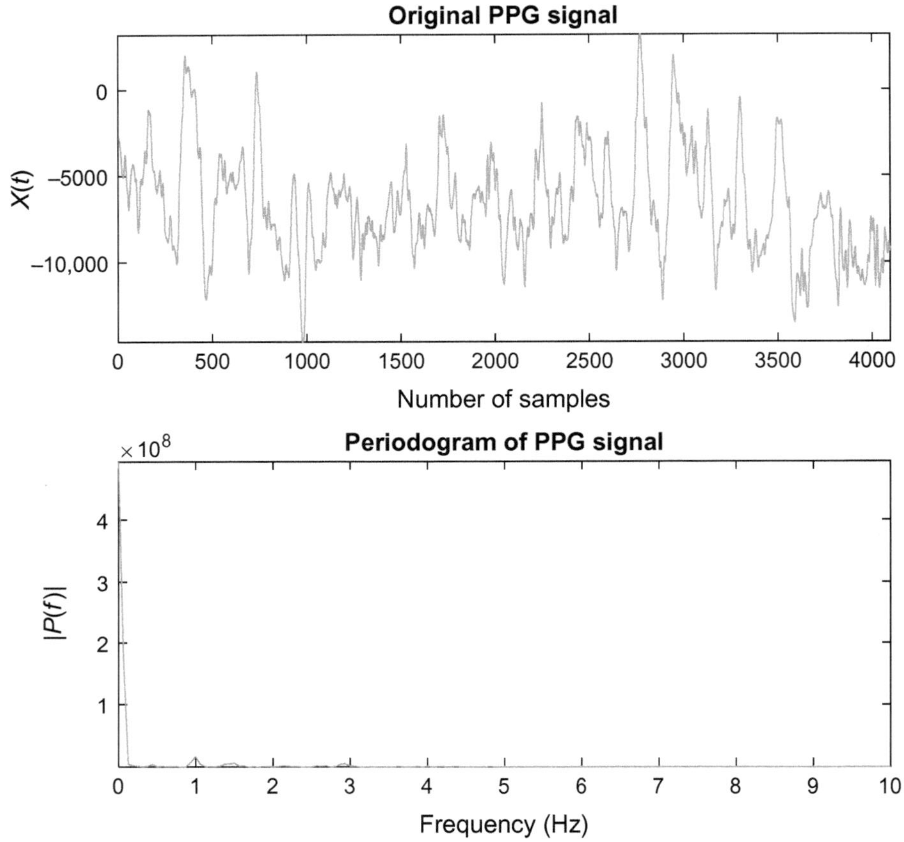

FIG. 3.10 Drawing of healthy PPG signal and its periodogram spectrum.

3.2.6 Welch Power Spectral Density

Welch's approach is one of the most popular techniques for estimating the averaged periodogram. Welch actually modified the averaging technique initially developed by Bartlett. He used overlapping sections and applied a window to every section. The advantage of using overlapping sections is the fact that more sections are averaged for a given length of the waveform. Spectrum acquired from noisy data is usually averaged from sections that overlap by 50%. In other applications, higher amounts of overlap have been suggested. Maximum overlap, shifting by only one sample to start the new section, is thus desirable if calculating time is not an issue (Semmlow, 2004).

If the periodogram is likely to be fairly flat, the application of a windowing technique for sidelobe control may be insignificant. Windows are actually claimed for favoring some data samples more than others and for only a minor reduction in estimation error at the cost of decreased frequency resolution. However, these statements are not correct for the Welch periodogram, because overlapping sections help in treating data samples equally whereas the estimation error decreases with increased number of sections. Furthermore, a windowing technique is recommended if the spectrum is likely to have large amplitude changes (Semmlow, 2004).

EXAMPLE 3.11. The following MATLAB code was used to plot the EEG signal and its Welch spectrum. You can download data from the following website: http://epileptologie-bonn.de/cms/front_content.php?idcat=193&lang=3&changelang=3

```
%Ch3_Ex11_EEG_Welch.m
%The following MATLAB code is used to plot
%the EEG signal and its Welch spectrum.
clc
clear
%Load Sample EEG Data downloaded from the web site
%http://epileptologie-bonn.de/cms/front_content.php?idcat=193&lang=3&changelang=3
load AS_BONN_ALL_EEG_DATA_4096.mat

%%
Fs = 173.6;             % Sampling frequency
T = 1/Fs;               % Sampling period
L = length(Normal_Eyes_Open(:,1));    % Length of signal
t = (0:L-1)*T;          % Time vector
segmentLength = L;
noverlap = L/4;

%%
%Plot the Original EEG Signal
subplot(2,1,1);
plot(Fs*t(1:L),Normal_Eyes_Open(:,1))
title('Original EEG Signal')
xlabel('Numeber of Samples')
ylabel('X(t)')
axis ([0 L -inf inf])

%%
% Obtain the Welch Spectrum of the signal using pwelch.
[Pxx,f] = pwelch(Normal_Eyes_Open(:,1),segmentLength,noverlap, L,Fs);
%Plot the Welch Spectrum.
subplot(2,1,2);
plot(f,Pxx)
title('Welch Spectrum of EEG Signal')
xlabel('Frequency (Hz)');
ylabel('|P(f)|')
axis ([0 50 -inf inf])
```

See Fig. 3.11.

3.3 PARAMETRIC MODEL-BASED METHODS

The methods for calculating the power spectra using the FT are denoted as classical methods, representing the most robust spectral estimators. Knowledge or assumptions concerning the nature of the waveform is not necessary but could be useful when selecting windows and averaging approaches. Additionally, the assumptions stating that the waveform outside the window equals zero may lead to the estimate distortion (Marple & Marple, 1987).

Modern methods of spectral analysis are intended to overcome some of the shortcomings and distortions produced by the classical techniques. They are mainly effective when short-duration signals are analyzed. Modern methods benefit from the fact that we know something about the waveform of interest, perhaps the process that generates it. In that case, these methods can make assumptions regarding the waveform beyond the window, enhancing spectral resolution for very noisy waveforms in particular. However, the suitability of the selected model will highly influence the spectral resolution (Marple & Marple, 1987).

Furthermore, the presence of computationally popular approaches for parameter evaluation partially describes the attractiveness of different linear model types,

FIG. 3.11 Drawing of healthy EEG signal and its Welch spectrum.

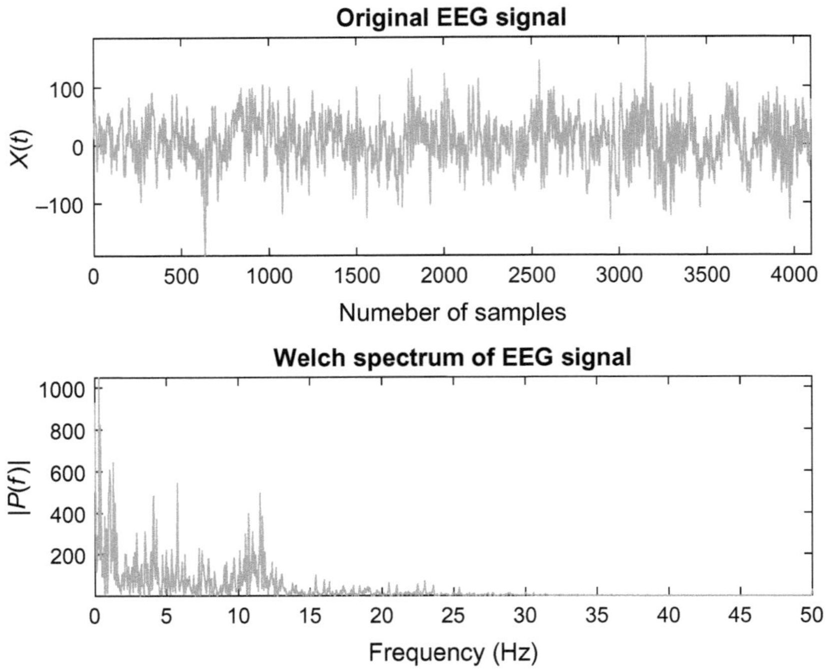

EXAMPLE 3.12. The following MATLAB code was used to plot the EMG signal and its Welch spectrum. For this example, EMGLAB data was used. You can download the data from the following website: http://www.emglab.net/emglab/Signals/signals.php

```
%Ch3_Ex12_EMG_Welch.m
%The following MATLAB code is used to plot
%the EMG signal and its Welch spectrum.
clc
clear
%Load Sample EMG Data downloaded from the web site
%http://www.emglab.net/emglab/Signals/signals.php
load EMGDAT.mat
%%
Fs = 10000;              % Sampling frequency
T = 1/Fs;                % Sampling period
L = length(CON(:,2));      % Length of signal
t = (0:L-1)*T;           % Time vector
segmentLength = L;
noverlap = L/4;
%%
%Plot the Original EMG Signal
subplot(2,1,1);
plot(Fs*t(1:L),CON(:,2))
%plot(t,CON(:,2))
title('Original EMG Signal')
xlabel('Number of Samples')
ylabel('X(t)')
axis ([0 L -inf inf])
%%
% Obtain the Welch Spectrum of the signal using pwelch.
[Pxx,f] = pwelch(CON(:,2),segmentLength,noverlap, L,Fs);
%Plot the Welch Spectrum.
subplot(2,1,2);
plot(f,Pxx)
title('Welch Spectrum of EMG Signal')
xlabel('Frequency (Hz)');
ylabel('|P(f)|')
axis ([0 1000 -inf inf])
```

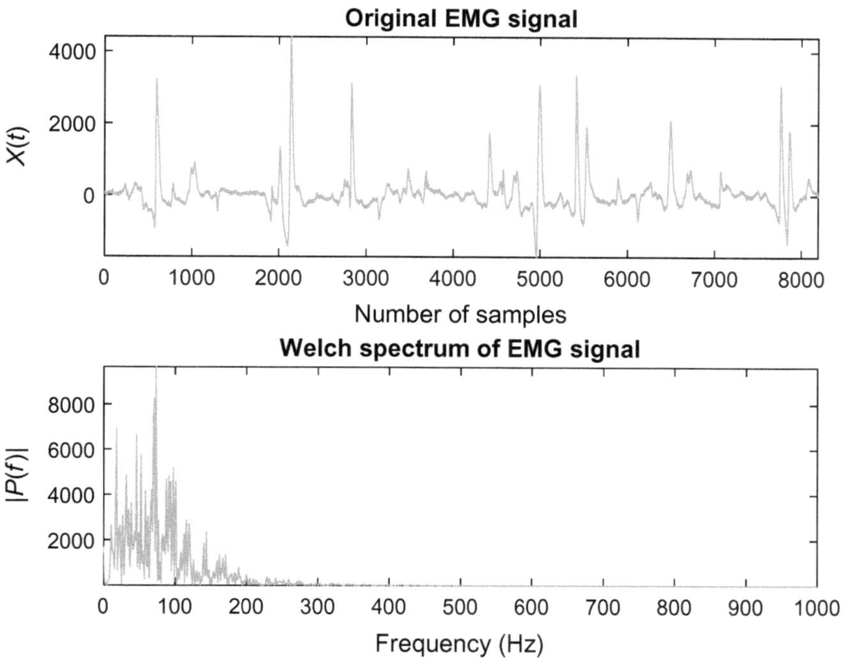

FIG. 3.12 Drawing of healthy EMG signal and its Welch spectrum.

EXAMPLE 3.13. The following MATLAB code was used to plot the ECG signal and its Welch spectrum. For this example, MIT-BIH Arrhythmia Database was used. You can download the data from the following website: https://www.physionet.org/physiobank/database/mitdb/

```
%Ch3_Ex13_ECG_Welch.m
%The following MATLAB code is used to plot
%the ECG signal and its Welch spectrum.
clc
clear
%Load Sample ECG Data downloaded from the web site
%https://www.physionet.org/physiobank/database/mitdb/
load MITBIH_ECG.mat
%%
Fs = 320;            % Sampling frequency
T = 1/Fs;             % Sampling period
L = length(ECGN(:,1));     % Length of signal
t = (0:L-1)*T;        % Time vector
segmentLength = L;
noverlap = L/4;
%%
%Plot NORMAL ECG Signal
subplot(2,1,1);
plot(ECGN(:,1));
title('Original ECG Signal')
xlabel('Number of Samples')
ylabel('Amplitude')
ylabel('X(t)')
axis ([0 L -inf inf])
%%
%Obtain the Welch Spectrum of the signal using pwelch.
[Pxx,f] = pwelch(ECGN(:,1),segmentLength,noverlap, L,Fs);
%Plot the Welch Spectrum.
subplot(2,1,2);
plot(f,Pxx)
title('Welch Spectrum of ECG Signal')
xlabel('Frequency (Hz)');
ylabel('|P(f)|')
axis ([0 100 -inf inf])
```

FIG. 3.13 Drawing of healthy ECG signal and its Welch spectrum.

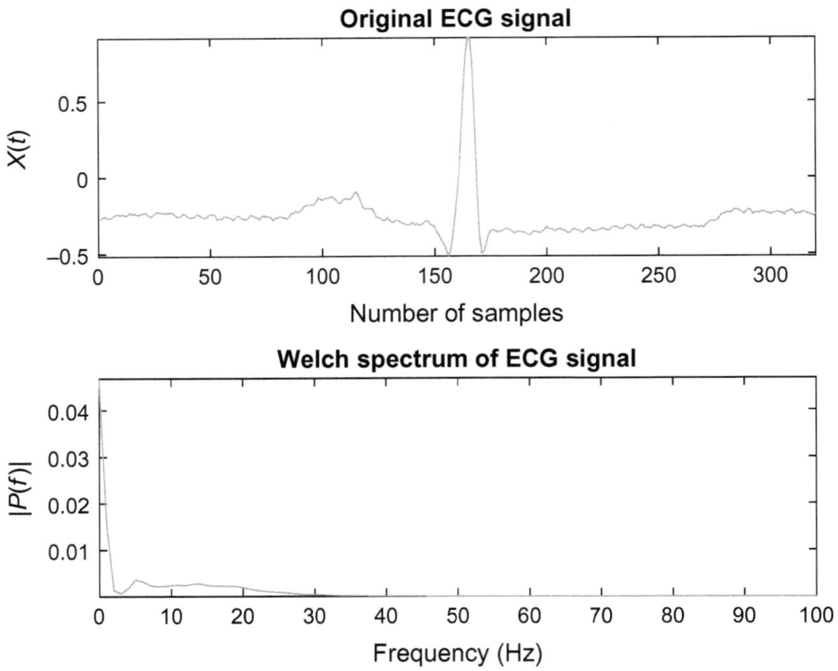

EXAMPLE 3.14. The following MATLAB code was used to plot the PCG signal and its Welch spectrum. For this example, PCG Phonogram Dataset was used. You can download the data from the following website: https://www.prweb.com/releases/2004/09/prweb162074.htm

```
%Ch3_Ex14_PCG_Welch.m
%The following MATLAB code is used to plot
%the PCG signal and its Welch spectrum.
clc
clear
%PCG Phonogram Data Set from BioSignetics Corporation
%Load Sample PCG Data downloaded from the web site
%https://www.prweb.com/releases/2004/09/prweb162074.htm
load PCG.mat
%%
Fs = 22000;              % Sampling frequency
T = 1/Fs;                % Sampling period
L = length(Aortic_Stenosis_2(1:40000,1)); % Length of signal
t = (0:L-1)*T;           % Time vector
segmentLength = L;
noverlap = L/4;
%%
%Plot Aortic_Stenosis PCG Signal
subplot(2,1,1);
plot(Aortic_Stenosis_2(1:40000,1));
title('Original Aortic Stenosis PCG Signal')
xlabel('Number of Samples')
ylabel('Amplitude')
ylabel('X(t)')
axis ([0 L -inf inf])
%Obtain the Welch Spectrum of the signal using pwelch.
[Pxx,f] = pwelch(Aortic_Stenosis_2(1:40000,1),segmentLength,noverlap, L,Fs);
%Plot the Welch Spectrum.
subplot(2,1,2);
plot(f,Pxx)
title('Welch Spectrum of Aortic Stenosis PCG Signal')
xlabel('Frequency (Hz)');
ylabel('|P(f)|')
axis ([0 500 -inf inf])
```

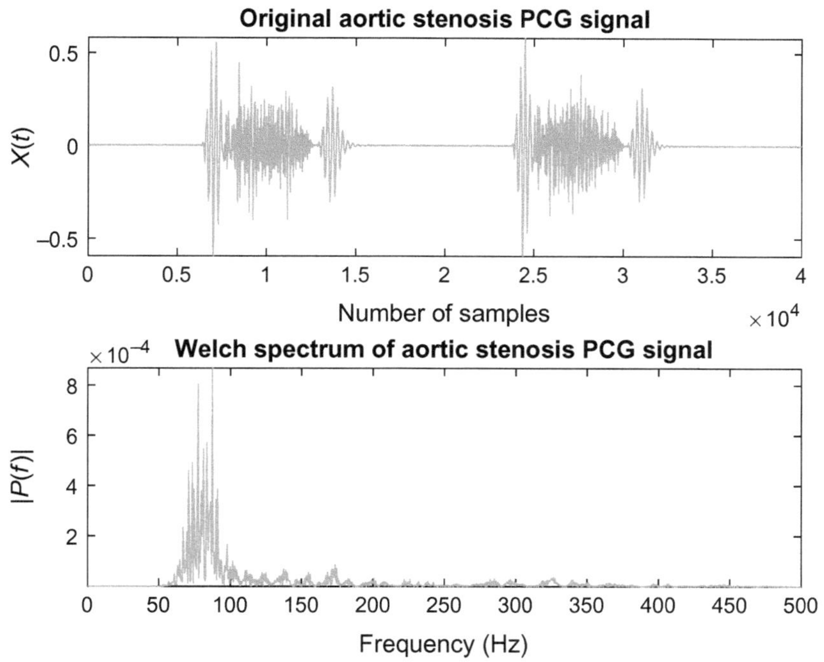

FIG. 3.14 Drawing of aortic stenosis PCG signal and its Welch spectrum.

EXAMPLE 3.15. The following MATLAB code was used to plot the PPG signal and its Welch spectrum. For this example, Wrist PPG During Exercise dataset was used. You can download the data from the following website: https://physionet.org/physiobank/database/wrist/

```
%Ch3_Ex15_PPG_Welch.m
%The following MATLAB code is used to plot
%the PPG signal and its Welch spectrum.
clc
clear
%Wrist PPG During Exercise
%Load Sample PPG Data downloaded from the web site
%https://physionet.org/physiobank/database/wrist/
load s1_high_resistance_bikem.mat
%%
Fs = 256;            % Sampling frequency
T = 1/Fs;             % Sampling period
L = length(val(2,4097:8192));    % Length of signal
t = (0:L-1)*T;        % Time vector
segmentLength = L;
noverlap = L/4;
%%
%Plot PPG Signal
subplot(2,1,1);
plot(val(2,4097:8192))
title('Original PPG Signal')
xlabel('Number of Samples')
ylabel('Amplitude')
ylabel('X(t)')
axis ([0 L -inf inf])

%Obtain the Welch Spectrum of the signal using pwelch.
[Pxx,f] = pwelch(val(2,4097:8192),segmentLength,noverlap, L,Fs);
%Plot the Welch Spectrum.
subplot(2,1,2);
plot(f,Pxx)
title('Welch Spectrum of PPG Signal')
xlabel('Frequency (Hz)');
ylabel('|P(f)|')
axis ([0 10 -inf inf])
```

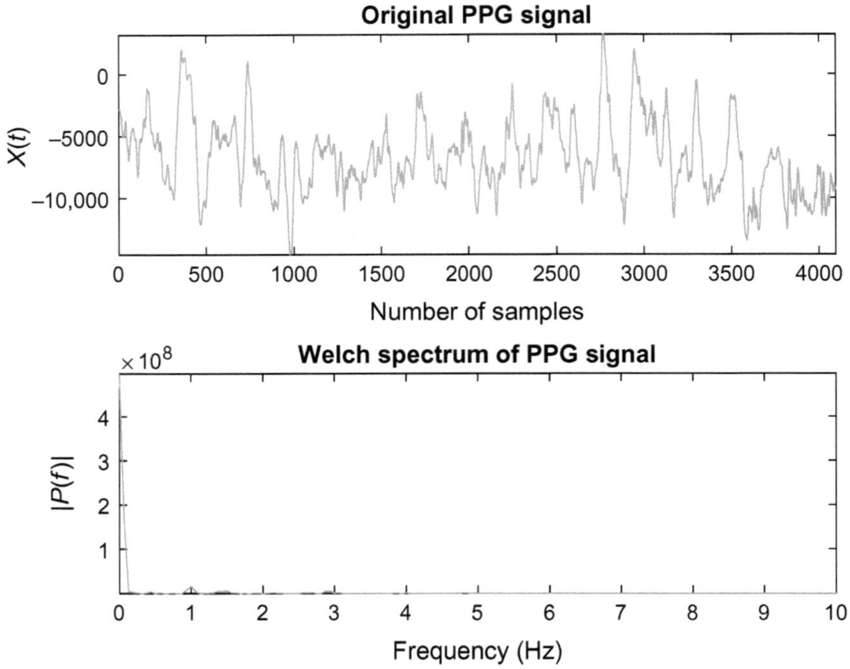

FIG. 3.15 Drawing of PPG signal and its Welch spectrum.

differentiated by the nature of their transfer functions. However, autoregressive (AR) modeling has received the most attention so far (Sörnmo & Laguna, 2005).

Some of the most popular model types are the following:

- AR,
- moving average (MA), and
- AR moving average (ARMA).

Some knowledge of the possible spectrum shapes is necessary for selecting the most suitable model. The equation for an AR model is,

$$y(n) = -\sum_{k=1}^{p} a(k)y(n-k) + v(n) \tag{3.19}$$

where $v(n)$ is the noise function fed as input to a filter of order p. White noise has the variance defined as $E[v^2(n)] = \sigma_v^2$. The output $y(n)$ is calculated by convolving the past versions of the output $y(n-k)$ with fixed model parameters $a(k)$.

A transfer function of the AR model contains a constant in the numerator and a polynomial in the denominator. Hence, the AR model is occasionally denoted as an all-pole model, if we neglect zeroes located at $z=0$. Using the Z-transform, the transfer function $H(z)$ of the AR model becomes:

$$H(z) = \frac{1}{A(z)} = \frac{1}{1 + a_1 z^{-1} + \cdots + a_p z^{-p}} \tag{3.20}$$

The model is characterized by the locations of the poles and zeroes. The poles are related with spectral peaks, while the zeroes are related with valleys in the spectrum. The AR model is thus primarily suitable for evaluating spectra with sharp peaks and no deep valleys.

On the contrary, the MA model is suitable for estimating spectra with the valleys and no sharp peaks. The MA model has a transfer function with just a numerator polynomial. Hence, it is occasionally denoted as an all-zero model. The MA model is defined by the following equation:

$$y(n) = -\sum_{k=1}^{q} b(k)v(n-k) \tag{3.21}$$

where $v(n)$ is the input noise function and q is the order of a model.

A model that combines properties of both the AR and MA models may be used if we expect the spectrum to have both sharp peaks and deep valleys. Likely, an ARMA model has a transfer function with both numerator and denominator polynomials, so it is occasionally denoted as a pole-zero model. The equation for an ARMA model is defined as:

$$y(n) = -\sum_{k=1}^{p} a(k)y(n-k) + \sum_{k=1}^{q} b(k)v(n-k) \tag{3.22}$$

The ARMA model can be equally well-defined by the rational transfer function,

$$H(z) = \frac{B(z)}{A(z)} = \frac{b_0 + b_1 z^{-1} + \cdots + b_q z^{-q}}{1 + a_1 z^{-1} + \cdots + a_p z^{-p}} \tag{3.23}$$

The evaluation of AR parameters normally requires solving a linear matrix equation, making the procedure quite efficient. On the contrary, evaluation techniques for computing the parameters of the ARMA and MA models involve considerably more calculations. Furthermore, these two algorithms may converge on the wrong solution or may not converge at all (Sörnmo & Laguna, 2005).

To find an optimal solution, the ARMA techniques should evaluate the AR and MA parameters together, which is not the case for most ARMA methods. In addition, the MA method does not represent a high-resolution spectral estimator because it cannot model narrowband spectra well. Therefore, it is not suitable for usage in power spectral estimation of biosignals (Semmlow, 2004).

Spectral estimation methods based on AR can be separated into two groups:

- Algorithms processing data sequentially.
- Algorithms processing block data.

The former is applicable if we need to quickly evaluate incoming data for real-time concerns, whereas the latter is suitable if we have the entire waveform stored in memory. Although the AR spectral technique is generally defined with respect to autocorrelation sequence estimation, superior results are achieved by algorithms that run directly on short data segments of the waveform, without estimating the autocorrelation sequence (Semmlow, 2004).

The most popular techniques for evaluating the AR model parameters and related power spectrum directly from the waveform are:

- Yule-Walker,
- Burg,
- Covariance, and
- Modified covariance methods.

All these methods for spectral estimation are implemented in the Signal Processing Toolbox within MATLAB, and they all produce similar spectra (Semmlow, 2004). Having a few options for AR spectral estimation to choose from, it is normal to consider which one of the aforementioned techniques is best for EEG analysis. The question of choosing the model order p should also be discussed. Among many design parameters, the sampling rate is also very important (Sörnmo & Laguna, 2005).

As different approaches highlight different spectral properties, the desired spectrum shape will determine the most suitable approach. For instance, the Yule-Walker technique yields spectra with the least resolution, hence providing the most smoothing. On the other hand, the modified covariance technique produces the sharpest spectral peaks, facilitating the detection of sinusoidal components in the waveform (Marple & Marple, 1987). The Burg and covariance techniques yield comparable spectra. However, the modified covariance technique may not result in a stable AR model with all poles located inside the unit circle (Therrien, 1992). Luckily, the stability plays no important role in spectral analysis but must be considered when designing a classifier based on AR modeling. Obviously, Burg's approach results in a stable AR model (Sörnmo & Laguna, 2005).

Besides choosing the type of model, we must also choose the order of model, denoted as p and/or q. Understanding the process of generating the data might be quite useful when selecting model order. There are a number of guidelines providing assistance regarding this matter (Semmlow, 2004). For example, selecting too large a p value results in fake spectral peaks, whereas a very small p value produces a desperately smooth spectrum having unsatisfactory resolution (Sörnmo & Laguna, 2005). The order of the model should be adequate to let the model spectrum fit the signal spectrum only, rejecting the noise (Semmlow, 2004). Actually, most studies have assumed a fixed model order. Even lower model orders guarantee some level of sturdiness against noise and artifacts while delivering enough spectral detail for classification of signal spectra (da Silva et al., 1987). The fifth- or sixth-order AR model has been successfully applied in some studies (Anderson, Stolz, & Shamsunder, 1998; Jansen, Bourne, & Ward, 1981). If we demand a more detailed representation of the power spectrum, a higher model order is a necessity. For that matter, every spectral peak of interest increases the model order by two, and this number defines a lower limit for p. An upper limit of the model order is freely defined with $p < N/3$ (Marple & Marple, 1987). However, the chosen model order is usually much lower than $N/3$ (Sörnmo & Laguna, 2005).

3.3.1 Autoregressive Model for Spectral Analysis

The model-based (parametric) methods are employed for modeling the time series data $x(n)$ as the output of a linear system represented by a rational assembly and the spectrum estimation process composed of two steps. In the first step, the model-based parameters are estimated from a given data sequence $x(n)$, $0 \leq n \leq N-1$. Then the PSD estimate is calculated from these estimations. The AR model is the frequently employed parametric method, as the estimation of AR parameters can be calculated simply by solving linear equations. In the AR model, the data can be modeled as an output of a causal, all-pole, discrete filter whose input is white noise. The AR model of order p can be expressed as:

$$x(n) = -\sum_{k=1}^{p} a(k)x(n-k) + w(n) \tag{3.24}$$

where $a(k)$ are the AR coefficients and $w(n)$ is white noise of variance equal to σ^2. The AR(p) model can be characterized by the AR parameters $\{a[1], a[2], \ldots, a[p], \sigma^2\}$. The PSD is

$$P_{AR}(f) = \frac{\sigma^2}{|A(f)|^2} \tag{3.25}$$

where $A(f) = 1 + a_1 e^{-j2\pi f} + \cdots + a_p e^{-j2\pi f p}$.

To find a suitable and stable AR model, different factors should be taken into consideration such as the model order selection, the signal length, and the level of stationary of the signal (Kay, 1993; Muthuswamy & Thakor, 1998; Pardey, Roberts, & Tarassenko, 1996).

Because the AR spectral estimation methods have computational efficiency and good performance, they are extensively utilized in practice. The AR spectral estimation techniques utilize the estimation of either the AR parameters or the reflection coefficients (Subasi, 2007).

3.3.2 Yule-Walker AR Modeling

Yule-Walker AR modeling parameters are estimated by minimizing an estimate of prediction error power,

$$\hat{\rho} = \frac{1}{N}\sum_{n=-\infty}^{\infty}\left|x(n) + \sum_{k=1}^{p} a(k)x(n-k)\right|^2 \quad (3.26)$$

The samples of the $x(n)$ process that are not observed in the range $0 \le n \le N-1$ are made equal to zero in Eq. (3.26). The estimated prediction error power is minimized by differentiating the equation with respect to the real and imaginary parts of the $a(k)$s. This can be achieved by utilizing the complex gradient to produce,

$$\frac{1}{N}\sum_{n=-\infty}^{\infty}\left(x(n) + \sum_{k=1}^{p} a(k)x(n-k)\right)x^*(n-l) = 0 \quad l = 1,2,\ldots,p \quad (3.27)$$

In matrix form, the set of equations becomes

$$\begin{bmatrix} \hat{r}(1) \\ \vdots \\ \hat{r}(p) \end{bmatrix} + \begin{bmatrix} \hat{r}(0) & \cdots & \hat{r}(-p+1) \\ \vdots & \ddots & \vdots \\ \hat{r}(p-1) & \cdots & \hat{r}(0) \end{bmatrix} \begin{bmatrix} \hat{a}(1) \\ \vdots \\ \hat{a}(p) \end{bmatrix} = \begin{bmatrix} 0 \\ \vdots \\ 0 \end{bmatrix}$$

or

$$\hat{r}_p + \hat{R}_p\hat{a} = 0 \quad (3.28)$$

where

$$\hat{r}(k) = \begin{cases} \frac{1}{N}\sum_{n=0}^{N-1-k} x^*(n)x(n+k) & k = 0,1,\ldots,p \\ \hat{r}^*(-k) & k = (-p+1),(-p+2),\ldots,-1 \end{cases}$$

From Eq. (3.7), the AR parameter estimates are found as,

$$\hat{a} = -\hat{R}_p^{-1}\hat{r}_p \quad (3.29)$$

The estimate of the white noise variance σ^2 is found as $\hat{\rho}_{\min}$, which is given by

$$\hat{\sigma}^2 = \hat{\rho}_{\min} = \frac{1}{N}\sum_{n=-\infty}^{\infty}\left|x(n) + \sum_{k=1}^{p}\hat{a}(k)x(n-k)\right|^2 \quad (3.30)$$

The final result is found by using Eq. (3.27),

$$\hat{\sigma}^2 = \hat{r}(0) + \sum_{k=1}^{p}\hat{a}(k)\hat{r}(-k) \quad (3.31)$$

From the estimates of the AR parameters, PSD estimation is formed as (Kay & Marple, 1981; Proakis & Manolakis, 2007; Stoica & Moses, 1997; Subasi, 2007):

$$\hat{P}_{YW}(f) = \frac{\hat{\sigma}^2}{\left|1 + \sum_{k=1}^{p}\hat{a}(k)e^{-j2\pi fk}\right|^2} \quad (3.32)$$

See Fig. 3.16.
See Fig. 3.17.
See Fig. 3.18.
See Fig. 3.19.

EXAMPLE 3.16. The following MATLAB code was used to plot the EEG signals and their Yule-Walker AR spectrum. You can download data from the following website: http://epileptologie-bonn.de/cms/front_content.php?idcat=193&lang=3&changelang=3

```
%Ch3_Ex16_EEG_Yule.m
%The following MATLAB code is used to plot
%the EEG signals and their Yule-Walker AR spectrum.
clc
clear
%Load Sample EEG Data downloaded from the web site
%http://epileptologie-bonn.de/cms/front_content.php?idcat=193&lang=3&changelang=3
load AS_BONN_ALL_EEG_DATA_4096.mat
%%
Fs = 173.6;              % Sampling frequency
T = 1/Fs;                % Sampling period
L = length(Normal_Eyes_Open(:,1));    % Length of signal
segmentLength = L;
order = 14;
%%
```

```
% Obtain the Yule-Walker AR Spectrum of the Normal EEG signal using pyulear.
[Pxx,F] = pyulear(Normal_Eyes_Open(:,1),order,segmentLength,Fs);
% Plot the Yule-Walker AR Spectrum of the signal
subplot(3,1,1)
plot(F,10*log10(Pxx))
ylabel('PSD (dB/Hz)');
xlabel('Frequency (Hz)');
legend('Normal Eyes Open')
title('Yule-Walker AR Spectrum of EEG Signals')
%%
% Obtain the Yule-Walker AR Spectrum of the Interictal EEG signal using pyulear.
[Pxx,F] = pyulear(Interictal(:,1),order,segmentLength,Fs);
% Plot the Yule-Walker AR Spectrum of the signal
subplot(3,1,2)
plot(F,10*log10(Pxx))
ylabel('PSD (dB/Hz)');
xlabel('Frequency (Hz)');
legend('Interictal')
%title('Yule-Walker AR')
%%
% Obtain the Yule-Walker AR Spectrum of the Ictal EEG signal using pyulear.
[Pxx,F] = pyulear(Ictal(:,1),order,segmentLength,Fs);
% Plot the Yule-Walker AR Spectrum of the signal
subplot(3,1,3)
plot(F,10*log10(Pxx))
ylabel('PSD (dB/Hz)');
xlabel('Frequency (Hz)');
legend('Ictal')
%title('Yule-Walker AR')
```

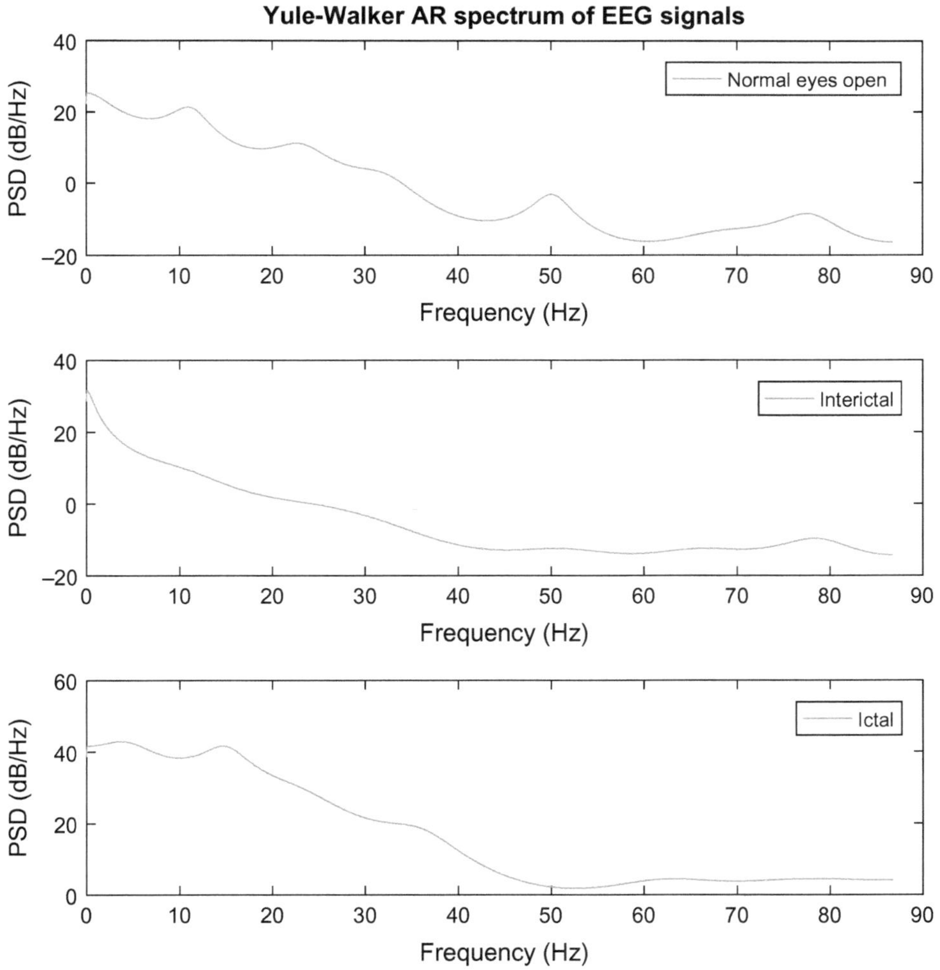

FIG. 3.16 Drawing of Yule-Walker AR spectrum of EEG signals.

EXAMPLE 3.17. The following MATLAB code was used to plot the EMG signals and their Yule-Walker AR spectrum. For this example, EMGLAB data was used. You can download the data from the following website: http://www.emglab.net/emglab/Signals/signals.php

```
%Ch3_Ex17_EMG_Yule.m
%The following MATLAB code is used to plot
%the EMG signals and their Yule-Walker AR spectrum.
clc
clear
%Load Sample EMG Data downloaded from the web site
%http://www.emglab.net/emglab/Signals/signals.php
load EMGDAT.mat
%%
Fs = 10000;            % Sampling frequency
T = 1/Fs;              % Sampling period
L = length(CON(:,2));    % Length of signal
segmentLength = L;
noverlap = L/4;
order = 34;
%%
% Obtain the Yule-Walker AR Spectrum of the Normal EEG signal using pyulear.
[Pxx,F] = pyulear(CON(:,1),order,segmentLength,Fs);
% Plot the Yule-Walker AR Spectrum of the signal
subplot(3,1,1)
plot(F,10*log10(Pxx))
ylabel('PSD (dB/Hz)');
xlabel('Frequency (Hz)');
legend('Control')
title('Yule-Walker AR Spectrum of EMG Signals')
%%
% Obtain the Yule-Walker AR Spectrum of the Interictal EEG signal using pyulear.
[Pxx,F] = pyulear(ALS(:,1),order,segmentLength,Fs);
% Plot the Yule-Walker AR Spectrum of the signal
subplot(3,1,2)
plot(F,10*log10(Pxx))
ylabel('PSD (dB/Hz)');
xlabel('Frequency (Hz)');
legend('ALS')
%title('Yule-Walker AR')
%%
% Obtain the Yule-Walker AR Spectrum of the Ictal EEG signal using pyulear.
[Pxx,F] = pyulear(MYO(:,1),order,segmentLength,Fs);
% Plot the Yule-Walker AR Spectrum of the signal
subplot(3,1,3)
plot(F,10*log10(Pxx))
ylabel('PSD (dB/Hz)');
xlabel('Frequency (Hz)');
legend('Myopathy')
%title('Yule-Walker AR')
```

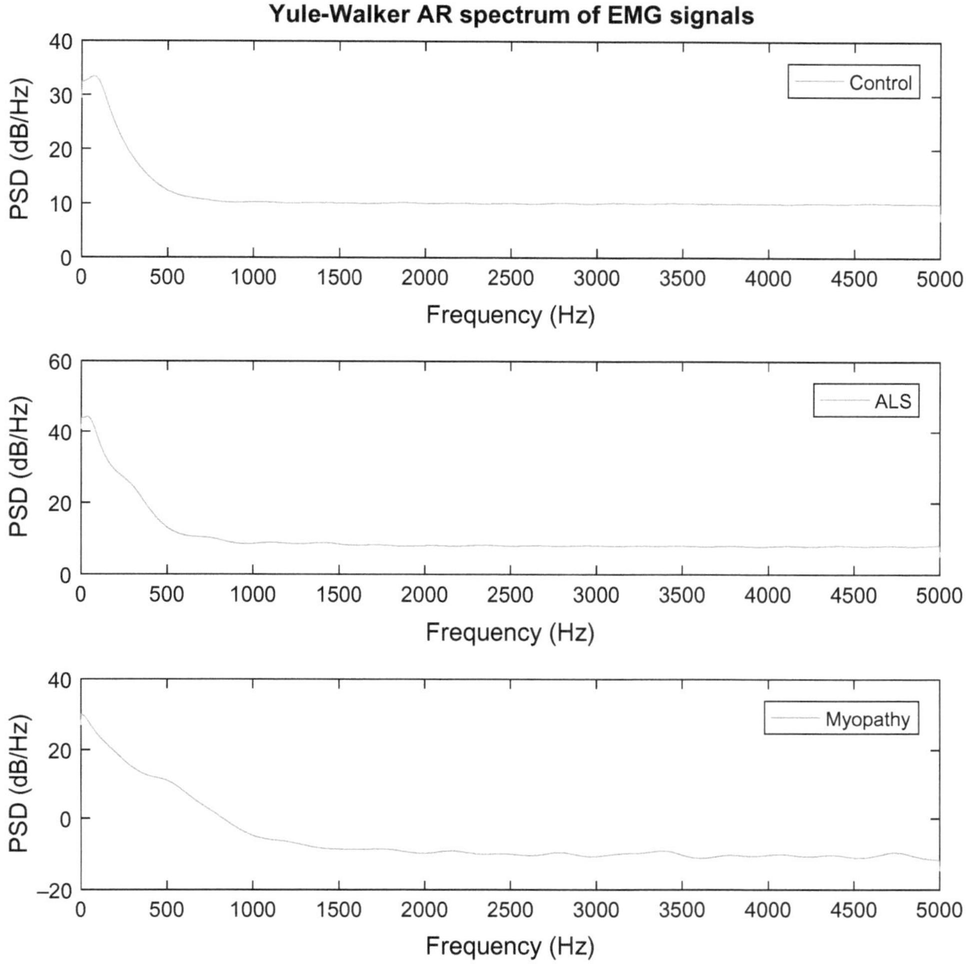

FIG. 3.17 Drawing of Yule-Walker AR spectrum of EMG signals.

EXAMPLE 3.18. The following MATLAB code was used to plot the ECG signals and their Yule-Walker AR spectrum. For this example, MIT-BIH Arrhythmia Database was used. You can download the data from the following website: https://www.physionet.org/physiobank/database/mitdb/

```
%Ch3_Ex18_ECG_Yule.m
%The following MATLAB code is used to plot
%the ECG signals and their Yule-Walker AR spectrum.
clc
clear
%Load Sample ECG Data downloaded from the web site
%https://www.physionet.org/physiobank/database/mitdb/
load MITBIH_ECG.mat
%%
Fs = 320;              % Sampling frequency
T = 1/Fs;              % Sampling period
L = length(ECGN(:,1));    % Length of signal
t = (0:L-1)*T;         % Time vector
```

```
segmentLength = L;
noverlap = L/4;
order = 34;
%%
% Obtain the Yule-Walker AR Spectrum of the Normal ECG signal using pyulear.
[Pxx,F] = pyulear(ECGN(:,1),order,segmentLength,Fs);
% Plot the Yule-Walker AR Spectrum of the signal
subplot(2,2,1)
plot(F,10*log10(Pxx))
ylabel('PSD (dB/Hz)');
xlabel('Frequency (Hz)');
legend('Normal ECG')
axis([0 L/2 -inf inf])
%title('Yule-Walker AR Spectrum of ECG Signals')
%%
% Obtain the Yule-Walker AR Spectrum of the ECG signal with APC using pyulear.
[Pxx,F] = pyulear(ECGAPC(:,1),order,segmentLength,Fs);
% Plot the Yule-Walker AR Spectrum of the signal
subplot(2,2,2)
plot(F,10*log10(Pxx))
ylabel('PSD (dB/Hz)');
xlabel('Frequency (Hz)');
legend('APC')
axis([0 L/2 -inf inf])
%title('Yule-Walker AR')
%%
% Obtain the Yule-Walker AR Spectrum of the ECG signal with LBBB using pyulear.
[Pxx,F] = pyulear(ECGLBBB(:,1),order,segmentLength,Fs);
% Plot the Yule-Walker AR Spectrum of the signal
subplot(2,2,3)
plot(F,10*log10(Pxx))
ylabel('PSD (dB/Hz)');
xlabel('Frequency (Hz)');
legend('LBBB')
axis([0 L/2 -inf inf])
%title('Yule-Walker AR')

%%
% Obtain the Yule-Walker AR Spectrum of the ECG signal with PVC using pyulear.
[Pxx,F] = pyulear(ECGPVC(:,1),order,segmentLength,Fs);
% Plot the Yule-Walker AR Spectrum of the signal
subplot(2,2,4)
plot(F,10*log10(Pxx))
ylabel('PSD (dB/Hz)');
xlabel('Frequency (Hz)');
legend('PVC')
axis([0 L/2 -inf inf])
%title('Yule-Walker AR')
```

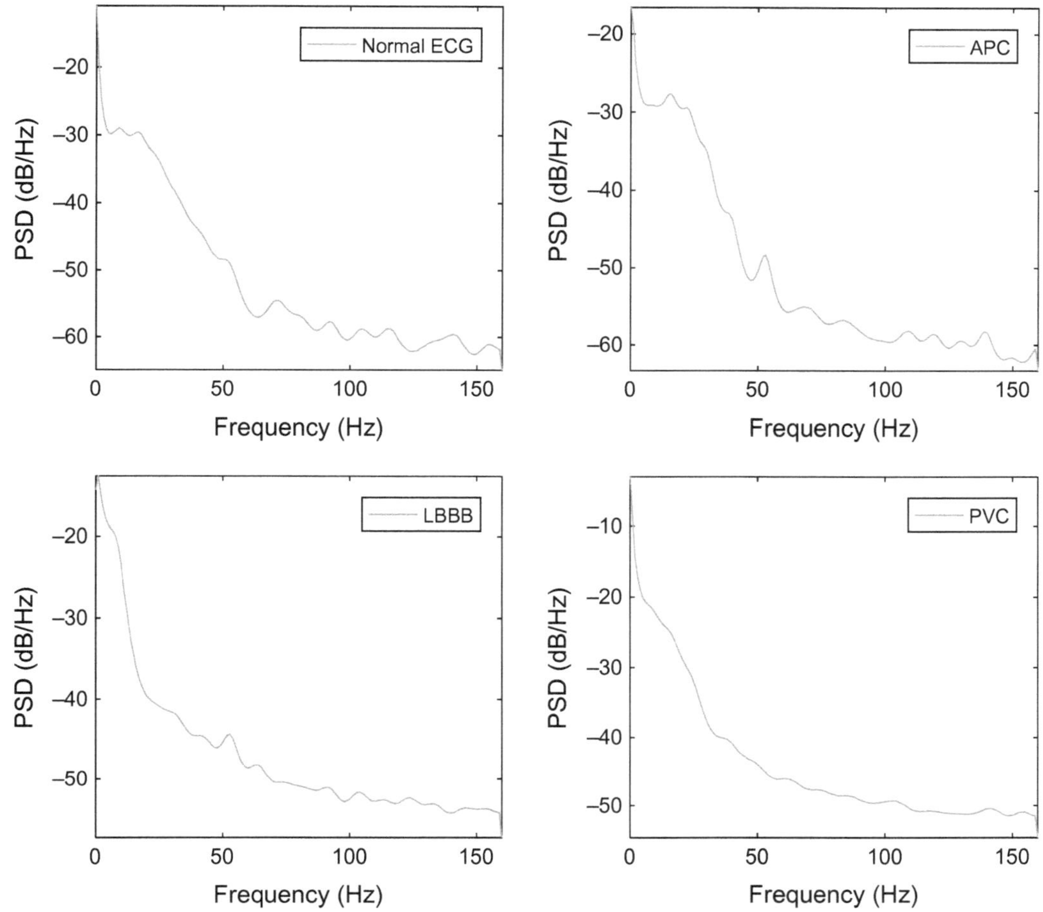

FIG. 3.18 Drawing of Yule-Walker AR spectrum of ECG signals.

EXAMPLE 3.19. The following MATLAB code was used to plot the Emotional EEG signals and their Yule-Walker AR spectrum. You can download data from the following website: http://bcmi.sjtu.edu.cn/~seed/index.html

```
%Ch3_Ex19_EEG_EMOTION_Yule.m
%The following MATLAB code is used to plot
%the EEG signals and their Yule-Walker AR spectrum.
clc
clear
%Load Sample EEG Data downloaded from the web site
%http://bcmi.sjtu.edu.cn/~seed/index.html
load EMOTIONSDAT.mat
%%
Fs = 200;            % Sampling frequency
T = 1/Fs;              % Sampling period
L = length(NEUTRAL(1,:));    % Length of signal
segmentLength = L;
order = 14;

%%
% Obtain the Yule-Walker AR Spectrum of the NEUTRAL EMOTIONAL EEG signal using pmusic.
[Pxx,F] = pyulear(NEUTRAL(1,:),order,segmentLength,Fs);
% Plot the Yule-Walker AR Spectrum of the signal
subplot(3,1,1);
plot(F,10*log10(Pxx))
```

```
ylabel('PSD (dB/Hz)');
xlabel('Frequency (Hz)');
legend('NEUTRAL')
title('Yule-Walker AR Spectrum of EEG Signals')
axis([0 100 -inf inf])

%%
% Obtain the Yule-Walker AR Spectrum of the NEGATIVE EMOTIONAL EEG signal using pmusic.
[Pxx,F] = pyulear(NEGATIVE(1,:),order,segmentLength,Fs);
% Plot the Yule-Walker AR Spectrum of the signal
subplot(3,1,2);
plot(F,10*log10(Pxx))
ylabel('PSD (dB/Hz)');
xlabel('Frequency (Hz)');
legend('NEGATIVE')
axis([0 100 -inf inf])

%%
% Obtain the Yule-Walker AR Spectrum of the POSITIVE EMOTIONAL EEG signal using pmusic.
[Pxx,F] = pyulear(POSITIVE(1,:),order,segmentLength,Fs);
% Plot the Yule-Walker AR Spectrum of the signal
subplot(3,1,3);
plot(F,10*log10(Pxx))
ylabel('PSD (dB/Hz)');
xlabel('Frequency (Hz)');
legend('POSITIVE')
axis([0 100 -inf inf])
```

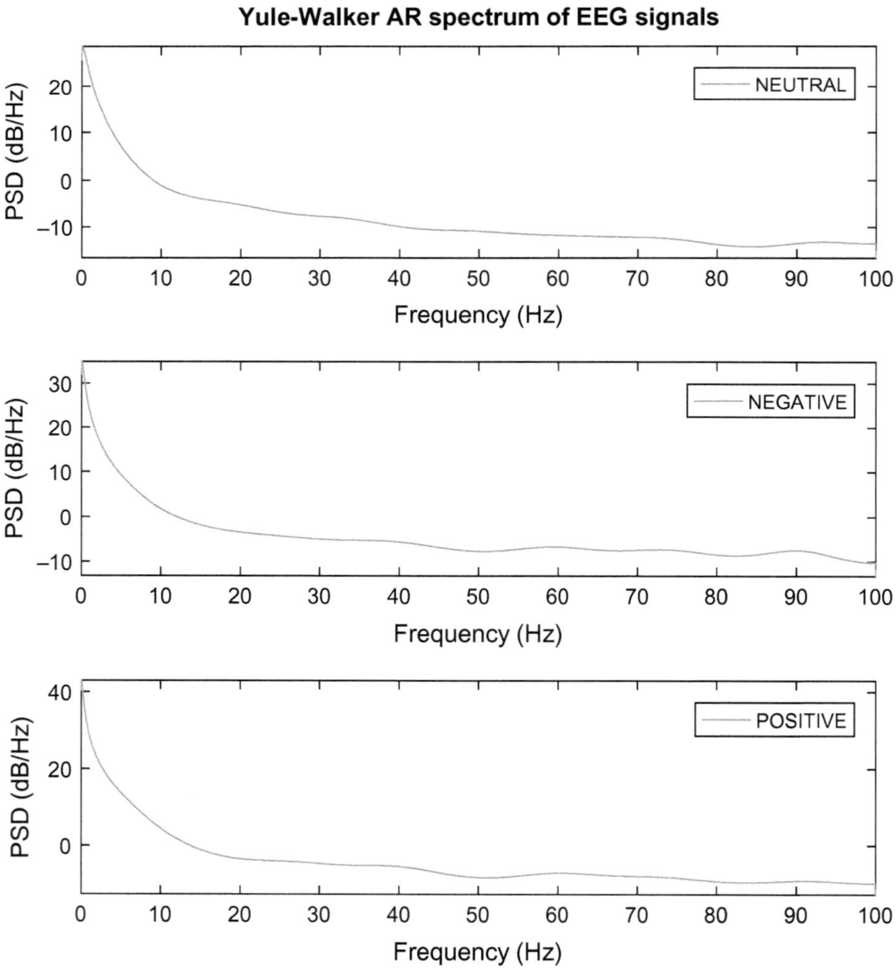

FIG. 3.19 Drawing of Yule-Walker AR spectrum of emotional EEG signals.

3.3.3 Covariance Method

For any complex time series, we can find a similar estimator by minimizing the estimate of the prediction error power,

$$\hat{\rho} = \frac{1}{N-p}\sum_{n=p}^{N-1}\left|x(n) + \sum_{k=1}^{p} a(k)x(n-k)\right|^2 \quad (3.33)$$

There is a difference between the covariance model and the autocorrelation model in the range of summation for the prediction error power estimate. In the covariance method, all the data points are calculated from the observed $\hat{\rho}$. It is not needed to make some part of the data equal to zero. The minimization of Eq. (3.33) can be affected by employing the complex gradient to achieve the AR parameter estimates as the solution of the equations,

$$\begin{bmatrix} c(1,0) \\ \vdots \\ c(p,0) \end{bmatrix} + \begin{bmatrix} c(1,1) & \cdots & c(1,p) \\ \vdots & \ddots & \vdots \\ c(p,1) & \cdots & c(p,p) \end{bmatrix} \begin{bmatrix} \hat{a}(1) \\ \vdots \\ \hat{a}(p) \end{bmatrix} = \begin{bmatrix} 0 \\ \vdots \\ 0 \end{bmatrix}$$

or

$$c_p + C_p\hat{a} = 0 \quad (3.34)$$

where $c(j,k) = \frac{1}{N-p}\sum_{n=p}^{N-1} x^*(n-j)x(n-k)$.

From Eq. (3.34), the AR parameter estimates are found as,

$$\hat{a} = -C_p^{-1}c_p \quad (3.35)$$

The white noise variance is estimated as

$$\hat{\sigma}^2 = \hat{\rho}_{\min} = c(0,0) + \sum_{k=1}^{p}\hat{a}(k)c(0,k) \quad (3.36)$$

PSD estimation can be found from the estimates of the AR parameters as (Kay & Marple, 1981; Proakis & Manolakis, 2007; Stoica & Moses, 1997; Subasi, 2007) (Fig. 3.20):

$$\hat{P}_{COV}(f) = \frac{\hat{\sigma}^2}{\left|1 + \sum_{k=1}^{p}\hat{a}(k)e^{-j2\pi fk}\right|^2} \quad (3.37)$$

EXAMPLE 3.20. The following MATLAB code was used to plot the EEG signals and their covariance spectrum. You can download data from the following website:
http://epileptologie-bonn.de/cms/front_content.php?idcat=193&lang=3&changelang=3

```
%Ch3_Ex21_EEG_Cov.m
%The following MATLAB code is used to plot
%the EEG signals and their Covariance spectrum.
clc
clear
%Load Sample EEG Data downloaded from the web site
%http://epileptologie-bonn.de/cms/front_content.php?idcat=193&lang=3&changelang=3
load AS_BONN_ALL_EEG_DATA_4096.mat
%%
Fs = 173.6;        % Sampling frequency
T = 1/Fs;          % Sampling period
L = length(Normal_Eyes_Open(:,1));    % Length of signal
segmentLength = L;
order = 14;
%%
% Obtain the Covariance Spectrum of the Normal EEG signal using pcov.
[Pxx,F] = pcov(Normal_Eyes_Open(:,1),order,segmentLength,Fs);
% Plot the Covariance Spectrum of the signal
subplot(3,1,1)
plot(F,10*log10(Pxx))
ylabel('PSD (dB/Hz)');
xlabel('Frequency (Hz)');
legend('Normal Eyes Open')
title('Covariance Spectrum of EEG Signals')

%%
% Obtain the Covariance Spectrum of the Interictal EEG signal using pcov.
```

```
[Pxx,F] = pcov(Interictal(:,1),order,segmentLength,Fs);
% Plot the Covariance Spectrum of the signal
subplot(3,1,2)
plot(F,10*log10(Pxx))
ylabel('PSD (dB/Hz)');
xlabel('Frequency (Hz)');
legend('Interictal')
%%
% Obtain the Covariance Spectrum of the Ictal EEG signal using pcov.
[Pxx,F] = pcov(Ictal(:,1),order,segmentLength,Fs);
% Plot the Covariance Spectrum of the signal
subplot(3,1,3)
plot(F,10*log10(Pxx))
ylabel('PSD (dB/Hz)');
xlabel('Frequency (Hz)');
legend('Ictal')
```

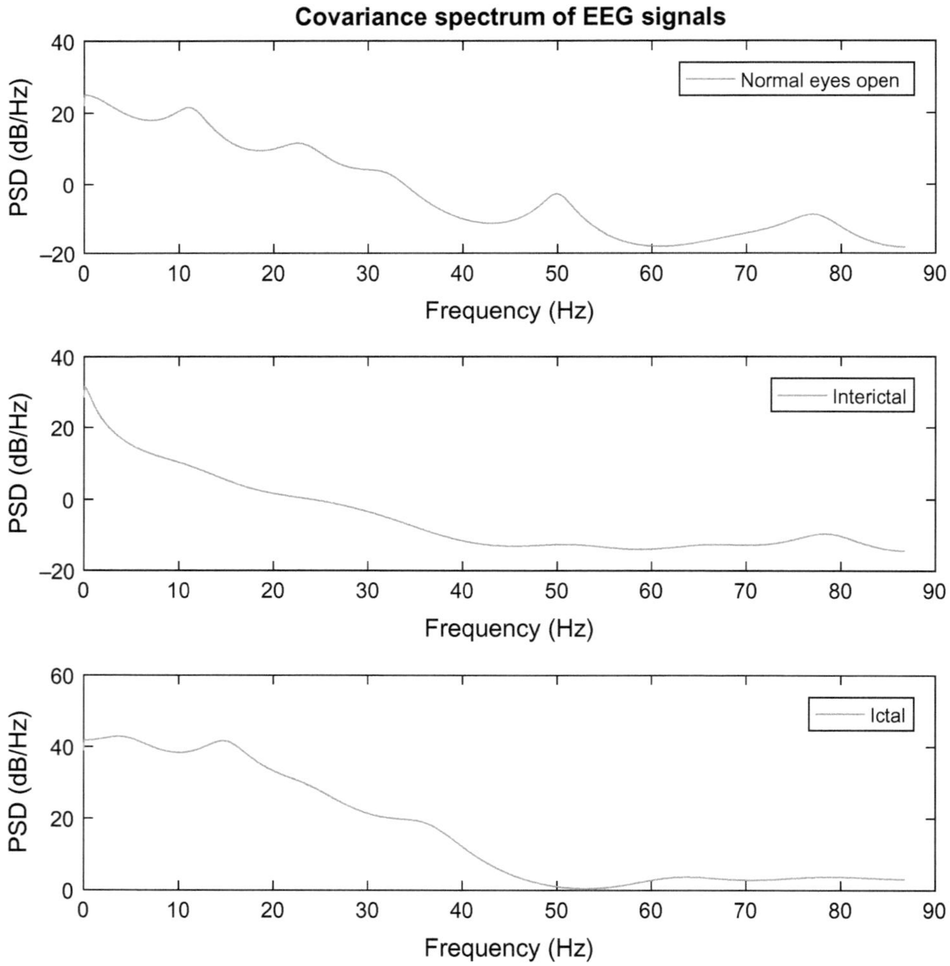

FIG. 3.20 Drawing of covariance spectrum of EEG signals.

EXAMPLE 3.21. The following MATLAB code was used to plot the EMG signals and their covariance spectrum. For this example, EMGLAB data was used. You can download the data from the following website: http://www.emglab.net/emglab/Signals/signals.php

```
%Ch3_Ex22_EMG_Cov.m
%The following MATLAB code is used to plot
%the EMG signals and their Covariance spectrum.
clc
clear
%Load Sample EMG Data downloaded from the web site
%http://www.emglab.net/emglab/Signals/signals.php
load EMGDAT.mat

%%
Fs = 10000;            % Sampling frequency
T = 1/Fs;              % Sampling period
L = length(CON(:,2));     % Length of signal
segmentLength = L;
noverlap = L/4;
order = 34;
%%
% Obtain the Covariance Spectrum of the Normal EMG signal using pcov.
[Pxx,F] = pcov(CON(:,2),order,segmentLength,Fs);
% Plot the Covariance Spectrum of the signal
subplot(3,1,1)
plot(F,10*log10(Pxx))
ylabel('PSD (dB/Hz)');
xlabel('Frequency (Hz)');
legend('Control')
title('Covariance Spectrum of EMG Signals')
%%
% Obtain the Covariance Spectrum of the ALS EMG signal using pcov.
[Pxx,F] = pcov(ALS(:,1),order,segmentLength,Fs);
% Plot the Covariance Spectrum of the signal
subplot(3,1,2)
plot(F,10*log10(Pxx))
ylabel('PSD (dB/Hz)');
xlabel('Frequency (Hz)');
legend('ALS')


%%
% Obtain the Covariance Spectrum of the Myopathic EMG signal using pcov.
[Pxx,F] = pcov(MYO(:,1),order,segmentLength,Fs);
% Plot the Covariance Spectrum of the signal
subplot(3,1,3)
plot(F,10*log10(Pxx))
ylabel('PSD (dB/Hz)');
xlabel('Frequency (Hz)');
legend('Myopathy')
```

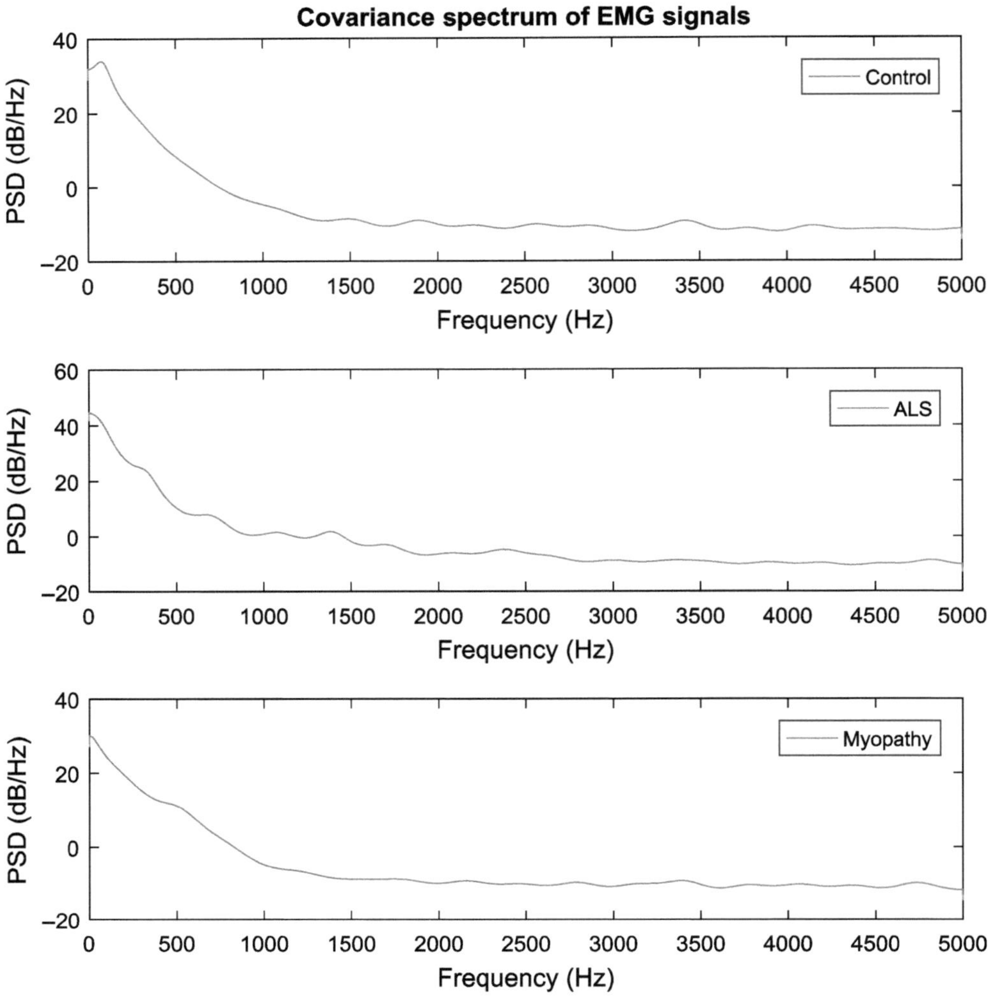

FIG. 3.21 Drawing of covariance spectrum of EMG signals.

EXAMPLE 3.22. The following MATLAB code was used to plot the surface EMG signals and their covariance spectrum. You can download data from the following website: https://archive.ics.uci.edu/ml/datasets/sEMG+for+Basic+Hand+movements#

```
%Ch3_Ex23_sEMG_Cov.m
%The following MATLAB code is used to plot
%the surface EMG signals and their Covariance spectrum.
clc
clear
%sEMG for Basic Hand movements Data Set
%Load Sample sEMG Data downloaded from the web site
%https://archive.ics.uci.edu/ml/datasets/sEMG+for+Basic+Hand+movements#
load sEMG_UCI_BHM.mat
%%
Fs = 500;              % Sampling frequency
T = 1/Fs;              % Sampling period
L = length(F1_spher_ch1(1,:));    % Length of signal
segmentLength = L;
noverlap = L/4;
order = 15;
%%
% Obtain the Covariance Spectrum of the Spherical sEMG signal using pcov.
[Pxx,F] = pcov(F1_spher_ch1(1,:),order,segmentLength,Fs);
```

```
% Plot the Covariance Spectrum of the signal
subplot(3,2,1);
plot(F,10*log10(Pxx))
ylabel('PSD (dB/Hz)');
xlabel('Frequency (Hz)');
title('Spherical')
axis ([0 250 -inf inf])

%%
% Obtain the Covariance Spectrum of the Tip sEMG signal using pcov.
[Pxx,F] = pcov(F1_tip_ch1(1,:),order,segmentLength,Fs);
% Plot the Covariance Spectrum of the signal
subplot(3,2,2);
plot(F,10*log10(Pxx))
ylabel('PSD (dB/Hz)');
xlabel('Frequency (Hz)');
title('Tip')
axis ([0 250 -inf inf])

%%
% Obtain the Covariance Spectrum of the Palmar sEMG signal using pcov.
[Pxx,F] = pcov(F1_palm_ch1(1,:),order,segmentLength,Fs);
% Plot the Covariance Spectrum of the signal
subplot(3,2,3);
plot(F,10*log10(Pxx))
ylabel('PSD (dB/Hz)');
xlabel('Frequency (Hz)');
title('Palmar')
axis ([0 250 -inf inf])

%%
% Obtain the Covariance Spectrum of the Lateral sEMG signal using pcov.
[Pxx,F] = pcov(F1_lat_ch1(1,:),order,segmentLength,Fs);
% Plot the Covariance Spectrum of the signal
subplot(3,2,4);
plot(F,10*log10(Pxx))
ylabel('PSD (dB/Hz)');
xlabel('Frequency (Hz)');
title('Lateral')
axis ([0 250 -inf inf])

%%
% Obtain the Covariance Spectrum of the Cylindrical sEMG signal using pcov.
[Pxx,F] = pcov(F1_cyl_ch1(1,:),order,segmentLength,Fs);
% Plot the Covariance Spectrum of the signal
subplot(3,2,5);
plot(F,10*log10(Pxx))
ylabel('PSD (dB/Hz)');
xlabel('Frequency (Hz)');
title('Cylindrical')
axis ([0 250 -inf inf])

%%
% Obtain the Covariance Spectrum of the Hook sEMG signal using pcov.
[Pxx,F] = pcov(F1_hook_ch1(1,:),order,segmentLength,Fs);
% Plot the Covariance Spectrum of the signal
subplot(3,2,6);
plot(F,10*log10(Pxx))
ylabel('PSD (dB/Hz)');
xlabel('Frequency (Hz)');
title('Hook')
axis ([0 250 -inf inf])
```

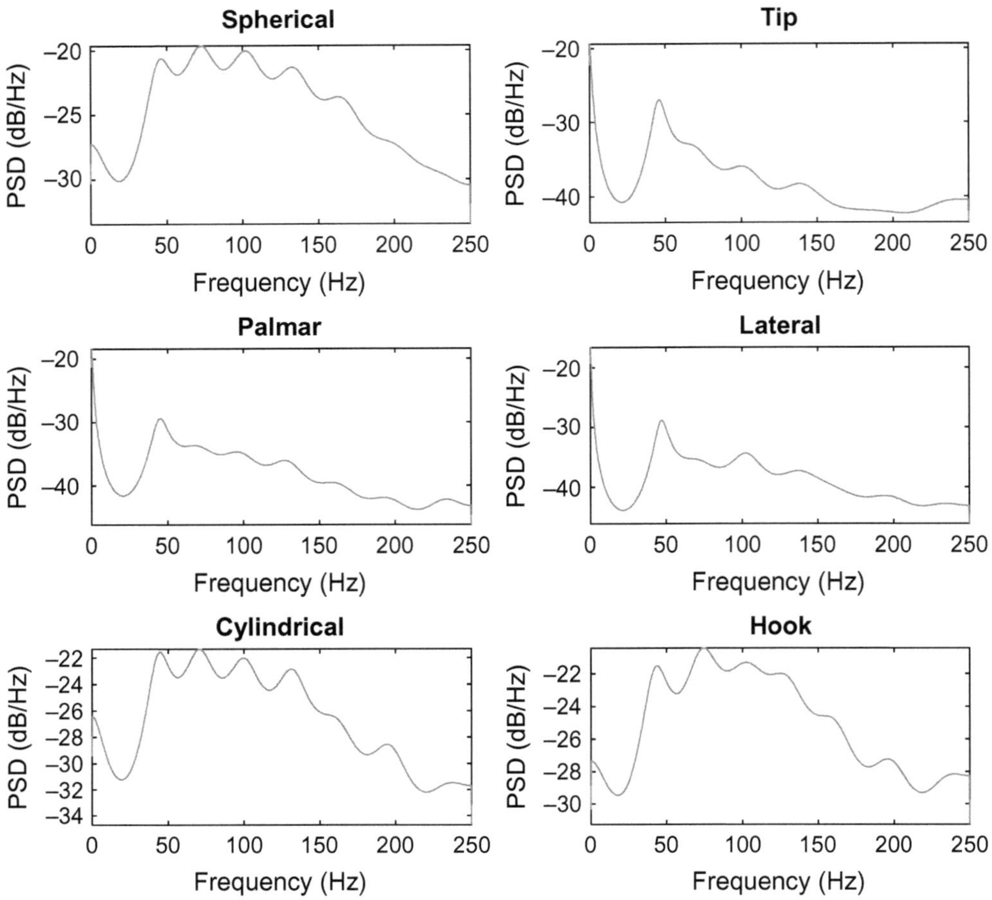

FIG. 3.22 Drawing of covariance spectrum of surface EMG signals.

EXAMPLE 3.23. The following MATLAB code was used to plot the ECG signals and their covariance spectrum. You can download data from the following website: https://www.physionet.org/physiobank/database/mitdb/

```
%Ch3_Ex24_ECG_Cov.m
%The following MATLAB code is used to plot
%the ECG signals and their Covariance spectrum.
clc
clear
%Load Sample ECG Data downloaded from the web site
%https://www.physionet.org/physiobank/database/mitdb/
load MITBIH_ECG.mat
%%
Fs = 320;              % Sampling frequency
T = 1/Fs;               % Sampling period
L = length(ECGN(:,1));    % Length of signal
t = (0:L-1)*T; % Time vector
segmentLength = L;
noverlap = L/4;
order=20;
%%
% Obtain the Covariance Spectrum of the Normal ECG signal using pcov.
[Pxx,F] = pcov(ECGN(:,1),order,segmentLength,Fs);
% Plot the Covariance Spectrum of the signal
subplot(2,2,1)
plot(F,10*log10(Pxx))
ylabel('PSD (dB/Hz)');
xlabel('Frequency (Hz)');
legend('Normal')
%title('Covariance Spectrum of ECG Signals')
axis([0 L/2 -inf inf])
```

```
%%
% Obtain the Covariance Spectrum of the APC ECG signal using pcov.
[Pxx,F] = pcov(ECGAPC(:,1),order,segmentLength,Fs);
% Plot the Covariance Spectrum of the signal
subplot(2,2,2)
plot(F,10*log10(Pxx))
ylabel('PSD (dB/Hz)');
xlabel('Frequency (Hz)');
legend('APC')
axis([0 L/2 -inf inf])

%%
% Obtain the Covariance Spectrum of the LBBB ECG signal using pcov.
[Pxx,F] = pcov(ECGLBBB(:,1),order,segmentLength,Fs);
% Plot the Covariance Spectrum of the signal
subplot(2,2,3)
plot(F,10*log10(Pxx))
ylabel('PSD (dB/Hz)');
xlabel('Frequency (Hz)');
legend('LBBB')
axis([0 L/2 -inf inf])
%%
% Obtain the Covariance Spectrum of the LBBB ECG signal using pcov.
[Pxx,F] = pcov(ECGRBBB(:,1),order,segmentLength,Fs);
% Plot the Covariance Spectrum of the signal
subplot(2,2,4)
plot(F,10*log10(Pxx))
ylabel('PSD (dB/Hz)');
xlabel('Frequency (Hz)');
legend('RBBB')
axis([0 L/2 -inf inf])
```

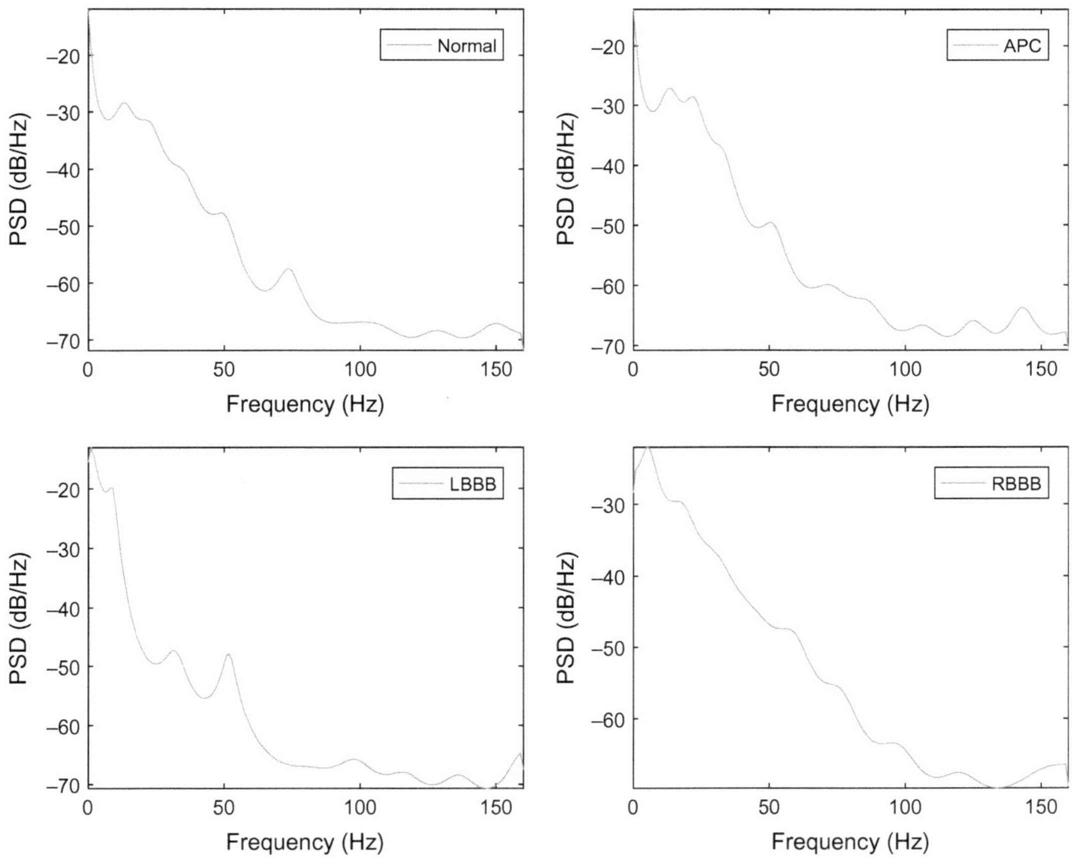

FIG. 3.23 Drawing of covariance spectrum of ECG signals.

3.3.4 Modified Covariance Method

For any AR(p) process, the optimal forward predictor is

$$\hat{x}(n) = -\sum_{k=1}^{p} a(k)x(n-k) \tag{3.38}$$

whereas the optimal backward predictor is

$$\hat{x}(n) = -\sum_{k=1}^{p} a^*(k)x(n+k) \tag{3.39}$$

where the $a(k)$s are the AR parameters. In both cases, the minimum prediction error power is taken as the white noise variance σ^2. The modified covariance method estimates the AR parameters by minimizing the average of the estimated forward and backward prediction error powers,

$$\hat{\rho} = \frac{1}{2}(\hat{\rho}^f + \hat{\rho}^b) \tag{3.40}$$

where $\hat{\rho}^f = \frac{1}{N-p}\sum_{n=p}^{N-1}\left|x(n) + \sum_{k=1}^{p} a(k)x(n-k)\right|^2$, $\hat{\rho}^b = \frac{1}{N-p}\sum_{n=0}^{N-1-p}\left|x(n) + \sum_{k=1}^{p} a^*(k)x(n+k)\right|^2$.

Similar to the covariance method, the summations are more than the prediction errors, which include observed time series samples. Eq. (3.40) can be minimized by employing the complex gradient to achieve

$$\frac{\partial\hat{\rho}}{\partial a(l)} = \frac{1}{N-p}\left[\sum_{n=p}^{N-1}\left(x(n) + \sum_{k=1}^{p} a(k)x(n-k)\right)x^*(n-l) + \sum_{n=0}^{N-1-p}\left(x^*(n) + \sum_{k=1}^{p} a(k)x^*(n+k)\right)x(n+l)\right] = 0$$

$$l = 1, 2, \ldots, p \tag{3.41}$$

After some simplification, the equation becomes

$$\sum_{k=1}^{p}\hat{a}(k)\left(\sum_{n=p}^{N-1}x(n-k)x^*(n-l) + \sum_{n=0}^{N-1-p}x^*(n+k)x(n+l)\right) = -\left(\sum_{n=p}^{N-1}x(n)x^*(n-l) + \sum_{n=0}^{N-1-p}x^*(n)x(n+l)\right) \quad l = 1, 2, \ldots, p \tag{3.42}$$

or in matrix form,

$$\begin{bmatrix} c(1,0) \\ \vdots \\ c(p,0) \end{bmatrix} + \begin{bmatrix} c(1,1) & \cdots & c(1,p) \\ \vdots & \ddots & \vdots \\ c(p,1) & \cdots & c(p,p) \end{bmatrix}\begin{bmatrix} \hat{a}(1) \\ \vdots \\ \hat{a}(p) \end{bmatrix} = \begin{bmatrix} 0 \\ \vdots \\ 0 \end{bmatrix}$$

$$c_p + C_p\hat{a} = 0 \tag{3.43}$$

where $c(j,k) = \frac{1}{2(N-p)}\left(\sum_{n=p}^{N-1}x^*(n-j)x(n-k) + \sum_{n=0}^{N-1-p}x(n+j)x^*(n+k)\right)$

From Eq. (3.43), the AR parameter estimates are found as,

$$\hat{a} = -C_p^{-1}c_p \tag{3.44}$$

The estimate of the white noise variance is

$$\hat{\sigma}^2 = \hat{\rho}_{\min} = \frac{1}{2(N-p)}\left[\sum_{n=p}^{N-1}\left(x(n) + \sum_{k=1}^{p}\hat{a}(k)x(n-k)\right)x^*(n) + \sum_{n=0}^{N-1-p}\left(x^*(n) + \sum_{k=1}^{p}\hat{a}(k)x^*(n+k)\right)x(n)\right]$$

where Eq. (3.41) has been employed, and finally,

$$\hat{\sigma}^2 = c(0,0) + \sum_{k=1}^{p}\hat{a}(k)c(0,k) \tag{3.45}$$

It is seen that the modified covariance method is similar to the covariance except for the definition of $c(j,k)$, the autocorrelation estimator. From the estimates of the AR parameters, PSD estimation is formed as (Kay & Marple, 1981; Proakis & Manolakis, 2007; Stoica & Moses, 1997; Subasi, 2007):

$$\hat{P}_{MCOV}(f) = \frac{\hat{\sigma}^2}{\left|1 + \sum_{k=1}^{p}\hat{a}(k)e^{-j2\pi fk}\right|^2} \tag{3.46}$$

EXAMPLE 3.24. The following MATLAB code was used to plot the EEG signals and their modified covariance spectrum. You can download data from the following website: https://www.physionet.org/pn6/chbmit/

```
%Ch3_Ex32_EEG_MCov.m
%The following MATLAB code is used to plot
%the EEG signals and their Modified Covariance spectrum.
clc
clear
%Load Sample EEG Data downloaded from the web site
```

```
%https://www.physionet.org/pn6/chbmit/
load AS_PHYSIONET_EEG_DATA.mat
%%
Fs = 256;           % Sampling frequency
T = 1/Fs;            % Sampling period
L = length(EEGICTAL(:,1));    % Length of signal
segmentLength = L;
order = 14;
%%
% Obtain the Modified Covariance Spectrum of the INTERICTAL EEG signal using pmcov.
[Pxx,F] = pmcov(EEGINTERICTAL(:,1),order,segmentLength,Fs);
% Plot the Modified Covariance Spectrum of the signal
subplot(3,1,1)
plot(F,10*log10(Pxx))
ylabel('PSD (dB/Hz)');
xlabel('Frequency (Hz)');
legend('INTERICTAL')
title('Modified Covariance Spectrum of EEG Signals')
axis([0 128 -inf inf])
%%
% Obtain the Modified Covariance Spectrum of the Preictal EEG signal using pmcov.
[Pxx,F] = pmcov(EEGPREICTAL(:,1),order,segmentLength,Fs);
% Plot the Modified Covariance Spectrum of the signal
subplot(3,1,2)
plot(F,10*log10(Pxx))
ylabel('PSD (dB/Hz)');
xlabel('Frequency (Hz)');
legend('PREICTAL')
%title('Modified Covariance')
axis([0 128 -inf inf])
%%
% Obtain the Modified Covariance Spectrum of the Ictal EEG signal using pmcov.
[Pxx,F] = pmcov(EEGICTAL(:,1),order,segmentLength,Fs);
% Plot the Covariance Spectrum of the signal
subplot(3,1,3)
plot(F,10*log10(Pxx))
ylabel('PSD (dB/Hz)');
xlabel('Frequency (Hz)');
legend('ICTAL')
%title('Modified Covariance')
axis([0 128 -inf inf])
```

3.3.5 Burg Method

The Burg AR model utilizes the minimization of the forward and backward prediction errors and estimation of the reflection coefficient. The forward and backward prediction errors for a pth-order model are defined as:

$$\hat{e}_{f,p}(n) = x(n) + \sum_{i=1}^{p} \hat{a}_{p,i} x(n-i) \quad n = p+1,\ldots,N \tag{3.47}$$

$$\hat{e}_{b,p}(n) = x(n-p) + \sum_{i=1}^{p} \hat{a}^{*}_{p,i} x(n-p+i) \quad n = p+1,\ldots,N \tag{3.48}$$

The AR parameters related to the reflection coefficient $\hat{k}_p$ can be denoted as

$$\hat{a}_{p,i} = \begin{cases} \hat{a}_{p-1,i} + \hat{k}_p \hat{a}^{*}_{p-1,p-i} & i = 1,\ldots,p-1 \\ \hat{k}_p & i = p \end{cases} \tag{3.49}$$

The Burg method reflects the recursive-in-order estimation of $\hat{k}_p$ given that the AR coefficients for order $p-1$ have been computed. The reflection coefficient estimate is given by

$$\hat{k}_p = \frac{-2 \sum_{n=p+1}^{N} \hat{e}_{f,p-1}(n) \hat{e}^{*}_{b,p-1}(n-1)}{\sum_{n=p+1}^{N} \left[\left| \hat{e}_{f,p-1}(n) \right|^2 + \left| \hat{e}_{b,p-1}(n-1) \right|^2 \right]} \tag{3.50}$$

The prediction errors satisfy the following recursive-in-order expressions,

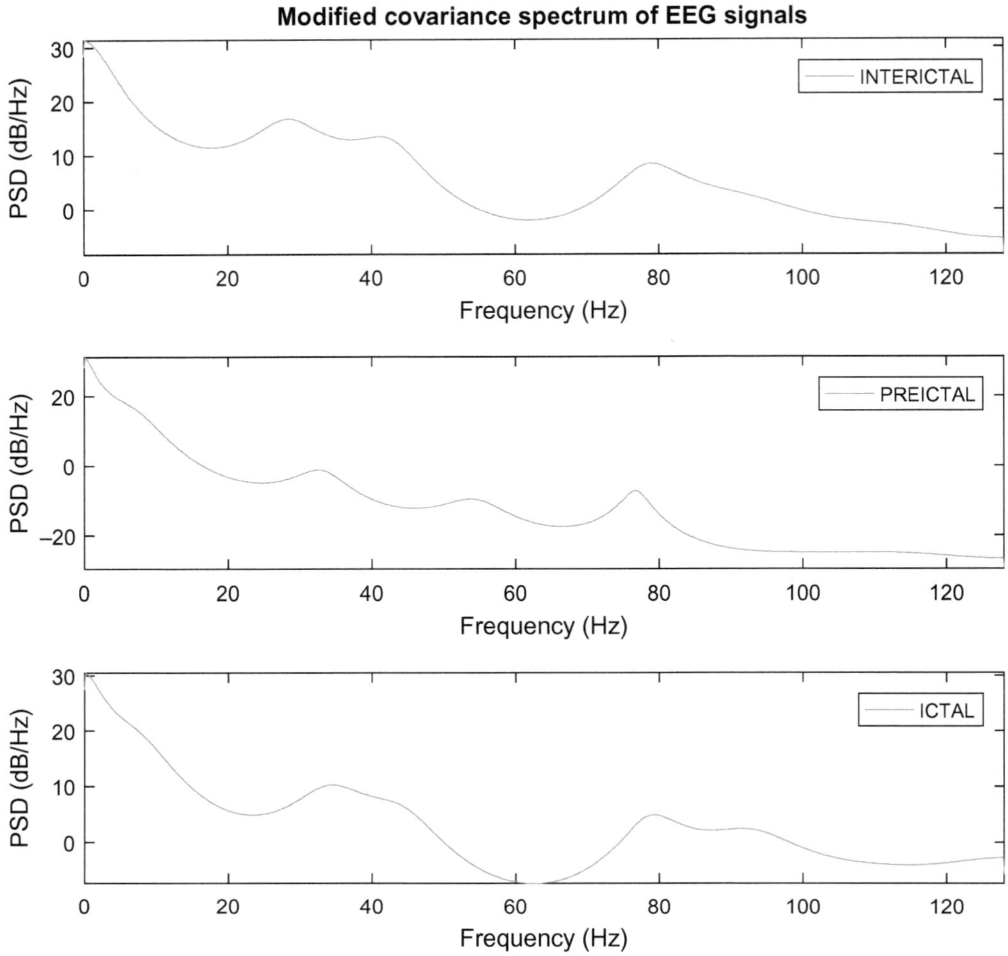

FIG. 3.24 Drawing of modified covariance spectrum of EEG signals.

$$\hat{e}_{f,p}(n) = \hat{e}_{f,p-1}(n) + \hat{k}_p \hat{e}_{b,p-1}(n-1) \quad (3.51)$$

$$\hat{e}_{b,p}(n) = \hat{e}_{b,p-1}(n-1) + \hat{k}_p^* \hat{e}_{f,p-1}(n) \quad (3.52)$$

and these expressions are employed to produce a recursive-in-order algorithm for estimating the AR coefficients. From the estimates of the AR parameters, PSD estimation is formed as (Kay & Marple, 1981; Proakis & Manolakis, 2007; Stoica & Moses, 1997):

$$\hat{P}_{BURG}(f) = \frac{\hat{e}_p}{\left|1 + \sum_{k=1}^{p} \hat{a}_p(k) e^{-j2\pi fk}\right|^2} \quad (3.53)$$

where $\hat{e}_p = \hat{e}_{f,p} + \hat{e}_{b,p}$ is the total least squares error (Subasi, 2007).

EXAMPLE 3.25. The following MATLAB code was used to plot the focal and nonfocal EEG signals and their AR Burg spectrum. You can download data from the following website: http://ntsa.upf.edu/downloads/andrzejak-rg-schindler-k-rummel-c-2012-nonrandomness-nonlinear-dependence-and

```
%Ch3_Ex36_EEG_Focal_Burg.m
%The following MATLAB code is used to plot
%the Focal and Non-Focal EEG signals and their AR Burg spectrum.
clc
clear
%Load Sample EEG Data downloaded from the web site
%http://ntsa.upf.edu/downloads/andrzejak-rg-schindler-k-rummel-c-2012-nonrandomness-nonlinear-
dependence-and
load FOCAL_NFOCAL.mat
```

```
%%
Fs = 512;           % Sampling frequency
T = 1/Fs;             % Sampling period
L = length(focal(:,1));    % Length of signal
segmentLength = L;
order = 14;
%%
% Obtain the Burg AR Spectrum of the FOCAL EEG signal using pburg.
[Pxx,F] = pburg(focal(:,1),order,segmentLength,Fs);
% Plot the Burg AR Spectrum of the signal
subplot(2,1,1);
plot(F,10*log10(Pxx))
ylabel('PSD (dB/Hz)');
xlabel('Frequency (Hz)');
legend('FOCAL')
title('Burg AR Spectrum of EEG Signals')
axis([0 256 -inf inf])
%%
% Obtain the Burg AR Spectrum of the NON-FOCAL EEG signal using pburg.
[Pxx,F] = pburg(nfocal(:,1),order,segmentLength,Fs);
% Plot the Burg AR Spectrum of the signal
subplot(2,1,2);
plot(F,10*log10(Pxx))
ylabel('PSD (dB/Hz)');
xlabel('Frequency (Hz)');
legend('NON-FOCAL')
axis([0 256 -inf inf])
```

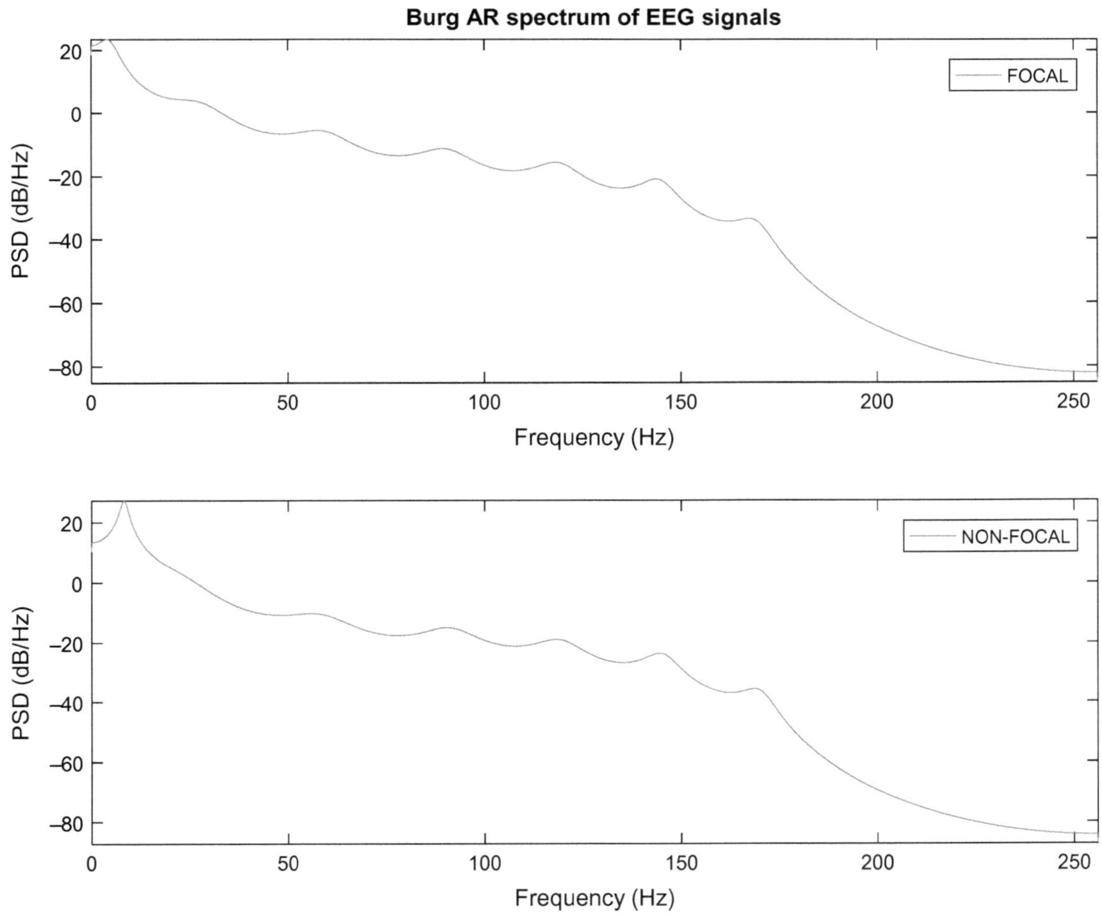

FIG. 3.25 Drawing of Burg AR spectrum of focal and nonfocal EEG signals.

Selecting a sampling rate that is fairly close to the Nyquist rate is usually recommended. This will ensure that the poles and zeroes are used on modeling signal components of interest instead of modeling high-frequency noise. Otherwise, the unchanged number of poles and zeroes must be considered and represent information in a considerably wider spectral interval, causing a certain loss of the spectral details (Sörnmo & Laguna, 2005). In conclusion, the AR model-based techniques can be suitable spectral estimators if we know the nature of the signal. However, we must pay attention to the selection of model type, model order, and sampling rate.

3.4 SUBSPACE-BASED METHODS FOR SPECTRAL ANALYSIS

Subspace-based methods are employed to estimate the powers and frequencies of signals from noise-corrupted measurements. These methods utilize the eigen-decomposition of the correlation matrix of the noise-corrupted signal, and they are best fitted to signals composed of several specific sinusoids buried in noise (Friedlander, 1990; Friedlander & Weiss, 1994; Swindlehurst & Kailath, 1992; Swindlehurst & Kailath, 1993; Viberg, 1995). The polynomial A(f) that includes zeroes on the unit circle are utilized to estimate the PSD:

$$A(f)=\sum_{k=0}^{m} a_k e^{-j2\pi fk} \tag{3.54}$$

where $A(f)$ represents the desired polynomial, a_k characterizes the desired polynomial coefficients, and m characterizes the order of the eigen-filter, $A(f)$. Moreover, the polynomial is expressed in terms of the autocorrelation matrix R of the input signal by assuming the noise is white:

$$R=E\left\{x(n)^{*}\cdot x(n)^{T}\right\}=SPS^{\#}+\sigma v^2 I \tag{3.55}$$

where $x(n)$ is observed signal, S characterizes the signal direction matrix of dimension $(m + 1) \times L$, L is the dimension of the signal subspace, R is the autocorrelation matrix of dimension $(m+1) \times (m+1)$, P is the signal power matrix of dimension $(L) \times (L)$, σv^2 characterizes the noise power, * represents the complex conjugate, I is the identity matrix, # represents the complex conjugate transposed, and T shows the transposed matrix. S is the signal direction matrix and denoted as

$$S=[Sw_1 Sw_2 \ldots Sw_L] \tag{3.56}$$

where $w_1, w_2, \ldots, w_L$ represent the signal frequencies:

$$Sw_i=\left[1\, e^{jwi} e^{j2wi} \ldots e^{jmwi}\right]^T \quad i=1,2,\ldots,L \tag{3.57}$$

In real-world applications, it is common to build the estimated autocorrelation matrix $\hat{R}$ from the autocorrelation lags:

$$\hat{R}(k)=\frac{1}{N}\sum_{n=0}^{N-1-k} x(n+k).x(n) \quad k=0,2,\ldots,m \tag{3.58}$$

where k is the autocorrelation lag index and N is the number of the signal samples. Then, the estimated autocorrelation matrix becomes

$$\hat{R}(k)=\begin{bmatrix} \hat{R}(0) & \hat{R}(1) & \hat{R}(2) & \ldots & .. & \hat{R}(m) \\ \hat{R}(1) & & & \ldots & .. & \hat{R}(m-1) \\ \hat{R}(2) & & . & & & \hat{R}(m-2) \\ & & . & . & & . \\ & & . & & . & . \\ \hat{R}(m) & \hat{R}(m-1) & \ldots & \ldots & \ldots & \hat{R}(0) \end{bmatrix} \tag{3.59}$$

Multiplying by the eigenvector of the autocorrelation matrix a, Eq. (3.55) can be denoted as

$$\hat{R}a=SPSa+\sigma v^2 a \tag{3.60}$$

where a characterizes the eigenvector of the estimated autocorrelation matrix $\hat{R}$ and a is represented as

$$[a_0, a_1, \ldots, a_m]^T$$

The subspace-based methods employ only the eigenvector corresponding to the minimum eigenvalue to construct the desired polynomial (Eq. 3.54) and to calculate the spectrum (Porat & Friedlander, 1988). Hence, it can be determined in a way that $S^{\#}a=0$. Then, the eigenvector $\boldsymbol{a}$ can be assumed in the noise subspace, and Eq. (3.60) becomes

$$\hat{R}a=\sigma v^2 a \tag{3.61}$$

under the constraint $a^{\#}a = 1$ where σv^2 is the noise power, which is the same as the minimum eigenvalue related to the eigenvector $\boldsymbol{a}$ in the subspace-based modeling. In principle, under the assumption of white noise, all noise subspace eigenvalues should be equal

$$\lambda_1=\lambda_2=\cdots=\lambda_k=\sigma v^2 \tag{3.62}$$

where λ_i represents the noise subspace eigenvalues, $i = 1, 2, \ldots, K$, and K represents the dimension of the noise subspace (Subasi, Erçelebi, Alkan, & Koklukaya, 2006).

3.4.1 MUSIC Modeling

The multiple signal classification (MUSIC) modeling is a noise subspace frequency estimator. The MUSIC modeling was proposed by Schmidt (1986) to remove the effects of spurious zeroes by employing the averaged spectra of all eigenvectors related to the noise subspace. The resulting PSD is calculated from

$$P_{\text{MUSIC}}(f) = \frac{1}{1/K \sum_{i=0}^{K-1} |A_i(f)|^2} \tag{3.63}$$

where K characterizes the dimension of noise subspace, and $A_i(f)$ characterizes the anticipated polynomial related to the eigenvectors of the noise subspace (Friedlander, 1990; Friedlander & Weiss, 1994; Schmidt, 1986; Subasi et al., 2006; Swindlehurst & Kailath, 1992; Swindlehurst & Kailath, 1993).

EXAMPLE 3.26. The following MATLAB code was used to plot the SEED EMOTION EEG signals and their MUSIC spectrum. You can download data from the following website: http://bcmi.sjtu.edu.cn/~seed/index.html

```
%Ch3_Ex41_EEG_EMOTION_MUSIC.m
%The following MATLAB code is used to plot
%the SEED EMOTION EEG signals and their MUSIC spectrum.
clc
clear
%Load Sample EEG Data downloaded from the web site
%http://bcmi.sjtu.edu.cn/~seed/index.html
load EMOTIONSDAT.mat
%%
Fs = 200;            % Sampling frequency
T = 1/Fs;              % Sampling period
L = length(NEUTRAL(1,:));    % Length of signal
NFFT = L;
order = 14;
%%
% Obtain the MUSIC Spectrum of the NEUTRAL EMOTIONAL EEG signal using pmusic.
[Pxx,F] = pmusic(NEUTRAL(1,:),order,NFFT,Fs);
% Plot the MUSIC Spectrum of the signal
subplot(3,1,1);
plot(F,10*log10(Pxx))
ylabel('PSD (dB/Hz)');
xlabel('Frequency (Hz)');
legend('NEUTRAL')
title('MUSIC Spectrum of Emotional EEG Signals')
axis([0 100 -inf inf])
%%
% Obtain the MUSIC Spectrum of the NEGATIVE EMOTIONAL EEG signal using pmusic.
[Pxx,F] = pmusic(NEGATIVE(1,:),order,NFFT,Fs);
% Plot the MUSIC Spectrum of the signal
subplot(3,1,2);
plot(F,10*log10(Pxx))
ylabel('PSD (dB/Hz)');
xlabel('Frequency (Hz)');
legend('NEGATIVE')
axis([0 100 -inf inf])

%%
% Obtain the MUSIC Spectrum of the POSITIVE EMOTIONAL EEG signal using pmusic.
[Pxx,F]=pmusic(POSITIVE(1,:),order,NFFT,Fs);
% Plot the MUSIC Spectrum of the signal
subplot(3,1,3);
plot(F,10*log10(Pxx))
ylabel('PSD (dB/Hz)');
xlabel('Frequency (Hz)');
legend('POSITIVE')
axis([0 100 -inf inf])
```

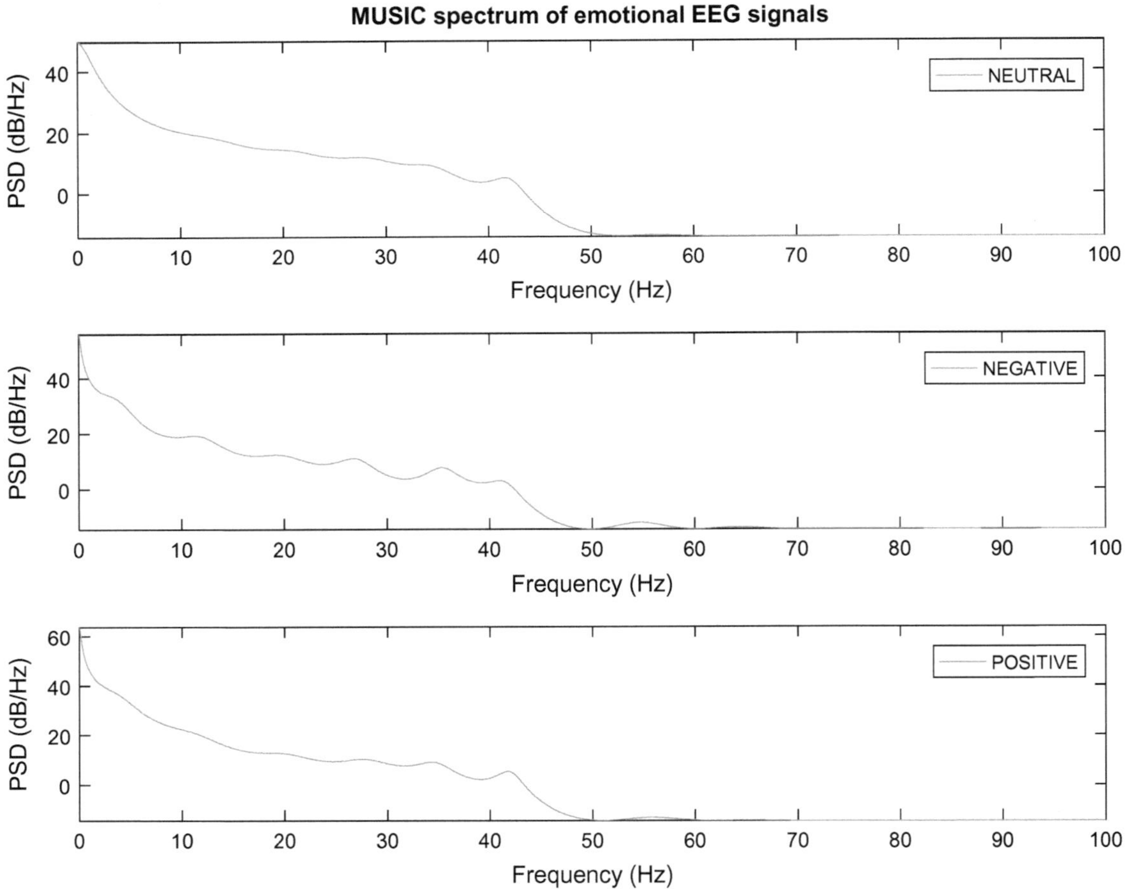

FIG. 3.26 Drawing of MUSIC spectrum of emotional EEG signals.

3.4.2 Eigenvector Modeling

Besides MUSIC modeling, the eigenvector modeling was examined (Proakis & Manolakis, 2007; Stoica & Moses, 1997; Viberg, 1995) to separate spurious zeroes from real zeroes. The eigenvector modeling forces spurious zeroes inside the unit circle and estimates the anticipated noise subspace vector $\boldsymbol{a}$ from either the signal subspace or noise eigenvectors. The PSD can be written by utilizing the eigenvector method as:

$$P_{ev}(f) = \frac{1}{\left(\sum_{i=0}^{K-1} |A_i(f)|^2/\lambda_i\right)} \tag{3.64}$$

To estimate the MUSIC and eigenvector PSD, the noise subspace dimension K should be determined by a technique such as the minimum description length (MDL) or AKAIKE (AIC) criteria (Proakis & Manolakis, 2007; Subasi et al., 2006).

EXAMPLE 3.27. The following MATLAB code was used to plot the sleep EEG signals and their eigen spectrum. You can download data from the following website: https://physionet.org/pn4/sleep-edfx/

```
%Ch3_Ex45_SLEEP_EEG_EIG.m
% The following MATLAB code is used to plot
%the SLEEP EEG signals and their EIGEN spectrum. clc
clear
%Load Sample EEG Data downloaded from the web site
%https://physionet.org/pn4/sleep-edfx/
load SC4001E0-PSG.mat
%%
Fs = 100;                 % Sampling frequency
T = 1/Fs;                 % Sampling period
L = length(sleepW(:)); % Length of signal
NFFT = L;
order = 35;
```

```
%%
% Obtain the EIGEN Spectrum of the AWAKE EEG signal using peig.
[Pxx,F] = peig(sleepW(:),order,NFFT,Fs);
% Plot the EIGEN Spectrum of the signal
subplot(3,2,1);
plot(F,10*log10(Pxx))
ylabel('PSD (dB/Hz)');
xlabel('Frequency (Hz)');
title('AWAKE')
axis([0 50 -inf inf])

%%
% Obtain the EIGEN Spectrum of the SLEEP STAGE I EEG signal using peig.
[Pxx,F] = peig(sleepI(:),order,NFFT,Fs);
% Plot the EIGEN Spectrum of the signal
subplot(3,2,2);
plot(F,10*log10(Pxx))
ylabel('PSD (dB/Hz)');
xlabel('Frequency (Hz)');
title('SLEEP STAGE I')
axis([0 50 -inf inf])

%%
% Obtain the EIGEN Spectrum of the SLEEP STAGE II EEG signal using peig.
[Pxx,F] = peig(sleepII(:),order,NFFT,Fs);
% Plot the EIGEN Spectrum of the signal
subplot(3,2,3);
plot(F,10*log10(Pxx))
ylabel('PSD (dB/Hz)');
xlabel('Frequency (Hz)');
title('SLEEP STAGE II')
axis([0 50 -inf inf])

%%
% Obtain the EIGEN Spectrum of the SLEEP STAGE III EEG signal using peig.
[Pxx,F] = peig(sleepIII(:),order,NFFT,Fs);
% Plot the EIGEN Spectrum of the signal
subplot(3,2,4);
plot(F,10*log10(Pxx))
ylabel('PSD (dB/Hz)');
xlabel('Frequency (Hz)');
title('SLEEP STAGE III')
axis([0 50 -inf inf])

%%
% Obtain the EIGEN Spectrum of the SLEEP STAGE IV EEG signal using peig.
[Pxx,F] = peig(sleepIV(:),order,NFFT,Fs);
% Plot the EIGEN Spectrum of the signal
subplot(3,2,5);
plot(F,10*log10(Pxx))
ylabel('PSD (dB/Hz)');
xlabel('Frequency (Hz)');
title('SLEEP STAGE IV')
axis([0 50 -inf inf])

%%
% Obtain the EIGEN Spectrum of the REM SLEEP EEG signal using peig.
[Pxx,F] = peig(sleepR(:),order,NFFT,Fs);
% Plot the EIGEN Spectrum of the signal
subplot(3,2,6);
plot(F,10*log10(Pxx))
ylabel('PSD (dB/Hz)');
xlabel('Frequency (Hz)');
title('REM SLEEP')
axis([0 50 -inf inf])
```

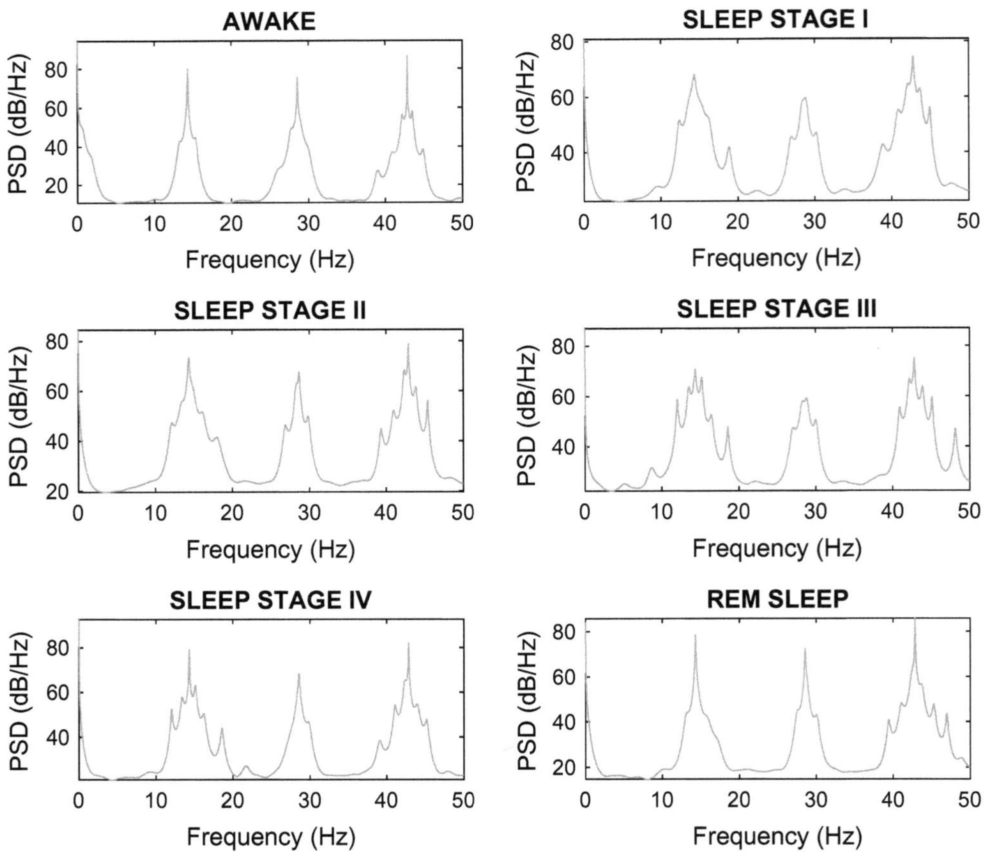

FIG. 3.27 Drawing of eigen spectrum of sleep EEG signals.

3.5 TIME-FREQUENCY ANALYSIS

Classical or modern spectral methods simply reveal which frequency components occur over the entire observation interval. If we do not require information at what time different signal frequencies exist, the spectral analysis represents a powerful signal-processing technique (Sörnmo & Laguna, 2005). The spectral analysis offers a complete and applicable solution for stationary waveforms. The statistical properties of these waveforms usually stay unaffected during the analysis, yet their frequencies are equally spread in time. However, many biomedical waveforms are not stationary, and their statistical properties vary significantly over time. The noticeable signal variations in EEG come as a result of different internal states of the subject such as sleep, eye movement, or meditation. Actually, these variations with time often draw the most attention and represent the main concern (Semmlow, 2004).

The FT "of a musical passage tells us what notes are played, but it is extremely different to figure out when they are played" (Hubbard, 1998). However, the frequency spectrum needs to hold the frequency timing information because we can uniquely recreate the waveform using the inverse FT. Actually, this information is embedded in the phase portion of the transform, which is hard to interpret and recover because particular time events are spread across all phase components, converting a local feature in time into a global feature in phase. The FT is thus not suitable for application to nonstationary signals upon which spectral components are time-dependent (Semmlow, 2004).

As a result, techniques to extract both time and frequency information from a waveform were developed, with the purpose of showing the frequencies at every time moment. Joint time-frequency information tremendously helped in the understanding of biomedical signals demonstrating nonstationary properties (Sörnmo & Laguna, 2005).

A number of methods developed for signal analysis with respect to both time and frequency can basically be separated into two categories: time-frequency methods and time-scale methods. The latter are sometimes also termed as wavelet analyzes (Semmlow, 2004). But we can also separate techniques that produce time-frequency representations into three main groups (Sörnmo & Laguna, 2005),

- *Linear, nonparametric methods'* time-frequency representations are acquired as a result of linear filtering. The short-time Fourier transform (STFT) is the typical representative, whereas the wavelet transform is a popular technique for describing transient signals.
- *Quadratic, nonparametric methods* have better time-frequency resolution. The Wigner-Ville distribution (WVD) and its variations are the most famous representatives. The first two groups make no specific assumptions about the signal, hence the name nonparametric.

- *Parametric methods* usually start with the AR model and assume that a statistical model with time-dependent parameters generates the signal of interest. Successive power spectra are then obtained from the resultant parameter estimates (Semmlow, 2004).

3.5.1 Short-Time Fourier Transform: The Spectrogram

The STFT is a commonly employed method for analyzing nonstationary signals as it is simple, computationally efficient, and achieves a consistent time-frequency spectrum. The main disadvantage is needing to making a compromise between time and frequency resolution of the decomposition. For transient signals, it might be impossible to competently handle this compromise (Marchant, 2003; Subasi & Kiymik, 2010).

The initial time-frequency approaches were based on cutting the waveform into several short, successive, possibly overlapping sections and applying the FT to all of them. The time-dependent nature of the signal was then represented by the resultant sequence of spectra. The process of cutting the waveform is performed using a window function, successfully isolating data sections from the waveform of interest. The FT is performed on data sections that are much shorter than the overall waveform; this technique is called STFT and has been effectively used in several biomedical applications. To define the equation for the STFT, we alter the equation for the FT by incorporating a sliding window function, $w(t)$, and get a two-dimensional function $X(t, f)$ defined by:

$$X(t,f) = \int_{-\infty}^{+\infty} x(\tau)w(\tau - t)e^{-j2\pi f\tau}d\tau \qquad (3.65)$$

where τ represents the variable that slides the window through the waveform $x(t)$.

The discrete form of the previous equation is defined as:

$$X(n, m) = \sum_{k=1}^{N} x(k)\left[W(k-n)e^{-j2\pi km/N}\right] \qquad (3.66)$$

The STFT produces a time-frequency representation/distribution of the waveform $x(t)$, whereas the spectrogram of the same waveform is equal to the squared magnitude of the STFT in Eq. (3.65),

$$P(t,f) = |X(t,f)|^2 \qquad (3.67)$$

Hence, the spectrogram represents a real-valued, nonnegative distribution (Semmlow, 2004).

There are two main difficulties with the spectrogram: (1) selecting an optimal window length for data segments, and (2) the time-frequency trade-off: shortening the data length, N, to achieve better time resolution will reduce frequency resolution. Shortening the data segment might also result in the loss of low frequencies, which are not completely involved in the data segment. Therefore, if the smaller window is employed to enhance the time resolution, then the frequency resolution is degraded and vice versa. This time-frequency trade-off has been equated to an uncertainty principle in which the product of time (T) and frequency resolution or bandwidth (B) must be:

$$BT \geq \frac{1}{4\pi} \qquad (3.68)$$

The trade-off between time and frequency resolution is essential in the STFT or spectrogram, which inspired a number of other time-frequency approaches as well as time–scale methods. In spite of these types of limitations, the STFT has been used effectively for a wide range of problems, mainly those where only high-frequency components are of interest and frequency resolution is not crucial (Semmlow, 2004).

EXAMPLE 3.28. The following MATLAB code was used to plot the ECG signals and their STFT spectrum. You can download data from the following website: https://www.physionet.org/pn3/incartdb/

```
%Ch3_Ex51_Spectrogram.m
%The following MATLAB code is used to plot
%the ECG signals and their Spectrogram (STFT spectrum).
clc
clear
%Load Sample ECG Data downloaded from the web site
%https://www.physionet.org/pn3/incartdb/
load PetersburgECG.mat
%%
Fs = 257;                % Sampling frequency
T = 1/Fs;                  % Sampling period
L = length(ECGN(:,1));     % Length of signal
segmentlen = L;
noverlap = 80;
NFFT = L/2;
```

```
%%
% Estimate the spectrum of the chirp using the short-time Fourier transform
% implemented in the |spectrogram| function. Divide the signal into
% sections of length L, windowed with a Hamming window. Specify 80
% samples of overlap between adjoining sections and evaluate the spectrum
% Suppress the default color bar.
subplot(2,2,1);
%Plot NORMAL ECG Signal
spectrogram(ECGN(:,1),segmentlen,noverlap,NFFT,Fs,'yaxis')
colorbar off
title('NORMAL ECG Signal')

%%
%Plot APC ECG Signal
subplot(2,2,2);
spectrogram(ECGAPC(:,1),segmentlen,noverlap,NFFT,Fs,'yaxis')
colorbar off
title('APC ECG Signal')
%%
%Plot PVC ECG Signal
subplot(2,2,3);
spectrogram(ECGPVC(:,1),segmentlen,noverlap,NFFT,Fs,'yaxis')
colorbar off
title('PVC ECG Signal')

%%
%Plot RBBB ECG Signal
subplot(2,2,4);
spectrogram(ECGRBBB(:,1),segmentlen,noverlap,NFFT,Fs,'yaxis')
colorbar off
title('RBBB ECG Signal')
```

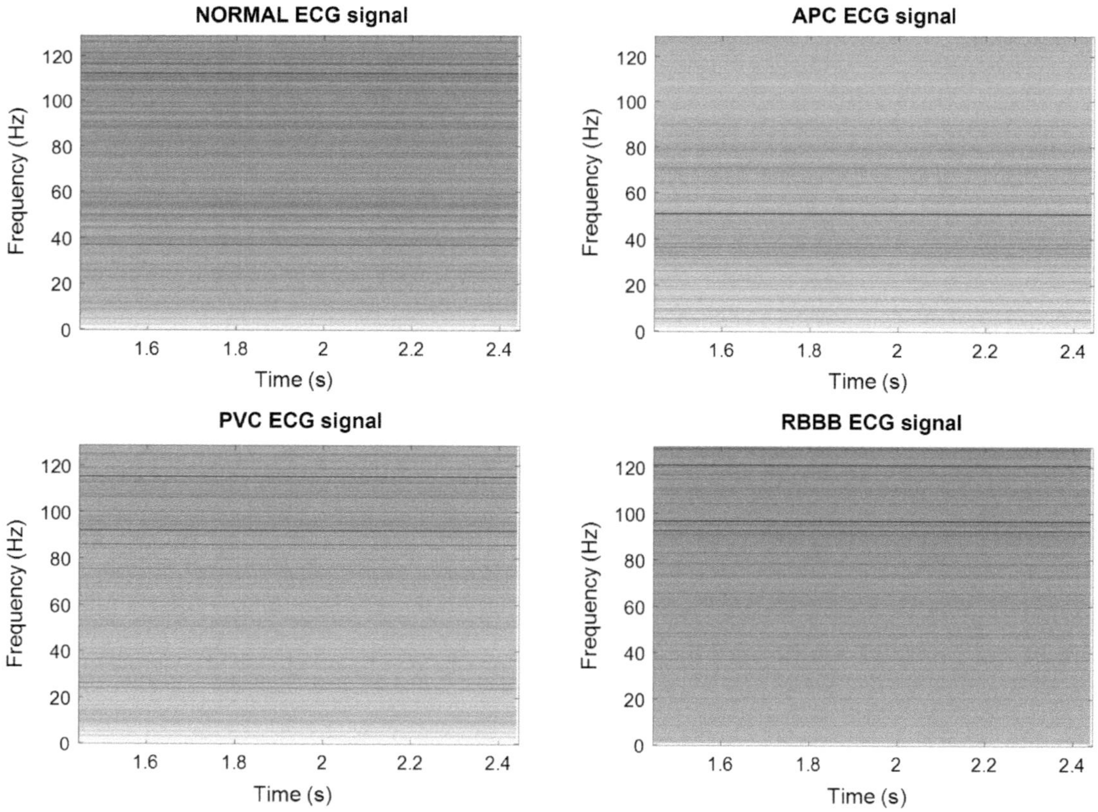

FIG. 3.28 Drawing of spectrogram of ECG signals.

3.5.2 Wigner-Ville Distribution

The first technique to overcome some limitations of the spectrogram was the WVD. It does not confine itself to the uncertainty principle and represents one of the most studied and best interpreted time-frequency approaches. The dual name is used because Wigner developed the method to be used in physics (Wigner, 1997), but Ville later applied it to signal processing (Ville, 1948). The original definition of the WVD that originated during the 1930s has been improved by a number of researchers with the purpose of overcoming its shortcomings (Cohen, 1995). In fact, the WVD now represents a special case of Cohen's class of distributions. The WVD and other distributions of Cohen's class make use of a variation of the autocorrelation function. In the WVD and all distributions of Cohen's class, we compare the waveform with itself for all potential shifts and values of time (Semmlow, 2004).

The WVD employs the autocorrelation function for calculating the power spectrum. Hence, the WVD can be defined from the time-domain representation of the signal:

$$WVD_X(t,f) = \int_{-\infty}^{-\infty} x\left(t+\frac{\tau}{2}\right) \cdot x^*\left(\tau-\frac{\tau}{2}\right) \cdot e^{-2j.\pi f.\tau} \cdot d\tau \quad (3.69)$$

where $R_x(t,\tau) = x\left(t+\frac{\tau}{2}\right) \cdot x^*\left(\tau-\frac{\tau}{2}\right)$ is the instantaneous autocorrelation function and * indicates conjugate operation (Semmlow, 2004). This distribution formula satisfies a large number of desirable mathematical properties. Principally, the WVD is mostly real-valued, and it conserves time and frequency shifts and satisfies the marginal properties. Although the WVD has the advantages of a high resolution in both time and frequency, the interfering terms produced by the interaction of two signal components make it tough to interpret the result. Therefore, it is better to diminish the number and the amplitude of the interference. If a time-smoothing window $g(t)$ and a frequency-smoothing window $h(t)$ are applied, then the WVD turns into the smoothed-pseudo-Wigner-Ville distribution (SPWVD), which is written as

$$SPWVD_x(t,f) = \iint h(t-\tau)g(f-\xi)W_x(\tau,\xi)^x d\tau d\xi \quad (3.70)$$

where W is the WVD. The previous trade-off of the spectrogram between time and frequency resolution can be replaced by a trade-off between the joint time-frequency resolution and the level of the interference terms (Boudreaux-Bartels & Murray, 1995; de Souza Neto et al., 2001; Marchant, 2003; Narasimhan & Nayak, 2003; Semmlow, 2004).

The WVD also possesses a number of weaknesses over the STFT. For instance, the WVD is not necessarily a positive-valued distribution like STFT, which does not represent a particularly serious issue in practical use, because the WVD's positive-valued areas match well with the time-frequency structure (Cohen, 1995). However, the manifestation of cross-products indicating energies where they do not exist represent the most severe drawback of WVD, restricting their use with multicomponent signals in practice. The development of other distributions has been highly driven by the presence of these fictional energies (Cohen, 1995). Consequently, different filters to the instantaneous autocorrelation function have been trialed to lessen the impact of the cross-products(Sörnmo & Laguna, 2005). The WVD also struggles with noise, which is basically spread across all time and frequency together with the cross-products. In certain situations, we can remove noise by applying a data window to the lag dimension of the instantaneous autocorrelation function. As known, windowing decreases frequency resolution, so a trade-off between the frequency resolution loss and the cross-product noise reduction is inevitable (Semmlow, 2004) (Figs. 3.29–3.31).

EXAMPLE 3.29. The following MATLAB code was used to plot the ECG signal and its Wigner-Ville time-frequency distributions. You can download data from the following website: https://www.physionet.org/pn3/incartdb/

Note that the Time-Frequency toolbox is used for running the following MATLAB code. You can download it from the following website: http://tftb.nongnu.org/

Do not forget to set the path for the Time-Frequency toolbox.

```
%Ch3_Ex52_WV.m
% Different Wigner-Ville time-frequency distribution
%Note that the Time-Frequency toolbox is used for running
%the following MATLAB code. You can download it from
%the following web site:
%http://tftb.nongnu.org/
%Set the path for the Time-Frequency toolbox

clc
clear
%Load Sample ECG Data downloaded from the web site
%https://www.physionet.org/pn3/incartdb/
load PetersburgECG.mat
```

```
%%
Fs = 257;           % Sampling frequency
T = 1/Fs;            % Sampling period
L = length(ECGN(:,1));    % Length of signal

%%
% Wigner-Ville time-frequency distribution
figure
[tfr,t,f] = tfrwv(ECGN(:,1));
mesh(t, f,abs(tfr(:,:)))
ylabel('Frequency(Hz) '); xlabel('Time(msec)');zlabel('dB');
title('Wigner Ville of ECG Signal')

%%
%Pseudo Wigner-Ville time-frequency distribution.
figure
[tfr,t,f] = tfrpwv(ECGN(:,1));
mesh(t, f,abs(tfr(:,:)))
ylabel('Frequency(Hz) '); xlabel('Time(sec)');zlabel('dB');
%axis([-inf inf 0 100 -inf inf])
title('Pseudo Wigner Ville of ECG Signal')

%%
%Smoothed Pseudo Wigner-Ville time-frequency distribution.
figure
[tfr,t,f] = tfrspwv(ECGN(:,1));
mesh(abs(tfr))
ylabel('Frequency(Hz) '); xlabel('Time(sec)');zlabel('dB');
title('Smoothed Pseudo Wigner Ville of ECG Signal')
```

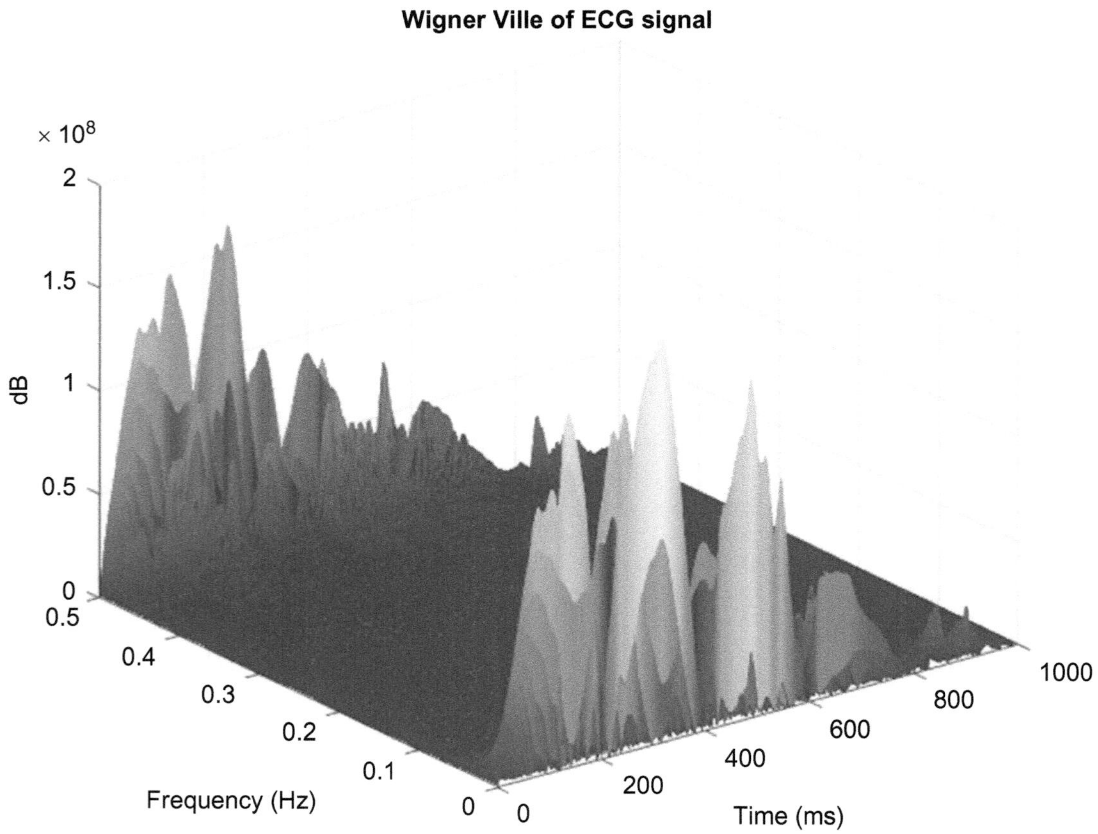

FIG. 3.29 The Wigner-Ville distribution of healthy ECG signal.

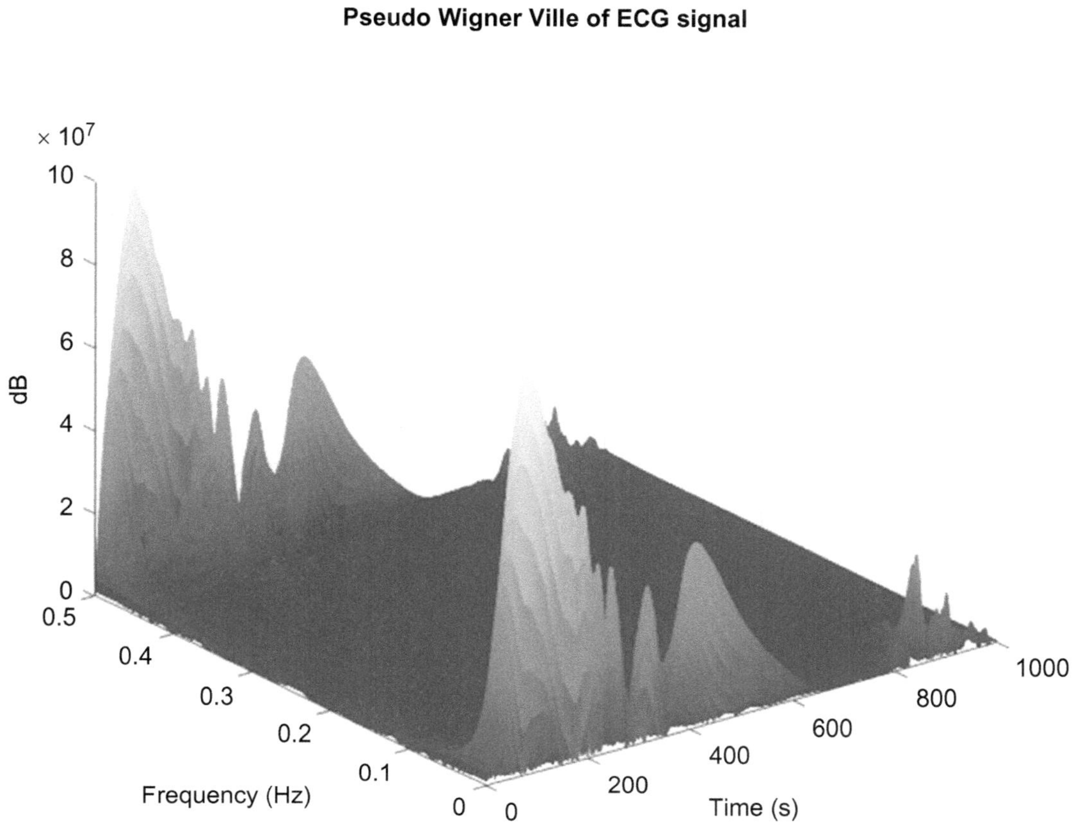

FIG. 3.30 The pseudo-Wigner-Ville distribution of healthy ECG signal.

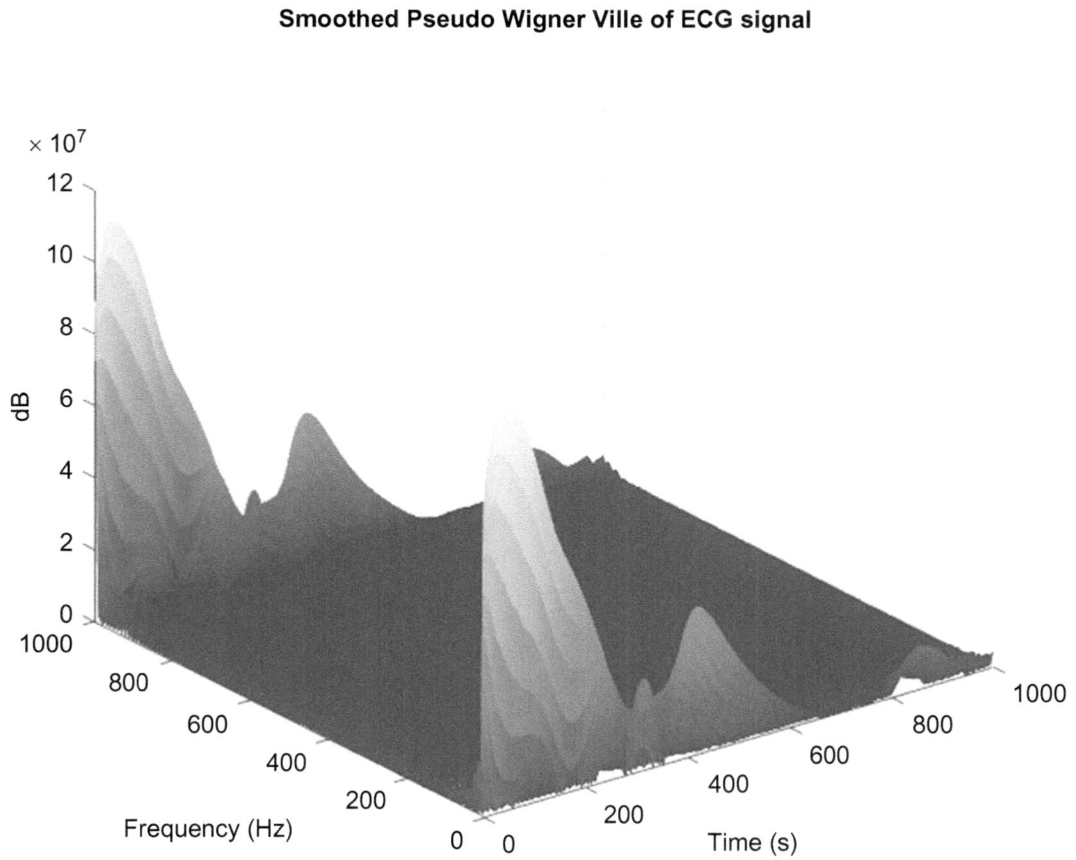

FIG. 3.31 The smoothed pseudo-Wigner-Ville distribution of healthy ECG signal.

3.5.3 Choi-Williams Distribution

The general equation of a time-frequency distribution from Cohen's class of distributions is defined as:

$$\rho(t,f) = \iiint g(\nu,\tau) R_{xx}(u,\tau) e^{j2\pi\nu u} e^{-j2\pi\nu t} e^{-j2\pi f\tau} d\nu du d\tau \tag{3.71}$$

where $g(\nu,\tau)$ represents the two-dimensional filter, which is also termed as a kernel. We can simplify Eq. (3.71) in two ways:

- Integrating over the variable ν because the rest of the integrand is not a function of ν, or
- Using the ambiguity function.

In the first approach, we multiply the kernel by the exponential function to produce a new function denoted as the determining function $G(u,\tau)$,

$$G(u,\tau) = \int_{-\infty}^{\infty} g(\nu,\tau) e^{j2\pi\nu u} d\nu \tag{3.72}$$

Now Eq. (3.71) becomes,

$$\rho(t,f) = \iint G(u-t,\tau) R_{xx}(u,\tau) e^{-j2\pi f\tau} du d\tau \tag{3.73}$$

In the second approach, we insert Eq. (3.72) into Eq. (3.71), so the general time-frequency distribution becomes:

$$\rho(t,f) = \iint g(\nu,\tau) A_x(\nu,\tau) e^{-j2\pi\nu t} e^{-j2\pi f\tau} d\nu d\tau \tag{3.74}$$

The task of a two-dimensional kernel function $g(\nu,\tau)$ is to shape the ambiguity function with the purpose of reducing phantom cross-terms positioned far away from the origin of $\nu-\tau$ domain, while preserving the auto-terms at the origin. The kernel is supposed to have its maximum at origin $g(0,0)=1$, and therefore can be looked at as a narrowband filter. This approach has its drawbacks because the kernel flattens out the ambiguity function, reducing the time-frequency resolution (Sörnmo & Laguna, 2005).

To make the resulting time-frequency distribution satisfying, the kernel must possess certain properties. If the kernel fulfils the condition $g(0,\tau)=1$, the frequency marginal condition is satisfied for time-frequency distributions of Cohen's class:

$$\int_{-\infty}^{\infty} \rho(t,f) dt = |X(f)|^2 \tag{3.75}$$

Likewise, the time marginal condition in Eq. (3.41) stays valid for $g(\nu,0)=1$. The condition $g(\nu,\tau)=g(-\nu,-\tau)$ is necessary to make the distribution real-valued (Sörnmo & Laguna, 2005).

We will also touch upon the favorite member amongst numerous Cohen's class distributions—the Choi-Williams distribution (CWD), widely used in biomedical signal processing. It is also denoted as an exponential distribution because of an exponential kernel function,

$$g(\nu,\tau) = e^{-\nu^2\tau^2/\sigma}, \quad \sigma > 0 \tag{3.76}$$

then, the determining function $G(t,\tau)$ becomes:

$$G(t,\tau) = \frac{\sqrt{\sigma/\pi}}{2\pi} e^{-\sigma t^2/4\tau^2} \tag{3.77}$$

where σ regulates the level of cross-term reduction and associated smoothing effect (Sörnmo & Laguna, 2005). Besides having cross-terms reduced, the CWD is also superior to WVD in terms of noise characteristics (Semmlow, 2004).

But there is a problem related to any kernel function that satisfies the marginal conditions, including the exponential kernel defined in Eq. (3.76). In fact, the CWD has vertical and horizontal ripple because the exponential kernel does not ensure the reduction of cross-products alongside the ν- and τ-axis. Auto-terms fixed around the same time, but existing at different frequencies, cause the vertical ripple, whereas auto-terms concentrated around the same frequency, but happening at different times, cause the horizontal ripple. Nonetheless, the overall understanding of the time-frequency analysis is barely influenced by the ripple manifestation (Sörnmo & Laguna, 2005).

The modified CWD integrates a data window and counts as a member of the class of reduced interference distributions (Semmlow & Griffel, 2014). We can further increase the performance of time-frequency distributions by applying a signal-dependent kernel instead of a fixed kernel, because the auto- and cross-term position depends on the examined signal. Signal-dependent kernel can diminish the ripple manifestation as well. Therefore, the selection of "optimal" signal-dependent kernel has been in focus of many important research studies (Nave & Cohen, 1993).

So, which type of numerous nonparametric time-frequency analysis techniques one should choose? If we need better time-frequency resolution, the signal-dependent kernel or CWD is usually the choice. Sometimes even the simple STFT may be suitable for certain biomedical signals analysis. Boashash and Sucic (2003) tried to define measures for comparing the performance of various nonparametric quadratic time-frequency methods. However, these findings should only be considered as a guide because they cannot be directly interpreted on EEG signals. Finally, the time-frequency technique producing the top performance in medical terms is eventually the favored one (Sörnmo & Laguna, 2005) (Fig. 3.32).

EXAMPLE 3.30. The following MATLAB code was used to plot the ECG signal and its Choi-Williams time-frequency distributions. You can download data from the following website: https://www.physionet.org/pn3/incartdb/

Note that the Time-Frequency toolbox is used for running the following MATLAB code. You can download it from the following website: http://tftb.nongnu.org/

Do not forget to set the path for the Time-Frequency toolbox.

```
%Ch3_Ex53_Choi_Williams.m
% Choi-Williams time-frequency distribution.
%Note that the Time-Frequency toolbox is used for running
%the following MATLAB code. You can download it from
%the following web site:
%http://tftb.nongnu.org/
%Set the path for the Time-Frequency toolbox

clc
clear
%Load Sample ECG Data downloaded from the web site
%https://www.physionet.org/pn3/incartdb/
load PetersburgECG.mat

%%
Fs = 257;              % Sampling frequency
T = 1/Fs;              % Sampling period
L = length(ECGN(:,1)); % Length of signal

%%
% Wigner Ville
figure
[tfr,t,f] = tfrcw(ECGN(:,1));
mesh(t, f,abs(tfr(:,:)))
ylabel('Frequency(Hz) '); xlabel('Time(msec)');zlabel('dB');
title('Choi-Williams time-frequency distribution of ECG Signal')
```

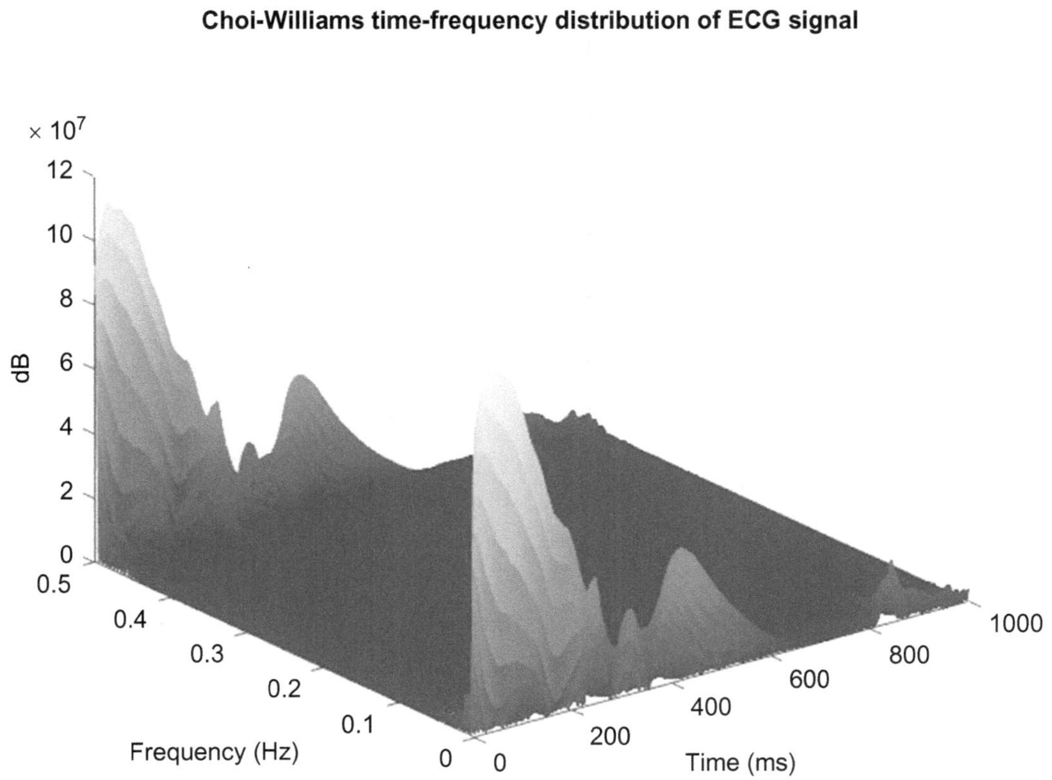

FIG. 3.32 The Choi-Williams time-frequency distribution of healthy ECG signal.

3.5.4 Analytic Signal

All distributions of Cohen's class perform better if applied to the analytic signal, which is a modified version of the real signal $x(t)$ having a number of advantages. As real signals contain both positive and negative frequencies generated from the FT, we only need to consider positive frequency components because negative ones are redundant and can be eliminated. Therefore, the analytic signal $x_A(t)$ can be defined in the frequency domain as:

$$X_A(f) = \begin{cases} 2X(f) & f \geq 0 \\ 0 & f < 0 \end{cases} \tag{3.78}$$

For the positive frequencies, the FT of the real signal $x(t)$ is multiplied by 2 to preserve the overall energy.

As negative frequency components will no longer create cross-terms, the analytic signal decreases their overall number. Furthermore, the sampling rate can be decreased if we use the analytic signal, which represents an additional advantage. Moreover, the instantaneous autocorrelation function is computed using evenly spaced values, hence it is undersampled by a factor of 2. Meanwhile, we must sample the data at twice the Nyquist frequency if the real signal is used (Semmlow, 2004).

The analytic signal can be created in several ways. Basically, only the imaginary part should be added to the real signal. The analytic signal defined in Eq. (3.78) has an imaginary component equal to the Hilbert Transform of the real signal (Cohen, 1995). In discretized form, this analytic signal is defined as,

$$x_A(n) = x(n) + jH[x(n)] = x(n) + j[x(n)*1/(\pi n)] \tag{3.79}$$

where H represents the Hilbert transform, and * stands for convolution operation.

The Hilbert transform can be implemented using the Hilbert transform filter in the form of a finite impulse response (FIR) filter having the following coefficients:

$$h(n) = \begin{cases} \dfrac{2\sin^2(\pi n/2)}{\pi n} & n \neq 0 \\ 0 & n = 0 \end{cases} \tag{3.80}$$

Although the Hilbert transform filter should basically be of an infinite impulse response (IIR) type, an FIR filter with the length of around 79 samples provides a satisfactory estimate in practical use (Boashash & Black, 1987).

3.5.5 Wavelet Analysis

Each transform should deliver supplementary information that usually opens up a new understanding about the original waveform. Numerous time-frequency techniques did not fully solve the time-frequency problem. The wavelet transform represents one more way of describing the time-frequency characteristics of a waveform. However, the waveform is now divided into scale segments rather than time sections (Semmlow, 2004).

In the FT, the waveform $x(t)$ was compared to a whole family of sinusoidal functions at harmonically related frequencies:

$$X(\omega_m) = \int_{-\infty}^{+\infty} x(t)e^{-j\omega_m t}dt \tag{3.81}$$

Sine functions are mainly favored because they have energy at only one particular frequency, making them a perfect choice for probing the frequency characteristic of a waveform called the frequency spectrum.

Nearly any family of functions could be used to evaluate or probe some specific characteristic or behavior of a waveform. The probing functions of finite length are suitable for sliding themselves over the waveform $x(t)$, which is done in the STFT utilizing sliding window function $w(t)$. In general, a transform including a sliding probing function can be defined as the following convolution:

$$X(t,m) = \int_{-\infty}^{+\infty} x(\tau)f(t-\tau)_m d\tau \tag{3.82}$$

where $f(t)_m$ represents a family of functions, whereas m is the family number. Note that the aforementioned comparisons in the last two equations were performed by multiplying the waveform with the probing functions followed by averaging. The family of functions (that is necessarily larg) should be capable of demonstrating all characteristics of the waveform $x(t)$ by establishing a redundant set of descriptions $X(t,m)$, more than enough to reconstruct $x(t)$, making this transform bilateral. This redundancy can occasionally be valuable for noise reduction, otherwise it is basically needless. Note that the FT is not redundant, whereas the STFT and all the distributions are (Semmlow, 2004).

Wavelets include two parameters: one is for scaling in time and another for sliding in time. A wavelet is an oscillating function with energy concentrated in time for improved characterization of transient signals. Band-pass filter characteristic is just one of several mathematical properties a wavelet function must have. Wavelet analysis tries to accomplish respectable localization in both frequency and time. Sliding and scaling, two new degrees of freedom, allow the analysis of the combined existence of fine structures and global waveforms in signals. A multiresolution analysis represents the essential idea of analyzing signals at different scales with a growing level of resolution (Sörnmo & Laguna, 2005).

Comprehensive explanations of wavelet analysis using mathematics can be found in several good books that also include the topic of filter banks, which is closely associated with wavelets (Basseville & Nikiforov, 1993; Gustafsson & Gustafsson, 2000).

3.5.6 Continuous Wavelet Transform

A family of wavelets $\Psi_{s,\tau}(t)$ is defined by sliding and scaling the mother wavelet $\Psi(t)$ with the continuous-valued parameters τ and s,

$$\Psi_{s,\tau}(t) = \frac{1}{\sqrt{s}}\Psi\left(\frac{t-\tau}{s}\right) \tag{3.83}$$

where the factor $1/\sqrt{s}$ ensures that all scaled functions contain the same energy. The wavelet expands for $s > 1$ and contracts for $0 < s < 1$.

The probing function $\Psi_{s,\tau}(t)$ always has an oscillatory form. For $s = 1$ and $\tau = 0$, it takes on its natural form called the mother wavelet, $\Psi_{1,0}(t) \equiv \Psi(t)$, together with several of its family members created by contraction and dilation. If a wavelet is contracted to a smaller time scale, that makes it more localized in time and less localized in frequency because the resultant band-pass frequency response has increased bandwidth and is moved to higher frequencies.

The continuous wavelet transform (CWT) $\omega(s,\tau)$ of a continuous-time signal $x(t)$ is defined by comparing the signal $x(t)$ to the probing function $\Psi_{s,\tau}(t)$:

$$\omega(s,\tau) = \int_{-\infty}^{+\infty} x(t)\Psi_{s,\tau}(t)dt \tag{3.84}$$

creating two-dimensional mapping onto the time-scale domain. As the last equation represents the convolution between the signal $x(t)$ and a filter with impulse response $\Psi(-t/s)/\sqrt{s}$, the CWT can be viewed as a linear filter.

As the CWT decomposes the waveform into coefficients of two variables s and τ, we need to perform a double integration to reconstruct the original waveform from the wavelet coefficients (Bodenstein, Schneider, & Malsburg, 1985):

$$x(t) = \frac{1}{C_\Psi}\int_{-\infty}^{\infty}\int_{0}^{\infty} \omega(s,\tau)\Psi_{s,\tau}(t)\frac{d\tau ds}{s^2} \tag{3.85}$$

where

$$C_\Psi = \int_{0}^{\infty} \frac{|\Psi(f)|^2}{|f|}df < \infty \tag{3.86}$$

and $\Psi(f)$ represents the FT of $\Psi(t)$.

The simplest wavelet is the Haar wavelet, a member of the Walsh basis functions. Another popular wavelet is the Mexican hat wavelet, defined by the equation:

$$\Psi(t) = \left(1 - 2t^2\right)e^{-t^2} \tag{3.87}$$

The Morlet wavelet, named after a pioneer of wavelet analysis, is given by the equation:

$$\Psi(t) = e^{-t^2}\cos\left(\pi\sqrt{\frac{2}{\ln 2}}t\right) \tag{3.88}$$

A wide variety of wavelets have been suggested, each one possessing some feature particularly suitable for certain applications.

Wavelets provide a trade-off between time and frequency localization. However, they do not occur at an exact time or frequency. More precisely, they are not exactly contained in either time or frequency but rather well contained in both. These ranges are also associated with the time and frequency resolution of the CWT. The shorter wavelet time range delivers an improved capability to isolate local time events but at the cost of frequency resolution because the wavelet only responds to high-frequency components. On the other hand, the CWT delivers enhanced frequency resolution for the longer wavelet length. This integrated compromise between time and frequency resolution makes CWT appropriate for analyzing signals that contain fast-changing (high-frequency) components covered by slowly changing (low-frequency) ones (Semmlow, 2004) (Fig. 3.33).

EXAMPLE 3.31. The following MATLAB code was used to plot the ECG signals and their continuous wavelet transform. You can download data from the following website: https://www.physionet.org/pn3/incartdb/

```
%Ch3_Ex54_CWT.m
% Time-Frequency Analysis with the Continuous Wavelet Transform
%
% This example shows how to use the continuous wavelet transform (CWT) to
% analyze ECG signals jointly in time and frequency. The example explains the
% relationship between scale and frequency.

%% Selecting a Wavelet and Scales
% Using the CWT for time-frequency analysis works best when the magnitude
% of the wavelet's Fourier transform is symmetric with respect to its
% center frequency. Additionally, use a complex-valued wavelet whose
% Fourier transform is nonzero only on the positive frequencies. Based on
% these criteria, the best wavelets for time-frequency analysis, in
% general, are analytic wavelets such as the bump and analytic Morlet
% wavelets supported by |cwtft|.
%
```

```
% After you select the wavelet, you must select the scales over which to
% compute the CWT.
%
% Both in time and frequency, the dilation (stretching and shrinking)
% operation on the wavelet is multiplicative. Therefore, the natural
% spacing of scales for the CWT is logarithmic where there is a constant
% ratio between successive elements. This differs from linearly-spaced
% scales where there is a constant difference between adjacent elements.

% The following examples illustrate applications of the CWT for
% time-frequency analysis.
clc
clear
%Load Sample ECG Data downloaded from the web site
%https://www.physionet.org/pn3/incartdb/
load PetersburgECG.mat

%%
Fs = 257;            % Sampling frequency
DT = 1/Fs;             % Sampling period
L = length(ECGN(:,1));    % Length of signal

%%
% Obtain and plot the CWT of Normal ECG Signal using the analytic Morlet wavelet.
t = 0:DT:(numel(ECGN(:,1))*DT)-DT;
f0 = 5/(2*pi);
scales = helperCWTTimeFreqVector(20,500,f0,0.001,32);
cwtecg = cwtft({ECGN(:,1),0.001},'wavelet','morl','scales',scales);
subplot(2,2,1);
helperCWTTimeFreqPlot(cwtecg.cfs,ECGN(:,1),cwtecg.frequencies,...
    'surf','CWT of Normal ECG Signal','Time (secs)','Frequency (Hz)')

colorbar off

%%
% Obtain and plot the CWT of APC ECG Signal using the analytic Morlet wavelet.
t = 0:DT:(numel(ECGAPC(:,1))*DT)-DT;
f0 = 5/(2*pi);
scales = helperCWTTimeFreqVector(20,500,f0,0.001,32);
cwtecg = cwtft({ECGAPC(:,1),0.001},'wavelet','morl','scales',scales);
subplot(2,2,2);
helperCWTTimeFreqPlot(cwtecg.cfs,ECGAPC(:,1),cwtecg.frequencies,...
    'surf','CWT of APC ECG Signal','Time (secs)','Frequency (Hz)')

colorbar off
%%
% Obtain and plot the CWT of PVC ECG Signal using the analytic Morlet wavelet.
t = 0:DT:(numel(ECGPVC(:,1))*DT)-DT;
f0 = 5/(2*pi);
scales = helperCWTTimeFreqVector(20,500,f0,0.001,32);
cwtecg = cwtft({ECGPVC(:,1),0.001},'wavelet','morl','scales',scales);
subplot(2,2,3);
helperCWTTimeFreqPlot(cwtecg.cfs,ECGPVC(:,1),cwtecg.frequencies,...
    'surf','CWT of PVC ECG Signal','Time (secs)','Frequency (Hz)')

colorbar off

%%
% Obtain and plot the CWT of RBBB ECG Signal using the analytic Morlet wavelet.
t = 0:DT:(numel(ECGRBBB(:,1))*DT)-DT;
f0 = 5/(2*pi);
scales = helperCWTTimeFreqVector(20,500,f0,0.001,32);
cwtecg = cwtft({ECGRBBB(:,1),0.001},'wavelet','morl','scales',scales);
subplot(2,2,4);
helperCWTTimeFreqPlot(cwtecg.cfs,ECGRBBB(:,1),cwtecg.frequencies,...
    'surf','CWT of RBBB ECG Signal','Time (secs)','Frequency (Hz)')

colorbar off
```

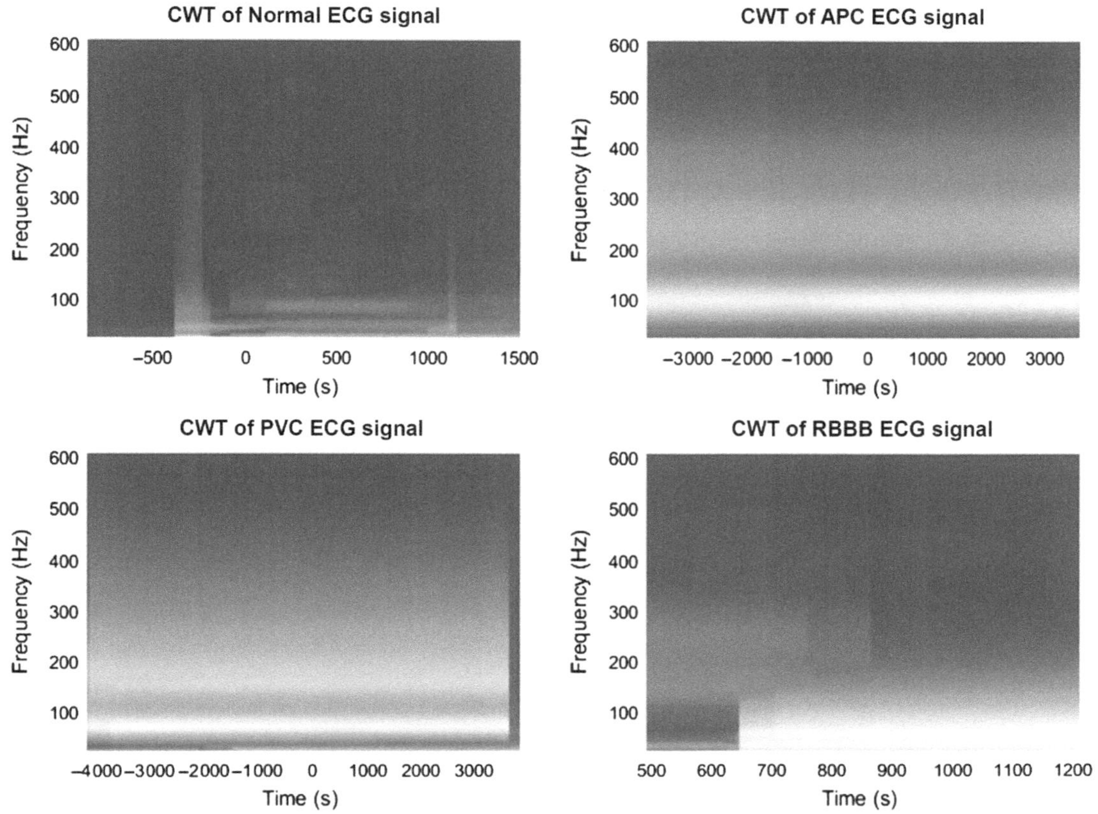

FIG. 3.33 The continuous wavelet transform of different ECG signals.

3.5.7 Discrete Wavelet Transform

The only major problem of CWT is its infinite redundancy, because it produces countless coefficients, quite more than is really required to accurately characterize the original signal. This redundancy becomes expensive only if we need to reconstruct the original signal, because all coefficients will be used making the calculation struggle rather unnecessary. The discrete wavelet transform (DWT) usually accomplishes the coefficient frugality by limiting the variation in scale and sliding to powers of 2; hence it is occasionally called the dyadic wavelet transform having the same abbreviation (DWT). However, we will still be able to accurately recover the original signal from the discrete coefficients of the dyadic wavelet transform (Bodenstein et al., 1985). If the selected wavelet belongs to an orthogonal family, the DWT even represents a nonredundant bilateral transform (Semmlow, 2004).

Dyadic sampling of the two wavelet parameters is defined as,

$$s = 2^{-j}, \quad \tau = k2^{-j} \tag{3.89}$$

where j and k are both integers. Therefore, the discretized probing function becomes:

$$\Psi_{j,k}(t) = 2^{j/2}\Psi\left(2^{j}t - k\right) \tag{3.90}$$

Inserting Eq. (3.90) into Eq. (3.84), we get the DWT:

$$\omega_{j,k} = \int_{-\infty}^{\infty} x(t)\Psi_{j,k}(t)dt \tag{3.91}$$

The original signal is recovered by the inverse DWT or the wavelet series expansion

$$x(t) = \sum_{j=-\infty}^{\infty}\sum_{k=-\infty}^{\infty} \omega_{j,k}\Psi_{j,k}(t) \tag{3.92}$$

where $\Psi_{j,k}(t)$ is a set of orthonormal basis functions. The wavelet series expansion represents the sum over two indices, j and k, that are associated with scaling and sliding of the basis functions $\Psi_{j,k}(t)$.

Here, we present a new concept, called the scaling function, which simplifies the application and calculation of the DWT. The calculation of the finest resolution is performed at first, followed by the calculation of the coarser resolutions by means of a smoothed form of the original waveform rather than the original waveform itself. This smoothed form is acquired using the scaling function that is occasionally termed the smoothing function (Semmlow, 2004) (Figs. 3.34–3.36).

EXAMPLE 3.32. The following MATLAB code was used to plot the EEG signals and their DWT. You can download data from the following website: http://epileptologie-bonn.de/cms/front_content.php?idcat=193&lang=3&changelang=3

```
%Ch3_Ex61_DWT_EEG.m
% DWT of NORMAL, INTERICTAL and ICTAL EEG signals
%Decompose EEG data using DWT coefficients A5, D4, D3, D2 and D1
clc
clear

%Load Sample EEG Data downloaded from the web site
%http://epileptologie-bonn.de/cms/front_content.php?idcat=193&lang=3&changelang=3
load AS_BONN_ALL_EEG_DATA_4096.mat
N=400;
wname = 'db4';
level=5;
%%
%%%%%%%%%%%%%%%%%%%%%%%%%%%%% A Z.ZIP EYES OPEN NORMAL SUBJECT
    [C,L]=wavedec(Normal_Eyes_Open(:,1),level,wname);
    cD1=detcoef(C,L,1);%
    cD2=detcoef(C,L,2);
    cD3=detcoef(C,L,3);
    cD4=detcoef(C,L,4);%
    cD5=detcoef(C,L,5);%
    cA5=appcoef(C,L,wname,level);

    subplot(4,2,1)
    plot(Normal_Eyes_Open(:,1))
    ylabel('EEG');
    title('Normal');
    axis([1 4000 -inf inf])

    subplot(4,2,3)
    plot(cD1)
    ylabel('cD1');
    axis([1 2000 -inf inf])

    subplot(4,2,4)
    plot(cD2)
    ylabel('cD2');
    axis([1 1000 -inf inf])

    subplot(4,2,5)
    plot(cD3)
    ylabel('cD3');
    axis([1 500 -inf inf])

    subplot(4,2,6)
    plot(cD4)
    ylabel('cD4');
    axis([1 260 -inf inf])

    subplot(4,2,7)
    plot(cD5)
    ylabel('cD5');
    axis([1 130 -inf inf])

    subplot(4,2,8)
    plot(cA5)
    ylabel('cA5');
    axis([1 130 -inf inf])

%%
% D F.ZIP EPILEPTIC SUBJECT (INTERICTAL) were recorded from within
% the epileptogenic zone during seizure free intervals
    [C,L]=wavedec(Interictal(:,1),level,wname);
    cD1=detcoef(C,L,1);            %
    cD2=detcoef(C,L,2);
    cD3=detcoef(C,L,3);
    cD4=detcoef(C,L,4);            %
```

```
    cD5=detcoef(C,L,5);              %
    cA5=appcoef(C,L,wname,level);

    figure
    subplot(4,2,1)
    plot(Interictal(:,1))
    ylabel('EEG');
    axis([1 4000 -inf inf])
    title('Interictal');

    subplot(4,2,3)
    plot(cD1)
    ylabel('cD1');
    axis([1 2000 -inf inf])

    subplot(4,2,4)
    plot(cD2)
    ylabel('cD2');
    axis([1 1000 -inf inf])

    subplot(4,2,5)
    plot(cD3)
    ylabel('cD3');
    axis([1 500 -inf inf])

    subplot(4,2,6)
    plot(cD4)
    ylabel('cD4');
    axis([1 260 -inf inf])

    subplot(4,2,7)
    plot(cD5)
    ylabel('cD5');
    axis([1 130 -inf inf])

    subplot(4,2,8)
    plot(cA5)
    label('cA5');
    axis([1 130 -inf inf])

%% E S.ZIP EPILEPTIC SUBJECT ICTAL DURING SEIZURE
%%%%%%%%%%%%%%%%%%%%%%%%%%%
  [C,L]=wavedec(Ictal(:,1),level,wname);
    cD1=detcoef(C,L,1);         %
    cD2=detcoef(C,L,2);
    cD3=detcoef(C,L,3);
    cD4=detcoef(C,L,4);         %
    cD5=detcoef(C,L,5);         %
    cA5=appcoef(C,L,wname,level);

    figure
    subplot(4,2,1)
    plot(Ictal(:,1))
    ylabel('EEG');
    axis([1 4000 -inf inf])
    title('Ictal');

    subplot(4,2,3)
    plot(cD1)
    ylabel('cD1');
    axis([1 2000 -inf inf])

    subplot(4,2,4)
    plot(cD2)
    ylabel('cD2');
    axis([1 1000 -inf inf])

    subplot(4,2,5)
    plot(cD3)
    ylabel('cD3');
    axis([1 500 -inf inf])
```

```
subplot(4,2,6)
plot(cD4)
ylabel('cD4');
axis([1 260 -inf inf])

subplot(4,2,7)
plot(cD5)
ylabel('cD5');
axis([1 130 -inf inf])

subplot(4,2,8)
plot(cA5)
ylabel('cA5');
axis([1 130 -inf inf])
```

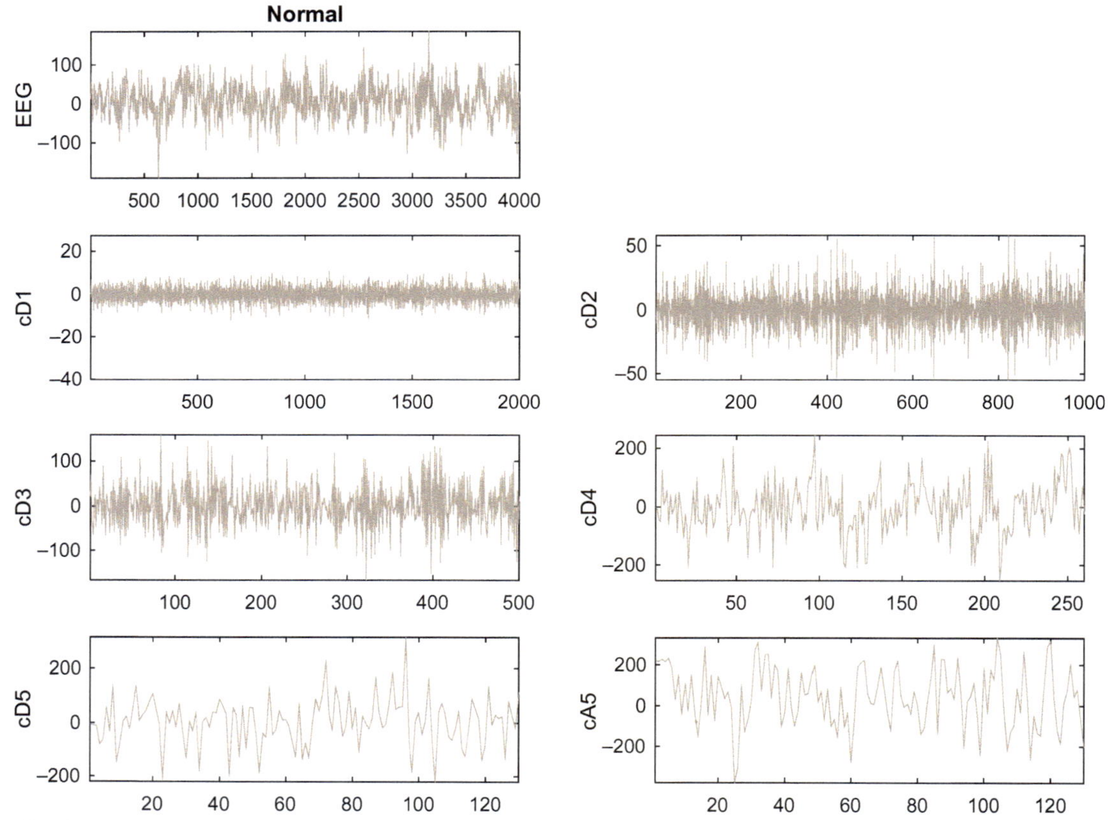

FIG. 3.34 The discrete wavelet transform of normal EEG signal.

3.5.8 Stationary Wavelet Transform

The stationary wavelet transform (SWT) computes all the decimated DWT for a given signal at one time. More specifically, for level 1, the SWT is attained by convolving the signal with the suitable filters as in the DWT but without employing downsampling. Then the detail and approximation coefficients at level 1 will be the same as the signal length. The general step j convolves the approximation coefficients at level $j-1$, with suitable filters without downsampling, to yield the detail and approximation coefficients at level j (Zhang, Dong, et al., 2015; Zhang, Li, et al., 2015).

DWT is a decomposition of the signal x(t) that can be considered as a successive band-pass filtering and downsampling. $x(t) = x_0(t)$ is decomposed into two parts: $y_1(t)$ presents the high-frequency parts, of $x_0(t)$ and $x_1(t)$ presents the low-frequency parts. The DWT is computationally faster and can be realized by consecutive filter banks. Unfortunately, the DWT is not shift-invariant when utilized to discrete time series x(t). If the input time series x(t) is shifted, the resulting coefficients may become totally different. The SWT (Nason & Silverman, 1995) has no such issues. Mainly, the SWT is DWT, but the downsampling step is replaced by an upsampling (Sudakov et al., 2017) (Figs. 3.37–3.39).

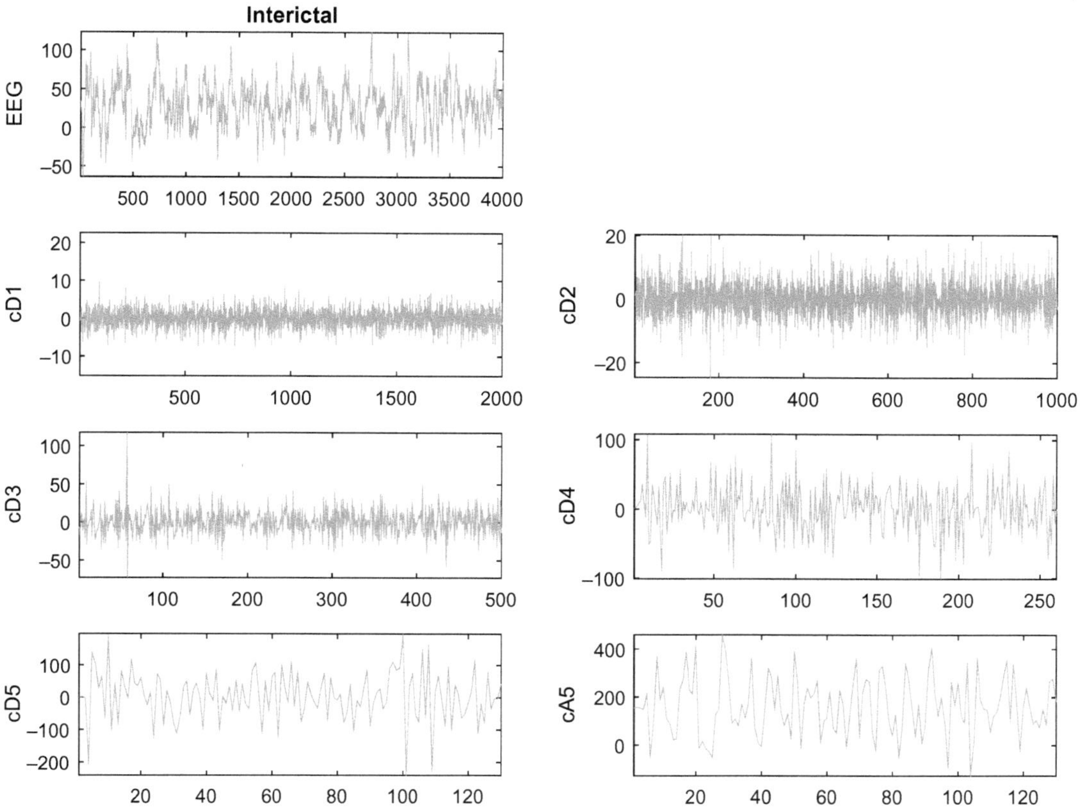

FIG. 3.35 The discrete wavelet transform of interictal EEG signal.

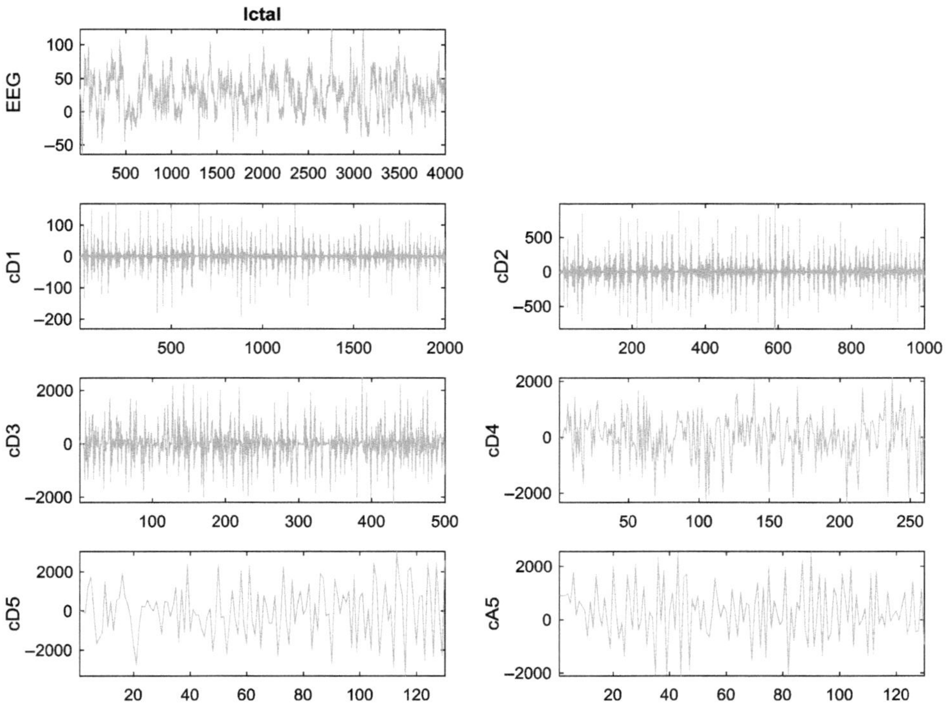

FIG. 3.36 The discrete wavelet transform of ictal EEG signal.

EXAMPLE 3.33. The following MATLAB code was used to plot the EEG signals and their SWT. You can download data from the following website: http://epileptologie-bonn.de/cms/front_content.php?idcat=193&lang=3&changelang=3

```
%Ch3_Ex66_SWT.m
%SWT of NORMAL, INTERICTAL and ICTAL EEG signals
clc
clear
%Load Sample EEG Data downloaded from the web site
%http://epileptologie-bonn.de/cms/front_content.php?idcat=193&lang=3&changelang=3
load AS_BONN_ALL_EEG_DATA_4096.mat
Level = 5;
wname = 'db4';
L=length(Normal_Eyes_Open(1:4096,1));
%%
% %%%%%%%%%%%%%%%%%%%%%%%%%%%%% A Z.ZIP EYES OPEN NORMAL SUBJECT
    [swa,swd] = swt(Normal_Eyes_Open(1:4096,1),Level,wname);
    subplot(4,2,1)
    plot(swa(1,:))
    ylabel('swa1');
    axis([1 L -inf inf])
    title('Normal');

    subplot(4,2,2)
    plot(swa(2,:))
    ylabel('swa2');
    axis([1 L -inf inf])

    subplot(4,2,3)
    plot(swa(2,:))
    ylabel('swa3');
    axis([1 L -inf inf])

    subplot(4,2,4)
    plot(swa(4,:))
    ylabel('swa4');
    axis([1 L -inf inf])

    subplot(4,2,5)
    plot(swd(1,:))
    ylabel('swd1');
    axis([1 L -inf inf])

    subplot(4,2,6)
        plot(swd(2,:))
    ylabel('swd2');
    axis([1 L -inf inf])

    subplot(4,2,7)
    plot(swd(3,:))
    ylabel('swd3');
    axis([1 L -inf inf])

    subplot(4,2,8)
    plot(swd(4,:))
    ylabel('swd4');
    axis([1 L -inf inf])

%%
%%%% D F.ZIP EPILEPTIC SUBJECT INTERICTAL were recorded from within
%%%%the epileptogenic zone during seizure free intervals %%%%
[swa,swd] = swt(Interictal(1:4096,1),Level,wname);
figure
    subplot(4,2,1)
    plot(swa(1,:))
    ylabel('swa1');
    axis([1 L -inf inf])
    title('Interictal');

    subplot(4,2,2)
    plot(swa(2,:))
    ylabel('swa2');
    axis([1 L -inf inf])
```

```
    subplot(4,2,3)
    plot(swa(2,:))
    ylabel('swa3');
    axis([1 L -inf inf])

    subplot(4,2,4)
    plot(swa(4,:))
    ylabel('swa4');
    axis([1 L -inf inf])

    subplot(4,2,5)
    plot(swd(1,:))
    ylabel('swd1');
    axis([1 L -inf inf])

    subplot(4,2,6)
    plot(swd(2,:))
    ylabel('swd2');
    axis([1 L -inf inf])

    subplot(4,2,7)
    plot(swd(3,:))
    ylabel('swd3');
    axis([1 L -inf inf])

    subplot(4,2,8)
    plot(swd(4,:))
    ylabel('swd4');
    axis([1 L -inf inf])
%% E S.ZIP EPILEPTIC SUBJECT ICTAL DURING SEIZURE
%%%%%%%%%%%%%%%%%%%%%%%%%%%%%

[swa,swd] = swt(Ictal(1:4096,1),Level,wname);
figure
    subplot(4,2,1)
    plot(swa(1,:))
    ylabel('swa1');
    axis([1 L -inf inf])
    title('Ictal');

    subplot(4,2,2)
    plot(swa(2,:))
    ylabel('swa2');
    axis([1 L -inf inf])

    subplot(4,2,3)
    plot(swa(2,:))
    ylabel('swa3');
    axis([1 L -inf inf])

    subplot(4,2,4)
    plot(swa(4,:))
    ylabel('swa4');
    axis([1 L -inf inf])

    subplot(4,2,5)
    plot(swd(1,:))
    ylabel('swd1');
    axis([1 L -inf inf])

    subplot(4,2,6)
    plot(swd(2,:))
    ylabel('swd2');
    axis([1 L -inf inf])

    subplot(4,2,7)
    plot(swd(3,:))
    ylabel('swd3');
    axis([1 L -inf inf])

    subplot(4,2,8)
    plot(swd(4,:))
    ylabel('swd4');
    axis([1 L -inf inf])
```

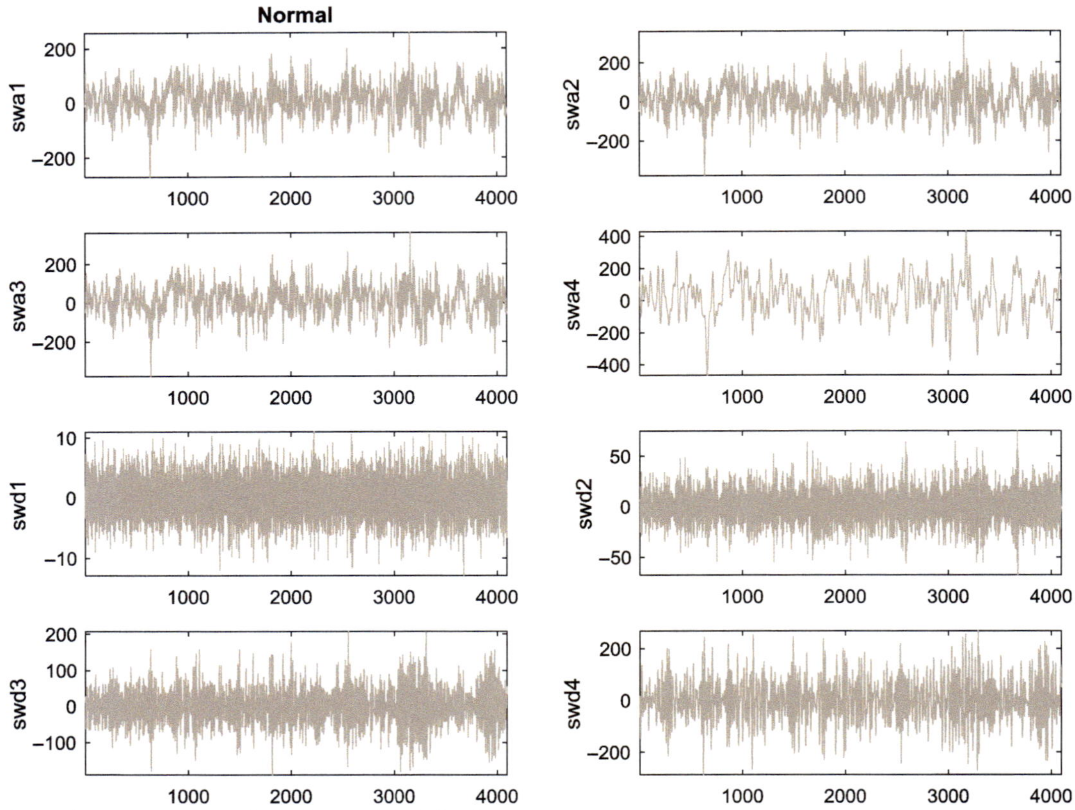

FIG. 3.37 The stationary wavelet transform of normal EEG signal.

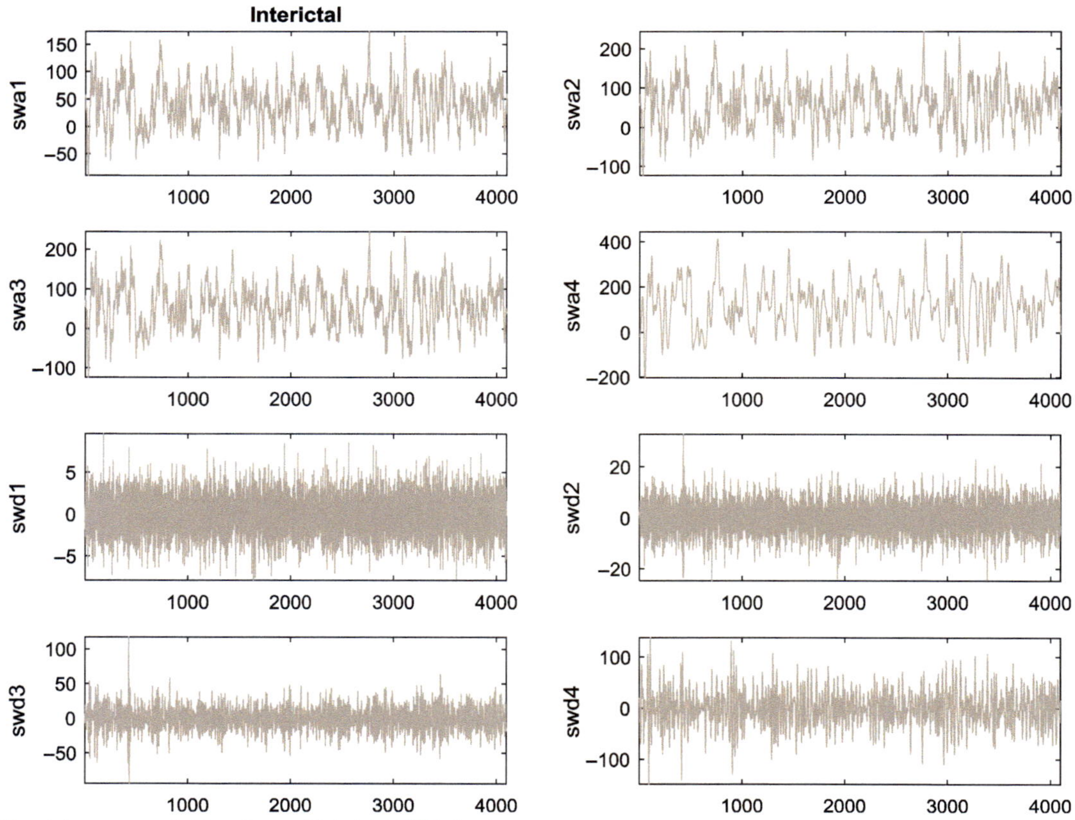

FIG. 3.38 The stationary wavelet transform of interictal EEG signal.

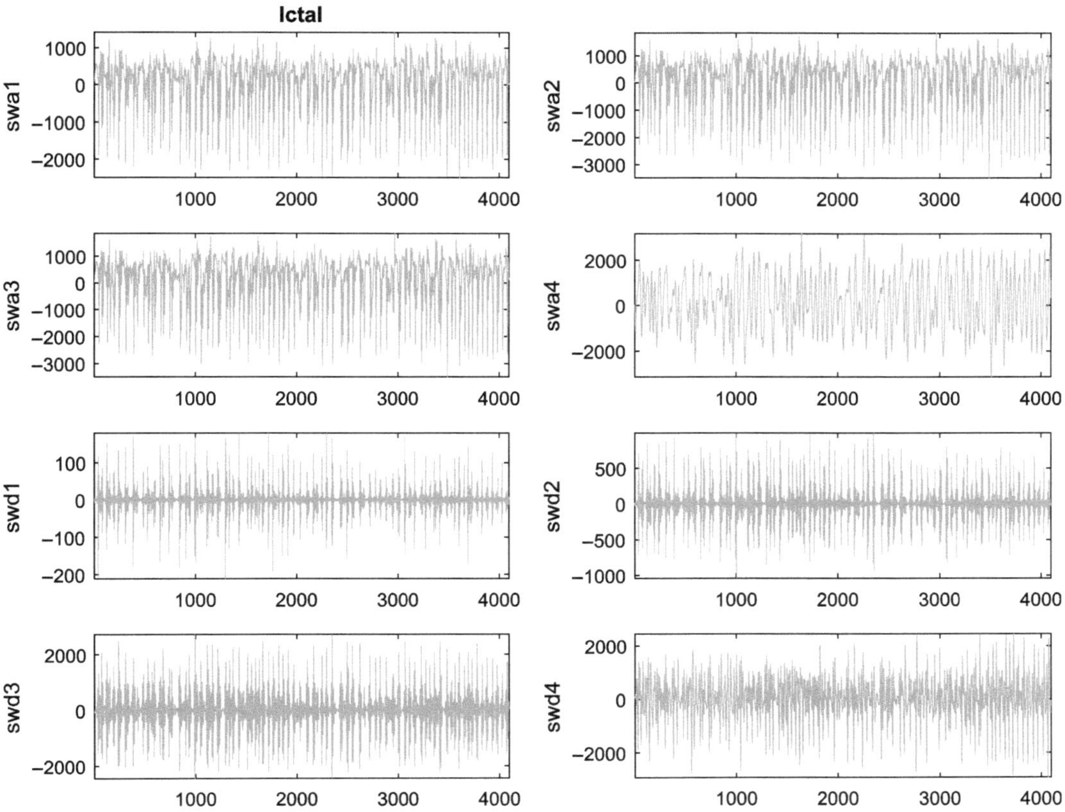

FIG. 3.39 The stationary wavelet transform of ictal EEG signal.

3.5.9 Wavelet Packet Decomposition

The wavelet transform accomplishes better time resolution of a signal by decomposing it into a set of basic functions. The wavelet packet decomposition (WPD) is known as the extension of the DWT where the low-frequency components, namely approximations, are decomposed. On the other hand, WPD utilizes both the approximations (low-frequency components) and the details (high-frequency components) (Daubechies, 1990; Learned & Willsky, 1995; Unser & Aldroubi, 1996). DWT and WPD differs from each other because WPD splits both the low- and high-frequency components into their sublevels. Accordingly, WPD yields an enhanced frequency resolution for a decomposed signal. WPD is considered as a continuous time wavelet transform, which is established at various frequencies at each scale or level. The WPD is beneficial to integrate different levels of decomposition for building the original signal (Kutlu & Kuntalp, 2012). The decomposition into WPD is realized in two steps. In the first step, the filter/downsampling cascade is modified. In the WPD structure at each level of the cascade, both branches (approximation and detailed coefficients) are further filtered and downsampled. In the second step, the tree is modified such that the most appropriate decomposition of a given signal is selected utilizing an entropy-based criterion. This procedure is known as pruning of a decomposition tree (Blinowska & Zygierewicz, 2011) (Figs. 3.40–3.42).

EXAMPLE 3.34. The following MATLAB code was used to plot the EEG signals and their WPD. You can download data from the following website: http://epileptologie-bonn.de/cms/front_content.php?idcat=193&lang=3&changelang=3

```
%Ch3_Ex71_WPD.m
% WPD of NORMAL, INTERICTAL and ICTAL EEG signals
clc
clear
%Load Sample EEG Data downloaded from the web site
%http://epileptologie-bonn.de/cms/front_content.php?idcat=193&lang=3&changelang=3
load AS_BONN_ALL_EEG_DATA_4096.mat
wname = 'db4';
L=length(Normal_Eyes_Open(1:4096,1));
%%
%%%%%%%%%%%%%%%%%%%%%%%%%%%% A Z.ZIP EYES OPEN NORMAL SUBJECT
    [wp10,wp11] = dwt(Normal_Eyes_Open(:,1),wname);
```

```
    [wp20,wp21] = dwt(wp10,wname);
    [wp22,wp23] = dwt(wp11,wname);

    [wp30,wp31] = dwt(wp20,wname);
    [wp32,wp33] = dwt(wp21,wname);
    [wp34,wp35] = dwt(wp22,wname);
    [wp36,wp37] = dwt(wp23,wname);
%%
    subplot(4,2,1)
    plot(wp30)
    ylabel('wp30');
    axis([1 500 -inf inf])
    title('Normal')

    subplot(4,2,2)
    plot(wp31)
    ylabel('wp31');
    axis([1 500 -inf inf])

    subplot(4,2,3)
    plot(wp32)
    ylabel('wp32');
    axis([1 500 -inf inf])

    subplot(4,2,4)
    plot(wp33)
    ylabel('wp33');
    axis([1 500 -inf inf])

    subplot(4,2,5)
    plot(wp34)
    ylabel('wp34');
    axis([1 500 -inf inf])

    subplot(4,2,6)
    plot(wp35)
    ylabel('wp35');
    axis([1 500 -inf inf])

    subplot(4,2,7)
    plot(wp36)
    ylabel('wp36');
    axis([1 500 -inf inf])

    subplot(4,2,8)
    plot(wp37)
    ylabel('wp37');
    axis([1 500 -inf inf])

%%
%%%% D F.ZIP EPILEPTIC SUBJECT INTERICTAL were recorded from within
%%%%the epileptogenic zone during seizure free intervals %%%%
    [wp10,wp11] = dwt(Interictal(:,1),wname);

    [wp20,wp21] = dwt(wp10,wname);
    [wp22,wp23] = dwt(wp11,wname);

    [wp30,wp31] = dwt(wp20,wname);
    [wp32,wp33] = dwt(wp21,wname);
    [wp34,wp35] = dwt(wp22,wname);
    [wp36,wp37] = dwt(wp23,wname);
%%
    figure
    subplot(4,2,1)
    plot(wp30)
    ylabel('wp30');
    axis([1 500 -inf inf])
    title('Interictal')

    subplot(4,2,2)
    plot(wp31)
```

```
    ylabel('wp31');
    axis([1 500 -inf inf])

    subplot(4,2,3)
    plot(wp32)
    ylabel('wp32');
    axis([1 500 -inf inf])

    subplot(4,2,4)
    plot(wp33)
    ylabel('wp33');
    axis([1 500 -inf inf])

    subplot(4,2,5)
    plot(wp34)
    ylabel('wp34');
    axis([1 500 -inf inf])

    subplot(4,2,6)
    plot(wp35)
    ylabel('wp35');
    axis([1 500 -inf inf])

    subplot(4,2,7)
    plot(wp36)
    ylabel('wp36');
    axis([1 500 -inf inf])

    subplot(4,2,8)
    plot(wp37)
    ylabel('wp37');
    axis([1 500 -inf inf])
%% E S.ZIP EPILEPTIC SUBJECT ICTAL DURING SEIZURE %%%%%%%%%%%%%%%%%%%%%%%
    [wp10,wp11] = dwt(Ictal(:,1),wname);

    [wp20,wp21] = dwt(wp10,wname);
    [wp22,wp23] = dwt(wp11,wname);

    [wp30,wp31] = dwt(wp20,wname);
    [wp32,wp33] = dwt(wp21,wname);
    [wp34,wp35] = dwt(wp22,wname);
    [wp36,wp37] = dwt(wp23,wname);
%%
    figure
    subplot(4,2,1)
    plot(wp30)
    ylabel('wp30');
    axis([1 500 -inf inf])
    title('Ictal')

    subplot(4,2,2)
    plot(wp31)
    ylabel('wp31');
    axis([1 500 -inf inf])

    subplot(4,2,3)
    plot(wp32)
    ylabel('wp32');
    axis([1 500 -inf inf])

    subplot(4,2,4)
    plot(wp33)
    ylabel('wp33');
    axis([1 500 -inf inf])

    subplot(4,2,5)
    plot(wp34)
    ylabel('wp34');
    axis([1 500 -inf inf])
```

```
subplot(4,2,6)
plot(wp35)
ylabel('wp35');
axis([1 500 -inf inf])

subplot(4,2,7)
plot(wp36)
ylabel('wp36');
axis([1 500 -inf inf])

subplot(4,2,8)
plot(wp37)
ylabel('wp37');
axis([1 500 -inf inf])
```

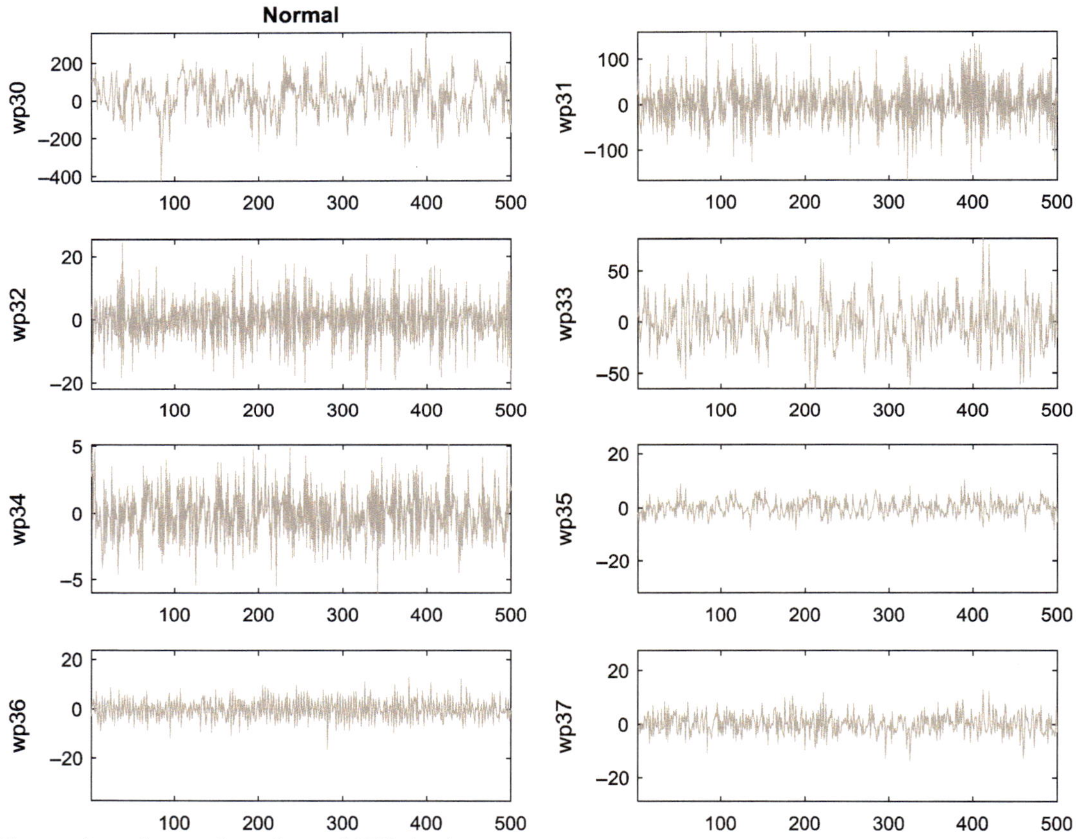

FIG. 3.40 The wavelet packet transform of normal EEG signal.

3.5.10 Dual Tree Complex Wavelet Transform

The dual tree complex wavelet transform (DTCWT) includes two real values. Because of the decimation process employed in the transform, the normal DWT is shift-variant. Hence, a fully diverse set of wavelet coefficients exist at the output and can produce with a small change in the input signal. For this reason, the DTCWT, a new wavelet transform was developed by Kingsbury (1998). Several software and hardware applications are realized for the application of the real DWT to create an effective wavelet-based algorithm, precisely for the applications when we can't use the DWT effectively. The DTCWT is a contemporary enhancement of the DWT having noteworthy new features. DTCWT is directionally selective in two dimensions and almost shift-invariant. Obviously, the multi-dimensional DTCWT is undividable yet it depends on a

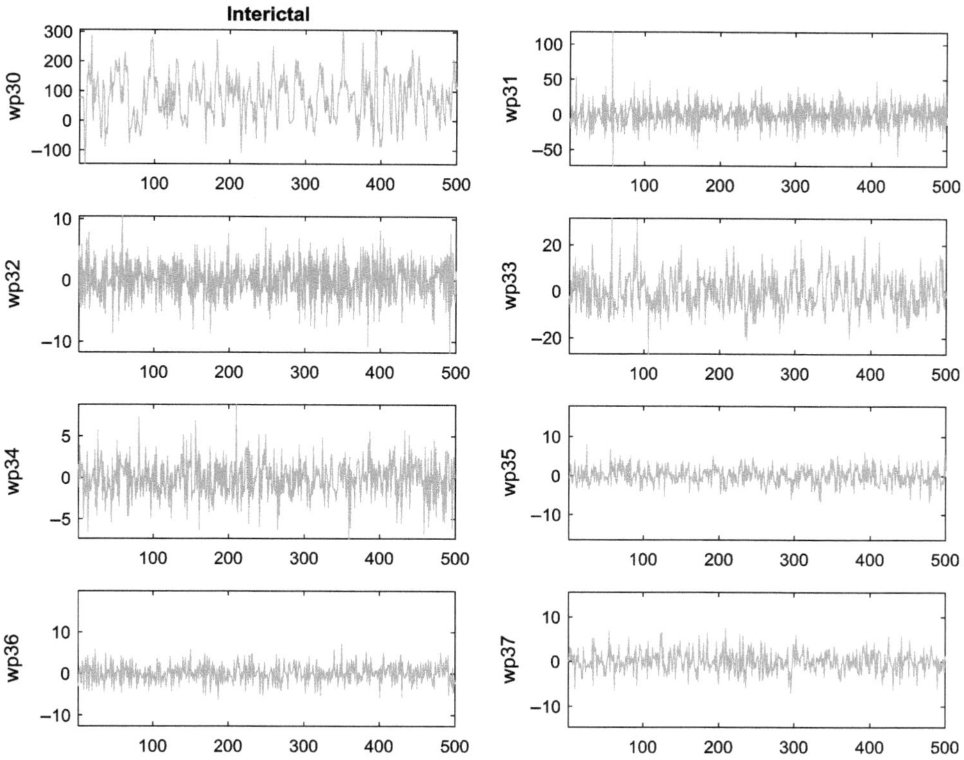

FIG. 3.41 The wavelet packet transform of interictal EEG signal.

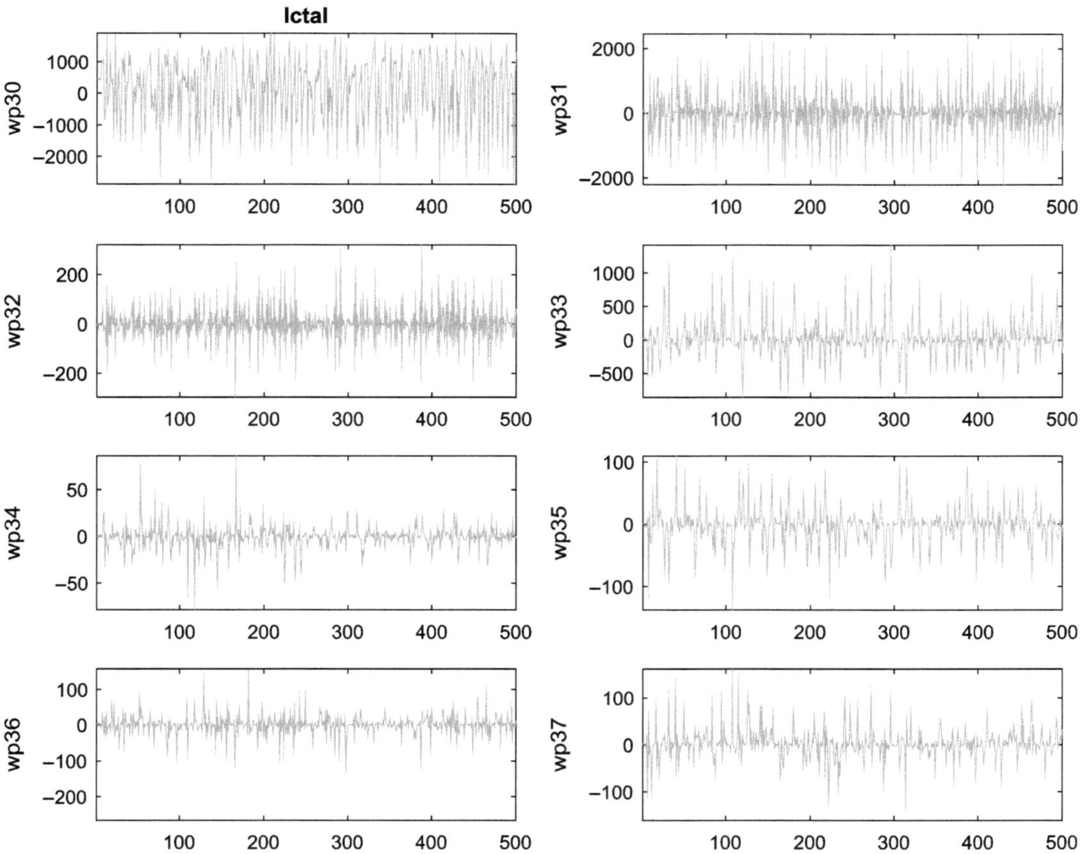

FIG. 3.42 The wavelet packet transform of ictal EEG signal.

filter bank that is computationally acceptable and distinguishable (Daubechies, 1992; Selesnick, Baraniuk, & Kingsbury, 2005). The DTCWT utilizes two real DWTs. One DWT characterizes the real part of the transform, whereas another DWT characterizes the imaginary part of the transform. The two sets of filters are developed to achieve a nearly analytic transform. The filters are real and do not need any complex arithmetic to realize the DTCWT. Moreover, the DTCWT is not a transform that is critically sampled, and it is twice as expensive in 1-D as the complete output signal rate is two times more than the input signal rate (Selesnick et al., 2005) (Figs. 3.43–3.45).

EXAMPLE 3.35. The following MATLAB code was used to plot the EEG signals and their DTCWT. You can download data from the following website: http://epileptologie-bonn.de/cms/front_content.php?idcat=193&lang=3&changelang=3

Note that the DTCWT toolbox is used for running the following MATLAB code. You can download it from the following website: http://eeweb.poly.edu/iselesni/WaveletSoftware/dt1D.html

Do not forget to set the path for the DTCWT toolbox.

```
%Ch3_Ex76_DTCWT.m
% DT-CWT of NORMAL, INTERICTAL and ICTAL EEG signals
%Note that the Dual-Tree Complex Wavelet Transform toolbox is used
%for running the following MATLAB code. You can download it from
%the following web site:
%http://eeweb.poly.edu/iselesni/WaveletSoftware/dt1D.html
%Set the path for the Dual-Tree Complex Wavelet Transform toolbox

clc
clear
%Load Sample EEG Data downloaded from the web site
%http://epileptologie-bonn.de/cms/front_content.php?idcat=193&lang=3&changelang=3
load AS_BONN_ALL_EEG_DATA_4096.mat
wname = 'db4';
L=1024; %Length of a signal
J = 4; %Number of level
[Faf, Fsf] = FSfarras; %Compute filters for stage 1
[af, sf] = dualfilt1;  %Compute filters for remaining stages

%%
% %%%%%%%%%%%%%%%%%%%%%%%%%%%%%% A Z.ZIP EYES OPEN NORMAL SUBJECT
w = dualtree(Normal_Eyes_Open(1:L,1), J, Faf, af); % Dual Tree CWT
%Signal length should be power of 2 (1024, 2048 etc)
    subplot(4,2,1)
    plot(w{1}{1})
    ylabel('w11');
    axis([1 L/2 -inf inf])
    title('Normal EEG')

    subplot(4,2,2)
    plot(w{1}{2})
    ylabel('w12');
    axis([1 L/2 -inf inf])

    subplot(4,2,3)
    plot(w{2}{1})
    ylabel('w21');
    axis([1 L/4 -inf inf])

    subplot(4,2,4)
    plot(w{2}{2})
    ylabel('w22');
    axis([1 L/4 -inf inf])

    subplot(4,2,5)
    plot(w{3}{1})
    ylabel('w31');
    axis([1 L/8 -inf inf])

    subplot(4,2,6)
    plot(w{3}{2})
```

```
    ylabel('w32');
    axis([1 L/8 -inf inf])

    subplot(4,2,7)
    plot(w{4}{1})
    ylabel('w41');
    axis([1 L/16 -inf inf])

    subplot(4,2,8)
    plot(w{2}{2})
    ylabel('w42');
    axis([1 L/16 -inf inf])

%%
%%%% D F.ZIP EPILEPTIC SUBJECT INTERICTAL were recorded from within
%%%%the epileptogenic zone during seizure free intervals %%%%
w = dualtree(Interictal(1:L,1), J, Faf, af); % Dual Tree CWT
%Signal length should be power of 2 (1024, 2048 etc)
figure
    subplot(4,2,1)
    plot(w{1}{1})
    ylabel('w11');
    axis([1 L/2 -inf inf])
    title('Interictal EEG')

    subplot(4,2,2)
    plot(w{1}{2})
    ylabel('w12');
    axis([1 L/2 -inf inf])

    subplot(4,2,3)
    plot(w{2}{1})
    ylabel('w21');
    axis([1 L/4 -inf inf])

    subplot(4,2,4)
    plot(w{2}{2})
    ylabel('w22');
    axis([1 L/4 -inf inf])

    subplot(4,2,5)
    plot(w{3}{1})
    ylabel('w31');
    axis([1 L/8 -inf inf])

    subplot(4,2,6)
    plot(w{3}{2})
    ylabel('w32');
    axis([1 L/8 -inf inf])

    subplot(4,2,7)
    plot(w{4}{1})
    ylabel('w41');
    axis([1 L/16 -inf inf])

    subplot(4,2,8)
    plot(w{2}{2})
    ylabel('w42');
    axis([1 L/16 -inf inf])

    %%
%%%% E S.ZIP EPILEPTIC SUBJECT ICTAL DURING SEIZURE %%%%%%%%%

w = dualtree(Ictal(1:L,1), J, Faf, af); % Dual Tree CWT
%Signal length should be power of 2 (1024, 2048 etc)
figure
    subplot(4,2,1)
    plot(w{1}{1})
    ylabel('w11');
    axis([1 L/2 -inf inf])
    title('Ictal EEG')

    subplot(4,2,2)
    plot(w{1}{2})
```

```
ylabel('w12');
axis([1 L/2 -inf inf])

subplot(4,2,3)
plot(w{2}{1})
ylabel('w21');
axis([1 L/4 -inf inf])

subplot(4,2,4)
plot(w{2}{2})
ylabel('w22');
axis([1 L/4 -inf inf])

subplot(4,2,5)
plot(w{3}{1})
ylabel('w31');
axis([1 L/8 -inf inf])

subplot(4,2,6)
plot(w{3}{2})
ylabel('w32');
axis([1 L/8 -inf inf])

subplot(4,2,7)
plot(w{4}{1})
ylabel('w41');
axis([1 L/16 -inf inf])

subplot(4,2,8)
plot(w{2}{2})
ylabel('w42');
axis([1 L/16 -inf inf])
```

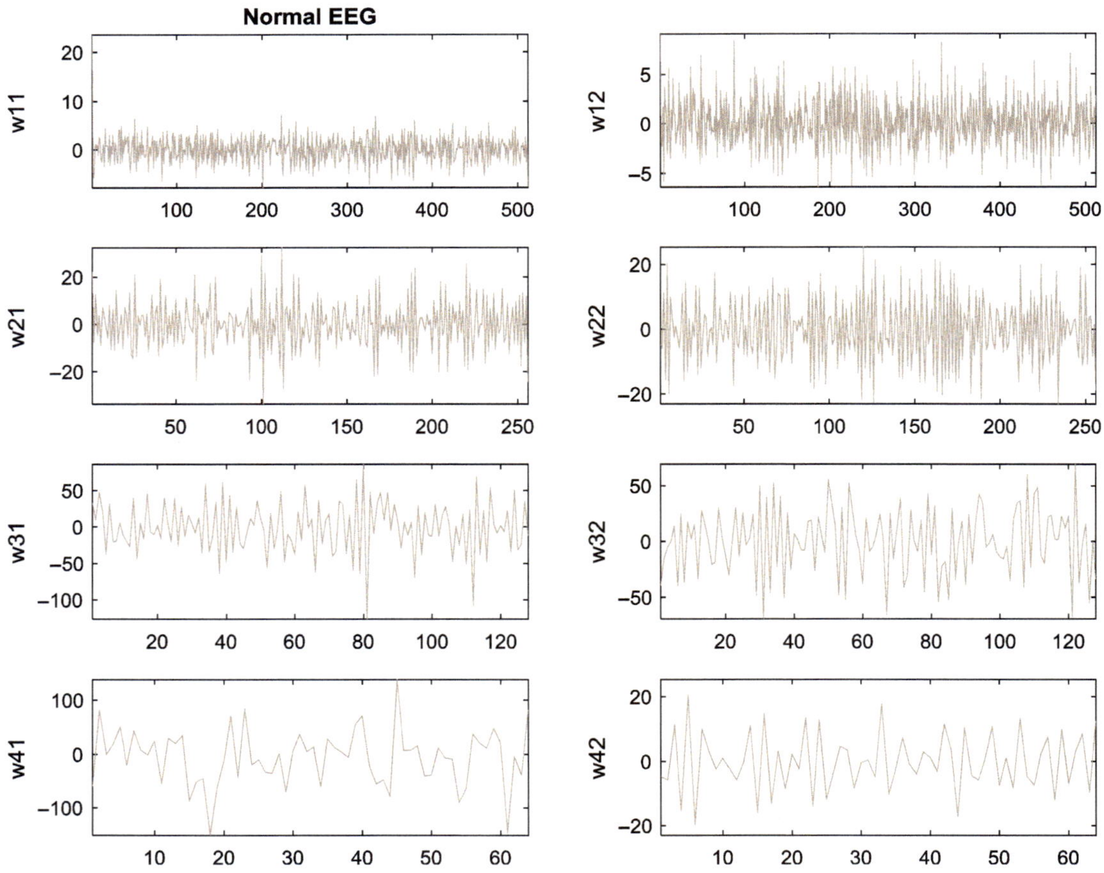

FIG. 3.43 The dual tree complex wavelet transform of normal EEG signal.

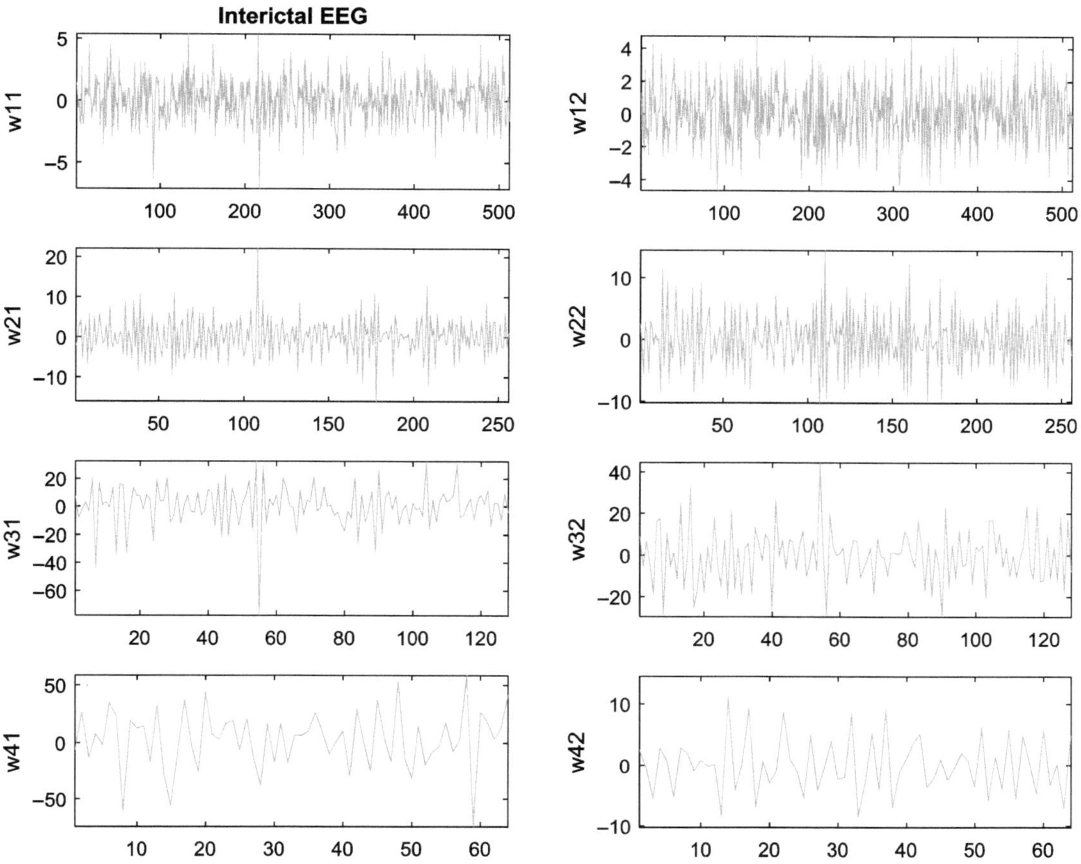

FIG. 3.44 The dual tree complex wavelet transform of interictal EEG signal.

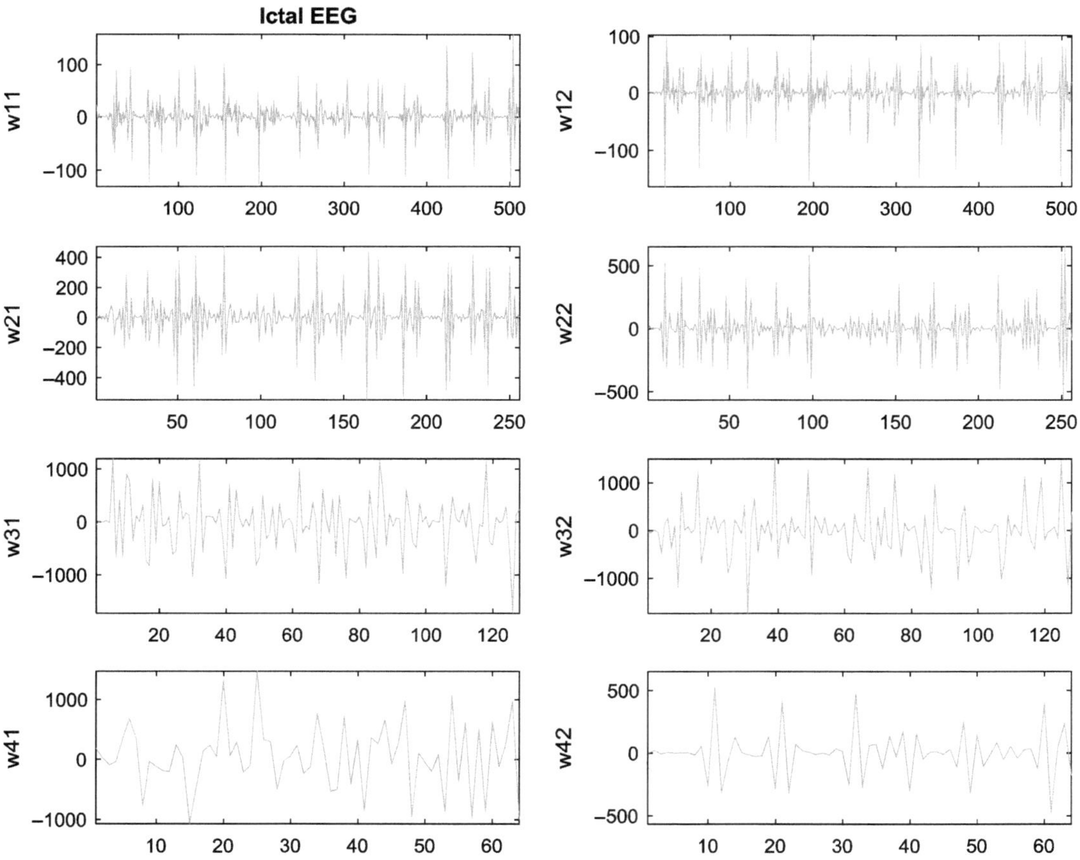

FIG. 3.45 The dual tree complex wavelet transform of ictal EEG signal.

3.5.11 Tunable *Q*-Factor Wavelet Transform

To investigate the oscillatory signals, the tunable Q-factor wavelet transform (TQWT) is considered as a robust transform. Optimally, according to the oscillatory nature of the signal, the Q-factor of a wavelet transform can be selected within that portion of signal. For example, a relatively high Q-factor should be used by the wavelet transform to analyze and process oscillatory signals such as EEG, ECG, EMG, and so on. However, a low Q-factor must be employed by the wavelet transform during the processing of signals that have little oscillatory behavior or do not have any oscillatory behavior. Most wavelet transforms do not possess much skill to adjust the wavelet's Q-factor rather than the continuous wavelet transform. The dyadic wavelet transform is proper for nonoscillatory such as piecewise-smooth because it includes a low Q-factor. Lastly, it is easy to tune discrete or approximate implementations of the continuous wavelet transform that might be formed to fulfill the perfect reconstruction property. The difference of the TQWT from the others is that it is considered as a result of the exploration of how well TQWT that depends on the structure of the discrete/dyadic wavelet transform might be enhanced (Daubechies, 1992; Selesnick, 2011) (Figs. 3.46–3.48).

EXAMPLE 3.36. The following MATLAB code is used to plot the EEG signals and their TQWT. You can download data from the following website: http://epileptologie-bonn.de/cms/front_content.php?idcat=193&lang=3&changelang=3

Note that the TQWT toolbox is used for running the following MATLAB code. You can download it from the following website: http://eeweb.poly.edu/iselesni/TQWT/index.html

Do not forget to set the path for the TQWT toolbox.

```
%Ch3_Ex81_TQWT.m
%  TQWT of NORMAL, INTERICTAL and ICTAL EEG signals
%Note that the Tunable Q-Factor Wavelet Transform (TQWT) toolbox is used
%for running the following MATLAB code. You can download it from
%the following web site:
%http://eeweb.poly.edu/iselesni/TQWT/index.html
%Set the path for the Tunable Q-Factor Wavelet Transform (TQWT) toolbox

% Q Q-factor Q>=1
% Use Q = 1 for non-oscillatory signals. A higher value of Q
% is appropriate for oscillatory signals.

% r Redundancy r > 1
% The total over-sampling rate when the TQWT is computed
% over many levels. It is recommended to select r >= 3 in
% order that the analysis/synthesis functions (wavelets) be
% well localized.

% J Number of stages (levels) of the TQWT J >= 1
% J is the number of times the two-channel filter bank is
% iterated. There will be a total of J + 1 subbands. (The
% last subband will be the low-pass subband.)

clc
clear
%Load Sample EEG Data downloaded from the web site
%http://epileptologie-bonn.de/cms/front_content.php?idcat=193&lang=3&changelang=3
load AS_BONN_ALL_EEG_DATA_4096.mat

Q = 2; r = 3; J = 7; % TQWT parameters % TQWT parameters
L=length(Normal_Eyes_Open(:,1));

%%
% %%%%%%%%%%%%%%%%%%%%%%%%%%%%%%% A Z.ZIP EYES OPEN NORMAL SUBJECT
w = tqwt_radix2(Normal_Eyes_Open(:,1),Q,r,J); % TQWT
    subplot(4,2,1)
    plot(w{1})
    ylabel('w1');
    axis([1 L/2 -inf inf])
    title('Normal EEG')
    subplot(4,2,2)
    plot(w{2})
    ylabel('w2');
    axis([1 L/2 -inf inf])
```

```
    subplot(4,2,3)
    plot(w{3})
    ylabel('w3');
    axis([1 L/2 -inf inf])

    subplot(4,2,4)
    plot(w{4})
    ylabel('w4');
    axis([1 L/2 -inf inf])

    subplot(4,2,5)
    plot(w{5})
    ylabel('w5');
    axis([1 L/4 -inf inf])

    subplot(4,2,6)
    plot(w{6})
    ylabel('w6');
    axis([1 L/4 -inf inf])

    subplot(4,2,7)
    plot(w{7})
    ylabel('w7');
    axis([1 L/4 -inf inf])

    subplot(4,2,8)
    plot(w{8})
    ylabel('w8');
    axis([1 L/4 -inf inf])

%%
%%%% D F.ZIP EPILEPTIC SUBJECT INTERICTAL were recorded from within
%%%%the epileptogenic zone during seizure free intervals %%%%
w = tqwt_radix2(Interictal(:,1),Q,r,J); % TQWT
figure
    subplot(4,2,1)
    plot(w{1})
    ylabel('w1');
    axis([1 L/2 -inf inf])
    title('Interictal EEG')

    subplot(4,2,2)
    plot(w{2})
    ylabel('w2');
    axis([1 L/2 -inf inf])

    subplot(4,2,3)
    plot(w{3})
    ylabel('w3');
    axis([1 L/2 -inf inf])

    subplot(4,2,4)
    plot(w{4})
    ylabel('w4');
    axis([1 L/2 -inf inf])

    subplot(4,2,5)
    plot(w{5})
    ylabel('w5');
    axis([1 L/4 -inf inf])

    subplot(4,2,6)
    plot(w{6})
    ylabel('w6');
    axis([1 L/4 -inf inf])

    subplot(4,2,7)
    plot(w{7})
    ylabel('w7');
    axis([1 L/4 -inf inf])
```

```
    subplot(4,2,8)
    plot(w{8})
    ylabel('w8');
    axis([1 L/4 -inf inf])

%% E S.ZIP EPILEPTIC SUBJECT ICTAL DURING SEIZURE
%%%%%%%%%%%%%%%%%%%%%%%%%%%%%
w = tqwt_radix2(Ictal(:,1),Q,r,J); % TQWT
figure
    subplot(4,2,1)
    plot(w{1})
    ylabel('w1');
    axis([1 L/2 -inf inf])
    title('Ictal EEG')

    subplot(4,2,2)
    plot(w{2})
    ylabel('w2');
    axis([1 L/2 -inf inf])

    subplot(4,2,3)
    plot(w{3})
    ylabel('w3');
    axis([1 L/2 -inf inf])

    subplot(4,2,4)
    plot(w{4})
    ylabel('w4');
    axis([1 L/2 -inf inf])

    subplot(4,2,5)
    plot(w{5})
    ylabel('w5');
    axis([1 L/4 -inf inf])

    subplot(4,2,6)
    plot(w{6})
    ylabel('w6');
    axis([1 L/4 -inf inf])

    subplot(4,2,7)
    plot(w{7})
    ylabel('w7');
    axis([1 L/4 -inf inf])

    subplot(4,2,8)
    plot(w{8})
    ylabel('w8');
    axis([1 L/4 -inf inf])
```

3.5.12 Flexible Analytic Wavelet Transform

Analytic wavelet transform with flexible time-frequency (FAWT) (Bayram, 2013; Kumar, Pachori, & Acharya, 2016; Kumar, Pachori, & Acharya, 2017;Zhang, Dong, et al., 2015 ; Zhang, Li, et al., 2015) is an efficient time-frequency method to analyze biomedical signals (Kumar et al., 2016; Kumar et al., 2017). The FAWT employs the Hilbert transform pairs of atoms and yields flexibility to regulate the quality factor (QF), redundancy, and dilation factor. Moreover, it is utilized in analyzing the signals with adjustable parameters. The *j*th level of FAWT can be achieved by employing iterative filter bank structure. The decomposition in this technique at each level delivers two channels, and one channel related to the low-pass and high-pass channels (Bayram, 2013; Gupta, Priya, Yadav, Pachori, & Acharya, 2017)

The FAWT is a rational-dilation wavelet transform, where the dilation factor, quality (Q)-factor, and redundancy (R) will be easily adjusted. The FAWT utilizes the fractional sampling rate in the low-pass and high-pass channels. Furthermore, it delivers analytic bases by dividing

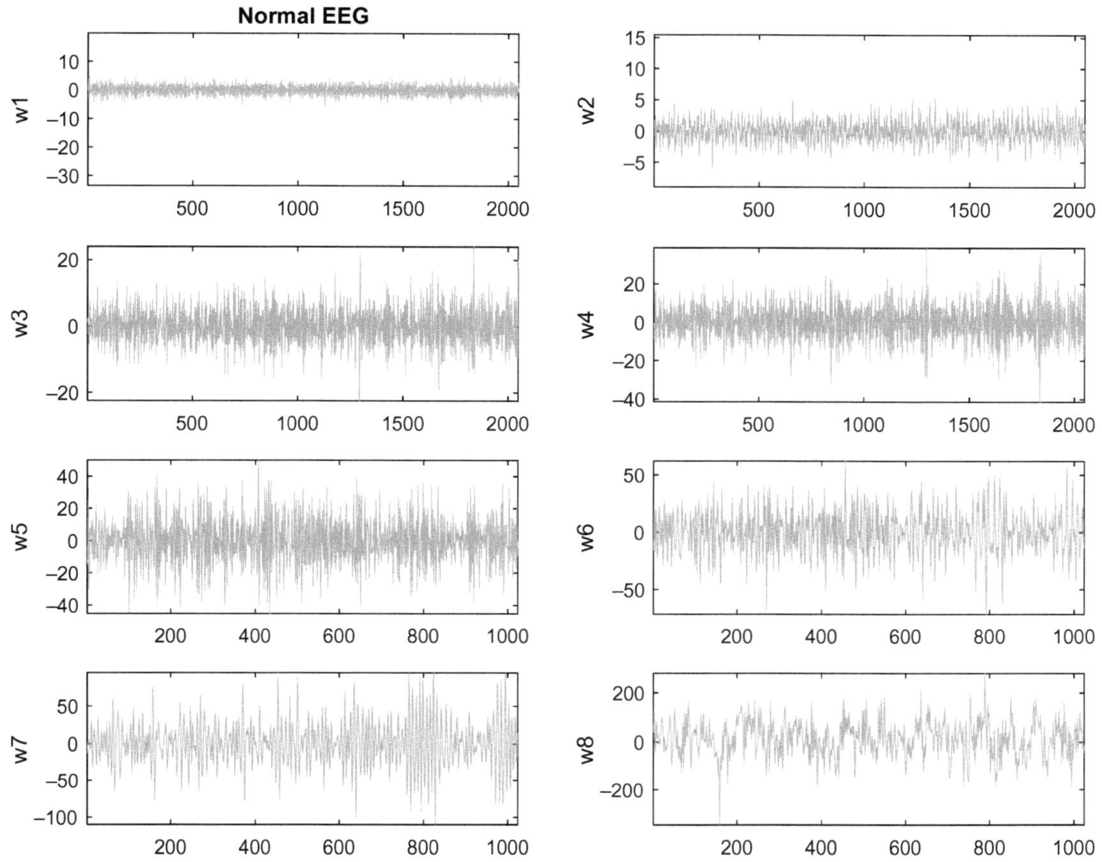

FIG. 3.46 The tunable Q-factor wavelet transform of normal EEG signal.

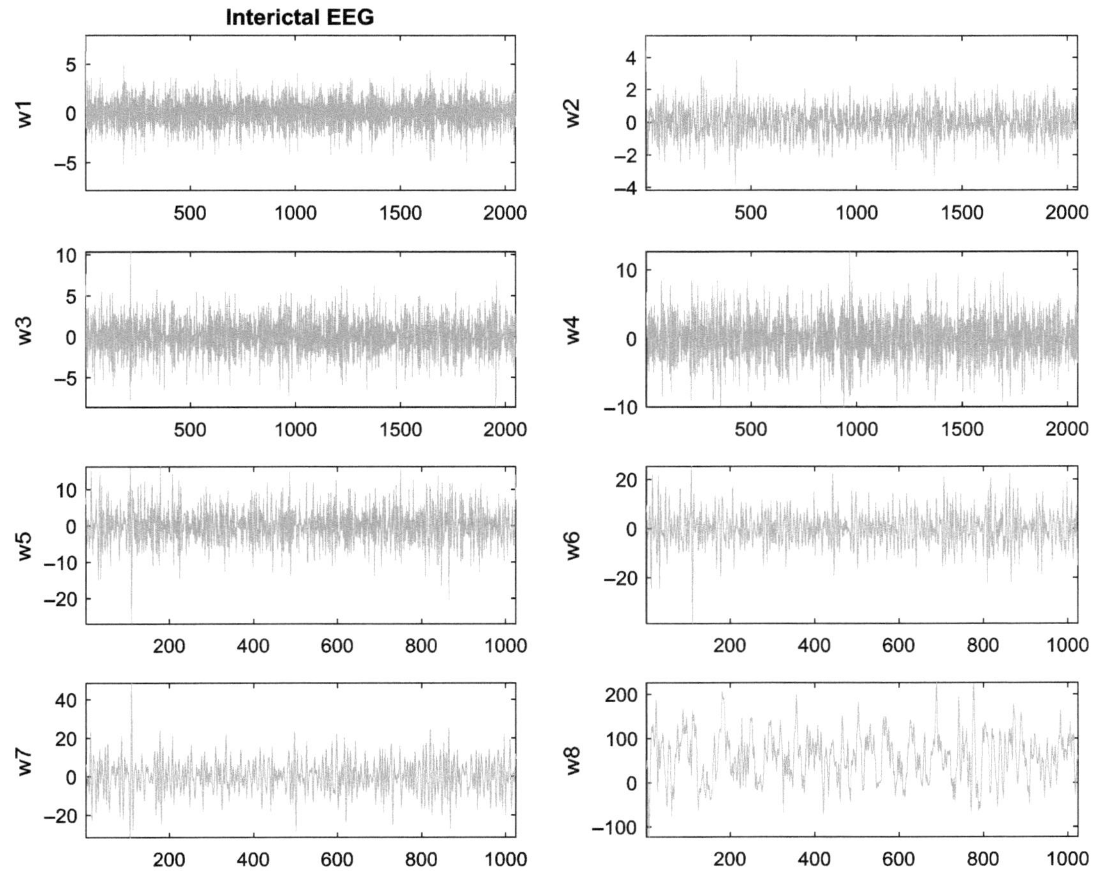

FIG. 3.47 The tunable Q-factor wavelet transform of interictal EEG signal.

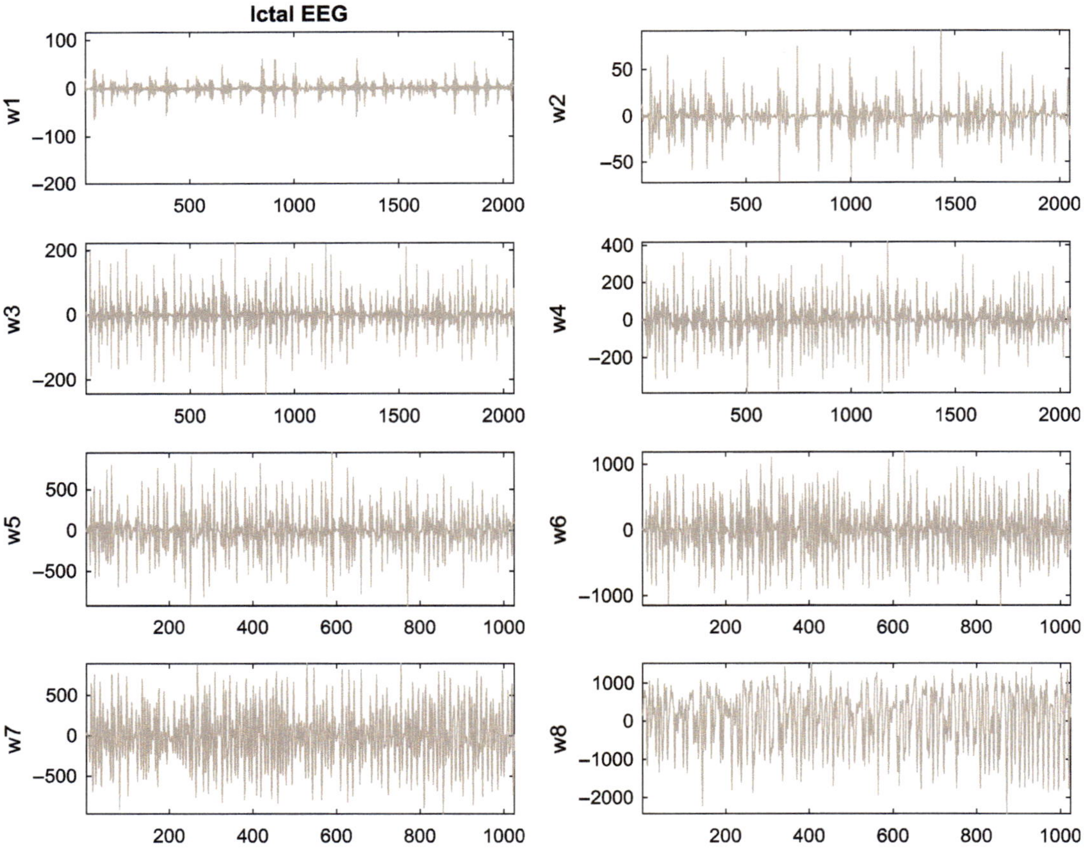

FIG. 3.48 The tunable Q-factor wavelet transform of ictal EEG signal.

the negative and positive frequencies in high-pass channels (Bayram, 2013). By utilizing analytic bases and fractional sampling rate, FAWT delivers tunable oscillatory bases, shift-invariance, and flexible time-frequency coverage (Zhang, Dong, et al., 2015; Zhang, Li, et al., 2015). These properties make FAWT appropriate to analyze the oscillatory and transient components of the biomedical signals. Q-factor adjusts the frequency resolution of FAWT. The high Q-factor delivers finer filter banks to analyze the biomedical signals in the frequency domain. For fixed dilation and Q-factors, the redundancy adjusts the position of the wavelet. The mathematical expressions of the filter banks for FAWT yields expression for perfect reconstruction condition of FAWT. The relation among the Q-factor and β is as follows (Bayram, 2013):

$$Q = \frac{2-\beta}{\beta} \tag{3.93}$$

Redundancy is represented as (Bayram, 2013):

$$R = \left(\frac{C}{D}\right)\frac{1}{(1-A/B)} \tag{3.94}$$

where A and B are used to adjust the sampling rate of the low-pass channel, and C and D are used to adjust the sampling rate of high-pass channels (Kumar et al., 2017) (Figs. 3.49–3.51).

EXAMPLE 3.37. The following MATLAB code was used to plot the EEG signals and their FAWT. You can download data from the following website: http://epileptologie-bonn.de/cms/front_content.php?idcat=193&lang=3&changelang=3

Note that the FAWT toolbox is used for running the following MATLAB code. You can download it from the following website: http://web.itu.edu.tr/ibayram/AnDWT/

Do not forget to set the path for the FAWT toolbox.

```
%Ch3_Ex86_FAWT.m
%Flexible analytic wavelet transform (FAWT) of NORMAL, INTERICTAL and ICTAL EEG signals
%Note that the Flexible analytic wavelet transform (FAWT) toolbox is used
%for running the following MATLAB code. You can download it from
%the following web site:
%http://web.itu.edu.tr/ibayram/AnDWT/
%Set the path for the Flexible analytic wavelet transform (FAWT) toolbox
clc
```

```
clear
%Load Sample EEG Data downloaded from the web site
%http://epileptologie-bonn.de/cms/front_content.php?idcat=193&lang=3&changelang=3
load AS_BONN_ALL_EEG_DATA_4096.mat

Level = 5;
wname = 'db4';
L=length(Normal_Eyes_Open(1:4096,1));
p = 5;
q = 6;
r = 1;
s = 2;
bet = 0.8*r/s;
J = 10;
Par = 200;

%%
% %%%%%%%%%%%%%%%%%%%%%%%%%%%%% A Z.ZIP EYES OPEN NORMAL SUBJECT

[Fp] = CreateFilters(L,p,q,r,s,bet,J,Par);%filters for the positive chirp parameter
w = RAnDwt(Normal_Eyes_Open(1:4096,1),p,q,r,s,J,Fp);
    subplot(4,2,1)
    plot(abs(w{1,1}))
    ylabel('w11');
    axis([1 L/4 -inf inf])
    title('Normal EEG')

    subplot(4,2,2)
    plot(abs(w{1,2}))
    ylabel('w12');
    axis([1 L/4 -inf inf])

    subplot(4,2,3)
    plot(abs(w{2,1}))
    ylabel('w21');
    axis([1 L/8 -inf inf])

    subplot(4,2,4)
    plot(abs(w{2,2}))
    ylabel('w22');
    axis([1 L/8 -inf inf])

    subplot(4,2,5)
    plot(abs(w{3,1}))
    ylabel('w31');
    axis([1 L/8 -inf inf])

    subplot(4,2,6)
    plot(abs(w{3,2}))
    ylabel('w32');
    axis([1 L/8 -inf inf])

    subplot(4,2,7)
    plot(abs(w{4,1}))
    ylabel('w41');
    axis([1 L/8 -inf inf])

    subplot(4,2,8)
    plot(abs(w{4,2}))
    ylabel('w42');
    axis([1 L/8 -inf inf])

%%
%%%% D F.ZIP EPILEPTIC SUBJECT INTERICTAL were recorded from within
%%%%the epileptogenic zone during seizure free intervals %%%%
figure
[Fp] = CreateFilters(L,p,q,r,s,bet,J,Par);%filters for the positive chirp parameter
w = RAnDwt(Interictal(1:4096,1),p,q,r,s,J,Fp);
    subplot(4,2,1)
    plot(abs(w{1,1}))
    ylabel('w11');
    axis([1 L/4 -inf inf])
    title('Interictal EEG')
```

```
    subplot(4,2,2)
    plot(abs(w{1,2}))
    ylabel('w12');
    axis([1 L/4 -inf inf])

    subplot(4,2,3)
    plot(abs(w{2,1}))
    ylabel('w21');
    axis([1 L/8 -inf inf])

    subplot(4,2,4)
    plot(abs(w{2,2}))
    ylabel('w22');
    axis([1 L/8 -inf inf])

    subplot(4,2,5)
    plot(abs(w{3,1}))
    ylabel('w31');
    axis([1 L/8 -inf inf])

    subplot(4,2,6)
    plot(abs(w{3,2}))
    ylabel('w32');
    axis([1 L/8 -inf inf])

    subplot(4,2,7)
    plot(abs(w{4,1}))
    ylabel('w41');
    axis([1 L/8 -inf inf])

    subplot(4,2,8)
    plot(abs(w{4,2}))
    ylabel('w42');
    axis([1 L/8 -inf inf])
%% E S.ZIP EPILEPTIC SUBJECT ICTAL DURING SEIZURE %%%%%%%%%%%%
figure
[Fp] = CreateFilters(L,p,q,r,s,bet,J,Par);%filters for the positive chirp parameter
w = RAnDwt(Ictal(1:4096,1),p,q,r,s,J,Fp);
    subplot(4,2,1)
    plot(abs(w{1,1}))
    ylabel('w11');
    axis([1 L/4 -inf inf])
    title('Ictal EEG')

    subplot(4,2,2)
    plot(abs(w{1,2}))
    ylabel('w12');
    axis([1 L/4 -inf inf])

    subplot(4,2,3)
    plot(abs(w{2,1}))
    ylabel('w21');
    axis([1 L/8 -inf inf])

    subplot(4,2,4)
    plot(abs(w{2,2}))
    ylabel('w22');
    axis([1 L/8 -inf inf])

    subplot(4,2,5)
    plot(abs(w{3,1}))
    ylabel('w31');
    axis([1 L/8 -inf inf])

    subplot(4,2,6)
    plot(abs(w{3,2}))
    ylabel('w32');
    axis([1 L/8 -inf inf])

    subplot(4,2,7)
    plot(abs(w{4,1}))
```

```
ylabel('w41');
axis([1 L/8 -inf inf])

subplot(4,2,8)
plot(abs(w{4,2}))
ylabel('w42');
axis([1 L/8 -inf inf])
```

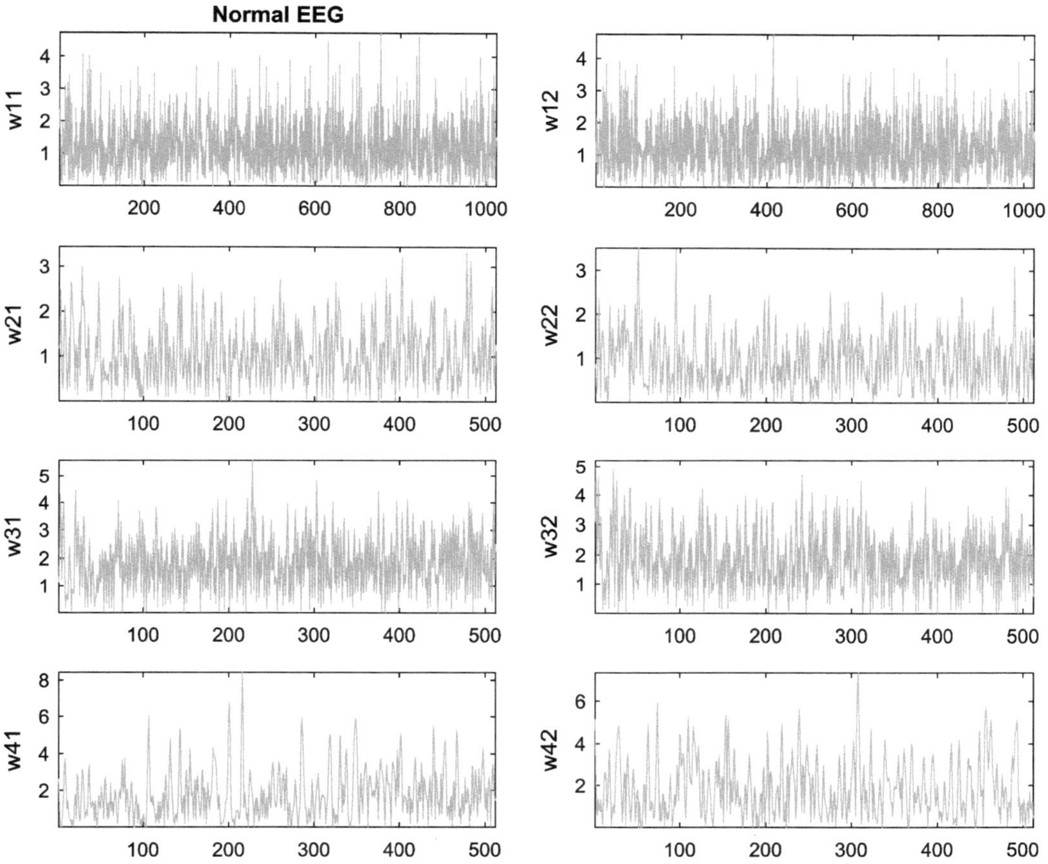

FIG. 3.49 The flexible analytic wavelet transform of normal EEG signal.

3.5.13 Empirical Wavelet Transform

Empirical wavelet transform (EWT) was developed by Gilles (2013) as a time-frequency technique to decompose signals using the adaptive wavelet based on information content. EWT first detects local maxima of the Fourier spectrum of the signal, then decomposes the spectrum based on the detected maxima, and lastly constructs the corresponding wavelet filter bank. It works in the following three steps:

1. Determine the frequency components of the applied signal using FFT.
2. The different modes are extracted by obtaining proper segmentation of the Fourier spectrum.
3. Apply scaling and wavelet functions corresponding to each detected segment. Segmentation of the Fourier spectrum is the most important step that provides the adaptability to this technique according to the analyzed signal.

The empirical wavelets are defined as band-pass filters. Littlewood-Paley and Mayer's wavelets are used as band-pass filters with the empirical wavelets $\Psi(w)$, and the empirical scaling function $\phi(w)$ can be expressed as follows (Gilles, 2013),

$$\Psi_n(w) = \begin{cases} 1 & \text{if } (1+\gamma)\Omega_n \le |w| \le (1-\gamma)\Omega_{n+1} \\ \cos\left[\frac{\pi}{2}\beta(\gamma, \Omega_{n+1})\right] & \text{if } (1-\gamma)\Omega_{n+1} \le |w| \le (1+\gamma)\Omega_{n+1} \\ \sin\left[\frac{\pi}{2}\beta(\gamma, \Omega_n)\right] & \text{if } (1-\gamma)\Omega_n \le |w| \le (1+\gamma)\Omega_n \\ 0 & \text{if otherwise} \end{cases} \tag{3.95}$$

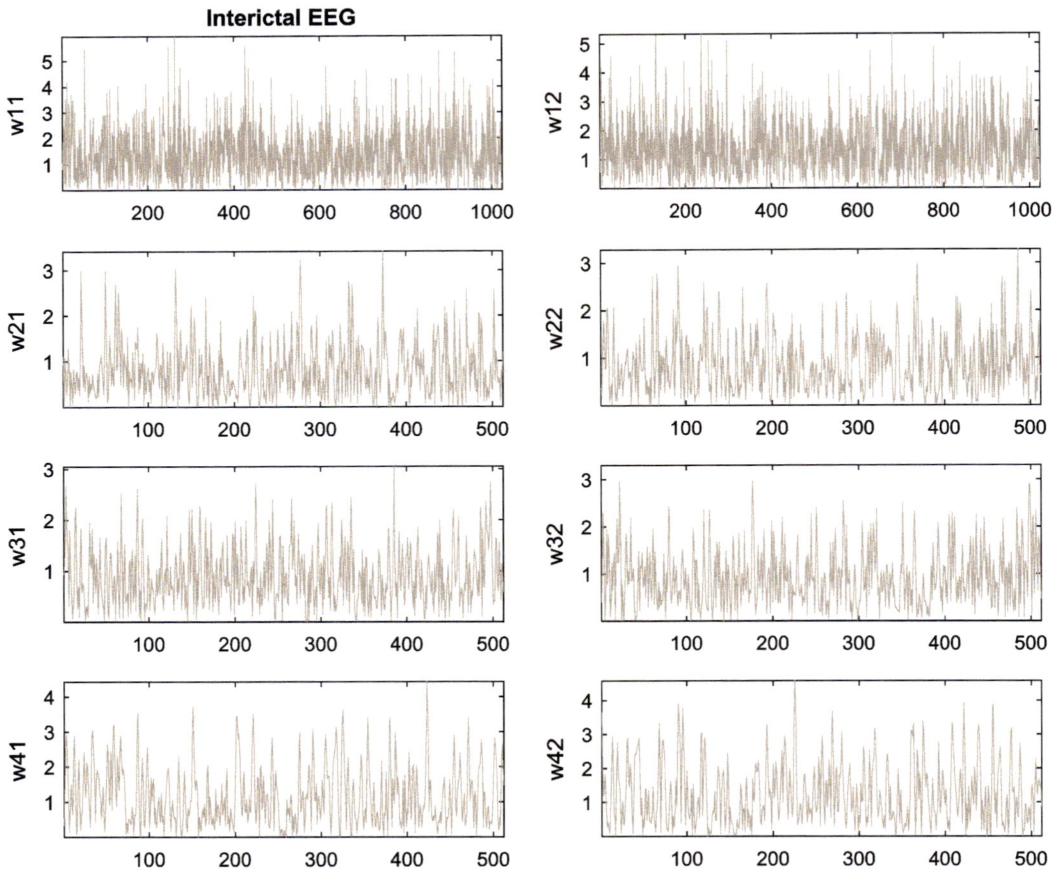

FIG. 3.50 The flexible analytic wavelet transform of interictal EEG signal.

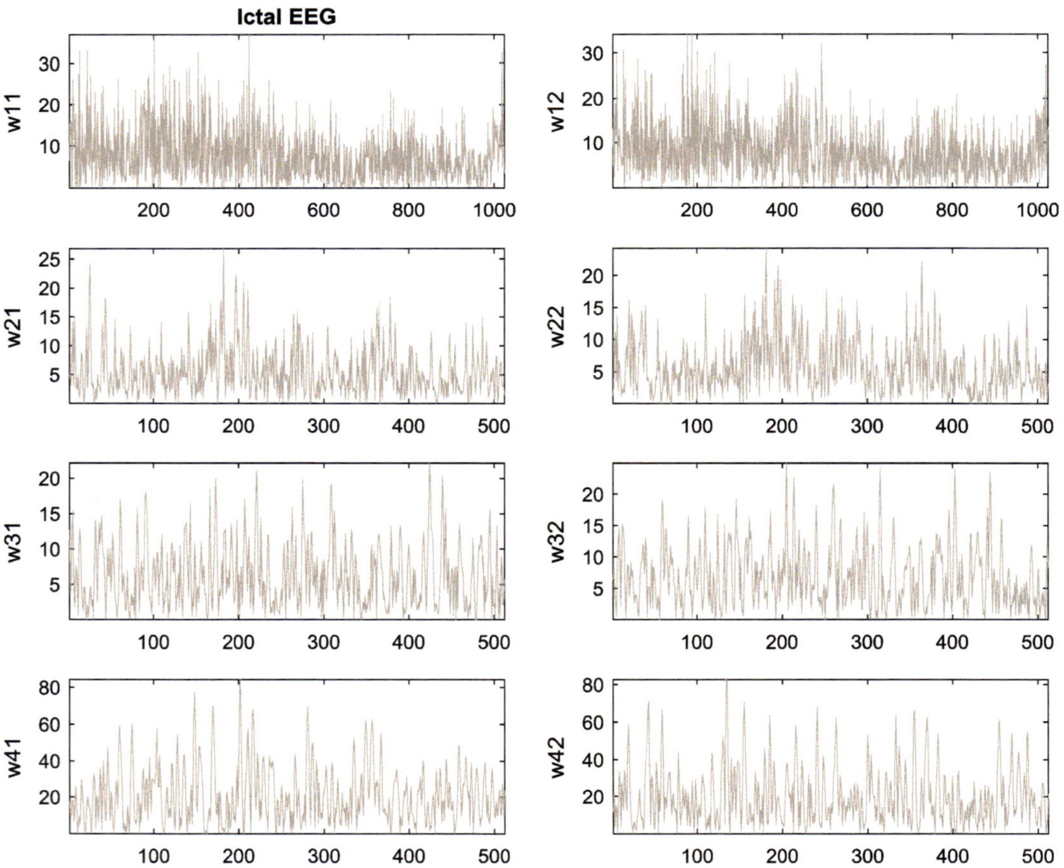

FIG. 3.51 The flexible analytic wavelet transform of interictal EEG signal.

and

$$\varphi_1(w) = \begin{cases} 1 & \text{if } |w| \le (1-\gamma)\Omega_1 \\ \cos\left[\frac{\pi}{2}\beta(\gamma, \Omega_1)\right] & \text{if } (1-\gamma)\Omega_1 \le |w| \le (1+\gamma)\Omega_1 \\ 0 & \text{if otherwise} \end{cases} \quad (3.96)$$

The function $\beta(x)$ is an arbitrary $C^k[(0,1)]$ function defined as

$$\beta(x) \begin{cases} 1 & \text{if } x \ge 1 \\ 0 & \text{if } x \le 0 \\ \beta(x) + \beta(1-x) = 1 & \text{if} \forall x \in [0, 1] \end{cases} \quad (3.97)$$

where $x = \frac{1}{2\gamma\Omega_n}(|\Omega| - (1-\gamma)\Omega_n)$, where, γ is a parameter to ensure no overlap between two consecutive transition areas. Many functions satisfy these properties; one widely used in the literature is

$$\beta(x) = x^4\left(35 - 84x + 70x^2 - 20x^3\right) \quad (3.98)$$

The EWT can be defined for the classical wavelet transform. The detailed coefficients are obtained by the inner product with the empirical scaling function as given here:

$$W_x(n,t) = x(t), \Psi_n = IFFT(X(w) \times \Psi_n(w)) \quad (3.99)$$

and the approximation coefficients are obtained by the inner product with the scaling function as given here:

$$W_x(1,t) = x(t), \varphi_1 = IFFT(X(w) \times \varphi_1(w)) \quad (3.100)$$

where $X(w)$ represents the FFT of the $x(t)$ signal (Figs. 3.52–3.55).

EXAMPLE 3.38. The following MATLAB code is used to plot the ECG signals and their EWT. You can download data from the following website: https://www.physionet.org/pn3/incartdb/Note that the EWT toolbox is used for running the following MATLAB code. You can download it from the following website: https://www.mathworks.com/matlabcentral/fileexchange/42141-empirical-wavelet-transforms

Do not forget to set the path for the EWT toolbox.

```
%Ch3_Ex93_EWT.m
%Empirical Wavelet Transform (EWT) of NORMAL, APC, PVC and RBBB ECG signals
%Note that the Empirical Wavelet Transform (EWT) toolbox is used
%for running the following MATLAB code. You can download it from
%the following web site:
%https://www.mathworks.com/matlabcentral/fileexchange/42141-empirical-wavelet-transforms
%Set the path for the Empirical Wavelet Transform (EWT) toolbox

clc
clear
%Load Sample ECG Data downloaded from the web site
%https://www.physionet.org/pn3/incartdb/
load PetersburgECG.mat

params.SamplingRate = 257; % sampling rate
% Choose the wanted global trend removal (none,plaw,poly,morpho,tophat)
params.globtrend = 'none';
params.degree=1; % degree for the polynomial interpolation
% Choose the wanted regularization (none,gaussian,avaerage,closing)
params.reg = 'none';
params.lengthFilter = 10;
params.sigmaFilter = 1.5;

% Choose the wanted detection method (locmax,locmaxmin,ftc,
% adaptive,adaptivereg,scalespace)
params.detect = 'scalespace';
params.typeDetect='otsu'; %for scalespace:otsu,halfnormal,empiricallaw,mean,kmeans
params.N = 3; % maximum number of bands
params.completion = 0; % choose if you want to force to have params.N modes
                       % in case the algorithm found less ones (0 or 1)

params.InitBounds = [2 25];

% Perform the detection on the log spectrum instead the spectrum
params.log=0;

% Choose the results you want to display (Show=1, Not Show=0)
Bound=0;   % Display the detected boundaries on the spectrum
Comp=0;    % Display the EWT components
Rec=0;     % Display the reconstructed signal
```

```
TFplane=0; % Display the time-frequency plane (by using the Hilbert
           % transform). You can decrease the frequency resolution by
           % changing the subresf variable below.
Demd=0;    % Display the Hilbert-Huang transform (YOU NEED TO HAVE
           % FLANDRIN'S EMD TOOLBOX)

subresf=2;
InitBounds = params.InitBounds;

%%
% %%%%%%%%%%%%%%%%%%%%%%%%%%%%%% NORMAL ECG Signal
[ewt,mfb,boundaries]=EWT1D(ECGN(:,1),params); % ewt
Hilb = EWT_InstantaneousComponents(ewt,boundaries); %hilbert transform
 %%
    subplot(4,2,1)
    plot(abs(Hilb{1}{1}))
    ylabel('EWT11');
    title('Normal ECG')

    subplot(4,2,2)
    plot(abs(Hilb{1}{2}))
    ylabel('EWT12');

    subplot(4,2,3)
    plot(abs(Hilb{2}{1}))
    ylabel('EWT21');

    subplot(4,2,4)
    plot(abs(Hilb{2}{2}))
    ylabel('EWT22');

    subplot(4,2,5)
    plot(abs(Hilb{3}{1}))
    ylabel('EWT31');

    subplot(4,2,6)
    plot(abs(Hilb{3}{2}))
    ylabel('EWT32');

    subplot(4,2,7)
    plot(abs(Hilb{4}{1}))
    ylabel('EWT41');

    subplot(4,2,8)
    plot(abs(Hilb{4}{2}))
    ylabel('EWT42');

%%
%%%% APC ECG Signal
[ewt,mfb,boundaries]=EWT1D(ECGAPC(:,1),params); % ewt
Hilb = EWT_InstantaneousComponents(ewt,boundaries); %hilbert transform
  %%
figure
    subplot(4,2,1)
    plot(abs(Hilb{1}{1}))
    ylabel('EWT11');
    title('APC ECG')

    subplot(4,2,2)
    plot(abs(Hilb{1}{2}))
    ylabel('EWT12');

    subplot(4,2,3)
    plot(abs(Hilb{2}{1}))
    ylabel('EWT21');

    subplot(4,2,4)
    plot(abs(Hilb{2}{2}))
    ylabel('EWT22');

    subplot(4,2,5)
    plot(abs(Hilb{3}{1}))
    ylabel('EWT31');
```

```
    subplot(4,2,6)
    plot(abs(Hilb{3}{2}))
    ylabel('EWT32');

    subplot(4,2,7)
    plot(abs(Hilb{4}{1}))
    ylabel('EWT41');

    subplot(4,2,8)
    plot(abs(Hilb{4}{2}))
    ylabel('EWT42');

   %%
%%%% PVC ECG Signal
[ewt,mfb,boundaries]=EWT1D(ECGPVC(:,1),params); % ewt
Hilb = EWT_InstantaneousComponents(ewt,boundaries); %hilbert transform
 %%
figure
    subplot(4,2,1)
    plot(abs(Hilb{1}{1}))
    ylabel('EWT11');
    title('PVC ECG')

    subplot(4,2,2)
    plot(abs(Hilb{1}{2}))
    ylabel('EWT12');

    subplot(4,2,3)
    plot(abs(Hilb{2}{1}))
    ylabel('EWT21');

    subplot(4,2,4)
    plot(abs(Hilb{2}{2}))
    ylabel('EWT22');

    subplot(4,2,5)
    plot(abs(Hilb{3}{1}))
    ylabel('EWT31');

    subplot(4,2,6)
    plot(abs(Hilb{3}{2}))
    ylabel('EWT32');

    subplot(4,2,7)
    plot(abs(Hilb{4}{1}))
    ylabel('EWT41');

    subplot(4,2,8)
    plot(abs(Hilb{4}{2}))
    ylabel('EWT42');

%%
%%%% RBBB ECG Signal
[ewt,mfb,boundaries]=EWT1D(ECGRBBB(:,1),params); % ewt
Hilb = EWT_InstantaneousComponents(ewt,boundaries); %hilbert transform
 %%
figure
    subplot(4,2,1)
    plot(abs(Hilb{1}{1}))
    ylabel('EWT11');
    title('RBBB ECG')

    subplot(4,2,2)
    plot(abs(Hilb{1}{2}))
    ylabel('EWT12');

    subplot(4,2,3)
    plot(abs(Hilb{2}{1}))
    ylabel('EWT21');

    subplot(4,2,4)
    plot(abs(Hilb{2}{2}))
    ylabel('EWT22');
```

```
subplot(4,2,5)
plot(abs(Hilb{3}{1}))
ylabel('EWT31');

subplot(4,2,6)
plot(abs(Hilb{3}{2}))
ylabel('EWT32');

subplot(4,2,7)
plot(abs(Hilb{4}{1}))
ylabel('EWT41');

subplot(4,2,8)
plot(abs(Hilb{4}{2}))
ylabel('EWT42');
```

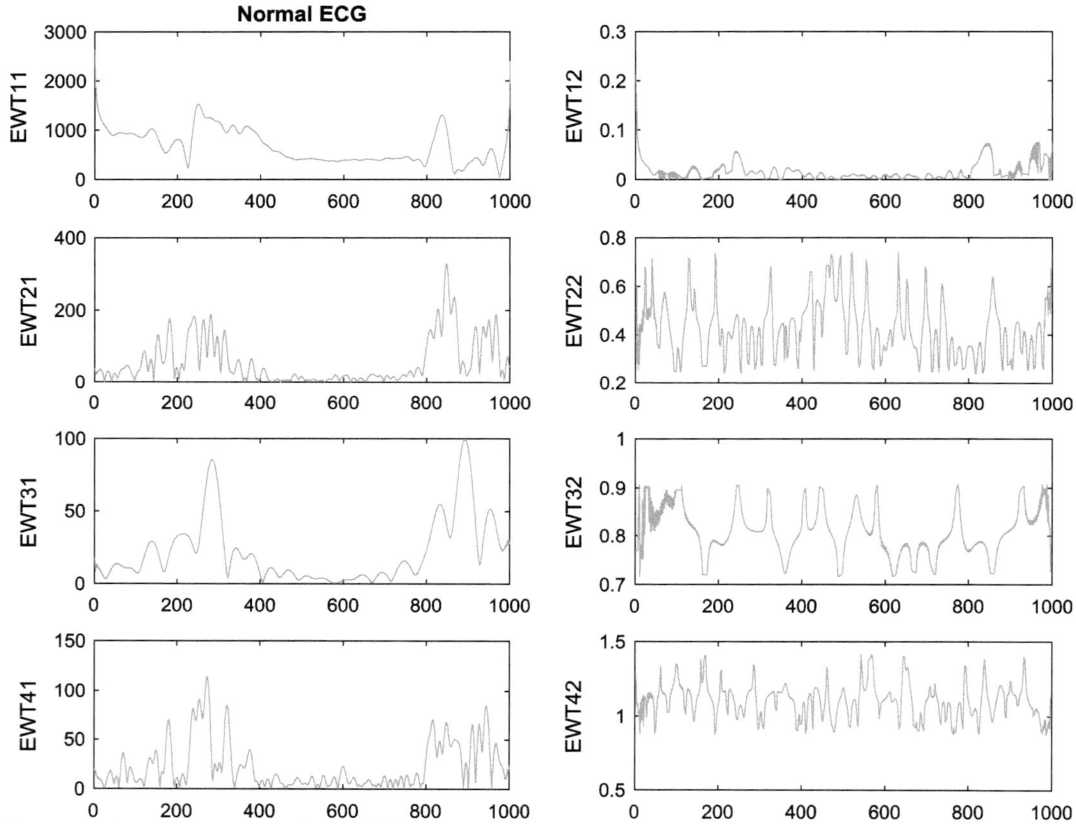

FIG. 3.52 The empirical wavelet transform of normal ECG signal.

3.5.14 Empirical Mode Decomposition

Empirical mode decomposition (EMD) was developed by Huang et al. (Huang et al., 1998) to characterize nonstationary signals as sums of zero-mean AM-FM components. Even though it proved remarkably effective (Flandrin, Rilling, & Goncalves, 2004), the algorithm has problems with the understanding of analytical formulation that allows for performance evaluation and theoretical analysis. Thus, Rilling, Flandrin, and Goncalves (2003) worked experimentally for a better understanding of the algorithm and suggested several improvements on the original formulation. The starting point of the EMD is to deliberate oscillations in signals at a very local level. Essentially, the signal $x(t)$ can be represented with a (local) high-frequency part or local detail associated with the oscillation terminating at the two minima and passing through the maximum that essentially exists among them, if it has two consecutive extrema. This can be achieved in a suitable way for all the oscillations establishing the whole signal, the method can then be

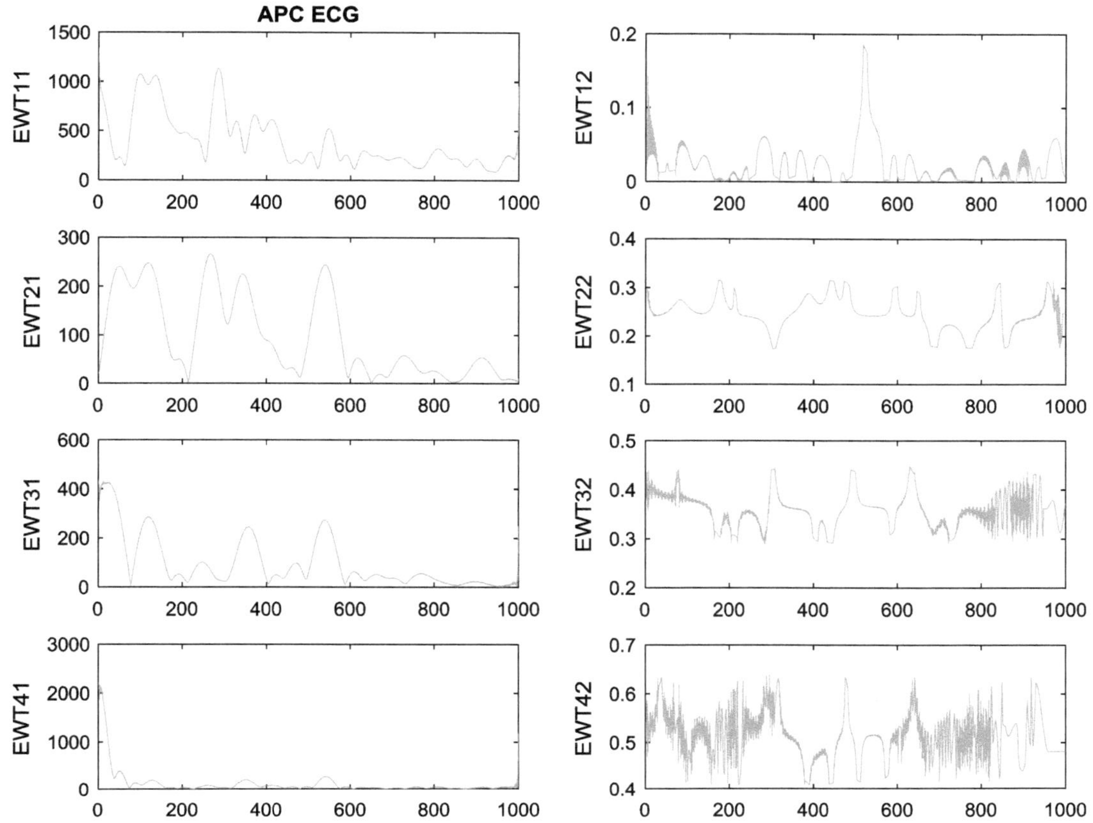

FIG. 3.53 The empirical wavelet transform of APC ECG signal.

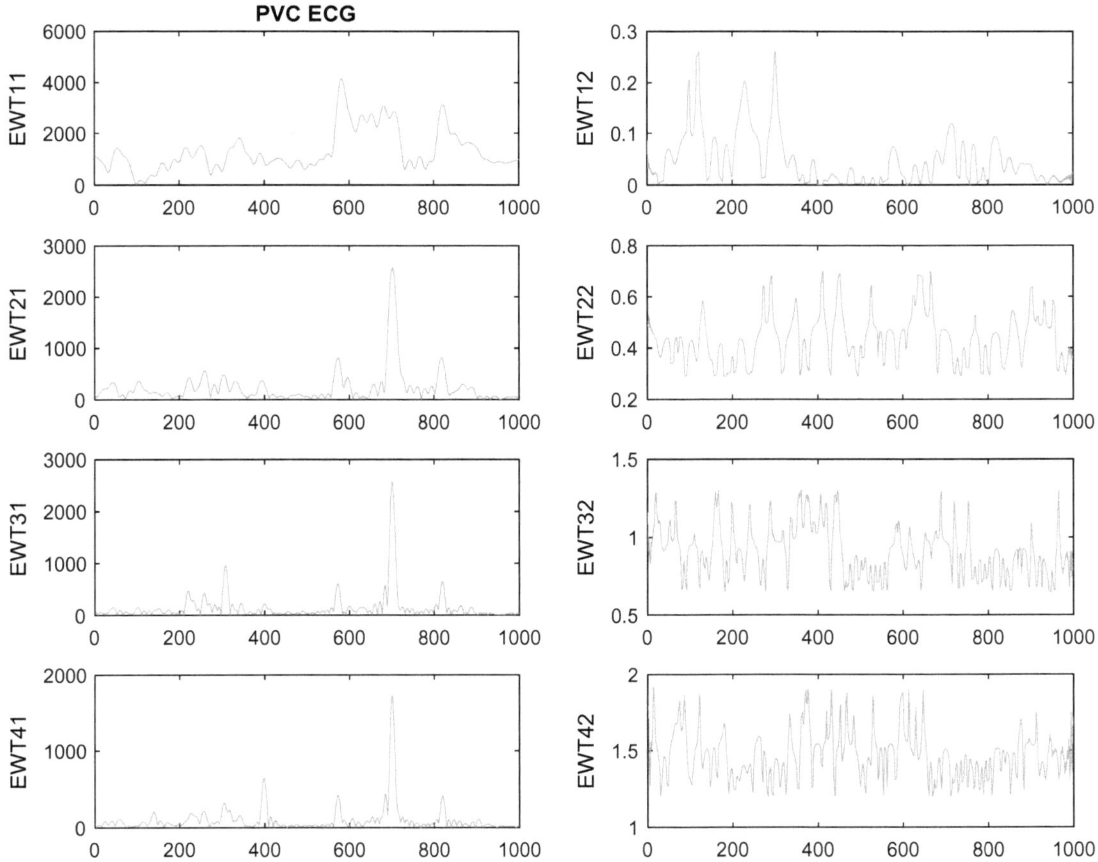

FIG. 3.54 The empirical wavelet transform of PVC ECG signal.

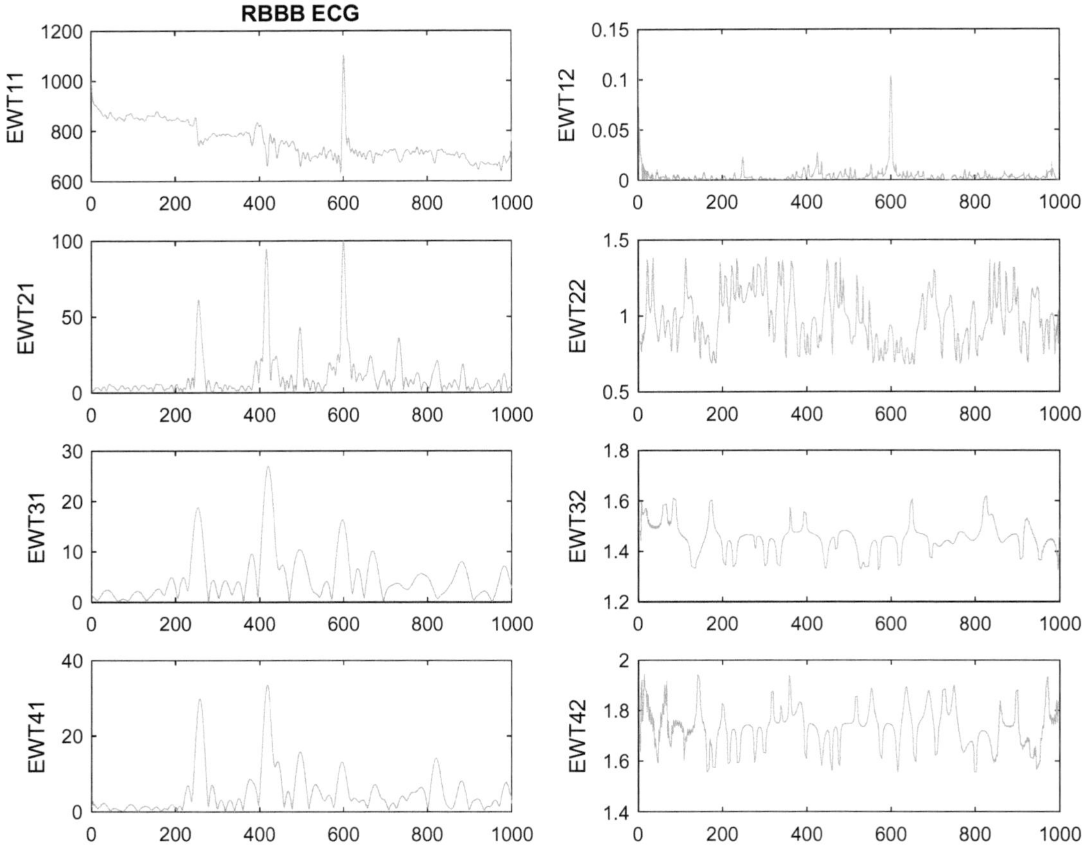

FIG. 3.55 The empirical wavelet transform of RBBB ECG signal.

employed on the residual comprising all local trends, and constitutive components of a signal can then be extracted iteratively. The efficient EMD algorithm can be summarized as follows:

1. identify all extrema of $x(t)$;
2. interpolate between minima (resp. maxima), ending up with envelope $e_{\min}(t)$ (resp. $e_{\max}(t)$);
3. compute the mean $m(t) = (e_{\min}(t) + e_{\max}(t))/2$;
4. extract the detail $d(t) = x(t) - m(t)$;
5. iterate on the residual $m(t)$.

In practice, this algorithm should be refined by a sifting process (Huang et al., 1998) by first iterating steps 1–4 upon the detail signal $d(t)$, until this latter can be considered as zero-mean according to some stopping condition. When this is realized, the detail is denoted as an intrinsic mode function (IMF), the related residual is computed, and step 5 is realized. Then, the number of extrema is decreased when working from one residual to the next, and the entire decomposition is defined to be completed with a finite number of modes. Modes and residuals are heuristically represented on "spectral" arguments. The high- vs. low-frequency differentiation, even with the harmonic oscillations, can be used only locally comparable to the wavelet transform. Modes are chosen with an adaptive and automatic time-variant filtering (Rilling et al., 2003) (Figs. 3.56–3.58).

EXAMPLE 3.39. The following MATLAB code was used to plot the EEG signals and their EMD. You can download data from the following website: http://epileptologie-bonn.de/cms/front_content.php?idcat=193&lang=3&changelang=3

Note that the EMD toolbox is used for running the following MATLAB code. You can download it from the following website: https://www.mathworks.com/matlabcentral/fileexchange/21409-empirical-mode-decomposition

Do not forget to set the path for the EMD toolbox.

```
%Ch3_Ex95_EMD.m
%Empirical Mode Decomposition (EMD) of NORMAL, INTERICTAL and ICTAL EEG signals
%Note that the their Empirical Mode Decomposition (EMD) toolbox is used
%for running the following MATLAB code. You can download it from
%the following web sites:
%http://perso.ens-lyon.fr/patrick.flandrin/emd.html
%https://www.mathworks.com/matlabcentral/fileexchange/21409-empirical-mode-decomposition
%Set the path for the their Empirical Mode Decomposition (EMD) toolbox
```

```
clc
clear
%Load Sample EEG Data downloaded from the web site
%http://epileptologie-bonn.de/cms/front_content.php?idcat=193&lang=3&changelang=3
load AS_BONN_ALL_EEG_DATA_4096.mat
L=length(Normal_Eyes_Open(1:4096,1));
%%
% %%%%%%%%%%%%%%%%%%%%%%%%%%%%% A Z.ZIP EYES OPEN NORMAL SUBJECT
imf=emd(Normal_Eyes_Open(1:4096,1));
    subplot(4,2,1)
    plot(imf(1,:))
    ylabel('IMF1');
    axis([1 L/4 -inf inf])
    title('Normal EEG')

    subplot(4,2,2)
    plot(imf(2,:))
    ylabel('IMF2');
    axis([1 L/4 -inf inf])

    subplot(4,2,3)
    plot(imf(3,:))
    ylabel('IMF3');
    axis([1 L/8 -inf inf])

    subplot(4,2,4)
    plot(imf(4,:))
    ylabel('IMF4');
    axis([1 L/8 -inf inf])

    subplot(4,2,5)
    plot(imf(5,:))
    ylabel('IMF5');
    axis([1 L/8 -inf inf])

    subplot(4,2,6)
    plot(imf(6,:))
    ylabel('IMF6');
    axis([1 L/8 -inf inf])

    subplot(4,2,7)
    plot(imf(7,:))
    ylabel('IMF7');
    axis([1 L/8 -inf inf])

    subplot(4,2,8)
    plot(imf(8,:))
    ylabel('IMF8');
    axis([1 L/8 -inf inf])

%%
%%%% D F.ZIP EPILEPTIC SUBJECT INTERICTAL were recorded from within
%%%%the epileptogenic zone during seizure free intervals %%%%
figure
imf=emd(Interictal(1:4096,1));
    subplot(4,2,1)
    plot(imf(1,:))
    ylabel('IMF1');
    axis([1 L/4 -inf inf])
    title('Interictal EEG')

    subplot(4,2,2)
    plot(imf(2,:))
    ylabel('IMF2');
    axis([1 L/4 -inf inf])

    subplot(4,2,3)
    plot(imf(3,:))
    ylabel('IMF3');
    axis([1 L/8 -inf inf])
```

```
    subplot(4,2,4)
    plot(imf(4,:))
    ylabel('IMF4');
    axis([1 L/8 -inf inf])

    subplot(4,2,5)
    plot(imf(5,:))
    ylabel('IMF5');
    axis([1 L/8 -inf inf])

    subplot(4,2,6)
    plot(imf(6,:))
    ylabel('IMF6');
    axis([1 L/8 -inf inf])

    subplot(4,2,7)
    plot(imf(7,:))
    ylabel('IMF7');
    axis([1 L/8 -inf inf])

    subplot(4,2,8)
    plot(imf(8,:))
    ylabel('IMF8');
    axis([1 L/8 -inf inf])

%% E S.ZIP EPILEPTIC SUBJECT ICTAL DURING SEIZURE %%%%%%%%%%%%
figure
imf=emd(Ictal(1:4096,1));
    subplot(4,2,1)
    plot(imf(1,:))
    ylabel('IMF1');
    axis([1 L/4 -inf inf])
    title('Ictal EEG')

    subplot(4,2,2)
    plot(imf(2,:))
    ylabel('IMF2');
    axis([1 L/4 -inf inf])

    subplot(4,2,3)
    plot(imf(3,:))
    ylabel('IMF3');
    axis([1 L/8 -inf inf])

    subplot(4,2,4)
    plot(imf(4,:))
    ylabel('IMF4');
    axis([1 L/8 -inf inf])

    subplot(4,2,5)
    plot(imf(5,:))
    ylabel('IMF5');
    axis([1 L/8 -inf inf])

    subplot(4,2,6)
    plot(imf(6,:))
    ylabel('IMF6');
    axis([1 L/8 -inf inf])

    subplot(4,2,7)
    plot(imf(7,:))
    ylabel('IMF7');
    axis([1 L/8 -inf inf])

    subplot(4,2,8)
    plot(imf(8,:))
    ylabel('IMF8');
    axis([1 L/8 -inf inf])
```

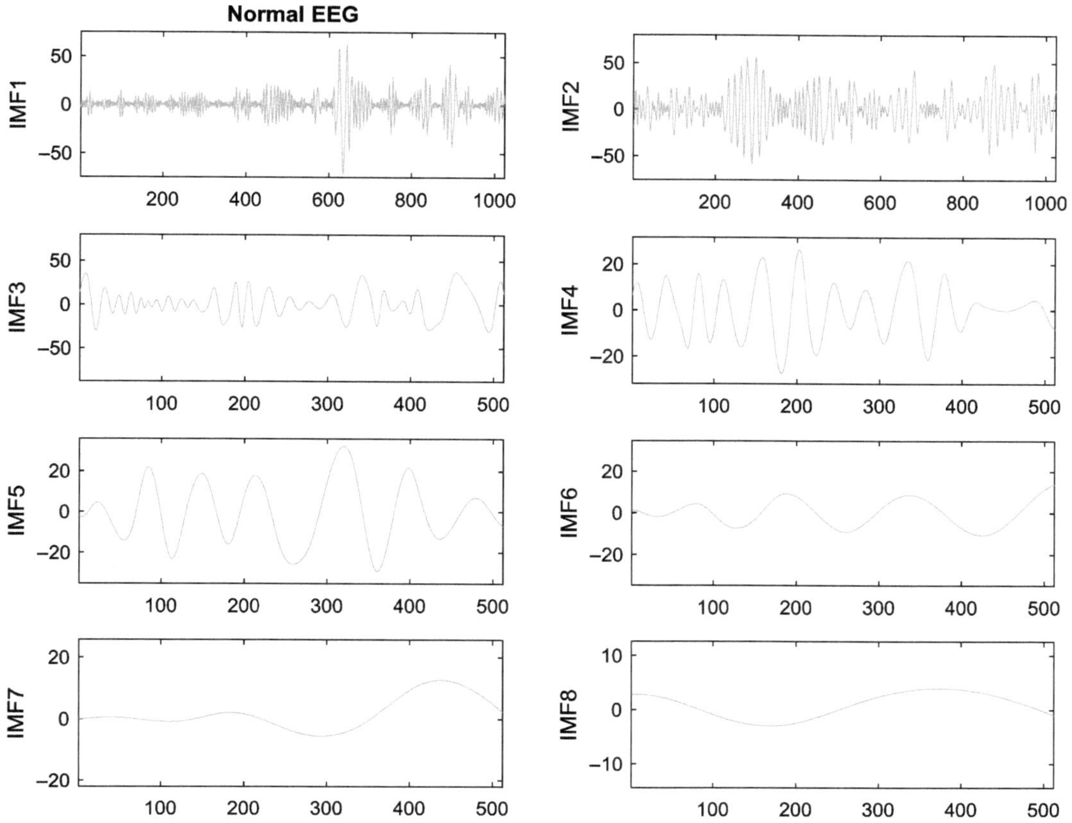

FIG. 3.56 The empirical mode decomposition of normal EEG signal.

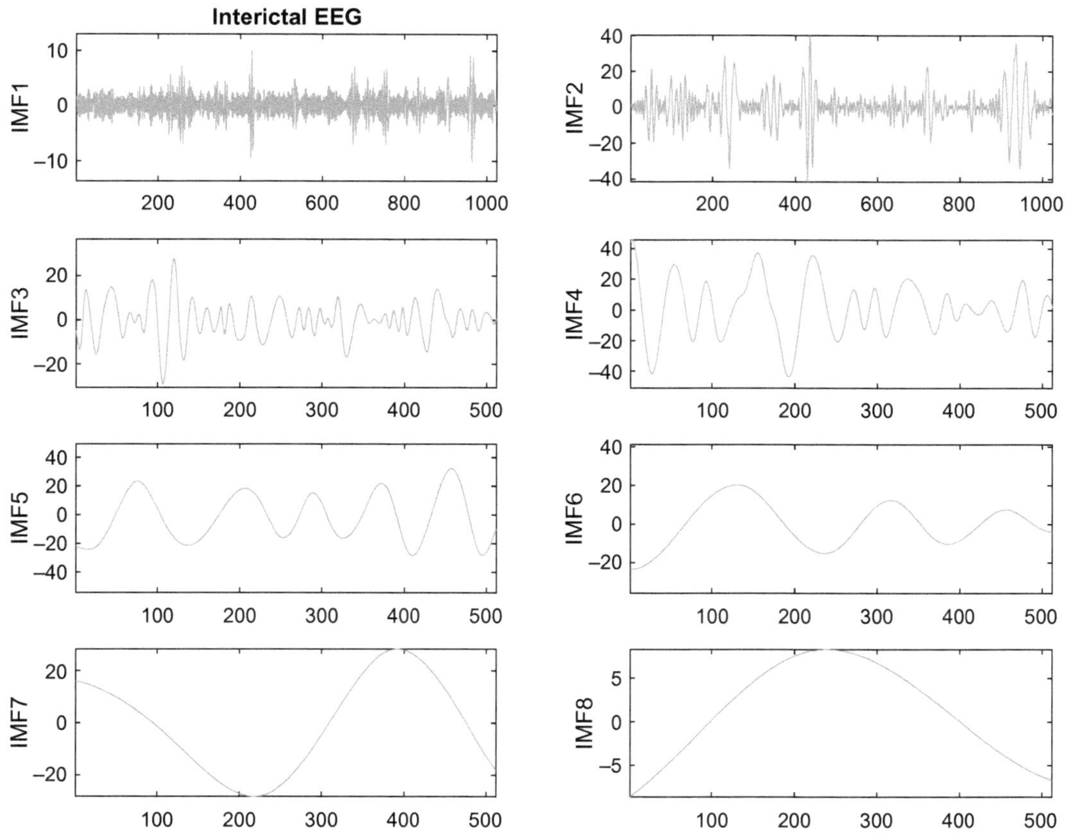

FIG. 3.57 The empirical mode decomposition of interictal EEG signal.

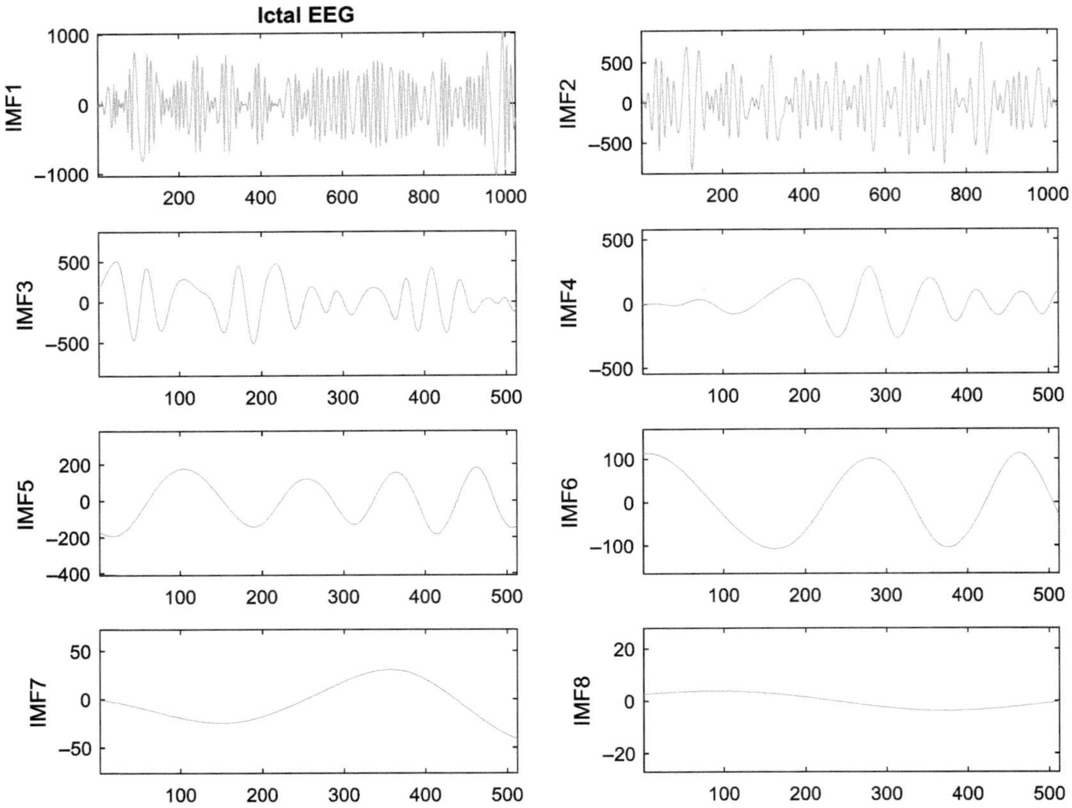

FIG. 3.58 The empirical mode decomposition of ictal EEG signal.

3.5.15 Ensemble Empirical Mode Decomposition

EMD has been recently developed (Huang et al., 1998) as an adaptive time-frequency signal analysis algorithm. It is verified to be pretty adaptable in several applications for the decomposition of signal where data is produced in noisy, nonlinear, nonstationary processes. But EMD still has some unresolved complications. One of the main disadvantages of the original EMD is the frequent presence of mode mixing presented as a single IMF either containing signals of broadly disparate scales, or a signal of a similar scale existing in different IMF components. Mode mixing is a result of signal intermittency. As studied by Huang et al. (1998), the intermittence could not only result in serious aliasing in the time-frequency distribution but also produce lack of physical meaning for the individual IMF. To eliminate this problem, Huang, Shen, and Long (1999) developed the intermittence test that can alleviate some of the complications. But the method itself has its own problems. To overcome these problems without introducing new ones, a new noise-assisted data analysis technique was adapted, called the ensemble EMD (EEMD), which describes the true IMF components as the mean of an ensemble of trials, each containing the signal plus a white noise of finite amplitude. With this ensemble technique, the scale can be clearly separated naturally without any a priori subjective criterion selection. This new technique is built on the understanding gathered from recent studies of the statistical properties of white noise (Flandrin et al., 2004; Wu & Huang, 2004) that presented the EMD successfully as an adaptive dyadic filter bank when applied to white noise. Their results confirmed that noise could support data analysis in the EMD.

The process of the EEMD is simply to add white noise to include the complete time-frequency space uniformly with the constituting components of several scales divided by the filter bank. Once the noise is added to the original signal by uniformly distributed white background, the different scale signals are automatically projected onto appropriate scales of reference created by the white noise in the background. Actually, every single trial might yield a very noisy signal for each of the noise-added decompositions composed of the signal and the added white noise. It can be cancelled out easily in the ensemble mean of adequate trials, as the noise in every trial is totally different in distinct trials. Hence, the ensemble mean is considered as the right solution for the persistent part of the signal, and several trials must be added in the ensemble (Wu & Huang, 2009).

Studies of Flandrin et al. (2004)) and Wu and Huang (2004) have explored the EMD works as a dyadic filter for different types of noise. This indicates that signals with a similar scale in a noisy dataset might be contained in one IMF component. It can be revealed that addition of noise with finite rather than infinitesimal amplitude to the data produce such a noisy dataset (Wu & Huang, 2009). Generally, all types of biomedical data consist of signal and noise, such that

$$x(t) = s(t) + n(t) \tag{3.101}$$

in which $x(t)$ is the recorded data, and $s(t)$ and $n(t)$ are the real signal and noise, respectively. Noise is omnipresent and characterizes an extremely unwanted portion of any dataset. Numerous data analysis approaches were mainly proposed to eliminate the noise and extract the actual signals from data. To enhance the exactness of measurements, the ensemble mean is an effective approach, in which data is gathered by separate observations that contain different noise. To generalize this ensemble idea, noise is added to the single dataset, $x(t)$, as if separate observations were made similar to a physical experiment that was repeated several times. The added white noise is considered as random noise that would be encountered during the measurement. Therefore, the ith "artificial" observation will be

$$x_i(t) = x(t) + w_i(t) \tag{3.102}$$

Even though noise addition might achive a smaller signal-to-noise ratio, the added white noise may produce a uniform reference scale distribution to smooth the EMD. Consequently, the low signal-noise ratio cannot disturb the decomposition process but essentially improves it to circumvent mode mixing. Moreover, a further step can be implemented by adding white noise to extract the real signals from the noisy data. This technique is called ensemble empirical mode decomposition (EEMD), which was developed as follows:

1. a white noise is added to the acquired data;
2. the data with added white noise is decomposed into IMFs;
3. step 1 and step 2 are repeated many times with different white noise series every time;
4. the (ensemble) means of corresponding IMFs of the decompositions are obtained at the end.

Because the added white noise data cancel each other in the EEMD decomposition, the mean IMFs stays within the natural dyadic filter windows by decreasing the chance of mode mixing and preserving the dyadic property (Wu & Huang, 2009) (Figs. 3.59–3.61).

EXAMPLE 3.40. The following MATLAB code was used to plot the EEG signals and their EEMD. You can download data from the following website: http://epileptologie-bonn.de/cms/front_content.php?idcat = 193&lang = 3&changelang = 3 Note that the EEMD toolbox is used for running the following MATLAB code. You can download it from the following website: http://perso.ens-lyon.fr/patrick.flandrin/emd.html

Do not forget to set the path for the EEMD toolbox.

```
%%Ch3_Ex101_EEMD.m
%Ensemble Empirical Mode Decomposition (EEMD) of NORMAL, INTERICTAL and ICTAL EEG signals
%Note that the Ensemble Empirical Mode Decomposition (EEMD) toolbox is used
%for running the following MATLAB code. You can download it from
%the following web site:
%http://perso.ens-lyon.fr/patrick.flandrin/emd.html
%Set the path for the Ensemble Empirical Mode Decomposition (EEMD) toolbox

clc
clear
%Load Sample EEG Data downloaded from the web site
%http://epileptologie-bonn.de/cms/front_content.php?idcat=193&lang=3&changelang=3
load AS_BONN_ALL_EEG_DATA_4096.mat

L=length(Normal_Eyes_Open(1:4096,1));
goal=7; %target mode numbers to decompose
ens=2;
nos=2;
%%
% %%%%%%%%%%%%%%%%%%%%%%%%%%%%%% A Z.ZIP EYES OPEN NORMAL SUBJECT
modes = eemd(Normal_Eyes_Open(1:1024,1), goal, ens, nos);
    subplot(4,2,1)
    plot(modes(1,:))
    ylabel('IMF1');
    axis([1 L/4 -inf inf])
    title('Normal EEG')

    subplot(4,2,2)
    plot(modes(2,:))
    ylabel('IMF2');
    axis([1 L/4 -inf inf])
```

```
    subplot(4,2,3)
    plot(modes(3,:))
    ylabel('IMF3');
    axis([1 L/4 -inf inf])

    subplot(4,2,4)
    plot(modes(4,:))
    ylabel('IMF4');
    axis([1 L/4 -inf inf])

    subplot(4,2,5)
    plot(modes(5,:))
    ylabel('IMF5');
    axis([1 L/4 -inf inf])

    subplot(4,2,6)
    plot(modes(6,:))
    ylabel('IMF6');
    axis([1 L/4 -inf inf])

    subplot(4,2,7)
    plot(modes(7,:))
    ylabel('IMF7');
    axis([1 L/4 -inf inf])

    subplot(4,2,8)
    plot(modes(8,:))
    ylabel('IMF8');
    axis([1 L/4 -inf inf])

%%
%%%% D F.ZIP EPILEPTIC SUBJECT INTERICTAL were recorded from within
%%%%the epileptogenic zone during seizure free intervals %%%%
figure
    modes = eemd(Interictal(1:1024,1), goal, ens, nos);
    subplot(4,2,1)
    plot(modes(1,:))
    ylabel('IMF1');
    axis([1 L/4 -inf inf])
    title('Interictal EEG')

    subplot(4,2,2)
    plot(modes(2,:))
    ylabel('IMF2');
    axis([1 L/4 -inf inf])

    subplot(4,2,3)
    plot(modes(3,:))
    ylabel('IMF3');
    axis([1 L/4 -inf inf])

    subplot(4,2,4)
    plot(modes(4,:))
    ylabel('IMF4');
    axis([1 L/4 -inf inf])

    subplot(4,2,5)
    plot(modes(5,:))
    ylabel('IMF5');
    axis([1 L/4 -inf inf])

    subplot(4,2,6)
    plot(modes(6,:))
    ylabel('IMF6');
    axis([1 L/4 -inf inf])

    subplot(4,2,7)
    plot(modes(7,:))
    ylabel('IMF7');
    axis([1 L/4 -inf inf])
```

```
    subplot(4,2,8)
    plot(modes(8,:))
    ylabel('IMF8');
    axis([1 L/4 -inf inf])

    %% E S.ZIP EPILEPTIC SUBJECT ICTAL DURING SEIZURE %%%%%%%%%%%%
figure
    modes = eemd(Ictal(1:1024,1), goal, ens, nos);
    subplot(4,2,1)
    plot(modes(1,:))
    ylabel('IMF1');
    axis([1 L/4 -inf inf])
    title('Ictal EEG')

    subplot(4,2,2)
    plot(modes(2,:))
    ylabel('IMF2');
    axis([1 L/4 -inf inf])

    subplot(4,2,3)
    plot(modes(3,:))
    ylabel('IMF3');
    axis([1 L/4 -inf inf])

    subplot(4,2,4)
    plot(modes(4,:))
    ylabel('IMF4');
    axis([1 L/4 -inf inf])

    subplot(4,2,5)
    plot(modes(5,:))
    ylabel('IMF5');
    axis([1 L/4 -inf inf])

    subplot(4,2,6)
    plot(modes(6,:))
    ylabel('IMF6');
    axis([1 L/4 -inf inf])

    subplot(4,2,7)
    plot(modes(7,:))
    ylabel('IMF7');
    axis([1 L/4 -inf inf])

    subplot(4,2,8)
    plot(modes(8,:))
    ylabel('IMF8');
    axis([1 L/4 -inf inf])
```

3.5.16 Complete Ensemble Empirical Mode Decomposition

Observe that, in EEMD, each $x^i[n]$ can be decomposed independently from each other to obtain a residue $r_k^i[n] = r_{k-1}^i[n] - IMF_k^i[n]$ for each one. Here, the decomposition modes are noted as $\widetilde{IMF}_k$ and a unique first residue can be calculated as:

$$r_1[n] = x[n] - \widetilde{IMF}_1\,[n] \tag{3.103}$$

where $\widetilde{IMF}_1\,[n]$ can be produced in the same way as EEMD. Then, the first EMD mode can be computed over an ensemble of $r_1[n]$ plus different realizations of a given noise obtaining $\widetilde{IMF}_2\,[n]$ by averaging. The next residue can be calculated as:

$$r_2[n] = r_1[n] - \widetilde{IMF}_2\,[n]$$

This process continues with the rest of the modes until the stopping criterion is reached. The operator $E_j(.)$ can be defined as to produce the jth mode taken by EMD for

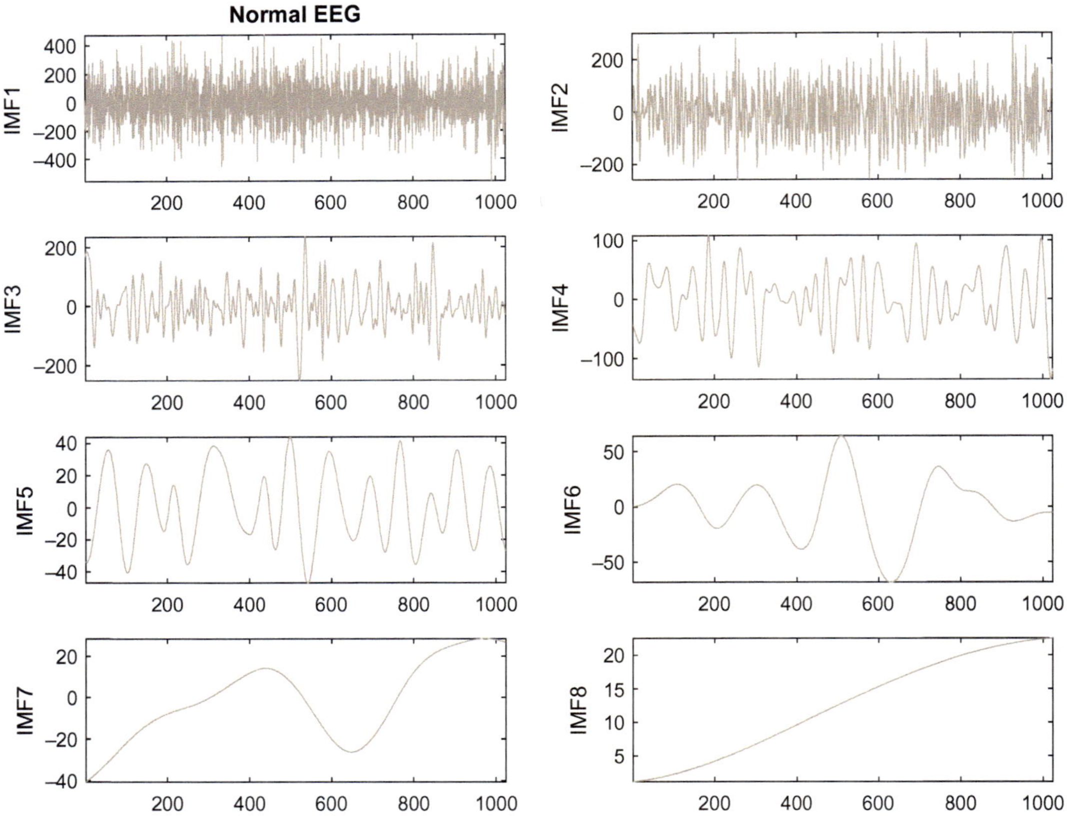

FIG. 3.59 The ensemble empirical mode decomposition of normal EEG signal.

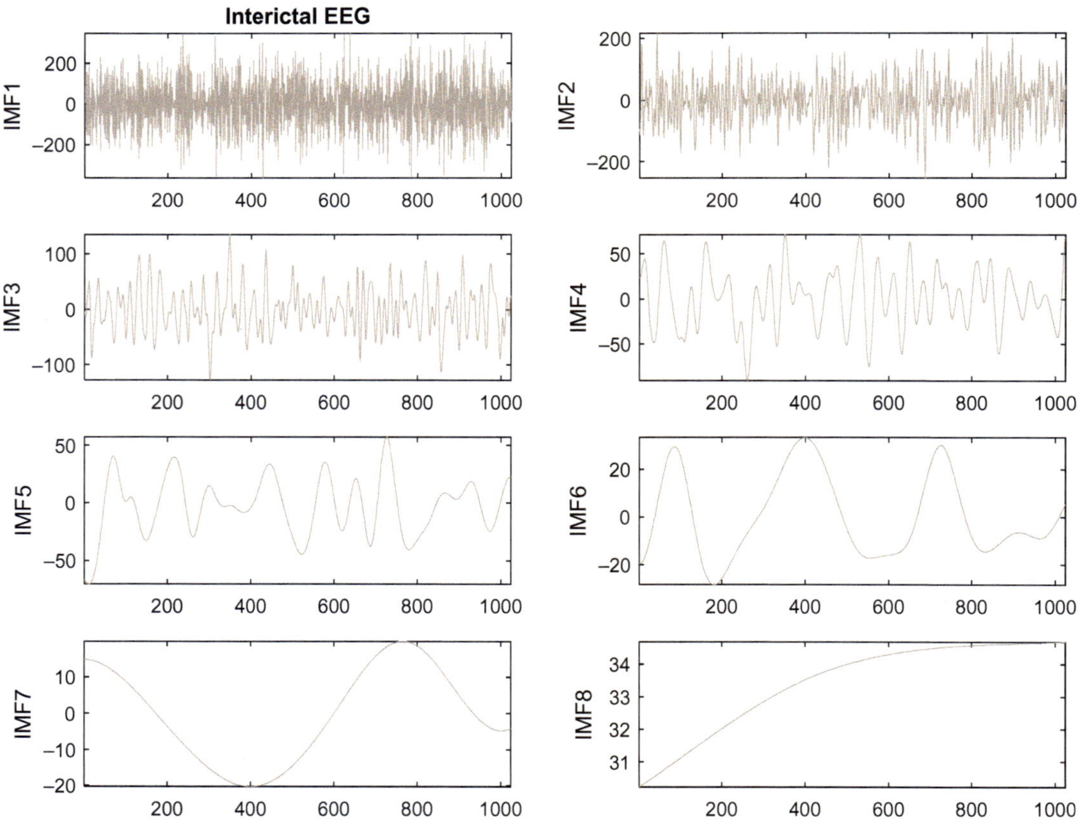

FIG. 3.60 The ensemble empirical mode decomposition of interictal EEG signal.

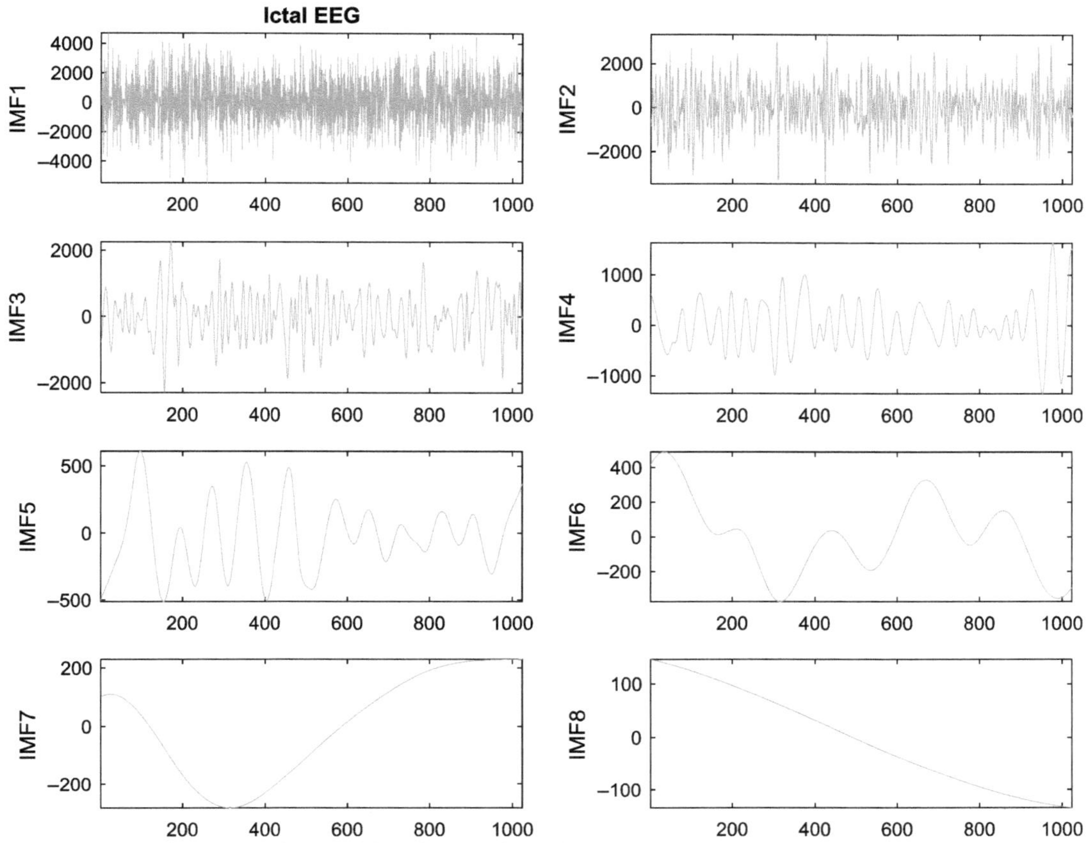

FIG. 3.61 The ensemble empirical mode decomposition of ictal EEG signal.

a given signal. If w^i is defined as a white noise with normal distribution $N(0;\ 1)$, then CEEMD can be described by the following algorithm:

1. I realizations $x[n] + \varepsilon_0\, w^i[n]$ is decomposed by EMD to get their first modes and computed as

$$\widetilde{IMF}_1\,[n] = \frac{1}{I}\sum_{i=1}^{I} IMF_1^i[n] = \overline{IMF_1}[n]$$

2. the first residue in $r_1[n] = x[n] - \widetilde{IMF}_1\,[n]$ is calculated at the first stage ($k = 1$)
3. realizations $r_1[n] + \varepsilon_1\, E_1(w^i[n])$ ($i = 1,\ldots, I$) are decomposed until their first EMD mode and then define the second mode:

$$\widetilde{IMF}_2\,[n] = \frac{1}{I}\sum_{i=1}^{I} E_1\left(r_1[n] + \varepsilon_1 E_1\left(w^i[n]\right)\right)$$

4. calculate the kth residue:

$$r_k[n] = r_{(k-1)}[n] - \widetilde{IMF}_1\,[n]\ (k = 2;\ \ldots;\ \boldsymbol{K}) \tag{3.104}$$

5. realizations $r_k[n] + \varepsilon_k\, E_k(w^i[n])$, ($i = 1,\ldots, \boldsymbol{I}$) are decomposed until their first EMD mode and then define the

$$\widetilde{IMF}_{(k+1)}\,[n] = \frac{1}{I}\sum_{i=1}^{I} E_1\left(r_k[n] + \varepsilon_k E_k\left(w^i[n]\right)\right) \tag{3.105}$$

6. go to step 4 for next k.

Steps 4–6 are carried out until the residue taken is no longer reasonable to be decomposed. The final residue satisfies:

$$R[n] = x[n] - \sum_{k=1}^{K} \widetilde{IMF}_k \tag{3.106}$$

where K is the total number of modes. Consequently, the original signal $x[n]$ can be represented as:

$$x[n] = \sum_{k=1}^{K} \widetilde{IMF}_k + R[n] \tag{3.107}$$

The last equation achieves a complete decomposition and delivers a precise reconstruction of the original data. Regarding to the amplitude of the added noise, Wu and Huang (2004) proposed to employ small amplitude values for data dominated by high-frequency signals and vice versa (Torres, Colominas, Schlotthauer, & Flandrin, 2011) (Figs. 3.62–3.64).

EXAMPLE 3.41. The following MATLAB code was used to plot the EEG signals and their CEEMD. You can download data from the following website: http://epileptologie-bonn.de/cms/front_content.php?idcat=193&lang=3&changelang=3
Note that the CEEMD toolbox is used for running the following MATLAB code. You can download it from the following website: http://perso.ens-lyon.fr/patrick.flandrin/emd.html

Do not forget to set the path for the CEEMD toolbox.

```
%Ch3_Ex106_CEEMD.m
% Example of the CEEMDAN performance, used in the work where CEEMDAN was first presented:
% Cite:
% M.E.TORRES, M.A. COLOMINAS, G. SCHLOTTHAUER, P. FLANDRIN,
% "A complete Ensemble Empirical Mode decomposition with adaptive noise,"
% IEEE Int. Conf. on Acoust., Speech and Signal Proc. ICASSP-11, pp. 4144-4147, Prague (CZ)
%
%Complete Ensemble Empirical Mode Decomposition (CEEMD) of NORMAL, INTERICTAL and ICTAL EEG
signals
% The code loads the signal EEG.mat
% It is an EEG from the epileptologie-bonn
%Load Sample EEG Data downloaded from the web site
%available at http://epileptologie-bonn.de/cms/front_content.php?idcat=193&lang=3&changelang=3

%Note that the Complete Ensemble Empirical Mode Decomposition (CEEMD) toolbox is used
%for running the following MATLAB code. You can download it from
%the following web site:
%http://perso.ens-lyon.fr/patrick.flandrin/emd.html
%http://bioingenieria.edu.ar/grupos/ldnlys/metorres/re_inter.htm#Codigos
%Set the path for the Complete Ensemble Empirical Mode Decomposition (CEEMD) toolbox

clc
clear
% The code loads the signal EEG.mat
% It is an EEG from the epileptologie-bonn
%Load Sample EEG Data downloaded from the web site
%available at http://epileptologie-bonn.de/cms/front_content.php?idcat=193&lang=3&changelang=3

clc
clear
load AS_BONN_ALL_EEG_DATA_4096.mat
L=length(Normal_Eyes_Open(1:1024,1));
Level=8; %target mode numbers to decompose 8 level
Nstd = 0.2;
NR = 500;
MaxIter = 5000;
%%
% %%%%%%%%%%%%%%%%%%%%%%%%%%%%%% A Z.ZIP EYES OPEN NORMAL SUBJECT
[modes its]=ceemdan(Normal_Eyes_Open(1:1024,1),Nstd,NR,MaxIter);
t=1:L;

[a b]=size(modes);
figure;
subplot(Level+1,1,1);
plot(t,Normal_Eyes_Open(1:1024,1));% the EEG signal is in the first row of the subplot
ylabel('EEG')
set(gca,'xtick',[])
axis tight;
title('Normal EEG')
```

```
for i=2:Level
    subplot(Level+1,1,i);
    plot(t,modes(i-1,:));
    ylabel (['IMF ' num2str(i-1)]);
    set(gca,'xtick',[])
    xlim([1 L])
end;

subplot(Level+1,1,Level+1)
plot(t,modes(Level,:))
ylabel(['IMF ' num2str(Level)])
xlim([1 L])

%%
%%%% D F.ZIP EPILEPTIC SUBJECT INTERICTAL were recorded from within
%%%%the epileptogenic zone during seizure free intervals %%%%
[modes its]=ceemdan(Interictal(1:1024,1),Nstd,NR,MaxIter);
t=1:L;

[a b]=size(modes);

figure;
subplot(Level+1,1,1);
plot(t,Interictal(1:1024,1));% the EEG signal is in the first row of the subplot
ylabel('EEG')
set(gca,'xtick',[])
axis tight;
title('Interictal EEG')

for i=2:Level
    subplot(Level+1,1,i);
    plot(t,modes(i-1,:));
    ylabel (['IMF ' num2str(i-1)]);
    set(gca,'xtick',[])
    xlim([1 L])
end;

subplot(Level+1,1,Level+1)
plot(t,modes(Level,:))
ylabel(['IMF ' num2str(Level)])
xlim([1 L])

%% E S.ZIP EPILEPTIC SUBJECT ICTAL DURING SEIZURE %%%%%%%%%%%%%
[modes its]=ceemdan(Ictal(1:1024,1),Nstd,NR,MaxIter);
t=1:L;

[a b]=size(modes);

figure;
subplot(Level+1,1,1);
plot(t,Ictal(1:1024,1));% the EEG signal is in the first row of the subplot
ylabel('EEG')
set(gca,'xtick',[])
axis tight;
title('Ictal EEG')

for i=2:Level
    subplot(Level+1,1,i);
    plot(t,modes(i-1,:));
```

```
    ylabel (['IMF ' num2str(i-1)]);
    set(gca,'xtick',[])
    xlim([1 L])
end;

subplot(Level+1,1,Level+1)
plot(t,modes(Level,:))
ylabel(['IMF ' num2str(Level)])
xlim([1 L])
```

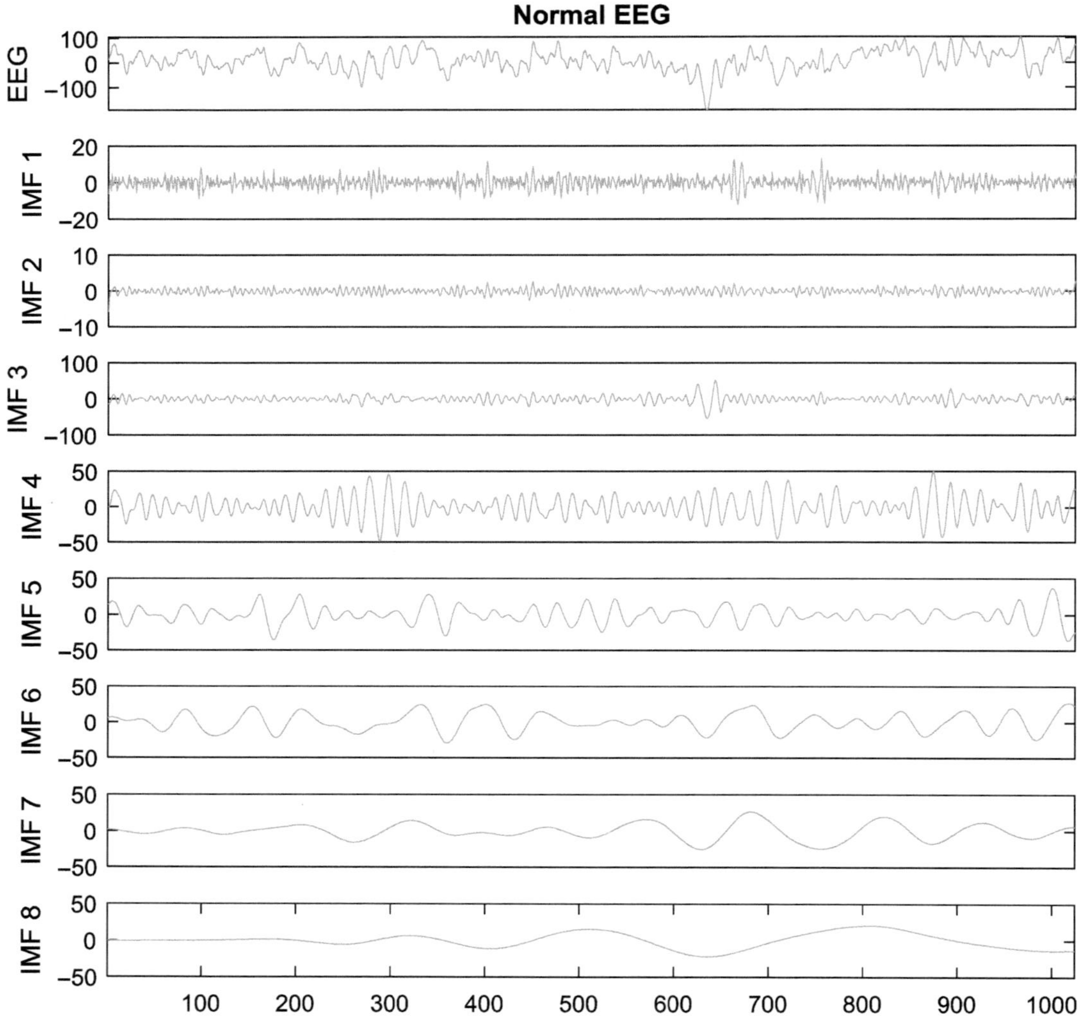

FIG. 3.62 The complete ensemble empirical mode decomposition of normal EEG signal.

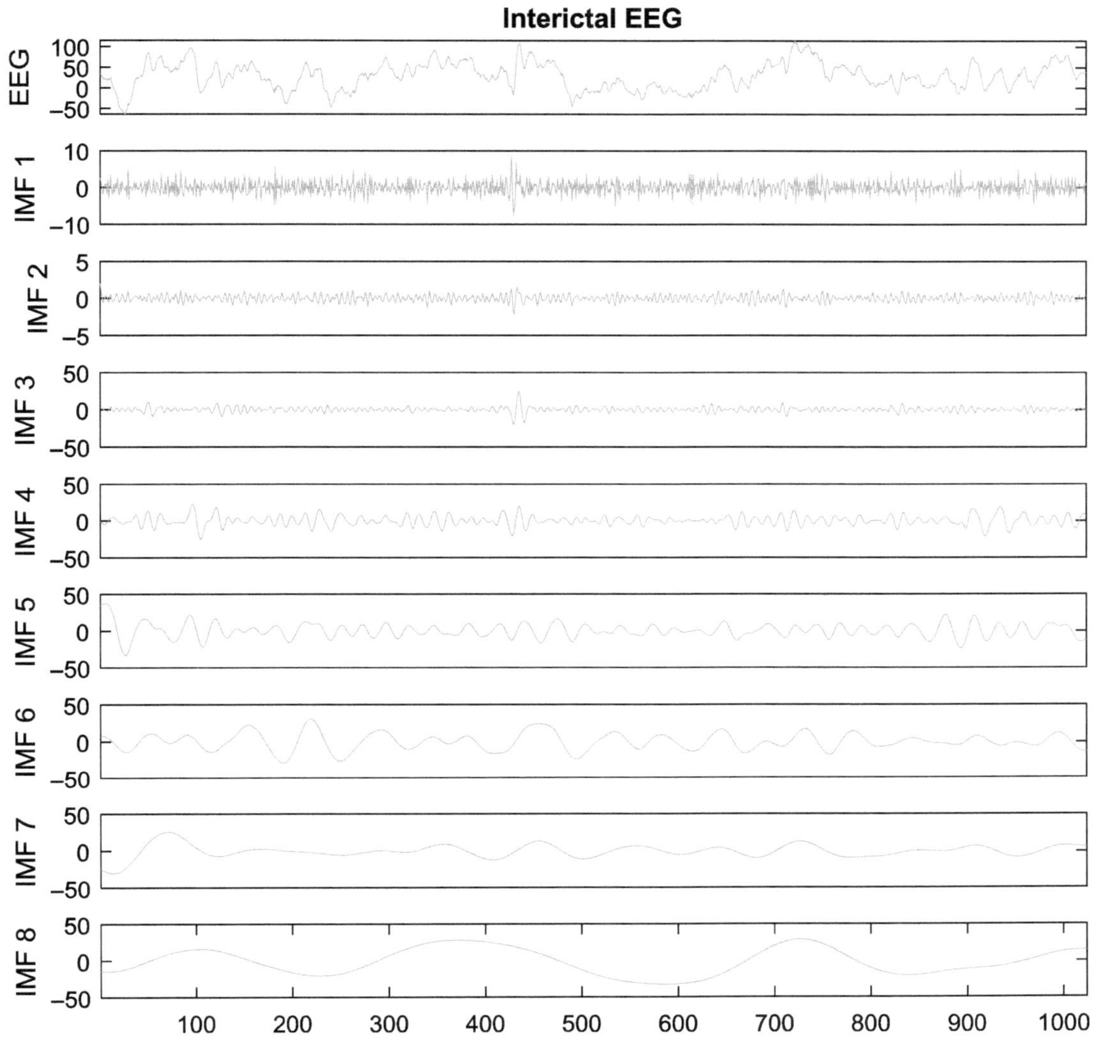

FIG. 3.63 The complete ensemble empirical mode decomposition of interictal EEG signal.

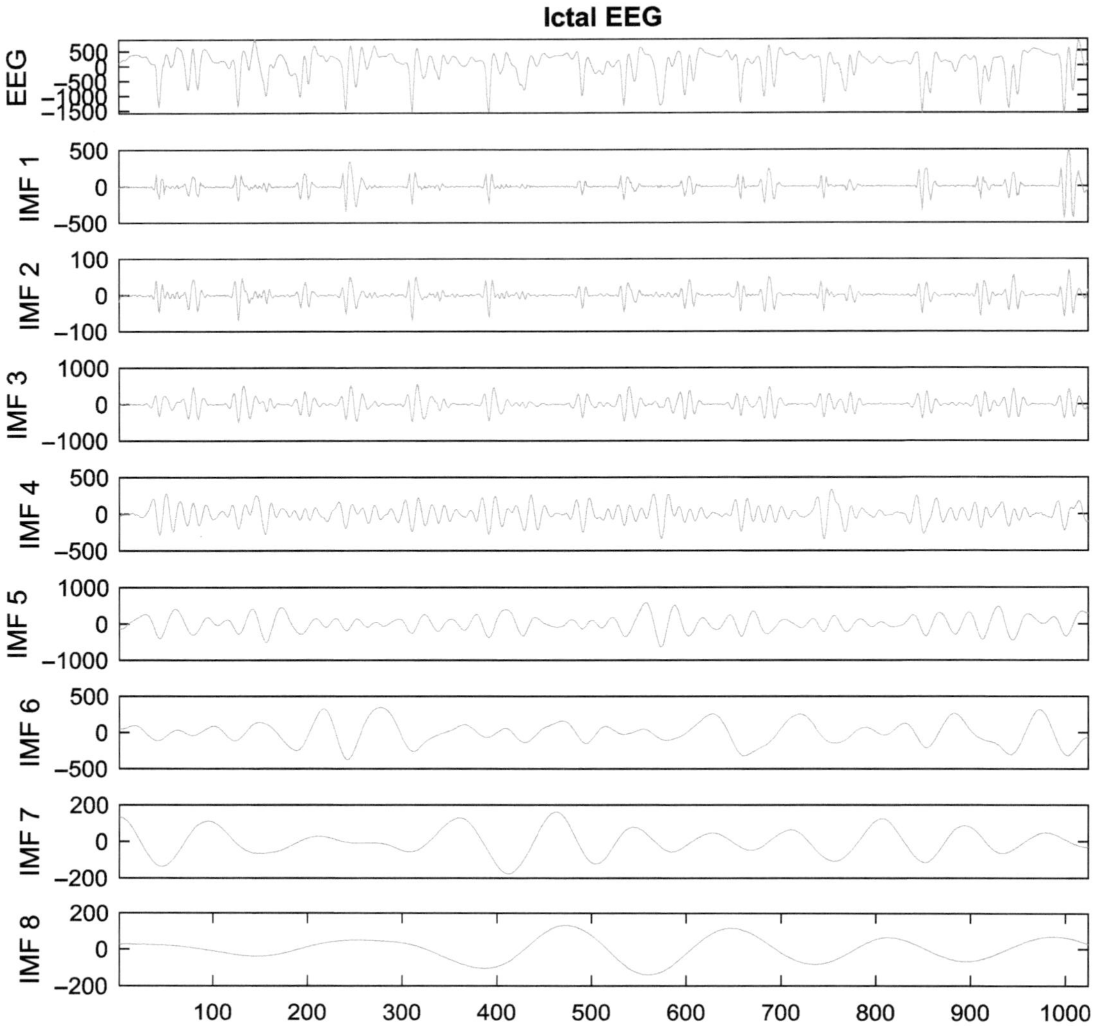

FIG. 3.64 The complete ensemble empirical mode decomposition of ictal EEG signal.

REFERENCES

Anderson, C. W., Stolz, E. A., & Shamsunder, S. (1998). Multivariate autoregressive models for classification of spontaneous electroencephalographic signals during mental tasks. *IEEE Transactions on Biomedical Engineering*, *45*(3), 277–286.

Basseville, M., & Nikiforov, I. V. (1993). *Detection of abrupt changes: Theory and application. Vol. 104*. Englewood Cliffs: Prentice Hall.

Bayram, İ. (2013). An analytic wavelet transform with a flexible time-frequency covering. *IEEE Transactions on Signal Processing*, *61*(5), 1131–1142.

Blinowska, K. J., & Zygierewicz, J. (2011). *Practical biomedical signal analysis using MATLAB®*. CRC Press.

Boashash, B., & Black, P. (1987). An efficient real-time implementation of the Wigner-Ville distribution. *IEEE Transactions on Acoustics, Speech, and Signal Processing*, *35*(11), 1611–1618.

Boashash, B., & Sucic, V. (2003). Resolution measure criteria for the objective assessment of the performance of quadratic time-frequency distributions. *IEEE Transactions on Signal Processing*, *51*(5), 1253–1263.

Bodenstein, G., Schneider, W., & Malsburg, C. (1985). Computerized EEG pattern classification by adaptive segmentation and probability-density-function classification. Description of the method. *Computers in Biology and Medicine*, *15*(5), 297–313.

Boudreaux-Bartels, G., & Murray, R. (1995). Time-frequency signal representations for biomedical signals. *The Biomedical Engineering Handbook*, 866–885.

Cohen, L. (1995). *Time-frequency analysis. 778*. Prentice hall.

da Silva, F. L., Niedermeyer, E., & da Silva, F. (1987). EEG analysis: theory and practice; computer-assisted EEG diagnosis: Pattern recognition techniques. In *Electroencephalography: Basic principles, clinical applications and related fields* (pp. 871–919).

Daubechies, I. (1990). The wavelet transform, time-frequency localization and signal analysis. *IEEE Transactions on Information Theory*, *36*(5), 961–1005.

Daubechies, I. (1992). *Ten lectures on wavelets*. SIAM.

de Souza Neto, E. P., Custaud, M.-A., Frutoso, J., Somody, L., Gharib, C., & Fortrat, J.-O. (2001). Smoothed pseudo Wigner-Ville distribution as an alternative to Fourier transform in rats. *Autonomic Neuroscience*, *87* (2–3), 258–267.

Dumermuth, G., & Flühler, H. (1967). Some modern aspects in numerical spectrum analysis of multichannel electroencephalographic data. *Medical and Biological Engineering*, *5*(4), 319–331.

Flandrin, P., Rilling, G., & Goncalves, P. (2004). Empirical mode decomposition as a filter bank. *IEEE Signal Processing Letters*, *11*(2), 112–114.

Friedlander, B. (1990). A sensitivity analysis of the MUSIC algorithm. *IEEE Transactions on Acoustics, Speech, and Signal Processing*, *38* (10), 1740–1751.

Friedlander, B., & Weiss, A. J. (1994). Effects of model errors on waveform estimation using the MUSIC algorithm. *IEEE Transactions on Signal Processing*, *42*(1), 147–155.

Gilles, J. (2013). Empirical wavelet transform. *IEEE Transactions on Signal Processing*, *61*(16), 3999–4010.

Gupta, V., Priya, T., Yadav, A. K., Pachori, R. B., & Acharya, U. R. (2017). Automated detection of focal EEG signals using features extracted from flexible analytic wavelet transform. *Pattern Recognition Letters*, *94*, 180–188.

Gustafsson, F., & Gustafsson, F. (2000). *Adaptive filtering and change detection. Vol. 1*. Citeseer.

Huang, N. E., Shen, Z., & Long, S. R. (1999). A new view of nonlinear water waves: the Hilbert spectrum. *Annual Review of Fluid Mechanics*, *31*(1), 417–457.

Huang, N. E., Shen, Z., Long, S. R., Wu, M. C., Shih, H. H., Zheng, Q., … Liu, H. H. (1998). The empirical mode decomposition and the Hilbert spectrum for nonlinear and non-stationary time series analysis. In *The Royal Society Presented at the Proceedings of the Royal Society of London A: Mathematical, physical and engineering sciences, Vol. 454* (pp. 903–995).

Hubbard, B. B. (1998). *The world according to wavelets: the story of a mathematical technique in the making*. AK Peters/CRC Press.

Jansen, B. H., Bourne, J. R., & Ward, J. W. (1981). Autoregressive estimation of short segment spectra for computerized EEG analysis. *IEEE Transactions on Biomedical Engineering*, (9), 630–638.

Proakis, J. G., & Manolakis, D. G. (2007). *Digital signal processing: principles, algorithms, and applications*. Pearson Prentice Hall.

Kay, S. M. (1993). *Fundamentals of statistical signal processing, volume I: Estimation theory*. .

Kay, S. M., & Marple, S. L. (1981). Spectrum analysis—a modern perspective. *Proceedings of the IEEE*, *69*(11), 1380–1419.

Kingsbury, N. G. (1998). The dual-tree complex wavelet transform: a new technique for shift invariance and directional filters. In *Presented at the Proc. 8th IEEE DSP workshop, Vol. 8, Utah* (p. 86).

Kumar, M., Pachori, R. B., & Acharya, U. R. (2016). An efficient automated technique for CAD diagnosis using flexible analytic wavelet transform and entropy features extracted from HRV signals. *Expert Systems With Applications*, *63*, 165–172.

Kumar, M., Pachori, R. B., & Acharya, U. R. (2017). Automated diagnosis of myocardial infarction ECG signals using sample entropy in flexible analytic wavelet transform framework. *Entropy*, *19*(9), 488.

Kutlu, Y., & Kuntalp, D. (2012). Feature extraction for ECG heartbeats using higher order statistics of WPD coefficients. *Computer Methods and Programs in Biomedicine*, *105*(3), 257–267.

Learned, R. E., & Willsky, A. S. (1995). A wavelet packet approach to transient signal classification. *Applied and Computational Harmonic Analysis*, *2*(3), 265–278.

Marchant, B. (2003). Time-frequency analysis for biosystems engineering. *Biosystems Engineering*, *85*(3), 261–281.

Marple, S. L., & Marple, S. L. (1987). *Digital spectral analysis: with applications. Vol. 5*. Englewood Cliffs, NJ: Prentice-Hall.

Muthuswamy, J., & Thakor, N. V. (1998). Spectral analysis methods for neurological signals. *Journal of Neuroscience Methods*, *83*(1), 1–14.

Narasimhan, S., & Nayak, M. B. (2003). Improved Wigner-Ville distribution performance by signal decomposition and modified group delay. *Signal Processing*, *83*(12), 2523–2538.

Nason, G. P., & Silverman, B. W. (1995). The stationary wavelet transform and some statistical applications. In *Wavelets and statistics* (pp. 281–299). Springer.

Nave, G., & Cohen, A. (1993). ECG compression using long-term prediction. *IEEE Transactions on Biomedical Engineering*, *40*(9), 877–885.

Pardey, J., Roberts, S., & Tarassenko, L. (1996). A review of parametric modeling techniques for EEG analysis. *Medical Engineering & Physics*, *18*(1), 2–11.

Porat, B., & Friedlander, B. (1988). Analysis of the asymptotic relative efficiency of the MUSIC algorithm. *IEEE Transactions on Acoustics, Speech, and Signal Processing*, *36*(4), 532–544.

Rilling, G., Flandrin, P., & Goncalves, P. (2003). On empirical mode decomposition and its algorithms. In *Presented at the IEEE-EURASIP workshop on nonlinear signal and image processing, NSIP-03, Grado (I), Vol. 3* (pp. 8–11).

Schmidt, R. (1986). Multiple emitter location and signal parameter estimation. *IEEE Transactions on Antennas and Propagation*, *34*(3), 276–280.

Selesnick, I. W. (2011). Wavelet transform with tunable Q-factor. *IEEE Transactions on Signal Processing*, *59*(8), 3560–3575.

Selesnick, I. W., Baraniuk, R. G., & Kingsbury, N. C. (2005). The dual-tree complex wavelet transform. *IEEE Signal Processing Magazine*, *22*(6), 123–151.

Semmlow, J. (2004). *Biosignal and biomedical image processing: MATLAB-based applications* (p. 2004).

Semmlow, J. L., & Griffel, B. (2014). *Biosignal and medical image processing*. CRC Press.

Sörnmo, L., & Laguna, P. (2005). *Bioelectrical signal processing in cardiac and neurological applications. Vol. 8*. Academic Press.

Stoica, P., & Moses, R. L. (1997). *Introduction to spectral analysis. Vol. 1*. Upper Saddle River, NJ: Prentice Hall.

Subasi, A. (2007). Selection of optimal AR spectral estimation method for EEG signals using Cramer-Rao bound. *Computers in Biology and Medicine*, *37*(2), 183–194.

Subasi, A., Erçelebi, E., Alkan, A., & Koklukaya, E. (2006). Comparison of subspace-based methods with AR parametric methods in epileptic seizure detection. *Computers in Biology and Medicine*, *36*(2), 195–208.

Subasi, A., & Kiymik, M. K. (2010). Muscle fatigue detection in EMG using time-frequency methods, ICA and neural networks. *Journal of Medical Systems*, *34*(4), 777–785.

Sudakov, O., Kriukova, G., Natarov, R., Gaidar, V., Maximyuk, O., Radchenko, S., & Isaev, D. (2017). Distributed system for sampling and analysis of electroencephalograms. In *Presented at the Intelligent Data Acquisition and Advanced Computing Systems: Technology and Applications (IDAACS), 2017 9th IEEE International Conference on IEEE, Vol. 1* (pp. 306–310).

Swindlehurst, A. L., & Kailath, T. (1992). A performance analysis of subspace-based methods in the presence of model errors. Part I. The MUSIC algorithm. *IEEE Transactions on Signal Processing*, *40*(7), 1758–1774.

Swindlehurst, A. L., & Kailath, T. (1993). A performance analysis of subspace-based methods in the presence of model errors: Part II-multidimensional algorithms. *IEEE Transactions on Signal Processing, 41*(9).

Therrien, C. W. (1992). *Discrete random signals and statistical signal processing*. Prentice Hall PTR.

Torres, M. E., Colominas, M. A., Schlotthauer, G., & Flandrin, P. (2011). A complete ensemble empirical mode decomposition with adaptive noise. In *Presented at the Acoustics, speech and signal processing (ICASSP), 2011 IEEE international conference on IEEE* (pp. 4144–4147).

Unser, M., & Aldroubi, A. (1996). A review of wavelets in biomedical applications. *Proceedings of the IEEE, 84*(4), 626–638.

Viberg, M. (1995). Subspace-based methods for the identification of linear time-invariant systems. *Automatica, 31*(12), 1835–1851.

Ville, J. (1948). Theorie et application dela notion de signal analytique. *Câbles et Transmissions, 2*(1), 61–74.

Wigner, E. P. (1997). *On the quantum correction for thermodynamic equilibrium. Part I. Physical Chemistry. Part II: Solid State Physics* (pp. 110–120). Springer.

Wu, Z., & Huang, N. E. (2004). A study of the characteristics of white noise using the empirical mode decomposition method. In *Presented at the Proceedings of the Royal Society of London A: Mathematical, Physical and Engineering Sciences, Vol. 460* (pp. 1597–1611): The Royal Society.

Wu, Z., & Huang, N. E. (2009). Ensemble empirical mode decomposition: a noise-assisted data analysis method. *Advances in Adaptive Data Analysis, 1*(01), 1–41.

Zhang, Y., Dong, Z., Liu, A., Wang, S., Ji, G., Zhang, Z., & Yang, J. (2015). Magnetic resonance brain image classification via stationary wavelet transform and generalized eigenvalue proximal support vector machine. *Journal of Medical Imaging and Health Informatics, 5*(7), 1395–1403.

Zhang, C., Li, B., Chen, B., Cao, H., Zi, Y., & He, Z. (2015). Weak fault signature extraction of rotating machinery using flexible analytic wavelet transform. *Mechanical Systems and Signal Processing, 64*, 162–187.

Chapter 4

Feature Extraction and Dimension Reduction

4.1 INTRODUCTION

Feature extraction deals with the problem of finding the most distinctive, informative, and reduced set of features to enhance the effectiveness of data storage and processing. Significant feature vectors remain the most common and appropriate means of signal depiction for classification problems. Many researchers in different areas, who are interested in data modeling and classification, are combining their efforts to solve the problems of feature extraction. Current advances in both biomedical signal processing and machine learning techniques have made it possible to develop recognition systems that can accomplish tasks that could not be achieved in the past. Feature extraction lies at the center of these developments with applications in biomedical signal analysis, as well as many other emerging applications (Guyon, Gunn, Nikravesh, & Zadeh, 2006).

Straightforward classifiers do not have a mechanism to deal with class boundaries. Hence, if the input variables (number of features) is large compared to the number of training data, class boundaries may not overlap. In such circumstances, the generalization ability of the classifier may not be good enough. Thus, to enhance the generalization ability, generally a small set of features from the original input variables are produced by feature extraction, dimension reduction, or feature selection. The most effective aspect in implementing a classifier with a high generalization ability is to employ a distinctive and informative set of features. But because there is no efficient way of finding an original set of features for a certain classification problem, it is required to find a set of original features by trial and error. If the number of features is very large, and every feature has little influence on the classification, it is more convenient to transform the set of features into a reduced set of features. In biomedical signal analysis, raw signal is transformed into a set of features by means of linear transformation. If each feature in the original set of features has an influence on the classification, the set is reduced by feature extraction, dimension reduction, or feature selection. By feature selection/reduction, meaningless or redundant features are eliminated such that a higher generalization performance and faster classification than by using the initial set of features can be achieved (Abe, 2010).

For effective classification, a precise feature extraction method is imperative to extract a distinctive and informative set of features from the original signals. Essentially, if the extracted features do not precisely designate the signals employed and are not relevant, a classification algorithm employing such features might have problems identifying the classes of the features (Siuly, Li, & Zhang, 2016). Consequently, classification accuracy might be lower. The feature extraction methods can be categorized into four groups: parametric methods, nonparametric methods, eigenvector methods, and time-frequency (TF) methods. The model-based or parametric methods produce a signal model with a known functional form, and then estimate the parameters in the produced model. The autoregressive (AR) model, moving average (MA) model, autoregressive-moving average (ARMA) model, and Lyapunov exponents are the popular parametric methods. The AR model is appropriate to present spectra with narrow peaks. The MA model achieves a good approximation for those spectra represented by broad peaks and sharp nulls. There is a slightly limited interest in employing the MA model for spectral estimation, because this type of spectra is found less frequently in applications than narrowband spectra. Spectra with both deep nulls and sharp peaks can be modeled by the ARMA model. The real-time ARMA estimators are computationally simple and often quite consistent, but their statistical accuracy may be poor in some cases (Kay, 1993; Kay & Marple, 1981; Proakis and Manolakis, 2007; Stoica & Moses, 1997). The nonparametric methods are based on the descriptions of power spectral density (PSD) to deliver spectral estimates. Two well-known nonparametric methods are the periodogram and the correlogram, providing a practically high resolution for adequately long data lengths, but they have poor spectral estimation because their variance is high and does not decrease with increasing data length. Eigenvector methods are employed to estimate the frequencies and powers of signals from noise-corrupted measurements. These methods are created by eigen decomposition of the

Practical Guide for Biomedical Signals Analysis Using Machine Learning Techniques. https://doi.org/10.1016/B978-0-12-817444-9.00004-0

correlation matrix of the noise-corrupted signal. The eigenvector methods such as multiple signal classification (MUSIC), Pisarenko, and minimum-norm are most appropriate for signals that are supposed to consist of numerous sinusoids suppressed with noise (Proakis and Manolakis, 2007; Stoica & Moses, 1997). The TF methods are extensively employed in biomedical signal processing and analysis (Siuly et al., 2016). Short-time Fourier transform (STFT), Wigner-Ville distribution, Cohen class kernel functions, wavelet transform (WT), discrete wavelet transform (DWT), wavelet packet decomposition (WPD), stationary wavelet transform (SWT), tunable Q-wavelet transform (TQWT), dual tree complex wavelet transform (DTCWT), empirical wavelet transform (EWT), empirical mode decomposition (EMD), ensemble EMD (EEMD), and complete EEMD (CEEMD) are widely known TF methods.

In the last decades, these feature extraction methods have been widely used in biomedical signal analysis. The performance of a classifier is contingent with the nature of the signal to be classified. There is no single classifier that operates best on all given problems. Numerous practical tests have been carried out to compare classifier performance and to recognize the characteristics of signals that determine the performance of the classifier. The total classification accuracy and confusion matrix are widely known methods to evaluate the performance of a classification system. Recently, receiver operating characteristic (ROC) curves have been employed to assess the trade-off between true- and false-positive rates of a given classification algorithm (Siuly et al., 2016).

This chapter explains different feature extraction methods for the classification of biomedical signals by giving different examples in MATLAB. Moreover, in this chapter, principal component analysis (PCA), independent component analysis (ICA), linear discriminant analysis (LDA), entropy, and statistical values are introduced as dimension reduction techniques.

4.2 FEATURE EXTRACTION METHODS

One of the critical stages in the classification of biomedical signals is feature extraction. Accordingly, the EEG signals consist of numerous data points, and distinctive and informative features can be extracted by using different feature extraction methods. These distinctive and informative parameters characterize the behavior of the biomedical signals that might indicate a specific action or activity. Emphasizing distinctive and informative features can characterize biomedical signals. The signal patterns employed in biomedical signal analysis can be represented by frequencies and amplitudes. These features can be extracted using different feature extraction algorithms, which is another step in signal processing to simplify the succeeding stage for classification (Graimann, Allison, & Pfurtscheller, 2009). Because all waveforms have limited duration and frequency, effective decomposition of the biomedical signals is needed for the integration of the time, frequency, and space dimensions. The biomedical signals can be decomposed using TF methods, which can detect changes in both time and frequency (Kevric & Subasi, 2017; Sanei, 2013).

It is essential to deal with smaller data that define appropriate features of the signals to achieve better performance. Features are usually accumulated into a feature vector by transforming signals into a relevant feature vector known as feature extraction. Distinctive features of a signal are analyzed by a signal classification framework, and depending on those distinctive features, a class of the signal is identified (Siuly et al., 2016). TF methods such as Wigner Ville transform, STFT, WT, DWT, WPD, SWT, TQWT, DTCWT, EWT, EMD, EEMD, and CEEMD decompose signals in both the time and frequency domains as explained in Chapter 3.

The aim of feature extraction is to extract features from the original signal to accomplish consistent classification. Feature extraction is the most crucial step of biomedical signal classification because the classification performance might be degraded if the features are not extracted in an appropriate way. The feature extraction step should reduce the original data to a lower dimension that includes most of the valuable information comprised in the original vector. Hence, it is essential to find the key features that characterize the whole dataset based on the nature of that dataset. Different statistical features can be extracted from each subsample data point because they are the most characteristic values to define the distribution of the biomedical signals. The features can be the minimum, maximum, mean, median, mode, standard deviation, variance, first quartile, third quartile, and inter-quartile range (IQR) of the biomedical signals (Siuly et al., 2016).

4.2.1 Examples for Feature Extraction

EXAMPLE 4.1. The following MATLAB code was used to extract features from the EEG signals using Yule-Walker AR spectrum. You can download data from the following website: http://epileptologie-bonn.de/cms/front_content.php?idcat=193&lang=3&changelang=3

```
%Ch4_Ex1_EEG_Yule.m
%The following MATLAB code is used to extract the features from
%the EEG signals using Yule-Walker AR spectrum.
```

```
clc
clear
%Load Sample EEG Data downloaded from the web site
%http://epileptologie-bonn.de/cms/front_content.php?idcat=193&lang=3&changelang=3
load AS_BONN_ALL_EEG_DATA_4096.mat

%%
Fs = 173.6;             % Sampling frequency
T = 1/Fs;               % Sampling period
segmentLength = 128;    % Length of a signal segment
Length=4096;% Length of signal
Nofsignal=100; %Number of Signal
order = 14;

%% A Z.ZIP EYES OPEN NORMAL SUBJECT

for i=1:Nofsignal
% Obtain the Yule-Walker AR Spectrum of the Normal EEG signal using pyulear.
    [Pxx,F] = pyulear(Normal_Eyes_Open(1:Length,i),order,segmentLength,Fs);
    ASData(i,:)=Pxx(:);
end
%%
% Obtain the Yule-Walker AR Spectrum of the Interictal EEG signal using pyulear.
[Pxx,F] = pyulear(Interictal(:,1),order,segmentLength,Fs);
for i=Nofsignal+1:2*Nofsignal
    [Pxx,F] = pyulear(Interictal(1:Length,i-Nofsignal),order,segmentLength,Fs);
    ASData(i,:)=Pxx(:);
end
%%
% Obtain the Yule-Walker AR Spectrum of the Ictal EEG signal using pyulear.
for i=2*Nofsignal+1:3*Nofsignal
[Pxx,F] = pyulear(Ictal(1:Length,i-2*Nofsignal),order,segmentLength,Fs);
ASData(i,:)=Pxx(:);
end
```

EXAMPLE 4.2 The following MATLAB code was used to extract features from the EEG signals using covariance spectrum. You can download data from the following website: http://epileptologie-bonn.de/cms/front_content.php?idcat = 193&lang = 3&changelang = 3

```
%Ch4_Ex6_EEG_Cov.m
%The following MATLAB code is used to extract the features from
%the EEG signals using Covariance.
clc
clear
%Load Sample EEG Data downloaded from the web site
%http://epileptologie-bonn.de/cms/front_content.php?idcat=193&lang=3&changelang=3
load AS_BONN_ALL_EEG_DATA_4096.mat
%%
Fs = 173.6;             % Sampling frequency
T = 1/Fs;               % Sampling period
segmentLength = 128;    % Length of a signal segment
Length=4096;% Length of signal
Nofsignal=100; %Number of Signal
order = 14;
```

```
%%
% Obtain the Covariance Spectrum of the Normal EEG signal using pcov.
for i=1:Nofsignal
    Pxx,F] = pcov(Normal_Eyes_Open(1:Length,i),order,segmentLength,Fs);
    ASData(i,:)=Pxx(:);
end
%%
% Obtain the Covariance Spectrum of the Interictal EEG signal using pcov.
for i=Nofsignal+1:2*Nofsignal
    [Pxx,F] = pcov(Interictal(1:Length,i-Nofsignal),order,segmentLength,Fs);
    ASData(i,:)=Pxx(:);
end
%%
% Obtain the Covariance Spectrum of the Ictal EEG signal using pcov.
for i=2*Nofsignal+1:3*Nofsignal
[Pxx,F] = pcov(Ictal(1:Length,i-2*Nofsignal),order,segmentLength,Fs);
ASData(i,:)=Pxx(:);
end
```

EXAMPLE 4.3 The following MATLAB code was used to extract features from the EEG signals using modified covariance spectrum. You can download data from the following website: http://epileptologie-bonn.de/cms/front_content.php?idcat=193&lang=3&changelang=3

```
%Ch4_Ex11_EEG_MCov.m
%The following MATLAB code is used to extract the features from
%the EEG signals using Modified Covariance.
clc
clear
%Load Sample EEG Data downloaded from the web site
%http://epileptologie-bonn.de/cms/front_content.php?idcat=193&lang=3&changelang=3
load AS_BONN_ALL_EEG_DATA_4096.mat
%%
Fs = 173.6;              % Sampling frequency
T = 1/Fs;                % Sampling period
segmentLength = 128;     % Length of a signal segment
Length=4096; % Length of signal
Nofsignal=100; %Number of Signal
order = 14;

%%
% Obtain the Modified Covariance Spectrum of the Normal EEG signal using pmcov.
for i=1:Nofsignal

    [Pxx,F] = pmcov(Normal_Eyes_Open(1:Length,i),order,segmentLength,Fs);
    ASData(i,:)=Pxx(:);
end
%%
% Obtain the Modified Covariance Spectrum of the Interictal EEG signal using pmcov.
for i=Nofsignal+1:2*Nofsignal
    [Pxx,F] = pmcov(Interictal(1:Length,i-Nofsignal),order,segmentLength,Fs);
    ASData(i,:)=Pxx(:);
end
%%
% Obtain the Modified Covariance Spectrum of the Ictal EEG signal using pmcov.
for i=2*Nofsignal+1:3*Nofsignal
[Pxx,F] = pmcov(Ictal(1:Length,i-2*Nofsignal),order,segmentLength,Fs);
ASData(i,:)=Pxx(:);
end
```

EXAMPLE 4.4 The following MATLAB code was used to extract features from the focal and nonfocal EEG signals using AR Burg spectrum. You can download data from the following website: http://ntsa.upf.edu/downloads/andrzejak-rg-schindler-k-rummel-c-2012-nonrandomness-nonlinear-dependence-and

```
%Ch4_Ex16_EEG_Focal_Burg.m
%The following MATLAB code is used to extract the features from
%the Focal and Non-Focal EEG signals using AR Burg spectrum.
clc
clear
%Load Sample EEG Data downloaded from the web site
%http://ntsa.upf.edu/downloads/andrzejak-rg-schindler-k-rummel-c-2012-nonrandomness-nonlinear-dependence-and
load FOCAL_NFOCAL.mat
%%
Fs = 512;              % Sampling frequency
T = 1/Fs;                % Sampling period
segmentLength = 128;    % Length of a signal segment
Length=4096;% Length of signal
Nofsignal=100; %Number of Signal
order = 14;
%%
% Obtain the Burg AR Spectrum of the FOCAL EEG signal using pburg.
for i=1:Nofsignal
    [Pxx,F] = pburg(focal(1:Length,i),order,segmentLength,Fs);
    ASData(i,:)=Pxx(:);
end
%%
% Obtain the Burg AR Spectrum of the NON-FOCAL EEG signal using pburg.
for i=Nofsignal+1:2*Nofsignal
    [Pxx,F] = pburg(nfocal(1:Length,i-Nofsignal),order,segmentLength,Fs);
    ASData(i,:)=Pxx(:);
end
```

EXAMPLE 4.5 The following MATLAB code was used to plot the SEED EMOTION EEG signals and their MUSIC spectrum. You can download data from the following website: http://bcmi.sjtu.edu.cn/~seed/index.html

```
%Ch4_Ex21_EEG_EMOTION_MUSIC.m
%The following MATLAB code is used to extract the features from
%the SEED EMOTION EEG signals and their MUSIC spectrum.
clc
clear
%Load Sample EEG Data downloaded from the web site
%http://bcmi.sjtu.edu.cn/~seed/index.html
load EMOTIONSDAT.mat
%%
Fs = 200;              % Sampling frequency
T = 1/Fs;                % Sampling period
NFFT = 128; %Number of FFT points
segmentLength = 128; % Length of a signal segment
Length=4096; % Length of signal
order = 14; %Order of the MUSIC Model
Nofsignal=100; %Number of Signal

%%
% Obtain the MUSIC Spectrum of the NEUTRAL EMOTIONAL EEG signal using pmusic.
```

```
for i=1:Nofsignal
[Pxx,F] = pmusic(NEUTRAL(i,1:Length),order,NFFT,Fs);
ASData(i,:)=Pxx(:);
end
%%
% Obtain the MUSIC Spectrum of the NEGATIVE EMOTIONAL EEG signal using pmusic.
for i=Nofsignal+1:2*Nofsignal
    [Pxx,F] = pmusic(NEGATIVE(i-Nofsignal,1:Length),order,NFFT,Fs);
    ASData(i,:)=Pxx(:);
end

%%
% Obtain the MUSIC Spectrum of the POSITIVE EMOTIONAL EEG signal using pmusic.
for i=2*Nofsignal+1:3*Nofsignal
[Pxx,F] = pmusic(POSITIVE(i-2*Nofsignal,1:Length),order,NFFT,Fs);
ASData(i,:)=Pxx(:);
end
```

EXAMPLE 4.6 The following MATLAB code was used to extract features from the EEG signals using EIGEN model. You can download data from the following website: http://epileptologie-bonn.de/cms/front_content.php?idcat=193&lang=3&changelang=3

```
%Ch4_Ex26_EEG_EIG.m
%The following MATLAB code is used to extract the features from
%the EEG signals using EIGEN Method.
clc
clear
%Load Sample EEG Data downloaded from the web site
%http://epileptologie-bonn.de/cms/front_content.php?idcat=193&lang=3&changelang=3
load AS_BONN_ALL_EEG_DATA_4096.mat
%%
Fs = 173.6;          % Sampling frequency
T = 1/Fs;            % Sampling period
segmentLength = 128; % Length of a signal segment
Length=4096; % Length of signal
Nofsignal=100; %Number of Signal
order = 34; %Order of the EIGEN Model
NFFT = 128; %Number of FFT points
Nofsignal=100; %Number of Signal

%%
% Obtain the EIGEN Spectrum of the NORMAL EEG signal using peig.
for i=1:Nofsignal
    [Pxx,F] = peig(Normal_Eyes_Open(1:Length,i),order,NFFT,Fs);
    ASData(i,:)=Pxx(:);
end
%%
% Obtain the EIGEN Spectrum of the INTERICTAL EEG signal using peig.
for i=Nofsignal+1:2*Nofsignal
    [Pxx,F] = peig(Interictal(1:Length,i-Nofsignal),order,NFFT,Fs);
    ASData(i,:)=Pxx(:);
end
%%
% Obtain the EIGEN Spectrum of the ICTAL EEG signal using peig.
for i=2*Nofsignal+1:3*Nofsignal
    [Pxx,F] = peig(Ictal(1:Length,i-2*Nofsignal),order,NFFT,Fs);
    ASData(i,:)=Pxx(:);
end
```

4.3 DIMENSION REDUCTION/FEATURE SELECTION METHODS

Dimension reduction is a procedure to decrease the dimension of the original feature vector, while retaining the most distinctive information and eliminating the remaining unrelated information, for decreasing the computational time in a classifier (Phinyomark et al., 2013). Most of the feature extraction methods yield redundant features. Actually, to improve the performance of a classifier and achieve minimum classification error, some types of feature selection/reduction methods that produce a new set of features must be applied. Several methods employed for dimension reduction and feature selection achieve better classification accuracy (Wołczowski & Zdunek, 2017). In feature selection and dimension reduction, the minimum subset of features should be selected from the original set of features that achieves the maximum generalization ability. To achieve this, the generalization ability of a subset of features are required to be estimated during the process of feature selection (Abe, 2010).

The dimension of biomedical signals should be reduced to analyze the data for achieving more accurate results. A small number of parameters are employed to reduce the dimension of the biomedical signals through different ways. Furthermore, the features or dimensions must be minimized to achieve better classification accuracy. For instance, the wavelet-based TF methods create wavelet coefficients to define the distribution of signal energy in both time and frequency domains, and they characterize the biomedical signals with a set of wavelet coefficients. Because wavelet-based feature extraction tools produce a feature vector that is too large to be used as an input to a classifier, a dimension reduction technique must be used to extract a smaller number of features from the wavelet coefficients. Recently, various dimension reduction methods such as Lyapunov exponents, low- or high-order statistics, and entropies have been employed for dimension reduction. Approximate entropy, which is a measure of complexity, can be applied to a noisy dataset and is superior than spectral entropy, Kolmogorov-Sinai entropy, and fractal dimension. Sample entropy presents less dependence on data length. Fuzzy entropy is another measure for complexity and originates from the fuzzy set theory. Another way of reducing dimension is to employ first-, second-, third-, and fourth-order statistics of the subbands, then the reduced feature set is calculated from the subbands of the TF decomposition.

Dimension reduction is another alternative to feature selection, for instance, PCA is employed effectively in diverse studies. Like feature selection, it produces a low-dimensional representation that assists to build lower capacity predictors to enhance generalization. But unlike feature selection, it may keep information from all the original input variables. Essentially, these methods are purely unsupervised, and they may eliminate low variance variations that are highly predictive of the target label or preserve some with high variance but are irrelevant for the classification task at hand. It is still possible to combine dimension reduction with a feature extraction algorithm to be utilized on the reduced dimensions to choose those most suitable for classification or vice versa (Bengio et al., 2003).

It is usual to determine that a subset of the feature set could achieve better classification accuracy than if all the features were used. The reason is that the presence of redundant features with reduced discriminative power may mystify the classifier. Therefore, the feature selection is of ultimate significance in generating a more accurate model for biomedical systems. There are several existing methods for feature selection of an optimal set of features to describe the problem (Begg, Lai, & Palaniswami, 2007).

Feature extraction transforms the data in the high-dimensional space to a low-dimensional space. The data transformation may be linear, as in PCA, but many non-linear dimension reduction techniques also exist. Dimension reduction can be done by LDA, PCA, and ICA. These three approaches will be defined to make our understanding clear about dimension reduction. As we know that PCA and ICA are used for two-dimensional non-Gaussian datasets, PCA extracts components with maximal variance and ICA extracts components with maximal independence. ICA considers signal elements as random variables with Gaussian distribution and minimized second-order statistics. Clearly, for any non-Gaussian distribution, the largest variances would not correspond to PCA basis vectors. ICA minimizes both second-order and higher-order dependencies in the input data and attempts to find the basis along which the data (when projected onto them) are statistically independent. LDA finds the vectors in the underlying space that best discriminate among classes (Delac, Grgic, & Grgic, 2005).

4.3.1 Statistical Features

Different statistical features can be extracted from each subsample data point because they are the most characteristic values to define the distribution of the biomedical signals. The features are the minimum, maximum, mean, median, mode, standard deviation, variance, first quartile, third quartile, and IQR of the biomedical signals (Siuly et al., 2016). The statistical values, namely mean, standard deviation, skewness, and kurtosis, are widely used to reduce the dimension of signal. The first- and second-order statistics are crucial in signal processing. On the other hand, when dealing with biomedical signals, many signals, particularly second-order statistics, are not enough. For this reason, higher-order statistics (HOS) should also be used for better characterization of the signals. Although the first- and second-order statistics characterize mean and variance, the HOS characterize higher-order moments (Kutlu & Kuntalp, 2012) (Mendel, 1991). If $X(n)$ is a random process, the moments of $X(n)$ can be defined as the coefficients in Taylor series expansion of the moment producing function (Kutlu & Kuntalp, 2012).

$$m_2(i) = E[X(n), X(n+i)] \tag{4.1}$$

$$\phi_x(w) = E[\exp(jwx)] \tag{4.2}$$

If discrete time signal has zero mean, then the moments are defined as:

$$m_3(i,j) = E[X(n), X(n+i) \cdot X(n+j)] \tag{4.3}$$

$$m_4(i,j,k) = E[X(n), X(n+i) \cdot X(n+j) \cdot X(n+k)] \tag{4.4}$$

where $E[\bullet]$ is the expected value of the random process $X(\bullet)$ (Kutlu & Kuntalp, 2012).

HOS refers to cumulants with three- and higher-order numbers that are linear combinations of lower-order moments and lower-order cumulants. The moments and cumulants are computed and then used as features under the supposition that the biomedical signal samples are random variables obtained from an unknown distribution, $P(x)$. The moment about zero, $\mu_n(0)$ (raw moment), of a continuous distribution, $P(x)$, is defined mathematically in following equation:

$$\mu_n(0) = \int x^n P(x) \tag{4.5}$$

In the case of biomedical signals, discrete distribution of $P(x)$ can be supposed and defines the moment as given in Eq. (4.6)

$$\mu_n(0) = \sum x^n P(x) \tag{4.6}$$

In Eq. (2.16), addition is taken over all existing discrete samples. Moments can be taken about a point in which case in Eq. (2.16) is calculated as given in Eq. (4.7)

$$\mu_n(a) = \sum (x-a)^n P(x) \tag{4.7}$$

The first moment, μ_1 represents the mean; the second moment, μ_2, the variance; and the third moment, μ_3, the skewness of the distribution that computes the asymmetry degree. The fourth-order moment is kurtosis measuring the distribution peakedness. Central moments are moments obtained about the distribution mean. The cumulants, k_n, of a distribution are the result obtained from the probability density function (PDF), $P(x)$. The higher-order cumulants are revealed to be linear combinations of the moments. Supposing raw moments, the first four cumulants can be expressed explicitly as given in Eq. (4.8)

$$k_1 = \mu_1$$

$$k_2 = \mu_2 - \mu_1^2$$

$$k_3 = 2\mu_1^3 - 3\mu_1\mu_2 + \mu_3$$

$$k_4 = -6\mu_1^4 + 12\mu_1^2\mu_2 - 3\mu_2^2 - 4\mu_1\mu_3 + \mu_4 \tag{4.8}$$

HOS methods for feature extraction are employed on biomedical signals to lessen the relative spread of properties (Begg et al., 2007).

Because TF-based feature extraction algorithms yield the feature vectors that are too large to be used as an input to a classifier, these types of dimension reduction techniques can be used to extract a smaller number of features from the coefficients. By employing first-, second-, third-, and fourth-order cumulants of each level of subband, the reduced new feature sets are calculated from the subbands of the signal decomposition. The capability of translating the set of coefficients into a reduced feature set designates one of the crucial steps in any classification task, because this reduced feature set characterizes the behavior of the biomedical signals in a better way. The following statistical features can be used for biomedical signal classification (Kevric & Subasi, 2017):

1. Mean absolute values of coefficients in each subband, $\mu = \frac{1}{M}\sum_{j=1}^{M} |y_j|$.
2. Average power of the coefficients in each subband, $\lambda = \sqrt{\frac{1}{M}\sum_{j=1}^{M} y_j^2}$.
3. Standard deviation of the coefficients in each subband, $\sigma = \sqrt{\frac{1}{M}\sum_{j=1}^{M} (y_j - \mu)^2}$.
4. Ratio of the absolute mean values of coefficients of adjacent subbands, $\chi = \frac{\sum_{j=1}^{M} |y_j|}{\sum_{j=1}^{M} |z_j|}$.
5. Skewness of the coefficients in each subband, $\varphi = \sqrt{\frac{1}{M}\sum_{j=1}^{M} \frac{(y_j - \mu)^3}{\sigma^3}}$.
6. Kurtosis of the coefficients in each subband, $\phi = \sqrt{\frac{1}{M}\sum_{j=1}^{M} \frac{(y_j - \mu)^4}{\sigma^4}}$.

4.3.2 Examples With Statistical Features

The following examples use different signal-processing techniques described in the previous chapter for feature extraction. The six statistical values utilized for the dimension reduction of biomedical signals are:

1. Mean absolute values of the signal coefficients in every subband.
2. Standard deviation of the signal coefficients in every subband.
3. Skewness of the signal coefficients in every subband.
4. Kurtosis of the signal coefficients in every subband.
5. Average power of the signal coefficients in every subband.
6. Ratio of the absolute mean values of signal coefficients of adjacent subbands.

EXAMPLE 4.7 The following MATLAB code was used to extract features from the EEG signals using DWT. You can download data from the following website:
http://epileptologie-bonn.de/cms/front_content.php?idcat=193&lang=3&changelang=3

```
%Ch4_Ex31_DWT_EEG.m
% DWT of NORMAL, INTERICTAL and ICTAL EEG signals
%%
%The following MATLAB code is used to Extract DWT features from the EEG signals
%Decompose EEG data using DWT coefficients
%Then uses Statistical features as:

%(1) Mean of the absolute values of the coefficients in each sub-band.
%(2) Standard deviation of the coefficients in each sub-band.
%(3) Skewness of the coefficients in each sub-band.
%(4) Kurtosis of the coefficients in each sub-band.
%(5) RMS power of the wavelet coefficients in each subband.
%(6) Ratio of the mean absolute values of adjacent subbands.

clc
clear
%Load Sample EEG Data downloaded from the web site
%http://epileptologie-bonn.de/cms/front_content.php?idcat=193&lang=3&changelang=3
load AS_BONN_ALL_EEG_DATA_4096.mat
wname = 'db4'; %Wavelet Name
level=8; %Decomposition Level
Length = 4096;     % Length of signal
Nofsignal=100; %Number of Signal
%%
%%%%%%%%%%%%%%%%%%%%%%%%%%%%%% A Z.ZIP EYES OPEN NORMAL SUBJECT

for i=1:Nofsignal;

[C,L]=wavedec(Normal_Eyes_Open(:,i),level,wname);

    cD1=detcoef(C,L,1);         %
    cD2=detcoef(C,L,2);
    cD3=detcoef(C,L,3);
    cD4=detcoef(C,L,4);         %
    cD5=detcoef(C,L,5);         %
    cD6=detcoef(C,L,6);         %
    cD7=detcoef(C,L,7);
    cD8=detcoef(C,L,8);         %
    cA8=appcoef(C,L,wname,level);

ASData(i,1)=mean(abs(cD1));
ASData(i,2)=mean(abs(cD2));
ASData(i,3)=mean(abs(cD3));
ASData(i,4)=mean(abs(cD4));
ASData(i,5)=mean(abs(cD5));
ASData(i,6)=mean(abs(cD6));
ASData(i,7)=mean(abs(cD7));
ASData(i,8)=mean(abs(cD8));
ASData(i,9)=mean(abs(cA8));

ASData(i,10)=std(cD1);
ASData(i,11)=std(cD2);
ASData(i,12)=std(cD3);
ASData(i,13)=std(cD4);
```

```
ASData(i,14)=std(cD5);
ASData(i,15)=std(cD6);
ASData(i,16)=std(cD7);
ASData(i,17)=std(cD8);
ASData(i,18)=std(cA8);

ASData(i,19)=skewness(cD1(:));
ASData(i,20)=skewness(cD2(:));
ASData(i,21)=skewness(cD3(:));
ASData(i,22)=skewness(cD4(:));
ASData(i,23)=skewness(cD5(:));
ASData(i,24)=skewness(cD6(:));
ASData(i,25)=skewness(cD7(:));
ASData(i,26)=skewness(cD8(:));
ASData(i,27)=skewness(cD8(:));

ASData(i,28)=kurtosis(cD1);
ASData(i,29)=kurtosis(cD2);
ASData(i,30)=kurtosis(cD3);
ASData(i,31)=kurtosis(cD4);
ASData(i,32)=kurtosis(cD5);
ASData(i,33)=kurtosis(cD6);
ASData(i,34)=kurtosis(cD7);
ASData(i,35)=kurtosis(cD8);
ASData(i,36)=kurtosis(cA8);

ASData(i,37)=rms(cD1(:));
ASData(i,38)=rms(cD2(:));
ASData(i,39)=rms(cD3(:));
ASData(i,40)=rms(cD4(:));
ASData(i,41)=rms(cD5(:));
ASData(i,42)=rms(cD6(:));
ASData(i,43)=rms(cD7(:));
ASData(i,44)=rms(cD8(:));
ASData(i,45)=rms(cA8(:));

ASData(i,46)=mean(abs(cD1))/mean(abs(cD2));
ASData(i,47)=mean(abs(cD2))/mean(abs(cD3));
ASData(i,48)=mean(abs(cD3))/mean(abs(cD4));
ASData(i,49)=mean(abs(cD4))/mean(abs(cD5));
ASData(i,50)=mean(abs(cD5))/mean(abs(cD6));
ASData(i,51)=mean(abs(cD6))/mean(abs(cD7));
ASData(i,52)=mean(abs(cD7))/mean(abs(cD8));
ASData(i,53)=mean(abs(cD8))/mean(abs(cA8));

end

%%
% D F.ZIP EPILEPTIC SUBJECT (INTERICTAL) were recorded from within
% the epileptogenic zone during seizure free intervals

for i=Nofsignal+1:2*Nofsignal;

 [C,L]=wavedec(Interictal(:,i-Nofsignal),level,wname);
     cD1=detcoef(C,L,1);         %
   cD2=detcoef(C,L,2);
   cD3=detcoef(C,L,3);
   cD4=detcoef(C,L,4);         %
```

```
    cD5=detcoef(C,L,5);         %
    cD6=detcoef(C,L,6);         %
    cD7=detcoef(C,L,7);
    cD8=detcoef(C,L,8);         %
    cA8=appcoef(C,L,wname,level);

ASData(i,1)=mean(abs(cD1));
ASData(i,2)=mean(abs(cD2));
ASData(i,3)=mean(abs(cD3));
ASData(i,4)=mean(abs(cD4));
ASData(i,5)=mean(abs(cD5));
ASData(i,6)=mean(abs(cD6));
ASData(i,7)=mean(abs(cD7));
ASData(i,8)=mean(abs(cD8));
ASData(i,9)=mean(abs(cA8));

ASData(i,10)=std(cD1);
ASData(i,11)=std(cD2);
ASData(i,12)=std(cD3);
ASData(i,13)=std(cD4);
ASData(i,14)=std(cD5);
ASData(i,15)=std(cD6);
ASData(i,16)=std(cD7);
ASData(i,17)=std(cD8);
ASData(i,18)=std(cA8);

ASData(i,19)=skewness(cD1(:));
ASData(i,20)=skewness(cD2(:));
ASData(i,21)=skewness(cD3(:));
ASData(i,22)=skewness(cD4(:));
ASData(i,23)=skewness(cD5(:));
ASData(i,24)=skewness(cD6(:));
ASData(i,25)=skewness(cD7(:));
ASData(i,26)=skewness(cD8(:));
ASData(i,27)=skewness(cD8(:));

ASData(i,28)=kurtosis(cD1);
ASData(i,29)=kurtosis(cD2);
ASData(i,30)=kurtosis(cD3);
ASData(i,31)=kurtosis(cD4);
ASData(i,32)=kurtosis(cD5);
ASData(i,33)=kurtosis(cD6);
ASData(i,34)=kurtosis(cD7);
ASData(i,35)=kurtosis(cD8);
ASData(i,36)=kurtosis(cA8);

ASData(i,37)=rms(cD1(:));
ASData(i,38)=rms(cD2(:));
ASData(i,39)=rms(cD3(:));
ASData(i,40)=rms(cD4(:));
ASData(i,41)=rms(cD5(:));
ASData(i,42)=rms(cD6(:));
ASData(i,43)=rms(cD7(:));
ASData(i,44)=rms(cD8(:));
ASData(i,45)=rms(cA8(:));

ASData(i,46)=mean(abs(cD1))/mean(abs(cD2));
ASData(i,47)=mean(abs(cD2))/mean(abs(cD3));
ASData(i,48)=mean(abs(cD3))/mean(abs(cD4));
```

```
ASData(i,49)=mean(abs(cD4))/mean(abs(cD5));
ASData(i,50)=mean(abs(cD5))/mean(abs(cD6));
ASData(i,51)=mean(abs(cD6))/mean(abs(cD7));
ASData(i,52)=mean(abs(cD7))/mean(abs(cD8));
ASData(i,53)=mean(abs(cD8))/mean(abs(cA8));

end

%% E S.ZIP EPILEPTIC SUBJECT ICTAL DURING SEIZURE
%%%%%%%%%%%%%%%%%%%%%%%%%%%%

for i=2*Nofsignal+1:3*Nofsignal;

[C,L]=wavedec(Ictal(:,i-2*Nofsignal),level,wname);

    cD1=detcoef(C,L,1);          %
    cD2=detcoef(C,L,2);
    cD3=detcoef(C,L,3);
    cD4=detcoef(C,L,4);          %
    cD5=detcoef(C,L,5);          %
    cD6=detcoef(C,L,6);          %
    cD7=detcoef(C,L,7);
    cD8=detcoef(C,L,8);          %
    cA8=appcoef(C,L,wname,level);


ASData(i,1)=mean(abs(cD1));
ASData(i,2)=mean(abs(cD2));
ASData(i,3)=mean(abs(cD3));
ASData(i,4)=mean(abs(cD4));
ASData(i,5)=mean(abs(cD5));
ASData(i,6)=mean(abs(cD6));
ASData(i,7)=mean(abs(cD7));
ASData(i,8)=mean(abs(cD8));
ASData(i,9)=mean(abs(cA8));

ASData(i,10)=std(cD1);
ASData(i,11)=std(cD2);
ASData(i,12)=std(cD3);
ASData(i,13)=std(cD4);
ASData(i,14)=std(cD5);
ASData(i,15)=std(cD6);
ASData(i,16)=std(cD7);
ASData(i,17)=std(cD8);
ASData(i,18)=std(cA8);

ASData(i,19)=skewness(cD1(:));
ASData(i,20)=skewness(cD2(:));
ASData(i,21)=skewness(cD3(:));
ASData(i,22)=skewness(cD4(:));
ASData(i,23)=skewness(cD5(:));
ASData(i,24)=skewness(cD6(:));
ASData(i,25)=skewness(cD7(:));
ASData(i,26)=skewness(cD8(:));
ASData(i,27)=skewness(cD8(:));

ASData(i,28)=kurtosis(cD1);
ASData(i,29)=kurtosis(cD2);
ASData(i,30)=kurtosis(cD3);
```

```
ASData(i,31)=kurtosis(cD4);
ASData(i,32)=kurtosis(cD5);
ASData(i,33)=kurtosis(cD6);
ASData(i,34)=kurtosis(cD7);
ASData(i,35)=kurtosis(cD8);
ASData(i,36)=kurtosis(cA8);

ASData(i,37)=rms(cD1(:));
ASData(i,38)=rms(cD2(:));
ASData(i,39)=rms(cD3(:));
ASData(i,40)=rms(cD4(:));
ASData(i,41)=rms(cD5(:));
ASData(i,42)=rms(cD6(:));
ASData(i,43)=rms(cD7(:));
ASData(i,44)=rms(cD8(:));
ASData(i,45)=rms(cA8(:));

ASData(i,46)=mean(abs(cD1))/mean(abs(cD2));
ASData(i,47)=mean(abs(cD2))/mean(abs(cD3));
ASData(i,48)=mean(abs(cD3))/mean(abs(cD4));
ASData(i,49)=mean(abs(cD4))/mean(abs(cD5));
ASData(i,50)=mean(abs(cD5))/mean(abs(cD6));
ASData(i,51)=mean(abs(cD6))/mean(abs(cD7));
ASData(i,52)=mean(abs(cD7))/mean(abs(cD8));
ASData(i,53)=mean(abs(cD8))/mean(abs(cA8));
end
```

EXAMPLE 4.8 The following MATLAB code was used to extract features from the EEG signals using SWT. You can download data from the following website:
http://epileptologie-bonn.de/cms/front_content.php?idcat=193&lang=3&changelang=3

```
%Ch4_Ex36_SWT_EEG.m
%SWT of NORMAL, INTERICTAL and ICTAL EEG signals
%The following MATLAB code is used to Extract SWT features from the EEG signals
%Decompose EEG data using SWT Transfrom
%Then uses Statistical features as:

%(1) Mean of the absolute values of the coefficients in each sub-band.
%(2) Standard deviation of the coefficients in each sub-band.
%(3) Skewness of the coefficients in each sub-band.
%(4) Kurtosis of the coefficients in each sub-band.
%(5) RMS power of the wavelet coefficients in each subband.
%(6) Ratio of the mean absolute values of adjacent subbands.

clc
clear
%Load Sample EEG Data downloaded from the web site
%http://epileptologie-bonn.de/cms/front_content.php?idcat=193&lang=3&changelang=3
load AS_BONN_ALL_EEG_DATA_4096.mat

wname = 'db4';
Level = 8;
Length = 4096; % Length of signal
Nofsignal=100; %Number of Signal
```

```
%%
%%%%%%%%%%%%%%%%%%%%%%%%%%%%%%% A Z.ZIP EYES OPEN NORMAL SUBJECT

for i=1:Nofsignal;

[swa,swd] = swt(Normal_Eyes_Open(1:Length,i),Level,wname);%SWT

ASData(i,1)=mean(abs(swa(1,:)));
ASData(i,2)=mean(abs(swa(2,:)));
ASData(i,3)=mean(abs(swa(3,:)));
ASData(i,4)=mean(abs(swa(4,:)));
ASData(i,5)=mean(abs(swa(5,:)));
ASData(i,6)=mean(abs(swa(6,:)));
ASData(i,7)=mean(abs(swa(7,:)));
ASData(i,8)=mean(abs(swa(8,:)));
ASData(i,9)=mean(abs(swd(1,:)));
ASData(i,10)=mean(abs(swd(2,:)));
ASData(i,11)=mean(abs(swd(3,:)));
ASData(i,12)=mean(abs(swd(4,:)));
ASData(i,13)=mean(abs(swd(5,:)));
ASData(i,14)=mean(abs(swd(6,:)));
ASData(i,15)=mean(abs(swd(7,:)));
ASData(i,16)=mean(abs(swd(8,:)));

ASData(i,17)=std(swa(1,:));
ASData(i,18)=std(swa(2,:));
ASData(i,19)=std(swa(3,:));
ASData(i,20)=std(swa(4,:));
ASData(i,21)=std(swa(5,:));
ASData(i,22)=std(swa(6,:));
ASData(i,23)=std(swa(7,:));
ASData(i,24)=std(swa(8,:));
ASData(i,25)=std(swd(1,:));
ASData(i,26)=std(swd(2,:));
ASData(i,27)=std(swd(3,:));
ASData(i,28)=std(swd(4,:));
ASData(i,29)=std(swd(5,:));
ASData(i,30)=std(swd(6,:));
ASData(i,31)=std(swd(7,:));
ASData(i,32)=std(swd(8,:));

ASData(i,33)=skewness(swa(1,:));
ASData(i,34)=skewness(swa(2,:));
ASData(i,35)=skewness(swa(3,:));
ASData(i,36)=skewness(swa(4,:));
ASData(i,37)=skewness(swa(5,:));
ASData(i,38)=skewness(swa(6,:));
ASData(i,39)=skewness(swa(7,:));
ASData(i,40)=skewness(swa(8,:));
ASData(i,41)=skewness(swd(1,:));
ASData(i,42)=skewness(swd(2,:));
ASData(i,43)=skewness(swd(3,:));
ASData(i,44)=skewness(swd(4,:));
ASData(i,45)=skewness(swd(5,:));
ASData(i,46)=skewness(swd(6,:));
ASData(i,47)=skewness(swd(7,:));
ASData(i,48)=skewness(swd(8,:));
```

```
ASData(i,49)=kurtosis(swa(1,:));
ASData(i,50)=kurtosis(swa(2,:));
ASData(i,51)=kurtosis(swa(3,:));
ASData(i,52)=kurtosis(swa(4,:));
ASData(i,53)=kurtosis(swa(5,:));
ASData(i,54)=kurtosis(swa(6,:));
ASData(i,55)=kurtosis(swa(7,:));
ASData(i,56)=kurtosis(swa(8,:));
ASData(i,57)=kurtosis(swd(1,:));
ASData(i,58)=kurtosis(swd(2,:));
ASData(i,59)=kurtosis(swd(3,:));
ASData(i,60)=kurtosis(swd(4,:));
ASData(i,61)=kurtosis(swd(5,:));
ASData(i,62)=kurtosis(swd(6,:));
ASData(i,63)=kurtosis(swd(7,:));
ASData(i,64)=kurtosis(swd(8,:));

        ASData(i,65)=rms(swa(1,:));
        ASData(i,66)=rms(swa(2,:));
        ASData(i,67)=rms(swa(3,:));
        ASData(i,68)=rms(swa(4,:));
        ASData(i,69)=rms(swa(5,:));
        ASData(i,70)=rms(swa(6,:));
        ASData(i,71)=rms(swa(7,:));
        ASData(i,72)=rms(swa(8,:));
        ASData(i,73)=rms(swd(1,:));
        ASData(i,74)=rms(swd(2,:));
        ASData(i,75)=rms(swd(3,:));
        ASData(i,76)=rms(swd(4,:));
        ASData(i,77)=rms(swd(5,:));
        ASData(i,78)=rms(swd(6,:));
        ASData(i,79)=rms(swd(7,:));
        ASData(i,80)=rms(swd(8,:));

        ASData(i,81)=mean(abs(swa(1,:)))/mean(abs(swa(2,:)));
        ASData(i,82)=mean(abs(swa(2,:)))/mean(abs(swa(3,:)));
        ASData(i,83)=mean(abs(swa(3,:)))/mean(abs(swa(4,:)));
        ASData(i,84)=mean(abs(swa(4,:)))/mean(abs(swa(5,:)));
        ASData(i,85)=mean(abs(swa(5,:)))/mean(abs(swa(6,:)));
        ASData(i,86)=mean(abs(swa(6,:)))/mean(abs(swa(7,:)));
        ASData(i,87)=mean(abs(swa(7,:)))/mean(abs(swa(8,:)));
        ASData(i,88)=mean(abs(swa(8,:)))/mean(abs(swd(1,:)));
        ASData(i,89)=mean(abs(swd(1,:)))/mean(abs(swd(2,:)));
        ASData(i,90)=mean(abs(swd(2,:)))/mean(abs(swd(3,:)));
        ASData(i,91)=mean(abs(swd(3,:)))/mean(abs(swd(4,:)));
        ASData(i,92)=mean(abs(swd(4,:)))/mean(abs(swd(5,:)));
        ASData(i,93)=mean(abs(swd(5,:)))/mean(abs(swd(6,:)));
        ASData(i,94)=mean(abs(swd(6,:)))/mean(abs(swd(7,:)));
        ASData(i,95)=mean(abs(swd(7,:)))/mean(abs(swd(8,:)));

end

%%
% D F.ZIP EPILEPTIC SUBJECT (INTERICTAL) were recorded from within
% the epileptogenic zone during seizure free intervals

for i=Nofsignal+1:2*Nofsignal;

[swa,swd] = swt(Interictal(1:Length,i-Nofsignal),Level,wname); %SWT
```

```
ASData(i,1)=mean(abs(swa(1,:)));
ASData(i,2)=mean(abs(swa(2,:)));
ASData(i,3)=mean(abs(swa(3,:)));
ASData(i,4)=mean(abs(swa(4,:)));
ASData(i,5)=mean(abs(swa(5,:)));
ASData(i,6)=mean(abs(swa(6,:)));
ASData(i,7)=mean(abs(swa(7,:)));
ASData(i,8)=mean(abs(swa(8,:)));
ASData(i,9)=mean(abs(swd(1,:)));
ASData(i,10)=mean(abs(swd(2,:)));
ASData(i,11)=mean(abs(swd(3,:)));
ASData(i,12)=mean(abs(swd(4,:)));
ASData(i,13)=mean(abs(swd(5,:)));
ASData(i,14)=mean(abs(swd(6,:)));
ASData(i,15)=mean(abs(swd(7,:)));
ASData(i,16)=mean(abs(swd(8,:)));

ASData(i,17)=std(swa(1,:));
ASData(i,18)=std(swa(2,:));
ASData(i,19)=std(swa(3,:));
ASData(i,20)=std(swa(4,:));
ASData(i,21)=std(swa(5,:));
ASData(i,22)=std(swa(6,:));
ASData(i,23)=std(swa(7,:));
ASData(i,24)=std(swa(8,:));
ASData(i,25)=std(swd(1,:));
ASData(i,26)=std(swd(2,:));
ASData(i,27)=std(swd(3,:));
ASData(i,28)=std(swd(4,:));
ASData(i,29)=std(swd(5,:));
ASData(i,30)=std(swd(6,:));
ASData(i,31)=std(swd(7,:));
ASData(i,32)=std(swd(8,:));

ASData(i,33)=skewness(swa(1,:));
ASData(i,34)=skewness(swa(2,:));
ASData(i,35)=skewness(swa(3,:));
ASData(i,36)=skewness(swa(4,:));
ASData(i,37)=skewness(swa(5,:));
ASData(i,38)=skewness(swa(6,:));
ASData(i,39)=skewness(swa(7,:));
ASData(i,40)=skewness(swa(8,:));
ASData(i,41)=skewness(swd(1,:));
ASData(i,42)=skewness(swd(2,:));
ASData(i,43)=skewness(swd(3,:));
ASData(i,44)=skewness(swd(4,:));
ASData(i,45)=skewness(swd(5,:));
ASData(i,46)=skewness(swd(6,:));
ASData(i,47)=skewness(swd(7,:));
ASData(i,48)=skewness(swd(8,:));

ASData(i,49)=kurtosis(swa(1,:));
ASData(i,50)=kurtosis(swa(2,:));
ASData(i,51)=kurtosis(swa(3,:));
ASData(i,52)=kurtosis(swa(4,:));
ASData(i,53)=kurtosis(swa(5,:));
ASData(i,54)=kurtosis(swa(6,:));
ASData(i,55)=kurtosis(swa(7,:));
```

```
ASData(i,56)=kurtosis(swa(8,:));
ASData(i,57)=kurtosis(swd(1,:));
ASData(i,58)=kurtosis(swd(2,:));
ASData(i,59)=kurtosis(swd(3,:));
ASData(i,60)=kurtosis(swd(4,:));
ASData(i,61)=kurtosis(swd(5,:));
ASData(i,62)=kurtosis(swd(6,:));
ASData(i,63)=kurtosis(swd(7,:));
ASData(i,64)=kurtosis(swd(8,:));

        ASData(i,65)=rms(swa(1,:));
        ASData(i,66)=rms(swa(2,:));
        ASData(i,67)=rms(swa(3,:));
        ASData(i,68)=rms(swa(4,:));
        ASData(i,69)=rms(swa(5,:));
        ASData(i,70)=rms(swa(6,:));
        ASData(i,71)=rms(swa(7,:));
        ASData(i,72)=rms(swa(8,:));
        ASData(i,73)=rms(swd(1,:));
        ASData(i,74)=rms(swd(2,:));
        ASData(i,75)=rms(swd(3,:));
        ASData(i,76)=rms(swd(4,:));
        ASData(i,77)=rms(swd(5,:));
        ASData(i,78)=rms(swd(6,:));
        ASData(i,79)=rms(swd(7,:));
        ASData(i,80)=rms(swd(8,:));

        ASData(i,81)=mean(abs(swa(1,:)))/mean(abs(swa(2,:)));
        ASData(i,82)=mean(abs(swa(2,:)))/mean(abs(swa(3,:)));
        ASData(i,83)=mean(abs(swa(3,:)))/mean(abs(swa(4,:)));
        ASData(i,84)=mean(abs(swa(4,:)))/mean(abs(swa(5,:)));
        ASData(i,85)=mean(abs(swa(5,:)))/mean(abs(swa(6,:)));
        ASData(i,86)=mean(abs(swa(6,:)))/mean(abs(swa(7,:)));
        ASData(i,87)=mean(abs(swa(7,:)))/mean(abs(swa(8,:)));
        ASData(i,88)=mean(abs(swa(8,:)))/mean(abs(swd(1,:)));
        ASData(i,89)=mean(abs(swd(1,:)))/mean(abs(swd(2,:)));
        ASData(i,90)=mean(abs(swd(2,:)))/mean(abs(swd(3,:)));
        ASData(i,91)=mean(abs(swd(3,:)))/mean(abs(swd(4,:)));
        ASData(i,92)=mean(abs(swd(4,:)))/mean(abs(swd(5,:)));
        ASData(i,93)=mean(abs(swd(5,:)))/mean(abs(swd(6,:)));
        ASData(i,94)=mean(abs(swd(6,:)))/mean(abs(swd(7,:)));
        ASData(i,95)=mean(abs(swd(7,:)))/mean(abs(swd(8,:)));

end

%% E S.ZIP EPILEPTIC SUBJECT ICTAL DURING SEIZURE
%%%%%%%%%%%%%%%%%%%%%%%%%%%%%%

for i=2*Nofsignal+1:3*Nofsignal;

[swa,swd] = swt(Ictal(1:Length,i-2*Nofsignal),Level,wname);%SWT

ASData(i,1)=mean(abs(swa(1,:)));
ASData(i,2)=mean(abs(swa(2,:)));
ASData(i,3)=mean(abs(swa(3,:)));
ASData(i,4)=mean(abs(swa(4,:)));
ASData(i,5)=mean(abs(swa(5,:)));
ASData(i,6)=mean(abs(swa(6,:)));
```

```
ASData(i,7)=mean(abs(swa(7,:)));
ASData(i,8)=mean(abs(swa(8,:)));
ASData(i,9)=mean(abs(swd(1,:)));
ASData(i,10)=mean(abs(swd(2,:)));
ASData(i,11)=mean(abs(swd(3,:)));
ASData(i,12)=mean(abs(swd(4,:)));
ASData(i,13)=mean(abs(swd(5,:)));
ASData(i,14)=mean(abs(swd(6,:)));
ASData(i,15)=mean(abs(swd(7,:)));
ASData(i,16)=mean(abs(swd(8,:)));

ASData(i,17)=std(swa(1,:));
ASData(i,18)=std(swa(2,:));
ASData(i,19)=std(swa(3,:));
ASData(i,20)=std(swa(4,:));
ASData(i,21)=std(swa(5,:));
ASData(i,22)=std(swa(6,:));
ASData(i,23)=std(swa(7,:));
ASData(i,24)=std(swa(8,:));
ASData(i,25)=std(swd(1,:));
ASData(i,26)=std(swd(2,:));
ASData(i,27)=std(swd(3,:));
ASData(i,28)=std(swd(4,:));
ASData(i,29)=std(swd(5,:));
ASData(i,30)=std(swd(6,:));
ASData(i,31)=std(swd(7,:));
ASData(i,32)=std(swd(8,:));

ASData(i,33)=skewness(swa(1,:));
ASData(i,34)=skewness(swa(2,:));
ASData(i,35)=skewness(swa(3,:));
ASData(i,36)=skewness(swa(4,:));
ASData(i,37)=skewness(swa(5,:));
ASData(i,38)=skewness(swa(6,:));
ASData(i,39)=skewness(swa(7,:));
ASData(i,40)=skewness(swa(8,:));
ASData(i,41)=skewness(swd(1,:));
ASData(i,42)=skewness(swd(2,:));
ASData(i,43)=skewness(swd(3,:));
ASData(i,44)=skewness(swd(4,:));
ASData(i,45)=skewness(swd(5,:));
ASData(i,46)=skewness(swd(6,:));
ASData(i,47)=skewness(swd(7,:));
ASData(i,48)=skewness(swd(8,:));

ASData(i,49)=kurtosis(swa(1,:));
ASData(i,50)=kurtosis(swa(2,:));
ASData(i,51)=kurtosis(swa(3,:));
ASData(i,52)=kurtosis(swa(4,:));
ASData(i,53)=kurtosis(swa(5,:));
ASData(i,54)=kurtosis(swa(6,:));
ASData(i,55)=kurtosis(swa(7,:));
ASData(i,56)=kurtosis(swa(8,:));
ASData(i,57)=kurtosis(swd(1,:));
ASData(i,58)=kurtosis(swd(2,:));
ASData(i,59)=kurtosis(swd(3,:));
ASData(i,60)=kurtosis(swd(4,:));
ASData(i,61)=kurtosis(swd(5,:));
```

```
ASData(i,62)=kurtosis(swd(6,:));
ASData(i,63)=kurtosis(swd(7,:));
ASData(i,64)=kurtosis(swd(8,:));

        ASData(i,65)=rms(swa(1,:));
        ASData(i,66)=rms(swa(2,:));
        ASData(i,67)=rms(swa(3,:));
        ASData(i,68)=rms(swa(4,:));
        ASData(i,69)=rms(swa(5,:));
        ASData(i,70)=rms(swa(6,:));
        ASData(i,71)=rms(swa(7,:));
        ASData(i,72)=rms(swa(8,:));
        ASData(i,73)=rms(swd(1,:));
        ASData(i,74)=rms(swd(2,:));
        ASData(i,75)=rms(swd(3,:));
        ASData(i,76)=rms(swd(4,:));
        ASData(i,77)=rms(swd(5,:));
        ASData(i,78)=rms(swd(6,:));
        ASData(i,79)=rms(swd(7,:));
        ASData(i,80)=rms(swd(8,:));

        ASData(i,81)=mean(abs(swa(1,:)))/mean(abs(swa(2,:)));
        ASData(i,82)=mean(abs(swa(2,:)))/mean(abs(swa(3,:)));
        ASData(i,83)=mean(abs(swa(3,:)))/mean(abs(swa(4,:)));
        ASData(i,84)=mean(abs(swa(4,:)))/mean(abs(swa(5,:)));
        ASData(i,85)=mean(abs(swa(5,:)))/mean(abs(swa(6,:)));
        ASData(i,86)=mean(abs(swa(6,:)))/mean(abs(swa(7,:)));
        ASData(i,87)=mean(abs(swa(7,:)))/mean(abs(swa(8,:)));
        ASData(i,88)=mean(abs(swa(8,:)))/mean(abs(swd(1,:)));
        ASData(i,89)=mean(abs(swd(1,:)))/mean(abs(swd(2,:)));
        ASData(i,90)=mean(abs(swd(2,:)))/mean(abs(swd(3,:)));
        ASData(i,91)=mean(abs(swd(3,:)))/mean(abs(swd(4,:)));
        ASData(i,92)=mean(abs(swd(4,:)))/mean(abs(swd(5,:)));
        ASData(i,93)=mean(abs(swd(5,:)))/mean(abs(swd(6,:)));
        ASData(i,94)=mean(abs(swd(6,:)))/mean(abs(swd(7,:)));
        ASData(i,95)=mean(abs(swd(7,:)))/mean(abs(swd(8,:)));
end
```

EXAMPLE 4.9 The following MATLAB code was used to extract the features from the EEG signals using the WPD. You can download data from the following website:
http://epileptologie-bonn.de/cms/front_content.php?idcat = 193&lang = 3&changelang = 3

```
%Ch4_Ex41_WPD_EEG.m
% WPD of NORMAL, INTERICTAL and ICTAL EEG signals
%The following MATLAB code is used to Extract WPD features from the EEG signals
%Decompose EEG data using WPD coefficients
%Then uses Statistical features as:

%(1) Mean of the absolute values of the coefficients in each sub-band.
%(2) Standard deviation of the coefficients in each sub-band.
%(3) Skewness of the coefficients in each sub-band.
%(4) Kurtosis of the coefficients in each sub-band.
%(5) RMS power of the wavelet coefficients in each subband.
%(6) Ratio of the mean absolute values of adjacent subbands.
```

```
clc
clear
%Load Sample EEG Data downloaded from the web site
%http://epileptologie-bonn.de/cms/front_content.php?idcat=193&lang=3&changelang=3
load AS_BONN_ALL_EEG_DATA_4096.mat
wname = 'db4';
Length = 4096;    % Length of signal
Nofsignal=100; %Number of Signal
%%
%%%%%%%%%%%%%%%%%%%%%%%%%%%%%% A Z.ZIP EYES OPEN NORMAL SUBJECT

for i=1:Nofsignal;

[wp10,wp11] = dwt(Normal_Eyes_Open(1:Length,i),wname);%WPD Decomposition
[wp20,wp21] = dwt(wp10,wname);
[wp22,wp23] = dwt(wp11,wname);

[wp30,wp31] = dwt(wp20,wname);
[wp32,wp33] = dwt(wp21,wname);
[wp34,wp35] = dwt(wp22,wname);
[wp36,wp37] = dwt(wp23,wname);

[wp40,wp41] = dwt(wp30,wname);
[wp42,wp43] = dwt(wp31,wname);
[wp44,wp45] = dwt(wp32,wname);
[wp46,wp47] = dwt(wp33,wname);
[wp48,wp49] = dwt(wp34,wname);
[wp4A,wp4B] = dwt(wp35,wname);
[wp4C,wp4D] = dwt(wp36,wname);
[wp4E,wp4F] = dwt(wp37,wname);

ASData(i,1)=mean(abs(wp40));
ASData(i,2)=mean(abs(wp41));
ASData(i,3)=mean(abs(wp42));
ASData(i,4)=mean(abs(wp43));
ASData(i,5)=mean(abs(wp44));
ASData(i,6)=mean(abs(wp45));
ASData(i,7)=mean(abs(wp46));
ASData(i,8)=mean(abs(wp47));
ASData(i,9)=mean(abs(wp48));
ASData(i,10)=mean(abs(wp49));
ASData(i,11)=mean(abs(wp4A));
ASData(i,12)=mean(abs(wp4B));
ASData(i,13)=mean(abs(wp4C));
ASData(i,14)=mean(abs(wp4D));
ASData(i,15)=mean(abs(wp4E));
ASData(i,16)=mean(abs(wp4F));

ASData(i,17)=std(wp40);
ASData(i,18)=std(wp41);
ASData(i,19)=std(wp42);
ASData(i,20)=std(wp43);
ASData(i,21)=std(wp44);
ASData(i,22)=std(wp45);
ASData(i,23)=std(wp46);
ASData(i,24)=std(wp47);
ASData(i,25)=std(wp48);
ASData(i,26)=std(wp49);
```

```
ASData(i,27)=std(wp4A);
ASData(i,28)=std(wp4B);
ASData(i,29)=std(wp4C);
ASData(i,30)=std(wp4D);
ASData(i,31)=std(wp4E);
ASData(i,32)=std(wp4F);

ASData(i,33)=skewness(wp40(:));
ASData(i,34)=skewness(wp41(:));
ASData(i,35)=skewness(wp42(:));
ASData(i,36)=skewness(wp43(:));
ASData(i,37)=skewness(wp44(:));
ASData(i,38)=skewness(wp45(:));
ASData(i,39)=skewness(wp46(:));
ASData(i,40)=skewness(wp47(:));
ASData(i,41)=skewness(wp48(:));
ASData(i,42)=skewness(wp49(:));
ASData(i,43)=skewness(wp4A(:));
ASData(i,44)=skewness(wp4B(:));
ASData(i,45)=skewness(wp4C(:));
ASData(i,46)=skewness(wp4D(:));
ASData(i,47)=skewness(wp4E(:));
ASData(i,48)=skewness(wp4F(:));

ASData(i,49)=kurtosis(wp40);
ASData(i,50)=kurtosis(wp41);
ASData(i,51)=kurtosis(wp42);
ASData(i,52)=kurtosis(wp43);
ASData(i,53)=kurtosis(wp44);
ASData(i,54)=kurtosis(wp45);
ASData(i,55)=kurtosis(wp46);
ASData(i,56)=kurtosis(wp47);
ASData(i,57)=kurtosis(wp48);
ASData(i,58)=kurtosis(wp49);
ASData(i,59)=kurtosis(wp4A);
ASData(i,60)=kurtosis(wp4B);
ASData(i,61)=kurtosis(wp4C);
ASData(i,62)=kurtosis(wp4D);
ASData(i,63)=kurtosis(wp4E);
ASData(i,64)=kurtosis(wp4F);

ASData(i,65)=rms(wp40(:));
ASData(i,66)=rms(wp41(:));
ASData(i,67)=rms(wp42(:));
ASData(i,68)=rms(wp43(:));
ASData(i,69)=rms(wp44(:));
ASData(i,70)=rms(wp45(:));
ASData(i,71)=rms(wp46(:));
ASData(i,72)=rms(wp47(:));
ASData(i,73)=rms(wp48(:));
ASData(i,74)=rms(wp49(:));
ASData(i,75)=rms(wp4A(:));
ASData(i,76)=rms(wp4B(:));
ASData(i,77)=rms(wp4C(:));
ASData(i,78)=rms(wp4D(:));
ASData(i,79)=rms(wp4E(:));
ASData(i,80)=rms(wp4F(:));
```

```
ASData(i,81)=mean(abs(wp40))/mean(abs(wp41));
ASData(i,82)=mean(abs(wp41))/mean(abs(wp42));
ASData(i,83)=mean(abs(wp42))/mean(abs(wp43));
ASData(i,84)=mean(abs(wp43))/mean(abs(wp44));
ASData(i,85)=mean(abs(wp44))/mean(abs(wp45));
ASData(i,86)=mean(abs(wp45))/mean(abs(wp46));
ASData(i,87)=mean(abs(wp46))/mean(abs(wp47));
ASData(i,88)=mean(abs(wp47))/mean(abs(wp48));
ASData(i,89)=mean(abs(wp48))/mean(abs(wp49));
ASData(i,90)=mean(abs(wp49))/mean(abs(wp4A));
ASData(i,91)=mean(abs(wp4A))/mean(abs(wp4B));
ASData(i,92)=mean(abs(wp4B))/mean(abs(wp4C));
ASData(i,93)=mean(abs(wp4C))/mean(abs(wp4D));
ASData(i,94)=mean(abs(wp4D))/mean(abs(wp4E));
ASData(i,95)=mean(abs(wp4E))/mean(abs(wp4F));

end

%%
% D F.ZIP EPILEPTIC SUBJECT (INTERICTAL) were recorded from within
% the epileptogenic zone during seizure free intervals

for i=Nofsignal+1:2*Nofsignal;

[wp10,wp11] = dwt(Interictal(1:Length,i-Nofsignal),wname);%WPD Decomposition
[wp20,wp21] = dwt(wp10,wname);
[wp22,wp23] = dwt(wp11,wname);

[wp30,wp31] = dwt(wp20,wname);
[wp32,wp33] = dwt(wp21,wname);
[wp34,wp35] = dwt(wp22,wname);
[wp36,wp37] = dwt(wp23,wname);

[wp40,wp41] = dwt(wp30,wname);
[wp42,wp43] = dwt(wp31,wname);
[wp44,wp45] = dwt(wp32,wname);
[wp46,wp47] = dwt(wp33,wname);
[wp48,wp49] = dwt(wp34,wname);
[wp4A,wp4B] = dwt(wp35,wname);
[wp4C,wp4D] = dwt(wp36,wname);
[wp4E,wp4F] = dwt(wp37,wname);

ASData(i,1)=mean(abs(wp40));
ASData(i,2)=mean(abs(wp41));
ASData(i,3)=mean(abs(wp42));
ASData(i,4)=mean(abs(wp43));
ASData(i,5)=mean(abs(wp44));
ASData(i,6)=mean(abs(wp45));
ASData(i,7)=mean(abs(wp46));
ASData(i,8)=mean(abs(wp47));
ASData(i,9)=mean(abs(wp48));
ASData(i,10)=mean(abs(wp49));
ASData(i,11)=mean(abs(wp4A));
ASData(i,12)=mean(abs(wp4B));
ASData(i,13)=mean(abs(wp4C));
ASData(i,14)=mean(abs(wp4D));
ASData(i,15)=mean(abs(wp4E));
ASData(i,16)=mean(abs(wp4F));
```

```
ASData(i,17)=std(wp40);
ASData(i,18)=std(wp41);
ASData(i,19)=std(wp42);
ASData(i,20)=std(wp43);
ASData(i,21)=std(wp44);
ASData(i,22)=std(wp45);
ASData(i,23)=std(wp46);
ASData(i,24)=std(wp47);
ASData(i,25)=std(wp48);
ASData(i,26)=std(wp49);
ASData(i,27)=std(wp4A);
ASData(i,28)=std(wp4B);
ASData(i,29)=std(wp4C);
ASData(i,30)=std(wp4D);
ASData(i,31)=std(wp4E);
ASData(i,32)=std(wp4F);

ASData(i,33)=skewness(wp40(:));
ASData(i,34)=skewness(wp41(:));
ASData(i,35)=skewness(wp42(:));
ASData(i,36)=skewness(wp43(:));
ASData(i,37)=skewness(wp44(:));
ASData(i,38)=skewness(wp45(:));
ASData(i,39)=skewness(wp46(:));
ASData(i,40)=skewness(wp47(:));
ASData(i,41)=skewness(wp48(:));
ASData(i,42)=skewness(wp49(:));
ASData(i,43)=skewness(wp4A(:));
ASData(i,44)=skewness(wp4B(:));
ASData(i,45)=skewness(wp4C(:));
ASData(i,46)=skewness(wp4D(:));
ASData(i,47)=skewness(wp4E(:));
ASData(i,48)=skewness(wp4F(:));

ASData(i,49)=kurtosis(wp40);
ASData(i,50)=kurtosis(wp41);
ASData(i,51)=kurtosis(wp42);
ASData(i,52)=kurtosis(wp43);
ASData(i,53)=kurtosis(wp44);
ASData(i,54)=kurtosis(wp45);
ASData(i,55)=kurtosis(wp46);
ASData(i,56)=kurtosis(wp47);
ASData(i,57)=kurtosis(wp48);
ASData(i,58)=kurtosis(wp49);
ASData(i,59)=kurtosis(wp4A);
ASData(i,60)=kurtosis(wp4B);
ASData(i,61)=kurtosis(wp4C);
ASData(i,62)=kurtosis(wp4D);
ASData(i,63)=kurtosis(wp4E);
ASData(i,64)=kurtosis(wp4F);

ASData(i,65)=rms(wp40(:));
ASData(i,66)=rms(wp41(:));
ASData(i,67)=rms(wp42(:));
ASData(i,68)=rms(wp43(:));
ASData(i,69)=rms(wp44(:));
ASData(i,70)=rms(wp45(:));
ASData(i,71)=rms(wp46(:));
```

```
ASData(i,72)=rms(wp47(:));
ASData(i,73)=rms(wp48(:));
ASData(i,74)=rms(wp49(:));
ASData(i,75)=rms(wp4A(:));
ASData(i,76)=rms(wp4B(:));
ASData(i,77)=rms(wp4C(:));
ASData(i,78)=rms(wp4D(:));
ASData(i,79)=rms(wp4E(:));
ASData(i,80)=rms(wp4F(:));

ASData(i,81)=mean(abs(wp40))/mean(abs(wp41));
ASData(i,82)=mean(abs(wp41))/mean(abs(wp42));
ASData(i,83)=mean(abs(wp42))/mean(abs(wp43));
ASData(i,84)=mean(abs(wp43))/mean(abs(wp44));
ASData(i,85)=mean(abs(wp44))/mean(abs(wp45));
ASData(i,86)=mean(abs(wp45))/mean(abs(wp46));
ASData(i,87)=mean(abs(wp46))/mean(abs(wp47));
ASData(i,88)=mean(abs(wp47))/mean(abs(wp48));
ASData(i,89)=mean(abs(wp48))/mean(abs(wp49));
ASData(i,90)=mean(abs(wp49))/mean(abs(wp4A));
ASData(i,91)=mean(abs(wp4A))/mean(abs(wp4B));
ASData(i,92)=mean(abs(wp4B))/mean(abs(wp4C));
ASData(i,93)=mean(abs(wp4C))/mean(abs(wp4D));
ASData(i,94)=mean(abs(wp4D))/mean(abs(wp4E));
ASData(i,95)=mean(abs(wp4E))/mean(abs(wp4F));
end

%% E S.ZIP EPILEPTIC SUBJECT ICTAL DURING SEIZURE
%%%%%%%%%%%%%%%%%%%%%%%%%%%%

for i=2*Nofsignal+1:3*Nofsignal;

[wp10,wp11] = dwt(Ictal(1:Length,i-2*Nofsignal),wname);%WPD Decomposition
[wp20,wp21] = dwt(wp10,wname);
[wp22,wp23] = dwt(wp11,wname);

[wp30,wp31] = dwt(wp20,wname);
[wp32,wp33] = dwt(wp21,wname);
[wp34,wp35] = dwt(wp22,wname);
[wp36,wp37] = dwt(wp23,wname);

[wp40,wp41] = dwt(wp30,wname);
[wp42,wp43] = dwt(wp31,wname);
[wp44,wp45] = dwt(wp32,wname);
[wp46,wp47] = dwt(wp33,wname);
[wp48,wp49] = dwt(wp34,wname);
[wp4A,wp4B] = dwt(wp35,wname);
[wp4C,wp4D] = dwt(wp36,wname);
[wp4E,wp4F] = dwt(wp37,wname);

ASData(i,1)=mean(abs(wp40));
ASData(i,2)=mean(abs(wp41));
ASData(i,3)=mean(abs(wp42));
ASData(i,4)=mean(abs(wp43));
ASData(i,5)=mean(abs(wp44));
ASData(i,6)=mean(abs(wp45));
ASData(i,7)=mean(abs(wp46));
ASData(i,8)=mean(abs(wp47));
```

```
ASData(i,9)=mean(abs(wp48));
ASData(i,10)=mean(abs(wp49));
ASData(i,11)=mean(abs(wp4A));
ASData(i,12)=mean(abs(wp4B));
ASData(i,13)=mean(abs(wp4C));
ASData(i,14)=mean(abs(wp4D));
ASData(i,15)=mean(abs(wp4E));
ASData(i,16)=mean(abs(wp4F));

ASData(i,17)=std(wp40);
ASData(i,18)=std(wp41);
ASData(i,19)=std(wp42);
ASData(i,20)=std(wp43);
ASData(i,21)=std(wp44);
ASData(i,22)=std(wp45);
ASData(i,23)=std(wp46);
ASData(i,24)=std(wp47);
ASData(i,25)=std(wp48);
ASData(i,26)=std(wp49);
ASData(i,27)=std(wp4A);
ASData(i,28)=std(wp4B);
ASData(i,29)=std(wp4C);
ASData(i,30)=std(wp4D);
ASData(i,31)=std(wp4E);
ASData(i,32)=std(wp4F);

ASData(i,33)=skewness(wp40(:));
ASData(i,34)=skewness(wp41(:));
ASData(i,35)=skewness(wp42(:));
ASData(i,36)=skewness(wp43(:));
ASData(i,37)=skewness(wp44(:));
ASData(i,38)=skewness(wp45(:));
ASData(i,39)=skewness(wp46(:));
ASData(i,40)=skewness(wp47(:));
ASData(i,41)=skewness(wp48(:));
ASData(i,42)=skewness(wp49(:));
ASData(i,43)=skewness(wp4A(:));
ASData(i,44)=skewness(wp4B(:));
ASData(i,45)=skewness(wp4C(:));
ASData(i,46)=skewness(wp4D(:));
ASData(i,47)=skewness(wp4E(:));
ASData(i,48)=skewness(wp4F(:));

ASData(i,49)=kurtosis(wp40);
ASData(i,50)=kurtosis(wp41);
ASData(i,51)=kurtosis(wp42);
ASData(i,52)=kurtosis(wp43);
ASData(i,53)=kurtosis(wp44);
ASData(i,54)=kurtosis(wp45);
ASData(i,55)=kurtosis(wp46);
ASData(i,56)=kurtosis(wp47);
ASData(i,57)=kurtosis(wp48);
ASData(i,58)=kurtosis(wp49);
ASData(i,59)=kurtosis(wp4A);
ASData(i,60)=kurtosis(wp4B);
ASData(i,61)=kurtosis(wp4C);
ASData(i,62)=kurtosis(wp4D);
ASData(i,63)=kurtosis(wp4E);
ASData(i,64)=kurtosis(wp4F);
```

```
ASData(i,65)=rms(wp40(:));
ASData(i,66)=rms(wp41(:));
ASData(i,67)=rms(wp42(:));
ASData(i,68)=rms(wp43(:));
ASData(i,69)=rms(wp44(:));
ASData(i,70)=rms(wp45(:));
ASData(i,71)=rms(wp46(:));
ASData(i,72)=rms(wp47(:));
ASData(i,73)=rms(wp48(:));
ASData(i,74)=rms(wp49(:));
ASData(i,75)=rms(wp4A(:));
ASData(i,76)=rms(wp4B(:));
ASData(i,77)=rms(wp4C(:));
ASData(i,78)=rms(wp4D(:));
ASData(i,79)=rms(wp4E(:));
ASData(i,80)=rms(wp4F(:));

ASData(i,81)=mean(abs(wp40))/mean(abs(wp41));
ASData(i,82)=mean(abs(wp41))/mean(abs(wp42));
ASData(i,83)=mean(abs(wp42))/mean(abs(wp43));
ASData(i,84)=mean(abs(wp43))/mean(abs(wp44));
ASData(i,85)=mean(abs(wp44))/mean(abs(wp45));
ASData(i,86)=mean(abs(wp45))/mean(abs(wp46));
ASData(i,87)=mean(abs(wp46))/mean(abs(wp47));
ASData(i,88)=mean(abs(wp47))/mean(abs(wp48));
ASData(i,89)=mean(abs(wp48))/mean(abs(wp49));
ASData(i,90)=mean(abs(wp49))/mean(abs(wp4A));
ASData(i,91)=mean(abs(wp4A))/mean(abs(wp4B));
ASData(i,92)=mean(abs(wp4B))/mean(abs(wp4C));
ASData(i,93)=mean(abs(wp4C))/mean(abs(wp4D));
ASData(i,94)=mean(abs(wp4D))/mean(abs(wp4E));
ASData(i,95)=mean(abs(wp4E))/mean(abs(wp4F));
end
```

EXAMPLE 4.10 The following MATLAB code was used to extract features from the EEG signals using the DTCWT. You can download data from the following website:
http://epileptologie-bonn.de/cms/front_content.php?idcat=193&lang=3&changelang=3

Note that the DTCWT toolbox is used for running the following MATLAB code. You can download it from the following website:

http://eeweb.poly.edu/iselesni/WaveletSoftware/dt1D.html

Do not forget to set the path for the DTCWT toolbox.

```
%Ch4_Ex46_DTCWT_EEG.m
% DTCWT of NORMAL, INTERICTAL and ICTAL EEG signals
%%
%The following MATLAB code is used to Extract DTCWT features from the EEG signals
%Decompose EEG data using DTCWT coefficients
%Then uses Statistical features as:

%(1) Mean of the absolute values of the coefficients in each sub-band.
%(2) Standard deviation of the coefficients in each sub-band.
%(3) Skewness of the coefficients in each sub-band.
%(4) Kurtosis of the coefficients in each sub-band.
%(5) RMS power of the wavelet coefficients in each subband.
%(6) Ratio of the mean absolute values of adjacent subbands.
```

```
clc
clear
%Load Sample EEG Data downloaded from the web site
%http://epileptologie-bonn.de/cms/front_content.php?idcat=193&lang=3&changelang=3
load AS_BONN_ALL_EEG_DATA_4096.mat
Length = 4096; % Length of signal
Nofsignal=100; %Number of Signal
J = 7; %Level of decomposition
[Faf, Fsf] = FSfarras; %Get PArameters
[af, sf] = dualfilt1;

%%
%%%%%%%%%%%%%%%%%%%%%%%%%% A Z.ZIP EYES OPEN NORMAL SUBJECT

for i=1:Nofsignal;

w = dualtree(Normal_Eyes_Open(1:Length,i), J, Faf, af); % Dual Tree CWT
%Signal length should be power of 2 (1024, 2048, 4096, 8192 etc)

ASData(i,1)=mean(abs(w{1}{1}));
ASData(i,2)=mean(abs(w{1}{2}));
ASData(i,3)=mean(abs(w{2}{1}));
ASData(i,4)=mean(abs(w{2}{2}));
ASData(i,5)=mean(abs(w{3}{1}));
ASData(i,6)=mean(abs(w{3}{2}));
ASData(i,7)=mean(abs(w{4}{1}));
ASData(i,8)=mean(abs(w{4}{2}));
ASData(i,9)=mean(abs(w{5}{1}));
ASData(i,10)=mean(abs(w{5}{2}));
ASData(i,11)=mean(abs(w{6}{1}));
ASData(i,12)=mean(abs(w{6}{2}));
ASData(i,13)=mean(abs(w{7}{1}));
ASData(i,14)=mean(abs(w{7}{2}));
ASData(i,15)=mean(abs(w{8}{1}));
ASData(i,16)=mean(abs(w{8}{1}));

ASData(i,17)=std(w{1}{1});
ASData(i,18)=std(w{1}{2});
ASData(i,19)=std(w{2}{1});
ASData(i,20)=std(w{2}{2});
ASData(i,21)=std(w{3}{1});
ASData(i,22)=std(w{3}{2});
ASData(i,23)=std(w{4}{1});
ASData(i,24)=std(w{4}{2});
ASData(i,25)=std(w{5}{1});
ASData(i,26)=std(w{5}{2});
ASData(i,27)=std(w{6}{1});
ASData(i,28)=std(w{6}{2});
ASData(i,29)=std(w{7}{1});
ASData(i,30)=std(w{7}{2});
ASData(i,31)=std(w{8}{1});
ASData(i,32)=std(w{8}{2});

ASData(i,33)=skewness(w{1}{1});
ASData(i,34)=skewness(w{1}{2});
ASData(i,35)=skewness(w{2}{1});
ASData(i,36)=skewness(w{2}{2});
ASData(i,37)=skewness(w{3}{1});
ASData(i,38)=skewness(w{3}{2});
ASData(i,39)=skewness(w{4}{1});
```

```
ASData(i,40)=skewness(w{4}{2});
ASData(i,41)=skewness(w{5}{1});
ASData(i,42)=skewness(w{5}{2});
ASData(i,43)=skewness(w{6}{1});
ASData(i,44)=skewness(w{6}{2});
ASData(i,45)=skewness(w{7}{1});
ASData(i,46)=skewness(w{7}{2});
ASData(i,47)=skewness(w{8}{1});
ASData(i,48)=skewness(w{8}{2});

ASData(i,49)=kurtosis(w{1}{1});
ASData(i,50)=kurtosis(w{1}{2});
ASData(i,51)=kurtosis(w{2}{1});
ASData(i,52)=kurtosis(w{2}{2});
ASData(i,53)=kurtosis(w{3}{1});
ASData(i,54)=kurtosis(w{3}{2});
ASData(i,55)=kurtosis(w{4}{1});
ASData(i,56)=kurtosis(w{4}{2});
ASData(i,57)=kurtosis(w{5}{1});
ASData(i,58)=kurtosis(w{5}{2});
ASData(i,59)=kurtosis(w{6}{1});
ASData(i,60)=kurtosis(w{6}{2});
ASData(i,61)=kurtosis(w{7}{1});
ASData(i,62)=kurtosis(w{7}{2});
ASData(i,63)=kurtosis(w{8}{1});
ASData(i,64)=kurtosis(w{8}{2});

        ASData(i,65)=rms(w{1}{1}(:));
        ASData(i,66)=rms(w{1}{2}(:));
        ASData(i,67)=rms(w{2}{1}(:));
        ASData(i,68)=rms(w{2}{2}(:));
        ASData(i,69)=rms(w{3}{1}(:));
        ASData(i,70)=rms(w{3}{2}(:));
        ASData(i,71)=rms(w{4}{1}(:));
        ASData(i,72)=rms(w{4}{2}(:));
        ASData(i,73)=rms(w{5}{1}(:));
        ASData(i,74)=rms(w{5}{2}(:));
        ASData(i,75)=rms(w{6}{1}(:));
        ASData(i,76)=rms(w{6}{2}(:));
        ASData(i,77)=rms(w{7}{1}(:));
        ASData(i,78)=rms(w{7}{2}(:));
        ASData(i,79)=rms(w{8}{1}(:));
        ASData(i,80)=rms(w{8}{2}(:));

        ASData(i,81)=mean(abs(w{1}{1}))/mean(abs(w{1}{2}));
        ASData(i,82)=mean(abs(w{1}{2}))/mean(abs(w{2}{1}));
        ASData(i,83)=mean(abs(w{2}{1}))/mean(abs(w{2}{2}));
        ASData(i,84)=mean(abs(w{2}{2}))/mean(abs(w{3}{1}));
        ASData(i,85)=mean(abs(w{3}{1}))/mean(abs(w{3}{2}));
        ASData(i,86)=mean(abs(w{3}{2}))/mean(abs(w{4}{1}));
        ASData(i,87)=mean(abs(w{4}{1}))/mean(abs(w{4}{2}));
        ASData(i,88)=mean(abs(w{4}{2}))/mean(abs(w{5}{1}));
        ASData(i,89)=mean(abs(w{5}{1}))/mean(abs(w{5}{2}));
        ASData(i,90)=mean(abs(w{5}{2}))/mean(abs(w{6}{1}));
        ASData(i,91)=mean(abs(w{6}{1}))/mean(abs(w{6}{2}));
        ASData(i,92)=mean(abs(w{6}{2}))/mean(abs(w{7}{1}));
        ASData(i,93)=mean(abs(w{7}{1}))/mean(abs(w{7}{2}));
        ASData(i,94)=mean(abs(w{7}{2}))/mean(abs(w{8}{1}));
        ASData(i,95)=mean(abs(w{8}{1}))/mean(abs(w{8}{2}));
end
```

```
%%
% D F.ZIP EPILEPTIC SUBJECT (INTERICTAL) were recorded from within
% the epileptogenic zone during seizure free intervals

for i=Nofsignal+1:2*Nofsignal;

w = dualtree(Interictal(1:Length,i-Nofsignal), J, Faf, af); % Dual Tree CWT
%Signal length should be power of 2 (1024, 2048, 4096, 8192 etc)

ASData(i,1)=mean(abs(w{1}{1}));
ASData(i,2)=mean(abs(w{1}{2}));
ASData(i,3)=mean(abs(w{2}{1}));
ASData(i,4)=mean(abs(w{2}{2}));
ASData(i,5)=mean(abs(w{3}{1}));
ASData(i,6)=mean(abs(w{3}{2}));
ASData(i,7)=mean(abs(w{4}{1}));
ASData(i,8)=mean(abs(w{4}{2}));
ASData(i,9)=mean(abs(w{5}{1}));
ASData(i,10)=mean(abs(w{5}{2}));
ASData(i,11)=mean(abs(w{6}{1}));
ASData(i,12)=mean(abs(w{6}{2}));
ASData(i,13)=mean(abs(w{7}{1}));
ASData(i,14)=mean(abs(w{7}{2}));
ASData(i,15)=mean(abs(w{8}{1}));
ASData(i,16)=mean(abs(w{8}{1}));

ASData(i,17)=std(w{1}{1});
ASData(i,18)=std(w{1}{2});
ASData(i,19)=std(w{2}{1});
ASData(i,20)=std(w{2}{2});
ASData(i,21)=std(w{3}{1});
ASData(i,22)=std(w{3}{2});
ASData(i,23)=std(w{4}{1});
ASData(i,24)=std(w{4}{2});
ASData(i,25)=std(w{5}{1});
ASData(i,26)=std(w{5}{2});
ASData(i,27)=std(w{6}{1});
ASData(i,28)=std(w{6}{2});
ASData(i,29)=std(w{7}{1});
ASData(i,30)=std(w{7}{2});
ASData(i,31)=std(w{8}{1});
ASData(i,32)=std(w{8}{2});

ASData(i,33)=skewness(w{1}{1});
ASData(i,34)=skewness(w{1}{2});
ASData(i,35)=skewness(w{2}{1});
ASData(i,36)=skewness(w{2}{2});
ASData(i,37)=skewness(w{3}{1});
ASData(i,38)=skewness(w{3}{2});
ASData(i,39)=skewness(w{4}{1});
ASData(i,40)=skewness(w{4}{2});
ASData(i,41)=skewness(w{5}{1});
ASData(i,42)=skewness(w{5}{2});
ASData(i,43)=skewness(w{6}{1});
ASData(i,44)=skewness(w{6}{2});
ASData(i,45)=skewness(w{7}{1});
ASData(i,46)=skewness(w{7}{2});
ASData(i,47)=skewness(w{8}{1});
ASData(i,48)=skewness(w{8}{2});
```

```
ASData(i,49)=kurtosis(w{1}{1});
ASData(i,50)=kurtosis(w{1}{2});
ASData(i,51)=kurtosis(w{2}{1});
ASData(i,52)=kurtosis(w{2}{2});
ASData(i,53)=kurtosis(w{3}{1});
ASData(i,54)=kurtosis(w{3}{2});
ASData(i,55)=kurtosis(w{4}{1});
ASData(i,56)=kurtosis(w{4}{2});
ASData(i,57)=kurtosis(w{5}{1});
ASData(i,58)=kurtosis(w{5}{2});
ASData(i,59)=kurtosis(w{6}{1});
ASData(i,60)=kurtosis(w{6}{2});
ASData(i,61)=kurtosis(w{7}{1});
ASData(i,62)=kurtosis(w{7}{2});
ASData(i,63)=kurtosis(w{8}{1});
ASData(i,64)=kurtosis(w{8}{2});

        ASData(i,65)=rms(w{1}{1}(:));
        ASData(i,66)=rms(w{1}{2}(:));
        ASData(i,67)=rms(w{2}{1}(:));
        ASData(i,68)=rms(w{2}{2}(:));
        ASData(i,69)=rms(w{3}{1}(:));
        ASData(i,70)=rms(w{3}{2}(:));
        ASData(i,71)=rms(w{4}{1}(:));
        ASData(i,72)=rms(w{4}{2}(:));
        ASData(i,73)=rms(w{5}{1}(:));
        ASData(i,74)=rms(w{5}{2}(:));
        ASData(i,75)=rms(w{6}{1}(:));
        ASData(i,76)=rms(w{6}{2}(:));
        ASData(i,77)=rms(w{7}{1}(:));
        ASData(i,78)=rms(w{7}{2}(:));
        ASData(i,79)=rms(w{8}{1}(:));
        ASData(i,80)=rms(w{8}{2}(:));

        ASData(i,81)=mean(abs(w{1}{1}))/mean(abs(w{1}{2}));
        ASData(i,82)=mean(abs(w{1}{2}))/mean(abs(w{2}{1}));
        ASData(i,83)=mean(abs(w{2}{1}))/mean(abs(w{2}{2}));
        ASData(i,84)=mean(abs(w{2}{2}))/mean(abs(w{3}{1}));
        ASData(i,85)=mean(abs(w{3}{1}))/mean(abs(w{3}{2}));
        ASData(i,86)=mean(abs(w{3}{2}))/mean(abs(w{4}{1}));
        ASData(i,87)=mean(abs(w{4}{1}))/mean(abs(w{4}{2}));
        ASData(i,88)=mean(abs(w{4}{2}))/mean(abs(w{5}{1}));
        ASData(i,89)=mean(abs(w{5}{1}))/mean(abs(w{5}{2}));
        ASData(i,90)=mean(abs(w{5}{2}))/mean(abs(w{6}{1}));
        ASData(i,91)=mean(abs(w{6}{1}))/mean(abs(w{6}{2}));
        ASData(i,92)=mean(abs(w{6}{2}))/mean(abs(w{7}{1}));
        ASData(i,93)=mean(abs(w{7}{1}))/mean(abs(w{7}{2}));
        ASData(i,94)=mean(abs(w{7}{2}))/mean(abs(w{8}{1}));
        ASData(i,95)=mean(abs(w{8}{1}))/mean(abs(w{8}{2}));
end

%% E S.ZIP EPILEPTIC SUBJECT ICTAL DURING SEIZURE
%%%%%%%%%%%%%%%%%%%%%%%%%%%%%%%%%%%%

for i=2*Nofsignal+1:3*Nofsignal;

w = dualtree(Ictal(1:Length,i-2*Nofsignal), J, Faf, af); % Dual Tree CWT
%Signal length should be power of 2 (1024, 2048, 4096, 8192 etc)
```

```
ASData(i,1)=mean(abs(w{1}{1}));
ASData(i,2)=mean(abs(w{1}{2}));
ASData(i,3)=mean(abs(w{2}{1}));
ASData(i,4)=mean(abs(w{2}{2}));
ASData(i,5)=mean(abs(w{3}{1}));
ASData(i,6)=mean(abs(w{3}{2}));
ASData(i,7)=mean(abs(w{4}{1}));
ASData(i,8)=mean(abs(w{4}{2}));
ASData(i,9)=mean(abs(w{5}{1}));
ASData(i,10)=mean(abs(w{5}{2}));
ASData(i,11)=mean(abs(w{6}{1}));
ASData(i,12)=mean(abs(w{6}{2}));
ASData(i,13)=mean(abs(w{7}{1}));
ASData(i,14)=mean(abs(w{7}{2}));
ASData(i,15)=mean(abs(w{8}{1}));
ASData(i,16)=mean(abs(w{8}{1}));

ASData(i,17)=std(w{1}{1});
ASData(i,18)=std(w{1}{2});
ASData(i,19)=std(w{2}{1});
ASData(i,20)=std(w{2}{2});
ASData(i,21)=std(w{3}{1});
ASData(i,22)=std(w{3}{2});
ASData(i,23)=std(w{4}{1});
ASData(i,24)=std(w{4}{2});
ASData(i,25)=std(w{5}{1});
ASData(i,26)=std(w{5}{2});
ASData(i,27)=std(w{6}{1});
ASData(i,28)=std(w{6}{2});
ASData(i,29)=std(w{7}{1});
ASData(i,30)=std(w{7}{2});
ASData(i,31)=std(w{8}{1});
ASData(i,32)=std(w{8}{2});

ASData(i,33)=skewness(w{1}{1});
ASData(i,34)=skewness(w{1}{2});
ASData(i,35)=skewness(w{2}{1});
ASData(i,36)=skewness(w{2}{2});
ASData(i,37)=skewness(w{3}{1});
ASData(i,38)=skewness(w{3}{2});
ASData(i,39)=skewness(w{4}{1});
ASData(i,40)=skewness(w{4}{2});
ASData(i,41)=skewness(w{5}{1});
ASData(i,42)=skewness(w{5}{2});
ASData(i,43)=skewness(w{6}{1});
ASData(i,44)=skewness(w{6}{2});
ASData(i,45)=skewness(w{7}{1});
ASData(i,46)=skewness(w{7}{2});
ASData(i,47)=skewness(w{8}{1});
ASData(i,48)=skewness(w{8}{2});

ASData(i,49)=kurtosis(w{1}{1});
ASData(i,50)=kurtosis(w{1}{2});
ASData(i,51)=kurtosis(w{2}{1});
ASData(i,52)=kurtosis(w{2}{2});
ASData(i,53)=kurtosis(w{3}{1});
ASData(i,54)=kurtosis(w{3}{2});
ASData(i,55)=kurtosis(w{4}{1});
```

```
ASData(i,56)=kurtosis(w{4}{2});
ASData(i,57)=kurtosis(w{5}{1});
ASData(i,58)=kurtosis(w{5}{2});
ASData(i,59)=kurtosis(w{6}{1});
ASData(i,60)=kurtosis(w{6}{2});
ASData(i,61)=kurtosis(w{7}{1});
ASData(i,62)=kurtosis(w{7}{2});
ASData(i,63)=kurtosis(w{8}{1});
ASData(i,64)=kurtosis(w{8}{2});

        ASData(i,65)=rms(w{1}{1}(:));
        ASData(i,66)=rms(w{1}{2}(:));
        ASData(i,67)=rms(w{2}{1}(:));
        ASData(i,68)=rms(w{2}{2}(:));
        ASData(i,69)=rms(w{3}{1}(:));
        ASData(i,70)=rms(w{3}{2}(:));
        ASData(i,71)=rms(w{4}{1}(:));
        ASData(i,72)=rms(w{4}{2}(:));
        ASData(i,73)=rms(w{5}{1}(:));
        ASData(i,74)=rms(w{5}{2}(:));
        ASData(i,75)=rms(w{6}{1}(:));
        ASData(i,76)=rms(w{6}{2}(:));
        ASData(i,77)=rms(w{7}{1}(:));
        ASData(i,78)=rms(w{7}{2}(:));
        ASData(i,79)=rms(w{8}{1}(:));
        ASData(i,80)=rms(w{8}{2}(:));

        ASData(i,81)=mean(abs(w{1}{1}))/mean(abs(w{1}{2}));
        ASData(i,82)=mean(abs(w{1}{2}))/mean(abs(w{2}{1}));
        ASData(i,83)=mean(abs(w{2}{1}))/mean(abs(w{2}{2}));
        ASData(i,84)=mean(abs(w{2}{2}))/mean(abs(w{3}{1}));
        ASData(i,85)=mean(abs(w{3}{1}))/mean(abs(w{3}{2}));
        ASData(i,86)=mean(abs(w{3}{2}))/mean(abs(w{4}{1}));
        ASData(i,87)=mean(abs(w{4}{1}))/mean(abs(w{4}{2}));
        ASData(i,88)=mean(abs(w{4}{2}))/mean(abs(w{5}{1}));
        ASData(i,89)=mean(abs(w{5}{1}))/mean(abs(w{5}{2}));
        ASData(i,90)=mean(abs(w{5}{2}))/mean(abs(w{6}{1}));
        ASData(i,91)=mean(abs(w{6}{1}))/mean(abs(w{6}{2}));
        ASData(i,92)=mean(abs(w{6}{2}))/mean(abs(w{7}{1}));
        ASData(i,93)=mean(abs(w{7}{1}))/mean(abs(w{7}{2}));
        ASData(i,94)=mean(abs(w{7}{2}))/mean(abs(w{8}{1}));
        ASData(i,95)=mean(abs(w{8}{1}))/mean(abs(w{8}{2}));
end
```

EXAMPLE 4.11 The following MATLAB code was used to extract features from the EEG signals using TQWT. You can download data from the following website:
http://epileptologie-bonn.de/cms/front_content.php?idcat=193&lang=3&changelang=3

Note that the TQWT toolbox is used for running the following MATLAB code. You can download it from the following website: http://eeweb.poly.edu/iselesni/TQWT/index.html

Do not forget to set the path for the TQWT toolbox.

```
%Ch4_Ex51_TQWT_EEG.m
%   TQWT of NORMAL, INTERICTAL and ICTAL EEG signals
```

```
%The following MATLAB code is used to Extract TQWT features from the EEG signals
%Decompose EEG data using TQWT Transfrom
%Then uses Statistical features as:

%(1) Mean of the absolute values of the coefficients in each sub-band.
%(2) Standard deviation of the coefficients in each sub-band.
%(3) Skewness of the coefficients in each sub-band.
%(4) Kurtosis of the coefficients in each sub-band.
%(5) RMS power of the wavelet coefficients in each subband.
%(6) Ratio of the mean absolute values of adjacent subbands.

%Note that the Tunable Q-Factor Wavelet Transform (TQWT) toolbox is used
%for running the following MATLAB code. You can download it from
%the following web site:
%http://eeweb.poly.edu/iselesni/TQWT/index.html
%Set the path for the Tunable Q-Factor Wavelet Transform (TQWT) toolbox

% Q Q-factor Q>=1
% Use Q = 1 for non-oscillatory signals. A higher value of Q
% is appropriate for oscillatory signals.

% r Redundancy r > 1
% The total over-sampling rate when the TQWT is computed
% over many levels. It is recommended to select r >= 3 in
% order that the analysis/synthesis functions (wavelets) be
% well localized.

% J Number of stages (levels) of the TQWT J >= 1
% J is the number of times the two-channel filter bank is
% iterated. There will be a total of J + 1 subbands. (The
% last subband will be the low-pass subband.)

clc
clear
%Load Sample EEG Data downloaded from the web site
%http://epileptologie-bonn.de/cms/front_content.php?idcat=193&lang=3&changelang=3
load AS_BONN_ALL_EEG_DATA_4096.mat

Q = 2; r = 3; % TQWT parameters
J = 15; %Number of Decomposition Level
Length = 4096;     % Length of signal
Nofsignal=100; %Number of Signal
%%
%%%%%%%%%%%%%%%%%%%%%%%%%%%%%% A Z.ZIP EYES OPEN NORMAL SUBJECT

for i=1:Nofsignal;

w = tqwt_radix2(Normal_Eyes_Open(1:Length,i),Q,r,J); % TQWT

ASData(i,1)=mean(abs(w{1}));
ASData(i,2)=mean(abs(w{2}));
ASData(i,3)=mean(abs(w{3}));
ASData(i,4)=mean(abs(w{4}));
ASData(i,5)=mean(abs(w{5}));
ASData(i,6)=mean(abs(w{6}));
ASData(i,7)=mean(abs(w{7}));
ASData(i,8)=mean(abs(w{8}));
ASData(i,9)=mean(abs(w{9}));
ASData(i,10)=mean(abs(w{10}));
```

```
ASData(i,11)=mean(abs(w{11}));
ASData(i,12)=mean(abs(w{12}));
ASData(i,13)=mean(abs(w{13}));
ASData(i,14)=mean(abs(w{14}));
ASData(i,15)=mean(abs(w{15}));
ASData(i,16)=mean(abs(w{16}));

ASData(i,17)=std(w{1});
ASData(i,18)=std(w{2});
ASData(i,19)=std(w{3});
ASData(i,20)=std(w{4});
ASData(i,21)=std(w{5});
ASData(i,22)=std(w{6});
ASData(i,23)=std(w{7});
ASData(i,24)=std(w{8});
ASData(i,25)=std(w{9});
ASData(i,26)=std(w{10});
ASData(i,27)=std(w{11});
ASData(i,28)=std(w{12});
ASData(i,29)=std(w{13});
ASData(i,30)=std(w{14});
ASData(i,31)=std(w{15});
ASData(i,32)=std(w{16});

ASData(i,33)=skewness(w{1});
ASData(i,34)=skewness(w{2});
ASData(i,35)=skewness(w{3});
ASData(i,36)=skewness(w{4});
ASData(i,37)=skewness(w{5});
ASData(i,38)=skewness(w{6});
ASData(i,39)=skewness(w{7});
ASData(i,40)=skewness(w{8});
ASData(i,41)=skewness(w{9});
ASData(i,42)=skewness(w{10});
ASData(i,43)=skewness(w{11});
ASData(i,44)=skewness(w{12});
ASData(i,45)=skewness(w{13});
ASData(i,46)=skewness(w{14});
ASData(i,47)=skewness(w{15});
ASData(i,48)=skewness(w{16});

ASData(i,49)=kurtosis(w{1});
ASData(i,50)=kurtosis(w{2});
ASData(i,51)=kurtosis(w{3});
ASData(i,52)=kurtosis(w{4});
ASData(i,53)=kurtosis(w{5});
ASData(i,54)=kurtosis(w{6});
ASData(i,55)=kurtosis(w{7});
ASData(i,56)=kurtosis(w{8});
ASData(i,57)=kurtosis(w{9});
ASData(i,58)=kurtosis(w{10});
ASData(i,59)=kurtosis(w{11});
ASData(i,60)=kurtosis(w{12});
ASData(i,61)=kurtosis(w{13});
ASData(i,62)=kurtosis(w{14});
ASData(i,63)=kurtosis(w{15});
ASData(i,64)=kurtosis(w{16});

        ASData(i,65)=rms(w{1}(:));
        ASData(i,66)=rms(w{2}(:));
```

```
        ASData(i,67)=rms(w{3}(:));
        ASData(i,68)=rms(w{4}(:));
        ASData(i,69)=rms(w{5}(:));
        ASData(i,70)=rms(w{6}(:));
        ASData(i,71)=rms(w{7}(:));
        ASData(i,72)=rms(w{8}(:));
        ASData(i,73)=rms(w{9}(:));
        ASData(i,74)=rms(w{10}(:));
        ASData(i,75)=rms(w{11}(:));
        ASData(i,76)=rms(w{12}(:));
        ASData(i,77)=rms(w{13}(:));
        ASData(i,78)=rms(w{14}(:));
        ASData(i,79)=rms(w{15}(:));
        ASData(i,80)=rms(w{16}(:));

        ASData(i,81)=mean(abs(w{1}))/mean(abs(w{2}));
        ASData(i,82)=mean(abs(w{2}))/mean(abs(w{3}));
        ASData(i,83)=mean(abs(w{3}))/mean(abs(w{4}));
        ASData(i,84)=mean(abs(w{4}))/mean(abs(w{5}));
        ASData(i,85)=mean(abs(w{5}))/mean(abs(w{6}));
        ASData(i,86)=mean(abs(w{6}))/mean(abs(w{7}));
        ASData(i,87)=mean(abs(w{7}))/mean(abs(w{8}));
        ASData(i,88)=mean(abs(w{8}))/mean(abs(w{9}));
        ASData(i,89)=mean(abs(w{9}))/mean(abs(w{10}));
        ASData(i,90)=mean(abs(w{10}))/mean(abs(w{11}));
        ASData(i,91)=mean(abs(w{11}))/mean(abs(w{12}));
        ASData(i,92)=mean(abs(w{12}))/mean(abs(w{13}));
        ASData(i,93)=mean(abs(w{13}))/mean(abs(w{14}));
        ASData(i,94)=mean(abs(w{14}))/mean(abs(w{15}));
        ASData(i,95)=mean(abs(w{15}))/mean(abs(w{16}));

end

%%
% D F.ZIP EPILEPTIC SUBJECT (INTERICTAL) were recorded from within
% the epileptogenic zone during seizure free intervals

for i=Nofsignal+1:2*Nofsignal;

w = tqwt_radix2(Interictal(1:Length,i-Nofsignal),Q,r,J); % TQWT

ASData(i,1)=mean(abs(w{1}));
ASData(i,2)=mean(abs(w{2}));
ASData(i,3)=mean(abs(w{3}));
ASData(i,4)=mean(abs(w{4}));
ASData(i,5)=mean(abs(w{5}));
ASData(i,6)=mean(abs(w{6}));
ASData(i,7)=mean(abs(w{7}));
ASData(i,8)=mean(abs(w{8}));
ASData(i,9)=mean(abs(w{9}));
ASData(i,10)=mean(abs(w{10}));
ASData(i,11)=mean(abs(w{11}));
ASData(i,12)=mean(abs(w{12}));
ASData(i,13)=mean(abs(w{13}));
ASData(i,14)=mean(abs(w{14}));
ASData(i,15)=mean(abs(w{15}));
ASData(i,16)=mean(abs(w{16}));
```

```
ASData(i,17)=std(w{1});
ASData(i,18)=std(w{2});
ASData(i,19)=std(w{3});
ASData(i,20)=std(w{4});
ASData(i,21)=std(w{5});
ASData(i,22)=std(w{6});
ASData(i,23)=std(w{7});
ASData(i,24)=std(w{8});
ASData(i,25)=std(w{9});
ASData(i,26)=std(w{10});
ASData(i,27)=std(w{11});
ASData(i,28)=std(w{12});
ASData(i,29)=std(w{13});
ASData(i,30)=std(w{14});
ASData(i,31)=std(w{15});
ASData(i,32)=std(w{16});

ASData(i,33)=skewness(w{1});
ASData(i,34)=skewness(w{2});
ASData(i,35)=skewness(w{3});
ASData(i,36)=skewness(w{4});
ASData(i,37)=skewness(w{5});
ASData(i,38)=skewness(w{6});
ASData(i,39)=skewness(w{7});
ASData(i,40)=skewness(w{8});
ASData(i,41)=skewness(w{9});
ASData(i,42)=skewness(w{10});
ASData(i,43)=skewness(w{11});
ASData(i,44)=skewness(w{12});
ASData(i,45)=skewness(w{13});
ASData(i,46)=skewness(w{14});
ASData(i,47)=skewness(w{15});
ASData(i,48)=skewness(w{16});

ASData(i,49)=kurtosis(w{1});
ASData(i,50)=kurtosis(w{2});
ASData(i,51)=kurtosis(w{3});
ASData(i,52)=kurtosis(w{4});
ASData(i,53)=kurtosis(w{5});
ASData(i,54)=kurtosis(w{6});
ASData(i,55)=kurtosis(w{7});
ASData(i,56)=kurtosis(w{8});
ASData(i,57)=kurtosis(w{9});
ASData(i,58)=kurtosis(w{10});
ASData(i,59)=kurtosis(w{11});
ASData(i,60)=kurtosis(w{12});
ASData(i,61)=kurtosis(w{13});
ASData(i,62)=kurtosis(w{14});
ASData(i,63)=kurtosis(w{15});
ASData(i,64)=kurtosis(w{16});

        ASData(i,65)=rms(w{1}(:));
        ASData(i,66)=rms(w{2}(:));
        ASData(i,67)=rms(w{3}(:));
        ASData(i,68)=rms(w{4}(:));
        ASData(i,69)=rms(w{5}(:));
        ASData(i,70)=rms(w{6}(:));
        ASData(i,71)=rms(w{7}(:));
```

```
        ASData(i,72)=rms(w{8}(:));
        ASData(i,73)=rms(w{9}(:));
        ASData(i,74)=rms(w{10}(:));
        ASData(i,75)=rms(w{11}(:));
        ASData(i,76)=rms(w{12}(:));
        ASData(i,77)=rms(w{13}(:));
        ASData(i,78)=rms(w{14}(:));
        ASData(i,79)=rms(w{15}(:));
        ASData(i,80)=rms(w{16}(:));

        ASData(i,81)=mean(abs(w{1}))/mean(abs(w{2}));
        ASData(i,82)=mean(abs(w{2}))/mean(abs(w{3}));
        ASData(i,83)=mean(abs(w{3}))/mean(abs(w{4}));
        ASData(i,84)=mean(abs(w{4}))/mean(abs(w{5}));
        ASData(i,85)=mean(abs(w{5}))/mean(abs(w{6}));
        ASData(i,86)=mean(abs(w{6}))/mean(abs(w{7}));
        ASData(i,87)=mean(abs(w{7}))/mean(abs(w{8}));
        ASData(i,88)=mean(abs(w{8}))/mean(abs(w{9}));
        ASData(i,89)=mean(abs(w{9}))/mean(abs(w{10}));
        ASData(i,90)=mean(abs(w{10}))/mean(abs(w{11}));
        ASData(i,91)=mean(abs(w{11}))/mean(abs(w{12}));
        ASData(i,92)=mean(abs(w{12}))/mean(abs(w{13}));
        ASData(i,93)=mean(abs(w{13}))/mean(abs(w{14}));
        ASData(i,94)=mean(abs(w{14}))/mean(abs(w{15}));
        ASData(i,95)=mean(abs(w{15}))/mean(abs(w{16}));

end

%% E S.ZIP EPILEPTIC SUBJECT ICTAL DURING SEIZURE
%%%%%%%%%%%%%%%%%%%%%%%%%%%%%

for i=2*Nofsignal+1:3*Nofsignal;

w = tqwt_radix2(Ictal(1:Length,i-2*Nofsignal),Q,r,J); % TQWT

ASData(i,1)=mean(abs(w{1}));
ASData(i,2)=mean(abs(w{2}));
ASData(i,3)=mean(abs(w{3}));
ASData(i,4)=mean(abs(w{4}));
ASData(i,5)=mean(abs(w{5}));
ASData(i,6)=mean(abs(w{6}));
ASData(i,7)=mean(abs(w{7}));
ASData(i,8)=mean(abs(w{8}));
ASData(i,9)=mean(abs(w{9}));
ASData(i,10)=mean(abs(w{10}));
ASData(i,11)=mean(abs(w{11}));
ASData(i,12)=mean(abs(w{12}));
ASData(i,13)=mean(abs(w{13}));
ASData(i,14)=mean(abs(w{14}));
ASData(i,15)=mean(abs(w{15}));
ASData(i,16)=mean(abs(w{16}));

ASData(i,17)=std(w{1});
ASData(i,18)=std(w{2});
ASData(i,19)=std(w{3});
ASData(i,20)=std(w{4});
ASData(i,21)=std(w{5});
ASData(i,22)=std(w{6});
ASData(i,23)=std(w{7});
```

```
ASData(i,24)=std(w{8});
ASData(i,25)=std(w{9});
ASData(i,26)=std(w{10});
ASData(i,27)=std(w{11});
ASData(i,28)=std(w{12});
ASData(i,29)=std(w{13});
ASData(i,30)=std(w{14});
ASData(i,31)=std(w{15});
ASData(i,32)=std(w{16});

ASData(i,33)=skewness(w{1});
ASData(i,34)=skewness(w{2});
ASData(i,35)=skewness(w{3});
ASData(i,36)=skewness(w{4});
ASData(i,37)=skewness(w{5});
ASData(i,38)=skewness(w{6});
ASData(i,39)=skewness(w{7});
ASData(i,40)=skewness(w{8});
ASData(i,41)=skewness(w{9});
ASData(i,42)=skewness(w{10});
ASData(i,43)=skewness(w{11});
ASData(i,44)=skewness(w{12});
ASData(i,45)=skewness(w{13});
ASData(i,46)=skewness(w{14});
ASData(i,47)=skewness(w{15});
ASData(i,48)=skewness(w{16});

ASData(i,49)=kurtosis(w{1});
ASData(i,50)=kurtosis(w{2});
ASData(i,51)=kurtosis(w{3});
ASData(i,52)=kurtosis(w{4});
ASData(i,53)=kurtosis(w{5});
ASData(i,54)=kurtosis(w{6});
ASData(i,55)=kurtosis(w{7});
ASData(i,56)=kurtosis(w{8});
ASData(i,57)=kurtosis(w{9});
ASData(i,58)=kurtosis(w{10});
ASData(i,59)=kurtosis(w{11});
ASData(i,60)=kurtosis(w{12});
ASData(i,61)=kurtosis(w{13});
ASData(i,62)=kurtosis(w{14});
ASData(i,63)=kurtosis(w{15});
ASData(i,64)=kurtosis(w{16});

       ASData(i,65)=rms(w{1}(:));
       ASData(i,66)=rms(w{2}(:));
       ASData(i,67)=rms(w{3}(:));
       ASData(i,68)=rms(w{4}(:));
       ASData(i,69)=rms(w{5}(:));
       ASData(i,70)=rms(w{6}(:));
       ASData(i,71)=rms(w{7}(:));
       ASData(i,72)=rms(w{8}(:));
       ASData(i,73)=rms(w{9}(:));
       ASData(i,74)=rms(w{10}(:));
       ASData(i,75)=rms(w{11}(:));
       ASData(i,76)=rms(w{12}(:));
       ASData(i,77)=rms(w{13}(:));
       ASData(i,78)=rms(w{14}(:));
       ASData(i,79)=rms(w{15}(:));
       ASData(i,80)=rms(w{16}(:));
```

```
        ASData(i,81)=mean(abs(w{1}))/mean(abs(w{2}));
        ASData(i,82)=mean(abs(w{2}))/mean(abs(w{3}));
        ASData(i,83)=mean(abs(w{3}))/mean(abs(w{4}));
        ASData(i,84)=mean(abs(w{4}))/mean(abs(w{5}));
        ASData(i,85)=mean(abs(w{5}))/mean(abs(w{6}));
        ASData(i,86)=mean(abs(w{6}))/mean(abs(w{7}));
        ASData(i,87)=mean(abs(w{7}))/mean(abs(w{8}));
        ASData(i,88)=mean(abs(w{8}))/mean(abs(w{9}));
        ASData(i,89)=mean(abs(w{9}))/mean(abs(w{10}));
        ASData(i,90)=mean(abs(w{10}))/mean(abs(w{11}));
        ASData(i,91)=mean(abs(w{11}))/mean(abs(w{12}));
        ASData(i,92)=mean(abs(w{12}))/mean(abs(w{13}));
        ASData(i,93)=mean(abs(w{13}))/mean(abs(w{14}));
        ASData(i,94)=mean(abs(w{14}))/mean(abs(w{15}));
        ASData(i,95)=mean(abs(w{15}))/mean(abs(w{16}));

end
```

EXAMPLE 4.12 The following MATLAB code was used to extract features from the EEG signals using FAWT. You can download data from the following website:

http://epileptologie-bonn.de/cms/front_content.php?idcat = 193&lang = 3&changelang = 3

Note that the FAWT toolbox is used for running the following MATLAB code. You can download it from the following website:

http://web.itu.edu.tr/ibayram/AnDWT/

Do not forget to set the path for the FAWT toolbox.

```
%Ch4_Ex56_FAWT_EEG.m
%   FAWT of NORMAL, INTERICTAL and ICTAL EEG signals
%The following MATLAB code is used to Extract FAWT features from the EEG signals
%Decompose EEG data using FAWT Transfrom
%Then uses Statistical features as:

%(1) Mean of the absolute values of the coefficients in each sub-band.
%(2) Standard deviation of the coefficients in each sub-band.
%(3) Skewness of the coefficients in each sub-band.
%(4) Kurtosis of the coefficients in each sub-band.
%(5) RMS power of the wavelet coefficients in each subband.
%(6) Ratio of the mean absolute values of adjacent subbands.

%Flexible analytic wavelet transform (FAWT) of NORMAL, INTERICTAL and ICTAL EEG signals
%Note that the Flexible analytic wavelet transform (FAWT) toolbox is used
%for running the following MATLAB code. You can download it from
%the following web site:
%http://web.itu.edu.tr/ibayram/AnDWT/
%Set the path for the Flexible analytic wavelet transform (FAWT) toolbox

clc
clear
%Load Sample EEG Data downloaded from the web site
%http://epileptologie-bonn.de/cms/front_content.php?idcat=193&lang=3&changelang=3
load AS_BONN_ALL_EEG_DATA_4096.mat

p = 5; %Set the FAWT parameters
q = 6;
```

```
r = 1;
s = 2;
bet = 0.8*r/s;
J = 10; %Number of Decomposition Level
Par = 200;

Length = 4096;    % Length of signal
Nofsignal=100; %Number of Signal
[Fp] = CreateFilters(Length,p,q,r,s,bet,J,Par);%filters for the positive
chirp parameter

%%
%%%%%%%%%%%%%%%%%%%%%%%%%%%%%%% A Z.ZIP EYES OPEN NORMAL SUBJECT

for i=1:Nofsignal;

w = RAnDwt(Normal_Eyes_Open(1:Length,i),p,q,r,s,J,Fp); %FAWT Transform

ASData(i,1)=mean(abs(abs(w{1,1})));
ASData(i,2)=mean(abs(abs(w{1,2})));
ASData(i,3)=mean(abs(abs(w{2,1})));
ASData(i,4)=mean(abs(abs(w{2,2})));
ASData(i,5)=mean(abs(abs(w{3,1})));
ASData(i,6)=mean(abs(abs(w{3,2})));
ASData(i,7)=mean(abs(abs(w{4,1})));
ASData(i,8)=mean(abs(abs(w{4,2})));
ASData(i,9)=mean(abs(abs(w{5,1})));
ASData(i,10)=mean(abs(abs(w{5,2})));
ASData(i,11)=mean(abs(abs(w{6,1})));
ASData(i,12)=mean(abs(abs(w{6,2})));
ASData(i,13)=mean(abs(abs(w{7,1})));
ASData(i,14)=mean(abs(abs(w{7,2})));
ASData(i,15)=mean(abs(abs(w{8,1})));
ASData(i,16)=mean(abs(abs(w{8,2})));

ASData(i,17)=std(abs(w{1,1}));
ASData(i,18)=std(abs(w{1,2}));
ASData(i,19)=std(abs(w{2,1}));
ASData(i,20)=std(abs(w{2,2}));
ASData(i,21)=std(abs(w{3,1}));
ASData(i,22)=std(abs(w{3,2}));
ASData(i,23)=std(abs(w{4,1}));
ASData(i,24)=std(abs(w{4,2}));
ASData(i,25)=std(abs(w{5,1}));
ASData(i,26)=std(abs(w{5,2}));
ASData(i,27)=std(abs(w{6,1}));
ASData(i,28)=std(abs(w{6,2}));
ASData(i,29)=std(abs(w{7,1}));
ASData(i,30)=std(abs(w{7,2}));
ASData(i,31)=std(abs(w{8,1}));
ASData(i,32)=std(abs(w{8,2}));

ASData(i,33)=skewness(abs(w{1,1}));
ASData(i,34)=skewness(abs(w{1,2}));
ASData(i,35)=skewness(abs(w{2,1}));
ASData(i,36)=skewness(abs(w{2,2}));
ASData(i,37)=skewness(abs(w{3,1}));
ASData(i,38)=skewness(abs(w{3,2}));
ASData(i,39)=skewness(abs(w{4,1}));
```

```
ASData(i,40)=skewness(abs(w{4,2}));
ASData(i,41)=skewness(abs(w{5,1}));
ASData(i,42)=skewness(abs(w{5,2}));
ASData(i,43)=skewness(abs(w{6,1}));
ASData(i,44)=skewness(abs(w{6,2}));
ASData(i,45)=skewness(abs(w{7,1}));
ASData(i,46)=skewness(abs(w{7,2}));
ASData(i,47)=skewness(abs(w{8,1}));
ASData(i,48)=skewness(abs(w{8,2}));

ASData(i,49)=kurtosis(abs(w{1,1}));
ASData(i,50)=kurtosis(abs(w{1,2}));
ASData(i,51)=kurtosis(abs(w{2,1}));
ASData(i,52)=kurtosis(abs(w{2,2}));
ASData(i,53)=kurtosis(abs(w{3,1}));
ASData(i,54)=kurtosis(abs(w{3,2}));
ASData(i,55)=kurtosis(abs(w{4,1}));
ASData(i,56)=kurtosis(abs(w{4,2}));
ASData(i,57)=kurtosis(abs(w{5,1}));
ASData(i,58)=kurtosis(abs(w{5,2}));
ASData(i,59)=kurtosis(abs(w{6,1}));
ASData(i,60)=kurtosis(abs(w{6,2}));
ASData(i,61)=kurtosis(abs(w{7,1}));
ASData(i,62)=kurtosis(abs(w{7,2}));
ASData(i,63)=kurtosis(abs(w{8,1}));
ASData(i,64)=kurtosis(abs(w{8,2}));

        ASData(i,65)=rms(abs(w{1,1}(:)));
        ASData(i,66)=rms(abs(w{1,2}(:)));
        ASData(i,67)=rms(abs(w{2,1}(:)));
        ASData(i,68)=rms(abs(w{2,2}(:)));
        ASData(i,69)=rms(abs(w{3,1}(:)));
        ASData(i,70)=rms(abs(w{3,2}(:)));
        ASData(i,71)=rms(abs(w{4,1}(:)));
        ASData(i,72)=rms(abs(w{4,2}(:)));
        ASData(i,73)=rms(abs(w{5,1}(:)));
        ASData(i,74)=rms(abs(w{5,2}(:)));
        ASData(i,75)=rms(abs(w{6,1}(:)));
        ASData(i,76)=rms(abs(w{6,2}(:)));
        ASData(i,77)=rms(abs(w{7,1}(:)));
        ASData(i,78)=rms(abs(w{7,2}(:)));
        ASData(i,79)=rms(abs(w{8,1}(:)));
        ASData(i,80)=rms(abs(w{8,2}(:)));

        ASData(i,81)=mean(abs(w{1,1}))/mean(abs(w{1,2}));
        ASData(i,82)=mean(abs(w{1,2}))/mean(abs(w{2,1}));
        ASData(i,83)=mean(abs(w{2,1}))/mean(abs(w{2,2}));
        ASData(i,84)=mean(abs(w{2,2}))/mean(abs(w{3,1}));
        ASData(i,85)=mean(abs(w{3,1}))/mean(abs(w{3,2}));
        ASData(i,86)=mean(abs(w{3,2}))/mean(abs(w{4,1}));
        ASData(i,87)=mean(abs(w{4,1}))/mean(abs(w{4,2}));
        ASData(i,88)=mean(abs(w{4,2}))/mean(abs(w{5,1}));
        ASData(i,89)=mean(abs(w{5,1}))/mean(abs(w{5,2}));
        ASData(i,90)=mean(abs(w{5,2}))/mean(abs(w{6,1}));
        ASData(i,91)=mean(abs(w{6,1}))/mean(abs(w{6,2}));
        ASData(i,92)=mean(abs(w{6,2}))/mean(abs(w{7,1}));
        ASData(i,93)=mean(abs(w{7,1}))/mean(abs(w{7,2}));
        ASData(i,94)=mean(abs(w{7,2}))/mean(abs(w{8,1}));
        ASData(i,95)=mean(abs(w{8,1}))/mean(abs(w{8,2}));
end
```

```
%%
% D F.ZIP EPILEPTIC SUBJECT (INTERICTAL) were recorded from within
% the epileptogenic zone during seizure free intervals

for i=Nofsignal+1:2*Nofsignal;

w = RAnDwt(Interictal(1:Length,i-Nofsignal),p,q,r,s,J,Fp); %FAWT Transform

ASData(i,1)=mean(abs(abs(w{1,1})));
ASData(i,2)=mean(abs(abs(w{1,2})));
ASData(i,3)=mean(abs(abs(w{2,1})));
ASData(i,4)=mean(abs(abs(w{2,2})));
ASData(i,5)=mean(abs(abs(w{3,1})));
ASData(i,6)=mean(abs(abs(w{3,2})));
ASData(i,7)=mean(abs(abs(w{4,1})));
ASData(i,8)=mean(abs(abs(w{4,2})));
ASData(i,9)=mean(abs(abs(w{5,1})));
ASData(i,10)=mean(abs(abs(w{5,2})));
ASData(i,11)=mean(abs(abs(w{6,1})));
ASData(i,12)=mean(abs(abs(w{6,2})));
ASData(i,13)=mean(abs(abs(w{7,1})));
ASData(i,14)=mean(abs(abs(w{7,2})));
ASData(i,15)=mean(abs(abs(w{8,1})));
ASData(i,16)=mean(abs(abs(w{8,2})));

ASData(i,17)=std(abs(w{1,1}));
ASData(i,18)=std(abs(w{1,2}));
ASData(i,19)=std(abs(w{2,1}));
ASData(i,20)=std(abs(w{2,2}));
ASData(i,21)=std(abs(w{3,1}));
ASData(i,22)=std(abs(w{3,2}));
ASData(i,23)=std(abs(w{4,1}));
ASData(i,24)=std(abs(w{4,2}));
ASData(i,25)=std(abs(w{5,1}));
ASData(i,26)=std(abs(w{5,2}));
ASData(i,27)=std(abs(w{6,1}));
ASData(i,28)=std(abs(w{6,2}));
ASData(i,29)=std(abs(w{7,1}));
ASData(i,30)=std(abs(w{7,2}));
ASData(i,31)=std(abs(w{8,1}));
ASData(i,32)=std(abs(w{8,2}));

ASData(i,33)=skewness(abs(w{1,1}));
ASData(i,34)=skewness(abs(w{1,2}));
ASData(i,35)=skewness(abs(w{2,1}));
ASData(i,36)=skewness(abs(w{2,2}));
ASData(i,37)=skewness(abs(w{3,1}));
ASData(i,38)=skewness(abs(w{3,2}));
ASData(i,39)=skewness(abs(w{4,1}));
ASData(i,40)=skewness(abs(w{4,2}));
ASData(i,41)=skewness(abs(w{5,1}));
ASData(i,42)=skewness(abs(w{5,2}));
ASData(i,43)=skewness(abs(w{6,1}));
ASData(i,44)=skewness(abs(w{6,2}));
ASData(i,45)=skewness(abs(w{7,1}));
ASData(i,46)=skewness(abs(w{7,2}));
ASData(i,47)=skewness(abs(w{8,1}));
ASData(i,48)=skewness(abs(w{8,2}));
```

```
ASData(i,49)=kurtosis(abs(w{1,1}));
ASData(i,50)=kurtosis(abs(w{1,2}));
ASData(i,51)=kurtosis(abs(w{2,1}));
ASData(i,52)=kurtosis(abs(w{2,2}));
ASData(i,53)=kurtosis(abs(w{3,1}));
ASData(i,54)=kurtosis(abs(w{3,2}));
ASData(i,55)=kurtosis(abs(w{4,1}));
ASData(i,56)=kurtosis(abs(w{4,2}));
ASData(i,57)=kurtosis(abs(w{5,1}));
ASData(i,58)=kurtosis(abs(w{5,2}));
ASData(i,59)=kurtosis(abs(w{6,1}));
ASData(i,60)=kurtosis(abs(w{6,2}));
ASData(i,61)=kurtosis(abs(w{7,1}));
ASData(i,62)=kurtosis(abs(w{7,2}));
ASData(i,63)=kurtosis(abs(w{8,1}));
ASData(i,64)=kurtosis(abs(w{8,2}));

        ASData(i,65)=rms(abs(w{1,1}(:)));
        ASData(i,66)=rms(abs(w{1,2}(:)));
        ASData(i,67)=rms(abs(w{2,1}(:)));
        ASData(i,68)=rms(abs(w{2,2}(:)));
        ASData(i,69)=rms(abs(w{3,1}(:)));
        ASData(i,70)=rms(abs(w{3,2}(:)));
        ASData(i,71)=rms(abs(w{4,1}(:)));
        ASData(i,72)=rms(abs(w{4,2}(:)));
        ASData(i,73)=rms(abs(w{5,1}(:)));
        ASData(i,74)=rms(abs(w{5,2}(:)));
        ASData(i,75)=rms(abs(w{6,1}(:)));
        ASData(i,76)=rms(abs(w{6,2}(:)));
        ASData(i,77)=rms(abs(w{7,1}(:)));
        ASData(i,78)=rms(abs(w{7,2}(:)));
        ASData(i,79)=rms(abs(w{8,1}(:)));
        ASData(i,80)=rms(abs(w{8,2}(:)));

        ASData(i,81)=mean(abs(w{1,1}))/mean(abs(w{1,2}));
        ASData(i,82)=mean(abs(w{1,2}))/mean(abs(w{2,1}));
        ASData(i,83)=mean(abs(w{2,1}))/mean(abs(w{2,2}));
        ASData(i,84)=mean(abs(w{2,2}))/mean(abs(w{3,1}));
        ASData(i,85)=mean(abs(w{3,1}))/mean(abs(w{3,2}));
        ASData(i,86)=mean(abs(w{3,2}))/mean(abs(w{4,1}));
        ASData(i,87)=mean(abs(w{4,1}))/mean(abs(w{4,2}));
        ASData(i,88)=mean(abs(w{4,2}))/mean(abs(w{5,1}));
        ASData(i,89)=mean(abs(w{5,1}))/mean(abs(w{5,2}));
        ASData(i,90)=mean(abs(w{5,2}))/mean(abs(w{6,1}));
        ASData(i,91)=mean(abs(w{6,1}))/mean(abs(w{6,2}));
        ASData(i,92)=mean(abs(w{6,2}))/mean(abs(w{7,1}));
        ASData(i,93)=mean(abs(w{7,1}))/mean(abs(w{7,2}));
        ASData(i,94)=mean(abs(w{7,2}))/mean(abs(w{8,1}));
        ASData(i,95)=mean(abs(w{8,1}))/mean(abs(w{8,2}));

end

%% E S.ZIP EPILEPTIC SUBJECT ICTAL DURING SEIZURE
%%%%%%%%%%%%%%%%%%%%%%%%%%%%%

for i=2*Nofsignal+1:3*Nofsignal;

w = RAnDwt(Ictal(1:Length,i-2*Nofsignal),p,q,r,s,J,Fp); %FAWT Transform
```

```
ASData(i,1)=mean(abs(abs(w{1,1})));
ASData(i,2)=mean(abs(abs(w{1,2})));
ASData(i,3)=mean(abs(abs(w{2,1})));
ASData(i,4)=mean(abs(abs(w{2,2})));
ASData(i,5)=mean(abs(abs(w{3,1})));
ASData(i,6)=mean(abs(abs(w{3,2})));
ASData(i,7)=mean(abs(abs(w{4,1})));
ASData(i,8)=mean(abs(abs(w{4,2})));
ASData(i,9)=mean(abs(abs(w{5,1})));
ASData(i,10)=mean(abs(abs(w{5,2})));
ASData(i,11)=mean(abs(abs(w{6,1})));
ASData(i,12)=mean(abs(abs(w{6,2})));
ASData(i,13)=mean(abs(abs(w{7,1})));
ASData(i,14)=mean(abs(abs(w{7,2})));
ASData(i,15)=mean(abs(abs(w{8,1})));
ASData(i,16)=mean(abs(abs(w{8,2})));

ASData(i,17)=std(abs(w{1,1}));
ASData(i,18)=std(abs(w{1,2}));
ASData(i,19)=std(abs(w{2,1}));
ASData(i,20)=std(abs(w{2,2}));
ASData(i,21)=std(abs(w{3,1}));
ASData(i,22)=std(abs(w{3,2}));
ASData(i,23)=std(abs(w{4,1}));
ASData(i,24)=std(abs(w{4,2}));
ASData(i,25)=std(abs(w{5,1}));
ASData(i,26)=std(abs(w{5,2}));
ASData(i,27)=std(abs(w{6,1}));
ASData(i,28)=std(abs(w{6,2}));
ASData(i,29)=std(abs(w{7,1}));
ASData(i,30)=std(abs(w{7,2}));
ASData(i,31)=std(abs(w{8,1}));
ASData(i,32)=std(abs(w{8,2}));

ASData(i,33)=skewness(abs(w{1,1}));
ASData(i,34)=skewness(abs(w{1,2}));
ASData(i,35)=skewness(abs(w{2,1}));
ASData(i,36)=skewness(abs(w{2,2}));
ASData(i,37)=skewness(abs(w{3,1}));
ASData(i,38)=skewness(abs(w{3,2}));
ASData(i,39)=skewness(abs(w{4,1}));
ASData(i,40)=skewness(abs(w{4,2}));
ASData(i,41)=skewness(abs(w{5,1}));
ASData(i,42)=skewness(abs(w{5,2}));
ASData(i,43)=skewness(abs(w{6,1}));
ASData(i,44)=skewness(abs(w{6,2}));
ASData(i,45)=skewness(abs(w{7,1}));
ASData(i,46)=skewness(abs(w{7,2}));
ASData(i,47)=skewness(abs(w{8,1}));
ASData(i,48)=skewness(abs(w{8,2}));

ASData(i,49)=kurtosis(abs(w{1,1}));
ASData(i,50)=kurtosis(abs(w{1,2}));
ASData(i,51)=kurtosis(abs(w{2,1}));
ASData(i,52)=kurtosis(abs(w{2,2}));
ASData(i,53)=kurtosis(abs(w{3,1}));
ASData(i,54)=kurtosis(abs(w{3,2}));
ASData(i,55)=kurtosis(abs(w{4,1}));
```

```
ASData(i,56)=kurtosis(abs(w{4,2}));
ASData(i,57)=kurtosis(abs(w{5,1}));
ASData(i,58)=kurtosis(abs(w{5,2}));
ASData(i,59)=kurtosis(abs(w{6,1}));
ASData(i,60)=kurtosis(abs(w{6,2}));
ASData(i,61)=kurtosis(abs(w{7,1}));
ASData(i,62)=kurtosis(abs(w{7,2}));
ASData(i,63)=kurtosis(abs(w{8,1}));
ASData(i,64)=kurtosis(abs(w{8,2}));

       ASData(i,65)=rms(abs(w{1,1}(:)));
       ASData(i,66)=rms(abs(w{1,2}(:)));
       ASData(i,67)=rms(abs(w{2,1}(:)));
       ASData(i,68)=rms(abs(w{2,2}(:)));
       ASData(i,69)=rms(abs(w{3,1}(:)));
       ASData(i,70)=rms(abs(w{3,2}(:)));
       ASData(i,71)=rms(abs(w{4,1}(:)));
       ASData(i,72)=rms(abs(w{4,2}(:)));
       ASData(i,73)=rms(abs(w{5,1}(:)));
       ASData(i,74)=rms(abs(w{5,2}(:)));
       ASData(i,75)=rms(abs(w{6,1}(:)));
       ASData(i,76)=rms(abs(w{6,2}(:)));
       ASData(i,77)=rms(abs(w{7,1}(:)));
       ASData(i,78)=rms(abs(w{7,2}(:)));
       ASData(i,79)=rms(abs(w{8,1}(:)));
       ASData(i,80)=rms(abs(w{8,2}(:)));

       ASData(i,81)=mean(abs(w{1,1}))/mean(abs(w{1,2}));
       ASData(i,82)=mean(abs(w{1,2}))/mean(abs(w{2,1}));
       ASData(i,83)=mean(abs(w{2,1}))/mean(abs(w{2,2}));
       ASData(i,84)=mean(abs(w{2,2}))/mean(abs(w{3,1}));
       ASData(i,85)=mean(abs(w{3,1}))/mean(abs(w{3,2}));
       ASData(i,86)=mean(abs(w{3,2}))/mean(abs(w{4,1}));
       ASData(i,87)=mean(abs(w{4,1}))/mean(abs(w{4,2}));
       ASData(i,88)=mean(abs(w{4,2}))/mean(abs(w{5,1}));
       ASData(i,89)=mean(abs(w{5,1}))/mean(abs(w{5,2}));
       ASData(i,90)=mean(abs(w{5,2}))/mean(abs(w{6,1}));
       ASData(i,91)=mean(abs(w{6,1}))/mean(abs(w{6,2}));
       ASData(i,92)=mean(abs(w{6,2}))/mean(abs(w{7,1}));
       ASData(i,93)=mean(abs(w{7,1}))/mean(abs(w{7,2}));
       ASData(i,94)=mean(abs(w{7,2}))/mean(abs(w{8,1}));
       ASData(i,95)=mean(abs(w{8,1}))/mean(abs(w{8,2}));
end
```

EXAMPLE 4.13 The following MATLAB code was used to extract features from the EEG signals EWT. You can download data from the following website:
http://epileptologie-bonn.de/cms/front_content.php?idcat=193&lang=3&changelang=3

Note that the EWT toolbox is used for running the following MATLAB code. You can download it from the following website: https://www.mathworks.com/matlabcentral/fileexchange/42141-empirical-wavelet-transforms

Do not forget to set the path for the EWT toolbox.

```
%Ch4_Ex61_EWT_EEG.m
%  EWT of NORMAL, INTERICTAL and ICTAL EEG signals
%The following MATLAB code is used to Extract EWT features from the EEG signals
%Decompose EEG data using EWT Transfrom
%Then uses Statistical features as:
```

```
%(1) Mean of the absolute values of the coefficients in each sub-band.
%(2) Standard deviation of the coefficients in each sub-band.
%(3) Skewness of the coefficients in each sub-band.
%(4) Kurtosis of the coefficients in each sub-band.
%(5) RMS power of the wavelet coefficients in each subband.
%(6) Ratio of the mean absolute values of adjacent subbands.

%Empirical Wavelet Transform (EWT) of NORMAL, INTERICTAL and ICTAL EEG signals
%Note that the Empirical Wavelet Transform (EWT) toolbox is used
%for running the following MATLAB code. You can download it from
%the following web site:
%https://www.mathworks.com/matlabcentral/fileexchange/42141-empirical-wavelet-transforms
%Set the path for the Empirical Wavelet Transform (EWT) toolbox

clc
clear
%Load Sample EEG Data downloaded from the web site
%http://epileptologie-bonn.de/cms/front_content.php?idcat=193&lang=3&changelang=3
load AS_BONN_ALL_EEG_DATA_4096.mat
Length = 4096;    % Length of signal
Nofsignal=100; %Number of Signal

%% User setup for filter parameters
params.SamplingRate = 173.6; % sampling rate
% Choose the wanted global trend removal (none,plaw,poly,morpho,tophat)
params.globtrend = 'none';
params.degree=1; % degree for the polynomial interpolation
% Choose the wanted regularization (none,gaussian,avaerage,closing)
params.reg = 'none';
params.lengthFilter = 10;
params.sigmaFilter = 1.5;

% Choose the wanted detection method (locmax,locmaxmin,ftc,
% adaptive,adaptivereg,scalespace)
params.detect = 'scalespace';
params.typeDetect='otsu'; %for scalespace:otsu,halfnormal,empiricallaw,mean,kmeans
params.N = 3; % maximum number of bands
params.completion = 0; % choose if you want to force to have params.N modes
                       % in case the algorithm found less ones (0 or 1)

params.InitBounds = [2 25];

% Perform the detection on the log spectrum instead the spectrum
params.log=0;

% Choose the results you want to display (Show=1, Not Show=0)
Bound=0;   % Display the detected boundaries on the spectrum
Comp=0;    % Display the EWT components
Rec=0;     % Display the reconstructed signal
TFplane=0; % Display the time-frequency plane (by using the Hilbert
           % transform). You can decrease the frequency resolution by
           % changing the subresf variable below.
Demd=0;    % Display the Hilbert-Huang transform (YOU NEED TO HAVE
           % FLANDRIN'S EMD TOOLBOX)

subresf=2;
InitBounds = params.InitBounds;

%%
%%%%%%%%%%%%%%%%%%%%%%%%%%%% A Z.ZIP EYES OPEN NORMAL SUBJECT
```

```
for i=1:Nofsignal;

% We perform the empirical transform and its inverse
% compute the EWT (and get the corresponding filter bank and list of
% boundaries)
[ewt,mfb,boundaries]=EWT1D(Normal_Eyes_Open(1:Length,i),params);
Hilb = EWT_InstantaneousComponents(ewt,boundaries);

ASData(i,1)=std(Hilb{1}{1});%Hilb{1}{1} amplitude
ASData(i,2)=std(Hilb{2}{1});
ASData(i,3)=std(Hilb{3}{1});
ASData(i,4)=std(Hilb{4}{1});
ASData(i,5)=std(Hilb{5}{1});
ASData(i,6)=std(Hilb{6}{1});
ASData(i,7)=std(Hilb{7}{1});
ASData(i,8)=std(Hilb{8}{1});

ASData(i,9)=std(Hilb{1}{2}); %Hilb{1}{2} frequency
ASData(i,10)=std(Hilb{2}{2});
ASData(i,11)=std(Hilb{3}{2});
ASData(i,12)=std(Hilb{4}{2});
ASData(i,13)=std(Hilb{5}{2});
ASData(i,14)=std(Hilb{6}{2});
ASData(i,15)=std(Hilb{7}{2});
ASData(i,16)=std(Hilb{8}{2});

ASData(i,17)=mean(Hilb{1}{1});%Hilb{1}{1} amplitude
ASData(i,18)=mean(Hilb{2}{1});
ASData(i,19)=mean(Hilb{3}{1});
ASData(i,20)=mean(Hilb{4}{1});
ASData(i,21)=mean(Hilb{5}{1});
ASData(i,22)=mean(Hilb{6}{1});
ASData(i,23)=mean(Hilb{7}{1});
ASData(i,24)=mean(Hilb{8}{1});

ASData(i,25)=mean(Hilb{1}{2});%Hilb{1}{2} frequency
ASData(i,26)=mean(Hilb{2}{2});
ASData(i,27)=mean(Hilb{3}{2});
ASData(i,28)=mean(Hilb{4}{2});
ASData(i,29)=mean(Hilb{5}{2});
ASData(i,30)=mean(Hilb{6}{2});
ASData(i,31)=mean(Hilb{6}{2});
ASData(i,32)=mean(Hilb{6}{2});

ASData(i,33)=skewness(Hilb{1}{1});%Hilb{1}{1} amplitude
ASData(i,34)=skewness(Hilb{2}{1});
ASData(i,35)=skewness(Hilb{3}{1});
ASData(i,36)=skewness(Hilb{4}{1});
ASData(i,37)=skewness(Hilb{5}{1});
ASData(i,38)=skewness(Hilb{6}{1});
ASData(i,39)=skewness(Hilb{7}{1});
ASData(i,40)=skewness(Hilb{8}{1});

ASData(i,41)=skewness(Hilb{1}{2});%Hilb{1}{2} frequency
ASData(i,42)=skewness(Hilb{2}{2});
ASData(i,43)=skewness(Hilb{3}{2});
ASData(i,44)=skewness(Hilb{4}{2});
ASData(i,45)=skewness(Hilb{5}{2});
```

```
ASData(i,46)=skewness(Hilb{6}{2});
ASData(i,47)=skewness(Hilb{7}{2});
ASData(i,48)=skewness(Hilb{8}{2});

ASData(i,49)=kurtosis(Hilb{1}{1});%Hilb{1}{1} amplitude
ASData(i,50)=kurtosis(Hilb{2}{1});
ASData(i,51)=kurtosis(Hilb{3}{1});
ASData(i,52)=kurtosis(Hilb{4}{1});
ASData(i,53)=kurtosis(Hilb{5}{1});
ASData(i,54)=kurtosis(Hilb{6}{1});
ASData(i,55)=kurtosis(Hilb{7}{1});
ASData(i,56)=kurtosis(Hilb{8}{1});

ASData(i,57)=kurtosis(Hilb{1}{2});%Hilb{1}{2} frequency
ASData(i,58)=kurtosis(Hilb{2}{2});
ASData(i,59)=kurtosis(Hilb{3}{2});
ASData(i,60)=kurtosis(Hilb{4}{2});
ASData(i,61)=kurtosis(Hilb{5}{2});
ASData(i,62)=kurtosis(Hilb{6}{2});
ASData(i,63)=kurtosis(Hilb{7}{2});
ASData(i,64)=kurtosis(Hilb{8}{2});

ASData(i,65)=rms(Hilb{1}{1});%Hilb{1}{1} amplitude
ASData(i,66)=rms(Hilb{2}{1});
ASData(i,67)=rms(Hilb{3}{1});
ASData(i,68)=rms(Hilb{4}{1});
ASData(i,69)=rms(Hilb{5}{1});
ASData(i,70)=rms(Hilb{6}{1});
ASData(i,71)=rms(Hilb{7}{1});
ASData(i,72)=rms(Hilb{8}{1});

ASData(i,73)=rms(Hilb{1}{2});%Hilb{1}{2} frequency
ASData(i,74)=rms(Hilb{2}{2});
ASData(i,75)=rms(Hilb{3}{2});
ASData(i,76)=rms(Hilb{4}{2});
ASData(i,77)=rms(Hilb{5}{2});
ASData(i,78)=rms(Hilb{6}{2});
ASData(i,79)=rms(Hilb{7}{2});
ASData(i,80)=rms(Hilb{8}{2});

ASData(i,81)=mean(abs(Hilb{1}{1}))/mean(abs(Hilb{2}{1})); %Hilb{1}{1} amplitude
ASData(i,82)=mean(abs(Hilb{2}{1}))/mean(abs(Hilb{3}{1}));
ASData(i,83)=mean(abs(Hilb{3}{1}))/mean(abs(Hilb{4}{1}));
ASData(i,84)=mean(abs(Hilb{4}{1}))/mean(abs(Hilb{5}{1}));
ASData(i,85)=mean(abs(Hilb{5}{1}))/mean(abs(Hilb{6}{1}));
ASData(i,86)=mean(abs(Hilb{6}{1}))/mean(abs(Hilb{7}{1}));
ASData(i,87)=mean(abs(Hilb{7}{1}))/mean(abs(Hilb{8}{1}));

ASData(i,88)=mean(abs(Hilb{1}{2}))/mean(abs(Hilb{2}{2}));%Hilb{1}{2} frequency
ASData(i,89)=mean(abs(Hilb{2}{2}))/mean(abs(Hilb{3}{2}));
ASData(i,90)=mean(abs(Hilb{3}{2}))/mean(abs(Hilb{4}{2}));
ASData(i,91)=mean(abs(Hilb{4}{2}))/mean(abs(Hilb{5}{2}));
ASData(i,92)=mean(abs(Hilb{5}{2}))/mean(abs(Hilb{6}{2}));
ASData(i,93)=mean(abs(Hilb{6}{2}))/mean(abs(Hilb{7}{2}));
ASData(i,94)=mean(abs(Hilb{7}{2}))/mean(abs(Hilb{8}{2}));

end
```

```
%%
% D F.ZIP EPILEPTIC SUBJECT (INTERICTAL) were recorded from within
% the epileptogenic zone during seizure free intervals

for i=Nofsignal+1:2*Nofsignal;

% We perform the empirical transform and its inverse
% compute the EWT (and get the corresponding filter bank and list of
% boundaries)
[ewt,mfb,boundaries]=EWT1D(Interictal(1:Length,i-Nofsignal),params);
Hilb = EWT_InstantaneousComponents(ewt,boundaries);

ASData(i,1)=std(Hilb{1}{1});%Hilb{1}{1} amplitude
ASData(i,2)=std(Hilb{2}{1});
ASData(i,3)=std(Hilb{3}{1});
ASData(i,4)=std(Hilb{4}{1});
ASData(i,5)=std(Hilb{5}{1});
ASData(i,6)=std(Hilb{6}{1});
ASData(i,7)=std(Hilb{7}{1});
ASData(i,8)=std(Hilb{8}{1});

ASData(i,9)=std(Hilb{1}{2}); %Hilb{1}{2} frequency
ASData(i,10)=std(Hilb{2}{2});
ASData(i,11)=std(Hilb{3}{2});
ASData(i,12)=std(Hilb{4}{2});
ASData(i,13)=std(Hilb{5}{2});
ASData(i,14)=std(Hilb{6}{2});
ASData(i,15)=std(Hilb{7}{2});
ASData(i,16)=std(Hilb{8}{2});

ASData(i,17)=mean(Hilb{1}{1});%Hilb{1}{1} amplitude
ASData(i,18)=mean(Hilb{2}{1});
ASData(i,19)=mean(Hilb{3}{1});
ASData(i,20)=mean(Hilb{4}{1});
ASData(i,21)=mean(Hilb{5}{1});
ASData(i,22)=mean(Hilb{6}{1});
ASData(i,23)=mean(Hilb{7}{1});
ASData(i,24)=mean(Hilb{8}{1});

ASData(i,25)=mean(Hilb{1}{2});%Hilb{1}{2} frequency
ASData(i,26)=mean(Hilb{2}{2});
ASData(i,27)=mean(Hilb{3}{2});
ASData(i,28)=mean(Hilb{4}{2});
ASData(i,29)=mean(Hilb{5}{2});
ASData(i,30)=mean(Hilb{6}{2});
ASData(i,31)=mean(Hilb{6}{2});
ASData(i,32)=mean(Hilb{6}{2});

ASData(i,33)=skewness(Hilb{1}{1});%Hilb{1}{1} amplitude
ASData(i,34)=skewness(Hilb{2}{1});
ASData(i,35)=skewness(Hilb{3}{1});
ASData(i,36)=skewness(Hilb{4}{1});
ASData(i,37)=skewness(Hilb{5}{1});
ASData(i,38)=skewness(Hilb{6}{1});
ASData(i,39)=skewness(Hilb{7}{1});
ASData(i,40)=skewness(Hilb{8}{1});

ASData(i,41)=skewness(Hilb{1}{2});%Hilb{1}{2} frequency
ASData(i,42)=skewness(Hilb{2}{2});
```

```
ASData(i,43)=skewness(Hilb{3}{2});
ASData(i,44)=skewness(Hilb{4}{2});
ASData(i,45)=skewness(Hilb{5}{2});
ASData(i,46)=skewness(Hilb{6}{2});
ASData(i,47)=skewness(Hilb{7}{2});
ASData(i,48)=skewness(Hilb{8}{2});

ASData(i,49)=kurtosis(Hilb{1}{1});%Hilb{1}{1} amplitude
ASData(i,50)=kurtosis(Hilb{2}{1});
ASData(i,51)=kurtosis(Hilb{3}{1});
ASData(i,52)=kurtosis(Hilb{4}{1});
ASData(i,53)=kurtosis(Hilb{5}{1});
ASData(i,54)=kurtosis(Hilb{6}{1});
ASData(i,55)=kurtosis(Hilb{7}{1});
ASData(i,56)=kurtosis(Hilb{8}{1});

ASData(i,57)=kurtosis(Hilb{1}{2});%Hilb{1}{2} frequency
ASData(i,58)=kurtosis(Hilb{2}{2});
ASData(i,59)=kurtosis(Hilb{3}{2});
ASData(i,60)=kurtosis(Hilb{4}{2});
ASData(i,61)=kurtosis(Hilb{5}{2});
ASData(i,62)=kurtosis(Hilb{6}{2});
ASData(i,63)=kurtosis(Hilb{7}{2});
ASData(i,64)=kurtosis(Hilb{8}{2});

ASData(i,65)=rms(Hilb{1}{1});%Hilb{1}{1} amplitude
ASData(i,66)=rms(Hilb{2}{1});
ASData(i,67)=rms(Hilb{3}{1});
ASData(i,68)=rms(Hilb{4}{1});
ASData(i,69)=rms(Hilb{5}{1});
ASData(i,70)=rms(Hilb{6}{1});
ASData(i,71)=rms(Hilb{7}{1});
ASData(i,72)=rms(Hilb{8}{1});

ASData(i,73)=rms(Hilb{1}{2});%Hilb{1}{2} frequency
ASData(i,74)=rms(Hilb{2}{2});
ASData(i,75)=rms(Hilb{3}{2});
ASData(i,76)=rms(Hilb{4}{2});
ASData(i,77)=rms(Hilb{5}{2});
ASData(i,78)=rms(Hilb{6}{2});
ASData(i,79)=rms(Hilb{7}{2});
ASData(i,80)=rms(Hilb{8}{2});

ASData(i,81)=mean(abs(Hilb{1}{1}))/mean(abs(Hilb{2}{1})); %Hilb{1}{1} amplitude
ASData(i,82)=mean(abs(Hilb{2}{1}))/mean(abs(Hilb{3}{1}));
ASData(i,83)=mean(abs(Hilb{3}{1}))/mean(abs(Hilb{4}{1}));
ASData(i,84)=mean(abs(Hilb{4}{1}))/mean(abs(Hilb{5}{1}));
ASData(i,85)=mean(abs(Hilb{5}{1}))/mean(abs(Hilb{6}{1}));
ASData(i,86)=mean(abs(Hilb{6}{1}))/mean(abs(Hilb{7}{1}));
ASData(i,87)=mean(abs(Hilb{7}{1}))/mean(abs(Hilb{8}{1}));

ASData(i,88)=mean(abs(Hilb{1}{2}))/mean(abs(Hilb{2}{2}));%Hilb{1}{2} frequency
ASData(i,89)=mean(abs(Hilb{2}{2}))/mean(abs(Hilb{3}{2}));
ASData(i,90)=mean(abs(Hilb{3}{2}))/mean(abs(Hilb{4}{2}));
ASData(i,91)=mean(abs(Hilb{4}{2}))/mean(abs(Hilb{5}{2}));
ASData(i,92)=mean(abs(Hilb{5}{2}))/mean(abs(Hilb{6}{2}));
ASData(i,93)=mean(abs(Hilb{6}{2}))/mean(abs(Hilb{7}{2}));
ASData(i,94)=mean(abs(Hilb{7}{2}))/mean(abs(Hilb{8}{2}));

end
```

```
%% E S.ZIP EPILEPTIC SUBJECT ICTAL DURING SEIZURE
%%%%%%%%%%%%%%%%%%%%%%%%%%%%%

for i=2*Nofsignal+1:3*Nofsignal;

% We perform the empirical transform and its inverse
% compute the EWT (and get the corresponding filter bank and list of
% boundaries)
[ewt,mfb,boundaries]=EWT1D(Ictal(1:Length,i-2*Nofsignal),params);
Hilb = EWT_InstantaneousComponents(ewt,boundaries);

ASData(i,1)=std(Hilb{1}{1});%Hilb{1}{1} amplitude
ASData(i,2)=std(Hilb{2}{1});
ASData(i,3)=std(Hilb{3}{1});
ASData(i,4)=std(Hilb{4}{1});
ASData(i,5)=std(Hilb{5}{1});
ASData(i,6)=std(Hilb{6}{1});
ASData(i,7)=std(Hilb{7}{1});
ASData(i,8)=std(Hilb{8}{1});

ASData(i,9)=std(Hilb{1}{2}); %Hilb{1}{2} frequency
ASData(i,10)=std(Hilb{2}{2});
ASData(i,11)=std(Hilb{3}{2});
ASData(i,12)=std(Hilb{4}{2});
ASData(i,13)=std(Hilb{5}{2});
ASData(i,14)=std(Hilb{6}{2});
ASData(i,15)=std(Hilb{7}{2});
ASData(i,16)=std(Hilb{8}{2});

ASData(i,17)=mean(Hilb{1}{1});%Hilb{1}{1} amplitude
ASData(i,18)=mean(Hilb{2}{1});
ASData(i,19)=mean(Hilb{3}{1});
ASData(i,20)=mean(Hilb{4}{1});
ASData(i,21)=mean(Hilb{5}{1});
ASData(i,22)=mean(Hilb{6}{1});
ASData(i,23)=mean(Hilb{7}{1});
ASData(i,24)=mean(Hilb{8}{1});

ASData(i,25)=mean(Hilb{1}{2});%Hilb{1}{2} frequency
ASData(i,26)=mean(Hilb{2}{2});
ASData(i,27)=mean(Hilb{3}{2});
ASData(i,28)=mean(Hilb{4}{2});
ASData(i,29)=mean(Hilb{5}{2});
ASData(i,30)=mean(Hilb{6}{2});
ASData(i,31)=mean(Hilb{6}{2});
ASData(i,32)=mean(Hilb{6}{2});


ASData(i,33)=skewness(Hilb{1}{1});%Hilb{1}{1} amplitude
ASData(i,34)=skewness(Hilb{2}{1});
ASData(i,35)=skewness(Hilb{3}{1});
ASData(i,36)=skewness(Hilb{4}{1});
ASData(i,37)=skewness(Hilb{5}{1});
ASData(i,38)=skewness(Hilb{6}{1});
ASData(i,39)=skewness(Hilb{7}{1});
ASData(i,40)=skewness(Hilb{8}{1});

ASData(i,41)=skewness(Hilb{1}{2});%Hilb{1}{2} frequency
ASData(i,42)=skewness(Hilb{2}{2});
```

```
ASData(i,43)=skewness(Hilb{3}{2});
ASData(i,44)=skewness(Hilb{4}{2});
ASData(i,45)=skewness(Hilb{5}{2});
ASData(i,46)=skewness(Hilb{6}{2});
ASData(i,47)=skewness(Hilb{7}{2});
ASData(i,48)=skewness(Hilb{8}{2});

ASData(i,49)=kurtosis(Hilb{1}{1});%Hilb{1}{1} amplitude
ASData(i,50)=kurtosis(Hilb{2}{1});
ASData(i,51)=kurtosis(Hilb{3}{1});
ASData(i,52)=kurtosis(Hilb{4}{1});
ASData(i,53)=kurtosis(Hilb{5}{1});
ASData(i,54)=kurtosis(Hilb{6}{1});
ASData(i,55)=kurtosis(Hilb{7}{1});
ASData(i,56)=kurtosis(Hilb{8}{1});

ASData(i,57)=kurtosis(Hilb{1}{2});%Hilb{1}{2} frequency
ASData(i,58)=kurtosis(Hilb{2}{2});
ASData(i,59)=kurtosis(Hilb{3}{2});
ASData(i,60)=kurtosis(Hilb{4}{2});
ASData(i,61)=kurtosis(Hilb{5}{2});
ASData(i,62)=kurtosis(Hilb{6}{2});
ASData(i,63)=kurtosis(Hilb{7}{2});
ASData(i,64)=kurtosis(Hilb{8}{2});


ASData(i,65)=rms(Hilb{1}{1});%Hilb{1}{1} amplitude
ASData(i,66)=rms(Hilb{2}{1});
ASData(i,67)=rms(Hilb{3}{1});
ASData(i,68)=rms(Hilb{4}{1});
ASData(i,69)=rms(Hilb{5}{1});
ASData(i,70)=rms(Hilb{6}{1});
ASData(i,71)=rms(Hilb{7}{1});
ASData(i,72)=rms(Hilb{8}{1});

ASData(i,73)=rms(Hilb{1}{2});%Hilb{1}{2} frequency
ASData(i,74)=rms(Hilb{2}{2});
ASData(i,75)=rms(Hilb{3}{2});
ASData(i,76)=rms(Hilb{4}{2});
ASData(i,77)=rms(Hilb{5}{2});
ASData(i,78)=rms(Hilb{6}{2});
ASData(i,79)=rms(Hilb{7}{2});
ASData(i,80)=rms(Hilb{8}{2});

ASData(i,81)=mean(abs(Hilb{1}{1}))/mean(abs(Hilb{2}{1})); %Hilb{1}{1} amplitude
ASData(i,82)=mean(abs(Hilb{2}{1}))/mean(abs(Hilb{3}{1}));
ASData(i,83)=mean(abs(Hilb{3}{1}))/mean(abs(Hilb{4}{1}));
ASData(i,84)=mean(abs(Hilb{4}{1}))/mean(abs(Hilb{5}{1}));
ASData(i,85)=mean(abs(Hilb{5}{1}))/mean(abs(Hilb{6}{1}));
ASData(i,86)=mean(abs(Hilb{6}{1}))/mean(abs(Hilb{7}{1}));
ASData(i,87)=mean(abs(Hilb{7}{1}))/mean(abs(Hilb{8}{1}));

ASData(i,88)=mean(abs(Hilb{1}{2}))/mean(abs(Hilb{2}{2}));%Hilb{1}{2} frequency
ASData(i,89)=mean(abs(Hilb{2}{2}))/mean(abs(Hilb{3}{2}));
ASData(i,90)=mean(abs(Hilb{3}{2}))/mean(abs(Hilb{4}{2}));
ASData(i,91)=mean(abs(Hilb{4}{2}))/mean(abs(Hilb{5}{2}));
ASData(i,92)=mean(abs(Hilb{5}{2}))/mean(abs(Hilb{6}{2}));
ASData(i,93)=mean(abs(Hilb{6}{2}))/mean(abs(Hilb{7}{2}));
ASData(i,94)=mean(abs(Hilb{7}{2}))/mean(abs(Hilb{8}{2}));
end
```

EXAMPLE 4.14 The following MATLAB code was used to extract features from the EEG signals using EMD. You can download data from the following website:
http://epileptologie-bonn.de/cms/front_content.php?idcat=193&lang=3&changelang=3

Note that the EMD toolbox is used for running the following MATLAB code. You can download it from the following website: https://www.mathworks.com/matlabcentral/fileexchange/21409-empirical-mode-decomposition

Do not forget to set the path for the EMD toolbox.

```
%Ch4_Ex66_EMD_EEG.m
%Empirical Mode Decomposition (EMD) of NORMAL, INTERICTAL and ICTAL EEG signals
%Note that the Empirical Mode Decomposition (EMD) toolbox is used
%for running the following MATLAB code. You can download it from
%the following web sites:
%http://perso.ens-lyon.fr/patrick.flandrin/emd.html
%https://www.mathworks.com/matlabcentral/fileexchange/21409-empirical-mode-decomposition
%Set the path for the Empirical Mode Decomposition (EMD) toolbox

%The following MATLAB code is used to Extract EMD features from the EEG signals
%Decompose EEG data using EMD
%Then uses Statistical features as:

%(1) Mean of the absolute values of the coefficients in each sub-band.
%(2) Standard deviation of the coefficients in each sub-band.
%(3) Skewness of the coefficients in each sub-band.
%(4) Kurtosis of the coefficients in each sub-band.
%(5) RMS power of the wavelet coefficients in each subband.
%(6) Ratio of the mean absolute values of adjacent subbands.

clc
clear
%Load Sample EEG Data downloaded from the web site
%http://epileptologie-bonn.de/cms/front_content.php?idcat=193&lang=3&changelang=3
load AS_BONN_ALL_EEG_DATA_4096.mat
Length = 4096;    % Length of signal
Nofsignal=100; %Number of Signal

%%
%%%%%%%%%%%%%%%%%%%%%%%%%%%%% A Z.ZIP EYES OPEN NORMAL SUBJECT

for i=1:Nofsignal;

imf=emd(Normal_Eyes_Open(1:Length,i));%Empirical Mode Decomposition

ASData(i,1)=mean(abs(imf(1,:)));
ASData(i,2)=mean(abs(imf(2,:)));
ASData(i,3)=mean(abs(imf(3,:)));
ASData(i,4)=mean(abs(imf(4,:)));
ASData(i,5)=mean(abs(imf(5,:)));
ASData(i,6)=mean(abs(imf(6,:)));
ASData(i,7)=mean(abs(imf(7,:)));
ASData(i,8)=mean(abs(imf(8,:)));
ASData(i,9)=mean(abs(imf(9,:)));
ASData(i,10)=mean(abs(imf(10,:)));
ASData(i,11)=mean(abs(imf(11,:)));
ASData(i,12)=mean(abs(imf(12,:)));

ASData(i,13)=std(imf(1,:));
ASData(i,14)=std(imf(2,:));
```

```
ASData(i,15)=std(imf(3,:));
ASData(i,16)=std(imf(4,:));
ASData(i,17)=std(imf(5,:));
ASData(i,18)=std(imf(6,:));
ASData(i,19)=std(imf(7,:));
ASData(i,20)=std(imf(8,:));
ASData(i,21)=std(imf(9,:));
ASData(i,22)=std(imf(10,:));
ASData(i,23)=std(imf(11,:));
ASData(i,24)=std(imf(12,:));

ASData(i,25)=skewness(imf(1,:));
ASData(i,26)=skewness(imf(2,:));
ASData(i,27)=skewness(imf(3,:));
ASData(i,28)=skewness(imf(4,:));
ASData(i,29)=skewness(imf(5,:));
ASData(i,30)=skewness(imf(6,:));
ASData(i,31)=skewness(imf(7,:));
ASData(i,32)=skewness(imf(8,:));
ASData(i,33)=skewness(imf(9,:));
ASData(i,34)=skewness(imf(10,:));
ASData(i,35)=skewness(imf(11,:));
ASData(i,36)=skewness(imf(12,:));

ASData(i,37)=kurtosis(imf(1,:));
ASData(i,38)=kurtosis(imf(2,:));
ASData(i,39)=kurtosis(imf(3,:));
ASData(i,40)=kurtosis(imf(4,:));
ASData(i,41)=kurtosis(imf(5,:));
ASData(i,42)=kurtosis(imf(6,:));
ASData(i,43)=kurtosis(imf(7,:));
ASData(i,44)=kurtosis(imf(8,:));
ASData(i,45)=kurtosis(imf(9,:));
ASData(i,46)=kurtosis(imf(10,:));
ASData(i,47)=kurtosis(imf(11,:));
ASData(i,48)=kurtosis(imf(12,:));
        ASData(i,49)=rms(imf(1,:));
        ASData(i,50)=rms(imf(2,:));
        ASData(i,51)=rms(imf(3,:));
        ASData(i,52)=rms(imf(4,:));
        ASData(i,53)=rms(imf(5,:));
        ASData(i,54)=rms(imf(6,:));
        ASData(i,55)=rms(imf(7,:));
        ASData(i,56)=rms(imf(8,:));
        ASData(i,56)=rms(imf(9,:));
        ASData(i,58)=rms(imf(10,:));
        ASData(i,59)=rms(imf(11,:));
        ASData(i,60)=rms(imf(12,:));

        ASData(i,61)=mean(abs(imf(1,:)))/mean(abs(imf(2,:)));
        ASData(i,62)=mean(abs(imf(2,:)))/mean(abs(imf(3,:)));
        ASData(i,63)=mean(abs(imf(3,:)))/mean(abs(imf(4,:)));
        ASData(i,64)=mean(abs(imf(4,:)))/mean(abs(imf(5,:)));
        ASData(i,65)=mean(abs(imf(5,:)))/mean(abs(imf(6,:)));
        ASData(i,66)=mean(abs(imf(6,:)))/mean(abs(imf(7,:)));
        ASData(i,67)=mean(abs(imf(7,:)))/mean(abs(imf(8,:)));
        ASData(i,68)=mean(abs(imf(8,:)))/mean(abs(imf(9,:)));
        ASData(i,69)=mean(abs(imf(9,:)))/mean(abs(imf(10,:)));
```

```
ASData(i,70)=mean(abs(imf(10,:)))/mean(abs(imf(11,:)));
ASData(i,71)=mean(abs(imf(11,:)))/mean(abs(imf(12,:)));

end

%%
% D F.ZIP EPILEPTIC SUBJECT (INTERICTAL) were recorded from within
% the epileptogenic zone during seizure free intervals

for i=Nofsignal+1:2*Nofsignal;
%Empirical Mode Decomposition
imf=emd(Interictal(1:Length,i-Nofsignal));

ASData(i,1)=mean(abs(imf(1,:)));
ASData(i,2)=mean(abs(imf(2,:)));
ASData(i,3)=mean(abs(imf(3,:)));
ASData(i,4)=mean(abs(imf(4,:)));
ASData(i,5)=mean(abs(imf(5,:)));
ASData(i,6)=mean(abs(imf(6,:)));
ASData(i,7)=mean(abs(imf(7,:)));
ASData(i,8)=mean(abs(imf(8,:)));
ASData(i,9)=mean(abs(imf(9,:)));
ASData(i,10)=mean(abs(imf(10,:)));
ASData(i,11)=mean(abs(imf(11,:)));
ASData(i,12)=mean(abs(imf(12,:)));

ASData(i,13)=std(imf(1,:));
ASData(i,14)=std(imf(2,:));
ASData(i,15)=std(imf(3,:));
ASData(i,16)=std(imf(4,:));
ASData(i,17)=std(imf(5,:));
ASData(i,18)=std(imf(6,:));
ASData(i,19)=std(imf(7,:));
ASData(i,20)=std(imf(8,:));
ASData(i,21)=std(imf(9,:));
ASData(i,22)=std(imf(10,:));
ASData(i,23)=std(imf(11,:));
ASData(i,24)=std(imf(12,:));

ASData(i,25)=skewness(imf(1,:));
ASData(i,26)=skewness(imf(2,:));
ASData(i,27)=skewness(imf(3,:));
ASData(i,28)=skewness(imf(4,:));
ASData(i,29)=skewness(imf(5,:));
ASData(i,30)=skewness(imf(6,:));
ASData(i,31)=skewness(imf(7,:));
ASData(i,32)=skewness(imf(8,:));
ASData(i,33)=skewness(imf(9,:));
ASData(i,34)=skewness(imf(10,:));
ASData(i,35)=skewness(imf(11,:));
ASData(i,36)=skewness(imf(12,:));

ASData(i,37)=kurtosis(imf(1,:));
ASData(i,38)=kurtosis(imf(2,:));
ASData(i,39)=kurtosis(imf(3,:));
ASData(i,40)=kurtosis(imf(4,:));
ASData(i,41)=kurtosis(imf(5,:));
ASData(i,42)=kurtosis(imf(6,:));
```

```
ASData(i,43)=kurtosis(imf(7,:));
ASData(i,44)=kurtosis(imf(8,:));
ASData(i,45)=kurtosis(imf(9,:));
ASData(i,46)=kurtosis(imf(10,:));
ASData(i,47)=kurtosis(imf(11,:));
ASData(i,48)=kurtosis(imf(12,:));
        ASData(i,49)=rms(imf(1,:));
        ASData(i,50)=rms(imf(2,:));
        ASData(i,51)=rms(imf(3,:));
        ASData(i,52)=rms(imf(4,:));
        ASData(i,53)=rms(imf(5,:));
        ASData(i,54)=rms(imf(6,:));
        ASData(i,55)=rms(imf(7,:));
        ASData(i,56)=rms(imf(8,:));
        ASData(i,56)=rms(imf(9,:));
        ASData(i,58)=rms(imf(10,:));
        ASData(i,59)=rms(imf(11,:));
        ASData(i,60)=rms(imf(12,:));

        ASData(i,61)=mean(abs(imf(1,:)))/mean(abs(imf(2,:)));
        ASData(i,62)=mean(abs(imf(2,:)))/mean(abs(imf(3,:)));
        ASData(i,63)=mean(abs(imf(3,:)))/mean(abs(imf(4,:)));
        ASData(i,64)=mean(abs(imf(4,:)))/mean(abs(imf(5,:)));
        ASData(i,65)=mean(abs(imf(5,:)))/mean(abs(imf(6,:)));
        ASData(i,66)=mean(abs(imf(6,:)))/mean(abs(imf(7,:)));
        ASData(i,67)=mean(abs(imf(7,:)))/mean(abs(imf(8,:)));
        ASData(i,68)=mean(abs(imf(8,:)))/mean(abs(imf(9,:)));
        ASData(i,69)=mean(abs(imf(9,:)))/mean(abs(imf(10,:)));
        ASData(i,70)=mean(abs(imf(10,:)))/mean(abs(imf(11,:)));
        ASData(i,71)=mean(abs(imf(11,:)))/mean(abs(imf(12,:)));

end

%% E S.ZIP EPILEPTIC SUBJECT ICTAL DURING SEIZURE
%%%%%%%%%%%%%%%%%%%%%%%%%%%%%

for i=2*Nofsignal+1:3*Nofsignal;
%Empirical Mode Decomposition
imf=emd(Ictal(1:Length,i-2*Nofsignal));

ASData(i,1)=mean(abs(imf(1,:)));
ASData(i,2)=mean(abs(imf(2,:)));
ASData(i,3)=mean(abs(imf(3,:)));
ASData(i,4)=mean(abs(imf(4,:)));
ASData(i,5)=mean(abs(imf(5,:)));
ASData(i,6)=mean(abs(imf(6,:)));
ASData(i,7)=mean(abs(imf(7,:)));
ASData(i,8)=mean(abs(imf(8,:)));
ASData(i,9)=mean(abs(imf(9,:)));
ASData(i,10)=mean(abs(imf(10,:)));
ASData(i,11)=mean(abs(imf(11,:)));
ASData(i,12)=mean(abs(imf(12,:)));

ASData(i,13)=std(imf(1,:));
ASData(i,14)=std(imf(2,:));
ASData(i,15)=std(imf(3,:));
ASData(i,16)=std(imf(4,:));
ASData(i,17)=std(imf(5,:));
```

```
ASData(i,18)=std(imf(6,:));
ASData(i,19)=std(imf(7,:));
ASData(i,20)=std(imf(8,:));
ASData(i,21)=std(imf(9,:));
ASData(i,22)=std(imf(10,:));
ASData(i,23)=std(imf(11,:));
ASData(i,24)=std(imf(12,:));

ASData(i,25)=skewness(imf(1,:));
ASData(i,26)=skewness(imf(2,:));
ASData(i,27)=skewness(imf(3,:));
ASData(i,28)=skewness(imf(4,:));
ASData(i,29)=skewness(imf(5,:));
ASData(i,30)=skewness(imf(6,:));
ASData(i,31)=skewness(imf(7,:));
ASData(i,32)=skewness(imf(8,:));
ASData(i,33)=skewness(imf(9,:));
ASData(i,34)=skewness(imf(10,:));
ASData(i,35)=skewness(imf(11,:));
ASData(i,36)=skewness(imf(12,:));

ASData(i,37)=kurtosis(imf(1,:));
ASData(i,38)=kurtosis(imf(2,:));
ASData(i,39)=kurtosis(imf(3,:));
ASData(i,40)=kurtosis(imf(4,:));
ASData(i,41)=kurtosis(imf(5,:));
ASData(i,42)=kurtosis(imf(6,:));
ASData(i,43)=kurtosis(imf(7,:));
ASData(i,44)=kurtosis(imf(8,:));
ASData(i,45)=kurtosis(imf(9,:));
ASData(i,46)=kurtosis(imf(10,:));
ASData(i,47)=kurtosis(imf(11,:));
ASData(i,48)=kurtosis(imf(12,:));
        ASData(i,49)=rms(imf(1,:));
        ASData(i,50)=rms(imf(2,:));
        ASData(i,51)=rms(imf(3,:));
        ASData(i,52)=rms(imf(4,:));
        ASData(i,53)=rms(imf(5,:));
        ASData(i,54)=rms(imf(6,:));
        ASData(i,55)=rms(imf(7,:));
        ASData(i,56)=rms(imf(8,:));
        ASData(i,56)=rms(imf(9,:));
        ASData(i,58)=rms(imf(10,:));
        ASData(i,59)=rms(imf(11,:));
        ASData(i,60)=rms(imf(12,:));

        ASData(i,61)=mean(abs(imf(1,:)))/mean(abs(imf(2,:)));
        ASData(i,62)=mean(abs(imf(2,:)))/mean(abs(imf(3,:)));
        ASData(i,63)=mean(abs(imf(3,:)))/mean(abs(imf(4,:)));
        ASData(i,64)=mean(abs(imf(4,:)))/mean(abs(imf(5,:)));
        ASData(i,65)=mean(abs(imf(5,:)))/mean(abs(imf(6,:)));
        ASData(i,66)=mean(abs(imf(6,:)))/mean(abs(imf(7,:)));
        ASData(i,67)=mean(abs(imf(7,:)))/mean(abs(imf(8,:)));
        ASData(i,68)=mean(abs(imf(8,:)))/mean(abs(imf(9,:)));
        ASData(i,69)=mean(abs(imf(9,:)))/mean(abs(imf(10,:)));
        ASData(i,70)=mean(abs(imf(10,:)))/mean(abs(imf(11,:)));
        ASData(i,71)=mean(abs(imf(11,:)))/mean(abs(imf(12,:)));
end
```

EXAMPLE 4.15 The following MATLAB code was used to extract features from the EEG signals using EEMD. You can download data from the following website:
http://epileptologie-bonn.de/cms/front_content.php?idcat=193&lang=3&changelang=3

Note that the EEMD toolbox is used for running the following MATLAB code. You can download it from the following website: http://perso.ens-lyon.fr/patrick.flandrin/emd.html

Do not forget to set the path for the EEMD toolbox.

```
%%Ch4_Ex71_EEMD.m
%Ensemble Empirical Mode Decomposition (EEMD) of NORMAL, INTERICTAL and ICTAL EEG signals
%Note that the Ensemble Empirical Mode Decomposition (EEMD) toolbox is used
%for running the following MATLAB code. You can download it from
%the following web site:
%http://perso.ens-lyon.fr/patrick.flandrin/emd.html
%Set the path for the Ensemble Empirical Mode Decomposition (EEMD) toolbox

%The following MATLAB code is used to Extract EEMD features from the EEG signals
%Decompose EEG data using EEMD
%you should adjust the level of decomposition
%depending on the length of the signal used

%Then uses Statistical features as:

%(1) Mean of the absolute values of the coefficients in each sub-band.
%(2) Standard deviation of the coefficients in each sub-band.
%(3) Skewness of the coefficients in each sub-band.
%(4) Kurtosis of the coefficients in each sub-band.
%(5) RMS power of the wavelet coefficients in each subband.
%(6) Ratio of the mean absolute values of adjacent subbands.

clc
clear
%Load Sample EEG Data downloaded from the web site
%http://epileptologie-bonn.de/cms/front_content.php?idcat=193&lang=3&changelang=3
load AS_BONN_ALL_EEG_DATA_4096.mat

goal=7; %target mode numbers to decompose
ens=2;
nos=2;

Length = 4096;    % Length of signal
Nofsignal=100; %Number of Signal

%%
%%%%%%%%%%%%%%%%%%%%%%%%%%%%%% A Z.ZIP EYES OPEN NORMAL SUBJECT

for i=1:Nofsignal;

%Decompose signals into modes
   modes= eemd(Normal_Eyes_Open(1:Length,i), goal, ens, nos);

ASData(i,1)=mean(abs(modes(1,:)));
ASData(i,2)=mean(abs(modes(2,:)));
ASData(i,3)=mean(abs(modes(3,:)));
ASData(i,4)=mean(abs(modes(4,:)));
ASData(i,5)=mean(abs(modes(5,:)));
ASData(i,6)=mean(abs(modes(6,:)));
ASData(i,7)=mean(abs(modes(7,:)));
ASData(i,8)=mean(abs(modes(8,:)));
```

```
ASData(i,9)=mean(abs(modes(9,:)));
ASData(i,10)=mean(abs(modes(10,:)));
ASData(i,11)=mean(abs(modes(11,:)));
ASData(i,12)=mean(abs(modes(12,:)));

ASData(i,13)=std(modes(1,:));
ASData(i,14)=std(modes(2,:));
ASData(i,15)=std(modes(3,:));
ASData(i,16)=std(modes(4,:));
ASData(i,17)=std(modes(5,:));
ASData(i,18)=std(modes(6,:));
ASData(i,19)=std(modes(7,:));
ASData(i,20)=std(modes(8,:));
ASData(i,21)=std(modes(9,:));
ASData(i,22)=std(modes(10,:));
ASData(i,23)=std(modes(11,:));
ASData(i,24)=std(modes(12,:));

ASData(i,25)=skewness(modes(1,:));
ASData(i,26)=skewness(modes(2,:));
ASData(i,27)=skewness(modes(3,:));
ASData(i,28)=skewness(modes(4,:));
ASData(i,29)=skewness(modes(5,:));
ASData(i,30)=skewness(modes(6,:));
ASData(i,31)=skewness(modes(7,:));
ASData(i,32)=skewness(modes(8,:));
ASData(i,33)=skewness(modes(9,:));
ASData(i,34)=skewness(modes(10,:));
ASData(i,35)=skewness(modes(11,:));
ASData(i,36)=skewness(modes(12,:));

ASData(i,37)=kurtosis(modes(1,:));
ASData(i,38)=kurtosis(modes(2,:));
ASData(i,39)=kurtosis(modes(3,:));
ASData(i,40)=kurtosis(modes(4,:));
ASData(i,41)=kurtosis(modes(5,:));
ASData(i,42)=kurtosis(modes(6,:));
ASData(i,43)=kurtosis(modes(7,:));
ASData(i,44)=kurtosis(modes(8,:));
ASData(i,45)=kurtosis(modes(9,:));
ASData(i,46)=kurtosis(modes(10,:));
ASData(i,47)=kurtosis(modes(11,:));
ASData(i,48)=kurtosis(modes(12,:));
        ASData(i,49)=rms(modes(1,:));
        ASData(i,50)=rms(modes(2,:));
        ASData(i,51)=rms(modes(3,:));
        ASData(i,52)=rms(modes(4,:));
        ASData(i,53)=rms(modes(5,:));
        ASData(i,54)=rms(modes(6,:));
        ASData(i,55)=rms(modes(7,:));
        ASData(i,56)=rms(modes(8,:));
        ASData(i,56)=rms(modes(9,:));
        ASData(i,58)=rms(modes(10,:));
        ASData(i,59)=rms(modes(11,:));
        ASData(i,60)=rms(modes(12,:));

        ASData(i,61)=mean(abs(modes(1,:)))/mean(abs(modes(2,:)));
        ASData(i,62)=mean(abs(modes(2,:)))/mean(abs(modes(3,:)));
        ASData(i,63)=mean(abs(modes(3,:)))/mean(abs(modes(4,:)));
        ASData(i,64)=mean(abs(modes(4,:)))/mean(abs(modes(5,:)));
```

```
        ASData(i,65)=mean(abs(modes(5,:)))/mean(abs(modes(6,:)));
        ASData(i,66)=mean(abs(modes(6,:)))/mean(abs(modes(7,:)));
        ASData(i,67)=mean(abs(modes(7,:)))/mean(abs(modes(8,:)));
        ASData(i,68)=mean(abs(modes(8,:)))/mean(abs(modes(9,:)));
        ASData(i,69)=mean(abs(modes(9,:)))/mean(abs(modes(10,:)));
        ASData(i,70)=mean(abs(modes(10,:)))/mean(abs(modes(11,:)));
        ASData(i,71)=mean(abs(modes(11,:)))/mean(abs(modes(12,:)));

end

%%
% D F.ZIP EPILEPTIC SUBJECT (INTERICTAL) were recorded from within
% the epileptogenic zone during seizure free intervals

for i=Nofsignal+1:2*Nofsignal;

%Decompose signals into modes
modes= eemd(Interictal(1:Length,i-Nofsignal), goal, ens, nos);

ASData(i,1)=mean(abs(modes(1,:)));
ASData(i,2)=mean(abs(modes(2,:)));
ASData(i,3)=mean(abs(modes(3,:)));
ASData(i,4)=mean(abs(modes(4,:)));
ASData(i,5)=mean(abs(modes(5,:)));
ASData(i,6)=mean(abs(modes(6,:)));
ASData(i,7)=mean(abs(modes(7,:)));
ASData(i,8)=mean(abs(modes(8,:)));
ASData(i,9)=mean(abs(modes(9,:)));
ASData(i,10)=mean(abs(modes(10,:)));
ASData(i,11)=mean(abs(modes(11,:)));
ASData(i,12)=mean(abs(modes(12,:)));

ASData(i,13)=std(modes(1,:));
ASData(i,14)=std(modes(2,:));
ASData(i,15)=std(modes(3,:));
ASData(i,16)=std(modes(4,:));
ASData(i,17)=std(modes(5,:));
ASData(i,18)=std(modes(6,:));
ASData(i,19)=std(modes(7,:));
ASData(i,20)=std(modes(8,:));
ASData(i,21)=std(modes(9,:));
ASData(i,22)=std(modes(10,:));
ASData(i,23)=std(modes(11,:));
ASData(i,24)=std(modes(12,:));

ASData(i,25)=skewness(modes(1,:));
ASData(i,26)=skewness(modes(2,:));
ASData(i,27)=skewness(modes(3,:));
ASData(i,28)=skewness(modes(4,:));
ASData(i,29)=skewness(modes(5,:));
ASData(i,30)=skewness(modes(6,:));
ASData(i,31)=skewness(modes(7,:));
ASData(i,32)=skewness(modes(8,:));
ASData(i,33)=skewness(modes(9,:));
ASData(i,34)=skewness(modes(10,:));
ASData(i,35)=skewness(modes(11,:));
ASData(i,36)=skewness(modes(12,:));

ASData(i,37)=kurtosis(modes(1,:));
ASData(i,38)=kurtosis(modes(2,:));
```

```
ASData(i,39)=kurtosis(modes(3,:));
ASData(i,40)=kurtosis(modes(4,:));
ASData(i,41)=kurtosis(modes(5,:));
ASData(i,42)=kurtosis(modes(6,:));
ASData(i,43)=kurtosis(modes(7,:));
ASData(i,44)=kurtosis(modes(8,:));
ASData(i,45)=kurtosis(modes(9,:));
ASData(i,46)=kurtosis(modes(10,:));
ASData(i,47)=kurtosis(modes(11,:));
ASData(i,48)=kurtosis(modes(12,:));
        ASData(i,49)=rms(modes(1,:));
        ASData(i,50)=rms(modes(2,:));
        ASData(i,51)=rms(modes(3,:));
        ASData(i,52)=rms(modes(4,:));
        ASData(i,53)=rms(modes(5,:));
        ASData(i,54)=rms(modes(6,:));
        ASData(i,55)=rms(modes(7,:));
        ASData(i,56)=rms(modes(8,:));
        ASData(i,56)=rms(modes(9,:));
        ASData(i,58)=rms(modes(10,:));
        ASData(i,59)=rms(modes(11,:));
        ASData(i,60)=rms(modes(12,:));

        ASData(i,61)=mean(abs(modes(1,:)))/mean(abs(modes(2,:)));
        ASData(i,62)=mean(abs(modes(2,:)))/mean(abs(modes(3,:)));
        ASData(i,63)=mean(abs(modes(3,:)))/mean(abs(modes(4,:)));
        ASData(i,64)=mean(abs(modes(4,:)))/mean(abs(modes(5,:)));
        ASData(i,65)=mean(abs(modes(5,:)))/mean(abs(modes(6,:)));
        ASData(i,66)=mean(abs(modes(6,:)))/mean(abs(modes(7,:)));
        ASData(i,67)=mean(abs(modes(7,:)))/mean(abs(modes(8,:)));
        ASData(i,68)=mean(abs(modes(8,:)))/mean(abs(modes(9,:)));
        ASData(i,69)=mean(abs(modes(9,:)))/mean(abs(modes(10,:)));
        ASData(i,70)=mean(abs(modes(10,:)))/mean(abs(modes(11,:)));
        ASData(i,71)=mean(abs(modes(11,:)))/mean(abs(modes(12,:)));

end

%% E S.ZIP EPILEPTIC SUBJECT ICTAL DURING SEIZURE
%%%%%%%%%%%%%%%%%%%%%%%%%%%%%

for i=2*Nofsignal+1:3*Nofsignal;

%Decompose signals into modes
modes= eemd(Ictal(1:Length,i-2*Nofsignal), goal, ens, nos);

ASData(i,1)=mean(abs(modes(1,:)));
ASData(i,2)=mean(abs(modes(2,:)));
ASData(i,3)=mean(abs(modes(3,:)));
ASData(i,4)=mean(abs(modes(4,:)));
ASData(i,5)=mean(abs(modes(5,:)));
ASData(i,6)=mean(abs(modes(6,:)));
ASData(i,7)=mean(abs(modes(7,:)));
ASData(i,8)=mean(abs(modes(8,:)));
ASData(i,9)=mean(abs(modes(9,:)));
ASData(i,10)=mean(abs(modes(10,:)));
ASData(i,11)=mean(abs(modes(11,:)));
ASData(i,12)=mean(abs(modes(12,:)));

ASData(i,13)=std(modes(1,:));
ASData(i,14)=std(modes(2,:));
ASData(i,15)=std(modes(3,:));
```

```
ASData(i,16)=std(modes(4,:));
ASData(i,17)=std(modes(5,:));
ASData(i,18)=std(modes(6,:));
ASData(i,19)=std(modes(7,:));
ASData(i,20)=std(modes(8,:));
ASData(i,21)=std(modes(9,:));
ASData(i,22)=std(modes(10,:));
ASData(i,23)=std(modes(11,:));
ASData(i,24)=std(modes(12,:));

ASData(i,25)=skewness(modes(1,:));
ASData(i,26)=skewness(modes(2,:));
ASData(i,27)=skewness(modes(3,:));
ASData(i,28)=skewness(modes(4,:));
ASData(i,29)=skewness(modes(5,:));
ASData(i,30)=skewness(modes(6,:));
ASData(i,31)=skewness(modes(7,:));
ASData(i,32)=skewness(modes(8,:));
ASData(i,33)=skewness(modes(9,:));
ASData(i,34)=skewness(modes(10,:));
ASData(i,35)=skewness(modes(11,:));
ASData(i,36)=skewness(modes(12,:));

ASData(i,37)=kurtosis(modes(1,:));
ASData(i,38)=kurtosis(modes(2,:));
ASData(i,39)=kurtosis(modes(3,:));
ASData(i,40)=kurtosis(modes(4,:));
ASData(i,41)=kurtosis(modes(5,:));
ASData(i,42)=kurtosis(modes(6,:));
ASData(i,43)=kurtosis(modes(7,:));
ASData(i,44)=kurtosis(modes(8,:));
ASData(i,45)=kurtosis(modes(9,:));
ASData(i,46)=kurtosis(modes(10,:));
ASData(i,47)=kurtosis(modes(11,:));
ASData(i,48)=kurtosis(modes(12,:));
        ASData(i,49)=rms(modes(1,:));
        ASData(i,50)=rms(modes(2,:));
        ASData(i,51)=rms(modes(3,:));
        ASData(i,52)=rms(modes(4,:));
        ASData(i,53)=rms(modes(5,:));
        ASData(i,54)=rms(modes(6,:));
        ASData(i,55)=rms(modes(7,:));
        ASData(i,56)=rms(modes(8,:));
        ASData(i,56)=rms(modes(9,:));
        ASData(i,58)=rms(modes(10,:));
        ASData(i,59)=rms(modes(11,:));
        ASData(i,60)=rms(modes(12,:));

        ASData(i,61)=mean(abs(modes(1,:)))/mean(abs(modes(2,:)));
        ASData(i,62)=mean(abs(modes(2,:)))/mean(abs(modes(3,:)));
        ASData(i,63)=mean(abs(modes(3,:)))/mean(abs(modes(4,:)));
        ASData(i,64)=mean(abs(modes(4,:)))/mean(abs(modes(5,:)));
        ASData(i,65)=mean(abs(modes(5,:)))/mean(abs(modes(6,:)));
        ASData(i,66)=mean(abs(modes(6,:)))/mean(abs(modes(7,:)));
        ASData(i,67)=mean(abs(modes(7,:)))/mean(abs(modes(8,:)));
        ASData(i,68)=mean(abs(modes(8,:)))/mean(abs(modes(9,:)));
        ASData(i,69)=mean(abs(modes(9,:)))/mean(abs(modes(10,:)));
        ASData(i,70)=mean(abs(modes(10,:)))/mean(abs(modes(11,:)));
        ASData(i,71)=mean(abs(modes(11,:)))/mean(abs(modes(12,:)));

end
```

EXAMPLE 4.16 The following MATLAB code was used to extract features from the EEG signals using CEEMD. You can download data from the following website: http://epileptologie-bonn.de/cms/front_content.php?idcat=193&lang=3&changelang=3

Note that the CEEMD toolbox is used for running the following MATLAB code. You can download it from the following website: http://perso.ens-lyon.fr/patrick.flandrin/emd.html

Do not forget to set the path for the CEEMD toolbox.

```
%Ch4_Ex76_CEEMD.m
%Complete Ensemble Empirical Mode Decomposition (CEEMD) of NORMAL,
INTERICTAL and ICTAL EEG signals

% Example of the CEEMDAN performance, used in the work where CEEMDAN was first presented:
% Cite:
% M.E.TORRES, M.A. COLOMINAS, G. SCHLOTTHAUER, P. FLANDRIN,
% "A complete Ensemble Empirical Mode decomposition with adaptive noise,"
% IEEE Int. Conf. on Acoust., Speech and Signal Proc. ICASSP-11, pp. 4144-4147, Prague (CZ)
%
%Note that the Complete Ensemble Empirical Mode Decomposition (CEEMD) toolbox is used
%for running the following MATLAB code. You can download it from
%the following web site:
%http://perso.ens-lyon.fr/patrick.flandrin/emd.html
%http://bioingenieria.edu.ar/grupos/ldnlys/metorres/re_inter.htm#Codigos
%Set the path for the Complete Ensemble Empirical Mode Decomposition (CEEMD) toolbox

%The following MATLAB code is used to Extract CEEMD features from the EEG signals
%Decompose EEG data using CEEMD
%you should adjust the level of decomposition
%depending on the length of the signal used

%Then uses Statistical features as:

%(1) Mean of the absolute values of the coefficients in each sub-band.
%(2) Standard deviation of the coefficients in each sub-band.
%(3) Skewness of the coefficients in each sub-band.
%(4) Kurtosis of the coefficients in each sub-band.
%(5) RMS power of the wavelet coefficients in each subband.
%(6) Ratio of the mean absolute values of adjacent subbands.

clc
clear
% The code loads the signal EEG.mat
% It is an EEG from the epileptologie-bonn
%Load Sample EEG Data downloaded from the web site
%available at http://epileptologie-bonn.de/cms/front_content.php?idcat=193&lang=3&changelang=3
load AS_BONN_ALL_EEG_DATA_4096.mat

Length = 4096; % Length of signal
Nofsignal=100; %Number of Signal
%% Define required parameters of CEEMD
Nstd = 0.2;
NR = 500;
MaxIter = 5000;

%%
%%%%%%%%%%%%%%%%%%%%%%%%%%%%%% A Z.ZIP EYES OPEN NORMAL SUBJECT

for i=1:Nofsignal;
```

```
%Decompose signals into modes
[modes its]=ceemdan(Normal_Eyes_Open(1:Length,i),Nstd,NR,MaxIter);

ASData(i,1)=mean(abs(modes(1,:)));
ASData(i,2)=mean(abs(modes(2,:)));
ASData(i,3)=mean(abs(modes(3,:)));
ASData(i,4)=mean(abs(modes(4,:)));
ASData(i,5)=mean(abs(modes(5,:)));
ASData(i,6)=mean(abs(modes(6,:)));
ASData(i,7)=mean(abs(modes(7,:)));
ASData(i,8)=mean(abs(modes(8,:)));
ASData(i,9)=mean(abs(modes(9,:)));
ASData(i,10)=mean(abs(modes(10,:)));
ASData(i,11)=mean(abs(modes(11,:)));
ASData(i,12)=mean(abs(modes(12,:)));

ASData(i,13)=std(modes(1,:));
ASData(i,14)=std(modes(2,:));
ASData(i,15)=std(modes(3,:));
ASData(i,16)=std(modes(4,:));
ASData(i,17)=std(modes(5,:));
ASData(i,18)=std(modes(6,:));
ASData(i,19)=std(modes(7,:));
ASData(i,20)=std(modes(8,:));
ASData(i,21)=std(modes(9,:));
ASData(i,22)=std(modes(10,:));
ASData(i,23)=std(modes(11,:));
ASData(i,24)=std(modes(12,:));

ASData(i,25)=skewness(modes(1,:));
ASData(i,26)=skewness(modes(2,:));
ASData(i,27)=skewness(modes(3,:));
ASData(i,28)=skewness(modes(4,:));
ASData(i,29)=skewness(modes(5,:));
ASData(i,30)=skewness(modes(6,:));
ASData(i,31)=skewness(modes(7,:));
ASData(i,32)=skewness(modes(8,:));
ASData(i,33)=skewness(modes(9,:));
ASData(i,34)=skewness(modes(10,:));
ASData(i,35)=skewness(modes(11,:));
ASData(i,36)=skewness(modes(12,:));

ASData(i,37)=kurtosis(modes(1,:));
ASData(i,38)=kurtosis(modes(2,:));
ASData(i,39)=kurtosis(modes(3,:));
ASData(i,40)=kurtosis(modes(4,:));
ASData(i,41)=kurtosis(modes(5,:));
ASData(i,42)=kurtosis(modes(6,:));
ASData(i,43)=kurtosis(modes(7,:));
ASData(i,44)=kurtosis(modes(8,:));
ASData(i,45)=kurtosis(modes(9,:));
ASData(i,46)=kurtosis(modes(10,:));
ASData(i,47)=kurtosis(modes(11,:));
ASData(i,48)=kurtosis(modes(12,:));
        ASData(i,49)=rms(modes(1,:));
        ASData(i,50)=rms(modes(2,:));
        ASData(i,51)=rms(modes(3,:));
        ASData(i,52)=rms(modes(4,:));
        ASData(i,53)=rms(modes(5,:));
```

```
        ASData(i,54)=rms(modes(6,:));
        ASData(i,55)=rms(modes(7,:));
        ASData(i,56)=rms(modes(8,:));
        ASData(i,56)=rms(modes(9,:));
        ASData(i,58)=rms(modes(10,:));
        ASData(i,59)=rms(modes(11,:));
        ASData(i,60)=rms(modes(12,:));

        ASData(i,61)=mean(abs(modes(1,:)))/mean(abs(modes(2,:)));
        ASData(i,62)=mean(abs(modes(2,:)))/mean(abs(modes(3,:)));
        ASData(i,63)=mean(abs(modes(3,:)))/mean(abs(modes(4,:)));
        ASData(i,64)=mean(abs(modes(4,:)))/mean(abs(modes(5,:)));
        ASData(i,65)=mean(abs(modes(5,:)))/mean(abs(modes(6,:)));
        ASData(i,66)=mean(abs(modes(6,:)))/mean(abs(modes(7,:)));
        ASData(i,67)=mean(abs(modes(7,:)))/mean(abs(modes(8,:)));
        ASData(i,68)=mean(abs(modes(8,:)))/mean(abs(modes(9,:)));
        ASData(i,69)=mean(abs(modes(9,:)))/mean(abs(modes(10,:)));
        ASData(i,70)=mean(abs(modes(10,:)))/mean(abs(modes(11,:)));
        ASData(i,71)=mean(abs(modes(11,:)))/mean(abs(modes(12,:)));

end

%%
% D F.ZIP EPILEPTIC SUBJECT (INTERICTAL) were recorded from within
% the epileptogenic zone during seizure free intervals

for i=Nofsignal+1:2*Nofsignal;

%Decompose signals into modes
[modes its]=ceemdan(Interictal(1:Length,i-Nofsignal),Nstd,NR,MaxIter);

ASData(i,1)=mean(abs(modes(1,:)));
ASData(i,2)=mean(abs(modes(2,:)));
ASData(i,3)=mean(abs(modes(3,:)));
ASData(i,4)=mean(abs(modes(4,:)));
ASData(i,5)=mean(abs(modes(5,:)));
ASData(i,6)=mean(abs(modes(6,:)));
ASData(i,7)=mean(abs(modes(7,:)));
ASData(i,8)=mean(abs(modes(8,:)));
ASData(i,9)=mean(abs(modes(9,:)));
ASData(i,10)=mean(abs(modes(10,:)));
ASData(i,11)=mean(abs(modes(11,:)));
ASData(i,12)=mean(abs(modes(12,:)));

ASData(i,13)=std(modes(1,:));
ASData(i,14)=std(modes(2,:));
ASData(i,15)=std(modes(3,:));
ASData(i,16)=std(modes(4,:));
ASData(i,17)=std(modes(5,:));
ASData(i,18)=std(modes(6,:));
ASData(i,19)=std(modes(7,:));
ASData(i,20)=std(modes(8,:));
ASData(i,21)=std(modes(9,:));
ASData(i,22)=std(modes(10,:));
ASData(i,23)=std(modes(11,:));
ASData(i,24)=std(modes(12,:));

ASData(i,25)=skewness(modes(1,:));
ASData(i,26)=skewness(modes(2,:));
```

```
ASData(i,27)=skewness(modes(3,:));
ASData(i,28)=skewness(modes(4,:));
ASData(i,29)=skewness(modes(5,:));
ASData(i,30)=skewness(modes(6,:));
ASData(i,31)=skewness(modes(7,:));
ASData(i,32)=skewness(modes(8,:));
ASData(i,33)=skewness(modes(9,:));
ASData(i,34)=skewness(modes(10,:));
ASData(i,35)=skewness(modes(11,:));
ASData(i,36)=skewness(modes(12,:));

ASData(i,37)=kurtosis(modes(1,:));
ASData(i,38)=kurtosis(modes(2,:));
ASData(i,39)=kurtosis(modes(3,:));
ASData(i,40)=kurtosis(modes(4,:));
ASData(i,41)=kurtosis(modes(5,:));
ASData(i,42)=kurtosis(modes(6,:));
ASData(i,43)=kurtosis(modes(7,:));
ASData(i,44)=kurtosis(modes(8,:));
ASData(i,45)=kurtosis(modes(9,:));
ASData(i,46)=kurtosis(modes(10,:));
ASData(i,47)=kurtosis(modes(11,:));
ASData(i,48)=kurtosis(modes(12,:));
        ASData(i,49)=rms(modes(1,:));
        ASData(i,50)=rms(modes(2,:));
        ASData(i,51)=rms(modes(3,:));
        ASData(i,52)=rms(modes(4,:));
        ASData(i,53)=rms(modes(5,:));
        ASData(i,54)=rms(modes(6,:));
        ASData(i,55)=rms(modes(7,:));
        ASData(i,56)=rms(modes(8,:));
        ASData(i,56)=rms(modes(9,:));
        ASData(i,58)=rms(modes(10,:));
        ASData(i,59)=rms(modes(11,:));
        ASData(i,60)=rms(modes(12,:));

        ASData(i,61)=mean(abs(modes(1,:)))/mean(abs(modes(2,:)));
        ASData(i,62)=mean(abs(modes(2,:)))/mean(abs(modes(3,:)));
        ASData(i,63)=mean(abs(modes(3,:)))/mean(abs(modes(4,:)));
        ASData(i,64)=mean(abs(modes(4,:)))/mean(abs(modes(5,:)));
        ASData(i,65)=mean(abs(modes(5,:)))/mean(abs(modes(6,:)));
        ASData(i,66)=mean(abs(modes(6,:)))/mean(abs(modes(7,:)));
        ASData(i,67)=mean(abs(modes(7,:)))/mean(abs(modes(8,:)));
        ASData(i,68)=mean(abs(modes(8,:)))/mean(abs(modes(9,:)));
        ASData(i,69)=mean(abs(modes(9,:)))/mean(abs(modes(10,:)));
        ASData(i,70)=mean(abs(modes(10,:)))/mean(abs(modes(11,:)));
        ASData(i,71)=mean(abs(modes(11,:)))/mean(abs(modes(12,:)));

end

%% E S.ZIP EPILEPTIC SUBJECT ICTAL DURING SEIZURE
%%%%%%%%%%%%%%%%%%%%%%%%%%%%%%

for i=2*Nofsignal+1:3*Nofsignal;

%Decompose signals into modes
[modes its]=ceemdan(Ictal(1:Length,i-2*Nofsignal),Nstd,NR,MaxIter);

ASData(i,1)=mean(abs(modes(1,:)));
ASData(i,2)=mean(abs(modes(2,:)));
```

```
ASData(i,3)=mean(abs(modes(3,:)));
ASData(i,4)=mean(abs(modes(4,:)));
ASData(i,5)=mean(abs(modes(5,:)));
ASData(i,6)=mean(abs(modes(6,:)));
ASData(i,7)=mean(abs(modes(7,:)));
ASData(i,8)=mean(abs(modes(8,:)));
ASData(i,9)=mean(abs(modes(9,:)));
ASData(i,10)=mean(abs(modes(10,:)));
ASData(i,11)=mean(abs(modes(11,:)));
ASData(i,12)=mean(abs(modes(12,:)));

ASData(i,13)=std(modes(1,:));
ASData(i,14)=std(modes(2,:));
ASData(i,15)=std(modes(3,:));
ASData(i,16)=std(modes(4,:));
ASData(i,17)=std(modes(5,:));
ASData(i,18)=std(modes(6,:));
ASData(i,19)=std(modes(7,:));
ASData(i,20)=std(modes(8,:));
ASData(i,21)=std(modes(9,:));
ASData(i,22)=std(modes(10,:));
ASData(i,23)=std(modes(11,:));
ASData(i,24)=std(modes(12,:));

ASData(i,25)=skewness(modes(1,:));
ASData(i,26)=skewness(modes(2,:));
ASData(i,27)=skewness(modes(3,:));
ASData(i,28)=skewness(modes(4,:));
ASData(i,29)=skewness(modes(5,:));
ASData(i,30)=skewness(modes(6,:));
ASData(i,31)=skewness(modes(7,:));
ASData(i,32)=skewness(modes(8,:));
ASData(i,33)=skewness(modes(9,:));
ASData(i,34)=skewness(modes(10,:));
ASData(i,35)=skewness(modes(11,:));
ASData(i,36)=skewness(modes(12,:));

ASData(i,37)=kurtosis(modes(1,:));
ASData(i,38)=kurtosis(modes(2,:));
ASData(i,39)=kurtosis(modes(3,:));
ASData(i,40)=kurtosis(modes(4,:));
ASData(i,41)=kurtosis(modes(5,:));
ASData(i,42)=kurtosis(modes(6,:));
ASData(i,43)=kurtosis(modes(7,:));
ASData(i,44)=kurtosis(modes(8,:));
ASData(i,45)=kurtosis(modes(9,:));
ASData(i,46)=kurtosis(modes(10,:));
ASData(i,47)=kurtosis(modes(11,:));
ASData(i,48)=kurtosis(modes(12,:));
        ASData(i,49)=rms(modes(1,:));
        ASData(i,50)=rms(modes(2,:));
        ASData(i,51)=rms(modes(3,:));
        ASData(i,52)=rms(modes(4,:));
        ASData(i,53)=rms(modes(5,:));
        ASData(i,54)=rms(modes(6,:));
        ASData(i,55)=rms(modes(7,:));
        ASData(i,56)=rms(modes(8,:));
        ASData(i,56)=rms(modes(9,:));
        ASData(i,58)=rms(modes(10,:));
```

```
        ASData(i,59)=rms(modes(11,:));
        ASData(i,60)=rms(modes(12,:));

        ASData(i,61)=mean(abs(modes(1,:)))/mean(abs(modes(2,:)));
        ASData(i,62)=mean(abs(modes(2,:)))/mean(abs(modes(3,:)));
        ASData(i,63)=mean(abs(modes(3,:)))/mean(abs(modes(4,:)));
        ASData(i,64)=mean(abs(modes(4,:)))/mean(abs(modes(5,:)));
        ASData(i,65)=mean(abs(modes(5,:)))/mean(abs(modes(6,:)));
        ASData(i,66)=mean(abs(modes(6,:)))/mean(abs(modes(7,:)));
        ASData(i,67)=mean(abs(modes(7,:)))/mean(abs(modes(8,:)));
        ASData(i,68)=mean(abs(modes(8,:)))/mean(abs(modes(9,:)));
        ASData(i,69)=mean(abs(modes(9,:)))/mean(abs(modes(10,:)));
        ASData(i,70)=mean(abs(modes(10,:)))/mean(abs(modes(11,:)));
        ASData(i,71)=mean(abs(modes(11,:)))/mean(abs(modes(12,:)));

end
```

4.3.3 Entropy

Entropy is a degree of uncertainty. The level of chaos in a biomedical signal can be measured using entropy of the system. Higher entropy signifies higher uncertainty and a more chaotic system. Entropy is given as

$$x(n) \int_{\min(x)}^{\max(x)} P_x \log\left(\frac{1}{p_x}\right) dx \tag{4.9}$$

where p_x is the PDF of signal $x(n)$. Usually, the distribution can be a joint PDF when the biomedical signal channels are jointly processed. Although this measure is utilized for single-channel biomedical signals, it can be simply extended to multichannel or even multidimensional signals. Thus, usually, the distribution can be a joint PDF when the multichannel signals are jointly processed. On the other hand, the PDF can be substituted by a conditional PDF in places where the incidence of an event is subject to another event. In this circumstance, the entropy is called conditional entropy. Entropy is very sensitive to noise. Noise increases the uncertainty, and noisy signals have higher entropy, even if the original signal is ordered. Entropy is employed in the calculation of many other useful parameters, such as mutual information, negentropy, nongaussianity and Kulback-Leibler divergence. These variables are extensively employed in the estimation of the degree of nonlinearity of the systems and correspondence of signals (Sanei, 2013). MATLAB code for calculating entropy is available at: https://www.mathworks.com/matlabcentral/fileexchange/28692-entropy

4.3.4 Kolmogorov Entropy

Kolmogorov entropy, also known as metric entropy, is an efficient way to determine the complexity of a system. This metric is frequently employed as a measure of chaos for dynamic systems. To calculate the Kolmogorov entropy, the phase space is separated into multidimensional hypercubes. The phase space is the space where all possible states of a system are denoted, each corresponding to one unique point in the phase space. In phase space, each parameter of the system or degree of freedom is denoted as an axis of a multidimensional space. A phase space may contain many dimensions. The hypercube is a generalization of a 3-cube to n-dimensions, also called an n-cube or measure polytope. It is a regular polytope with mutually perpendicular sides and is, then, an orthotope. But the estimation of these joint probabilities for large dimensional data is computationally costly. On the other hand, in practice, long data sequences are normally required to perform a precise estimation of Kolmogorov entropy. (Sanei, 2013). MATLAB code for calculating Kolmogorov entropy is available at:

https://www.mathworks.com/matlabcentral/fileexchange/6886-kolmogorov-complexity

4.3.5 Approximate and Sample Entropy

Another widespread used metric for chaos is approximate entropy (AE). AE can be calculated from the discrete time signals, particularly for real-time applications. This measure can enumerate the complexity or irregularity of the system. AE is less sensitive to noise and can be employed for short-length data. Moreover, it is resilient to short strong transient interferences (outliers) such as spikes (Sanei, 2013). MATLAB code for calculating AE is available at: https://www.mathworks.com/matlabcentral/fileexchange/26546-approximate-entropy or https://www.mathworks.com/matlabcentral/fileexchange/32427-fast-approximate-entropy

Even for low-dimensional chaotic systems, a huge number of points is essential to accomplish convergence of the algorithms assessing dimension or entropy of the process. To eliminate this problem and deliver a measure capable of measuring the changes of process complexity, modifications of the entropy measure were utilized. AE measures the (logarithmic) likelihood of trajectories that are close to each other and remain close on the next incremental comparison. However, methodological pitfalls in AE were eliminated by another measure called sample entropy (SaEn). MATLAB code for calculating SaEn is available at: http://www.physionet.org/physiotools/sampen/matlab/ or https://www.mathworks.com/matlabcentral/fileexchange/35784-sample-entropy

Because SaEn cannot describe the complexity of the signal completely, another modification called multiscale entropy (MSE) is proposed to involve partitioning the signal into disjoined windows, and the data are averaged inside each time window, which in fact is equivalent to downsampling, or low-pass filtering (Blinowska & Zygierewicz, 2011).

4.3.6 Detrended Fluctuation Analysis

Detrended fluctuation analysis (DFA) measures intrinsic fractal-like correlation properties of dynamic systems. The ultimate feature of a fractal system is scale invariance or self-similarity in different scales. In DFA, the variability of the signal is analyzed with respect to local trends in data windows. The method permits distinguishing long-range correlations embedded in a nonstationary signal. This detrending procedure followed by the fluctuation measurement process is repeated over the whole signal over all time scales. As with most approaches that rely on line fitting, it is always possible to find a number by the DFA method, but this does not essentially mean that the signal is self-similar (Blinowska & Zygierewicz, 2011). MATLAB code for calculating DFA is available at: https://www.mathworks.com/matlabcentral/fileexchange/19795-detrended-fluctuation-analysis or https://www.mathworks.com/matlabcentral/fileexchange/67889-detrended-fluctuation-analysis-dfa

EXAMPLE 4.17 The following MATLAB code was used to extract features from the EEG signals using WPD. Then it used different entropy measures for dimension reduction. You can download data from the following website: http://epileptologie-bonn.de/cms/front_content.php?idcat=193&lang=3&changelang=3

```
%Ch4_Ex111_WPD_Entropy_EEG.m
% WPD of NORMAL, INTERICTAL and ICTAL EEG signals
%The following MATLAB code is used to Extract WPD features from the EEG signals
%Decompose EEG data using WPD coefficients
%Then Entropy values as:

%(1) Entropy the absolute values of the coefficients in each sub-band.
%(2) Approximate Entropy of the coefficients in each sub-band.
%(3) Sample Entropy of the coefficients in each sub-band.
%(4) Kolmogorov Entropy of the coefficients in each sub-band.
%(5) Shannon Entropy of the wavelet coefficients in each subband.

clc
clear
%Load Sample EEG Data downloaded from the web site
%http://epileptologie-bonn.de/cms/front_content.php?idcat=193&lang=3&changelang=3
load AS_BONN_ALL_EEG_DATA_4096.mat
wname = 'db4';
Length = 4096;     % Length of signal
Nofsignal=100; %Number of Signal
%%
%%%%%%%%%%%%%%%%%%%%%%%%%%%%% A Z.ZIP EYES OPEN NORMAL SUBJECT

for i=1:Nofsignal;

[wp10,wp11] = dwt(Normal_Eyes_Open(1:Length,i),wname);%WPD Decomposition
[wp20,wp21] = dwt(wp10,wname);
[wp22,wp23] = dwt(wp11,wname);

[wp30,wp31] = dwt(wp20,wname);
[wp32,wp33] = dwt(wp21,wname);
[wp34,wp35] = dwt(wp22,wname);
[wp36,wp37] = dwt(wp23,wname);
```

```
[wp40,wp41] = dwt(wp30,wname);
[wp42,wp43] = dwt(wp31,wname);
[wp44,wp45] = dwt(wp32,wname);
[wp46,wp47] = dwt(wp33,wname);
[wp48,wp49] = dwt(wp34,wname);
[wp4A,wp4B] = dwt(wp35,wname);
[wp4C,wp4D] = dwt(wp36,wname);
[wp4E,wp4F] = dwt(wp37,wname);

ASData(i,1)=Entropy(abs(wp40));
ASData(i,2)=Entropy(abs(wp41));
ASData(i,3)=Entropy(abs(wp42));
ASData(i,4)=Entropy(abs(wp43));
ASData(i,5)=Entropy(abs(wp44));
ASData(i,6)=Entropy(abs(wp45));
ASData(i,7)=Entropy(abs(wp46));
ASData(i,8)=Entropy(abs(wp47));
ASData(i,9)=Entropy(abs(wp48));
ASData(i,10)=Entropy(abs(wp49));
ASData(i,11)=Entropy(abs(wp4A));
ASData(i,12)=Entropy(abs(wp4B));
ASData(i,13)=Entropy(abs(wp4C));
ASData(i,14)=Entropy(abs(wp4D));
ASData(i,15)=Entropy(abs(wp4E));
ASData(i,16)=Entropy(abs(wp4F));

ASData(i,17)=approx_entropy(64,0.5,abs(wp40));
ASData(i,18)=approx_entropy(64,0.5,abs(wp41));
ASData(i,19)=approx_entropy(64,0.5,abs(wp42));
ASData(i,20)=approx_entropy(64,0.5,abs(wp43));
ASData(i,21)=approx_entropy(64,0.5,abs(wp44));
ASData(i,22)=approx_entropy(64,0.5,abs(wp45));
ASData(i,23)=approx_entropy(64,0.5,abs(wp46));
ASData(i,24)=approx_entropy(64,0.5,abs(wp47));
ASData(i,25)=approx_entropy(64,0.5,abs(wp48));
ASData(i,26)=approx_entropy(64,0.5,abs(wp49));
ASData(i,27)=approx_entropy(64,0.5,abs(wp4A));
ASData(i,28)=approx_entropy(64,0.5,abs(wp4B));
ASData(i,29)=approx_entropy(64,0.5,abs(wp4C));
ASData(i,30)=approx_entropy(64,0.5,abs(wp4D));
ASData(i,31)=approx_entropy(64,0.5,abs(wp4E));
ASData(i,32)=approx_entropy(64,0.5,abs(wp4F));

ASData(i,33)=SampEn(5, 0.7,abs(wp40));
ASData(i,34)=SampEn(5, 0.7,abs(wp41));
ASData(i,35)=SampEn(5, 0.7,abs(wp42));
ASData(i,36)=SampEn(5, 0.7,abs(wp43));
ASData(i,37)=SampEn(5, 0.7,abs(wp44));
ASData(i,38)=SampEn(5, 0.7,abs(wp45));
ASData(i,39)=SampEn(5, 0.7,abs(wp46));
ASData(i,40)=SampEn(5, 0.7,abs(wp47));
ASData(i,41)=SampEn(5, 0.7,abs(wp48));
ASData(i,42)=SampEn(5, 0.7,abs(wp49));
ASData(i,43)=SampEn(5, 0.7,abs(wp4A));
ASData(i,44)=SampEn(5, 0.7,abs(wp4B));
ASData(i,45)=SampEn(5, 0.7,abs(wp4C));
ASData(i,46)=SampEn(5, 0.7,abs(wp4D));
ASData(i,47)=SampEn(5, 0.7,abs(wp4E));
ASData(i,48)=SampEn(5, 0.7,abs(wp4F));
```

```
ASData(i,49)=kolmogorov(abs(wp40));
ASData(i,50)=kolmogorov(abs(wp41));
ASData(i,51)=kolmogorov(abs(wp42));
ASData(i,52)=kolmogorov(abs(wp43));
ASData(i,53)=kolmogorov(abs(wp44));
ASData(i,54)=kolmogorov(abs(wp45));
ASData(i,55)=kolmogorov(abs(wp46));
ASData(i,56)=kolmogorov(abs(wp47));
ASData(i,57)=kolmogorov(abs(wp48));
ASData(i,58)=kolmogorov(abs(wp49));
ASData(i,59)=kolmogorov(abs(wp4A));
ASData(i,60)=kolmogorov(abs(wp4B));
ASData(i,61)=kolmogorov(abs(wp4C));
ASData(i,62)=kolmogorov(abs(wp4D));
ASData(i,63)=kolmogorov(abs(wp4E));
ASData(i,64)=kolmogorov(abs(wp4F));


ASData(i,65)=shannon_entro(abs(wp40(:)));
ASData(i,66)=shannon_entro(abs(wp41(:)));
ASData(i,67)=shannon_entro(abs(wp42(:)));
ASData(i,68)=shannon_entro(abs(wp43(:)));
ASData(i,69)=shannon_entro(abs(wp44(:)));
ASData(i,70)=shannon_entro(abs(wp45(:)));
ASData(i,71)=shannon_entro(abs(wp46(:)));
ASData(i,72)=shannon_entro(abs(wp47(:)));
ASData(i,73)=shannon_entro(abs(wp48(:)));
ASData(i,74)=shannon_entro(abs(wp49(:)));
ASData(i,75)=shannon_entro(abs(wp4A(:)));
ASData(i,76)=shannon_entro(abs(wp4B(:)));
ASData(i,77)=shannon_entro(abs(wp4C(:)));
ASData(i,78)=shannon_entro(abs(wp4D(:)));
ASData(i,79)=shannon_entro(abs(wp4E(:)));
ASData(i,80)=shannon_entro(abs(wp4F(:)));
end

%%
% D F.ZIP EPILEPTIC SUBJECT (INTERICTAL) were recorded from within
% the epileptogenic zone during seizure free intervals

for i=Nofsignal+1:2*Nofsignal;

[wp10,wp11] = dwt(Interictal(1:Length,i-Nofsignal),wname);%WPD Decomposition
[wp20,wp21] = dwt(wp10,wname);
[wp22,wp23] = dwt(wp11,wname);

[wp30,wp31] = dwt(wp20,wname);
[wp32,wp33] = dwt(wp21,wname);
[wp34,wp35] = dwt(wp22,wname);
[wp36,wp37] = dwt(wp23,wname);

[wp40,wp41] = dwt(wp30,wname);
[wp42,wp43] = dwt(wp31,wname);
[wp44,wp45] = dwt(wp32,wname);
[wp46,wp47] = dwt(wp33,wname);
[wp48,wp49] = dwt(wp34,wname);
[wp4A,wp4B] = dwt(wp35,wname);
[wp4C,wp4D] = dwt(wp36,wname);
[wp4E,wp4F] = dwt(wp37,wname);
```

```
ASData(i,1)=Entropy(abs(wp40));
ASData(i,2)=Entropy(abs(wp41));
ASData(i,3)=Entropy(abs(wp42));
ASData(i,4)=Entropy(abs(wp43));
ASData(i,5)=Entropy(abs(wp44));
ASData(i,6)=Entropy(abs(wp45));
ASData(i,7)=Entropy(abs(wp46));
ASData(i,8)=Entropy(abs(wp47));
ASData(i,9)=Entropy(abs(wp48));
ASData(i,10)=Entropy(abs(wp49));
ASData(i,11)=Entropy(abs(wp4A));
ASData(i,12)=Entropy(abs(wp4B));
ASData(i,13)=Entropy(abs(wp4C));
ASData(i,14)=Entropy(abs(wp4D));
ASData(i,15)=Entropy(abs(wp4E));
ASData(i,16)=Entropy(abs(wp4F));

ASData(i,17)=approx_entropy(64,0.5,abs(wp40));
ASData(i,18)=approx_entropy(64,0.5,abs(wp41));
ASData(i,19)=approx_entropy(64,0.5,abs(wp42));
ASData(i,20)=approx_entropy(64,0.5,abs(wp43));
ASData(i,21)=approx_entropy(64,0.5,abs(wp44));
ASData(i,22)=approx_entropy(64,0.5,abs(wp45));
ASData(i,23)=approx_entropy(64,0.5,abs(wp46));
ASData(i,24)=approx_entropy(64,0.5,abs(wp47));
ASData(i,25)=approx_entropy(64,0.5,abs(wp48));
ASData(i,26)=approx_entropy(64,0.5,abs(wp49));
ASData(i,27)=approx_entropy(64,0.5,abs(wp4A));
ASData(i,28)=approx_entropy(64,0.5,abs(wp4B));
ASData(i,29)=approx_entropy(64,0.5,abs(wp4C));
ASData(i,30)=approx_entropy(64,0.5,abs(wp4D));
ASData(i,31)=approx_entropy(64,0.5,abs(wp4E));
ASData(i,32)=approx_entropy(64,0.5,abs(wp4F));

ASData(i,33)=SampEn(5, 0.7,abs(wp40));
ASData(i,34)=SampEn(5, 0.7,abs(wp41));
ASData(i,35)=SampEn(5, 0.7,abs(wp42));
ASData(i,36)=SampEn(5, 0.7,abs(wp43));
ASData(i,37)=SampEn(5, 0.7,abs(wp44));
ASData(i,38)=SampEn(5, 0.7,abs(wp45));
ASData(i,39)=SampEn(5, 0.7,abs(wp46));
ASData(i,40)=SampEn(5, 0.7,abs(wp47));
ASData(i,41)=SampEn(5, 0.7,abs(wp48));
ASData(i,42)=SampEn(5, 0.7,abs(wp49));
ASData(i,43)=SampEn(5, 0.7,abs(wp4A));
ASData(i,44)=SampEn(5, 0.7,abs(wp4B));
ASData(i,45)=SampEn(5, 0.7,abs(wp4C));
ASData(i,46)=SampEn(5, 0.7,abs(wp4D));
ASData(i,47)=SampEn(5, 0.7,abs(wp4E));
ASData(i,48)=SampEn(5, 0.7,abs(wp4F));

ASData(i,49)=kolmogorov(abs(wp40));
ASData(i,50)=kolmogorov(abs(wp41));
ASData(i,51)=kolmogorov(abs(wp42));
ASData(i,52)=kolmogorov(abs(wp43));
ASData(i,53)=kolmogorov(abs(wp44));
ASData(i,54)=kolmogorov(abs(wp45));
ASData(i,55)=kolmogorov(abs(wp46));
ASData(i,56)=kolmogorov(abs(wp47));
```

```
ASData(i,57)=kolmogorov(abs(wp48));
ASData(i,58)=kolmogorov(abs(wp49));
ASData(i,59)=kolmogorov(abs(wp4A));
ASData(i,60)=kolmogorov(abs(wp4B));
ASData(i,61)=kolmogorov(abs(wp4C));
ASData(i,62)=kolmogorov(abs(wp4D));
ASData(i,63)=kolmogorov(abs(wp4E));
ASData(i,64)=kolmogorov(abs(wp4F));


ASData(i,65)=shannon_entro(abs(wp40(:)));
ASData(i,66)=shannon_entro(abs(wp41(:)));
ASData(i,67)=shannon_entro(abs(wp42(:)));
ASData(i,68)=shannon_entro(abs(wp43(:)));
ASData(i,69)=shannon_entro(abs(wp44(:)));
ASData(i,70)=shannon_entro(abs(wp45(:)));
ASData(i,71)=shannon_entro(abs(wp46(:)));
ASData(i,72)=shannon_entro(abs(wp47(:)));
ASData(i,73)=shannon_entro(abs(wp48(:)));
ASData(i,74)=shannon_entro(abs(wp49(:)));
ASData(i,75)=shannon_entro(abs(wp4A(:)));
ASData(i,76)=shannon_entro(abs(wp4B(:)));
ASData(i,77)=shannon_entro(abs(wp4C(:)));
ASData(i,78)=shannon_entro(abs(wp4D(:)));
ASData(i,79)=shannon_entro(abs(wp4E(:)));
ASData(i,80)=shannon_entro(abs(wp4F(:)));

end

%% E S.ZIP EPILEPTIC SUBJECT ICTAL DURING SEIZURE
%%%%%%%%%%%%%%%%%%%%%%%%%%%%%

for i=2*Nofsignal+1:3*Nofsignal;

[wp10,wp11] = dwt(Ictal(1:Length,i-2*Nofsignal),wname);%WPD Decomposition
[wp20,wp21] = dwt(wp10,wname);
[wp22,wp23] = dwt(wp11,wname);

[wp30,wp31] = dwt(wp20,wname);
[wp32,wp33] = dwt(wp21,wname);
[wp34,wp35] = dwt(wp22,wname);
[wp36,wp37] = dwt(wp23,wname);

[wp40,wp41] = dwt(wp30,wname);
[wp42,wp43] = dwt(wp31,wname);
[wp44,wp45] = dwt(wp32,wname);
[wp46,wp47] = dwt(wp33,wname);
[wp48,wp49] = dwt(wp34,wname);
[wp4A,wp4B] = dwt(wp35,wname);
[wp4C,wp4D] = dwt(wp36,wname);
[wp4E,wp4F] = dwt(wp37,wname);


ASData(i,1)=Entropy(abs(wp40));
ASData(i,2)=Entropy(abs(wp41));
ASData(i,3)=Entropy(abs(wp42));
ASData(i,4)=Entropy(abs(wp43));
ASData(i,5)=Entropy(abs(wp44));
ASData(i,6)=Entropy(abs(wp45));
```

```
ASData(i,7)=Entropy(abs(wp46));
ASData(i,8)=Entropy(abs(wp47));
ASData(i,9)=Entropy(abs(wp48));
ASData(i,10)=Entropy(abs(wp49));
ASData(i,11)=Entropy(abs(wp4A));
ASData(i,12)=Entropy(abs(wp4B));
ASData(i,13)=Entropy(abs(wp4C));
ASData(i,14)=Entropy(abs(wp4D));
ASData(i,15)=Entropy(abs(wp4E));
ASData(i,16)=Entropy(abs(wp4F));

ASData(i,17)=approx_entropy(64,0.5,abs(wp40));
ASData(i,18)=approx_entropy(64,0.5,abs(wp41));
ASData(i,19)=approx_entropy(64,0.5,abs(wp42));
ASData(i,20)=approx_entropy(64,0.5,abs(wp43));
ASData(i,21)=approx_entropy(64,0.5,abs(wp44));
ASData(i,22)=approx_entropy(64,0.5,abs(wp45));
ASData(i,23)=approx_entropy(64,0.5,abs(wp46));
ASData(i,24)=approx_entropy(64,0.5,abs(wp47));
ASData(i,25)=approx_entropy(64,0.5,abs(wp48));
ASData(i,26)=approx_entropy(64,0.5,abs(wp49));
ASData(i,27)=approx_entropy(64,0.5,abs(wp4A));
ASData(i,28)=approx_entropy(64,0.5,abs(wp4B));
ASData(i,29)=approx_entropy(64,0.5,abs(wp4C));
ASData(i,30)=approx_entropy(64,0.5,abs(wp4D));
ASData(i,31)=approx_entropy(64,0.5,abs(wp4E));
ASData(i,32)=approx_entropy(64,0.5,abs(wp4F));

ASData(i,33)=SampEn(5, 0.7,abs(wp40));
ASData(i,34)=SampEn(5, 0.7,abs(wp41));
ASData(i,35)=SampEn(5, 0.7,abs(wp42));
ASData(i,36)=SampEn(5, 0.7,abs(wp43));
ASData(i,37)=SampEn(5, 0.7,abs(wp44));
ASData(i,38)=SampEn(5, 0.7,abs(wp45));
ASData(i,39)=SampEn(5, 0.7,abs(wp46));
ASData(i,40)=SampEn(5, 0.7,abs(wp47));
ASData(i,41)=SampEn(5, 0.7,abs(wp48));
ASData(i,42)=SampEn(5, 0.7,abs(wp49));
ASData(i,43)=SampEn(5, 0.7,abs(wp4A));
ASData(i,44)=SampEn(5, 0.7,abs(wp4B));
ASData(i,45)=SampEn(5, 0.7,abs(wp4C));
ASData(i,46)=SampEn(5, 0.7,abs(wp4D));
ASData(i,47)=SampEn(5, 0.7,abs(wp4E));
ASData(i,48)=SampEn(5, 0.7,abs(wp4F));

ASData(i,49)=kolmogorov(abs(wp40));
ASData(i,50)=kolmogorov(abs(wp41));
ASData(i,51)=kolmogorov(abs(wp42));
ASData(i,52)=kolmogorov(abs(wp43));
ASData(i,53)=kolmogorov(abs(wp44));
ASData(i,54)=kolmogorov(abs(wp45));
ASData(i,55)=kolmogorov(abs(wp46));
ASData(i,56)=kolmogorov(abs(wp47));
ASData(i,57)=kolmogorov(abs(wp48));
ASData(i,58)=kolmogorov(abs(wp49));
ASData(i,59)=kolmogorov(abs(wp4A));
ASData(i,60)=kolmogorov(abs(wp4B));
ASData(i,61)=kolmogorov(abs(wp4C));
ASData(i,62)=kolmogorov(abs(wp4D));
```

```
ASData(i,63)=kolmogorov(abs(wp4E));
ASData(i,64)=kolmogorov(abs(wp4F));

ASData(i,65)=shannon_entro(abs(wp40(:)));
ASData(i,66)=shannon_entro(abs(wp41(:)));
ASData(i,67)=shannon_entro(abs(wp42(:)));
ASData(i,68)=shannon_entro(abs(wp43(:)));
ASData(i,69)=shannon_entro(abs(wp44(:)));
ASData(i,70)=shannon_entro(abs(wp45(:)));
ASData(i,71)=shannon_entro(abs(wp46(:)));
ASData(i,72)=shannon_entro(abs(wp47(:)));
ASData(i,73)=shannon_entro(abs(wp48(:)));
ASData(i,74)=shannon_entro(abs(wp49(:)));
ASData(i,75)=shannon_entro(abs(wp4A(:)));
ASData(i,76)=shannon_entro(abs(wp4B(:)));
ASData(i,77)=shannon_entro(abs(wp4C(:)));
ASData(i,78)=shannon_entro(abs(wp4D(:)));
ASData(i,79)=shannon_entro(abs(wp4E(:)));
ASData(i,80)=shannon_entro(abs(wp4F(:)));

end
```

4.3.7 Principal Component Analysis

PCA is a branch of statistics known as multivariate analysis. As the name indicates, multivariate analysis deals with the analysis of multiple variables or measurements. Multivariate data can be presented in M-dimensional space, in which every spatial dimension includes one signal. Mostly, multivariate analysis attempts to yield results that considers the relationship between multiple variables, as well as within the variables, and employs tools that work on all data. A main issue of multivariate analysis is to invent transformations of the multivariate data that yields a smaller dataset. For instance, it might contain relevant information in a multidimensional variable that can be represented by employing fewer dimensions (i.e., variables), and the reduced set of variables might be more meaningful than the original dataset. In transformations that decrease the dimension of a multivariable dataset, the idea is to transform one set of variables into a new set to produce new variables that are quite small compared to the original data. As the values of these variables are comparatively small, they might not contain much useful information to the overall dataset and, thus, can be removed. With a suitable transformation, a large number of variables that contribute only marginally to the total information can be eliminated. The data transformation employed to achive a new set of variables is a linear function, as linear transformations are easier to calculate. As a linear transformation, this process can be interpreted as a rotation and possibly scaling of the original dataset in M-dimensional space. In PCA, the aim is to transform the dataset to yield a new set of variables (principal components) that are uncorrelated. The objective is to reduce the dimension of the data, not necessarily to yield more meaningful variables. PCA can reduce the number of variables in a dataset without loss of information and identify new variables with better meaning. It transforms a set of correlated variables into a new set of uncorrelated variables. If the variables in a dataset are already uncorrelated, PCA is useless. Moreover, the principal components are orthogonal and are ordered in terms of the variability they characterize (Semmlow, 2004).

In big datasets with lots of features, it may be more convenient to determine a smaller and more compact feature representation employing a feature transformation. One technique is to employ PCA, which uses a projection to the features to find a reduced representation. Assume that the training set contains n training examples, that is, $X=\{x_1, x_2, \ldots, x_j\}$, then the algorithm makes principal components, P_k, which are linear combinations of the original features X (Begg et al., 2007). This can be written as

$$Pk = a_{k1}x_1 + a_{k2}x_2 + \cdots + a_n x_{kn} \tag{4.10}$$

where $\sum_i a_{ki}^2 = 1$.

The principal component vectors are produced in a way that they are orthogonal to each other and, for this reason, have the highest variance (principal components). Generally, the training data is first normalized to zero mean and unit variance before employing the PCA algorithm (Begg et al., 2007).

Preprocessing generally transforms a high-dimensional space to a lower-dimensional space with minimum information loss. The procedure is known as feature extraction. The PCA is a well-known feature extraction technique and permits the removal of the second-order correlation between given random processes. The PCA linearly transforms a high-dimensional input vector into a lower-dimensional one whose components are uncorrelated, by computing the eigenvectors of the covariance matrix of the input signal. The PCA normally utilizes the optimization of some information criterion, such as the maximization of the variance of the projected signal or the minimization of the reconstruction error. The goal of the PCA is to extract M-orthonormal directions $\mathbf{w}_i \in R^n$, $i = 1, 2, \ldots, M$, in the input space that interpret for the mininmum variance of the signal. Afterward, an input vector $\mathbf{x} \in R^n$ is transformed into a lower M-dimensional space without losing vital intrinsic information. The vector x can be characterized by being projected onto the M-dimensional subspace spanned by $\mathbf{w}_i$ utilizing the inner products $\mathbf{x}^T\mathbf{w}_i$. This yields the dimension reduction (Du & Swamy, 2006).

PCA is a technique to decompose a multichannel signal into components that are linearly independent, that is, temporally and spatially uncorrelated. The samples from all channels are treated at a given time interval as a point in the space of dimension equal to the number of channels. The original time series can be improved as a linear combination of these components to yield a reduction of data dimension by neglecting the smallest variance. PCA can be realized by means of the singular value decomposition (SVD) algorithm (Blinowska & Zygierewicz, 2011). The SVD algorithm is implemented in MATLAB as a function svd; to find the decomposition of x execute:

```
[P,A,M] = svd(x);
```

At this step, the PCA is used to reduce the dimensionality of the biomedical dataset and to produce fewer numbers of uncorrelated variables, which are employed as features for better classification of biomedical signals. Usually, the recorded multichannel biomedical signals are huge in number, encompassing a large amount of redundant information that are highly correlated. The PCA is a useful method to transform a number of correlated variables into a smaller number of uncorrelated variables called principal components. The principal components characterize the most informative data carried by the original signals to deliver discriminative information about those signals. Thus, the PCA features work better in different biomedical signal classification (Siuly et al., 2016). PCA can be performed in MATLAB using function `pca` from the Statistics and Machine Learning Toolbox.

EXAMPLE 4.18 The following MATLAB code is used to extract features from the EEG signals using AR Burg spectrum. Then it uses PCA for dimension reduction. You can download data from the following website: http://epileptologie-bonn.de/cms/front_content.php?idcat=193&lang=3&changelang=3

```
%Ch4_Ex86_EEG_Burg_PCA.m
%The following MATLAB code is used to extract the features from
%the EEG signals using Burg AR.
%then it uses PCA for dimension reduction
clc
clear
%Load Sample EEG Data downloaded from the web site
%http://epileptologie-bonn.de/cms/front_content.php?idcat=193&lang=3&changelang=3
load AS_BONN_ALL_EEG_DATA_4096.mat
%%
Fs = 173.6;              % Sampling frequency
T = 1/Fs;                % Sampling period
segmentLength = 128; % Length of a signal segment
Length=4096;             % Length of signal
Nofsignal=100;           %Number of Signal
order = 14;
%% FEATURE EXTRACTION
% Obtain the Burg AR Spectrum of the Normal EEG signal using pburg.
for i=1:Nofsignal
[Pxx,F] = pburg(Normal_Eyes_Open(1:Length,i),order,segmentLength,Fs);
  ASData(i,:)=Pxx(:);
end
%%
% Obtain the Burg AR Spectrum of the Interictal EEG signal using pburg.
for i=Nofsignal+1:2*Nofsignal
[Pxx,F] = pburg(Interictal(1:Length,i-Nofsignal),order,segmentLength,Fs);
     ASData(i,:)=Pxx(:);
```

```
end
%%
% Obtain the Burg AR Spectrum of the Ictal EEG signal using pburg.
for i=2*Nofsignal+1:3*Nofsignal
[Pxx,F] = pburg(Ictal(1:Length,i-2*Nofsignal),order,segmentLength,Fs);
    %[Pxx,F] = pmcov(Ictal(1:Length,i-2*Nofsignal),order,segmentLength,Fs);
ASData(i,:)=Pxx(:);
end

%% Dimension Reduction with PCA
ASDataPCA=pca(ASData);
```

4.3.8 Independent Component Analysis

ICA is a recently developed statistical model as an extension of the PCA. The ICA was initially presented for blind source separation (BSS), and then adapted for feature extraction and dimension reduction. The ICA can be used for BSS, feature extraction, dimension reduction, and signal detection. Although the undercomplete ICA can be utilized for feature extraction, the overcomplete ICA can be utilized for dimension reduction of biomedical signals based on multiscale and redundant basis sets (Du & Swamy, 2006). The idea of ICA is to decompose the signals into their constituent independent components. The joint source signals are thought to be independent from each other, and this notion plays a critical role in denoising and separation of the signals (Sanei & Chambers, 2013).

ICA is a statistical signal-processing technique that decomposes a multichannel signal into components that are statistically independent. The objective of the ICA is to estimate

$$\mathbf{y} = \mathbf{W}^T\mathbf{x} \tag{4.11}$$

such that the components of y must be statistically independent. The HOS of the original signals are essential to estimate the signal, rather than the second-order moment or covariance of the samples as utilized in the PCA. The Cramer-Rao bound is employed to estimate the source signals in the ICA, based on the assumption that all independent components have finite variance (Du & Swamy, 2006). To find the independent components in the light of the central limit theorem, components of least Gaussian distributions should be found. To realize this method, the heuristic, which assumes that the required independent components must have identical distributions, must be followed (Hyvärinen & Oja, 2000). There are several tools for performing ICA in MATLAB. To use the fastica algorithm, one must install the fastica toolbox https://research.ics.aalto.fi/ica/fastica/code/dlcode.shtml or https://github.com/davidkun/FastICA

and include it in the MATLAB path. Usually, the physiological implication of any differences in the results of dissimilar algorithms have not been systematically verified for biomedical signals. The most frequent applications of ICA in the field of biomedical signal analysis are related to EEG and MEG artifact reduction, feature extraction, and dimension reduction in studies of EEG signal classification, and as a preprocessing step in the search of sources of EEG and MEG activity localization. ICA is also utilized in analysis of heart and muscle signals (Blinowska & Zygierewicz, 2011).

EXAMPLE 4.19 The following MATLAB code is used to extract features from the EEG signals using AR Burg spectrum. Then it uses ICA for dimension reduction. You can download fastICA from the website: https://research.ics.aalto.fi/ica/fastica/

You can download data from the following website:
http://epileptologie-bonn.de/cms/front_content.php?idcat=193&lang=3&changelang=3

```
%Ch4_Ex91_EEG_Burg_fastICA.m
%The following MATLAB code is used to extract the features from
%the EEG signals using Burg AR.
%then it uses fastICA for dimension reduction
%You can download fastICA from the web site
%https://research.ics.aalto.fi/ica/fastica/

clc
clear
```

```
%Load Sample EEG Data downloaded from the web site
%http://epileptologie-bonn.de/cms/front_content.php?idcat=193&lang=3&changelang=3
load AS_BONN_ALL_EEG_DATA_4096.mat
%%
Fs = 173.6;          % Sampling frequency
T = 1/Fs;            % Sampling period
segmentLength = 128; % Length of a signal segment
Length=4096;% Length of signal
Nofsignal=100; %Number of Signal
order = 14;
%% FEATURE EXTRACTION
% Obtain the Burg AR Spectrum of the Normal EEG signal using pburg.
for i=1:Nofsignal
[Pxx,F] = pburg(Normal_Eyes_Open(1:Length,i),order,segmentLength,Fs);
ASData(i,:)=Pxx(:);
end
%%
% Obtain the Burg AR Spectrum of the Interictal EEG signal using pburg.
for i=Nofsignal+1:2*Nofsignal
[Pxx,F] = pburg(Interictal(1:Length,i-Nofsignal),order,segmentLength,Fs);
ASData(i,:)=Pxx(:);
end
%%
% Obtain the Burg AR Spectrum of the Ictal EEG signal using pburg.
for i=2*Nofsignal+1:3*Nofsignal
[Pxx,F] = pburg(Ictal(1:Length,i-2*Nofsignal),order,segmentLength,Fs);
  ASData(i,:)=Pxx(:);
end

%% Dimension Reduction with fast ICA
ASDataFastICA=fastica(ASData');
```

4.3.9 Linear Discriminant Analysis

PCA is widely used for feature extraction. As a variant of PCA, kernel PCA has widespread acceptance. Because PCA does not use class information, the first principal component is not essentially vital for class separation. On the other hand, LDA looks for the axis that maximally separates training data projected on this axis into two classes. Applications of LDA are limited to a case in which every class is composed of one cluster, and they are not heavily overlapped. By suitable selection of kernels and their parameter values, kernel discriminant analysis (KDA) eliminates the limitation of LDA. It can be extended to multiclass problems as well. KDA is employed as criteria for feature selection and kernel selection as well as feature extraction (Abe, 2010).

Among dimension reduction methods, LDA is a widespread one that has been widely employed. LDA aims to maximize the ratio of the between-class scatter and total data scatter in projected space, and the label of each data is required. However, in real applications, the labeled data are rare and unlabeled data are in large quantity, so LDA is difficult to use under such cases (Wang, Lu, Gu, Du, & Yang, 2016).

EXAMPLE 4.20 The following MATLAB code is used to extract features from the EEG signals using AR Burg spectrum. Then it uses LDA for dimension reduction. You can download drtoolbox from the website: https://lvdmaaten.github.io/drtoolbox/

You can download data from the following website:
http://epileptologie-bonn.de/cms/front_content.php?idcat=193&lang=3&changelang=3

```
%Ch4_Ex96_EEG_Burg_LDA.m
%The following MATLAB code is used to extract the features from
%the EEG signals using Burg AR.
```

```
%then it uses LDA of drtoolbox for dimension reduction
%https://lvdmaaten.github.io/drtoolbox/
clc
clear
%Load Sample EEG Data downloaded from the web site
%http://epileptologie-bonn.de/cms/front_content.php?idcat=193&lang=3&changelang=3
load AS_BONN_ALL_EEG_DATA_4096.mat
%%
Fs = 173.6;          % Sampling frequency
T = 1/Fs;            % Sampling period
segmentLength = 128; % Length of a signal segment
Length=4096;% Length of signal
Nofsignal=100; %Number of Signal
order = 14;
%% FEATURE EXTRACTION
% Obtain the Burg AR Spectrum of the Normal EEG signal using pburg.
for i=1:Nofsignal
[Pxx,F] = pburg(Normal_Eyes_Open(1:Length,i),order,segmentLength,Fs);
  %[Pxx,F] = pmcov(Normal_Eyes_Open(1:Length,i),order,segmentLength,Fs);
    ASData(i,:)=Pxx(:);
end
%%
% Obtain the Burg AR Spectrum of the Interictal EEG signal using pburg.
for i=Nofsignal+1:2*Nofsignal
[Pxx,F] = pburg(Interictal(1:Length,i-Nofsignal),order,segmentLength,Fs);

% [Pxx,F] = pmcov(Interictal(1:Length,i-Nofsignal),order,segmentLength,Fs);
    ASData(i,:)=Pxx(:);
end
%%
% Obtain the Modified Covariance Spectrum of the Ictal EEG signal using pburg.
for i=2*Nofsignal+1:3*Nofsignal
[Pxx,F] = pburg(Ictal(1:Length,i-2*Nofsignal),order,segmentLength,Fs);
    %[Pxx,F] = pmcov(Ictal(1:Length,i-2*Nofsignal),order,segmentLength,Fs);
ASData(i,:)=Pxx(:);
end

%% Create Target Vector
yapp=[1*ones(1,Nofsignal) 2*ones(1,Nofsignal) 3*ones(1,Nofsignal)];

%% Dimension Reduction with fast LCA
ASDataLDA=lda(ASData,yapp,3);
```

4.4 ELECTROCARDIOGRAM SIGNAL PREPROCESSING

One of the first applications of computer processing was analysis of ECG. This early work was done using decision tree logic. Very soon, it became obvious that computer interpretation and accuracy of the measurement were highly interconnected. As a consequence of this, signal processing became crucial for accurate diagnosis, particularly when analyzing ECG records under ambulatory or tiring circumstances. Furthermore, theoretical improvements in signal processing contributed considerably to a dynamic property of the ECG signal. It is, as a result, expected to contribute to systems for ECG interpretation, intensive care monitoring, ambulatory ECG monitoring, stress testing, and so on. Generally, a group of algorithms extract fundamental ECG measurements of wave durations and amplitudes to compress the data for efficient transmission or storage. Although these algorithms are often designed to work in chronological order, data on the incidence time of a P-QRS-T complex, as realized by the P-QRS-T detector, is occasionally included into several algorithms to increase efficiency. The complexity of every algorithm differs from application to application, for instance, noise filtering employed in ambulatory monitoring is considerably more advanced than filtering required in resting ECG analysis. Significant focus is on the design of filters, having both linear and nonlinear structures, for the elimination of baseline wander and powerline interference. Both kinds of interruption require the design of a narrowband filter. Elimination of noise caused by muscle activity illustrates an additional significant filtering difficulty that is much more complicated to handle. Muscle noise in the ECG can be decreased each time, because it is suitable to use methods that gain from the fact that the ECG is a recurrent signal (Sörnmo & Laguna, 2005).

During ECG recording, different origins can possibly append noise to the recorded signal and damage the output signal. The occurrence of noise in signal systems can have a harsh effect on understanding of the data and lead to an incorrect diagnosis. Noteworthy origins of noise (Friesen et al., 1990) include the following (Begg et al., 2007):

- **a.** Electrical interference from powerlines at 50 or 60 Hz powerline frequency. These consist of fundamental signals plus their harmonics with up to a maximum of 50% peak-to-peak ECG amplitude.
- **b.** Muscle contraction and muscle activity can produce high-frequency electromyography (EMG) noise.
- **c.** Motion artifacts, for example, movement of the electrode over the skin surface.
- **d.** Impedance alterations at the skin/electrode interface because of momentary loss of contact or loose electrodes.
- **e.** Baseline drifts due to respiration.
- **f.** Noise appearing because of instrumentation or electronic devices.
- **g.** Electrosurgical noise.

For this reason, preprocessing of the ECG signals is required to decrease the different noise components and raise the signal-to-noise ratio (SNR) to a great degree. The ECG signal is filtered with a single band-pass filter with a 10- to 25-Hz passband. Zigel, Cohen, & Katz (2000) developed filtering by employing two split infinite impulse response filters to unlike parts of the ECG waveform. In their scheme, the P- and T-wave segments were filtered with a 0.1- to 50-Hz passband while the QRS complex was filtered using a 0.1- to 100-Hz passband (Begg et al., 2007).

The filtering techniques are widely employed for signal denoising, including for analysis of ECG signals. The ECG filtering must be done when the preferred data remains exact. A major consideration while denoising is the level to which the P-QRS-T complexes control the output of the filter. The existence of a P-QRS-T complex and its occurrence time is a fundamental requirement in ECG signal processing. The implementation of a P-QRS-T detector is significant because poor detection efficiency might limit the general system performance. When the P-QRS-T complex is detected, important waves can be analyzed to classify signals correctly (Sörnmo & Laguna, 2005).

4.4.1 QRS Detection Algorithms

Typical characteristics of the QRS complex, for example, as physical morphology, its time, and frequency domain characteristics, offer a great amount of data regarding cardiac health. There have been a significant number of researches dedicated to the design of QRS detection algorithms over the last decades. In a review made by Friesen et al. (1990), they provided a number of references to QRS detection and gave an evaluation of nine QRS detection algorithms classified with regard to the method employed to discover the QRS complex. Their preface test results discovered that none of the algorithms were able to detect the QRS complexes without false positives for all the noise types at the highest noise levels. It was discovered that algorithms based on amplitude and slope showed the highest efficiency for EMG-corrupted ECG. An algorithm utilizing a digital filter had the best performance for the composite-noise-corrupted data. After the QRS complexes have been determined, you can extract further data from the physical features such as highest amplitude, length of QT segment, and RT segment duration (Begg et al., 2007).

EXAMPLE 4.21 The following MATLAB code was to denoise and detect P-QRS-T segments of a given ECG signal as a feature. This example is taken from the signal processing toolbox of the MATLAB (Krauss, Shure, & Little, 2017).

```
% Measuring Amplitudes of Peaks
% This example shows peak analysis in an ECG (Electro-cardiogram) signal. ECG is a
% measure of electrical activity of the heart over time. The signal is measured by electrodes
% attached to the skin and is sensitive to disturbances such as power source interference
% and noises due to movement artifacts.
load noisyecg.mat
t = 1:length(noisyECG_withTrend);
figure
plot(t,noisyECG_withTrend)
title('Signal with a Trend')
```

```
xlabel('Samples');
ylabel('Voltage(mV)')
legend('Noisy ECG Signal')
grid on

%%
% Detrending Data
% The above signal shows a baseline shift and therefore does not represent the true
% amplitude. In order to remove the trend, fit a low order polynomial to the signal and use
% the polynomial to detrend it.
[p,s,mu] = polyfit((1:numel(noisyECG_withTrend))',noisyECG_withTrend,6);
f_y = polyval(p,(1:numel(noisyECG_withTrend))',[],mu);
ECG_data = noisyECG_withTrend - f_y; % Detrend data
figure
plot(t,ECG_data); grid on
ax = axis; axis([ax(1:2) -1.2 1.2])
title('Detrended ECG Signal')
xlabel('Samples'); ylabel('Voltage(mV)')
legend('Detrended ECG Signal')

%%
% After detrending, find the QRS-complex which is the most prominent repeating peak
% in the ECG signal. The QRS-complex corresponds to the depolarization of the right and
% left ventricles of the human heart. It can be used to determine a patient's cardiac rate or
% predict abnormalities in heart function. The following figure shows the shape of the QRScomplex
% in an ECG signal.
%
% Thresholding to Find Peaks of Interest
% The QRS-complex consists of three major components: Q-wave, R-wave, S-wave. The
% R-waves can be detected by thresholding peaks above 0.5mV. Notice that the R-waves are
% separated by more than 200 samples. Use this information to remove unwanted peaks by
% specifying a 'MinPeakDistance'.

[~,locs_Rwave] = findpeaks(ECG_data,'MinPeakHeight',0.5,...
'MinPeakDistance',200);

%%
% For detection of the S-waves, find the local minima in the signal and apply thresholds
% appropriately.
% Finding Local Minima in Signal
% Local minima can be detected by finding peaks on an inverted version of the original
% signal.
ECG_inverted = -ECG_data;
[~,locs_Swave] = findpeaks(ECG_inverted,'MinPeakHeight',0.5,...
'MinPeakDistance',200);
% The following plot shows the R-waves and S-waves detected in the signal.
figure
hold on
plot(t,ECG_data);
plot(locs_Rwave,ECG_data(locs_Rwave),'rv','MarkerFaceColor','r');
plot(locs_Swave,ECG_data(locs_Swave),'rs','MarkerFaceColor','b');
axis([0 1850 -1.1 1.1]); grid on;
legend('ECG Signal','R-waves','S-waves');
xlabel('Samples'); ylabel('Voltage(mV)')
title('R-wave and S-wave in Noisy ECG Signal')

%%
% Next, we try and determine the locations of the Q-waves. Thresholding the peaks to
% locate the Q-waves results in detection of unwanted peaks as the Q-waves are buried in
```

```
% noise. We filter the signal first and then find the peaks. Savitzky-Golay filtering is used
% to remove noise in the signal.
smoothECG = sgolayfilt(ECG_data,7,21);
figure
plot(t,ECG_data,'b',t,smoothECG,'r'); grid on
axis tight;
xlabel('Samples'); ylabel('Voltage(mV)');
legend('Noisy ECG Signal','Filtered Signal')
title('Filtering Noisy ECG Signal')

%%
% We perform peak detection on the smooth signal and use logical indexing to find the
% locations of the Q-waves.
[~,min_locs] = findpeaks(-smoothECG,'MinPeakDistance',40);
% Peaks between -0.2mV and -0.5mV
locs_Qwave = min_locs(smoothECG(min_locs)>-0.5 & smoothECG(min_locs)<-0.2);
figure
hold on
plot(t,smoothECG);
plot(locs_Qwave,smoothECG(locs_Qwave),'rs','MarkerFaceColor','g');
plot(locs_Rwave,smoothECG(locs_Rwave),'rv','MarkerFaceColor','r');
plot(locs_Swave,smoothECG(locs_Swave),'rs','MarkerFaceColor','b');
grid on
title('Thresholding Peaks in Signal')
xlabel('Samples'); ylabel('Voltage(mV)')
ax = axis; axis([0 1850 -1.1 1.1])
legend('Smooth ECG signal','Q-wave','R-wave','S-wave');

%%
% The above figure shows that the QRS-complex successfully detected in the noisy ECG
% signal.
% Error Between Noisy and Smooth Signal
% Notice the average difference between the QRS-complex in the raw and the detrended
% filtered signal.
% Values of the Extrema
[val_Qwave, val_Rwave, val_Swave] = deal(smoothECG(locs_Qwave), smoothECG(locs_Rwave),smoothECG
(locs_Swave));
meanError_Qwave = mean((noisyECG_withTrend(locs_Qwave) - val_Qwave));
meanError_Rwave = mean((noisyECG_withTrend(locs_Rwave) - val_Rwave));
meanError_Swave = mean((noisyECG_withTrend(locs_Swave) - val_Swave));
%%
% This demonstrates that it is essential to detrend a noisy signal for efficient peak analysis.
% Peak Properties
% Some important peak properties involve rise time, fall time, rise level, and fall level.
% These properties are computed for each of the QRS-complexes in the ECG signal. The
% average values for these properties are displayed on the figure below.
avg_riseTime = mean(locs_Rwave-locs_Qwave); % Average Rise time
avg_fallTime = mean(locs_Swave-locs_Rwave); % Average Fall time
avg_riseLevel = mean(val_Rwave-val_Qwave); % Average Rise Level
avg_fallLevel = mean(val_Rwave-val_Swave); % Average Fall Level
helperPeakAnalysisPlot(t,smoothECG,...
locs_Qwave,locs_Rwave,locs_Swave,...
val_Qwave,val_Rwave,val_Swave,...
avg_riseTime,avg_fallTime,...
avg_riseLevel,avg_fallLevel)
```

REFERENCES

Abe, S. (2010). Feature selection and extraction. In *Support vector machines for pattern classification* (pp. 331–341). Springer.

Begg, R., Lai, D. T., & Palaniswami, M. (2007). *Computational intelligence in biomedical engineering*. CRC Press.

Bengio, Y., Delalleau, O., Roux, N., Paiement, J.-F., Vincent, P., & Ouimet, M. (2003). *Feature extraction: Foundations and applications, chapter spectral dimensionality reduction.*

Blinowska, K. J., & Zygierewicz, J. (2011). *Practical biomedical signal analysis using MATLAB®*. CRC Press.

Delac, K., Grgic, M., & Grgic, S. (2005). A comparative study of PCA, ICA, and LDA. In *Presented at the Proc. of the 5th EURASIP conference focused on speech and image processing, multimedia communications and services*, (pp. 99–106).

Du, K.-L., & Swamy, M. (2006). Principal component analysis networks. *Neural Networks in a Softcomputing Framework*, 295–351.

Friesen, G. M., Jannett, T. C., Jadallah, M. A., Yates, S. L., Quint, S. R., & Nagle, H. T. (1990). A comparison of the noise sensitivity of nine QRS detection algorithms. *IEEE Transactions on Biomedical Engineering*, *37*(1), 85–98.

Graimann, B., Allison, B., & Pfurtscheller, G. (2009). Brain-computer interfaces: a gentle introduction. In *Brain-computer interfaces* (pp. 1–27): Springer.

Guyon, I., Gunn, S., Nikravesh, M., & Zadeh, L. (2006). *Feature extraction: Foundations and applications (studies in fuzziness and soft computing)*. Secaucus, NJ, USA: Springer.

Hyvärinen, A., & Oja, E. (2000). Independent component analysis: algorithms and applications. *Neural Networks*, *13*(4–5), 411–430.

Proakis, J. G., & Manolakis, D. G. (2007). *Digital signal processing: Principles, algorithms, and applications*. Pearson Prentice Hall.

Kay, S. M. (1993). *Fundamentals of statistical signal processing, volume I: Estimation theory*.

Kay, S. M., & Marple, S. L. (1981). Spectrum analysis—a modern perspective. *Proceedings of the IEEE*, *69*(11), 1380–1419.

Kevric, J., & Subasi, A. (2017). Comparison of signal decomposition methods in classification of EEG signals for motor-imagery BCI system. *Biomedical Signal Processing and Control*, *31*, 398–406.

Krauss, T. P., Shure, L., & Little, J. N. (2017). *Signal processing toolbox user's guide*. The Math Works.

Kutlu, Y., & Kuntalp, D. (2012). Feature extraction for ECG heartbeats using higher order statistics of WPD coefficients. *Computer Methods and Programs in Biomedicine*, *105*(3), 257–267.

Mendel, J. M. (1991). Tutorial on higher-order statistics (spectra) in signal processing and system theory: theoretical results and some applications. *Proceedings of the IEEE*, *79*(3), 278–305.

Phinyomark, A., Quaine, F., Charbonnier, S., Serviere, C., Tarpin-Bernard, F., & Laurillau, Y. (2013). EMG feature evaluation for improving myoelectric pattern recognition robustness. *Expert Systems With Applications*, *40*(12), 4832–4840.

Sanei, S. (2013). *Adaptive processing of brain signals*. John Wiley & Sons.

Sanei, S., & Chambers, J. A. (2013). *EEG signal processing*. John Wiley & Sons.

Semmlow, J. (2004). *Biosignal and biomedical image processing: MATLAB-based applications*. Marcel Dekker.

Siuly, S., Li, Y., & Zhang, Y. (2016). *EEG signal analysis and classification*. Springer.

Sörnmo, L., & Laguna, P. (2005). *Bioelectrical signal processing in cardiac and neurological applications. Vol. 8*. Academic Press.

Stoica, P., & Moses, R. L. (1997). *Introduction to spectral analysis. Vol. 1*. Upper Saddle River, NJ: Prentice Hall.

Wang, S., Lu, J., Gu, X., Du, H., & Yang, J. (2016). Semi-supervised linear discriminant analysis for dimension reduction and classification. *Pattern Recognition*, *57*, 179–189.

Wołczowski, A., & Zdunek, R. (2017). Electromyography and mechanomyography signal recognition: experimental analysis using multi-way array decomposition methods. *Biocybernetics and Biomedical Engineering*, *37*(1), 103–113. https://doi.org/10.1016/j.bbe.2016.09.004.

Zigel, Y., Cohen, A., & Katz, A. (2000). ECG signal compression using analysis by synthesis coding. *IEEE Transactions on Biomedical Engineering*, *47*(10), 1308–1316.

Chapter 5

Biomedical Signal Classification Methods

5.1 INTRODUCTION

A machine learning model can be defined by using some parameters implemented by a computer program that optimizes those parameters of the model by employing training data or past experience. Machine learning utilizes the theory of statistics in creating models, as the main task is to make inference using a sample. In some applications, the efficiency of the learning algorithm might be as significant as its classification accuracy. Machine learning methods are employed as a decision support system in medicine for medical diagnoses (Alpaydin, 2014).

Clustering methods are used to separate or divide data or samples into a number of classes, and only the number of clusters can be recognized and fed into the clustering algorithm by the user. Classification is similar to clustering except the classifier is trained by utilizing a set of previously labeled data. Hence, the testing data is classified somehow using the similarity of a category of labeled data. In practice, the aim is to find a boundary between two or more classes and label them based on their measured features. There is always an uncertainty in clustering or classification regarding which features should be used and how those features are extracted or enhanced. In the context of biomedical signal analysis, the classification of the data in feature spaces is generally essential. There have been plenty of classification methods developed within the last four decades. The most popular ones are linear discriminant analysis (LDA), Naïve Bayes (NB), *k*-nearest neighbor (*k*-NN), artificial neural networks (ANNs), support vector machines (SVMs), and decision tree (DT) algorithms (Sanei, 2013). The utilization of all these methods in biomedical signal analysis is the objective of this chapter.

The research community has a great deal of interest in machine learning techniques for recognition, classification, and diagnosis of diseases. Several examples, including diagnosis of cardiac diseases, and brain or muscular disorders, usually lead to life-threatening circumstances. Early and precise diagnoses are the primary implication so that preventive measures or treatment can be administered. The analysis of biomedical signals is crucial for monitoring abnormalities in the human body. The diagnostic procedure encompasses the elimination of features from biomedical signals and succeeding comparisons with known illnesses to discover any differences from the normal characteristics of the signal. Such a monitoring system must be able to find abnormalities represented by signal shape changes. Machine learning methods are capable of automating the process of biomedical signal analysis and classification among normal and pathological patterns by creating decision surfaces classifying these patterns. Automatic detection and classification of biomedical signals using several signal-processing techniques have developed into a critical aspect of clinical monitoring (Begg, Lai, & Palaniswami, 2007). The aim of this chapter is to present how to design an efficient system to carry out real-time monitoring and alert clinicians when life-threatening conditions start to surface.

Generally, the biomedical signal classification process can be divided into four stages, namely (1) signal acquisition and segmentation, (2) signal denoising, (3) feature extraction/dimension reduction, and (4) recognition and classification. As seen in Fig. 5.1, the biomedical signals are recorded from the human body and then denoised to reduce the noise detected by other electrical activities of the body or other types of artifact. Then, the features are extracted from the biomedical signal acquired from the previous stage and converted into a feature vector. The relevant structure in the raw data is characterized by the feature vector. In the third step, a dimension reduction is applied to remove redundant information from the feature vector, producing a reduced feature vector. In the fourth stage, a classifier categorizes the reduced feature vector.

5.2 PERFORMANCE EVALUATION METRICS

To obtain a reliable assessment of the target approximation quality represented by the model, the evaluation of the classification model is utilized. Depending on the application area, different performance measures can be employed. Because the model is created based on a training dataset, which is usually a small subset of the domain, the generalization property of the model is essential for quality. Hence, it is crucial to discriminate the value of a particular dataset, the value of the training set (training performance), and its expected performance on the whole domain (true performance). Training performance of the model is determined by an assessment of the model on the training set used

Practical Guide for Biomedical Signals Analysis Using Machine Learning Techniques. https://doi.org/10.1016/B978-0-12-817444-9.00005-2

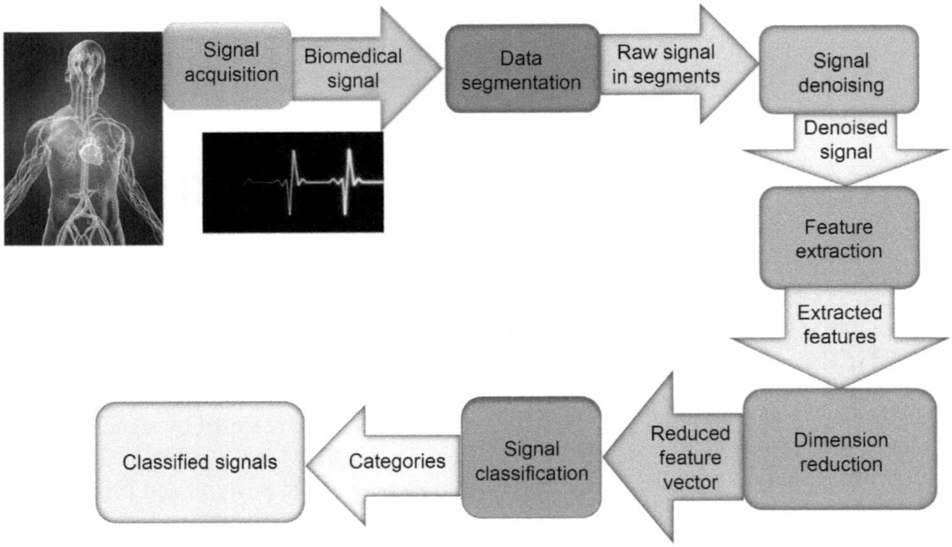

FIG. 5.1 A general framework for biomedical signal classification.

to build the model. Although this performance is useful for better understanding of the model, it is not of significant interest as the classification of the training data is not the purpose of classification models. Expected performance of the model, on the entire domain, represents its true performance. True performance of the model shows its ability to correctly classify new instances from the given domain. It always remains unknown and can only be estimated by dataset performance, as the true class labels are usually unavailable for the domain. To assess the true performance, that is, to consistently estimate the unknown values of the adopted performance measures on the entire domain that are comprised of generally unseen instances, appropriate evaluation procedures are needed (Cichosz, 2014).

The measures used for assessing the performance of a methodology are a crucial part of its design. In machine learning, there are numerous kinds of performance evaluation measures. In this book, the performance of a classifier is measured on the basis of standard criteria employed in biomedical signal analysis. These include classification accuracy, sensitivity (or true positive rate [TPR] or recall), specificity, false alarm rate (FAR), F-measure, and receiving operating characteristic (ROC) curve. These measures are employed for the estimation of the behavior of the classifiers on the extracted feature data (Siuly, Li, & Zhang, 2016).

k-Fold cross-validation is a sophisticated assessment technique that handles the trade-off between bias and variance. It randomly divides the dataset into k subsets of the same size and then repeats over these subsets. After all k repetitions are accomplished, the model built without specific instance in the training set is utilized to generate a predicted class label for every instance in the dataset. The resulting vector of classifications can be compared to the true class labels employing one or more selected measures of the performance. A single repetition of k-fold cross-validation is similar to the hold-out procedure, with $(k-1)/k$ of data chosen for training and $1/k$ of data chosen for assessment. For sufficiently large k, this does not diminish the training set size to an extent that would have impact on model quality, as the validation set is small. As in k repetitions where all existing instances are employed for the model assessment, the variance of performance evaluations is not increased. The k-fold cross-validation technique efficiently virtualizes the training and validation of test sets. All existing instances from the set can be utilized for both model creation and assessment but not at the same time. The difference of this procedure from the hold-out procedure repeated k times is that all the validation sets from successive repetitions are disjointed and together cover the whole existing dataset (Cichosz, 2014).

The leave-one-out is the validation process that utilizes the idea of k-fold cross-validation to the extreme, as it employs the number of instances in the dataset as the value of k. A model built on the dataset with one instance removed to classify for this instance is employed in the process that iterates over all instances. Leave-one-out is a form of cross-validation with no compromise, as its major advantage is large number of k, that is, it has no pessimistic bias. In practice, even the variance increases due to large k and cannot be reduced by multiple runs, therefore, the leave-one-out assessment process occasionally produces overoptimistic estimates. The reason is that, for especially larger datasets, the distinct models formed on succeeding iterations are rather unlikely to differ considerably from one another, as well as from the model that can be created employing the whole dataset, as their training sets differ just in a single instance. However, during building classification models based on small datasets in which a single instance still substantially matters, the leave-one-out represents a reasonable evaluation procedure. Due to the computational expense of creating as many models as instances available, it cannot be applied easily to larger datasets. In contrast to hold-out and cross-validation with k less than the dataset size, leave-one-out is perfectly deterministic and reproducible, and leaves no space for randomness in the evaluation process (Cichosz, 2014).

Performance measures of a classifier can be achieved by comparing the true class labels of the instances from the dataset and the predictions produced by the classifier on the same dataset. This subset is usually separated from the training set and called as the validation set or test set. For the performance evaluation which can affect the final model by choosing a classification algorithm, tuning its parameters and choosing attributes called intermediate evaluation and generally validation set is used. When the performance of the eventually formed model is to be evaluated, test set is used. Intermediate evaluation associated with the model selection is based on evaluation results (Cichosz, 2014).

Many applications show that it is not enough to know how often the evaluation model is incorrect or even the average costs of its mistakes. It is crucial to know how often the model fails to correctly predict particular classes in those cases when the classes of the target concept have different predictability or different occurrence rates. This may also match with nonuniform misclassification costs. In those cases, performance of the model can be deeply evaluated using a confusion matrix. Confusion matrix provides a useful insight of the capability of the model to predict particular classes and also its generalization properties. However, it does not often give the required ability of ranking and comparing different models based on their performance to assist in selecting the best of the several candidate models. Different measures of performance may be derived from the confusion matrix, but they can be only applied to two-class models. The convention used in a confusion matrix is that "positive" or "negative" refers to the class labels predicted by the model, and "true" and "false" refers to the accuracy of the prediction (Cichosz, 2014). The most popular measures of the performance calculated for a 2 x 2 confusion matrix are shown in Fig. 5.2.

	PREDICTED CLASS		
		Class = Yes	Class = No
ACTUAL CLASS	Class = Yes	(TP)	(FN)
	Class = No	(FP)	(TN)

FIG. 5.2 Representation of confusion matrix.

Misclassification error is represented as the ratio of incorrectly classified instances to all instances:

$$\text{Classification Error} = \frac{FP + FN}{TP + TN + FP + FN} \tag{5.1}$$

Accuracy is given as the ratio of correctly classified instances to all instances:

$$\text{Accuracy} = \frac{TP + TN}{TP + FP + FN + TN} \tag{5.2}$$

TPR or sensitivity is the ratio of instances correctly classified as positive to all positive instances:

$$\text{TPR} = \text{sensitivity} = \frac{TP}{TP + FN} \tag{5.3}$$

False positive rate (specificity) is represented as the ratio of instances incorrectly classified as positive to all negative instances:

$$\text{FPR} = \text{specificity} = \frac{TN}{FP + TN} \tag{5.4}$$

Sensitivity and specificity are the measures employed in biomedical applications and in studies involving image and visual data. In such application areas, the number of examples in one class is significantly lower than the overall number of examples. The experimental setting is represented as follows: there is a class of special interest (usually positive) within the set of classes. The rest of the classes are either left, as is in the case of multiclass classification, or combined into one, as in binary classification. The measures of choice are obtained for the positive class (Sokolova, Japkowicz, & Szpakowicz, 2006).

Precision is the ratio of instances correctly classified as positive to all instances classified as positive:

$$\text{precision} = \frac{TP}{TP + FP} \tag{5.5}$$

A relation between properly classified instances, that is, true positives, and misclassified instances, that is, false negatives is called recall.

$$\text{recall} = \frac{TP}{TP + FN} \tag{5.6}$$

To have reasonable measure of the performance of the classifier based on its confusion matrix, a comprehensive pair of balancing indicators must be utilized. In this way, the model selection process is much more difficult, as there is no single criterion to rank candidate models from which the best could be chosen. Some measures to facilitate this task have tried to fold two complementary indicators into a single one. One well-known example is the F-measure, which is defined as the harmonic mean of the precision and recall indicators (Cichosz, 2014):

$$F - \text{measure} = 2 \times \frac{\text{Precision} \times \text{Recall}}{\text{Precision} + \text{Recall}} = \frac{2 \times TP}{2 \times TP + FP + FN} \tag{5.7}$$

The misclassification error and accuracy represent the same class-insensitive performance measures. The remaining measures of performance are much more interesting class-sensitive indicators that designate the level of success of a classifier to correctly detect the positive class. It does not make sense to use them all simultaneously, as they would result in considerable informational redundancy. However, it is not sufficient to use just one indicator from this set. That is the reason why performance measures are usually considered in pairs such as TPR and false positive rate, precision and recall, or sensitivity and specificity. An important member of all these pairs is the TPR under different names. It characterizes the ratio of positive instances that the classifier correctly detects. TPR should be maximized, and it can be represented by 1 for a trivial classifier that always predicts the positive class. This is the reason why it has to be used together with a complementary indicator (Cichosz, 2014).

As one of its complementary indicators, false positive rate can be used as it represents the ratio of negative instances incorrectly classified as positive. This indicator should be minimized, and a trivial classifier using a perfect 0 false positive rate would be one that always predicts the negative class. Another indicator used together with TPR is precision. It is given as the ratio of all instances classified as positive that are truly positive, and this ratio should be maximized. Precision can be maximized by the impractical classifier, which never predicts the positive class to avoid false alarms. The last complementary pair uses the specificity, which is represented as 1's complement of the false positive rate. The indicators in the mentioned pairs are complementary, as one of them represents the capability to detect positive instances, and the other avoids misdetection of negative instances. Each indicator in a complementary pair can be individually optimized by a trivial and useless classifier. Furthermore, there is a trade-off between the indicators in the same pair as improvement of one is likely to make the other worse or may, at best, leave it unchanged (Cichosz, 2014).

The previously mentioned measures of performance use comparisons of predicted class labels of the model with true classes. Different performance measures need to be defined for the classifiers that can predict different class labels at different operating points depending on the cut-off value. Receiver operating characteristic (ROC) analysis is one of the convenient tools that facilitates classifier performance evaluation in multiple operating points, operating point comparison, and operating point selection. ROC is suitable for classifier evaluation. The ROC uses a Cartesian coordinate system where its y-axis represents the TPR, whereas the x-axis represents the false positive rate. These axes makes the ROC plane. A single point on the ROC plane that visualizes the underlying trade-off between true positives and false positives represents the performance of a discrete classifier. In a similar way, a single operating point of a scoring classifier is also given as a point on the ROC plane. The point, with a TPR of 1 and a false positive rate of 0, that is, (0, 1), represents the perfect operating point, with all instances classified correctly. The (1, 0) point, with a TPR of 0 and a false positive rate of 1, is the worst operating point, with all instances classified incorrectly. The (0, 0) point correlates to a classifier that always predicts class 0, yielding no positives, and the (1, 1) point corresponds to a classifier that always predicts class 1 (Cichosz, 2014).

The ROC curve is obtained by combining all operating points of a scoring classifier on the ROC plane with line segments. Thus, a visual representation of performance of the classifier that is independent of the cut-off value is obtained. The ROC curve demonstrates the whole range of different operating points, with the corresponding different levels of the trade-off of the true positives and false positives, in a single plot. Performance of a scoring classifier that depends only on its scoring function component can be graphically indicated using a ROC curve. The scoring function captures the relationship between classes and attribute values. It is necessary to find all possible operating points of a scoring classifier to create the ROC curve because it is based on its scores produced for a dataset. As several operating points can be realized whenever the predicted class label changes for at least one instance, ROC can identify all possible operating points by considering all such cut-off values that produce several class predictions for at least one instance. After the sorting instances with respect to their scores, there is exactly one cut-off value separating two consecutive scores for instances of several classes, which produces a distinct operating point (Cichosz, 2014).

Beside the usage of ROC analysis when comparing different operating points and identifying the best operating point, it can also be employed to compare scoring classifiers irrespective of their labeling functions, that is, the scoring functions alone. This process requires comparing different ROC curves. When one curve is entirely above another, comparison is quite trivial as it can be immediately concluded that the former is better. This is not always the case because, for each operating point of the worse curve, there can be a superior operating point on the second curve; therefore, such an operating point does not have to be attainable on the better curve, but a superior operating point can be obtained by interpolation, in those cases (Cichosz, 2014).

In the case where ROC curves intersect, the situation is no longer so clear. When curves intersect, some parts of one curve are above the other and some parts are below. That means that, in some ranges of the false positive rate, one model accomplishes a higher TPR, and in other ranges, the other model has more true positives. Depending on the range of concern and where the ultimate desired operating point is most likely to lie, one curve might be preferable to the other. Occasionally a simple comparison criterion is required even in more complex cases. In the case when there are a variety of models produced using different algorithms or parameter settings, the quick and easy way of ranking them with respect to their predictive utility without considering any particular operating points is needed. Such commonly used criterion is the area under the ROC curve (AUC). While performing comparison of models, the model that has a greater AUC value can be roughly considered superior with respect to its overall predictive performance potential, even if another model with a lower AUC value can actually produce a more preferable operating point than any point attainable by the former model. This measure of performance is employed to evaluate the effect of several parameter settings for classification algorithms and to select a subset of the most promising models before the selection of an operating point. This measure is also useful when the best scoring model needs to be chosen to be subsequently employed at several different operating points (Cichosz, 2014).

Kappa statistic (κ) is a measure that takes an expected figure into account by deducting it from the predictor's successes. It expresses the result as a proportion of the total for a perfect predictor. The Kappa's maximum value is 100%, and the expected value for a random predictor with the same column total is 0. The measurement of the agreement between predicted and observed categorizations of a dataset, and correcting the agreement that occurs by chance, is performed by κ. Similarly to the plain success rate, it does not take into account the costs (Hall, Witten, & Frank, 2011). κ uses the fact that the classifier will agree or disagree simply by chance. κ is the most frequently used statistic for the evaluation of categorical data when there is no independent means of assessing the probability of chance agreement between two or more observers. Cohen (1960) defined κ as an agreement index and defined as the following:

$$K = \frac{P_0 - P_e}{1 - P_e} \tag{5.8}$$

where P_0 is an observed agreement and defined as

$$P_0 = \frac{TP + TN}{TP + TN + FP + FN} \tag{5.9}$$

and P_e measures the probability of random agreement (Yang & Zhou, 2015). Overall random agreement probability is the probability that they agreed on, either Yes or No, that is:

$$P_e = P_{\text{YES}} + P_{\text{NO}} \tag{5.10}$$

where

$$P_{\text{YES}} = \frac{TP + FP}{TP + TN + FP + FN} \times \frac{TP + FN}{TP + TN + FP + FN} \tag{5.11}$$

$$P_{\text{NO}} = \frac{FN + TN}{TP + TN + FP + FN} \times \frac{FP + TN}{TP + TN + FP + FN} \tag{5.12}$$

In the last decade, several feature extraction methods have been combined with different types of classifiers. The performance of a classifier is contingent with the characteristics of the data to be classified. There is no single classifier that operates best on all given problems. Numerous practical tests have been completed to compare classifier performance and to recognize the characteristics of data that determine the classifier performance. The measures of accuracy and confusion matrix are widespread approaches for assessing the quality of a classification algorithm. Recently, ROC curves have been employed to assess the trade-off between true- and false-positive rates of classification algorithms (Siuly et al., 2016). This book primarilly uses accuracy to assess the performance of classification algorithms. The confusion matrix and ROC curves are also utilized to assess the performance of the classifiers.

5.3 LINEAR DISCRIMINANT ANALYSIS

Linear discriminant analysis (LDA) is a classification technique employed to find a linear combination of features that represents or divides two or more classes of data. The subsequent combination can be employed as a linear classifier. In LDA, the classes are assumed to be normally distributed. Similar to principal component analysis (PCA), LDA can be utilized for both data classification and dimension reduction. In a two-class dataset, given the a priori probabilities for class 1 and class 2 are p_1 and p_2, respectively, and class means and overall mean as μ_1, μ_2, and μ, and the class variances as cov_1 and cov_2:

$$\mu = p_1 \times \mu_1 + p_2 \times \mu_2 \tag{5.13}$$

Then, within-class and between-class scatters are employed to express the required criteria for class separability. The scatter measures for multiclass situation are calculated as:

$$S_w = \sum_{j=1}^{C} p_j x \text{cov}_j \tag{5.14}$$

where C refers to the number of classes and

$$\text{cov}_j = (\mathbf{x}_j - \mu_j)(\mathbf{x}_j - \mu_j)^T \tag{5.15}$$

The between-class scatter is calculated as:

$$S_b = \frac{1}{C} \sum_{j=1}^{C} (\mu_j - \mu)(\mu_j - \mu)^T \tag{5.16}$$

Then, the aim is to find a discriminant plane to maximize the ratio of between-class to within-class scatters (variances):

$$J_{\text{LDA}} = \frac{w S_b w^T}{w S_w w^T} \tag{5.17}$$

In practical cases, the class covariances and means are not known. But they can be estimated from the training set. Either the maximum a posteriori estimate or the maximum likelihood estimate can be employed in place of the exact value in the previously discussed equations (Sanei, 2013).

EXAMPLE 5.1. The following MATLAB code was used to extract features from the electroencephalogram (EEG) signals using AR Burg method. Then it classified these data using LDA. You can download data from the following website: http://epileptologie-bonn.de/cms/front_content.php?idcat=193&lang=3&changelang=3

```
%% Ch5_Ex1_EEG_ARBURG_LDA.m
%The following MATLAB code is used to extract the features from
%the EEG signals using AR Burg.
%Then it classifies these data using Linear Discriminant Analysis (LDA)
clc
clear
%Load Sample EEG Data downloaded from the web site
%http://epileptologie-bonn.de/cms/front_content.php?idcat=193&lang=3&changelang=3
load AS_BONN_ALL_EEG_DATA_4096.mat
%%
Fs = 173.6;             % Sampling frequency
T = 1/Fs;               % Sampling period
segmentLength = 128;  % Length of a signal segment
Length=4096;% Length of signal
Nofsignal=100; %Number of Signal
order = 14;

%%
% Obtain the AR Burg Spectrum of the Normal EEG signal using pburg.
for i=1:Nofsignal
    [Pxx,F] = pburg(Normal_Eyes_Open(1:Length,i),order,segmentLength,Fs);
    ASData(i,:)=Pxx(:);
end
%%
% Obtain the AR Burg Spectrum of the Interictal EEG signal using pburg.
for i=Nofsignal+1:2*Nofsignal
    [Pxx,F] = pburg(Interictal(1:Length,i-Nofsignal),order,segmentLength,Fs);
    ASData(i,:)=Pxx(:);
end
%%
% Obtain the AR Burg Spectrum of the Ictal EEG signal using pburg.
for i=2*Nofsignal+1:3*Nofsignal
[Pxx,F] = pburg(Ictal(1:Length,i-2*Nofsignal),order,segmentLength,Fs);
ASData(i,:)=Pxx(:);
end

%Make input as the Data Matrix
Inputs=ASData;

%%
%You can create target using excel and the import it from the MATLAB HOME
%menu and Import data and then save it as EEGTargets
%Load Target
load EEGTargets

%% Classification
% This example shows how to perform classification using discriminant
% analysis. Suppose you have a
% data set containing observations with measurements on different variables
% (called predictors) and their known class labels. If you obtain predictor
% values for new observations, could you determine to which classes those
% observations probably belong? This is the problem of classification.
```

```
%%
% Suppose you have Normal, Interictal and Ictal EEG data, and you need to
% determine their classes using discriminant analysis.

%% Linear Discriminant Analysis
% The |fitcdiscr| function can perform classification using different types
% of discriminant analysis. First classify the data using the default
% linear discriminant analysis (LDA).

lda = fitcdiscr(Inputs,Targets);
ldaClass = resubPredict(lda);
%%
% The observations with known class labels are usually called the training data.
% Now compute the resubstitution error, which is the misclassification error
% (the proportion of misclassified observations) on the training set.

ldaResubErr = resubLoss(lda)
%%
% You can also compute the confusion matrix on the training set. A
% confusion matrix contains information about known class labels and
% predicted class labels. Generally speaking, the (i,j) element in the
% confusion matrix is the number of samples whose known class label is
% class i and whose predicted class is j. The diagonal elements represent
% correctly classified observations.

[ldaResubCM,grpOrder]=confusionmat(Targets,ldaClass)

% Calculate the Total Classification Accuracy
TotalAccuracy=(ldaResubCM(1,1)+ldaResubCM(2,2)+ldaResubCM(3,3))/(3*Nofsignal)

disp('Total Classification Accuracy=')
disp(TotalAccuracy)

%%
% You have computed the resubstitution error. Usually people are more
% interested in the test error (also referred to as generalization error),
% which is the expected prediction error on an independent set. In fact,
% the resubstitution error will likely under-estimate the test error.
%
% In this case you don't have another labeled data set, but you can
% simulate one by doing cross-validation. A stratified 10-fold
% cross-validation is a popular choice for estimating the test error on
% classification algorithms. It randomly divides the training set into 10
% disjoint subsets. Each subset has roughly equal size and roughly the same
% class proportions as in the training set. Remove one subset, train the
% classification model using the other nine subsets, and use the trained
% model to classify the removed subset. You could repeat this by removing
% each of the ten subsets one at a time.
%
% Because cross-validation randomly divides data, its outcome depends on
% the initial random seed. To reproduce the exact results in this example,
% execute the following command:
rng(0,'twister');

%%
% First use |cvpartition| to generate 10 disjoint stratified subsets.
cp = cvpartition(Targets,'KFold',10)
%%
```

```
% The |crossval| and |kfoldLoss| methods can estimate the misclassification
% error for LDA using the given data partition |cp|.
%
% Estimate the true test error for LDA using 10-fold stratified
% cross-validation.

cvlda = crossval(lda,'CVPartition',cp);
ldaCVErr = kfoldLoss(cvlda)
%%
% The LDA cross-validation error has the same value as the LDA
% resubstitution error on this data.

%% Conclusions
% This example shows how to perform classification with LDA in MATLAB(R)
%using Statistics and Machine Learning Toolbox(TM) functions.
```

EXAMPLE 5.2. The following MATLAB code was used to extract features from the electrocardiogram (ECG) signals using the covariance method. Then it classified these data using the LDA classifier. You can download data from the following website: https://www.physionet.org/physiobank/database/mitdb/

```
%% Ch5_Ex2_ECG_COV_LDA.m
%The following MATLAB code is used to extract the features from
%the ECG signals using Covariance Method.
%Then it classifies ECG Signals Using LDA
clc
clear
%Load Sample ECG Data downloaded from the web site
%https://www.physionet.org/physiobank/database/mitdb/
load MITBIH_ECG.mat
%%
Fs = 320;              % Sampling frequency
T = 1/Fs;                % Sampling period
segmentLength = 128;    % Length of a signal segment
Length=320;% Length of signal
Nofsignal=300; %Number of Signal
order = 14;
%%
% Obtain the Covariance Spectrum of the Normal ECG signal using pcov.
for i=1:Nofsignal
    [Pxx,F] = pcov(ECGN(1:Length,i),order,segmentLength,Fs);
    ASData(i,:)=Pxx(:);
end
%%
% Obtain the Covariance Spectrum of the ECG signal with APC using pcov.
[Pxx,F] = pcov(ECGAPC(:,1),order,segmentLength,Fs);
for i=Nofsignal+1:2*Nofsignal
    [Pxx,F] = pcov(ECGAPC(1:Length,i-Nofsignal),order,segmentLength,Fs);
    ASData(i,:)=Pxx(:);
end
%%
% Obtain the Covariance Spectrum of the ECG signal with LBBB using pcov.
for i=2*Nofsignal+1:3*Nofsignal
[Pxx,F]=pcov(ECGPVC(1:Length,i-2*Nofsignal),order,segmentLength,Fs);
ASData(i,:)=Pxx(:);
end

%%
% Obtain the Covariance Spectrum of the ECG signal with PVC using pcov.
```

```
for i=3*Nofsignal+1:4*Nofsignal
[Pxx,F]=pcov(ECGLBBB(1:Length,i-3*Nofsignal),order,segmentLength,Fs);
ASData(i,:)=Pxx(:);
end

%%
% Obtain the Covariance Spectrum of the ECG signal with RBBB using pcov.

for i=4*Nofsignal+1:5*Nofsignal
[Pxx,F]=pcov(ECGRBBB(1:Length,i-4*Nofsignal),order,segmentLength,Fs);
ASData(i,:)=Pxx(:);
end

%Create the Data Matrix as Inputs
Inputs=ASData;

%%
%You can create target using excel and the import it from the MATLAB HOME
%menu and Import data and then save it as EEGTargets
%Load Target
load ECGTargets

%% Classification
% This example shows how to perform classification using discriminant
% analysis. Suppose you have a
% data set containing observations with measurements on different variables
% (called predictors) and their known class labels. If you obtain predictor
% values for new observations, could you determine to which classes those
% observations probably belong? This is the problem of classification.

%%
% Suppose you have Normal, Interictal and Ictal EEG data, and you need to
% determine their classes using discriminant analysis.

%% Linear Discriminant Analysis
% The |fitcdiscr| function can perform classification using different types
% of discriminant analysis. First classify the data using the default
% linear discriminant analysis (LDA).

lda = fitcdiscr(Inputs,Targets);
ldaClass = resubPredict(lda);
%%
% The observations with known class labels are usually called the training data.
% Now compute the resubstitution error, which is the misclassification error
% (the proportion of misclassified observations) on the training set.

ldaResubErr = resubLoss(lda)
%%
% You can also compute the confusion matrix on the training set. A
% confusion matrix contains information about known class labels and
% predicted class labels. Generally speaking, the (i,j) element in the
% confusion matrix is the number of samples whose known class label is
% class i and whose predicted class is j. The diagonal elements represent
% correctly classified observations.

[ldaResubCM,grpOrder] = confusionmat(Targets,ldaClass)

% Calculate the Total Classification Accuracy
TotalAccuracy=(ldaResubCM(1,1)+ldaResubCM(2,2)+ldaResubCM(3,3))/(3*Nofsignal)
```

```
disp('Total Classification Accuracy=')
disp(TotalAccuracy)

%%
% You have computed the resubstitution error. Usually people are more
% interested in the test error (also referred to as generalization error),
% which is the expected prediction error on an independent set. In fact,
% the resubstitution error will likely under-estimate the test error.
%
% In this case you don't have another labeled data set, but you can
% simulate one by doing cross-validation. A stratified 10-fold
% cross-validation is a popular choice for estimating the test error on
% classification algorithms. It randomly divides the training set into 10
% disjoint subsets. Each subset has roughly equal size and roughly the same
% class proportions as in the training set. Remove one subset, train the
% classification model using the other nine subsets, and use the trained
% model to classify the removed subset. You could repeat this by removing
% each of the ten subsets one at a time.
%
% Because cross-validation randomly divides data, its outcome depends on
% the initial random seed. To reproduce the exact results in this example,
% execute the following command:

rng(0,'twister');

%%
% First use |cvpartition| to generate 10 disjoint stratified subsets.
cp = cvpartition(Targets,'KFold',10)
%%
% The |crossval| and |kfoldLoss| methods can estimate the misclassification
% error for LDA using the given data partition |cp|.
%
% Estimate the true test error for LDA using 10-fold stratified
% cross-validation.

cvlda = crossval(lda,'CVPartition',cp);
ldaCVErr = kfoldLoss(cvlda)
%%
% The LDA cross-validation error has the same value as the LDA
% resubstitution error on this data.

%% Conclusions
% This example shows how to perform classification using LDA in MATLAB(R)
%using Statistics and Machine Learning Toolbox(TM) functions.
```

EXAMPLE 5.3. The following MATLAB code is used to extract features from the ECG signals using MUSIC method. Then it classifies these data LDA classifier. You can download data from the following website: https://www.physionet.org/physiobank/database/mitdb/

```
%% Ch5_Ex3_ECG_COV_LDA.m
%The following MATLAB code is used to extract the features from
%the ECG signals using MUSIC Method.
%Then it classifies ECG Signals Using LDA
clc
clear
%Load Sample ECG Data downloaded from the web site
%https://www.physionet.org/physiobank/database/mitdb/
load MITBIH_ECG.mat
```

```
%%
Fs = 320;             % Sampling frequency
T = 1/Fs;              % Sampling period
segmentLength = 128;   % Length of a signal segment
Length=320;% Length of signal
Nofsignal=300; %Number of Signal
order = 34;
%%
% Obtain the MUSIC Spectrum of the Normal ECG signal using pmusic.
for i=1:Nofsignal
    [Pxx,F] = pmusic(ECGN(1:Length,i),order,segmentLength,Fs);
    ASData(i,:)=Pxx(:);
end
%%
% Obtain the MUSIC Spectrum of the ECG signal with APC using pmusic.
[Pxx,F] = pmusic(ECGAPC(:,1),order,segmentLength,Fs);
for i=Nofsignal+1:2*Nofsignal
    [Pxx,F]=pmusic(ECGAPC(1:Length,i-Nofsignal),order,segmentLength,Fs);
    ASData(i,:)=Pxx(:);
end
%%
% Obtain the MUSIC Spectrum of the ECG signal with LBBB using pmusic.
for i=2*Nofsignal+1:3*Nofsignal
[Pxx,F]=pmusic(ECGPVC(1:Length,i-2*Nofsignal),order,segmentLength,Fs);
ASData(i,:)=Pxx(:);
end

%%
% Obtain the MUSIC Spectrum of the ECG signal with PVC using pmusic.

for i=3*Nofsignal+1:4*Nofsignal
[Pxx,F] = pmusic(ECGLBBB(1:Length,i-3*Nofsignal),order,segmentLength,Fs);
ASData(i,:)=Pxx(:);
end

%%
% Obtain the MUSIC Spectrum of the ECG signal with RBBB using pmusic.

for i=4*Nofsignal+1:5*Nofsignal
[Pxx,F]=pmusic(ECGRBBB(1:Length,i-4*Nofsignal),order,segmentLength,Fs);
ASData(i,:)=Pxx(:);
end

%Create the Data Matrix as Inputs
Inputs=ASData;

%%
%You can create target using excel and the import it from the MATLAB HOME
%menu and Import data and then save it as EEGTargets
%Load Target
load ECGTargets

%% Classification
% This example shows how to perform classification using discriminant
% analysis. Suppose you have a
% data set containing observations with measurements on different variables
% (called predictors) and their known class labels. If you obtain predictor
% values for new observations, could you determine to which classes those
% observations probably belong? This is the problem of classification.
```

```
%%
% Suppose you have Normal, Interictal and Ictal EEG data, and you need to
% determine their classes using discriminant analysis.

%% Linear Discriminant Analysis
% The |fitcdiscr| function can perform classification using different types
% of discriminant analysis. First classify the data using the default
% linear discriminant analysis (LDA).

lda = fitcdiscr(Inputs,Targets);
ldaClass = resubPredict(lda);
%%
% The observations with known class labels are usually called the training data.
% Now compute the resubstitution error, which is the misclassification error
% (the proportion of misclassified observations) on the training set.

ldaResubErr = resubLoss(lda)
%%
% You can also compute the confusion matrix on the training set. A
% confusion matrix contains information about known class labels and
% predicted class labels. Generally speaking, the (i,j) element in the
% confusion matrix is the number of samples whose known class label is
% class i and whose predicted class is j. The diagonal elements represent
% correctly classified observations.

[ldaResubCM,grpOrder] = confusionmat(Targets,ldaClass)

% Calculate the Total Classification Accuracy
TotalAccuracy=(ldaResubCM(1,1)+ldaResubCM(2,2)+ldaResubCM(3,3))/(3*Nofsignal)

disp('Total Classification Accuracy=')
disp(TotalAccuracy)

%%
% You have computed the resubstitution error. Usually people are more
% interested in the test error (also referred to as generalization error),
% which is the expected prediction error on an independent set. In fact,
% the resubstitution error will likely under-estimate the test error.
%
% In this case you don't have another labeled data set, but you can
% simulate one by doing cross-validation. A stratified 10-fold
% cross-validation is a popular choice for estimating the test error on
% classification algorithms. It randomly divides the training set into 10
% disjoint subsets. Each subset has roughly equal size and roughly the same
% class proportions as in the training set. Remove one subset, train the
% classification model using the other nine subsets, and use the trained
% model to classify the removed subset. You could repeat this by removing
% each of the ten subsets one at a time.
%
% Because cross-validation randomly divides data, its outcome depends on
% the initial random seed. To reproduce the exact results in this example,
% execute the following command:

rng(0,'twister');

%%
% First use |cvpartition| to generate 10 disjoint stratified subsets.
cp = cvpartition(Targets,'KFold',10)
%%
% The |crossval| and |kfoldLoss| methods can estimate the misclassification
% error for LDA using the given data partition |cp|.
```

```
%
% Estimate the true test error for LDA using 10-fold stratified
% cross-validation.

cvlda = crossval(lda,'CVPartition',cp);
ldaCVErr = kfoldLoss(cvlda)
%%
% The LDA cross-validation error has the same value as the LDA
% resubstitution error on this data.

%% Conclusions
% This example shows how to perform classification using LDA in MATLAB(R)
%using Statistics and Machine Learning Toolbox(TM) functions.
```

EXAMPLE 5.4. The following MATLAB code was used to extract features from the EEG signals using modified covariance method. Then it classified these data using LDA. You can download data from the following website: http://epileptologie-bonn.de/cms/front_content.php?idcat=193&lang=3&changelang=3

```
%% Ch5_Ex4_EEG_MCOV_LDA.m
%The following MATLAB code is used to extract the features from
%the EEG signals using Modified Covariance.
%Then it classifies EEG Signals Using Discriminant Analysis

clc
clear
%Load Sample EEG Data downloaded from the web site
%http://epileptologie-bonn.de/cms/front_content.php?idcat=193&lang=3&changelang=3
load AS_BONN_ALL_EEG_DATA_4096.mat
%%
Fs = 173.6;           % Sampling frequency
T = 1/Fs;             % Sampling period
segmentLength = 128; % Length of a signal segment
Length=4096;% Length of signal
Nofsignal=100; %Number of Signal
order = 33;

%%
% Obtain the Modified Covariance Spectrum of the Normal EEG signal using pmcov.

for i=1:Nofsignal
    [Pxx,F] = pmcov(Normal_Eyes_Open(1:Length,i),order,segmentLength,Fs);
    ASData(i,:)=Pxx(:);
end
%%
% Obtain the Modified Covariance Spectrum of the Interictal EEG signal using pmcov.
for i=Nofsignal+1:2*Nofsignal
    [Pxx,F] = pmcov(Interictal(1:Length,i-Nofsignal),order,segmentLength,Fs);
    ASData(i,:)=Pxx(:);
end
%%
% Obtain the Modified Covariance Spectrum of the Ictal EEG signal using pmcov.
for i=2*Nofsignal+1:3*Nofsignal
[Pxx,F] = pmcov(Ictal(1:Length,i-2*Nofsignal),order,segmentLength,Fs);
ASData(i,:)=Pxx(:);
end
```

```
%Make Inputs as the Data Matrix
Inputs=ASData;

%%
%You can create target using excel and the import it from the MATLAB HOME
%menu and Import data and then save it as EEGTargets
load EEGTargets

%% Classification
% This example shows how to perform classification using discriminant
% analysis. Suppose you have a
% data set containing observations with measurements on different variables
% (called predictors) and their known class labels. If you obtain predictor
% values for new observations, could you determine to which classes those
% observations probably belong? This is the problem of classification.

%%
% Suppose you have Normal, Interictal and Ictal EEG data, and you need to
% determine their classes using discriminant analysis.

%% Linear Discriminant Analysis
% The |fitcdiscr| function can perform classification using different types
% of discriminant analysis. First classify the data using the default
% linear discriminant analysis (LDA).

lda = fitcdiscr(Inputs,Targets);
ldaClass = resubPredict(lda);
%%
% The observations with known class labels are usually called the training data.
% Now compute the resubstitution error, which is the misclassification error
% (the proportion of misclassified observations) on the training set.

ldaResubErr = resubLoss(lda)
%%
% You can also compute the confusion matrix on the training set. A
% confusion matrix contains information about known class labels and
% predicted class labels. Generally speaking, the (i,j) element in the
% confusion matrix is the number of samples whose known class label is
% class i and whose predicted class is j. The diagonal elements represent
% correctly classified observations.

[ldaResubCM,grpOrder] = confusionmat(Targets,ldaClass)

% Calculate the Total Classification Accuracy
TotalAccuracy=(ldaResubCM(1,1)+ldaResubCM(2,2)+ldaResubCM(3,3))/(3*Nofsignal)

disp('Total Classification Accuracy=')
disp(TotalAccuracy)

%%
% You have computed the resubstitution error. Usually people are more
% interested in the test error (also referred to as generalization error),
% which is the expected prediction error on an independent set. In fact,
% the resubstitution error will likely under-estimate the test error.
%
% In this case you don't have another labeled data set, but you can
% simulate one by doing cross-validation. A stratified 10-fold
% cross-validation is a popular choice for estimating the test error on
```

```
% classification algorithms. It randomly divides the training set into 10
% disjoint subsets. Each subset has roughly equal size and roughly the same
% class proportions as in the training set. Remove one subset, train the
% classification model using the other nine subsets, and use the trained
% model to classify the removed subset. You could repeat this by removing
% each of the ten subsets one at a time.
%
% Because cross-validation randomly divides data, its outcome depends on
% the initial random seed. To reproduce the exact results in this example,
% execute the following command:

rng(0,'twister');

%%
% First use |cvpartition| to generate 10 disjoint stratified subsets.
cp = cvpartition(Targets,'KFold',10)
%%
% The |crossval| and |kfoldLoss| methods can estimate the misclassification
% error for LDA using the given data partition |cp|.
%
% Estimate the true test error for LDA using 10-fold stratified
% cross-validation.

cvlda = crossval(lda,'CVPartition',cp);
ldaCVErr = kfoldLoss(cvlda)
%%
% The LDA cross-validation error has the same value as the LDA
% resubstitution error on this data.

%% Conclusions
% This example shows how to perform classification using LDA in MATLAB(R)
%using Statistics and Machine Learning Toolbox(TM) functions.
```

EXAMPLE 5.5. The following MATLAB code was used to extract features from the EEG signals using wavelet packet decomposition (WPD). Then it used statistical values of WPD subbands. Then it classified these data using LDA. You can download data from the following website:
http://epileptologie-bonn.de/cms/front_content.php?idcat = 193&lang = 3&changelang = 3

```
%% Ch5_Ex5_EEG_WPD_LDA.m
%WPD of NORMAL, INTERICTAL and ICTAL EEG signals
%The following MATLAB code is used to Extract WPD features from the EEG signals
%Decompose EEG data using WPD coefficients
%Then uses Statistical features as:

%(1) Mean of the absolute values of the coefficients in each sub-band.
%(2) Standard deviation of the coefficients in each sub-band.
%(3) Skewness of the coefficients in each sub-band.
%(4) Kurtosis of the coefficients in each sub-band.
%(5) RMS power of the wavelet coefficients in each subband.
%(6) Ratio of the mean absolute values of adjacent subbands.

%Then it classifies EEG Signals Using Linear Discriminant Analysis
clc
clear
%Load Sample EEG Data downloaded from the web site
%http://epileptologie-bonn.de/cms/front_content.php?idcat=193&lang=3&changelang=3
```

```
load AS_BONN_ALL_EEG_DATA_4096.mat
wname = 'db4';
Length = 4096;      % Length of signal
Nofsignal=100; %Number of Signal
%%
%%%%%%%%%%%%%%%%%%%%%%%%%%%%%% A Z.ZIP EYES OPEN NORMAL SUBJECT

for i=1:Nofsignal;

[wp10,wp11] = dwt(Normal_Eyes_Open(1:Length,i),wname);%WPD Decomposition
[wp20,wp21] = dwt(wp10,wname);
[wp22,wp23] = dwt(wp11,wname);

[wp30,wp31] = dwt(wp20,wname);
[wp32,wp33] = dwt(wp21,wname);
[wp34,wp35] = dwt(wp22,wname);
[wp36,wp37] = dwt(wp23,wname);

[wp40,wp41] = dwt(wp30,wname);
[wp42,wp43] = dwt(wp31,wname);
[wp44,wp45] = dwt(wp32,wname);
[wp46,wp47] = dwt(wp33,wname);
[wp48,wp49] = dwt(wp34,wname);
[wp4A,wp4B] = dwt(wp35,wname);
[wp4C,wp4D] = dwt(wp36,wname);
[wp4E,wp4F] = dwt(wp37,wname);

ASData(i,1)=mean(abs(wp40));
ASData(i,2)=mean(abs(wp41));
ASData(i,3)=mean(abs(wp42));
ASData(i,4)=mean(abs(wp43));
ASData(i,5)=mean(abs(wp44));
ASData(i,6)=mean(abs(wp45));
ASData(i,7)=mean(abs(wp46));
ASData(i,8)=mean(abs(wp47));
ASData(i,9)=mean(abs(wp48));
ASData(i,10)=mean(abs(wp49));
ASData(i,11)=mean(abs(wp4A));
ASData(i,12)=mean(abs(wp4B));
ASData(i,13)=mean(abs(wp4C));
ASData(i,14)=mean(abs(wp4D));
ASData(i,15)=mean(abs(wp4E));
ASData(i,16)=mean(abs(wp4F));

ASData(i,17)=std(wp40);
ASData(i,18)=std(wp41);
ASData(i,19)=std(wp42);
ASData(i,20)=std(wp43);
ASData(i,21)=std(wp44);
ASData(i,22)=std(wp45);
ASData(i,23)=std(wp46);
ASData(i,24)=std(wp47);
ASData(i,25)=std(wp48);
ASData(i,26)=std(wp49);
ASData(i,27)=std(wp4A);
ASData(i,28)=std(wp4B);
ASData(i,29)=std(wp4C);
ASData(i,30)=std(wp4D);
ASData(i,31)=std(wp4E);
ASData(i,32)=std(wp4F);
```

```
ASData(i,33)=skewness(wp40(:));
ASData(i,34)=skewness(wp41(:));
ASData(i,35)=skewness(wp42(:));
ASData(i,36)=skewness(wp43(:));
ASData(i,37)=skewness(wp44(:));
ASData(i,38)=skewness(wp45(:));
ASData(i,39)=skewness(wp46(:));
ASData(i,40)=skewness(wp47(:));
ASData(i,41)=skewness(wp48(:));
ASData(i,42)=skewness(wp49(:));
ASData(i,43)=skewness(wp4A(:));
ASData(i,44)=skewness(wp4B(:));
ASData(i,45)=skewness(wp4C(:));
ASData(i,46)=skewness(wp4D(:));
ASData(i,47)=skewness(wp4E(:));
ASData(i,48)=skewness(wp4F(:));

ASData(i,49)=kurtosis(wp40);
ASData(i,50)=kurtosis(wp41);
ASData(i,51)=kurtosis(wp42);
ASData(i,52)=kurtosis(wp43);
ASData(i,53)=kurtosis(wp44);
ASData(i,54)=kurtosis(wp45);
ASData(i,55)=kurtosis(wp46);
ASData(i,56)=kurtosis(wp47);
ASData(i,57)=kurtosis(wp48);
ASData(i,58)=kurtosis(wp49);
ASData(i,59)=kurtosis(wp4A);
ASData(i,60)=kurtosis(wp4B);
ASData(i,61)=kurtosis(wp4C);
ASData(i,62)=kurtosis(wp4D);
ASData(i,63)=kurtosis(wp4E);
ASData(i,64)=kurtosis(wp4F);

ASData(i,65)=rms(wp40(:));
ASData(i,66)=rms(wp41(:));
ASData(i,67)=rms(wp42(:));
ASData(i,68)=rms(wp43(:));
ASData(i,69)=rms(wp44(:));
ASData(i,70)=rms(wp45(:));
ASData(i,71)=rms(wp46(:));
ASData(i,72)=rms(wp47(:));
ASData(i,73)=rms(wp48(:));
ASData(i,74)=rms(wp49(:));
ASData(i,75)=rms(wp4A(:));
ASData(i,76)=rms(wp4B(:));
ASData(i,77)=rms(wp4C(:));
ASData(i,78)=rms(wp4D(:));
ASData(i,79)=rms(wp4E(:));
ASData(i,80)=rms(wp4F(:));

ASData(i,81)=mean(abs(wp40))/mean(abs(wp41));
ASData(i,82)=mean(abs(wp41))/mean(abs(wp42));
ASData(i,83)=mean(abs(wp42))/mean(abs(wp43));
ASData(i,84)=mean(abs(wp43))/mean(abs(wp44));
ASData(i,85)=mean(abs(wp44))/mean(abs(wp45));
ASData(i,86)=mean(abs(wp45))/mean(abs(wp46));
ASData(i,87)=mean(abs(wp46))/mean(abs(wp47));
ASData(i,88)=mean(abs(wp47))/mean(abs(wp48));
ASData(i,89)=mean(abs(wp48))/mean(abs(wp49));
```

```
ASData(i,90)=mean(abs(wp49))/mean(abs(wp4A));
ASData(i,91)=mean(abs(wp4A))/mean(abs(wp4B));
ASData(i,92)=mean(abs(wp4B))/mean(abs(wp4C));
ASData(i,93)=mean(abs(wp4C))/mean(abs(wp4D));
ASData(i,94)=mean(abs(wp4D))/mean(abs(wp4E));
ASData(i,95)=mean(abs(wp4E))/mean(abs(wp4F));

end

%%
% D F.ZIP EPILEPTIC SUBJECT (INTERICTAL) were recorded from within
% the epileptogenic zone during seizure free intervals

for i=Nofsignal+1:2*Nofsignal;

[wp10,wp11] = dwt(Interictal(1:Length,i-Nofsignal),wname);%WPD Decomposition
[wp20,wp21] = dwt(wp10,wname);
[wp22,wp23] = dwt(wp11,wname);

[wp30,wp31] = dwt(wp20,wname);
[wp32,wp33] = dwt(wp21,wname);
[wp34,wp35] = dwt(wp22,wname);
[wp36,wp37] = dwt(wp23,wname);

[wp40,wp41] = dwt(wp30,wname);
[wp42,wp43] = dwt(wp31,wname);
[wp44,wp45] = dwt(wp32,wname);
[wp46,wp47] = dwt(wp33,wname);
[wp48,wp49] = dwt(wp34,wname);
[wp4A,wp4B] = dwt(wp35,wname);
[wp4C,wp4D] = dwt(wp36,wname);
[wp4E,wp4F] = dwt(wp37,wname);

ASData(i,1)=mean(abs(wp40));
ASData(i,2)=mean(abs(wp41));
ASData(i,3)=mean(abs(wp42));
ASData(i,4)=mean(abs(wp43));
ASData(i,5)=mean(abs(wp44));
ASData(i,6)=mean(abs(wp45));
ASData(i,7)=mean(abs(wp46));
ASData(i,8)=mean(abs(wp47));
ASData(i,9)=mean(abs(wp48));
ASData(i,10)=mean(abs(wp49));
ASData(i,11)=mean(abs(wp4A));
ASData(i,12)=mean(abs(wp4B));
ASData(i,13)=mean(abs(wp4C));
ASData(i,14)=mean(abs(wp4D));
ASData(i,15)=mean(abs(wp4E));
ASData(i,16)=mean(abs(wp4F));

ASData(i,17)=std(wp40);
ASData(i,18)=std(wp41);
ASData(i,19)=std(wp42);
ASData(i,20)=std(wp43);
ASData(i,21)=std(wp44);
ASData(i,22)=std(wp45);
ASData(i,23)=std(wp46);
ASData(i,24)=std(wp47);
ASData(i,25)=std(wp48);
ASData(i,26)=std(wp49);
```

```
ASData(i,27)=std(wp4A);
ASData(i,28)=std(wp4B);
ASData(i,29)=std(wp4C);
ASData(i,30)=std(wp4D);
ASData(i,31)=std(wp4E);
ASData(i,32)=std(wp4F);

ASData(i,33)=skewness(wp40(:));
ASData(i,34)=skewness(wp41(:));
ASData(i,35)=skewness(wp42(:));
ASData(i,36)=skewness(wp43(:));
ASData(i,37)=skewness(wp44(:));
ASData(i,38)=skewness(wp45(:));
ASData(i,39)=skewness(wp46(:));
ASData(i,40)=skewness(wp47(:));
ASData(i,41)=skewness(wp48(:));
ASData(i,42)=skewness(wp49(:));
ASData(i,43)=skewness(wp4A(:));
ASData(i,44)=skewness(wp4B(:));
ASData(i,45)=skewness(wp4C(:));
ASData(i,46)=skewness(wp4D(:));
ASData(i,47)=skewness(wp4E(:));
ASData(i,48)=skewness(wp4F(:));

ASData(i,49)=kurtosis(wp40);
ASData(i,50)=kurtosis(wp41);
ASData(i,51)=kurtosis(wp42);
ASData(i,52)=kurtosis(wp43);
ASData(i,53)=kurtosis(wp44);
ASData(i,54)=kurtosis(wp45);
ASData(i,55)=kurtosis(wp46);
ASData(i,56)=kurtosis(wp47);
ASData(i,57)=kurtosis(wp48);
ASData(i,58)=kurtosis(wp49);
ASData(i,59)=kurtosis(wp4A);
ASData(i,60)=kurtosis(wp4B);
ASData(i,61)=kurtosis(wp4C);
ASData(i,62)=kurtosis(wp4D);
ASData(i,63)=kurtosis(wp4E);
ASData(i,64)=kurtosis(wp4F);

ASData(i,65)=rms(wp40(:));
ASData(i,66)=rms(wp41(:));
ASData(i,67)=rms(wp42(:));
ASData(i,68)=rms(wp43(:));
ASData(i,69)=rms(wp44(:));
ASData(i,70)=rms(wp45(:));
ASData(i,71)=rms(wp46(:));
ASData(i,72)=rms(wp47(:));
ASData(i,73)=rms(wp48(:));
ASData(i,74)=rms(wp49(:));
ASData(i,75)=rms(wp4A(:));
ASData(i,76)=rms(wp4B(:));
ASData(i,77)=rms(wp4C(:));
ASData(i,78)=rms(wp4D(:));
ASData(i,79)=rms(wp4E(:));
ASData(i,80)=rms(wp4F(:));
```

```
ASData(i,81)=mean(abs(wp40))/mean(abs(wp41));
ASData(i,82)=mean(abs(wp41))/mean(abs(wp42));
ASData(i,83)=mean(abs(wp42))/mean(abs(wp43));
ASData(i,84)=mean(abs(wp43))/mean(abs(wp44));
ASData(i,85)=mean(abs(wp44))/mean(abs(wp45));
ASData(i,86)=mean(abs(wp45))/mean(abs(wp46));
ASData(i,87)=mean(abs(wp46))/mean(abs(wp47));
ASData(i,88)=mean(abs(wp47))/mean(abs(wp48));
ASData(i,89)=mean(abs(wp48))/mean(abs(wp49));
ASData(i,90)=mean(abs(wp49))/mean(abs(wp4A));
ASData(i,91)=mean(abs(wp4A))/mean(abs(wp4B));
ASData(i,92)=mean(abs(wp4B))/mean(abs(wp4C));
ASData(i,93)=mean(abs(wp4C))/mean(abs(wp4D));
ASData(i,94)=mean(abs(wp4D))/mean(abs(wp4E));
ASData(i,95)=mean(abs(wp4E))/mean(abs(wp4F));
end

%% E S.ZIP EPILEPTIC SUBJECT ICTAL DURING SEIZURE
%%%%%%%%%%%%%%%%%%%%%%%%%%%%

for i=2*Nofsignal+1:3*Nofsignal;

[wp10,wp11] = dwt(Ictal(1:Length,i-2*Nofsignal),wname);%WPD Decomposition
[wp20,wp21] = dwt(wp10,wname);
[wp22,wp23] = dwt(wp11,wname);

[wp30,wp31] = dwt(wp20,wname);
[wp32,wp33] = dwt(wp21,wname);
[wp34,wp35] = dwt(wp22,wname);
[wp36,wp37] = dwt(wp23,wname);

[wp40,wp41] = dwt(wp30,wname);
[wp42,wp43] = dwt(wp31,wname);
[wp44,wp45] = dwt(wp32,wname);
[wp46,wp47] = dwt(wp33,wname);
[wp48,wp49] = dwt(wp34,wname);
[wp4A,wp4B] = dwt(wp35,wname);
[wp4C,wp4D] = dwt(wp36,wname);
[wp4E,wp4F] = dwt(wp37,wname);

ASData(i,1)=mean(abs(wp40));
ASData(i,2)=mean(abs(wp41));
ASData(i,3)=mean(abs(wp42));
ASData(i,4)=mean(abs(wp43));
ASData(i,5)=mean(abs(wp44));
ASData(i,6)=mean(abs(wp45));
ASData(i,7)=mean(abs(wp46));
ASData(i,8)=mean(abs(wp47));
ASData(i,9)=mean(abs(wp48));
ASData(i,10)=mean(abs(wp49));
ASData(i,11)=mean(abs(wp4A));
ASData(i,12)=mean(abs(wp4B));
ASData(i,13)=mean(abs(wp4C));
ASData(i,14)=mean(abs(wp4D));
ASData(i,15)=mean(abs(wp4E));
ASData(i,16)=mean(abs(wp4F));
```

```
ASData(i,17)=std(wp40);
ASData(i,18)=std(wp41);
ASData(i,19)=std(wp42);
ASData(i,20)=std(wp43);
ASData(i,21)=std(wp44);
ASData(i,22)=std(wp45);
ASData(i,23)=std(wp46);
ASData(i,24)=std(wp47);
ASData(i,25)=std(wp48);
ASData(i,26)=std(wp49);
ASData(i,27)=std(wp4A);
ASData(i,28)=std(wp4B);
ASData(i,29)=std(wp4C);
ASData(i,30)=std(wp4D);
ASData(i,31)=std(wp4E);
ASData(i,32)=std(wp4F);

ASData(i,33)=skewness(wp40(:));
ASData(i,34)=skewness(wp41(:));
ASData(i,35)=skewness(wp42(:));
ASData(i,36)=skewness(wp43(:));
ASData(i,37)=skewness(wp44(:));
ASData(i,38)=skewness(wp45(:));
ASData(i,39)=skewness(wp46(:));
ASData(i,40)=skewness(wp47(:));
ASData(i,41)=skewness(wp48(:));
ASData(i,42)=skewness(wp49(:));
ASData(i,43)=skewness(wp4A(:));
ASData(i,44)=skewness(wp4B(:));
ASData(i,45)=skewness(wp4C(:));
ASData(i,46)=skewness(wp4D(:));
ASData(i,47)=skewness(wp4E(:));
ASData(i,48)=skewness(wp4F(:));

ASData(i,49)=kurtosis(wp40);
ASData(i,50)=kurtosis(wp41);
ASData(i,51)=kurtosis(wp42);
ASData(i,52)=kurtosis(wp43);
ASData(i,53)=kurtosis(wp44);
ASData(i,54)=kurtosis(wp45);
ASData(i,55)=kurtosis(wp46);
ASData(i,56)=kurtosis(wp47);
ASData(i,57)=kurtosis(wp48);
ASData(i,58)=kurtosis(wp49);
ASData(i,59)=kurtosis(wp4A);
ASData(i,60)=kurtosis(wp4B);
ASData(i,61)=kurtosis(wp4C);
ASData(i,62)=kurtosis(wp4D);
ASData(i,63)=kurtosis(wp4E);
ASData(i,64)=kurtosis(wp4F);

ASData(i,65)=rms(wp40(:));
ASData(i,66)=rms(wp41(:));
ASData(i,67)=rms(wp42(:));
ASData(i,68)=rms(wp43(:));
ASData(i,69)=rms(wp44(:));
ASData(i,70)=rms(wp45(:));
ASData(i,71)=rms(wp46(:));
```

```
ASData(i,72)=rms(wp47(:));
ASData(i,73)=rms(wp48(:));
ASData(i,74)=rms(wp49(:));
ASData(i,75)=rms(wp4A(:));
ASData(i,76)=rms(wp4B(:));
ASData(i,77)=rms(wp4C(:));
ASData(i,78)=rms(wp4D(:));
ASData(i,79)=rms(wp4E(:));
ASData(i,80)=rms(wp4F(:));

ASData(i,81) = mean(abs(wp40))/mean(abs(wp41));
ASData(i,82) = mean(abs(wp41))/mean(abs(wp42));
ASData(i,83) = mean(abs(wp42))/mean(abs(wp43));
ASData(i,84) = mean(abs(wp43))/mean(abs(wp44));
ASData(i,85) = mean(abs(wp44))/mean(abs(wp45));
ASData(i,86) = mean(abs(wp45))/mean(abs(wp46));
ASData(i,87) = mean(abs(wp46))/mean(abs(wp47));
ASData(i,88) = mean(abs(wp47))/mean(abs(wp48));
ASData(i,89) = mean(abs(wp48))/mean(abs(wp49));
ASData(i,90) = mean(abs(wp49))/mean(abs(wp4A));
ASData(i,91) = mean(abs(wp4A))/mean(abs(wp4B));
ASData(i,92) = mean(abs(wp4B))/mean(abs(wp4C));
ASData(i,93) = mean(abs(wp4C))/mean(abs(wp4D));
ASData(i,94) = mean(abs(wp4D))/mean(abs(wp4E));
ASData(i,95) = mean(abs(wp4E))/mean(abs(wp4F));

end

%Make input the Data Matrix
Inputs=ASData;

%%
%You can create target using excel and the import it from the MATLAB HOME
%menu and Import data and then save it as EEGTargets
%Load Targets
load EEGTargets

%% Classification
% This example shows how to perform classification using discriminant
% analysis. Suppose you have a
% data set containing observations with measurements on different variables
% (called predictors) and their known class labels. If you obtain predictor
% values for new observations, could you determine to which classes those
% observations probably belong? This is the problem of classification.

% Copyright 2002-2014 The MathWorks, Inc.

%%
% Suppose you have Normal, Interictal and Ictal EEG data, and you need to
% determine their classes using discriminant analysis.
%% Linear Discriminant Analysis
% The |fitcdiscr| function can perform classification using different types
% of discriminant analysis. First classify the data using the default
% linear discriminant analysis (LDA).

lda = fitcdiscr(Inputs,Targets);
ldaClass = resubPredict(lda);
%%
% The observations with known class labels are usually called the training data.
% Now compute the resubstitution error, which is the misclassification error
% (the proportion of misclassified observations) on the training set.
```

```
ldaResubErr = resubLoss(lda)
%%
% You can also compute the confusion matrix on the training set. A
% confusion matrix contains information about known class labels and
% predicted class labels. Generally speaking, the (i,j) element in the
% confusion matrix is the number of samples whose known class label is
% class i and whose predicted class is j. The diagonal elements represent
% correctly classified observations.

[ldaResubCM,grpOrder] = confusionmat(Targets,ldaClass)

% Calculate the Total Classification Accuracy
TotalAccuracy=(ldaResubCM(1,1)+ldaResubCM(2,2)+ldaResubCM(3,3))/(3*Nofsignal)

disp('Total Classification Accuracy=')
disp(TotalAccuracy)

%%
% You have computed the resubstitution error. Usually people are more
% interested in the test error (also referred to as generalization error),
% which is the expected prediction error on an independent set. In fact,
% the resubstitution error will likely under-estimate the test error.
%
% In this case you don't have another labeled data set, but you can
% simulate one by doing cross-validation. A stratified 10-fold
% cross-validation is a popular choice for estimating the test error on
% classification algorithms. It randomly divides the training set into 10
% disjoint subsets. Each subset has roughly equal size and roughly the same
% class proportions as in the training set. Remove one subset, train the
% classification model using the other nine subsets, and use the trained
% model to classify the removed subset. You could repeat this by removing
% each of the ten subsets one at a time.
%
% Because cross-validation randomly divides data, its outcome depends on
% the initial random seed. To reproduce the exact results in this example,
% execute the following command:

rng(0,'twister');

%%
% First use |cvpartition| to generate 10 disjoint stratified subsets.
cp = cvpartition(Targets,'KFold',10)
%%
% The |crossval| and |kfoldLoss| methods can estimate the misclassification
% error for LDA using the given data partition |cp|.
%
% Estimate the true test error for LDA using 10-fold stratified
% cross-validation.

cvlda = crossval(lda,'CVPartition',cp);
ldaCVErr = kfoldLoss(cvlda)
%%
% The LDA cross-validation error has the same value as the LDA
% resubstitution error on this data.

%% Conclusions
% This example shows how to perform classification using LDA in MATLAB(R)
% using Statistics and Machine Learning Toolbox(TM) functions.
```

5.4 NAÏVE BAYES

When given a set of objects where each belongs to a known class and has a known vector of variables, the aim is to create a rule that will assign a future object to a class, given only the vectors of variables describing the future object. Because the NB method does not need a complicated iterative parameter estimation scheme, it is very easy to build. In other words, NB can be easily applied to huge datasets, and users who are unskilled in classification can understand why it makes specific classifications. Performance of the algorithm is quite good. In any particular application, it may not be the best possible classifier, but it is robust and can perform quite well. There are several factors that should be considered, which can make assumptions that are not as detrimental as they may appear. A selection step of an a prior variable has often already taken place, in which variables that are highly correlated can be eliminated on the basis that they are likely to contribute in a similar way to the separation of classes. This presents the conclusion that the relationship between the remaining variables might be approximated by their independence. If interactions of those variables are assumed to be zero, the algorithm delivers an implicit regularization step that results in decreasing the variance of the model and achieving more precise classification. There are cases when the variables are correlated, and the optimal decision surface matches the surface formed under the independence assumption. This shows that the assumption is not at all detrimental to performance. The decision surface produced by the NB model can be in a complicated nonlinear shape. The surface is linear but highly nonlinear in the original variables; in this way, it can fit quite complicated surfaces (Wu et al., 2008).

EXAMPLE 5.6. The following MATLAB code was used to extract features from the EEG signals using AR Burg method. Then it classified these data using NB classifier. You can download data from the following website: http://epileptologie-bonn.de/cms/front_content.php?idcat = 193&lang = 3&changelang = 3

```
%% Ch5_Ex11_EEG_ARBURG_NB.m
%The following MATLAB code is used to extract the features from
%the EEG signals using AR Burg.
%Then it classifies EEG Signals Using Naive Bayes Classifier
clc
clear
%Load Sample EEG Data downloaded from the web site
%http://epileptologie-bonn.de/cms/front_content.php?idcat=193&lang=3&changelang=3
load AS_BONN_ALL_EEG_DATA_4096.mat
%%
Fs = 173.6;          % Sampling frequency
T = 1/Fs;            % Sampling period
segmentLength = 128; % Length of a signal segment
Length=4096;% Length of signal
Nofsignal=100; %Number of Signal
order = 14;

%%
% Obtain the AR Burg Spectrum of the Normal EEG signal using pburg.
for i=1:Nofsignal
    [Pxx,F] = pburg(Normal_Eyes_Open(1:Length,i),order,segmentLength,Fs);
    ASData(i,:)=Pxx(:);
end
%%
% Obtain the AR Burg Spectrum of the Interictal EEG signal using pburg.
for i=Nofsignal+1:2*Nofsignal
    [Pxx,F] = pburg(Interictal(1:Length,i-Nofsignal),order,segmentLength,Fs);
    ASData(i,:)=Pxx(:);
end
%%
% Obtain the AR Burg Spectrum of the Ictal EEG signal using pburg.
for i=2*Nofsignal+1:3*Nofsignal
[Pxx,F] = pburg(Ictal(1:Length,i-2*Nofsignal),order,segmentLength,Fs);
ASData(i,:)=Pxx(:);
end
```

```
%Make input the Data Matrix
Inputs=ASData;

%%
%You can create target using excel and the import it from the MATLAB HOME
%menu and Import data and then save it as EEGTargets
%Load Target
load EEGTargets

%% Classification
% This example shows how to perform classification using Naive Bayes
%Suppose you have a
% data set containing observations with measurements on different variables
% (called predictors) and their known class labels. If you obtain predictor
% values for new observations, could you determine to which classes those
% observations probably belong? This is the problem of classification.

%%
% Suppose you have Normal, Interictal and Ictal EEG data, and you need to
% determine their classes using Naive Bayes.

%% Naive Bayes Classifiers
% The |fitcdiscr| function has other two other types, 'DiagLinear' and
% 'DiagQuadratic'. They are similar to 'linear' and 'quadratic', but with
% diagonal covariance matrix estimates. These diagonal choices are specific
% examples of a naive Bayes classifier, because they assume the variables are
% conditionally independent given the class label. Naive Bayes classifiers are
% among the most popular classifiers. While the assumption of class-conditional
% independence between variables is not true in general, naive Bayes classifiers
% have been found to work well in practice on many data sets.
%
% The |fitcnb| function can be used to create a more general type of naive Bayes
% classifier.

%%
% You have computed the resubstitution error. Usually people are more
% interested in the test error (also referred to as generalization error),
% which is the expected prediction error on an independent set. In fact,
% the resubstitution error will likely under-estimate the test error.
%
% In this case you don't have another labeled data set, but you can
% simulate one by doing cross-validation. A stratified 10-fold
% cross-validation is a popular choice for estimating the test error on
% classification algorithms. It randomly divides the training set into 10
% disjoint subsets. Each subset has roughly equal size and roughly the same
% class proportions as in the training set. Remove one subset, train the
% classification model using the other nine subsets, and use the trained
% model to classify the removed subset. You could repeat this by removing
% each of the ten subsets one at a time.
%
% Because cross-validation randomly divides data, its outcome depends on
% the initial random seed. To reproduce the exact results in this example,
% execute the following command:

rng(0,'twister');

%%
% First use |cvpartition| to generate 10 disjoint stratified subsets.
cp = cvpartition(Targets,'KFold',10)
```

```
%%
% First model each variable in each class using a Gaussian distribution.
% You can compute the resubstitution error and the cross-validation error.

nbGau = fitcnb(Inputs, Targets);
nbGauResubErr = resubLoss(nbGau)
nbGauCV = crossval(nbGau, 'CVPartition',cp);
nbGauCVErr = kfoldLoss(nbGauCV)

%%
% You can also compute the confusion matrix on the training set. A
% confusion matrix contains information about known class labels and
% predicted class labels. Generally speaking, the (i,j) element in the
% confusion matrix is the number of samples whose known class label is
% class i and whose predicted class is j. The diagonal elements represent
% correctly classified observations.
nbClass = resubPredict(nbGau);
[nbResubCM,grpOrder] = confusionmat(Targets,nbClass)

% Calculate the Total Classification Accuracy
TotalAccuracy = (nbResubCM(1,1)+nbResubCM(2,2)+nbResubCM(3,3))/(3*Nofsignal)

disp('Total Classification Accuracy=')
disp(TotalAccuracy)
%%
% So far you have assumed the variables from each class have a multivariate
% normal distribution. Often that is a reasonable assumption, but sometimes
% you may not be willing to make that assumption or you may see clearly
% that it is not valid. Now try to model each variable in each class using
% a kernel density estimation, which is a more flexible nonparametric
% technique. Here we set the kernel to |box|.

nbKD = fitcnb(Inputs, Targets, 'DistributionNames','kernel', 'Kernel','box');
nbKDResubErr = resubLoss(nbKD)
nbKDCV = crossval(nbKD, 'CVPartition',cp);
nbKDCVErr = kfoldLoss(nbKDCV)

%%
% For this data set, the naive Bayes classifier with kernel density
% estimation gets smaller resubstitution error and cross-validation error
% than the naive Bayes classifier with a Gaussian distribution.

%% Conclusions
% This example shows how to perform classification using Naive Bayes in MATLAB(R)
% using Statistics and Machine Learning Toolbox(TM) functions.
```

EXAMPLE 5.7. The following MATLAB code is used to extract features from the ECG signals using modified covariance method. Then it classifies these data using NB classifier. You can download data from the following website: https://www.physionet.org/physiobank/database/mitdb/

```
%% Ch5_Ex12_ECG_MCOV_NB.m
%The following MATLAB code is used to extract the features from
%the ECG signals using Modified Covariance spectrum.
%Then it classifies EEG Signals Using Naive Bayes Classifier
clc
clear
```

```
%Load Sample ECG Data downloaded from the web site
%https://www.physionet.org/physiobank/database/mitdb/
load MITBIH_ECG.mat
%%
Fs = 320;            % Sampling frequency
T = 1/Fs;             % Sampling period
segmentLength = 128;   % Length of a signal segment
Length=320;% Length of signal
Nofsignal=300; %Number of Signal
order = 34;
%%
% Obtain the Modified Covariance Spectrum of the Normal ECG signal using pmcov.
for i=1:Nofsignal
    [Pxx,F] = pmcov(ECGN(1:Length,i),order,segmentLength,Fs);
    ASData(i,:)=Pxx(:);
end
%%
% Obtain the Modified Covariance Spectrum of the ECG signal with APC using pmcov.
[Pxx,F] = pmcov(ECGAPC(:,1),order,segmentLength,Fs);
for i=Nofsignal+1:2*Nofsignal
    [Pxx,F] = pmcov(ECGAPC(1:Length,i-Nofsignal),order,segmentLength,Fs);
    ASData(i,:)=Pxx(:);
end
%%
% Obtain the Modified Covariance Spectrum of the ECG signal with LBBB using pmcov.
for i=2*Nofsignal+1:3*Nofsignal
[Pxx,F]=pmcov(ECGPVC(1:Length,i-2*Nofsignal),order,segmentLength,Fs);
ASData(i,:)=Pxx(:);
end

%%
% Obtain the Modified Covariance Spectrum of the ECG signal with PVC using pmcov.
for i=3*Nofsignal+1:4*Nofsignal
[Pxx,F] = pmcov(ECGLBBB(1:Length,i-3*Nofsignal),order,segmentLength,Fs);
ASData(i,:)=Pxx(:);
end

%%
% Obtain the Modified Covariance Spectrum of the ECG signal with RBBB using pmcov.
for i=4*Nofsignal+1:5*Nofsignal
[Pxx,F] = pmcov(ECGRBBB(1:Length,i-4*Nofsignal),order,segmentLength,Fs);
ASData(i,:)=Pxx(:);
end

%Make Inputs as the Data Matrix
Inputs=ASData;
%%
%You can create target using excel and the import it from the MATLAB HOME
%menu and Import data and then save it as EEGTargets
%Load Target
load ECGTargets

%% Classification
% This example shows how to perform classification using Naive Bayes
%Suppose you have a
% data set containing observations with measurements on different variables
% (called predictors) and their known class labels. If you obtain predictor
% values for new observations, could you determine to which classes those
% observations probably belong? This is the problem of classification.
```

```
%%
% Suppose you have Normal, Interictal and Ictal EEG data, and you need to
% determine their classes using Naive Bayes.

%% Naive Bayes Classifiers
% The |fitcdiscr| function has other two other types, 'DiagLinear' and
% 'DiagQuadratic'. They are similar to 'linear' and 'quadratic', but with
% diagonal covariance matrix estimates. These diagonal choices are specific
% examples of a naive Bayes classifier, because they assume the variables are
% conditionally independent given the class label. Naive Bayes classifiers are
% among the most popular classifiers. While the assumption of class-conditional
% independence between variables is not true in general, naive Bayes classifiers
% have been found to work well in practice on many data sets.
%
% The |fitcnb| function can be used to create a more general type of naive Bayes
% classifier.

%%
% You have computed the resubstitution error. Usually people are more
% interested in the test error (also referred to as generalization error),
% which is the expected prediction error on an independent set. In fact,
% the resubstitution error will likely under-estimate the test error.
%
% In this case you don't have another labeled data set, but you can
% simulate one by doing cross-validation. A stratified 10-fold
% cross-validation is a popular choice for estimating the test error on
% classification algorithms. It randomly divides the training set into 10
% disjoint subsets. Each subset has roughly equal size and roughly the same
% class proportions as in the training set. Remove one subset, train the
% classification model using the other nine subsets, and use the trained
% model to classify the removed subset. You could repeat this by removing
% each of the ten subsets one at a time.
%
% Because cross-validation randomly divides data, its outcome depends on
% the initial random seed. To reproduce the exact results in this example,
% execute the following command:

rng(0,'twister');

%%
% First use |cvpartition| to generate 10 disjoint stratified subsets.
cp = cvpartition(Targets,'KFold',10)

%%
% First model each variable in each class using a Gaussian distribution.
% You can compute the resubstitution error and the cross-validation error.

nbGau = fitcnb(Inputs, Targets);
nbGauResubErr = resubLoss(nbGau)
nbGauCV = crossval(nbGau, 'CVPartition',cp);
nbGauCVErr = kfoldLoss(nbGauCV)

%%
% You can also compute the confusion matrix on the training set. A
% confusion matrix contains information about known class labels and
% predicted class labels. Generally speaking, the (i,j) element in the
% confusion matrix is the number of samples whose known class label is
% class i and whose predicted class is j. The diagonal elements represent
% correctly classified observations.
```

```
nbClass = resubPredict(nbGau);
[nbResubCM,grpOrder] = confusionmat(Targets,nbClass)

% Calculate the Total Classification Accuracy
TotalAccuracy=(nbResubCM(1,1)+nbResubCM(2,2)+nbResubCM(3,3))/(3*Nofsignal)

disp('Total Classification Accuracy=')
disp(TotalAccuracy)
%%
% So far you have assumed the variables from each class have a multivariate
% normal distribution. Often that is a reasonable assumption, but sometimes
% you may not be willing to make that assumption or you may see clearly
% that it is not valid. Now try to model each variable in each class using
% a kernel density estimation, which is a more flexible nonparametric
% technique. Here we set the kernel to |box|.

nbKD = fitcnb(Inputs, Targets, 'DistributionNames','kernel', 'Kernel','box');
nbKDResubErr = resubLoss(nbKD)
nbKDCV = crossval(nbKD, 'CVPartition',cp);
nbKDCVErr = kfoldLoss(nbKDCV)

%%
% For this data set, the naive Bayes classifier with kernel density
% estimation gets smaller resubstitution error and cross-validation error
% than the naive Bayes classifier with a Gaussian distribution.

%% Conclusions
% This example shows how to perform classification using Naive Bayes in MATLAB(R)
% using Statistics and Machine Learning Toolbox(TM) functions.
```

EXAMPLE 5.8. The following MATLAB code was used to extract features from the EEG signals using WPD. Then it used statistical values of WPD subbands. Then it classified these data using NB classifier. You can download data from the following website:
http://epileptologie-bonn.de/cms/front_content.php?idcat=193&lang=3&changelang=3

```
%Ch5_Ex15_EEG_WPD_NB.m
% WPD of NORMAL, INTERICTAL and ICTAL EEG signals
%The following MATLAB code is used to Extract WPD features from the EEG signals
%Decompose EEG data using WPD coefficients
%Then uses Statistical features as:

%(1) Mean of the absolute values of the coefficients in each sub-band.
%(2) Standard deviation of the coefficients in each sub-band.
%(3) Skewness of the coefficients in each sub-band.
%(4) Kurtosis of the coefficients in each sub-band.
%(5) RMS power of the wavelet coefficients in each subband.
%(6) Ratio of the mean absolute values of adjacent subbands.

%Then it classifies EEG Signals Using Naive Bayes Classifier
clc
clear
%Load Sample EEG Data downloaded from the web site
%http://epileptologie-bonn.de/cms/front_content.php?idcat=193&lang=3&changelang=3
load AS_BONN_ALL_EEG_DATA_4096.mat
wname = 'db4';
```

```
Length = 4096; % Length of signal
Nofsignal=100; %Number of Signal
%%
%%%%%%%%%%%%%%%%%%%%%%%%%%%%% A Z.ZIP EYES OPEN NORMAL SUBJECT

for i=1:Nofsignal;

[wp10,wp11] = dwt(Normal_Eyes_Open(1:Length,i),wname);%WPD Decomposition
[wp20,wp21] = dwt(wp10,wname);
[wp22,wp23] = dwt(wp11,wname);

[wp30,wp31] = dwt(wp20,wname);
[wp32,wp33] = dwt(wp21,wname);
[wp34,wp35] = dwt(wp22,wname);
[wp36,wp37] = dwt(wp23,wname);

[wp40,wp41] = dwt(wp30,wname);
[wp42,wp43] = dwt(wp31,wname);
[wp44,wp45] = dwt(wp32,wname);
[wp46,wp47] = dwt(wp33,wname);
[wp48,wp49] = dwt(wp34,wname);
[wp4A,wp4B] = dwt(wp35,wname);
[wp4C,wp4D] = dwt(wp36,wname);
[wp4E,wp4F] = dwt(wp37,wname);

ASData(i,1)=mean(abs(wp40));
ASData(i,2)=mean(abs(wp41));
ASData(i,3)=mean(abs(wp42));
ASData(i,4)=mean(abs(wp43));
ASData(i,5)=mean(abs(wp44));
ASData(i,6)=mean(abs(wp45));
ASData(i,7)=mean(abs(wp46));
ASData(i,8)=mean(abs(wp47));
ASData(i,9)=mean(abs(wp48));
ASData(i,10)=mean(abs(wp49));
ASData(i,11)=mean(abs(wp4A));
ASData(i,12)=mean(abs(wp4B));
ASData(i,13)=mean(abs(wp4C));
ASData(i,14)=mean(abs(wp4D));
ASData(i,15)=mean(abs(wp4E));
ASData(i,16)=mean(abs(wp4F));

ASData(i,17)=std(wp40);
ASData(i,18)=std(wp41);
ASData(i,19)=std(wp42);
ASData(i,20)=std(wp43);
ASData(i,21)=std(wp44);
ASData(i,22)=std(wp45);
ASData(i,23)=std(wp46);
ASData(i,24)=std(wp47);
ASData(i,25)=std(wp48);
ASData(i,26)=std(wp49);
ASData(i,27)=std(wp4A);
ASData(i,28)=std(wp4B);
ASData(i,29)=std(wp4C);
ASData(i,30)=std(wp4D);
ASData(i,31)=std(wp4E);
ASData(i,32)=std(wp4F);
```

```
ASData(i,33)=skewness(wp40(:));
ASData(i,34)=skewness(wp41(:));
ASData(i,35)=skewness(wp42(:));
ASData(i,36)=skewness(wp43(:));
ASData(i,37)=skewness(wp44(:));
ASData(i,38)=skewness(wp45(:));
ASData(i,39)=skewness(wp46(:));
ASData(i,40)=skewness(wp47(:));
ASData(i,41)=skewness(wp48(:));
ASData(i,42)=skewness(wp49(:));
ASData(i,43)=skewness(wp4A(:));
ASData(i,44)=skewness(wp4B(:));
ASData(i,45)=skewness(wp4C(:));
ASData(i,46)=skewness(wp4D(:));
ASData(i,47)=skewness(wp4E(:));
ASData(i,48)=skewness(wp4F(:));

ASData(i,49)=kurtosis(wp40);
ASData(i,50)=kurtosis(wp41);
ASData(i,51)=kurtosis(wp42);
ASData(i,52)=kurtosis(wp43);
ASData(i,53)=kurtosis(wp44);
ASData(i,54)=kurtosis(wp45);
ASData(i,55)=kurtosis(wp46);
ASData(i,56)=kurtosis(wp47);
ASData(i,57)=kurtosis(wp48);
ASData(i,58)=kurtosis(wp49);
ASData(i,59)=kurtosis(wp4A);
ASData(i,60)=kurtosis(wp4B);
ASData(i,61)=kurtosis(wp4C);
ASData(i,62)=kurtosis(wp4D);
ASData(i,63)=kurtosis(wp4E);
ASData(i,64)=kurtosis(wp4F);

ASData(i,65)=rms(wp40(:));
ASData(i,66)=rms(wp41(:));
ASData(i,67)=rms(wp42(:));
ASData(i,68)=rms(wp43(:));
ASData(i,69)=rms(wp44(:));
ASData(i,70)=rms(wp45(:));
ASData(i,71)=rms(wp46(:));
ASData(i,72)=rms(wp47(:));
ASData(i,73)=rms(wp48(:));
ASData(i,74)=rms(wp49(:));
ASData(i,75)=rms(wp4A(:));
ASData(i,76)=rms(wp4B(:));
ASData(i,77)=rms(wp4C(:));
ASData(i,78)=rms(wp4D(:));
ASData(i,79)=rms(wp4E(:));
ASData(i,80)=rms(wp4F(:));

ASData(i,81) = mean(abs(wp40))/mean(abs(wp41));
ASData(i,82) = mean(abs(wp41))/mean(abs(wp42));
ASData(i,83) = mean(abs(wp42))/mean(abs(wp43));
ASData(i,84) = mean(abs(wp43))/mean(abs(wp44));
ASData(i,85) = mean(abs(wp44))/mean(abs(wp45));
ASData(i,86) = mean(abs(wp45))/mean(abs(wp46));
ASData(i,87) = mean(abs(wp46))/mean(abs(wp47));
```

```
ASData(i,88) = mean(abs(wp47))/mean(abs(wp48));
ASData(i,89) = mean(abs(wp48))/mean(abs(wp49));
ASData(i,90) = mean(abs(wp49))/mean(abs(wp4A));
ASData(i,91) = mean(abs(wp4A))/mean(abs(wp4B));
ASData(i,92) = mean(abs(wp4B))/mean(abs(wp4C));
ASData(i,93) = mean(abs(wp4C))/mean(abs(wp4D));
ASData(i,94) = mean(abs(wp4D))/mean(abs(wp4E));
ASData(i,95) = mean(abs(wp4E))/mean(abs(wp4F));

end

%%
% D F.ZIP EPILEPTIC SUBJECT (INTERICTAL) were recorded from within
% the epileptogenic zone during seizure free intervals

for i=Nofsignal+1:2*Nofsignal;

[wp10,wp11] = dwt(Interictal(1:Length,i-Nofsignal),wname);%WPD Decomposition
[wp20,wp21] = dwt(wp10,wname);
[wp22,wp23] = dwt(wp11,wname);

[wp30,wp31] = dwt(wp20,wname);
[wp32,wp33] = dwt(wp21,wname);
[wp34,wp35] = dwt(wp22,wname);
[wp36,wp37] = dwt(wp23,wname);

[wp40,wp41] = dwt(wp30,wname);
[wp42,wp43] = dwt(wp31,wname);
[wp44,wp45] = dwt(wp32,wname);

[wp46,wp47] = dwt(wp33,wname);
[wp48,wp49] = dwt(wp34,wname);
[wp4A,wp4B] = dwt(wp35,wname);
[wp4C,wp4D] = dwt(wp36,wname);
[wp4E,wp4F] = dwt(wp37,wname);

ASData(i,1)=mean(abs(wp40));
ASData(i,2)=mean(abs(wp41));
ASData(i,3)=mean(abs(wp42));
ASData(i,4)=mean(abs(wp43));
ASData(i,5)=mean(abs(wp44));
ASData(i,6)=mean(abs(wp45));
ASData(i,7)=mean(abs(wp46));
ASData(i,8)=mean(abs(wp47));
ASData(i,9)=mean(abs(wp48));
ASData(i,10)=mean(abs(wp49));
ASData(i,11)=mean(abs(wp4A));
ASData(i,12)=mean(abs(wp4B));
ASData(i,13)=mean(abs(wp4C));
ASData(i,14)=mean(abs(wp4D));
ASData(i,15)=mean(abs(wp4E));
ASData(i,16)=mean(abs(wp4F));

ASData(i,17)=std(wp40);
ASData(i,18)=std(wp41);
ASData(i,19)=std(wp42);
ASData(i,20)=std(wp43);
ASData(i,21)=std(wp44);
```

```
ASData(i,22)=std(wp45);
ASData(i,23)=std(wp46);
ASData(i,24)=std(wp47);
ASData(i,25)=std(wp48);
ASData(i,26)=std(wp49);
ASData(i,27)=std(wp4A);
ASData(i,28)=std(wp4B);
ASData(i,29)=std(wp4C);
ASData(i,30)=std(wp4D);
ASData(i,31)=std(wp4E);
ASData(i,32)=std(wp4F);

ASData(i,33)=skewness(wp40(:));
ASData(i,34)=skewness(wp41(:));
ASData(i,35)=skewness(wp42(:));
ASData(i,36)=skewness(wp43(:));
ASData(i,37)=skewness(wp44(:));
ASData(i,38)=skewness(wp45(:));
ASData(i,39)=skewness(wp46(:));
ASData(i,40)=skewness(wp47(:));
ASData(i,41)=skewness(wp48(:));
ASData(i,42)=skewness(wp49(:));
ASData(i,43)=skewness(wp4A(:));
ASData(i,44)=skewness(wp4B(:));
ASData(i,45)=skewness(wp4C(:));
ASData(i,46)=skewness(wp4D(:));
ASData(i,47)=skewness(wp4E(:));
ASData(i,48)=skewness(wp4F(:));

ASData(i,49)=kurtosis(wp40);
ASData(i,50)=kurtosis(wp41);
ASData(i,51)=kurtosis(wp42);
ASData(i,52)=kurtosis(wp43);
ASData(i,53)=kurtosis(wp44);
ASData(i,54)=kurtosis(wp45);
ASData(i,55)=kurtosis(wp46);
ASData(i,56)=kurtosis(wp47);
ASData(i,57)=kurtosis(wp48);
ASData(i,58)=kurtosis(wp49);
ASData(i,59)=kurtosis(wp4A);
ASData(i,60)=kurtosis(wp4B);
ASData(i,61)=kurtosis(wp4C);
ASData(i,62)=kurtosis(wp4D);
ASData(i,63)=kurtosis(wp4E);
ASData(i,64)=kurtosis(wp4F);

ASData(i,65)=rms(wp40(:));
ASData(i,66)=rms(wp41(:));
ASData(i,67)=rms(wp42(:));
ASData(i,68)=rms(wp43(:));
ASData(i,69)=rms(wp44(:));
ASData(i,70)=rms(wp45(:));
ASData(i,71)=rms(wp46(:));
ASData(i,72)=rms(wp47(:));
ASData(i,73)=rms(wp48(:));
ASData(i,74)=rms(wp49(:));
ASData(i,75)=rms(wp4A(:));
ASData(i,76)=rms(wp4B(:));
```

```
ASData(i,77)=rms(wp4C(:));
ASData(i,78)=rms(wp4D(:));
ASData(i,79)=rms(wp4E(:));
ASData(i,80)=rms(wp4F(:));

ASData(i,81)=mean(abs(wp40))/mean(abs(wp41));
ASData(i,82)=mean(abs(wp41))/mean(abs(wp42));
ASData(i,83)=mean(abs(wp42))/mean(abs(wp43));
ASData(i,84)=mean(abs(wp43))/mean(abs(wp44));
ASData(i,85)=mean(abs(wp44))/mean(abs(wp45));
ASData(i,86)=mean(abs(wp45))/mean(abs(wp46));
ASData(i,87)=mean(abs(wp46))/mean(abs(wp47));
ASData(i,88)=mean(abs(wp47))/mean(abs(wp48));
ASData(i,89)=mean(abs(wp48))/mean(abs(wp49));
ASData(i,90)=mean(abs(wp49))/mean(abs(wp4A));
ASData(i,91)=mean(abs(wp4A))/mean(abs(wp4B));
ASData(i,92)=mean(abs(wp4B))/mean(abs(wp4C));
ASData(i,93)=mean(abs(wp4C))/mean(abs(wp4D));
ASData(i,94)=mean(abs(wp4D))/mean(abs(wp4E));
ASData(i,95)=mean(abs(wp4E))/mean(abs(wp4F));
end

%% E S.ZIP EPILEPTIC SUBJECT ICTAL DURING SEIZURE
%%%%%%%%%%%%%%%%%%%%%%%%%%%%%

for i=2*Nofsignal+1:3*Nofsignal;

[wp10,wp11] = dwt(Ictal(1:Length,i-2*Nofsignal),wname);%WPD Decomposition
[wp20,wp21] = dwt(wp10,wname);
[wp22,wp23] = dwt(wp11,wname);

[wp30,wp31] = dwt(wp20,wname);
[wp32,wp33] = dwt(wp21,wname);
[wp34,wp35] = dwt(wp22,wname);
[wp36,wp37] = dwt(wp23,wname);
[wp40,wp41] = dwt(wp30,wname);
[wp42,wp43] = dwt(wp31,wname);
[wp44,wp45] = dwt(wp32,wname);
[wp46,wp47] = dwt(wp33,wname);
[wp48,wp49] = dwt(wp34,wname);
[wp4A,wp4B] = dwt(wp35,wname);
[wp4C,wp4D] = dwt(wp36,wname);
[wp4E,wp4F] = dwt(wp37,wname);

ASData(i,1)=mean(abs(wp40));
ASData(i,2)=mean(abs(wp41));
ASData(i,3)=mean(abs(wp42));
ASData(i,4)=mean(abs(wp43));
ASData(i,5)=mean(abs(wp44));
ASData(i,6)=mean(abs(wp45));
ASData(i,7)=mean(abs(wp46));
ASData(i,8)=mean(abs(wp47));
ASData(i,9)=mean(abs(wp48));
ASData(i,10)=mean(abs(wp49));
ASData(i,11)=mean(abs(wp4A));
ASData(i,12)=mean(abs(wp4B));
ASData(i,13)=mean(abs(wp4C));
ASData(i,14)=mean(abs(wp4D));
```

```
ASData(i,15)=mean(abs(wp4E));
ASData(i,16)=mean(abs(wp4F));

ASData(i,17)=std(wp40);
ASData(i,18)=std(wp41);
ASData(i,19)=std(wp42);
ASData(i,20)=std(wp43);
ASData(i,21)=std(wp44);
ASData(i,22)=std(wp45);
ASData(i,23)=std(wp46);
ASData(i,24)=std(wp47);
ASData(i,25)=std(wp48);
ASData(i,26)=std(wp49);
ASData(i,27)=std(wp4A);
ASData(i,28)=std(wp4B);
ASData(i,29)=std(wp4C);
ASData(i,30)=std(wp4D);
ASData(i,31)=std(wp4E);
ASData(i,32)=std(wp4F);

ASData(i,33)=skewness(wp40(:));
ASData(i,34)=skewness(wp41(:));
ASData(i,35)=skewness(wp42(:));
ASData(i,36)=skewness(wp43(:));
ASData(i,37)=skewness(wp44(:));
ASData(i,38)=skewness(wp45(:));
ASData(i,39)=skewness(wp46(:));
ASData(i,40)=skewness(wp47(:));
ASData(i,41)=skewness(wp48(:));
ASData(i,42)=skewness(wp49(:));
ASData(i,43)=skewness(wp4A(:));
ASData(i,44)=skewness(wp4B(:));
ASData(i,45)=skewness(wp4C(:));
ASData(i,46)=skewness(wp4D(:));
ASData(i,47)=skewness(wp4E(:));
ASData(i,48)=skewness(wp4F(:));

ASData(i,49)=kurtosis(wp40);
ASData(i,50)=kurtosis(wp41);
ASData(i,51)=kurtosis(wp42);
ASData(i,52)=kurtosis(wp43);
ASData(i,53)=kurtosis(wp44);
ASData(i,54)=kurtosis(wp45);
ASData(i,55)=kurtosis(wp46);
ASData(i,56)=kurtosis(wp47);
ASData(i,57)=kurtosis(wp48);
ASData(i,58)=kurtosis(wp49);
ASData(i,59)=kurtosis(wp4A);
ASData(i,60)=kurtosis(wp4B);
ASData(i,61)=kurtosis(wp4C);
ASData(i,62)=kurtosis(wp4D);
ASData(i,63)=kurtosis(wp4E);
ASData(i,64)=kurtosis(wp4F);

ASData(i,65)=rms(wp40(:));
ASData(i,66)=rms(wp41(:));
ASData(i,67)=rms(wp42(:));
ASData(i,68)=rms(wp43(:));
```

```
ASData(i,69)=rms(wp44(:));
ASData(i,70)=rms(wp45(:));
ASData(i,71)=rms(wp46(:));
ASData(i,72)=rms(wp47(:));
ASData(i,73)=rms(wp48(:));
ASData(i,74)=rms(wp49(:));
ASData(i,75)=rms(wp4A(:));
ASData(i,76)=rms(wp4B(:));
ASData(i,77)=rms(wp4C(:));
ASData(i,78)=rms(wp4D(:));
ASData(i,79)=rms(wp4E(:));
ASData(i,80)=rms(wp4F(:));

ASData(i,81)=mean(abs(wp40))/mean(abs(wp41));
ASData(i,82)=mean(abs(wp41))/mean(abs(wp42));
ASData(i,83)=mean(abs(wp42))/mean(abs(wp43));
ASData(i,84)=mean(abs(wp43))/mean(abs(wp44));
ASData(i,85)=mean(abs(wp44))/mean(abs(wp45));
ASData(i,86)=mean(abs(wp45))/mean(abs(wp46));
ASData(i,87)=mean(abs(wp46))/mean(abs(wp47));
ASData(i,88)=mean(abs(wp47))/mean(abs(wp48));
ASData(i,89)=mean(abs(wp48))/mean(abs(wp49));
ASData(i,90)=mean(abs(wp49))/mean(abs(wp4A));
ASData(i,91)=mean(abs(wp4A))/mean(abs(wp4B));
ASData(i,92)=mean(abs(wp4B))/mean(abs(wp4C));
ASData(i,93)=mean(abs(wp4C))/mean(abs(wp4D));
ASData(i,94)=mean(abs(wp4D))/mean(abs(wp4E));
ASData(i,95)=mean(abs(wp4E))/mean(abs(wp4F));
end

%Make input the Data Matrix
Inputs=ASData;

%%
%You can create target using excel and the import it from the MATLAB HOME
%menu and Import data and then save it as EEGTargets
%Load Target
load EEGTargets

%%
%% Classification
% This example shows how to perform classification using discriminant
% analysis. Suppose you have a
% data set containing observations with measurements on different variables
% (called predictors) and their known class labels. If you obtain predictor
% values for new observations, could you determine to which classes those
% observations probably belong? This is the problem of classification.

% Copyright 2002-2014 The MathWorks, Inc.

%%
% Suppose you have Normal, Interictal and Ictal EEG data, and you need to
% determine their classes using discriminant analysis.

%% Naive Bayes Classifiers
% The |fitcdiscr| function has other two other types, 'DiagLinear' and
% 'DiagQuadratic'. They are similar to 'linear' and 'quadratic', but with
% diagonal covariance matrix estimates. These diagonal choices are specific
```

```
% examples of a naive Bayes classifier, because they assume the variables are
% conditionally independent given the class label. Naive Bayes classifiers are
% among the most popular classifiers. While the assumption of class-conditional
% independence between variables is not true in general, naive Bayes classifiers
% have been found to work well in practice on many data sets.
%
% The |fitcnb| function can be used to create a more general type of naive Bayes
% classifier.

%%
% You have computed the resubstitution error. Usually people are more
% interested in the test error (also referred to as generalization error),
% which is the expected prediction error on an independent set. In fact,
% the resubstitution error will likely under-estimate the test error.
%
% In this case you don't have another labeled data set, but you can
% simulate one by doing cross-validation. A stratified 10-fold
% cross-validation is a popular choice for estimating the test error on
% classification algorithms. It randomly divides the training set into 10
% disjoint subsets. Each subset has roughly equal size and roughly the same
% class proportions as in the training set. Remove one subset, train the
% classification model using the other nine subsets, and use the trained
% model to classify the removed subset. You could repeat this by removing
% each of the ten subsets one at a time.
%
% Because cross-validation randomly divides data, its outcome depends on
% the initial random seed. To reproduce the exact results in this example,
% execute the following command:

rng(0,'twister');

%%
% First use |cvpartition| to generate 10 disjoint stratified subsets.
cp = cvpartition(Targets,'KFold',10)
%%
% First model each variable in each class using a Gaussian distribution.
% You can compute the resubstitution error and the cross-validation error.

nbGau = fitcnb(Inputs, Targets);
nbGauResubErr = resubLoss(nbGau)
nbGauCV = crossval(nbGau, 'CVPartition',cp);
nbGauCVErr = kfoldLoss(nbGauCV)

%%
% You can also compute the confusion matrix on the training set. A
% confusion matrix contains information about known class labels and
% predicted class labels. Generally speaking, the (i,j) element in the
% confusion matrix is the number of samples whose known class label is
% class i and whose predicted class is j. The diagonal elements represent
% correctly classified observations.
nbClass = resubPredict(nbGau);
[nbResubCM,grpOrder] = confusionmat(Targets,nbClass)

% Calculate the Total Classification Accuracy
TotalAccuracy=(nbResubCM(1,1)+nbResubCM(2,2)+nbResubCM(3,3))/(3*Nofsignal)

disp('Total Classification Accuracy=')
disp(TotalAccuracy)
```

```
%%
% So far you have assumed the variables from each class have a multivariate
% normal distribution. Often that is a reasonable assumption, but sometimes
% you may not be willing to make that assumption or you may see clearly
% that it is not valid. Now try to model each variable in each class using
% a kernel density estimation, which is a more flexible nonparametric
% technique. Here we set the kernel to |box|.

nbKD = fitcnb(Inputs, Targets, 'DistributionNames','kernel', 'Kernel','box');
nbKDResubErr = resubLoss(nbKD)
nbKDCV = crossval(nbKD, 'CVPartition',cp);
nbKDCVErr = kfoldLoss(nbKDCV)

%%
% For this data set, the naive Bayes classifier with kernel density
% estimation gets smaller resubstitution error and cross-validation error
% than the naive Bayes classifier with a Gaussian distribution.

%% Conclusions
% This example shows how to perform classification in MATLAB(R) using
% Statistics and Machine Learning Toolbox(TM) functions.
```

5.5 *K*-NEAREST NEIGHBOR

In many cases, the *k*-NN is a simple but effective nonparametric method (Hand, Mannila, & Smyth, 2001). For a data record *t* to be classified, its *k* nearest neighbors form a *neighborhood of t*. Generally, majority voting among the data records in the neighborhood is employed to decide the classification for *t* with or without considering the distance-based weighing. To use *k*-NN, a proper value for *k* is needed to choose, and the achievement of the classifier is mostly dependent on this value. In a sense, the *k*-NN classifier is biased by *k*. There are several ways to select the *k* value, but a simple one is to execute the classifier numerous times with different *k* values and then select the one with the highest accuracy. *k*-NN has a high cost of classifying new instances because nearly all computation takes place at the classification time rather than the first encountered training examples. There are methods (Mitchell, 1997) (Bishop, 2007) that are employed to decrease the necessary computation at query time, such as indexing training examples (Guo, Wang, Bell, Bi, & Greer, 2003).

k-NN is a case-based learning algorithm that keeps all the training data for classification. Because *k*-NN is a lazy learning method, it cannot be employed in many applications with a large repository. To enhance its efficiency, some samples should be found to represent the complete training data for classification, and then build an inductive learning model from the training dataset employing this model for classification. Because *k*-NN is a simple but efficient technique for classification, and it is substantial as one of the most effective methods, it stimulates creation of a *k*-NN model to enhance its competence whilst conserving its classification accuracy as well. If the Euclidean distance is used as a similarity measure, many data points with the same class label are close to each other according to distance measure in many local areas. If these samples are taken as a model to represent the complete training dataset, it will meaningfully decrease the number of data points for classification, and as a result its efficiency will improve. Apparently, if a sample covers a new data point, the class label of this representative will classify it. If not, the distance of the new data point to each sample's nearest boundary must be calculated, and each sample's nearest boundary as a data point should be taken. The new data point in the spirit of *k*-NN can then be classified (Guo et al., 2003).

In model creation, every data point has its largest local neighborhood covering the maximal number of data points with the same class label. Based on these local neighborhoods, the largest local neighborhood might be attained in every cycle. This largest global neighborhood may be realized as a sample to characterize all the data points covered by it. For data points not covered by any samples, the previous operation must be repeated until selected samples have covered all the data points (Guo et al., 2003).

EXAMPLE 5.9. The following MATLAB code was used to extract features from the EEG signals using AR Burg method. Then it classified these data using *k*-NN. You can download data from the following website: http://epileptologie-bonn.de/cms/front_content.php?idcat=193&lang=3&changelang=3

```
%% Ch5_Ex21_EEG_ARBURG_kNN.m
%The following MATLAB code is used to extract the features from
%the EEG signals using AR Burg.
%Then it classifies EEG Signals Using k-NN Classifier
%Cross-Validate k-NN Classifier
clc
clear
%Load Sample EEG Data downloaded from the web site
%http://epileptologie-bonn.de/cms/front_content.php?idcat=193&lang=3&changelang=3
load AS_BONN_ALL_EEG_DATA_4096.mat
%%
Fs = 173.6;          % Sampling frequency
T = 1/Fs;            % Sampling period
segmentLength = 128; % Length of a signal segment
Length=4096;% Length of signal
Nofsignal=100; %Number of Signal
order = 14;

%%
% Obtain the AR Burg Spectrum of the Normal EEG signal using pburg.
for i=1:Nofsignal
    [Pxx,F] = pburg(Normal_Eyes_Open(1:Length,i),order,segmentLength,Fs);
    ASData(i,:)=Pxx(:);
end
%%
% Obtain the AR Burg Spectrum of the Interictal EEG signal using pburg.
for i=Nofsignal+1:2*Nofsignal
    [Pxx,F] = pburg(Interictal(1:Length,i-Nofsignal),order,segmentLength,Fs);
    ASData(i,:)=Pxx(:);
end

%%
% Obtain the AR Burg Spectrum of the Ictal EEG signal using pburg.
for i=2*Nofsignal+1:3*Nofsignal
[Pxx,F] = pburg(Ictal(1:Length,i-2*Nofsignal),order,segmentLength,Fs);
ASData(i,:)=Pxx(:);
end

%Make input Data Matrix
Inputs=ASData;

%%
%You can create target using excel and the import it from the MATLAB HOME
%menu and Import data and then save it as EEGTargets
%Load Target
load EEGTargets

%% Examine Quality of K-NN Classifier
% This example shows how to examine the quality of a k-nearest neighbor
% classifier using resubstitution and cross validation.
%%
% Construct a K-NN classifier

rng(10); % For reproducibility
Mdl = fitcknn(Inputs,Targets,'NumNeighbors',4);
```

```
%%
% Examine the resubstitution loss, which, by default, is the fraction of
% misclassifications from the predictions of |Mdl|.
disp('Resubstitution loss')
rloss = resubLoss(Mdl)
%%
% The classifier predicts incorrectly for 3.67% of the training data.
%%
% Construct a cross-validated classifier from the model.
CVMdl = crossval(Mdl);
%%
% Examine the cross-validation loss, which is the average loss of each
% cross-validation model when predicting on data that is not used for
% training.
disp('Cross-validation loss')
kloss = kfoldLoss(CVMdl)
%%
% The cross-validated classification accuracy resembles the resubstitution
% accuracy. Therefore, you can expect |Mdl| to misclassify approximately 3.67%
% of new data, assuming that the new data has about the same distribution
% as the training data.

%% Display the Confusion Matrix
kNNMdl = fitcknn(Inputs,Targets);
%%
% The observations with known class labels are usually called the training data.
% Now compute the resubstitution error, which is the misclassification error
% (the proportion of misclassified observations) on the training set.

kNNResubErr = resubLoss(kNNMdl)
%%
% You can also compute the confusion matrix on the training set. A
% confusion matrix contains information about known class labels and
% predicted class labels. Generally speaking, the (i,j) element in the
% confusion matrix is the number of samples whose known class label is
% class i and whose predicted class is j. The diagonal elements represent
% correctly classified observations.

kNNClass = resubPredict(kNNMdl);
[kNNResubCM,grpOrder] = confusionmat(Targets,kNNClass)

% Calculate the Total Classification Accuracy
TotalAccuracy=(kNNResubCM(1,1)+kNNResubCM(2,2)+kNNResubCM(3,3))/(3*Nofsignal)

disp('Total Classification Accuracy=')
disp(TotalAccuracy)
```

EXAMPLE 5.10. The following MATLAB code is used to extract features from the ECG signals using modified covariance method. Then it classifies these data using k-NN. You can download data from the following website: https://www.physionet.org/physiobank/database/mitdb/

```
%% Ch5_Ex22_ECG_MCOV_kNN.m
%The following MATLAB code is used to extract the features from
%the ECG signals using Modified Covariance spectrum.
%Then it classifies ECG Signals Using k-NN Classifier
% Cross-Validate k-NN Classifier
```

```
clc
clear
%Load Sample ECG Data downloaded from the web site
%https://www.physionet.org/physiobank/database/mitdb/
load MITBIH_ECG.mat
%%
Fs = 320;             % Sampling frequency
T = 1/Fs;              % Sampling period
segmentLength = 128; % Length of a signal segment
Length=320;% Length of signal
Nofsignal=300; %Number of Signal
order = 34;
%%
% Obtain the Modified Covariance Spectrum of the Normal ECG signal using pmcov.
for i=1:Nofsignal
    [Pxx,F] = pmcov(ECGN(1:Length,i),order,segmentLength,Fs);
    ASData(i,:)=Pxx(:);
end
%%
% Obtain the Modified Covariance Spectrum of the ECG signal with APC using pmcov.
[Pxx,F] = pmcov(ECGAPC(:,1),order,segmentLength,Fs);
for i=Nofsignal+1:2*Nofsignal
    [Pxx,F] = pmcov(ECGAPC(1:Length,i-Nofsignal),order,segmentLength,Fs);
    ASData(i,:)=Pxx(:);
end
%%
% Obtain the Modified Covariance Spectrum of the ECG signal with PVC using pmcov.
for i=2*Nofsignal+1:3*Nofsignal
[Pxx,F] = pmcov(ECGPVC(1:Length,i-2*Nofsignal),order,segmentLength,Fs);
ASData(i,:)=Pxx(:);
end
%%
% Obtain the Modified Covariance Spectrum of the ECG signal with LBBB using pmcov.
for i=3*Nofsignal+1:4*Nofsignal
[Pxx,F] = pmcov(ECGLBBB(1:Length,i-3*Nofsignal),order,segmentLength,Fs);
ASData(i,:)=Pxx(:);
end

%%
% Obtain the Modified Covariance Spectrum of the ECG signal with RBBB using pmcov.
for i=4*Nofsignal+1:5*Nofsignal
[Pxx,F] = pmcov(ECGRBBB(1:Length,i-4*Nofsignal),order,segmentLength,Fs);
ASData(i,:)=Pxx(:);
end

%Make input the Data Matrix
Inputs=ASData;
%%
%You can create target using excel and the import it from the MATLAB HOME
%menu and Import data and then save it as EEGTargets
%Load Target
load ECGTargets

%% Examine Quality of K-NN Classifier
% This example shows how to examine the quality of a k-nearest neighbor
% classifier using resubstitution and cross validation.
%%
% Construct a K-NN classifier
```

```
rng(10); % For reproducibility
Mdl = fitcknn(Inputs,Targets,'NumNeighbors',4);
%%
% Examine the resubstitution loss, which, by default, is the fraction of
% misclassifications from the predictions of |Mdl|.
disp('Resubstitution loss')
rloss = resubLoss(Mdl)

%%
% Construct a cross-validated classifier from the model.
CVMdl = crossval(Mdl);
%%
% Examine the cross-validation loss, which is the average loss of each
% cross-validation model when predicting on data that is not used for
% training.
disp('Cross-validation loss')
kloss = kfoldLoss(CVMdl)
%%
% The cross-validated classification accuracy resembles the resubstitution
% accuracy. Therefore, you can expect |Mdl| to misclassify approximately 3.67%
% of new data, assuming that the new data has about the same distribution
% as the training data.

%% Display the Confusion Matrix
kNNMdl = fitcknn(Inputs,Targets);
%%
% The observations with known class labels are usually called the training data.
% Now compute the resubstitution error, which is the misclassification error
% (the proportion of misclassified observations) on the training set.

kNNResubErr = resubLoss(kNNMdl)
%%
% You can also compute the confusion matrix on the training set. A
% confusion matrix contains information about known class labels and
% predicted class labels. Generally speaking, the (i,j) element in the
% confusion matrix is the number of samples whose known class label is
% class i and whose predicted class is j. The diagonal elements represent
% correctly classified observations.

kNNClass = resubPredict(kNNMdl);
[kNNResubCM,grpOrder] = confusionmat(Targets,kNNClass)

% Calculate the Total Classification Accuracy
TotalAccuracy=(kNNResubCM(1,1)+kNNResubCM(2,2)+kNNResubCM(3,3))/(3*Nofsignal)

disp('Total Classification Accuracy=')
disp(TotalAccuracy)
```

EXAMPLE 5.11. The following MATLAB code is used to extract features from the EEG signals using Wavelet Packet Decomposition (WPD). Then it uses statistical values of WPD subbands. Then it classifies these data using k-NN classifier. You can download data from the following web site: http://epileptologie-bonn.de/cms/front_content.php?idcat=193&lang=3&changelang=3

```
%Ch5_Ex25_EEG_WPD_kNN.m

% WPD of NORMAL, INTERICTAL and ICTAL EEG signals
%The following MATLAB code is used to Extract WPD features from the EEG signals
%Decompose EEG data using WPD coefficients
%Then uses Statistical features as:
```

```
%(1) Mean of the absolute values of the coefficients in each sub-band.
%(2) Standard deviation of the coefficients in each sub-band.
%(3) Skewness of the coefficients in each sub-band.
%(4) Kurtosis of the coefficients in each sub-band.
%(5) RMS power of the wavelet coefficients in each subband.
%(6) Ratio of the mean absolute values of adjacent subbands.

%Then it classifies EEG Signals Using k-NN Classifier
% Cross-Validate k-NN Classifier
clc
clear
%Load Sample EEG Data downloaded from the web site
%http://epileptologie-bonn.de/cms/front_content.php?idcat=193&lang=3&changelang=3
load AS_BONN_ALL_EEG_DATA_4096.mat
wname = 'db4';
Length = 4096;      % Length of signal
Nofsignal=100; %Number of Signal
%%
%%%%%%%%%%%%%%%%%%%%%%%%%%%%%% A Z.ZIP EYES OPEN NORMAL SUBJECT

for i=1:Nofsignal;
[wp10,wp11] = dwt(Normal_Eyes_Open(1:Length,i),wname);%WPD Decomposition
[wp20,wp21] = dwt(wp10,wname);
[wp22,wp23] = dwt(wp11,wname);

[wp30,wp31] = dwt(wp20,wname);
[wp32,wp33] = dwt(wp21,wname);
[wp34,wp35] = dwt(wp22,wname);
[wp36,wp37] = dwt(wp23,wname);

[wp40,wp41] = dwt(wp30,wname);
[wp42,wp43] = dwt(wp31,wname);
[wp44,wp45] = dwt(wp32,wname);
[wp46,wp47] = dwt(wp33,wname);
[wp48,wp49] = dwt(wp34,wname);
[wp4A,wp4B] = dwt(wp35,wname);
[wp4C,wp4D] = dwt(wp36,wname);
[wp4E,wp4F] = dwt(wp37,wname);

ASData(i,1)=mean(abs(wp40));
ASData(i,2)=mean(abs(wp41));
ASData(i,3)=mean(abs(wp42));
ASData(i,4)=mean(abs(wp43));
ASData(i,5)=mean(abs(wp44));
ASData(i,6)=mean(abs(wp45));
ASData(i,7)=mean(abs(wp46));
ASData(i,8)=mean(abs(wp47));
ASData(i,9)=mean(abs(wp48));
ASData(i,10)=mean(abs(wp49));
ASData(i,11)=mean(abs(wp4A));
ASData(i,12)=mean(abs(wp4B));
ASData(i,13)=mean(abs(wp4C));
ASData(i,14)=mean(abs(wp4D));
ASData(i,15)=mean(abs(wp4E));
ASData(i,16)=mean(abs(wp4F));

ASData(i,17)=std(wp40);
ASData(i,18)=std(wp41);
ASData(i,19)=std(wp42);
```

```
ASData(i,20)=std(wp43);
ASData(i,21)=std(wp44);
ASData(i,22)=std(wp45);
ASData(i,23)=std(wp46);
ASData(i,24)=std(wp47);
ASData(i,25)=std(wp48);
ASData(i,26)=std(wp49);
ASData(i,27)=std(wp4A);
ASData(i,28)=std(wp4B);
ASData(i,29)=std(wp4C);
ASData(i,30)=std(wp4D);
ASData(i,31)=std(wp4E);
ASData(i,32)=std(wp4F);

ASData(i,33)=skewness(wp40(:));
ASData(i,34)=skewness(wp41(:));
ASData(i,35)=skewness(wp42(:));
ASData(i,36)=skewness(wp43(:));
ASData(i,37)=skewness(wp44(:));
ASData(i,38)=skewness(wp45(:));
ASData(i,39)=skewness(wp46(:));
ASData(i,40)=skewness(wp47(:));
ASData(i,41)=skewness(wp48(:));
ASData(i,42)=skewness(wp49(:));
ASData(i,43)=skewness(wp4A(:));
ASData(i,44)=skewness(wp4B(:));
ASData(i,45)=skewness(wp4C(:));
ASData(i,46)=skewness(wp4D(:));
ASData(i,47)=skewness(wp4E(:));
ASData(i,48)=skewness(wp4F(:));

ASData(i,49)=kurtosis(wp40);
ASData(i,50)=kurtosis(wp41);
ASData(i,51)=kurtosis(wp42);
ASData(i,52)=kurtosis(wp43);
ASData(i,53)=kurtosis(wp44);
ASData(i,54)=kurtosis(wp45);
ASData(i,55)=kurtosis(wp46);
ASData(i,56)=kurtosis(wp47);
ASData(i,57)=kurtosis(wp48);
ASData(i,58)=kurtosis(wp49);
ASData(i,59)=kurtosis(wp4A);
ASData(i,60)=kurtosis(wp4B);
ASData(i,61)=kurtosis(wp4C);
ASData(i,62)=kurtosis(wp4D);
ASData(i,63)=kurtosis(wp4E);
ASData(i,64)=kurtosis(wp4F);

ASData(i,65)=rms(wp40(:));
ASData(i,66)=rms(wp41(:));
ASData(i,67)=rms(wp42(:));
ASData(i,68)=rms(wp43(:));
ASData(i,69)=rms(wp44(:));
ASData(i,70)=rms(wp45(:));
ASData(i,71)=rms(wp46(:));
ASData(i,72)=rms(wp47(:));
ASData(i,73)=rms(wp48(:));
ASData(i,74)=rms(wp49(:));
```

```
ASData(i,75)=rms(wp4A(:));
ASData(i,76)=rms(wp4B(:));
ASData(i,77)=rms(wp4C(:));
ASData(i,78)=rms(wp4D(:));
ASData(i,79)=rms(wp4E(:));
ASData(i,80)=rms(wp4F(:));

ASData(i,81)=mean(abs(wp40))/mean(abs(wp41));
ASData(i,82)=mean(abs(wp41))/mean(abs(wp42));
ASData(i,83)=mean(abs(wp42))/mean(abs(wp43));
ASData(i,84)=mean(abs(wp43))/mean(abs(wp44));
ASData(i,85)=mean(abs(wp44))/mean(abs(wp45));
ASData(i,86)=mean(abs(wp45))/mean(abs(wp46));
ASData(i,87)=mean(abs(wp46))/mean(abs(wp47));
ASData(i,88)=mean(abs(wp47))/mean(abs(wp48));
ASData(i,89)=mean(abs(wp48))/mean(abs(wp49));
ASData(i,90)=mean(abs(wp49))/mean(abs(wp4A));
ASData(i,91)=mean(abs(wp4A))/mean(abs(wp4B));
ASData(i,92)=mean(abs(wp4B))/mean(abs(wp4C));
ASData(i,93)=mean(abs(wp4C))/mean(abs(wp4D));
ASData(i,94)=mean(abs(wp4D))/mean(abs(wp4E));
ASData(i,95)=mean(abs(wp4E))/mean(abs(wp4F));

end
%%
% D F.ZIP EPILEPTIC SUBJECT (INTERICTAL) were recorded from within
% the epileptogenic zone during seizure free intervals

for i=Nofsignal+1:2*Nofsignal;

[wp10,wp11] = dwt(Interictal(1:Length,i-Nofsignal),wname);%WPD Decomposition
[wp20,wp21] = dwt(wp10,wname);
[wp22,wp23] = dwt(wp11,wname);

[wp30,wp31] = dwt(wp20,wname);
[wp32,wp33] = dwt(wp21,wname);
[wp34,wp35] = dwt(wp22,wname);
[wp36,wp37] = dwt(wp23,wname);

[wp40,wp41] = dwt(wp30,wname);
[wp42,wp43] = dwt(wp31,wname);
[wp44,wp45] = dwt(wp32,wname);
[wp46,wp47] = dwt(wp33,wname);
[wp48,wp49] = dwt(wp34,wname);
[wp4A,wp4B] = dwt(wp35,wname);
[wp4C,wp4D] = dwt(wp36,wname);
[wp4E,wp4F] = dwt(wp37,wname);

ASData(i,1)=mean(abs(wp40));
ASData(i,2)=mean(abs(wp41));
ASData(i,3)=mean(abs(wp42));
ASData(i,4)=mean(abs(wp43));
ASData(i,5)=mean(abs(wp44));
ASData(i,6)=mean(abs(wp45));
ASData(i,7)=mean(abs(wp46));
ASData(i,8)=mean(abs(wp47));
ASData(i,9)=mean(abs(wp48));
ASData(i,10)=mean(abs(wp49));
```

```
ASData(i,11)=mean(abs(wp4A));
ASData(i,12)=mean(abs(wp4B));
ASData(i,13)=mean(abs(wp4C));
ASData(i,14)=mean(abs(wp4D));
ASData(i,15)=mean(abs(wp4E));
ASData(i,16)=mean(abs(wp4F));

ASData(i,17)=std(wp40);
ASData(i,18)=std(wp41);
ASData(i,19)=std(wp42);
ASData(i,20)=std(wp43);
ASData(i,21)=std(wp44);
ASData(i,22)=std(wp45);
ASData(i,23)=std(wp46);
ASData(i,24)=std(wp47);
ASData(i,25)=std(wp48);
ASData(i,26)=std(wp49);
ASData(i,27)=std(wp4A);
ASData(i,28)=std(wp4B);
ASData(i,29)=std(wp4C);
ASData(i,30)=std(wp4D);
ASData(i,31)=std(wp4E);
ASData(i,32)=std(wp4F);

ASData(i,33)=skewness(wp40(:));
ASData(i,34)=skewness(wp41(:));
ASData(i,35)=skewness(wp42(:));
ASData(i,36)=skewness(wp43(:));
ASData(i,37)=skewness(wp44(:));
ASData(i,38)=skewness(wp45(:));
ASData(i,39)=skewness(wp46(:));
ASData(i,40)=skewness(wp47(:));
ASData(i,41)=skewness(wp48(:));
ASData(i,42)=skewness(wp49(:));
ASData(i,43)=skewness(wp4A(:));
ASData(i,44)=skewness(wp4B(:));
ASData(i,45)=skewness(wp4C(:));
ASData(i,46)=skewness(wp4D(:));
ASData(i,47)=skewness(wp4E(:));
ASData(i,48)=skewness(wp4F(:));

ASData(i,49)=kurtosis(wp40);
ASData(i,50)=kurtosis(wp41);
ASData(i,51)=kurtosis(wp42);
ASData(i,52)=kurtosis(wp43);
ASData(i,53)=kurtosis(wp44);
ASData(i,54)=kurtosis(wp45);
ASData(i,55)=kurtosis(wp46);
ASData(i,56)=kurtosis(wp47);
ASData(i,57)=kurtosis(wp48);
ASData(i,58)=kurtosis(wp49);
ASData(i,59)=kurtosis(wp4A);
ASData(i,60)=kurtosis(wp4B);
ASData(i,61)=kurtosis(wp4C);
ASData(i,62)=kurtosis(wp4D);
ASData(i,63)=kurtosis(wp4E);
ASData(i,64)=kurtosis(wp4F);
```

```
ASData(i,65)=rms(wp40(:));
ASData(i,66)=rms(wp41(:));
ASData(i,67)=rms(wp42(:));
ASData(i,68)=rms(wp43(:));
ASData(i,69)=rms(wp44(:));
ASData(i,70)=rms(wp45(:));
ASData(i,71)=rms(wp46(:));
ASData(i,72)=rms(wp47(:));
ASData(i,73)=rms(wp48(:));
ASData(i,74)=rms(wp49(:));
ASData(i,75)=rms(wp4A(:));
ASData(i,76)=rms(wp4B(:));
ASData(i,77)=rms(wp4C(:));
ASData(i,78)=rms(wp4D(:));
ASData(i,79)=rms(wp4E(:));
ASData(i,80)=rms(wp4F(:));

ASData(i,81)=mean(abs(wp40))/mean(abs(wp41));
ASData(i,82)=mean(abs(wp41))/mean(abs(wp42));
ASData(i,83)=mean(abs(wp42))/mean(abs(wp43));
ASData(i,84)=mean(abs(wp43))/mean(abs(wp44));
ASData(i,85)=mean(abs(wp44))/mean(abs(wp45));
ASData(i,86)=mean(abs(wp45))/mean(abs(wp46));
ASData(i,87)=mean(abs(wp46))/mean(abs(wp47));
ASData(i,88)=mean(abs(wp47))/mean(abs(wp48));
ASData(i,89)=mean(abs(wp48))/mean(abs(wp49));
ASData(i,90)=mean(abs(wp49))/mean(abs(wp4A));
ASData(i,91)=mean(abs(wp4A))/mean(abs(wp4B));
ASData(i,92)=mean(abs(wp4B))/mean(abs(wp4C));
ASData(i,93)=mean(abs(wp4C))/mean(abs(wp4D));
ASData(i,94)=mean(abs(wp4D))/mean(abs(wp4E));
ASData(i,95)=mean(abs(wp4E))/mean(abs(wp4F));
end

%% E S.ZIP EPILEPTIC SUBJECT ICTAL DURING SEIZURE
%%%%%%%%%%%%%%%%%%%%%%%%%%%

for i=2*Nofsignal+1:3*Nofsignal;

[wp10,wp11] = dwt(Ictal(1:Length,i-2*Nofsignal),wname);%WPD Decomposition
[wp20,wp21] = dwt(wp10,wname);
[wp22,wp23] = dwt(wp11,wname);

[wp30,wp31] = dwt(wp20,wname);
[wp32,wp33] = dwt(wp21,wname);
[wp34,wp35] = dwt(wp22,wname);
[wp36,wp37] = dwt(wp23,wname);

[wp40,wp41] = dwt(wp30,wname);
[wp42,wp43] = dwt(wp31,wname);
[wp44,wp45] = dwt(wp32,wname);
[wp46,wp47] = dwt(wp33,wname);
[wp48,wp49] = dwt(wp34,wname);
[wp4A,wp4B] = dwt(wp35,wname);
[wp4C,wp4D] = dwt(wp36,wname);
[wp4E,wp4F] = dwt(wp37,wname);

ASData(i,1)=mean(abs(wp40));
ASData(i,2)=mean(abs(wp41));
```

```
ASData(i,3)=mean(abs(wp42));
ASData(i,4)=mean(abs(wp43));
ASData(i,5)=mean(abs(wp44));
ASData(i,6)=mean(abs(wp45));
ASData(i,7)=mean(abs(wp46));
ASData(i,8)=mean(abs(wp47));
ASData(i,9)=mean(abs(wp48));
ASData(i,10)=mean(abs(wp49));
ASData(i,11)=mean(abs(wp4A));
ASData(i,12)=mean(abs(wp4B));
ASData(i,13)=mean(abs(wp4C));
ASData(i,14)=mean(abs(wp4D));
ASData(i,15)=mean(abs(wp4E));
ASData(i,16)=mean(abs(wp4F));

ASData(i,17)=std(wp40);
ASData(i,18)=std(wp41);
ASData(i,19)=std(wp42);
ASData(i,20)=std(wp43);
ASData(i,21)=std(wp44);
ASData(i,22)=std(wp45);
ASData(i,23)=std(wp46);
ASData(i,24)=std(wp47);
ASData(i,25)=std(wp48);
ASData(i,26)=std(wp49);
ASData(i,27)=std(wp4A);
ASData(i,28)=std(wp4B);
ASData(i,29)=std(wp4C);
ASData(i,30)=std(wp4D);
ASData(i,31)=std(wp4E);
ASData(i,32)=std(wp4F);

ASData(i,33)=skewness(wp40(:));
ASData(i,34)=skewness(wp41(:));
ASData(i,35)=skewness(wp42(:));
ASData(i,36)=skewness(wp43(:));
ASData(i,37)=skewness(wp44(:));
ASData(i,38)=skewness(wp45(:));
ASData(i,39)=skewness(wp46(:));
ASData(i,40)=skewness(wp47(:));
ASData(i,41)=skewness(wp48(:));
ASData(i,42)=skewness(wp49(:));
ASData(i,43)=skewness(wp4A(:));
ASData(i,44)=skewness(wp4B(:));
ASData(i,45)=skewness(wp4C(:));
ASData(i,46)=skewness(wp4D(:));
ASData(i,47)=skewness(wp4E(:));
ASData(i,48)=skewness(wp4F(:));

ASData(i,49)=kurtosis(wp40);
ASData(i,50)=kurtosis(wp41);
ASData(i,51)=kurtosis(wp42);
ASData(i,52)=kurtosis(wp43);
ASData(i,53)=kurtosis(wp44);
ASData(i,54)=kurtosis(wp45);
ASData(i,55)=kurtosis(wp46);
ASData(i,56)=kurtosis(wp47);
ASData(i,57)=kurtosis(wp48);
```

```
ASData(i,58)=kurtosis(wp49);
ASData(i,59)=kurtosis(wp4A);
ASData(i,60)=kurtosis(wp4B);
ASData(i,61)=kurtosis(wp4C);
ASData(i,62)=kurtosis(wp4D);
ASData(i,63)=kurtosis(wp4E);
ASData(i,64)=kurtosis(wp4F);

ASData(i,65)=rms(wp40(:));
ASData(i,66)=rms(wp41(:));
ASData(i,67)=rms(wp42(:));
ASData(i,68)=rms(wp43(:));
ASData(i,69)=rms(wp44(:));
ASData(i,70)=rms(wp45(:));
ASData(i,71)=rms(wp46(:));
ASData(i,72)=rms(wp47(:));
ASData(i,73)=rms(wp48(:));
ASData(i,74)=rms(wp49(:));
ASData(i,75)=rms(wp4A(:));
ASData(i,76)=rms(wp4B(:));
ASData(i,77)=rms(wp4C(:));
ASData(i,78)=rms(wp4D(:));
ASData(i,79)=rms(wp4E(:));
ASData(i,80)=rms(wp4F(:));

ASData(i,81)=mean(abs(wp40))/mean(abs(wp41));
ASData(i,82)=mean(abs(wp41))/mean(abs(wp42));
ASData(i,83)=mean(abs(wp42))/mean(abs(wp43));
ASData(i,84)=mean(abs(wp43))/mean(abs(wp44));
ASData(i,85)=mean(abs(wp44))/mean(abs(wp45));
ASData(i,86)=mean(abs(wp45))/mean(abs(wp46));
ASData(i,87)=mean(abs(wp46))/mean(abs(wp47));
ASData(i,88)=mean(abs(wp47))/mean(abs(wp48));
ASData(i,89)=mean(abs(wp48))/mean(abs(wp49));
ASData(i,90)=mean(abs(wp49))/mean(abs(wp4A));
ASData(i,91)=mean(abs(wp4A))/mean(abs(wp4B));
ASData(i,92)=mean(abs(wp4B))/mean(abs(wp4C));
ASData(i,93)=mean(abs(wp4C))/mean(abs(wp4D));
ASData(i,94)=mean(abs(wp4D))/mean(abs(wp4E));
ASData(i,95)=mean(abs(wp4E))/mean(abs(wp4F));
end

%Make input the Data Matrix
Inputs=ASData;

%%
%You can create target using excel and the import it from the MATLAB HOME
%menu and Import data and then save it as EEGTargets
%Load Target
load EEGTargets

%% Examine Quality of KNN Classifier
% This example shows how to examine the quality of a _k_-nearest neighbor
% classifier using resubstitution and cross validation.
%%
% Construct a KNN classifier

rng(10); % For reproducibility
Mdl = fitcknn(Inputs,Targets,'NumNeighbors',4);
```

```
%%
% Examine the resubstitution loss, which, by default, is the fraction of
% misclassifications from the predictions of |Mdl|.
disp('Resubstitution loss')
rloss = resubLoss(Mdl)
%%
% The classifier predicts incorrectly for 0.036% of the training data.
%%
% Construct a cross-validated classifier from the model.
CVMdl = crossval(Mdl);
%%
% Examine the cross-validation loss, which is the average loss of each
% cross-validation model when predicting on data that is not used for
% training.
disp('Cross-validation loss')
kloss = kfoldLoss(CVMdl)
%%
% The cross-validated classification accuracy resembles the resubstitution
% accuracy. Therefore, you can expect |Mdl| to misclassify approximately 4%
% of new data, assuming that the new data has about the same distribution
% as the training data.

%% Display the Confusion Matrix
kNNMdl = fitcknn(Inputs,Targets);
%%
% The observations with known class labels are usually called the training data.
% Now compute the resubstitution error, which is the misclassification error
% (the proportion of misclassified observations) on the training set.

kNNResubErr = resubLoss(kNNMdl)
%%
% You can also compute the confusion matrix on the training set. A
% confusion matrix contains information about known class labels and
% predicted class labels. Generally speaking, the (i,j) element in the
% confusion matrix is the number of samples whose known class label is
% class i and whose predicted class is j. The diagonal elements represent
% correctly classified observations.

kNNClass = resubPredict(kNNMdl);
[kNNResubCM,grpOrder] = confusionmat(Targets,kNNClass)

% Calculate the Total Classification Accuracy
TotalAccuracy=(kNNResubCM(1,1)+kNNResubCM(2,2)+kNNResubCM(3,3))/(3*Nofsignal)

disp('Total Classification Accuracy=')
disp(TotalAccuracy)
```

5.6 ARTIFICIAL NEURAL NETWORKS

The brain serves as an inspiration for the ANN models. It represents a device that processes information and has some extraordinary abilities that outreach current biomedical applications. If it was possible to understand the way in which brain operates and performs these functions, algorithmic solutions to these tasks could be defined and eventually implemented on computers. There is quite a difference between a computer and the human brain. Although a computer has only one processor, the human brain consists of an enormous number of processing units, specifically the neurons. It is believed that those processing units are much simpler and slower than the processor of the computer. However, neurons in the brain have connections, that is, synapses, with other neurons, and all of them operate in parallel. This enormous connectivity of the brain provides its computational power. In computers, memory represents a separate part and is passive, whereas the processor is active. In the human brain, both processing units and memory are distributed together; neurons perform processing, and memory is found in the synapses between neurons (Alpaydin, 2014).

A perceptron with a single layer of weights can only be used to approximate linear functions of the input and cannot give a solution to problems where the discriminant to be estimated is nonlinear. On the other hand, multilayer perceptron (MLP) can implement nonlinear discriminants, if used for classification. The output of the MLP is the linear combination of the nonlinear basis function values given by the hidden units. Hidden units perform a nonlinear transformation from the d-dimensional input space to the H-dimensional space spanned by the hidden units, and, in this space, the second output layer implements a linear function. Because MLP is not limited to one hidden layer, more hidden layers with their own weights can be placed after the first layer with sigmoid hidden units to calculate nonlinear functions of the first layer of hidden units, implementing more complex functions of the inputs. People rarely use more than one hidden layer in practice, because the analysis of a network including many hidden layers is complicated. Training of the MLP is identical to the training of the perceptron. In contrary to the perceptron, the output of the MLP is a nonlinear function of the input because of the nonlinear basis function in the hidden units. Considering the hidden units as inputs, the second layer is a perceptron and the way of updating parameters. In other words, the error propagates from the output back to the inputs, and hence this algorithm is called backpropagation algorithm (Rumelhart, Hinton, & Williams, 1986).(Alpaydin, 2014)

EXAMPLE 5.12. The following MATLAB code was used to extract features from the EEG signals using AR Burg method. Then it classified these data using ANN. You can download data from the following website: http://epileptologie-bonn.de/cms/front_content.php?idcat=193&lang=3&changelang=3

```
%% Ch5_Ex31_EEG_ARBURG_ANN.m
%The following MATLAB code is used to extract the features from
%the EEG signals using AR Burg.
%Then it classifies EEG Signals Using ANN Classifier
clc
clear
%Load Sample EEG Data downloaded from the web site
%http://epileptologie-bonn.de/cms/front_content.php?idcat=193&lang=3&changelang=3
load AS_BONN_ALL_EEG_DATA_4096.mat
%%
Fs = 173.6;             % Sampling frequency
T = 1/Fs;               % Sampling period
segmentLength = 128; % Length of a signal segment
Length=4096;            % Length of signal
Nofsignal=100;          % Number of Signal
order = 14;

%%
% Obtain the AR Burg Spectrum of the Normal EEG signal using pburg.
for i=1:Nofsignal
    [Pxx,F] = pburg(Normal_Eyes_Open(1:Length,i),order,segmentLength,Fs);
    ASData(i,:)=Pxx(:);
end
%%
% Obtain the AR Burg Spectrum of the Interictal EEG signal using pburg.
for i=Nofsignal+1:2*Nofsignal
    [Pxx,F] = pburg(Interictal(1:Length,i-Nofsignal),order,segmentLength,Fs);
    ASData(i,:)=Pxx(:);
end
%%
% Obtain the AR Burg Spectrum of the Ictal EEG signal using pburg.
for i=2*Nofsignal+1:3*Nofsignal
[Pxx,F] = pburg(Ictal(1:Length,i-2*Nofsignal),order,segmentLength,Fs);
ASData(i,:)=Pxx(:);
end
```

```
%Transpose the Data Matrix
Inputs=ASData';
%% TARGET GENERATION
% Your target set must have one label for each sample,
%so it must contains 300 elements. Here is an easy way to build one.
%Start by numbering your classes from 1 to 3.
%Make sure that your X Array is sorted by class,
%meaning all the samples from the first class,
%then all the samples from the second class, and so on. Then:

 % Classes from 1 to 3
 y = 1:3;
 % 100 samples per class
 y = repmat(y, 100, 1);
 % Reshape to obtain a vector
 y = reshape(y, 1, numel(y));
% At this stage, you should have a 1-by-300 vector with numeric class labels from 1 to 3.
%This is what you want if you use the Statistics and Machine Learning Toolbox.
%
% Now, I suppose you intend to use the Neural Network Toolbox.
%In that case the syntax is a bit different: the class labels are in a 3-by-300 matrix,
%each column being a sample, with the row corresponding to the class getting value '1' and the
rest '0'.
%To build it from the previous syntax, type:

Targets = full(ind2vec(y));

%% CLASSIFACATION
% Solve a Pattern Recognition Problem with a Neural Network
% Script generated by NPRTOOL
%
% Create a Pattern Recognition Network
hiddenLayerSize = 33;
net = patternnet(hiddenLayerSize);
% Set up Division of Data for Training, Validation, Testing
net.divideParam.trainRatio = 50/100;
net.divideParam.valRatio = 25/100;
net.divideParam.testRatio = 25/100;
% Train the Network
[net,tr] = train(net,Inputs,Targets);
% Test the Network
outputs = net(Inputs);
errors = gsubtract(Targets,outputs);
performance = perform(net,Targets,outputs)
% View the Network
view(net)
% Plots
figure, plotperform(tr)
figure, plottrainstate(tr)
figure, ploterrhist(errors)
%%
% One measure of how well the neural network has fit the data is the
% confusion plot. Here the confusion matrix is plotted across all samples.
%
% The confusion matrix shows the percentages of correct and incorrect
% classifications. Correct classifications are the green squares on the
% matrices diagonal. Incorrect classifications form the red squares.
%
% If the network has learned to classify properly, the percentages in the
% red squares should be very small, indicating few misclassifications.
%
```

```
% If this is not the case then further training, or training a network
% with more hidden neurons, would be advisable.

figure, plotconfusion(Targets,outputs)
%%
% Another measure of how well the neural network has fit data is the
% receiver operating characteristic plot. This shows how the false
% positive and true positive rates relate as the thresholding of outputs
% is varied from 0 to 1.
%
% The farther left and up the line is, the fewer false positives need to
% be accepted in order to get a high true positive rate. The best
% classifiers will have a line going from the bottom left corner, to the
% top left corner, to the top right corner, or close to that.

figure, plotroc(Targets,outputs)
```

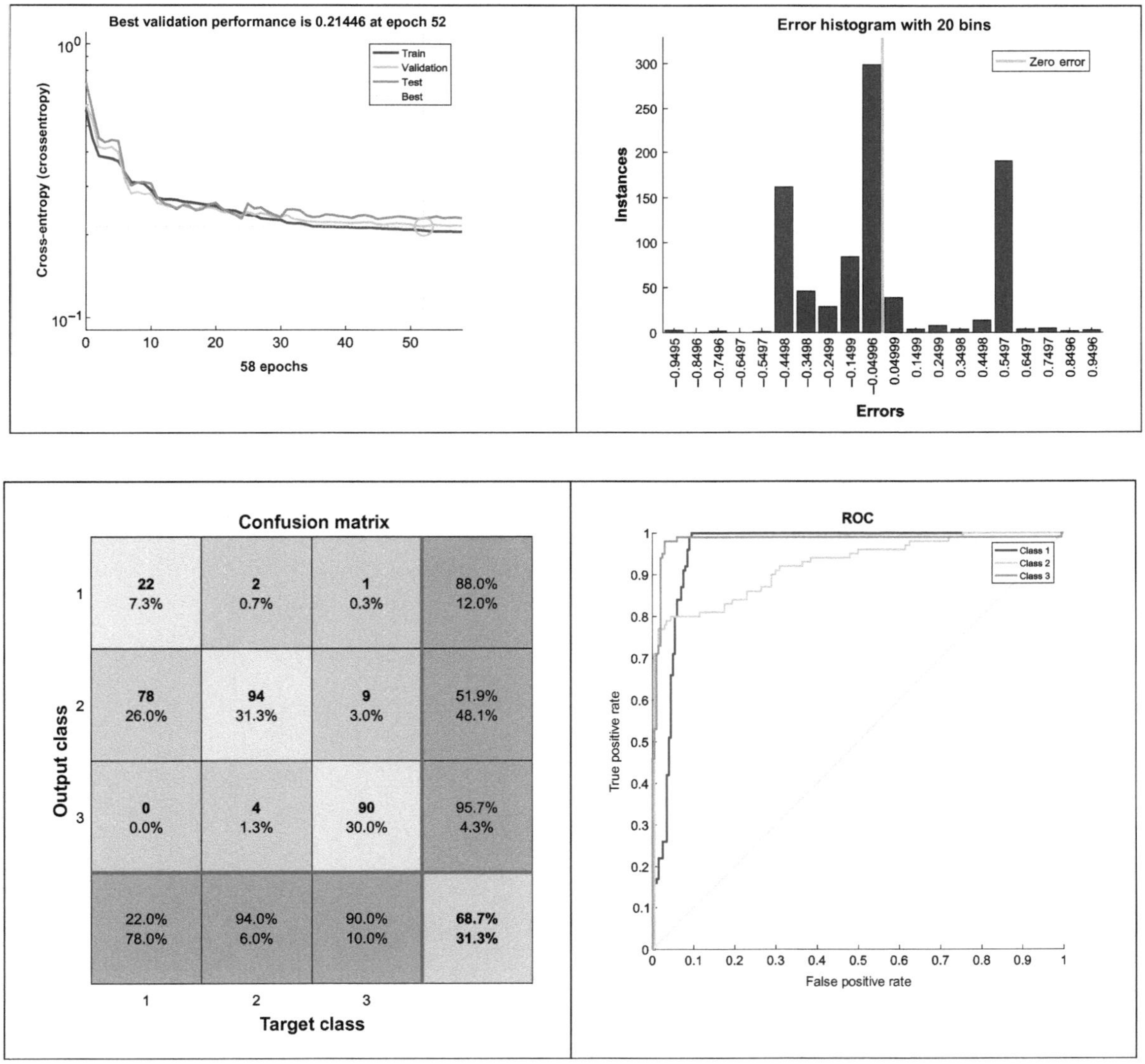

FIG. 5.3 Representation of performance of the ANN classifier for EEG signals features extracted using AR Burg method.

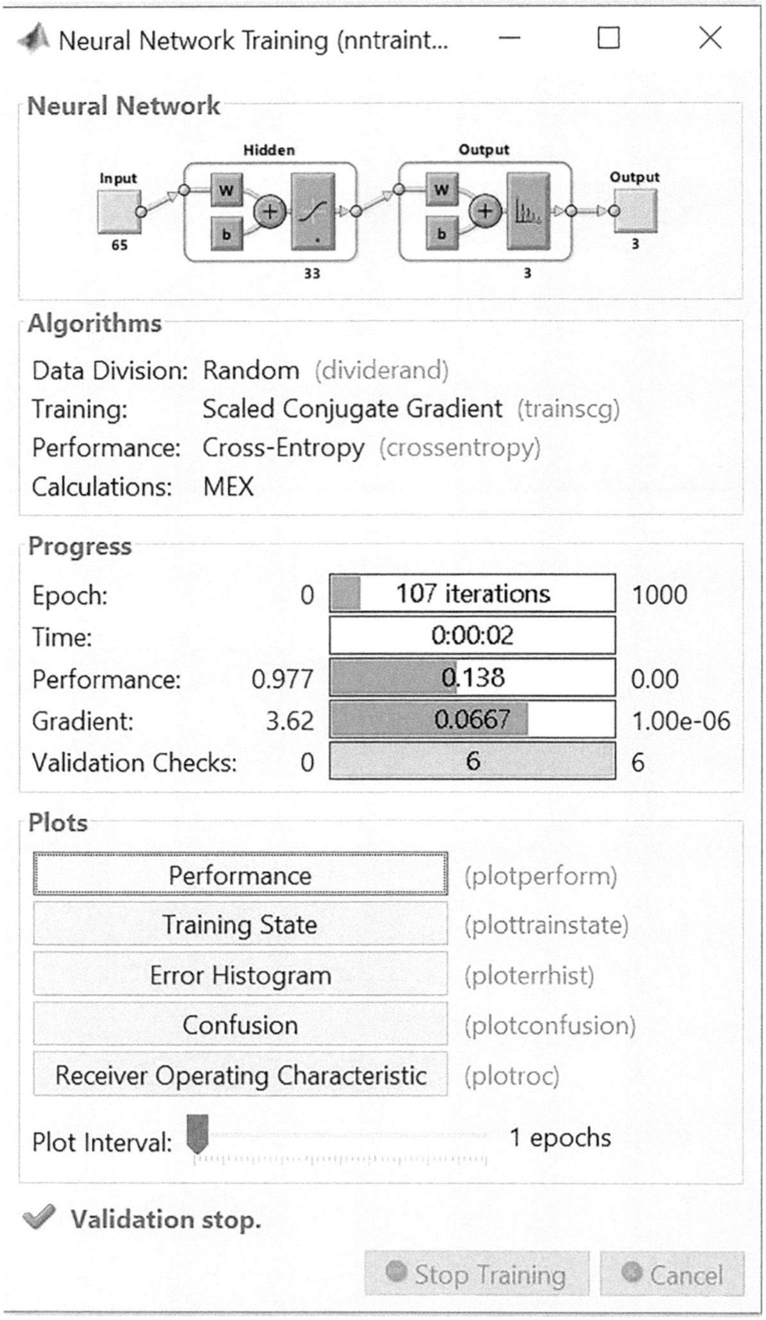

FIG. 5.4 The Neural Network Training tool of the ANN classifier for EEG signals features extracted using AR Burg method.

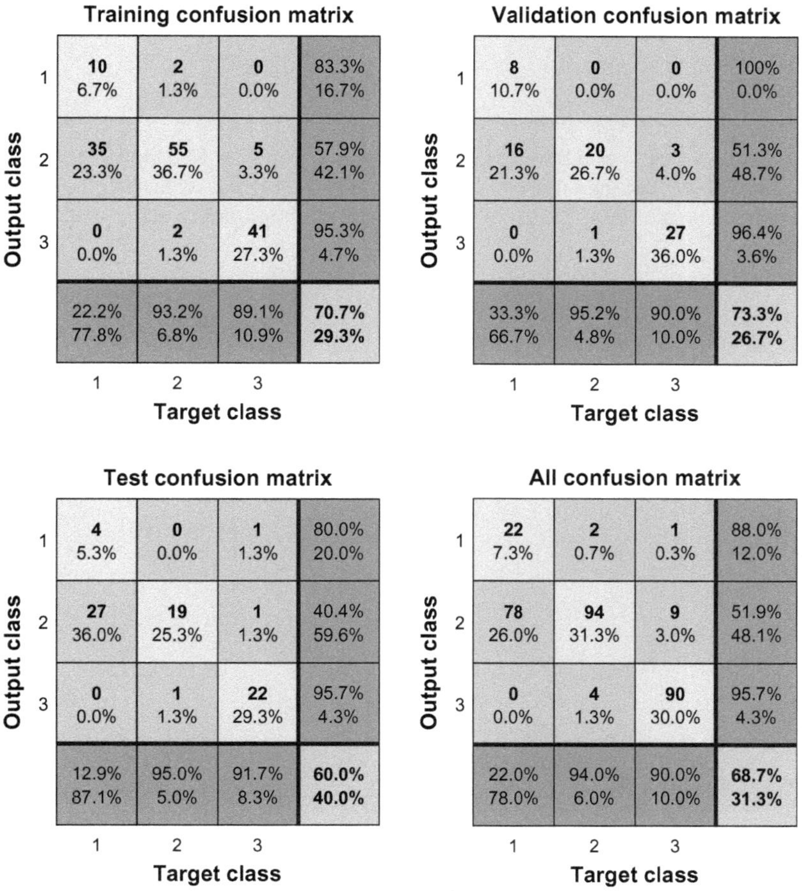

FIG. 5.5 Confusion matrices of train, validation, test, and all sets of the ANN classifier for EEG signals features extracted using AR Burg method.

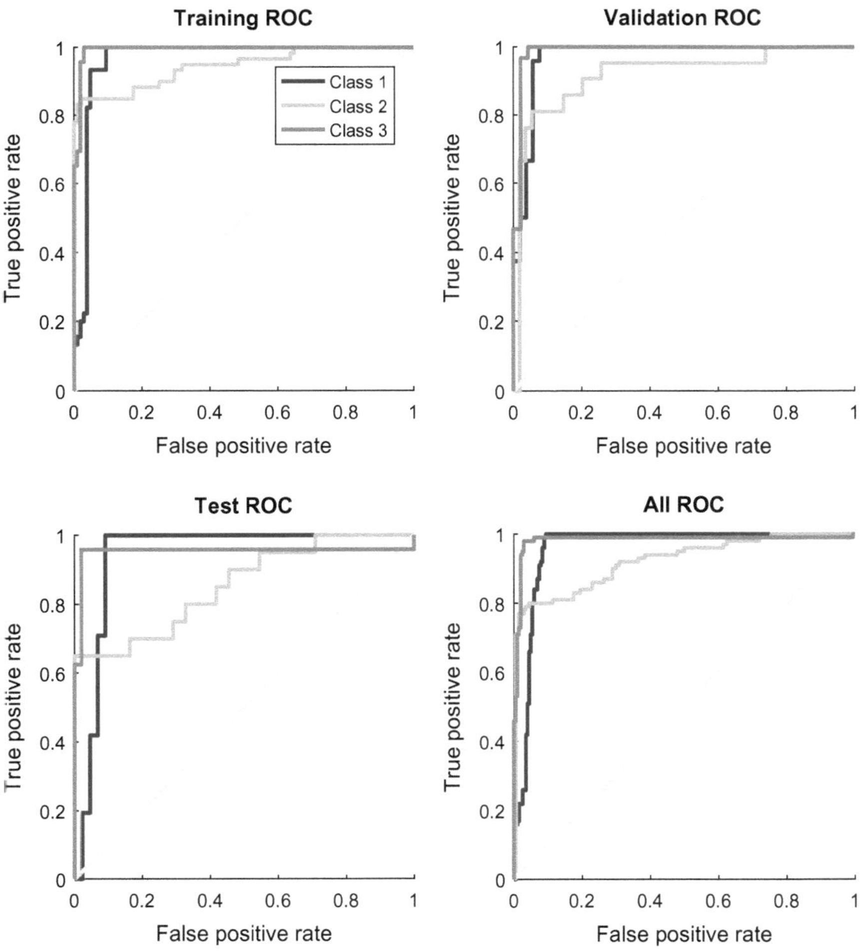

FIG. 5.6 ROC area of train, validation, test, and all sets of the ANN Classifier for EEG signals features extracted using AR Burg method.

EXAMPLE 5.13. The following MATLAB code was used to extract features from the focal and nonfocal EEG signals using AR Burg method. Then it classified these data using ANN. You can download data from the following website: http://ntsa.upf.edu/downloads/andrzejak-rg-schindler-k-rummel-c-2012-nonrandomness-nonlinear-dependence-and

```
%% Ch5_Ex32_EEG_Focal_ARBURG_ANN.m
%Classify Focal and Non-Focal EEG Signals Using ANN
%The following MATLAB code is used to extract the features from
%the Focal and Non-Focal EEG signals using AR Burg spectrum.
clc
clear
%Load Sample EEG Data downloaded from the web site
%http://ntsa.upf.edu/downloads/andrzejak-rg-schindler-k-rummel-c-2012-nonrandomness-nonlinear-
dependence-and
load FOCAL_NFOCAL.mat
%%
Fs = 512;           % Sampling frequency
T = 1/Fs;            % Sampling period
segmentLength = 128; % Length of a signal segment
Length=4096;% Length of signal
Nofsignal=1000; %Number of Signal
order = 14;
%%
% Obtain the AR Burg Spectrum of the FOCAL EEG signal using pburg.
for i=1:Nofsignal

    [Pxx,F] = pburg(focal(1:Length,i),order,segmentLength,Fs);
    ASData(i,:)=Pxx(:);
end
%%
% Obtain the AR Burg Spectrum of the NON-FOCAL EEG signal using pburg.

for i=Nofsignal+1:2*Nofsignal
    [Pxx,F] = pburg(nfocal(1:Length,i-Nofsignal),order,segmentLength,Fs);
    ASData(i,:)=Pxx(:);
end

%Transpose the Data Matrix
Inputs=ASData';
%% TARGET GENERATION
% Your target set must have one label for each sample,
%so it must contains 2000 elements. Here is an easy way to build one.
%Start by numbering your classes from 1 to 2.
%Make sure that your X Array is sorted by class,
%meaning all the samples from the first class,
%then all the samples from the second class, and so on. Then:

 % Classes from 1 to 2
 y = 1:2;
 % 1000 samples per class
 y = repmat(y, Nofsignal, 1);
 % Reshape to obtain a vector
 y = reshape(y, 1, numel(y));
% At this stage, you should have a 1-by-2000 vector with numeric class labels from 1 to 3.
%This is what you want if you use the Statistics and Machine Learning Toolbox.
%
% Now, I suppose you intend to use the Neural Network Toolbox.
%In that case the syntax is a bit different: the class labels are in a 3-by-300 matrix,
```

```
%each column being a sample, with the row corresponding to the class getting value '1' and the
rest '0'.
%To build it from the previous syntax, type:

Targets = full(ind2vec(y));

%% CLASSIFICATION
% Solve a Pattern Recognition Problem with a Neural Network
% Script generated by NPRTOOL
%
% Create a Pattern Recognition Network
hiddenLayerSize = 15;
net = patternnet(hiddenLayerSize);
% Set up Division of Data for Training, Validation, Testing
net.divideParam.trainRatio = 70/100;
net.divideParam.valRatio = 15/100;
net.divideParam.testRatio = 15/100;
% Train the Network
[net,tr] = train(net,Inputs,Targets);
% Test the Network
outputs = net(Inputs);
errors = gsubtract(Targets,outputs);
performance = perform(net,Targets,outputs);
% View the Network
view(net)
% Plots
figure, plotperform(tr)
figure, plottrainstate(tr)
figure, ploterrhist(errors)

%%
% One measure of how well the neural network has fit the data is the
% confusion plot. Here the confusion matrix is plotted across all samples.
%
% The confusion matrix shows the percentages of correct and incorrect
% classifications. Correct classifications are the green squares on the
% matrices diagonal. Incorrect classifications form the red squares.
%
% If the network has learned to classify properly, the percentages in the
% red squares should be very small, indicating few misclassifications.
%
% If this is not the case then further training, or training a network
% with more hidden neurons, would be advisable.
figure, plotconfusion(Targets,outputs)
%%
% Another measure of how well the neural network has fit data is the
% receiver operating characteristic plot. This shows how the false
% positive and true positive rates relate as the thresholding of outputs
% is varied from 0 to 1.
%
% The farther left and up the line is, the fewer false positives need to
% be accepted in order to get a high true positive rate. The best
% classifiers will have a line going from the bottom left corner, to the
% top left corner, to the top right corner, or close to that.

figure, plotroc(Targets,outputs)
```

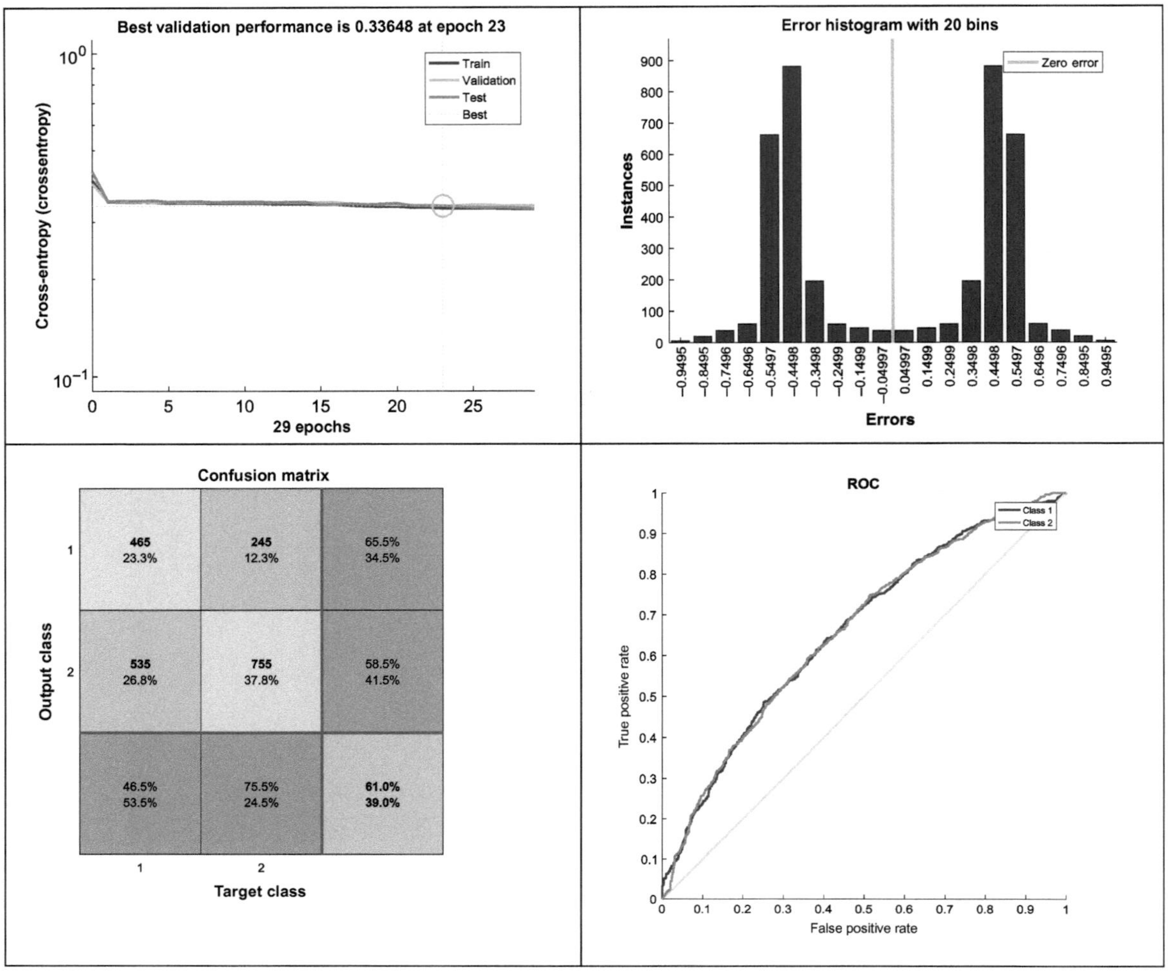

FIG. 5.7 Representation of performance of the ANN classifier for the focal and nonfocal EEG signals features extracted using AR Burg method.

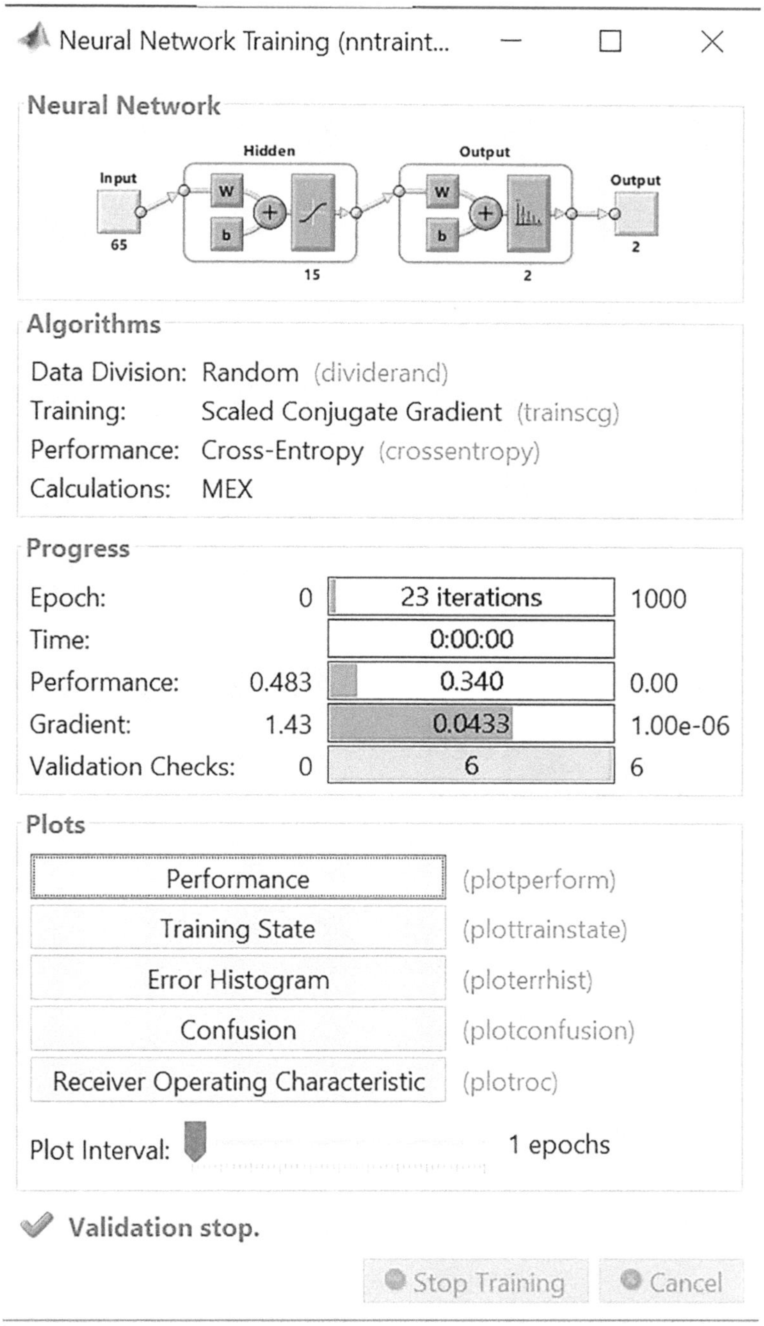

FIG. 5.8 The Neural Network Training tool of the ANN classifier for the focal and nonfocal EEG signals features extracted using AR Burg method.

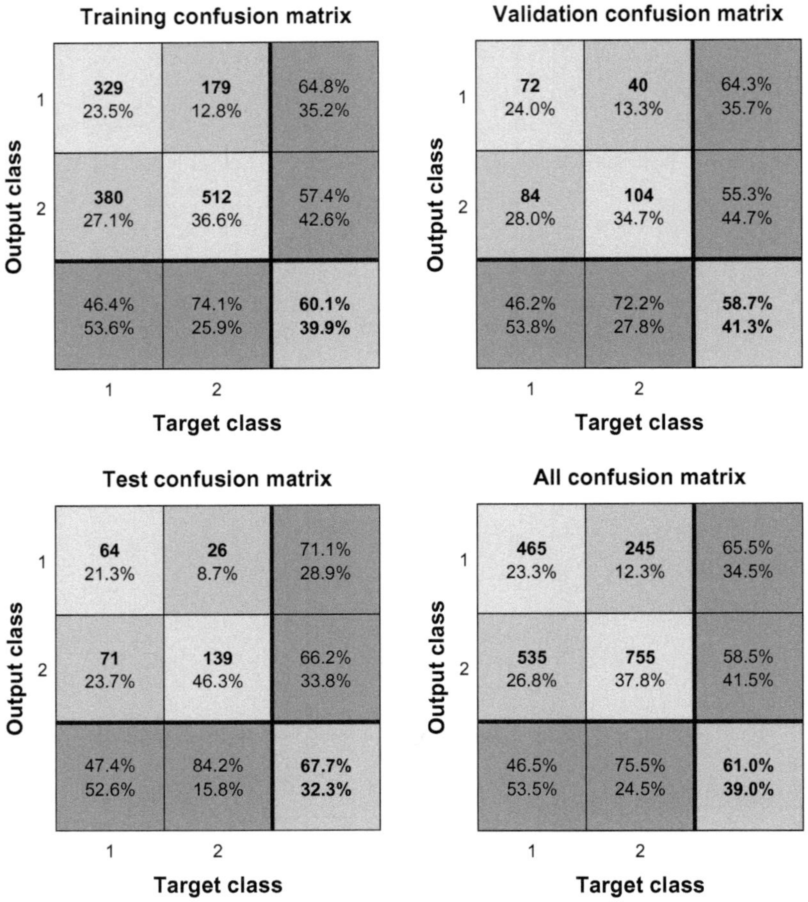

FIG. 5.9 Confusion matrices of train, validation, test, and all sets of the ANN classifier for the focal and nonfocal EEG signals features extracted using AR Burg method.

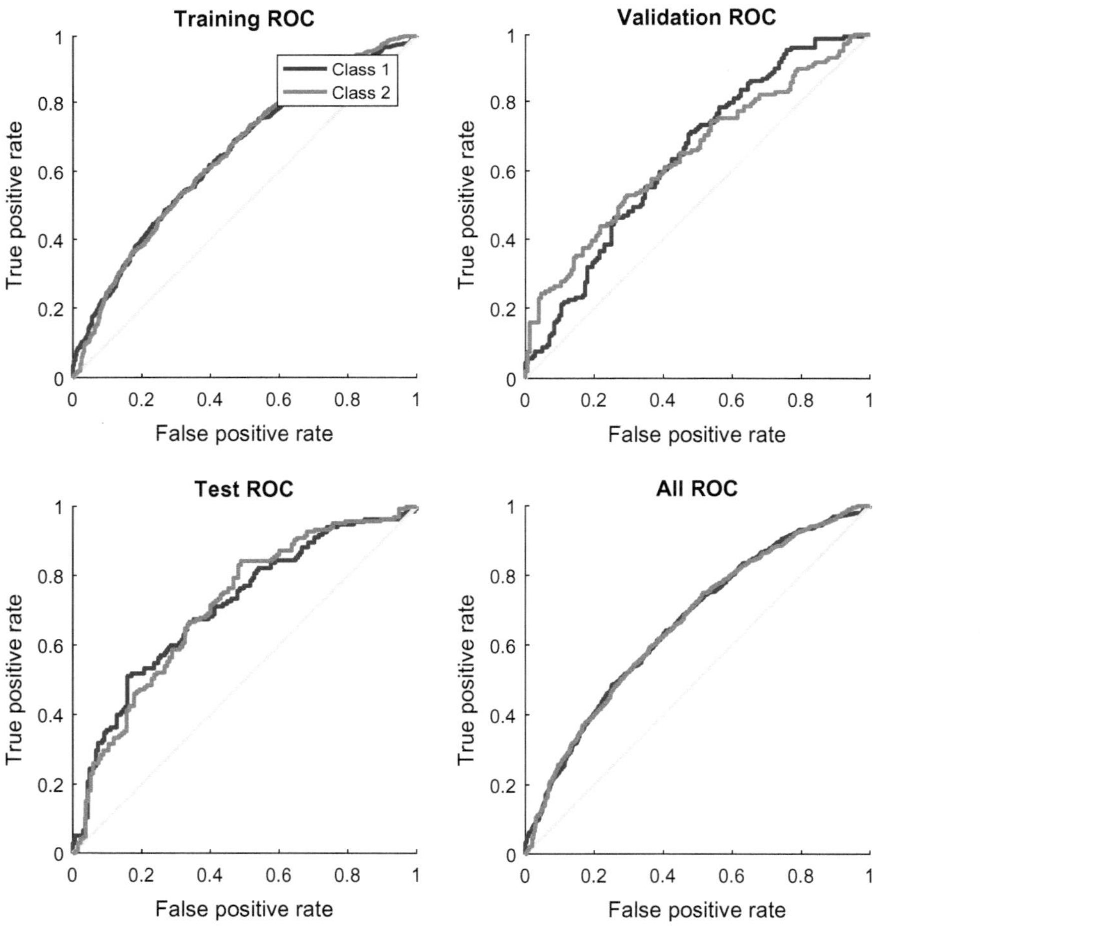

FIG. 5.10 ROC area of train, validation, test, and all sets of the ANN classifier for the focal and nonfocal EEG signals features extracted using AR Burg method.

EXAMPLE 5.14. The following MATLAB code is used to extract features from the ECG signals using covariance method. Then it classifies these data ANN classifier. You can download data from the following website: https://www.physionet.org/physiobank/database/mitdb/

```
%% Ch5_Ex33_ECG_COV_ANN.m
%The following MATLAB code is used to extract the features from
%the ECG signals using Covariance Method.
%Then it classifies ECG Signals Using ANN
clc
clear
%Load Sample ECG Data downloaded from the web site
%https://www.physionet.org/physiobank/database/mitdb/
load MITBIH_ECG.mat
%%
Fs = 320;           % Sampling frequency
T = 1/Fs;            % Sampling period
segmentLength = 128; % Length of a signal segment
Length=320;% Length of signal
Nofsignal=300; %Number of Signal
order = 14;
%%
% Obtain the Covariance Spectrum of the Normal ECG signal using pcov.
for i=1:Nofsignal
    [Pxx,F] = pcov(ECGN(1:Length,i),order,segmentLength,Fs);
    ASData(i,:)=Pxx(:);
end
%%
% Obtain the Covariance Spectrum of the ECG signal with APC using pcov.
[Pxx,F] = pcov(ECGAPC(:,1),order,segmentLength,Fs);
for i=Nofsignal+1:2*Nofsignal
    [Pxx,F] = pcov(ECGAPC(1:Length,i-Nofsignal),order,segmentLength,Fs);
    ASData(i,:)=Pxx(:);
end
%%
% Obtain the Covariance Spectrum of the ECG signal with PVC using pcov.
for i=2*Nofsignal+1:3*Nofsignal
[Pxx,F] = pcov(ECGPVC(1:Length,i-2*Nofsignal),order,segmentLength,Fs);
ASData(i,:)=Pxx(:);
end

%%
% Obtain the Covariance Spectrum of the ECG signal with LBBB using pcov.
for i=3*Nofsignal+1:4*Nofsignal
[Pxx,F] = pcov(ECGLBBB(1:Length,i-3*Nofsignal),order,segmentLength,Fs);
ASData(i,:)=Pxx(:);
end

%%
% Obtain the Covariance Spectrum of the ECG signal with RBBB using pcov.
for i=4*Nofsignal+1:5*Nofsignal
[Pxx,F] = pcov(ECGRBBB(1:Length,i-4*Nofsignal),order,segmentLength,Fs);
ASData(i,:)=Pxx(:);
end

%Transpose the Data Matrix
Inputs=ASData';
%% TARGET GENERATION
% Your target set must have one label for each sample,
```

```
%so it must contains 300 elements. Here is an easy way to build one.
%Start by numbering your classes from 1 to 5.
%Make sure that your X Array is sorted by class,
%meaning all the samples from the first class,
%then all the samples from the second class, and so on. Then:

 % Classes from 1 to 5
 y = 1:5;
 % 300 samples per class
 y = repmat(y, Nofsignal, 1);
 % Reshape to obtain a vector
 y = reshape(y, 1, numel(y));
% At this stage, you should have a 1-by-1500 vector with numeric class labels from 1 to 3.
%This is what you want if you use the Statistics and Machine Learning Toolbox.
%
% Now, I suppose you intend to use the Neural Network Toolbox.
%In that case the syntax is a bit different: the class labels are in a 3-by-300 matrix,
%each column being a sample, with the row corresponding to the class getting value '1' and the
rest '0'.
%To build it from the previous syntax, type:

Targets = full(ind2vec(y));

%% CLASSIFACATION
% Solve a Pattern Recognition Problem with a Neural Network
% Script generated by NPRTOOL
%
% Create a Pattern Recognition Network
hiddenLayerSize = 35;
net = patternnet(hiddenLayerSize);
% Set up Division of Data for Training, Validation, Testing
net.divideParam.trainRatio = 70/100;
net.divideParam.valRatio = 15/100;
net.divideParam.testRatio = 15/100;
% Train the Network
[net,tr] = train(net,Inputs,Targets);
% Test the Network
outputs = net(Inputs);
errors = gsubtract(Targets,outputs);
performance = perform(net,Targets,outputs);
% View the Network
view(net)
% Plots
figure, plotperform(tr)
figure, plottrainstate(tr)
figure, ploterrhist(errors)
%%
% One measure of how well the neural network has fit the data is the
% confusion plot. Here the confusion matrix is plotted across all samples.
%
% The confusion matrix shows the percentages of correct and incorrect
% classifications. Correct classifications are the green squares on the
% matrices diagonal. Incorrect classifications form the red squares.
%
% If the network has learned to classify properly, the percentages in the
% red squares should be very small, indicating few misclassifications.
%
```

```
% If this is not the case then further training, or training a network
% with more hidden neurons, would be advisable.

figure, plotconfusion(Targets,outputs)
%%
% Another measure of how well the neural network has fit data is the
% receiver operating characteristic plot. This shows how the false
% positive and true positive rates relate as the thresholding of outputs
% is varied from 0 to 1.
%
% The farther left and up the line is, the fewer false positives need to
% be accepted in order to get a high true positive rate. The best
% classifiers will have a line going from the bottom left corner, to the
% top left corner, to the top right corner, or close to that.

figure, plotroc(Targets,outputs)
```

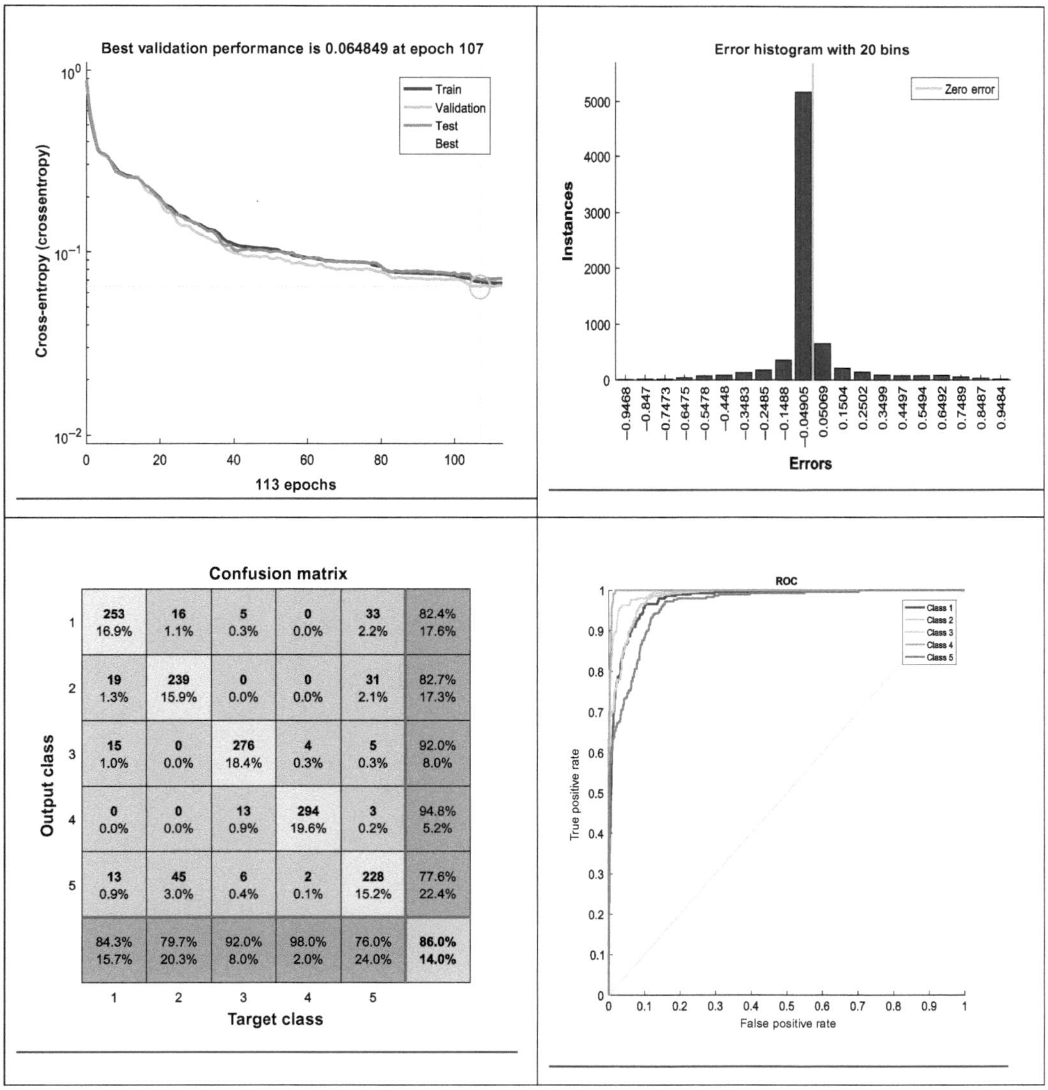

FIG. 5.11 Representation of performance of the ANN classifier for the ECG signals features extracted using covariance method.

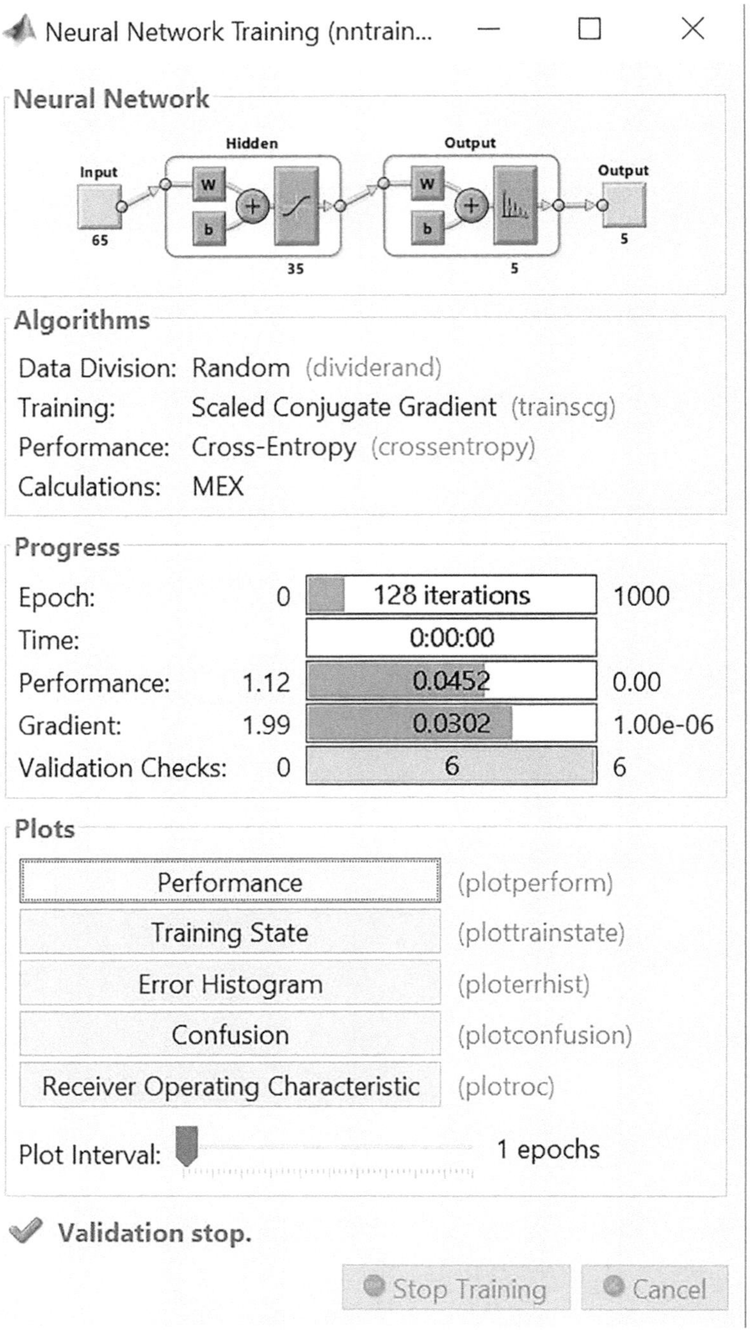

FIG. 5.12 The Neural Network Training tool of the ANN classifier for the ECG signals features extracted using covariance method.

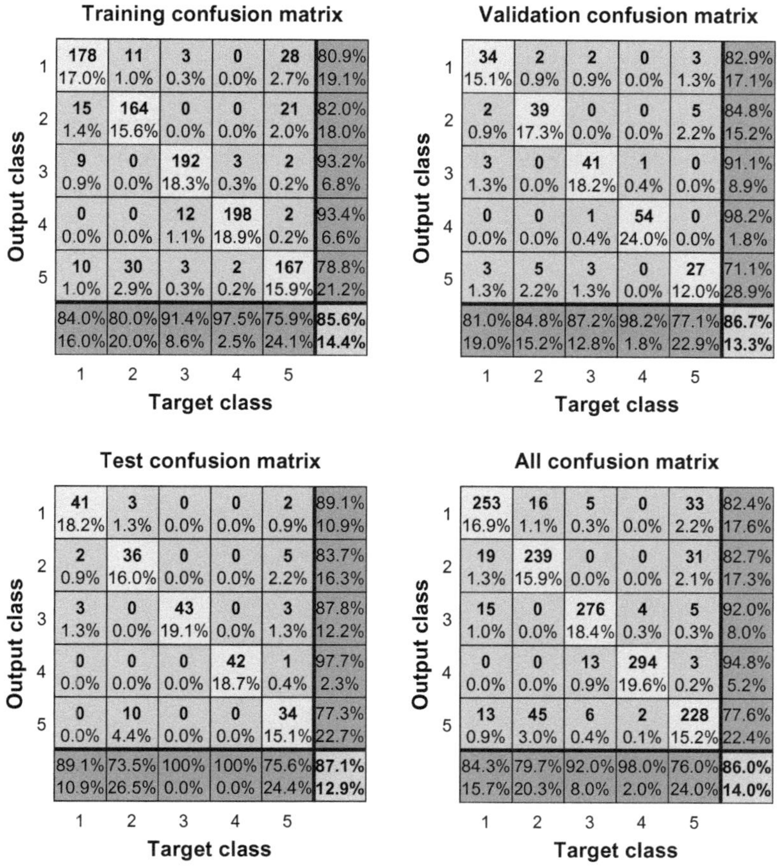

FIG. 5.13 Confusion matrices of train, validation, test, and all sets of the ANN classifier for the ECG signals features extracted using covariance method.

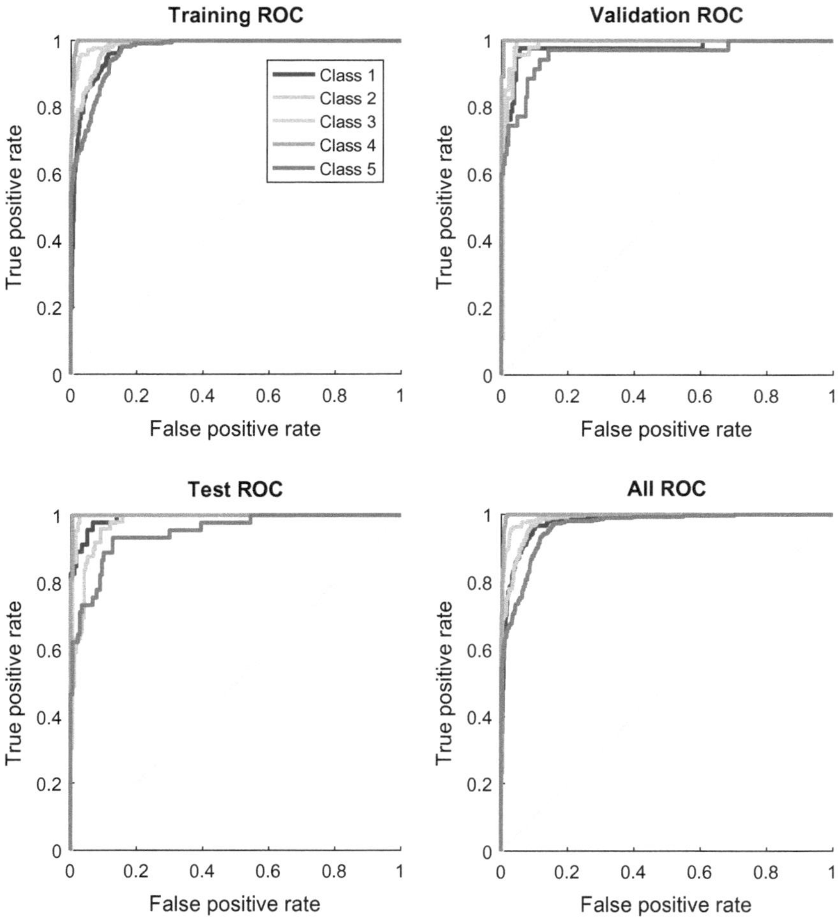

FIG. 5.14 ROC area of train, validation, test, and all sets of the ANN classifier for the ECG signals features extracted using covariance method.

EXAMPLE 5.15. The following MATLAB code was used to extract features from the EEG signals using EIGEN method. Then it classified these data using ANN. You can download data from the following website: http://epileptologie-bonn.de/cms/front_content.php?idcat=193&lang=3&changelang=3

```
%% Ch5_Ex34_EEG_EIG_ANN.m
%The following MATLAB code is used to extract the features from
%the EEG signals using EIGEN Method.
%Then it classifies EEG Signals Using ANN Classifier
clc
clear
%Load Sample EEG Data downloaded from the web site
%http://epileptologie-bonn.de/cms/front_content.php?idcat=193&lang=3&changelang=3
load AS_BONN_ALL_EEG_DATA_4096.mat
%%
Fs = 173.6;          % Sampling frequency
T = 1/Fs;            % Sampling period
segmentLength = 128; % Length of a signal segment
Length=4096;% Length of signal
order = 34; %Order of the EIGEN Model
NFFT = 128; %Number of FFT points
Nofsignal=100; %Number of Signal

%% Preparing the Data
% Data for classification problems are set up for a neural network by
% organizing the data into two matrices, the input matrix X and the target
% matrix T.
%
% Each ith column of the input matrix will have 65 elements from EIGEN
% spectrum
%
% Each corresponding column of the target matrix will have three elements.
%
% Here such the dataset is created.

%%
% Obtain the EIGEN Spectrum of the NORMAL EEG signal using peig.
for i=1:Nofsignal
    [Pxx,F] = peig(Normal_Eyes_Open(1:Length,i),order,NFFT,Fs);
    ASData(i,:)=Pxx(:);
end
%%
% Obtain the EIGEN Spectrum of the INTERICTAL EEG signal using peig.
for i=Nofsignal+1:2*Nofsignal
    [Pxx,F] = peig(Interictal(1:Length,i-Nofsignal),order,NFFT,Fs);
    ASData(i,:)=Pxx(:);
end
%%
% Obtain the EIGEN Spectrum of the ICTAL EEG signal using peig.
for i=2*Nofsignal+1:3*Nofsignal
    [Pxx,F] = peig(Ictal(1:Length,i-2*Nofsignal),order,NFFT,Fs);
    ASData(i,:)=Pxx(:);
end

%Transpose the Data Matrix
Inputs=ASData';

%% TARGET GENERATION
% Your target set must have one label for each sample,
```

```
%so it must contains 300 elements. Here is an easy way to build one.
%Start by numbering your classes from 1 to 3.
%Make sure that your X Array is sorted by class,
%meaning all the samples from the first class,
%then all the samples from the second class, and so on. Then:

 % Classes from 1 to 3
 y = 1:3;
 % Nofsignal samples per class
 y = repmat(y, Nofsignal, 1);
 % Reshape to obtain a vector
 y = reshape(y, 1, numel(y));
% At this stage, you should have a 1-by-300 vector with numeric class labels from 1 to 3.
%This is what you want if you use the Statistics and Machine Learning Toolbox.
%
% Now, I suppose you intend to use the Neural Network Toolbox.
%In that case the syntax is a bit different: the class labels are in a 3-by-300 matrix,
%each column being a sample, with the row corresponding to the class getting value '1' and the
rest '0'.
%To build it from the previous syntax, type:

Targets = full(ind2vec(y));

%% CLASSIFICATION
%% Building the Neural Network Classifier
% The next step is to create a neural network that will learn to identify
% the EEG Signals.
%
% Since the neural network starts with random initial weights, the results
% of this example will differ slightly every time it is run. The random seed
% is set to avoid this randomness. However this is not necessary for your
% own applications.

%%
% Two-layer (i.e. one-hidden-layer) feed forward neural networks can learn
% any input-output relationship given enough neurons in the hidden layer.
% Layers which are not output layers are called hidden layers.
%
% We will try a single hidden layer of 50 neurons for this example. In
% general, more difficult problems require more neurons, and perhaps more
% layers. Simpler problems require fewer neurons.
%
% The input and output have sizes of 0 because the network has not yet
% been configured to match our input and target data. This will happen
% when the network is trained.

net = patternnet(50);
view(net)

%%
% Now the network is ready to be trained. The samples are automatically
% divided into training, validation and test sets. The training set is
% used to teach the network. Training continues as long as the network
% continues improving on the validation set. The test set provides a
% completely independent measure of network accuracy.

[net,tr] = train(net,Inputs,Targets);
nntraintool
```

```
%%
% To see how the network's performance improved during training, either
% click the "Performance" button in the training tool, or call PLOTPERFORM.
%
% Performance is measured in terms of mean squared error, and shown in
% log scale. It rapidly decreased as the network was trained.
%
% Performance is shown for each of the training, validation and test sets.
% The version of the network that did best on the validation set is
% was after training.

figure, plotperform(tr)

%% Testing the Classifier
% The trained neural network can now be tested with the testing samples
% This will give us a sense of how well the network will do when applied
% to data from the real world.
%
% The network outputs will be in the range 0 to 1, so we can use *vec2ind*
% function to get the class indices as the position of the highest element
% in each output vector.

testX = Inputs(:,tr.testInd);
testT = Targets(:,tr.testInd);

testY = net(testX);
testIndices = vec2ind(testY)

%%
% One measure of how well the neural network has fit the data is the
% confusion plot. Here the confusion matrix is plotted across all samples.
%
% The confusion matrix shows the percentages of correct and incorrect
% classifications. Correct classifications are the green squares on the
% matrices diagonal. Incorrect classifications form the red squares.
%
% If the network has learned to classify properly, the percentages in the
% red squares should be very small, indicating few misclassifications.
%
% If this is not the case then further training, or training a network
% with more hidden neurons, would be advisable.
figure, plotconfusion(testT,testY)

%%
% Here are the overall percentages of correct and incorrect classification.

[c,cm] = confusion(testT,testY)

fprintf('Percentage Correct Classification : %f%%\n', 100*(1-c));
fprintf('Percentage Incorrect Classification : %f%%\n', 100*c);

%%
% Another measure of how well the neural network has fit data is the
% receiver operating characteristic plot. This shows how the false
% positive and true positive rates relate as the thresholding of outputs
% is varied from 0 to 1.
%
```

```
% The farther left and up the line is, the fewer false positives need to
% be accepted in order to get a high true positive rate. The best
% classifiers will have a line going from the bottom left corner, to the
% top left corner, to the top right corner, or close to that.
figure, plotroc(testT,testY)

%%
% This example illustrated using a neural network to classify EEG Signals.
%
% Explore other examples and the documentation for more insight into neural
% networks and its applications.

displayEndOfDemoMessage(mfilename)
```

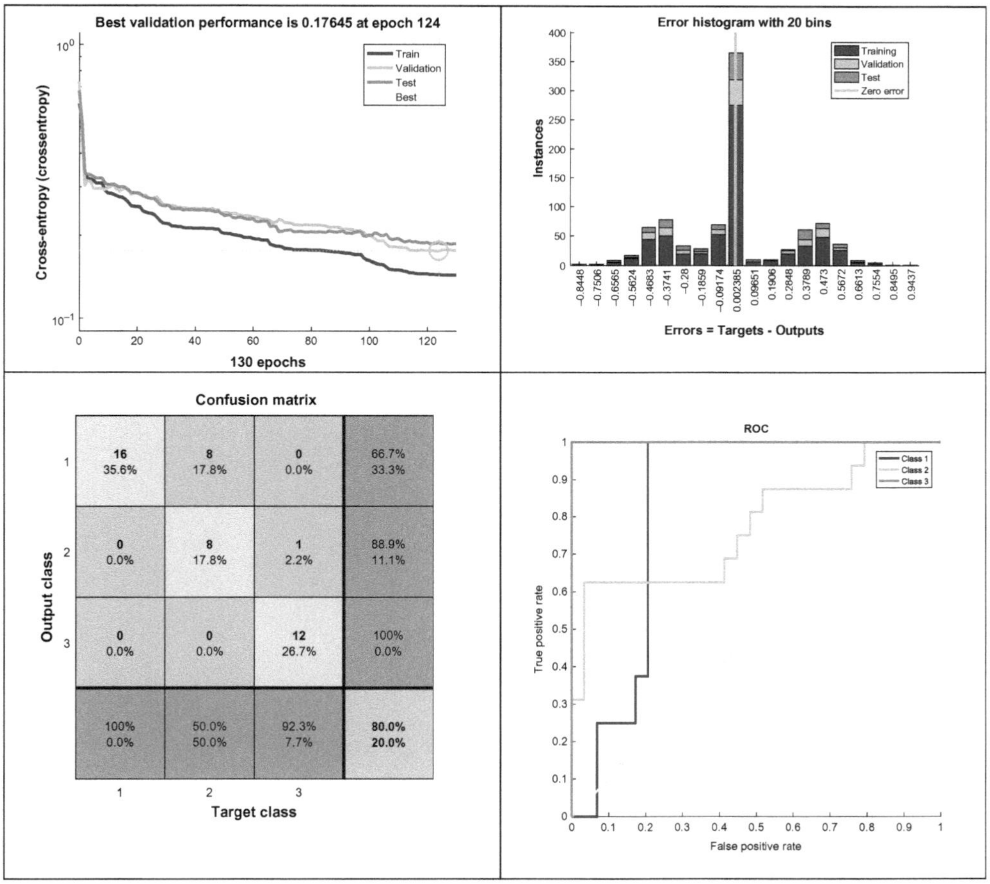

FIG. 5.15 Representation of performance of the ANN classifier for the EEG signals features extracted using Eigen method.

FIG. 5.16 The Neural Network Training tool of the ANN classifier for the EEG signals features extracted using Eigen method.

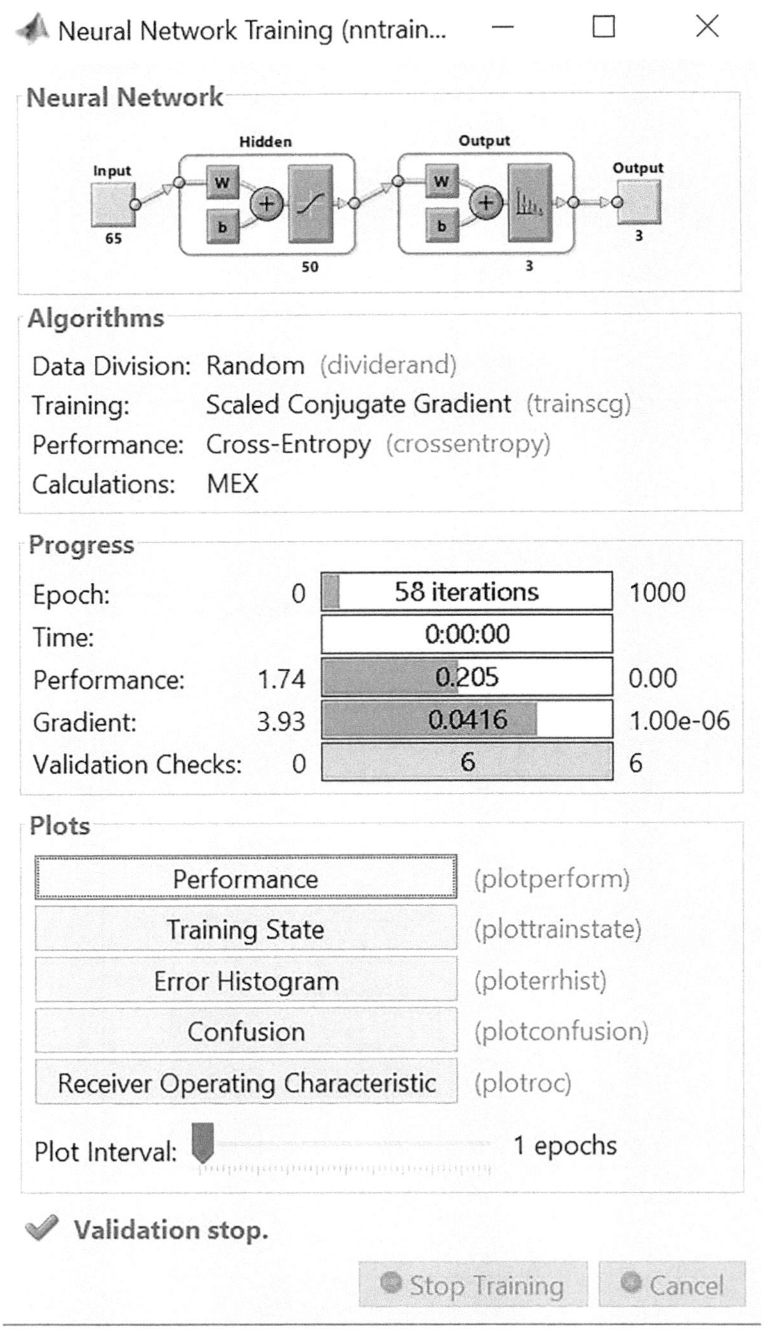

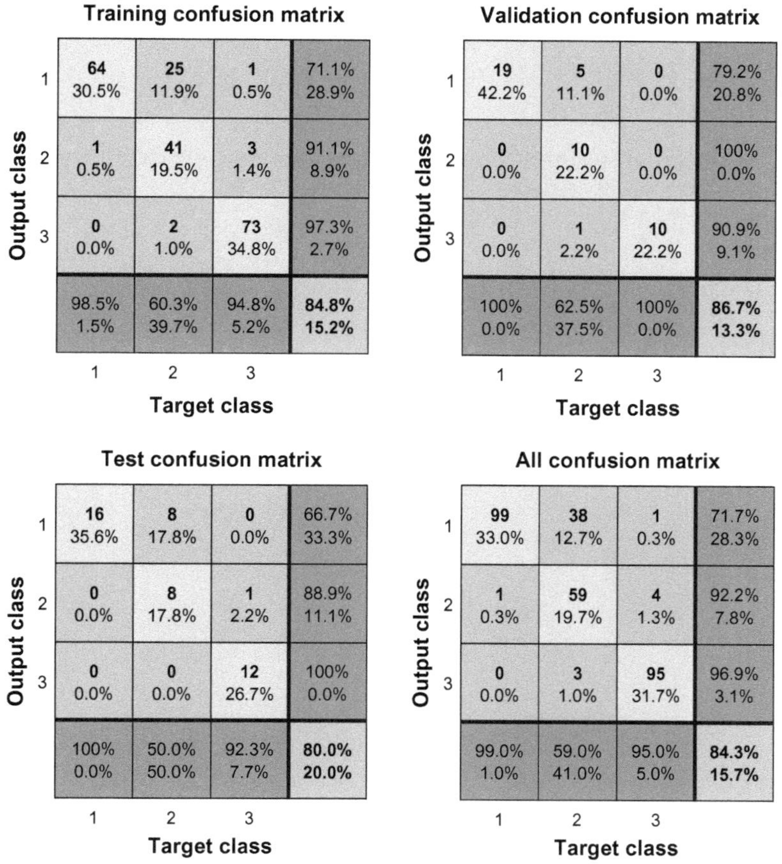

FIG. 5.17 Confusion matrices of train, validation, test, and all sets of the ANN classifier for EEG signals features extracted using Eigen method.

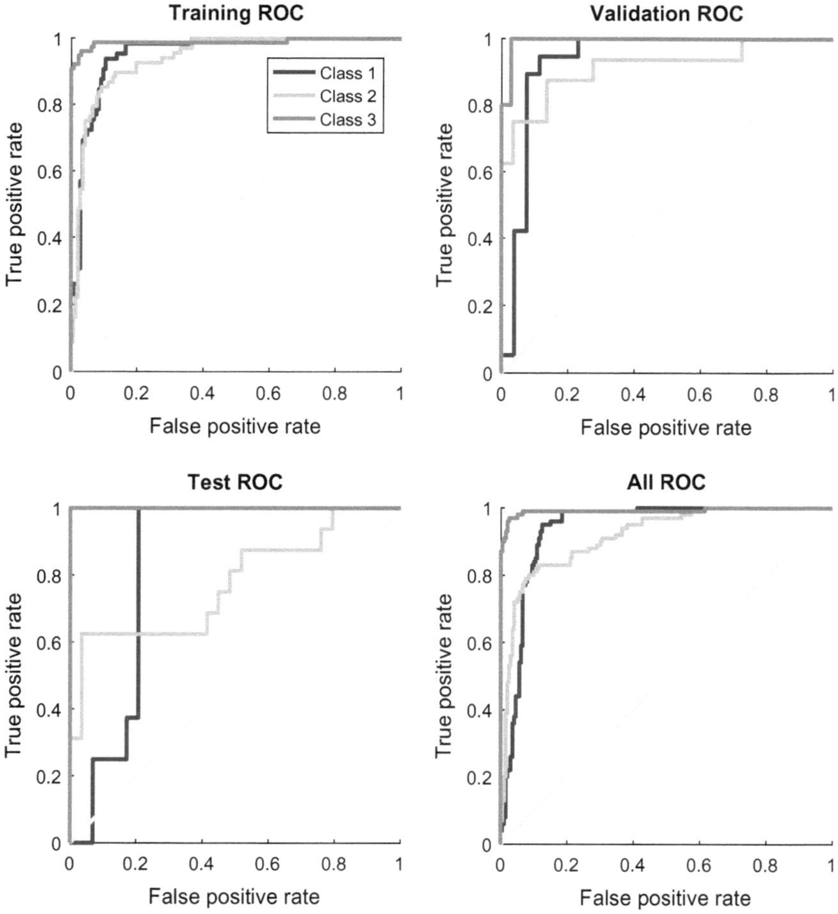

FIG. 5.18 ROC area of train, validation, test, and all sets of the ANN classifier for EEG signals features extracted using Eigen method.

EXAMPLE 5.16. The following MATLAB code was used to extract features from the EEG signals using modified covariance method. Then it classified these data using ANN. You can download data from the following website: http://epileptologie-bonn.de/cms/front_content.php?idcat=193&lang=3&changelang=3

```
%% Ch5_Ex35_EEG_MCOV_ANN.m
%The following MATLAB code is used to extract the features from
%the EEG signals using Modified Covariance.
%Then it classifies EEG Signals Using ANN
clc
clear
%Load Sample EEG Data downloaded from the web site
%http://epileptologie-bonn.de/cms/front_content.php?idcat=193&lang=3&changelang=3
load AS_BONN_ALL_EEG_DATA_4096.mat
%%
Fs = 173.6;            % Sampling frequency
T = 1/Fs;              % Sampling period
segmentLength = 128; % Length of a signal segment
Length=4096;% Length of signal
Nofsignal=100; %Number of Signal
order = 14;

%%
% Obtain the Modified Covariance Spectrum of the Normal EEG signal using pmcov.
for i=1:Nofsignal
    [Pxx,F] = pmcov(Normal_Eyes_Open(1:Length,i),order,segmentLength,Fs);
    ASData(i,:)=Pxx(:);
end
%%
% Obtain the Modified Covariance Spectrum of the Interictal EEG signal using pmcov.
for i=Nofsignal+1:2*Nofsignal
    [Pxx,F] = pmcov(Interictal(1:Length,i-Nofsignal),order,segmentLength,Fs);
    ASData(i,:)=Pxx(:);
end
%%
% Obtain the Modified Covariance Spectrum of the Ictal EEG signal using pmcov.
for i=2*Nofsignal+1:3*Nofsignal
[Pxx,F] = pmcov(Ictal(1:Length,i-2*Nofsignal),order,segmentLength,Fs);
ASData(i,:)=Pxx(:);
end

%Transpose the Data Matrix
Inputs=ASData';
%% TARGET GENERATION
% Your target set must have one label for each sample,
%so it must contains 300 elements. Here is an easy way to build one.
%Start by numbering your classes from 1 to 3.
%Make sure that your X Array is sorted by class,
%meaning all the samples from the first class,
%then all the samples from the second class, and so on. Then:

 % Classes from 1 to 3
 y = 1:3;
 % 100 samples per class
 y = repmat(y, 100, 1);
 % Reshape to obtain a vector
 y = reshape(y, 1, numel(y));
% At this stage, you should have a 1-by-300 vector with numeric class labels from 1 to 3.
%This is what you want if you use the Statistics and Machine Learning Toolbox.
%
```

```
% Now, I suppose you intend to use the Neural Network Toolbox.
%In that case the syntax is a bit different: the class labels are in a 3-by-300 matrix,
%each column being a sample, with the row corresponding to the class getting value '1' and the
rest '0'.
%To build it from the previous syntax, type:

Targets = full(ind2vec(y));

%% CLASSIFACATION
% Solve a Pattern Recognition Problem with a Neural Network
% Script generated by NPRTOOL
%
% Create a Pattern Recognition Network
hiddenLayerSize = 50;
net = patternnet(hiddenLayerSize);
% Set up Division of Data for Training, Validation, Testing
net.divideParam.trainRatio = 70/100;
net.divideParam.valRatio = 15/100;
net.divideParam.testRatio = 15/100;
% Train the Network
[net,tr] = train(net,Inputs,Targets);
% Test the Network
outputs = net(Inputs);
errors = gsubtract(Targets,outputs);
performance = perform(net,Targets,outputs);
% View the Network
view(net)
% Plots
figure, plotperform(tr)
figure, plottrainstate(tr)
figure, ploterrhist(errors)
%%
% One measure of how well the neural network has fit the data is the
% confusion plot. Here the confusion matrix is plotted across all samples.
%
% The confusion matrix shows the percentages of correct and incorrect
% classifications. Correct classifications are the green squares on the
% matrices diagonal. Incorrect classifications form the red squares.
%
% If the network has learned to classify properly, the percentages in the
% red squares should be very small, indicating few misclassifications.
%
% If this is not the case then further training, or training a network
% with more hidden neurons, would be advisable.
figure, plotconfusion(Targets,outputs)
%%
% Another measure of how well the neural network has fit data is the
% receiver operating characteristic plot. This shows how the false
% positive and true positive rates relate as the thresholding of outputs
% is varied from 0 to 1.
%
% The farther left and up the line is, the fewer false positives need to
% be accepted in order to get a high true positive rate. The best
% classifiers will have a line going from the bottom left corner, to the
% top left corner, to the top right corner, or close to that.

figure, plotroc(Targets,outputs)
```

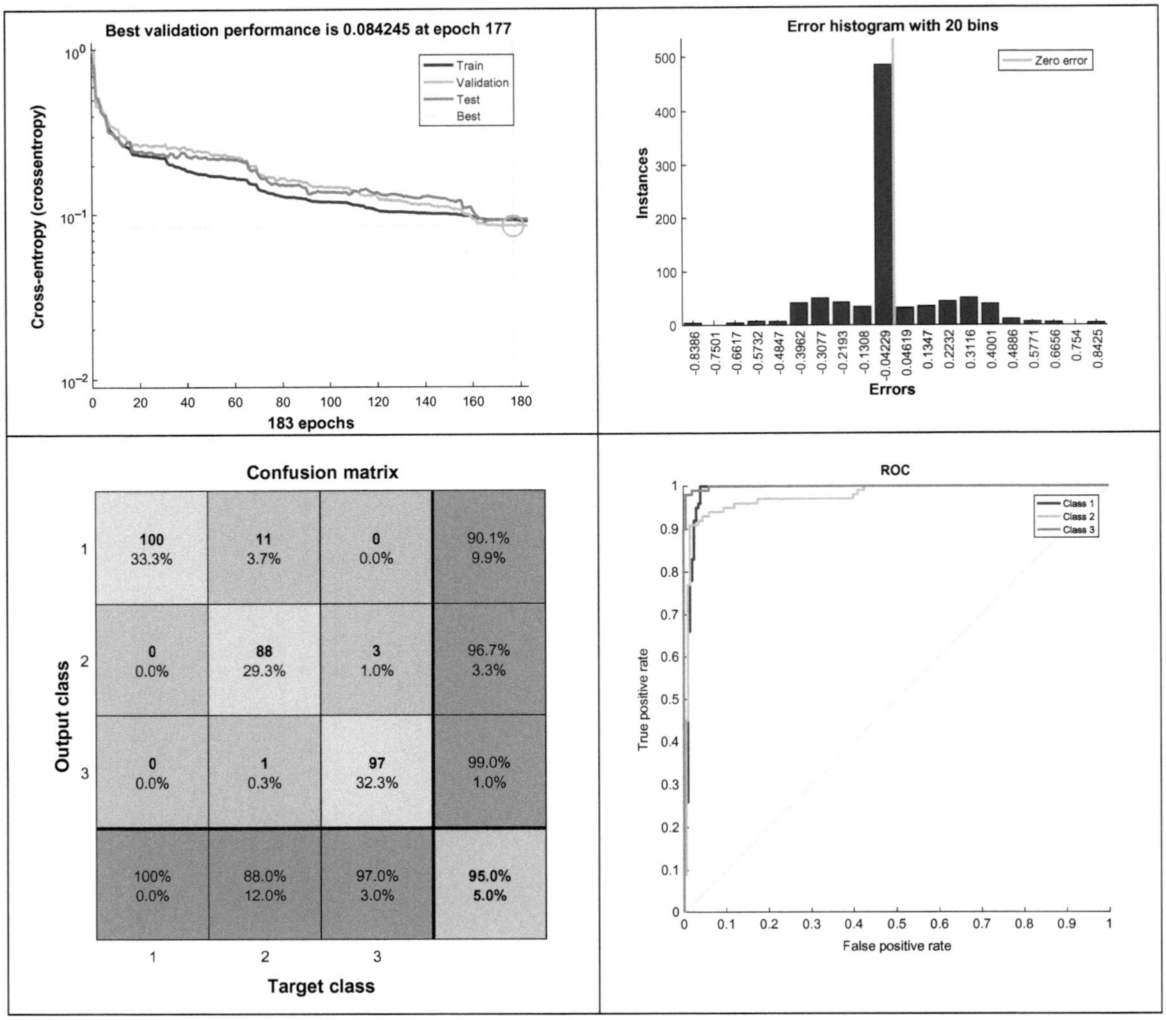

FIG. 5.19 Representation of performance of the ANN classifier for the EEG signals features extracted using modified covariance method.

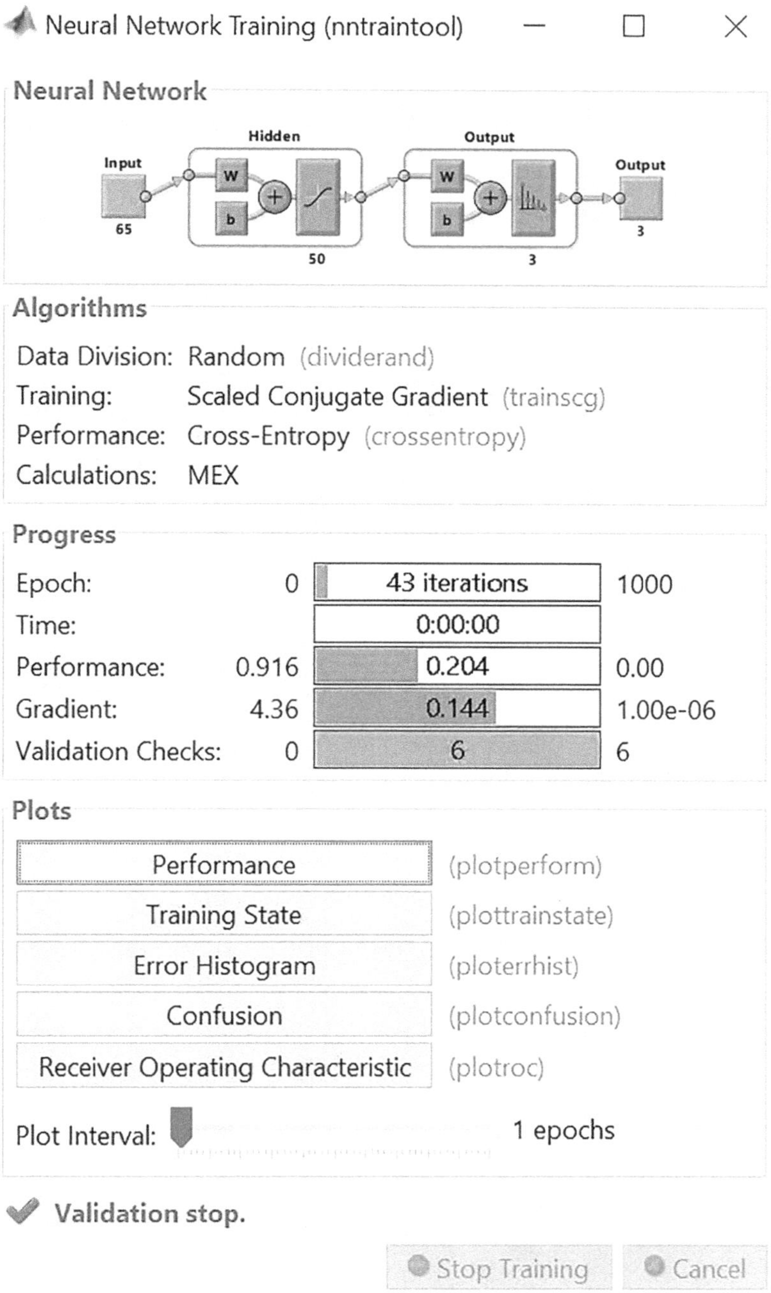

FIG. 5.20 The Neural Network Training tool of the ANN classifier for the EEG signals features extracted using modified covariance method.

FIG. 5.21 Confusion matrices of train, validation, test, and all sets of the ANN classifier for EEG signals features extracted using modified covariance method.

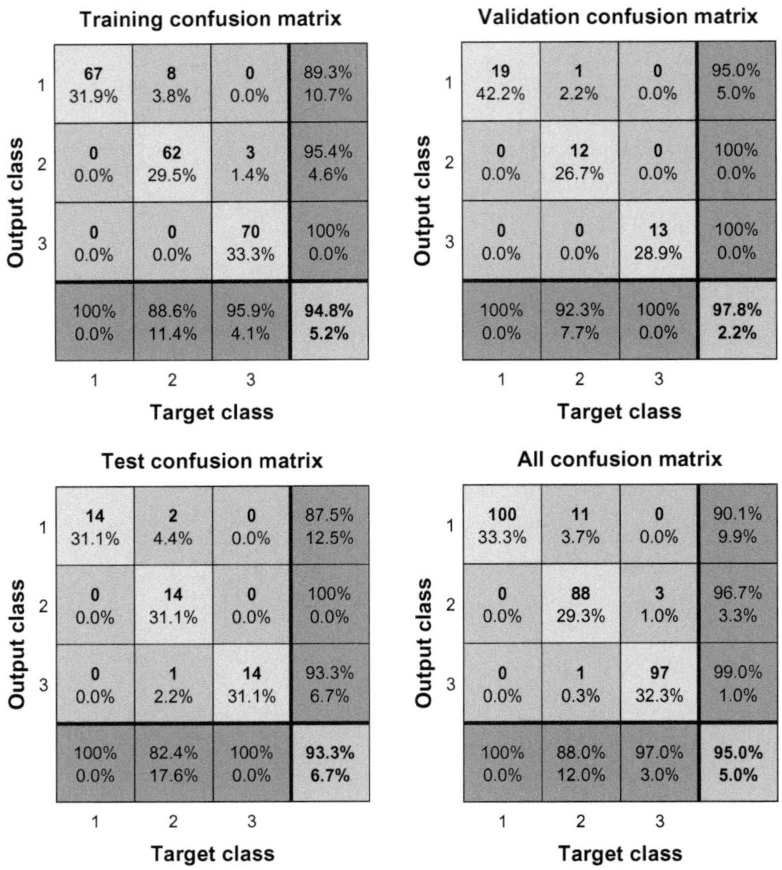

FIG. 5.22 ROC area of train, validation, test, and all sets of the ANN classifier for EEG signals features extracted using modified covariance method.

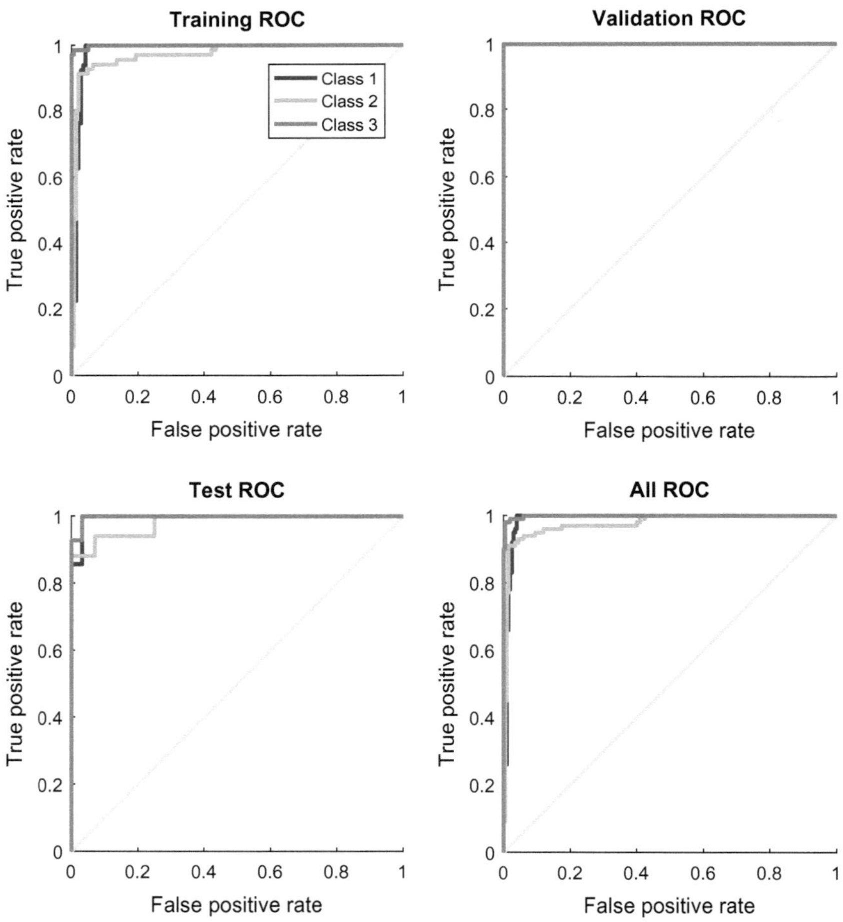

EXAMPLE 5.17 The following MATLAB code was used to extract features from the EEG signals using WPD. Then it used statistical values of WPD subbands. Then it classified these data using ANN classifier. You can download data from the following website:
http://epileptologie-bonn.de/cms/front_content.php?idcat=193&lang=3&changelang=3

```
% %% Ch5_Ex37_EEG_WPD_ANN.m
% WPD of NORMAL, INTERICTAL and ICTAL EEG signals
%The following MATLAB code is used to Extract WPD features from the EEG signals
%Decompose EEG data using WPD coefficients
%Then uses Statistical features as:

%(1) Mean of the absolute values of the coefficients in each sub-band.
%(2) Standard deviation of the coefficients in each sub-band.
%(3) Skewness of the coefficients in each sub-band.
%(4) Kurtosis of the coefficients in each sub-band.
%(5) RMS power of the wavelet coefficients in each subband.
%(6) Ratio of the mean absolute values of adjacent subbands.

%Then it classifies EEG Signals Using ANN Classifier
clc
clear
%Load Sample EEG Data downloaded from the web site
%http://epileptologie-bonn.de/cms/front_content.php?idcat=193&lang=3&changelang=3
load AS_BONN_ALL_EEG_DATA_4096.mat
wname = 'db4';
Length = 4096;     % Length of signal
Nofsignal=100; %Number of Signal
%%
%%%%%%%%%%%%%%%%%%%%%%%%%%%%%% A Z.ZIP EYES OPEN NORMAL SUBJECT

for i=1:Nofsignal;

[wp10,wp11] = dwt(Normal_Eyes_Open(1:Length,i),wname);%WPD Decomposition
[wp20,wp21] = dwt(wp10,wname);
[wp22,wp23] = dwt(wp11,wname);

[wp30,wp31] = dwt(wp20,wname);
[wp32,wp33] = dwt(wp21,wname);
[wp34,wp35] = dwt(wp22,wname);
[wp36,wp37] = dwt(wp23,wname);

[wp40,wp41] = dwt(wp30,wname);
[wp42,wp43] = dwt(wp31,wname);
[wp44,wp45] = dwt(wp32,wname);
[wp46,wp47] = dwt(wp33,wname);
[wp48,wp49] = dwt(wp34,wname);
[wp4A,wp4B] = dwt(wp35,wname);
[wp4C,wp4D] = dwt(wp36,wname);
[wp4E,wp4F] = dwt(wp37,wname);

ASData(i,1)=mean(abs(wp40));
ASData(i,2)=mean(abs(wp41));
ASData(i,3)=mean(abs(wp42));
ASData(i,4)=mean(abs(wp43));
ASData(i,5)=mean(abs(wp44));
ASData(i,6)=mean(abs(wp45));
ASData(i,7)=mean(abs(wp46));
ASData(i,8)=mean(abs(wp47));
ASData(i,9)=mean(abs(wp48));
```

```
ASData(i,10)=mean(abs(wp49));
ASData(i,11)=mean(abs(wp4A));
ASData(i,12)=mean(abs(wp4B));
ASData(i,13)=mean(abs(wp4C));
ASData(i,14)=mean(abs(wp4D));
ASData(i,15)=mean(abs(wp4E));
ASData(i,16)=mean(abs(wp4F));

ASData(i,17)=std(wp40);
ASData(i,18)=std(wp41);
ASData(i,19)=std(wp42);
ASData(i,20)=std(wp43);
ASData(i,21)=std(wp44);
ASData(i,22)=std(wp45);
ASData(i,23)=std(wp46);
ASData(i,24)=std(wp47);
ASData(i,25)=std(wp48);
ASData(i,26)=std(wp49);
ASData(i,27)=std(wp4A);
ASData(i,28)=std(wp4B);
ASData(i,29)=std(wp4C);
ASData(i,30)=std(wp4D);
ASData(i,31)=std(wp4E);
ASData(i,32)=std(wp4F);

ASData(i,33)=skewness(wp40(:));
ASData(i,34)=skewness(wp41(:));
ASData(i,35)=skewness(wp42(:));
ASData(i,36)=skewness(wp43(:));
ASData(i,37)=skewness(wp44(:));
ASData(i,38)=skewness(wp45(:));
ASData(i,39)=skewness(wp46(:));
ASData(i,40)=skewness(wp47(:));
ASData(i,41)=skewness(wp48(:));
ASData(i,42)=skewness(wp49(:));
ASData(i,43)=skewness(wp4A(:));
ASData(i,44)=skewness(wp4B(:));
ASData(i,45)=skewness(wp4C(:));
ASData(i,46)=skewness(wp4D(:));
ASData(i,47)=skewness(wp4E(:));
ASData(i,48)=skewness(wp4F(:));

ASData(i,49)=kurtosis(wp40);
ASData(i,50)=kurtosis(wp41);
ASData(i,51)=kurtosis(wp42);
ASData(i,52)=kurtosis(wp43);
ASData(i,53)=kurtosis(wp44);
ASData(i,54)=kurtosis(wp45);
ASData(i,55)=kurtosis(wp46);
ASData(i,56)=kurtosis(wp47);
ASData(i,57)=kurtosis(wp48);
ASData(i,58)=kurtosis(wp49);
ASData(i,59)=kurtosis(wp4A);
ASData(i,60)=kurtosis(wp4B);
ASData(i,61)=kurtosis(wp4C);
ASData(i,62)=kurtosis(wp4D);
ASData(i,63)=kurtosis(wp4E);
ASData(i,64)=kurtosis(wp4F);
```

```
ASData(i,65)=rms(wp40(:));
ASData(i,66)=rms(wp41(:));
ASData(i,67)=rms(wp42(:));
ASData(i,68)=rms(wp43(:));
ASData(i,69)=rms(wp44(:));
ASData(i,70)=rms(wp45(:));
ASData(i,71)=rms(wp46(:));
ASData(i,72)=rms(wp47(:));
ASData(i,73)=rms(wp48(:));
ASData(i,74)=rms(wp49(:));
ASData(i,75)=rms(wp4A(:));
ASData(i,76)=rms(wp4B(:));
ASData(i,77)=rms(wp4C(:));
ASData(i,78)=rms(wp4D(:));
ASData(i,79)=rms(wp4E(:));
ASData(i,80)=rms(wp4F(:));

ASData(i,81)=mean(abs(wp40))/mean(abs(wp41));
ASData(i,82)=mean(abs(wp41))/mean(abs(wp42));
ASData(i,83)=mean(abs(wp42))/mean(abs(wp43));
ASData(i,84)=mean(abs(wp43))/mean(abs(wp44));
ASData(i,85)=mean(abs(wp44))/mean(abs(wp45));
ASData(i,86)=mean(abs(wp45))/mean(abs(wp46));
ASData(i,87)=mean(abs(wp46))/mean(abs(wp47));
ASData(i,88)=mean(abs(wp47))/mean(abs(wp48));
ASData(i,89)=mean(abs(wp48))/mean(abs(wp49));
ASData(i,90)=mean(abs(wp49))/mean(abs(wp4A));
ASData(i,91)=mean(abs(wp4A))/mean(abs(wp4B));
ASData(i,92)=mean(abs(wp4B))/mean(abs(wp4C));
ASData(i,93)=mean(abs(wp4C))/mean(abs(wp4D));
ASData(i,94)=mean(abs(wp4D))/mean(abs(wp4E));
ASData(i,95)=mean(abs(wp4E))/mean(abs(wp4F));

end

%%
% D F.ZIP EPILEPTIC SUBJECT (INTERICTAL) were recorded from within
% the epileptogenic zone during seizure free intervals

for i=Nofsignal+1:2*Nofsignal;

[wp10,wp11] = dwt(Interictal(1:Length,i-Nofsignal),wname);%WPD Decomposition
[wp20,wp21] = dwt(wp10,wname);
[wp22,wp23] = dwt(wp11,wname);

[wp30,wp31] = dwt(wp20,wname);
[wp32,wp33] = dwt(wp21,wname);
[wp34,wp35] = dwt(wp22,wname);
[wp36,wp37] = dwt(wp23,wname);

[wp40,wp41] = dwt(wp30,wname);
[wp42,wp43] = dwt(wp31,wname);
[wp44,wp45] = dwt(wp32,wname);
[wp46,wp47] = dwt(wp33,wname);
[wp48,wp49] = dwt(wp34,wname);
[wp4A,wp4B] = dwt(wp35,wname);
[wp4C,wp4D] = dwt(wp36,wname);
[wp4E,wp4F] = dwt(wp37,wname);
```

```
ASData(i,1)=mean(abs(wp40));
ASData(i,2)=mean(abs(wp41));
ASData(i,3)=mean(abs(wp42));
ASData(i,4)=mean(abs(wp43));
ASData(i,5)=mean(abs(wp44));
ASData(i,6)=mean(abs(wp45));
ASData(i,7)=mean(abs(wp46));
ASData(i,8)=mean(abs(wp47));
ASData(i,9)=mean(abs(wp48));
ASData(i,10)=mean(abs(wp49));
ASData(i,11)=mean(abs(wp4A));
ASData(i,12)=mean(abs(wp4B));
ASData(i,13)=mean(abs(wp4C));
ASData(i,14)=mean(abs(wp4D));
ASData(i,15)=mean(abs(wp4E));
ASData(i,16)=mean(abs(wp4F));

ASData(i,17)=std(wp40);
ASData(i,18)=std(wp41);
ASData(i,19)=std(wp42);
ASData(i,20)=std(wp43);
ASData(i,21)=std(wp44);
ASData(i,22)=std(wp45);
ASData(i,23)=std(wp46);
ASData(i,24)=std(wp47);
ASData(i,25)=std(wp48);
ASData(i,26)=std(wp49);
ASData(i,27)=std(wp4A);
ASData(i,28)=std(wp4B);
ASData(i,29)=std(wp4C);
ASData(i,30)=std(wp4D);
ASData(i,31)=std(wp4E);
ASData(i,32)=std(wp4F);

ASData(i,33)=skewness(wp40(:));
ASData(i,34)=skewness(wp41(:));
ASData(i,35)=skewness(wp42(:));
ASData(i,36)=skewness(wp43(:));
ASData(i,37)=skewness(wp44(:));
ASData(i,38)=skewness(wp45(:));
ASData(i,39)=skewness(wp46(:));
ASData(i,40)=skewness(wp47(:));
ASData(i,41)=skewness(wp48(:));
ASData(i,42)=skewness(wp49(:));
ASData(i,43)=skewness(wp4A(:));
ASData(i,44)=skewness(wp4B(:));
ASData(i,45)=skewness(wp4C(:));
ASData(i,46)=skewness(wp4D(:));
ASData(i,47)=skewness(wp4E(:));
ASData(i,48)=skewness(wp4F(:));

ASData(i,49)=kurtosis(wp40);
ASData(i,50)=kurtosis(wp41);
ASData(i,51)=kurtosis(wp42);
ASData(i,52)=kurtosis(wp43);
ASData(i,53)=kurtosis(wp44);
ASData(i,54)=kurtosis(wp45);
ASData(i,55)=kurtosis(wp46);
ASData(i,56)=kurtosis(wp47);
ASData(i,57)=kurtosis(wp48);
```

```
ASData(i,58)=kurtosis(wp49);
ASData(i,59)=kurtosis(wp4A);
ASData(i,60)=kurtosis(wp4B);
ASData(i,61)=kurtosis(wp4C);
ASData(i,62)=kurtosis(wp4D);
ASData(i,63)=kurtosis(wp4E);
ASData(i,64)=kurtosis(wp4F);

ASData(i,65)=rms(wp40(:));
ASData(i,66)=rms(wp41(:));
ASData(i,67)=rms(wp42(:));
ASData(i,68)=rms(wp43(:));
ASData(i,69)=rms(wp44(:));
ASData(i,70)=rms(wp45(:));
ASData(i,71)=rms(wp46(:));
ASData(i,72)=rms(wp47(:));
ASData(i,73)=rms(wp48(:));
ASData(i,74)=rms(wp49(:));
ASData(i,75)=rms(wp4A(:));
ASData(i,76)=rms(wp4B(:));
ASData(i,77)=rms(wp4C(:));
ASData(i,78)=rms(wp4D(:));
ASData(i,79)=rms(wp4E(:));
ASData(i,80)=rms(wp4F(:));

ASData(i,81)=mean(abs(wp40))/mean(abs(wp41));
ASData(i,82)=mean(abs(wp41))/mean(abs(wp42));
ASData(i,83)=mean(abs(wp42))/mean(abs(wp43));
ASData(i,84)=mean(abs(wp43))/mean(abs(wp44));
ASData(i,85)=mean(abs(wp44))/mean(abs(wp45));
ASData(i,86)=mean(abs(wp45))/mean(abs(wp46));
ASData(i,87)=mean(abs(wp46))/mean(abs(wp47));
ASData(i,88)=mean(abs(wp47))/mean(abs(wp48));
ASData(i,89)=mean(abs(wp48))/mean(abs(wp49));
ASData(i,90)=mean(abs(wp49))/mean(abs(wp4A));
ASData(i,91)=mean(abs(wp4A))/mean(abs(wp4B));
ASData(i,92)=mean(abs(wp4B))/mean(abs(wp4C));
ASData(i,93)=mean(abs(wp4C))/mean(abs(wp4D));
ASData(i,94)=mean(abs(wp4D))/mean(abs(wp4E));
ASData(i,95)=mean(abs(wp4E))/mean(abs(wp4F));
end

%% E S.ZIP EPILEPTIC SUBJECT ICTAL DURING SEIZURE
%%%%%%%%%%%%%%%%%%%%%%%%%%%%

for i=2*Nofsignal+1:3*Nofsignal;

[wp10,wp11] = dwt(Ictal(1:Length,i-2*Nofsignal),wname);%WPD Decomposition
[wp20,wp21] = dwt(wp10,wname);
[wp22,wp23] = dwt(wp11,wname);

[wp30,wp31] = dwt(wp20,wname);
[wp32,wp33] = dwt(wp21,wname);
[wp34,wp35] = dwt(wp22,wname);
[wp36,wp37] = dwt(wp23,wname);

[wp40,wp41] = dwt(wp30,wname);
[wp42,wp43] = dwt(wp31,wname);
[wp44,wp45] = dwt(wp32,wname);
```

```
[wp46,wp47] = dwt(wp33,wname);
[wp48,wp49] = dwt(wp34,wname);
[wp4A,wp4B] = dwt(wp35,wname);
[wp4C,wp4D] = dwt(wp36,wname);
[wp4E,wp4F] = dwt(wp37,wname);

ASData(i,1)=mean(abs(wp40));
ASData(i,2)=mean(abs(wp41));
ASData(i,3)=mean(abs(wp42));
ASData(i,4)=mean(abs(wp43));
ASData(i,5)=mean(abs(wp44));
ASData(i,6)=mean(abs(wp45));
ASData(i,7)=mean(abs(wp46));
ASData(i,8)=mean(abs(wp47));
ASData(i,9)=mean(abs(wp48));
ASData(i,10)=mean(abs(wp49));
ASData(i,11)=mean(abs(wp4A));
ASData(i,12)=mean(abs(wp4B));
ASData(i,13)=mean(abs(wp4C));
ASData(i,14)=mean(abs(wp4D));
ASData(i,15)=mean(abs(wp4E));
ASData(i,16)=mean(abs(wp4F));

ASData(i,17)=std(wp40);
ASData(i,18)=std(wp41);
ASData(i,19)=std(wp42);
ASData(i,20)=std(wp43);
ASData(i,21)=std(wp44);
ASData(i,22)=std(wp45);
ASData(i,23)=std(wp46);
ASData(i,24)=std(wp47);
ASData(i,25)=std(wp48);
ASData(i,26)=std(wp49);
ASData(i,27)=std(wp4A);
ASData(i,28)=std(wp4B);
ASData(i,29)=std(wp4C);
ASData(i,30)=std(wp4D);
ASData(i,31)=std(wp4E);
ASData(i,32)=std(wp4F);

ASData(i,33)=skewness(wp40(:));
ASData(i,34)=skewness(wp41(:));
ASData(i,35)=skewness(wp42(:));
ASData(i,36)=skewness(wp43(:));
ASData(i,37)=skewness(wp44(:));
ASData(i,38)=skewness(wp45(:));
ASData(i,39)=skewness(wp46(:));
ASData(i,40)=skewness(wp47(:));
ASData(i,41)=skewness(wp48(:));
ASData(i,42)=skewness(wp49(:));
ASData(i,43)=skewness(wp4A(:));
ASData(i,44)=skewness(wp4B(:));
ASData(i,45)=skewness(wp4C(:));
ASData(i,46)=skewness(wp4D(:));
ASData(i,47)=skewness(wp4E(:));
ASData(i,48)=skewness(wp4F(:));

ASData(i,49)=kurtosis(wp40);
ASData(i,50)=kurtosis(wp41);
```

```
ASData(i,51)=kurtosis(wp42);
ASData(i,52)=kurtosis(wp43);
ASData(i,53)=kurtosis(wp44);
ASData(i,54)=kurtosis(wp45);
ASData(i,55)=kurtosis(wp46);
ASData(i,56)=kurtosis(wp47);
ASData(i,57)=kurtosis(wp48);
ASData(i,58)=kurtosis(wp49);
ASData(i,59)=kurtosis(wp4A);
ASData(i,60)=kurtosis(wp4B);
ASData(i,61)=kurtosis(wp4C);
ASData(i,62)=kurtosis(wp4D);
ASData(i,63)=kurtosis(wp4E);
ASData(i,64)=kurtosis(wp4F);

ASData(i,65)=rms(wp40(:));
ASData(i,66)=rms(wp41(:));
ASData(i,67)=rms(wp42(:));
ASData(i,68)=rms(wp43(:));
ASData(i,69)=rms(wp44(:));
ASData(i,70)=rms(wp45(:));
ASData(i,71)=rms(wp46(:));
ASData(i,72)=rms(wp47(:));
ASData(i,73)=rms(wp48(:));
ASData(i,74)=rms(wp49(:));
ASData(i,75)=rms(wp4A(:));
ASData(i,76)=rms(wp4B(:));
ASData(i,77)=rms(wp4C(:));
ASData(i,78)=rms(wp4D(:));
ASData(i,79)=rms(wp4E(:));
ASData(i,80)=rms(wp4F(:));

ASData(i,81)=mean(abs(wp40))/mean(abs(wp41));
ASData(i,82)=mean(abs(wp41))/mean(abs(wp42));
ASData(i,83)=mean(abs(wp42))/mean(abs(wp43));
ASData(i,84)=mean(abs(wp43))/mean(abs(wp44));
ASData(i,85)=mean(abs(wp44))/mean(abs(wp45));
ASData(i,86)=mean(abs(wp45))/mean(abs(wp46));
ASData(i,87)=mean(abs(wp46))/mean(abs(wp47));
ASData(i,88)=mean(abs(wp47))/mean(abs(wp48));
ASData(i,89)=mean(abs(wp48))/mean(abs(wp49));
ASData(i,90)=mean(abs(wp49))/mean(abs(wp4A));
ASData(i,91)=mean(abs(wp4A))/mean(abs(wp4B));
ASData(i,92)=mean(abs(wp4B))/mean(abs(wp4C));
ASData(i,93)=mean(abs(wp4C))/mean(abs(wp4D));
ASData(i,94)=mean(abs(wp4D))/mean(abs(wp4E));
ASData(i,95)=mean(abs(wp4E))/mean(abs(wp4F));
end

%Transpose the Data Matrix
Inputs=ASData';

%% TARGET GENERATION
% Your target set must have one label for each sample,
%so it must contains 300 elements. Here is an easy way to build one.
%Start by numbering your classes from 1 to 3.
%Make sure that your X Array is sorted by class,
%meaning all the samples from the first class,
%then all the samples from the second class, and so on. Then:
```

```
 % Classes from 1 to 3
 y = 1:3;
 % Nofsignal samples per class
 y = repmat(y, Nofsignal, 1);
 % Reshape to obtain a vector
 y = reshape(y, 1, numel(y));
% At this stage, you should have a 1-by-300 vector with numeric class labels from 1 to 3.
%This is what you want if you use the Statistics and Machine Learning Toolbox.
%
% Now, I suppose you intend to use the Neural Network Toolbox.
%In that case the syntax is a bit different: the class labels are in a 3-by-300 matrix,
%each column being a sample, with the row corresponding to the class getting value '1' and the
rest '0'.
%To build it from the previous syntax, type:

Targets = full(ind2vec(y));

%% CLASSIFACATION
% Solve a Pattern Recognition Problem with a Neural Network
% Script generated by NPRTOOL
%
% Create a Pattern Recognition Network
hiddenLayerSize = 50;
net = patternnet(hiddenLayerSize);

% Train the network. The network uses the default Levenberg-Marquardt algorithm
% (trainlm) for training. For problems in which Levenberg-Marquardt does not
% produce as accurate results as desired, or for large data problems, consider setting
% the network training function to Bayesian Regularization (trainbr) or Scaled
% Conjugate Gradient (trainscg), respectively, with either
net.trainFcn = 'trainbr';
net.trainFcn = 'trainscg';

% Set up Division of Data for Training, Validation, Testing
net.divideParam.trainRatio = 70/100;
net.divideParam.valRatio = 15/100;
net.divideParam.testRatio = 15/100;
% Train the Network
[net,tr] = train(net,Inputs,Targets);
% Test the Network
outputs = net(Inputs);
errors = gsubtract(Targets,outputs);
performance = perform(net,Targets,outputs)
% View the Network
view(net)
% Plots
% Uncomment these lines to enable various plots.
figure, plotperform(tr)
figure, plottrainstate(tr)
figure, ploterrhist(errors)
%%
% One measure of how well the neural network has fit the data is the
% confusion plot. Here the confusion matrix is plotted across all samples.
%
% The confusion matrix shows the percentages of correct and incorrect
% classifications. Correct classifications are the green squares on the
% matrices diagonal. Incorrect classifications form the red squares.
%
% If the network has learned to classify properly, the percentages in the
```

```
% red squares should be very small, indicating few misclassifications.
%
% If this is not the case then further training, or training a network
% with more hidden neurons, would be advisable.

figure, plotconfusion(Targets,outputs)
%%
% Another measure of how well the neural network has fit data is the
% receiver operating characteristic plot. This shows how the false
% positive and true positive rates relate as the thresholding of outputs
% is varied from 0 to 1.
%
% The farther left and up the line is, the fewer false positives need to
% be accepted in order to get a high true positive rate. The best
% classifiers will have a line going from the bottom left corner, to the
% top left corner, to the top right corner, or close to that.

figure, plotroc(Targets,outputs)
```

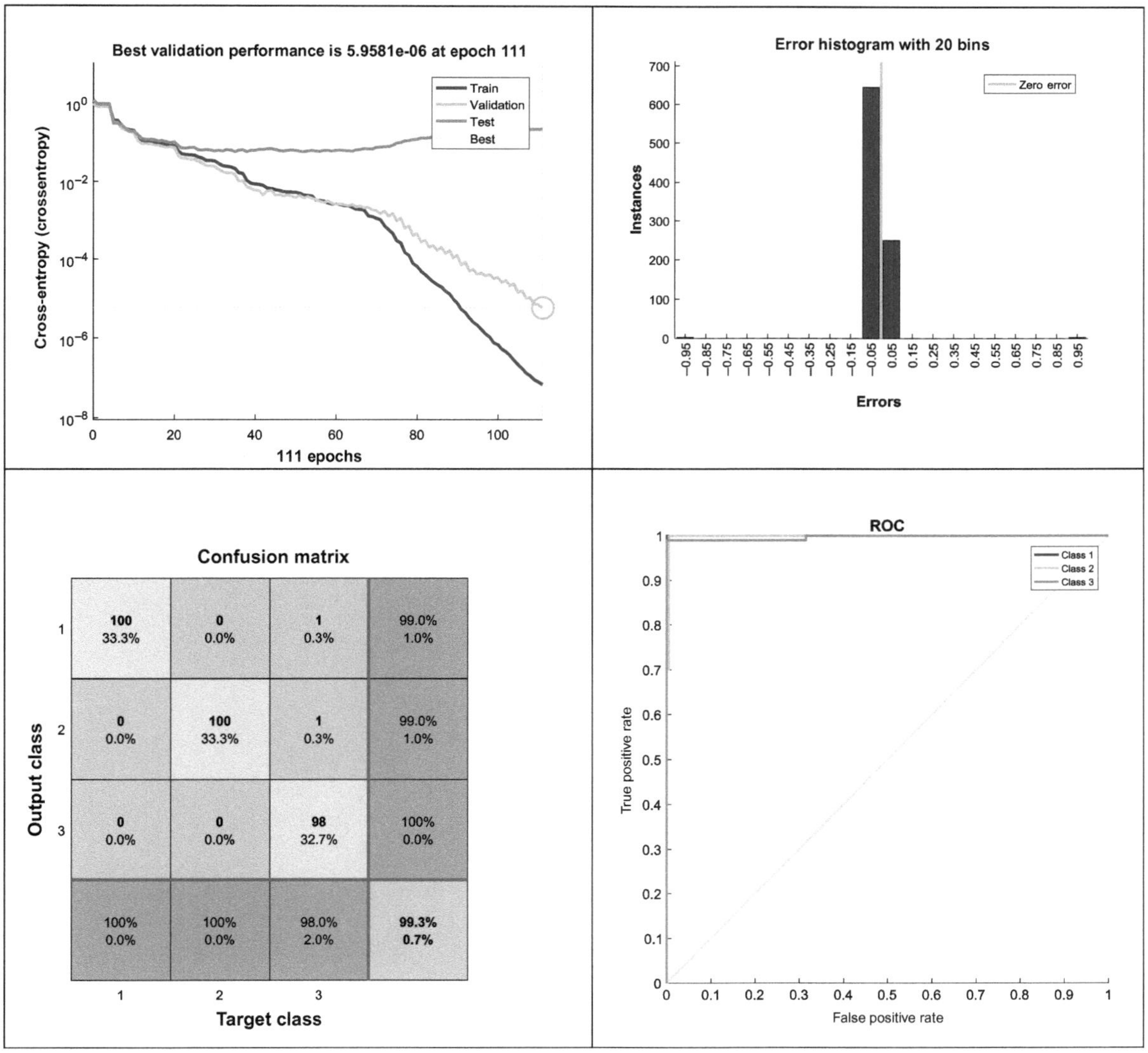

FIG. 5.23 Representation of performance of the ANN classifier for the EEG signals features extracted using wavelet packet transform (WPT).

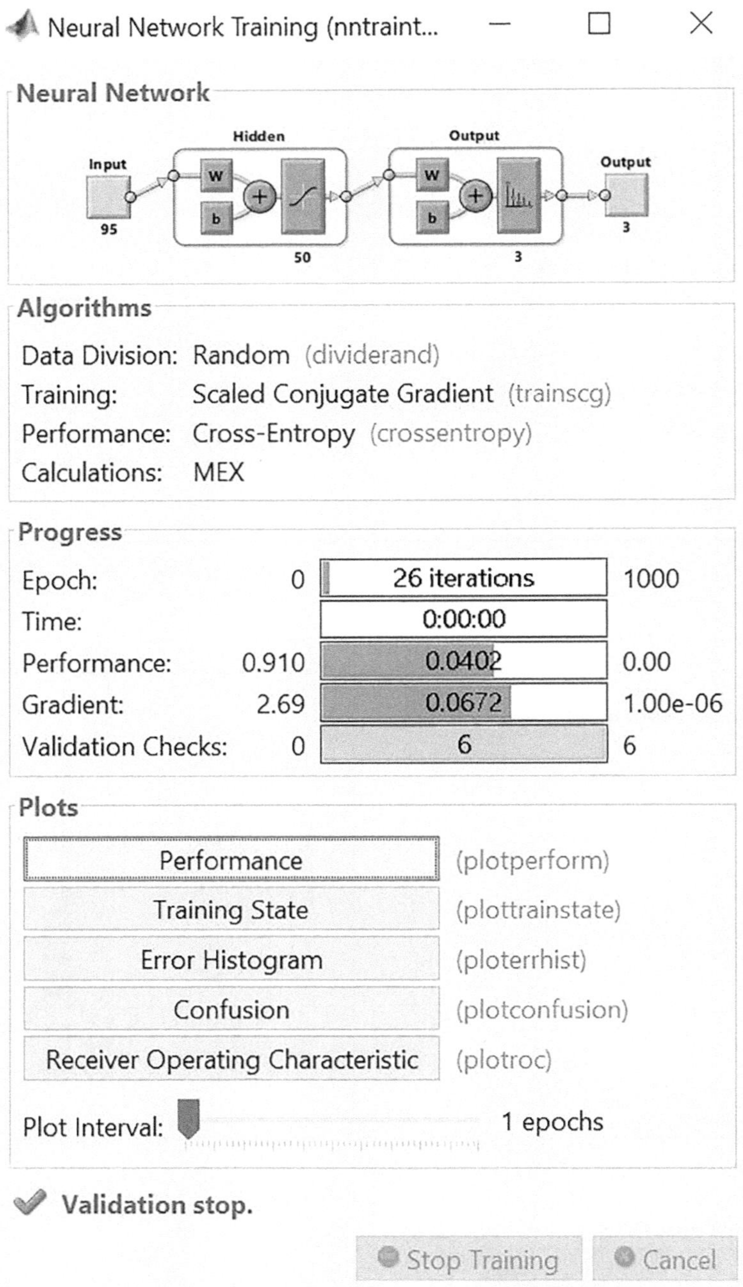

FIG. 5.24 The Neural Network Training tool of the ANN classifier for the EEG signals features extracted using WPT.

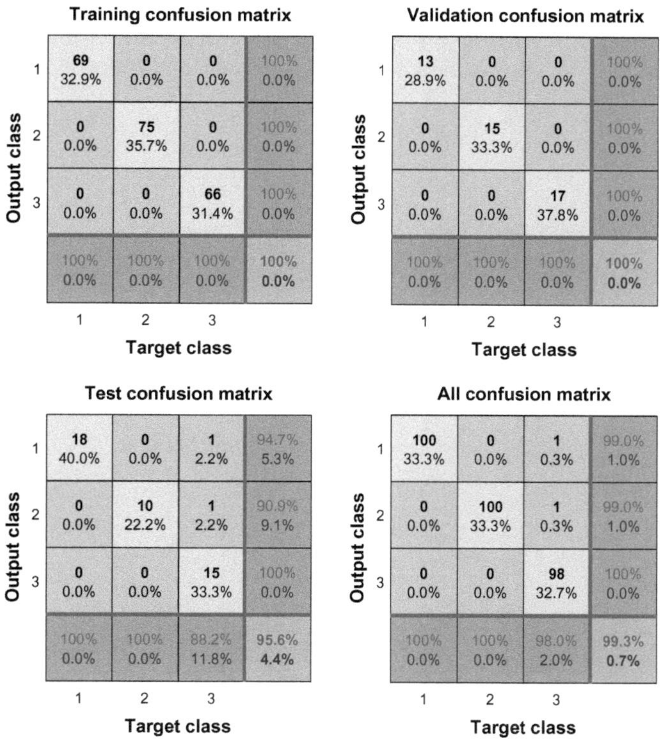

FIG. 5.25 Confusion matrices of train, validation, test, and all sets of the ANN classifier for EEG signals features extracted using WPT.

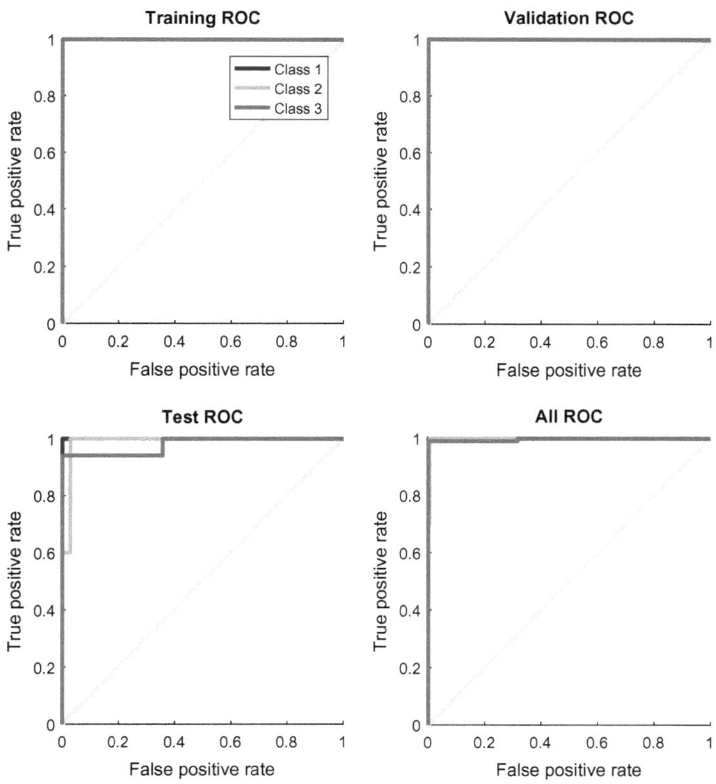

FIG. 5.26 ROC area of train, validation, test, and all sets of the ANN classifier for EEG signals features extracted using WPT.

EXAMPLE 5.18. The following MATLAB code was used to extract features from the EEG signals using stationary wavelet transform (SWT). Then it used statistical values of SWT subbands. Then it classified these data using ANN classifier. You can download data from the following website: http://epileptologie-bonn.de/cms/front_content.php?idcat=193&lang=3&changelang=3

```
%% Ch5_Ex38_EEG_SWT_ANN.m
%SWT of NORMAL, INTERICTAL and ICTAL EEG signals
%The following MATLAB code is used to Extract SWT features from the EEG signals
%Decompose EEG data using SWT Transfrom
%Then uses Statistical features as:

%(1) Mean of the absolute values of the coefficients in each sub-band.
%(2) Standard deviation of the coefficients in each sub-band.
%(3) Skewness of the coefficients in each sub-band.
%(4) Kurtosis of the coefficients in each sub-band.
%(5) RMS power of the wavelet coefficients in each subband.
%(6) Ratio of the mean absolute values of adjacent subbands.

%Then it classifies EEG Signals Using ANN Classifier
clc
clear
%Load Sample EEG Data downloaded from the web site
%http://epileptologie-bonn.de/cms/front_content.php?idcat=193&lang=3&changelang=3
load AS_BONN_ALL_EEG_DATA_4096.mat

wname = 'db4';
Level = 8;
Length = 4096;      % Length of signal
Nofsignal=100; %Number of Signal
%%
%%%%%%%%%%%%%%%%%%%%%%%%%%%%% A Z.ZIP EYES OPEN NORMAL SUBJECT

for i=1:Nofsignal;

[swa,swd] = swt(Normal_Eyes_Open(1:Length,i),Level,wname);%SWT

ASData(i,1)=mean(abs(swa(1,:)));
ASData(i,2)=mean(abs(swa(2,:)));
ASData(i,3)=mean(abs(swa(3,:)));
ASData(i,4)=mean(abs(swa(4,:)));
ASData(i,5)=mean(abs(swa(5,:)));
ASData(i,6)=mean(abs(swa(6,:)));
ASData(i,7)=mean(abs(swa(7,:)));
ASData(i,8)=mean(abs(swa(8,:)));
ASData(i,9)=mean(abs(swd(1,:)));
ASData(i,10)=mean(abs(swd(2,:)));
ASData(i,11)=mean(abs(swd(3,:)));
ASData(i,12)=mean(abs(swd(4,:)));
ASData(i,13)=mean(abs(swd(5,:)));
ASData(i,14)=mean(abs(swd(6,:)));
ASData(i,15)=mean(abs(swd(7,:)));
ASData(i,16)=mean(abs(swd(8,:)));

ASData(i,17)=std(swa(1,:));
ASData(i,18)=std(swa(2,:));
ASData(i,19)=std(swa(3,:));
ASData(i,20)=std(swa(4,:));
ASData(i,21)=std(swa(5,:));
ASData(i,22)=std(swa(6,:));
ASData(i,23)=std(swa(7,:));
```

```
ASData(i,24)=std(swa(8,:));
ASData(i,25)=std(swd(1,:));
ASData(i,26)=std(swd(2,:));
ASData(i,27)=std(swd(3,:));
ASData(i,28)=std(swd(4,:));
ASData(i,29)=std(swd(5,:));
ASData(i,30)=std(swd(6,:));
ASData(i,31)=std(swd(7,:));
ASData(i,32)=std(swd(8,:));

ASData(i,33)=skewness(swa(1,:));
ASData(i,34)=skewness(swa(2,:));
ASData(i,35)=skewness(swa(3,:));
ASData(i,36)=skewness(swa(4,:));
ASData(i,37)=skewness(swa(5,:));
ASData(i,38)=skewness(swa(6,:));
ASData(i,39)=skewness(swa(7,:));
ASData(i,40)=skewness(swa(8,:));
ASData(i,41)=skewness(swd(1,:));
ASData(i,42)=skewness(swd(2,:));
ASData(i,43)=skewness(swd(3,:));
ASData(i,44)=skewness(swd(4,:));
ASData(i,45)=skewness(swd(5,:));
ASData(i,46)=skewness(swd(6,:));
ASData(i,47)=skewness(swd(7,:));
ASData(i,48)=skewness(swd(8,:));

ASData(i,49)=kurtosis(swa(1,:));
ASData(i,50)=kurtosis(swa(2,:));
ASData(i,51)=kurtosis(swa(3,:));
ASData(i,52)=kurtosis(swa(4,:));
ASData(i,53)=kurtosis(swa(5,:));
ASData(i,54)=kurtosis(swa(6,:));
ASData(i,55)=kurtosis(swa(7,:));
ASData(i,56)=kurtosis(swa(8,:));
ASData(i,57)=kurtosis(swd(1,:));
ASData(i,58)=kurtosis(swd(2,:));
ASData(i,59)=kurtosis(swd(3,:));
ASData(i,60)=kurtosis(swd(4,:));
ASData(i,61)=kurtosis(swd(5,:));
ASData(i,62)=kurtosis(swd(6,:));
ASData(i,63)=kurtosis(swd(7,:));
ASData(i,64)=kurtosis(swd(8,:));

        ASData(i,65)=rms(swa(1,:));
        ASData(i,66)=rms(swa(2,:));
        ASData(i,67)=rms(swa(3,:));
        ASData(i,68)=rms(swa(4,:));
        ASData(i,69)=rms(swa(5,:));
        ASData(i,70)=rms(swa(6,:));
        ASData(i,71)=rms(swa(7,:));
        ASData(i,72)=rms(swa(8,:));
        ASData(i,73)=rms(swd(1,:));
        ASData(i,74)=rms(swd(2,:));
        ASData(i,75)=rms(swd(3,:));
        ASData(i,76)=rms(swd(4,:));
        ASData(i,77)=rms(swd(5,:));
        ASData(i,78)=rms(swd(6,:));
        ASData(i,79)=rms(swd(7,:));
        ASData(i,80)=rms(swd(8,:));
```

```
        ASData(i,81)=mean(abs(swa(1,:)))/mean(abs(swa(2,:)));
        ASData(i,82)=mean(abs(swa(2,:)))/mean(abs(swa(3,:)));
        ASData(i,83)=mean(abs(swa(3,:)))/mean(abs(swa(4,:)));
        ASData(i,84)=mean(abs(swa(4,:)))/mean(abs(swa(5,:)));
        ASData(i,85)=mean(abs(swa(5,:)))/mean(abs(swa(6,:)));
        ASData(i,86)=mean(abs(swa(6,:)))/mean(abs(swa(7,:)));
        ASData(i,87)=mean(abs(swa(7,:)))/mean(abs(swa(8,:)));
        ASData(i,88)=mean(abs(swa(8,:)))/mean(abs(swd(1,:)));
        ASData(i,89)=mean(abs(swd(1,:)))/mean(abs(swd(2,:)));
        ASData(i,90)=mean(abs(swd(2,:)))/mean(abs(swd(3,:)));
        ASData(i,91)=mean(abs(swd(3,:)))/mean(abs(swd(4,:)));
        ASData(i,92)=mean(abs(swd(4,:)))/mean(abs(swd(5,:)));
        ASData(i,93)=mean(abs(swd(5,:)))/mean(abs(swd(6,:)));
        ASData(i,94)=mean(abs(swd(6,:)))/mean(abs(swd(7,:)));
        ASData(i,95)=mean(abs(swd(7,:)))/mean(abs(swd(8,:)));

end

%%
% D F.ZIP EPILEPTIC SUBJECT (INTERICTAL) were recorded from within
% the epileptogenic zone during seizure free intervals

for i=Nofsignal+1:2*Nofsignal;

[swa,swd] = swt(Interictal(1:Length,i-Nofsignal),Level,wname); %SWT

ASData(i,1)=mean(abs(swa(1,:)));
ASData(i,2)=mean(abs(swa(2,:)));
ASData(i,3)=mean(abs(swa(3,:)));
ASData(i,4)=mean(abs(swa(4,:)));
ASData(i,5)=mean(abs(swa(5,:)));
ASData(i,6)=mean(abs(swa(6,:)));
ASData(i,7)=mean(abs(swa(7,:)));
ASData(i,8)=mean(abs(swa(8,:)));
ASData(i,9)=mean(abs(swd(1,:)));
ASData(i,10)=mean(abs(swd(2,:)));
ASData(i,11)=mean(abs(swd(3,:)));
ASData(i,12)=mean(abs(swd(4,:)));
ASData(i,13)=mean(abs(swd(5,:)));
ASData(i,14)=mean(abs(swd(6,:)));
ASData(i,15)=mean(abs(swd(7,:)));
ASData(i,16)=mean(abs(swd(8,:)));

ASData(i,17)=std(swa(1,:));
ASData(i,18)=std(swa(2,:));
ASData(i,19)=std(swa(3,:));
ASData(i,20)=std(swa(4,:));
ASData(i,21)=std(swa(5,:));
ASData(i,22)=std(swa(6,:));
ASData(i,23)=std(swa(7,:));
ASData(i,24)=std(swa(8,:));
ASData(i,25)=std(swd(1,:));
ASData(i,26)=std(swd(2,:));
ASData(i,27)=std(swd(3,:));
ASData(i,28)=std(swd(4,:));
ASData(i,29)=std(swd(5,:));
ASData(i,30)=std(swd(6,:));
ASData(i,31)=std(swd(7,:));
ASData(i,32)=std(swd(8,:));
```

```
ASData(i,33)=skewness(swa(1,:));
ASData(i,34)=skewness(swa(2,:));
ASData(i,35)=skewness(swa(3,:));
ASData(i,36)=skewness(swa(4,:));
ASData(i,37)=skewness(swa(5,:));
ASData(i,38)=skewness(swa(6,:));
ASData(i,39)=skewness(swa(7,:));
ASData(i,40)=skewness(swa(8,:));
ASData(i,41)=skewness(swd(1,:));
ASData(i,42)=skewness(swd(2,:));
ASData(i,43)=skewness(swd(3,:));
ASData(i,44)=skewness(swd(4,:));
ASData(i,45)=skewness(swd(5,:));
ASData(i,46)=skewness(swd(6,:));
ASData(i,47)=skewness(swd(7,:));
ASData(i,48)=skewness(swd(8,:));

ASData(i,49)=kurtosis(swa(1,:));
ASData(i,50)=kurtosis(swa(2,:));
ASData(i,51)=kurtosis(swa(3,:));
ASData(i,52)=kurtosis(swa(4,:));
ASData(i,53)=kurtosis(swa(5,:));
ASData(i,54)=kurtosis(swa(6,:));
ASData(i,55)=kurtosis(swa(7,:));
ASData(i,56)=kurtosis(swa(8,:));
ASData(i,57)=kurtosis(swd(1,:));
ASData(i,58)=kurtosis(swd(2,:));
ASData(i,59)=kurtosis(swd(3,:));
ASData(i,60)=kurtosis(swd(4,:));
ASData(i,61)=kurtosis(swd(5,:));
ASData(i,62)=kurtosis(swd(6,:));
ASData(i,63)=kurtosis(swd(7,:));
ASData(i,64)=kurtosis(swd(8,:));

        ASData(i,65)=rms(swa(1,:));
        ASData(i,66)=rms(swa(2,:));
        ASData(i,67)=rms(swa(3,:));
        ASData(i,68)=rms(swa(4,:));
        ASData(i,69)=rms(swa(5,:));
        ASData(i,70)=rms(swa(6,:));
        ASData(i,71)=rms(swa(7,:));
        ASData(i,72)=rms(swa(8,:));
        ASData(i,73)=rms(swd(1,:));
        ASData(i,74)=rms(swd(2,:));
        ASData(i,75)=rms(swd(3,:));
        ASData(i,76)=rms(swd(4,:));
        ASData(i,77)=rms(swd(5,:));
        ASData(i,78)=rms(swd(6,:));
        ASData(i,79)=rms(swd(7,:));
        ASData(i,80)=rms(swd(8,:));

        ASData(i,81)=mean(abs(swa(1,:)))/mean(abs(swa(2,:)));
        ASData(i,82)=mean(abs(swa(2,:)))/mean(abs(swa(3,:)));
        ASData(i,83)=mean(abs(swa(3,:)))/mean(abs(swa(4,:)));
        ASData(i,84)=mean(abs(swa(4,:)))/mean(abs(swa(5,:)));
        ASData(i,85)=mean(abs(swa(5,:)))/mean(abs(swa(6,:)));
        ASData(i,86)=mean(abs(swa(6,:)))/mean(abs(swa(7,:)));
        ASData(i,87)=mean(abs(swa(7,:)))/mean(abs(swa(8,:)));
        ASData(i,88)=mean(abs(swa(8,:)))/mean(abs(swd(1,:)));
        ASData(i,89)=mean(abs(swd(1,:)))/mean(abs(swd(2,:)));
```

```
        ASData(i,90)=mean(abs(swd(2,:)))/mean(abs(swd(3,:)));
        ASData(i,91)=mean(abs(swd(3,:)))/mean(abs(swd(4,:)));
        ASData(i,92)=mean(abs(swd(4,:)))/mean(abs(swd(5,:)));
        ASData(i,93)=mean(abs(swd(5,:)))/mean(abs(swd(6,:)));
        ASData(i,94)=mean(abs(swd(6,:)))/mean(abs(swd(7,:)));
        ASData(i,95)=mean(abs(swd(7,:)))/mean(abs(swd(8,:)));

end

%% E S.ZIP EPILEPTIC SUBJECT ICTAL DURING SEIZURE
%%%%%%%%%%%%%%%%%%%%%%%%%%%%

for i=2*Nofsignal+1:3*Nofsignal;

[swa,swd] = swt(Ictal(1:Length,i-2*Nofsignal),Level,wname);%SWT

ASData(i,1)=mean(abs(swa(1,:)));
ASData(i,2)=mean(abs(swa(2,:)));
ASData(i,3)=mean(abs(swa(3,:)));
ASData(i,4)=mean(abs(swa(4,:)));
ASData(i,5)=mean(abs(swa(5,:)));
ASData(i,6)=mean(abs(swa(6,:)));
ASData(i,7)=mean(abs(swa(7,:)));
ASData(i,8)=mean(abs(swa(8,:)));
ASData(i,9)=mean(abs(swd(1,:)));
ASData(i,10)=mean(abs(swd(2,:)));
ASData(i,11)=mean(abs(swd(3,:)));
ASData(i,12)=mean(abs(swd(4,:)));
ASData(i,13)=mean(abs(swd(5,:)));
ASData(i,14)=mean(abs(swd(6,:)));
ASData(i,15)=mean(abs(swd(7,:)));
ASData(i,16)=mean(abs(swd(8,:)));

ASData(i,17)=std(swa(1,:));
ASData(i,18)=std(swa(2,:));
ASData(i,19)=std(swa(3,:));
ASData(i,20)=std(swa(4,:));
ASData(i,21)=std(swa(5,:));
ASData(i,22)=std(swa(6,:));
ASData(i,23)=std(swa(7,:));
ASData(i,24)=std(swa(8,:));
ASData(i,25)=std(swd(1,:));
ASData(i,26)=std(swd(2,:));
ASData(i,27)=std(swd(3,:));
ASData(i,28)=std(swd(4,:));
ASData(i,29)=std(swd(5,:));
ASData(i,30)=std(swd(6,:));
ASData(i,31)=std(swd(7,:));
ASData(i,32)=std(swd(8,:));

ASData(i,33)=skewness(swa(1,:));
ASData(i,34)=skewness(swa(2,:));
ASData(i,35)=skewness(swa(3,:));
ASData(i,36)=skewness(swa(4,:));
ASData(i,37)=skewness(swa(5,:));
ASData(i,38)=skewness(swa(6,:));
ASData(i,39)=skewness(swa(7,:));
ASData(i,40)=skewness(swa(8,:));
ASData(i,41)=skewness(swd(1,:));
ASData(i,42)=skewness(swd(2,:));
```

```
ASData(i,43)=skewness(swd(3,:));
ASData(i,44)=skewness(swd(4,:));
ASData(i,45)=skewness(swd(5,:));
ASData(i,46)=skewness(swd(6,:));
ASData(i,47)=skewness(swd(7,:));
ASData(i,48)=skewness(swd(8,:));

ASData(i,49)=kurtosis(swa(1,:));
ASData(i,50)=kurtosis(swa(2,:));
ASData(i,51)=kurtosis(swa(3,:));
ASData(i,52)=kurtosis(swa(4,:));
ASData(i,53)=kurtosis(swa(5,:));
ASData(i,54)=kurtosis(swa(6,:));
ASData(i,55)=kurtosis(swa(7,:));
ASData(i,56)=kurtosis(swa(8,:));
ASData(i,57)=kurtosis(swd(1,:));
ASData(i,58)=kurtosis(swd(2,:));
ASData(i,59)=kurtosis(swd(3,:));
ASData(i,60)=kurtosis(swd(4,:));
ASData(i,61)=kurtosis(swd(5,:));
ASData(i,62)=kurtosis(swd(6,:));
ASData(i,63)=kurtosis(swd(7,:));
ASData(i,64)=kurtosis(swd(8,:));

        ASData(i,65)=rms(swa(1,:));
        ASData(i,66)=rms(swa(2,:));
        ASData(i,67)=rms(swa(3,:));
        ASData(i,68)=rms(swa(4,:));
        ASData(i,69)=rms(swa(5,:));
        ASData(i,70)=rms(swa(6,:));
        ASData(i,71)=rms(swa(7,:));
        ASData(i,72)=rms(swa(8,:));
        ASData(i,73)=rms(swd(1,:));
        ASData(i,74)=rms(swd(2,:));
        ASData(i,75)=rms(swd(3,:));
        ASData(i,76)=rms(swd(4,:));
        ASData(i,77)=rms(swd(5,:));
        ASData(i,78)=rms(swd(6,:));
        ASData(i,79)=rms(swd(7,:));
        ASData(i,80)=rms(swd(8,:));

        ASData(i,81)=mean(abs(swa(1,:)))/mean(abs(swa(2,:)));
        ASData(i,82)=mean(abs(swa(2,:)))/mean(abs(swa(3,:)));
        ASData(i,83)=mean(abs(swa(3,:)))/mean(abs(swa(4,:)));
        ASData(i,84)=mean(abs(swa(4,:)))/mean(abs(swa(5,:)));
        ASData(i,85)=mean(abs(swa(5,:)))/mean(abs(swa(6,:)));
        ASData(i,86)=mean(abs(swa(6,:)))/mean(abs(swa(7,:)));
        ASData(i,87)=mean(abs(swa(7,:)))/mean(abs(swa(8,:)));
        ASData(i,88)=mean(abs(swa(8,:)))/mean(abs(swd(1,:)));
        ASData(i,89)=mean(abs(swd(1,:)))/mean(abs(swd(2,:)));
        ASData(i,90)=mean(abs(swd(2,:)))/mean(abs(swd(3,:)));
        ASData(i,91)=mean(abs(swd(3,:)))/mean(abs(swd(4,:)));
        ASData(i,92)=mean(abs(swd(4,:)))/mean(abs(swd(5,:)));
        ASData(i,93)=mean(abs(swd(5,:)))/mean(abs(swd(6,:)));
        ASData(i,94)=mean(abs(swd(6,:)))/mean(abs(swd(7,:)));
        ASData(i,95)=mean(abs(swd(7,:)))/mean(abs(swd(8,:)));
end

%Transpose the Data Matrix
Inputs=ASData';
```

```
%% TARGET GENERATION
% Your target set must have one label for each sample,
%so it must contains 300 elements. Here is an easy way to build one.
%Start by numbering your classes from 1 to 3.
%Make sure that your X Array is sorted by class,
%meaning all the samples from the first class,
%then all the samples from the second class, and so on. Then:

 % Classes from 1 to 3
 y = 1:3;
 % Nofsignal samples per class
 y = repmat(y, Nofsignal, 1);
 % Reshape to obtain a vector
 y = reshape(y, 1, numel(y));
% At this stage, you should have a 1-by-300 vector with numeric class labels from 1 to 3.
%This is what you want if you use the Statistics and Machine Learning Toolbox.
%
% Now, I suppose you intend to use the Neural Network Toolbox.
%In that case the syntax is a bit different: the class labels are in a 3-by-300 matrix,
%each column being a sample, with the row corresponding to the class getting value '1' and the
rest '0'.
%To build it from the previous syntax, type:

Targets = full(ind2vec(y));

%% CLASSIFICATION
% Solve a Pattern Recognition Problem with a Neural Network
% Script generated by NPRTOOL
%
% Create a Pattern Recognition Network
hiddenLayerSize = 50;
net = patternnet(hiddenLayerSize);
% Set up Division of Data for Training, Validation, Testing
net.divideParam.trainRatio = 70/100;
net.divideParam.valRatio = 15/100;
net.divideParam.testRatio = 15/100;
% Train the Network
[net,tr] = train(net,Inputs,Targets);
% Test the Network
outputs = net(Inputs);
errors = gsubtract(Targets,outputs);
performance = perform(net,Targets,outputs)
% View the Network
view(net)
% Performance Plots
% Comment/Uncomment these lines to disable/enable various plots.
figure, plotperform(tr)
figure, plottrainstate(tr)
figure, ploterrhist(errors)

%%
% One measure of how well the neural network has fit the data is the
% confusion plot. Here the confusion matrix is plotted across all samples.
%
% The confusion matrix shows the percentages of correct and incorrect
% classifications. Correct classifications are the green squares on the
% matrices diagonal. Incorrect classifications form the red squares.
%
% If the network has learned to classify properly, the percentages in the
```

```
% red squares should be very small, indicating few misclassifications.
%
% If this is not the case then further training, or training a network
% with more hidden neurons, would be advisable.

figure, plotconfusion(Targets,outputs)
%%
% Another measure of how well the neural network has fit data is the
% receiver operating characteristic plot. This shows how the false
% positive and true positive rates relate as the thresholding of outputs
% is varied from 0 to 1.
%
% The farther left and up the line is, the fewer false positives need to
% be accepted in order to get a high true positive rate. The best
% classifiers will have a line going from the bottom left corner, to the
% top left corner, to the top right corner, or close to that.

figure, plotroc(Targets,outputs)
```

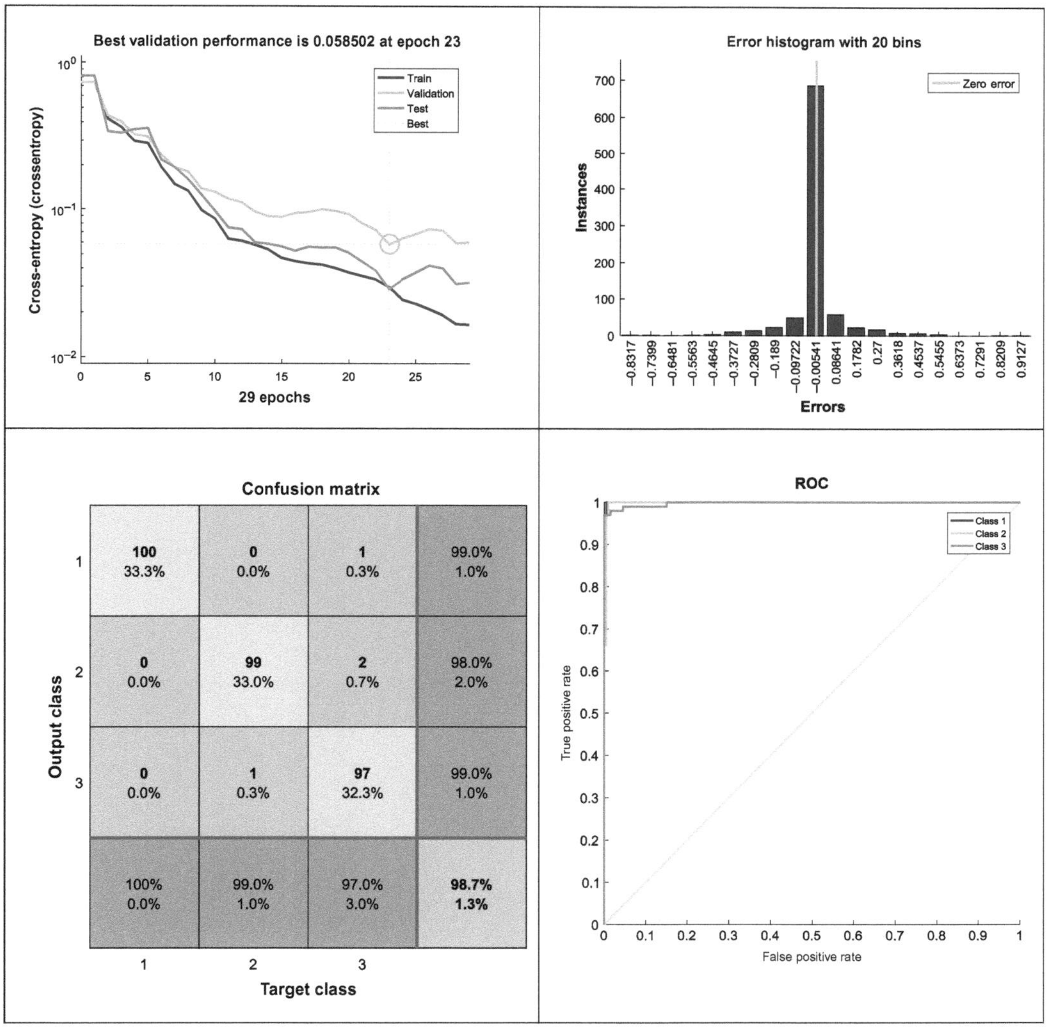

FIG. 5.27 Representation of performance of the ANN classifier for the EEG signals features extracted using SWT.

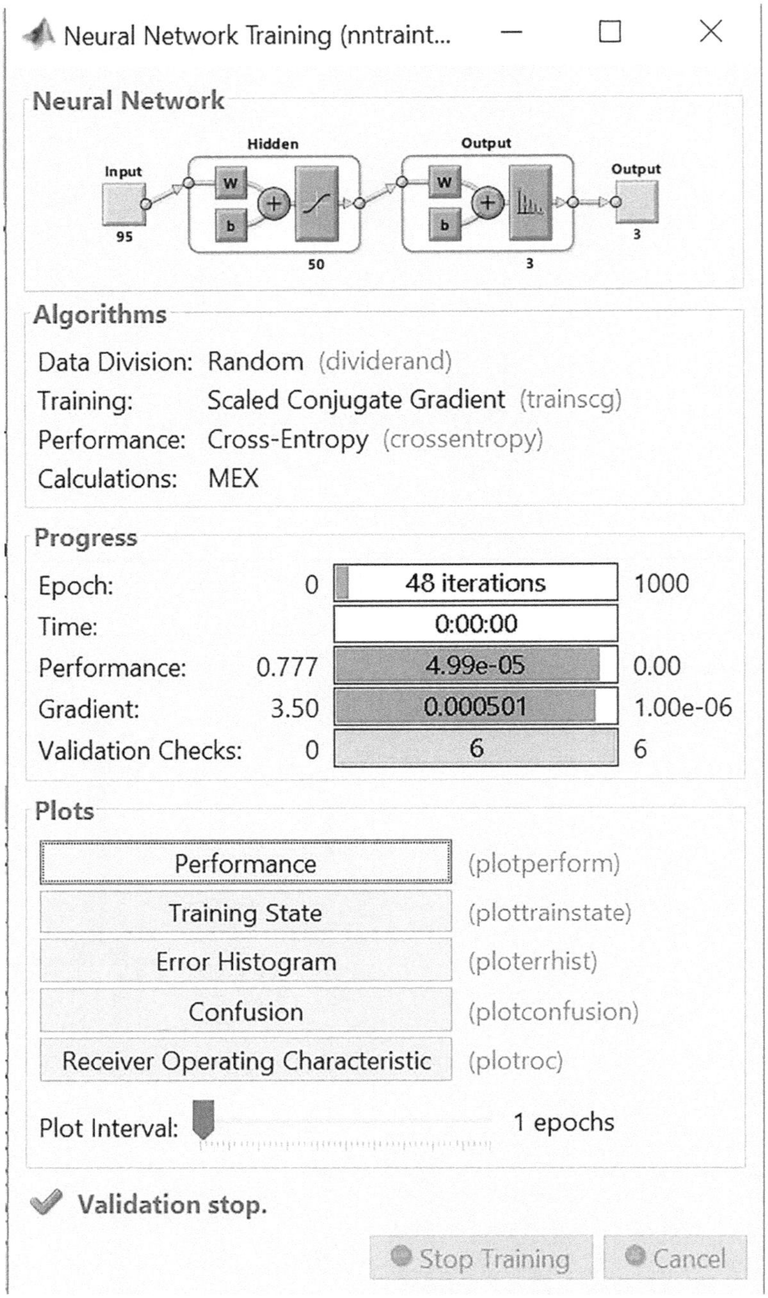

FIG. 5.28 The Neural Network Training tool of the ANN classifier for the EEG signals features extracted using SWT.

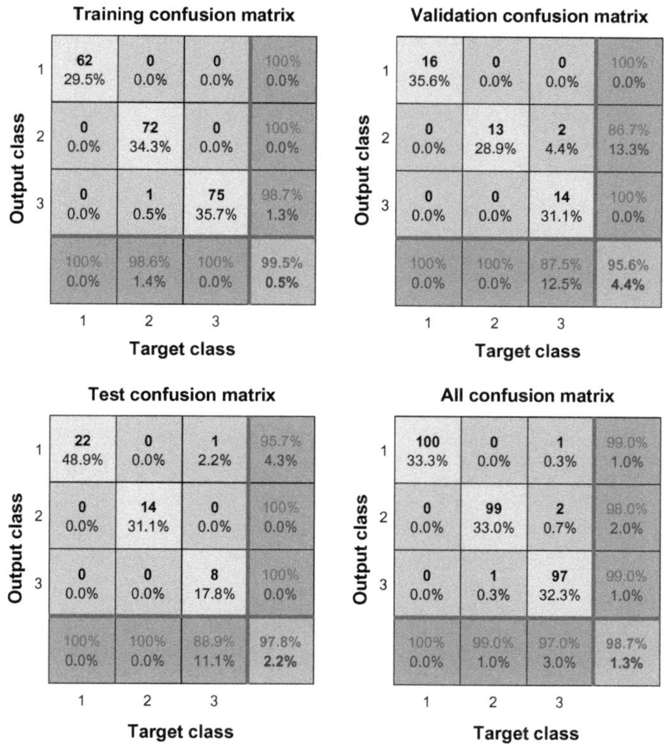

FIG. 5.29 Confusion matrices of train, validation, test, and all sets of the ANN classifier for EEG signals features extracted using SWT.

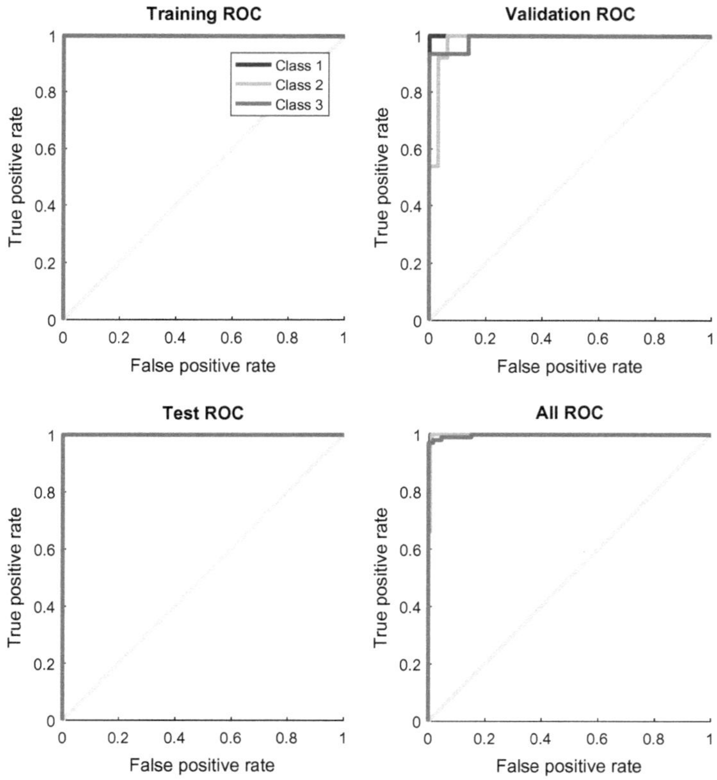

FIG. 5.30 ROC area of train, validation, test, and all sets of the ANN classifier for EEG signals features extracted using SWT.

5.7 SUPPORT VECTOR MACHINES

SVMs were proposed by Vapnik as a supervised learning method. SVM can classify both linear and nonlinear data. SVM is an algorithm that uses nonlinear mapping to transform the original training data into a higher dimension. Within this new dimension, it searches for the linear optimal separating hyperplane. With appropriate nonlinear mapping to a sufficiently high dimension, data from two classes can always be separated by a hyperplane. The SVM finds this hyperplane using support vectors and margins. Although the training time of even the fastest SVMs can be extremely slow, they are highly accurate, owing to their ability to model complex nonlinear decision boundaries. They are much less prone to overfitting than other methods. If the data are not linearly separable, no straight line can be found that would separate the classes. The linear SVMs we studied would not be able to find a feasible solution here. The approach described for linear SVMs can be extended to create nonlinear SVMs for the classification of linearly inseparable data. Such SVMs are capable of finding nonlinear decision boundaries in input space. We obtain a nonlinear SVM by extending the approach for linear SVMs with two main steps. In the first step, we transform the original input data into a higher dimensional space using nonlinear mapping. Once the data have been transformed into the new higher space, the second step searches for a linear separating hyperplane in the new space. We again end up with a quadratic optimization problem that can be solved using the linear SVM formulation. The maximal marginal hyperplane found in the new space corresponds to a nonlinear separating hypersurface in the original space (Han, Pei, & Kamber, 2011).

There is no modification in the formulation of the SVM for the multidimensional cases. The dimension of the hyperplane changes depending on the number of feature types. In such nonseparable cases, utilization of a nonlinear function may benefit to separate the datasets. Kernel mapping provides an alternative solution by nonlinearly mapping the data into a higher dimensional feature space to separate such cases. A highly nonlinear classification function, such as a polynomial or a radial basis function or even a sigmoidal neural network, can be trained by employing a strong and effective algorithm that does not suffer from local minima. Radial basis function is the most widely known nonlinear kernel employed in SVM for nonseparable classes (Sanei, 2013).

EXAMPLE 5.19. The following MATLAB code was used to extract features from the EEG signals using AR Burg method. Then it classified these data using SVMs. You can download data from the following website: http://epileptologie-bonn.de/cms/front_content.php?idcat=193&lang=3&changelang=3

```
%% Ch5_Ex41_EEG_ARBURG_SVM.m
%The following MATLAB code is used to extract the features from
%the EEG signals using AR Burg.
%Then it classifies EEG Signals Using SVM Classifier
% Cross-Validate ECOC SVM Classifier

clc
clear
%Load Sample EEG Data downloaded from the web site
%http://epileptologie-bonn.de/cms/front_content.php?idcat=193&lang=3&changelang=3
load AS_BONN_ALL_EEG_DATA_4096.mat
%%
Fs = 173.6;           % Sampling frequency
T = 1/Fs;             % Sampling period
segmentLength = 128; % Length of a signal segment
Length=4096;% Length of signal
Nofsignal=100; %Number of Signal
order = 14;

%%
% Obtain the AR Burg Spectrum of the Normal EEG signal using pburg.
for i=1:Nofsignal
    [Pxx,F] = pburg(Normal_Eyes_Open(1:Length,i),order,segmentLength,Fs);
    ASData(i,:)=Pxx(:);
end
%%
% Obtain the AR Burg Spectrum of the Interictal EEG signal using pburg.
for i=Nofsignal+1:2*Nofsignal
```

```
    [Pxx,F] = pburg(Interictal(1:Length,i-Nofsignal),order,segmentLength,Fs);
    ASData(i,:)=Pxx(:);
end
%%
% Obtain the AR Burg Spectrum of the Ictal EEG signal using pburg.
for i=2*Nofsignal+1:3*Nofsignal
[Pxx,F] = pburg(Ictal(1:Length,i-2*Nofsignal),order,segmentLength,Fs);
ASData(i,:)=Pxx(:);
end

%Make input the Data Matrix
Inputs=ASData;

%%
%You can create target using excel and the import it from the MATLAB HOME
%menu and Import data and then save it as EEGTargets
%Load Target
load EEGTargets

%% CLASSIFACATION
rng(1); % For reproducibility
% Create an SVM template. It is good practice to standardize the predictors.
t = templateSVM('Standardize',1)
% Train the ECOC classifier. It is good practice to specify the class order.
Mdl = fitcecoc(Inputs,Targets,'Learners',t,...
'ClassNames',{'NORMAL','INTERICTAL','ICTAL'});
% % Mdl is a ClassificationECOC classifier. You can access its properties using dot
% % notation.
% % Cross-validate Mdl using 10-fold cross-validation.
CVMdl = crossval(Mdl);
% CVMdl is a ClassificationPartitionedECOC cross-validated ECOC classifier.
% Estimate the generalization error.
oosLoss = kfoldLoss(CVMdl)

%% Display the Confusion Matrix

MdlConf = fitcecoc(Inputs,Targets,'Learners',t,...
'ClassNames',{'NORMAL','INTERICTAL','ICTAL'});
svmClass = resubPredict(MdlConf);
[svmResubCM,grpOrder] = confusionmat(Targets,svmClass)

% Calculate the Total Classification Accuracy
TotalAccuracy=(svmResubCM(1,1)+svmResubCM(2,2)+svmResubCM(3,3))/(3*Nofsignal)

disp('Total Classification Accuracy=')
disp(TotalAccuracy)
```

EXAMPLE 5.20 The following MATLAB code was used to extract features from the ECG signals using AR Burg method. Then it classified these data using SVMs. You can download data from the following website: https://www.physionet.org/physiobank/database/mitdb/

```
%% Ch5_Ex42_ECG_ARBURG_SVM.m
%The following MATLAB code is used to extract the features from
%the ECG signals using AR Burg spectrum.
%Then it classifies ECG Signals Using SVM Classifier
```

```
% Cross-Validate ECOC SVM Classifier
clc
clear
%Load Sample ECG Data downloaded from the web site
%https://www.physionet.org/physiobank/database/mitdb/
load MITBIH_ECG.mat
%%
Fs = 320;          % Sampling frequency
T = 1/Fs;            % Sampling period
segmentLength = 128; % Length of a signal segment
Length=320;% Length of signal
Nofsignal=300; %Number of Signal
order = 34;
%%
% Obtain the AR Burg Spectrum of the Normal ECG signal using pburg.
for i=1:Nofsignal
    [Pxx,F] = pburg(ECGN(1:Length,i),order,segmentLength,Fs);
    ASData(i,:)=Pxx(:);
end
%%
% Obtain the AR Burg Spectrum of the ECG signal with APC using pburg.
[Pxx,F] = pburg(ECGAPC(:,1),order,segmentLength,Fs);
for i=Nofsignal+1:2*Nofsignal
    [Pxx,F] = pburg(ECGAPC(1:Length,i-Nofsignal),order,segmentLength,Fs);
    ASData(i,:)=Pxx(:);
end
%%
% Obtain the AR Burg Spectrum of the ECG signal with PVC using pburg.
for i=2*Nofsignal+1:3*Nofsignal
[Pxx,F] = pburg(ECGPVC(1:Length,i-2*Nofsignal),order,segmentLength,Fs);
ASData(i,:)=Pxx(:);
end

%%
% Obtain the AR Burg Spectrum of the ECG signal with LBBB using pburg.
for i=3*Nofsignal+1:4*Nofsignal
[Pxx,F] = pburg(ECGLBBB(1:Length,i-3*Nofsignal),order,segmentLength,Fs);
ASData(i,:)=Pxx(:);
end

%%
% Obtain the Covariance Spectrum of the ECG signal with RBBB using pburg.
for i=4*Nofsignal+1:5*Nofsignal
[Pxx,F] = pburg(ECGRBBB(1:Length,i-4*Nofsignal),order,segmentLength,Fs);
ASData(i,:)=Pxx(:);
end

%Create the Data Matrix as Inputs
Inputs=ASData;
%%
%You can create target using excel and the import it from the MATLAB HOME
%menu and Import data and then save it as EEGTargets
%Load Target
load ECGTargets

%% CLASSIFACATION
rng(1); % For reproducibility
% Create an SVM template. It is good practice to standardize the predictors.
```

```
t = templateSVM('Standardize',1)
% Train the ECOC classifier. It is good practice to specify the class order.
Mdl = fitcecoc(Inputs,Targets,'Learners',t,...
'ClassNames',{'NORMAL','APC','PVC','LBBB','RBBB'});
% % Mdl is a ClassificationECOC classifier. You can access its properties using dot
% % notation.
% % Cross-validate Mdl using 10-fold cross-validation.
CVMdl = crossval(Mdl);
% CVMdl is a ClassificationPartitionedECOC cross-validated ECOC classifier.
% Estimate the generalization error.
disp('Generalization error=')
oosLoss = kfoldLoss(CVMdl)

%% Display the Confusion Matrix

MdlConf = fitcecoc(Inputs,Targets,'Learners',t,...
'ClassNames',{'NORMAL','APC','PVC','LBBB','RBBB'});
svmClass = resubPredict(MdlConf);
[svmResubCM,grpOrder] = confusionmat(Targets,svmClass)

% Calculate the Total Classification Accuracy
TotalAccuracy=(svmResubCM(1,1)+svmResubCM(2,2)+svmResubCM(3,3))/(3*Nofsignal)

disp('Total Classification Accuracy=')
disp(TotalAccuracy)
```

EXAMPLE 5.21. The following MATLAB code was used to extract features from the ECG signals using covariance method. Then it classified these data using SVMs. You can download data from the following website: https://www.physionet.org/physiobank/database/mitdb/

```
%% Ch5_Ex43_ECG_COV_SVM.m

%The following MATLAB code is used to extract the features from
%the ECG signals using Covariance spectrum.
%Then it classifies ECG Signals Using SVM Classifier
% Cross-Validate ECOC SVM Classifier

clc
clear
%Load Sample ECG Data downloaded from the web site
%https://www.physionet.org/physiobank/database/mitdb/
load MITBIH_ECG.mat
%%
Fs = 320;        % Sampling frequency
T = 1/Fs;            % Sampling period
segmentLength = 128; % Length of a signal segment
Length=320;% Length of signal
Nofsignal=300; %Number of Signal
order = 34;
%%
% Obtain the Covariance Spectrum of the Normal ECG signal using pcov.
for i=1:Nofsignal
```

```
    [Pxx,F] = pcov(ECGN(1:Length,i),order,segmentLength,Fs);
    ASData(i,:)=Pxx(:);
end
%%
% Obtain the Covariance Spectrum of the ECG signal with APC using pcov.
[Pxx,F] = pcov(ECGAPC(:,1),order,segmentLength,Fs);
for i=Nofsignal+1:2*Nofsignal
    [Pxx,F] = pcov(ECGAPC(1:Length,i-Nofsignal),order,segmentLength,Fs);
    ASData(i,:)=Pxx(:);
end
%%
% Obtain the Covariance Spectrum of the ECG signal with PVC using pcov.
for i=2*Nofsignal+1:3*Nofsignal
[Pxx,F] = pcov(ECGPVC(1:Length,i-2*Nofsignal),order,segmentLength,Fs);
ASData(i,:)=Pxx(:);
end

%%
% Obtain the Covariance Spectrum of the ECG signal with LBBB using pcov.
for i=3*Nofsignal+1:4*Nofsignal
[Pxx,F] = pcov(ECGLBBB(1:Length,i-3*Nofsignal),order,segmentLength,Fs);
ASData(i,:)=Pxx(:);
end

%%
% Obtain the Covariance Spectrum of the ECG signal with RBBB using pcov.
for i=4*Nofsignal+1:5*Nofsignal
[Pxx,F] = pcov(ECGRBBB(1:Length,i-4*Nofsignal),order,segmentLength,Fs);
ASData(i,:)=Pxx(:);
end

%Create the Data Matrix as Inputs
Inputs=ASData;
%%
%You can create target using excel and the import it from the MATLAB HOME
%menu and Import data and then save it as EEGTargets
%Load Target
load ECGTargets

%% CLASSIFACATION
rng(1); % For reproducibility
% Create an SVM template. It is good practice to standardize the predictors.
t = templateSVM('Standardize',1)
% Train the ECOC classifier. It is good practice to specify the class order.
Mdl = fitcecoc(Inputs,Targets,'Learners',t,...
'ClassNames',{'NORMAL','APC','PVC','LBBB','RBBB'});
% % Mdl is a ClassificationECOC classifier. You can access its properties using dot
% % notation.
% % Cross-validate Mdl using 10-fold cross-validation.
CVMdl = crossval(Mdl);
% CVMdl is a ClassificationPartitionedECOC cross-validated ECOC classifier.
% Estimate the generalization error.
disp('Generalization error=')
oosLoss = kfoldLoss(CVMdl)

%% Display the Confusion Matrix

MdlConf = fitcecoc(Inputs,Targets,'Learners',t,...
```

```
'ClassNames',{'NORMAL','APC','PVC','LBBB','RBBB'});
svmClass = resubPredict(MdlConf);
[svmResubCM,grpOrder] = confusionmat(Targets,svmClass)

% Calculate the Total Classification Accuracy
TotalAccuracy=(svmResubCM(1,1)+svmResubCM(2,2)+svmResubCM(3,3))/(3*Nofsignal)

disp('Total Classification Accuracy=')
disp(TotalAccuracy)
```

EXAMPLE 5.22 The following MATLAB code was used to extract features from the EEG signals using modified covariance method. Then it classified these data using SVMs. You can download data from the following website: http://epileptologie-bonn.de/cms/front_content.php?idcat=193&lang=3&changelang=3

```
%% Ch5_Ex44_EEG_MCOV_SVM.m

%The following MATLAB code is used to extract the features from
%the EEG signals using Modified Covariance.
%Then it classifies EEG Signals Using SVM Classifier
% Cross-Validate ECOC SVM Classifier
clc
clear
%Load Sample EEG Data downloaded from the web site
%http://epileptologie-bonn.de/cms/front_content.php?idcat=193&lang=3&changelang=3
load AS_BONN_ALL_EEG_DATA_4096.mat
%%
Fs = 173.6;          % Sampling frequency
T = 1/Fs;            % Sampling period
segmentLength = 128; % Length of a signal segment
Length=4096;% Length of signal
Nofsignal=100; %Number of Signal
order = 34;

%%
% Obtain the Modified Covariance Spectrum of the Normal EEG signal using pmcov.
for i=1:Nofsignal

    [Pxx,F] = pmcov(Normal_Eyes_Open(1:Length,i),order,segmentLength,Fs);
    ASData(i,:)=Pxx(:);
end
%%
% Obtain the Modified Covariance Spectrum of the Interictal EEG signal using pmcov.
for i=Nofsignal+1:2*Nofsignal
    [Pxx,F] = pmcov(Interictal(1:Length,i-Nofsignal),order,segmentLength,Fs);
    ASData(i,:)=Pxx(:);
end
%%
% Obtain the Modified Covariance Spectrum of the Ictal EEG signal using pmcov.
for i=2*Nofsignal+1:3*Nofsignal
[Pxx,F] = pmcov(Ictal(1:Length,i-2*Nofsignal),order,segmentLength,Fs);
ASData(i,:)=Pxx(:);
end

%Create the Data Matrix as Inputs
```

```
Inputs=ASData;
%%
%You can create target using excel and the import it from the MATLAB HOME
%menu and Import data and then save it as EEGTargets
%Load Target
load EEGTargets

%% CLASSIFACATION
rng(1); % For reproducibility
% Create an SVM template. It is good practice to standardize the predictors.
t = templateSVM('Standardize',1)
% Train the ECOC classifier. It is good practice to specify the class order.
Mdl = fitcecoc(Inputs,Targets,'Learners',t,...
'ClassNames',{'NORMAL','INTERICTAL','ICTAL'});
% % Mdl is a ClassificationECOC classifier. You can access its properties using dot
% % notation.
% % Cross-validate Mdl using 10-fold cross-validation.
CVMdl = crossval(Mdl);
% CVMdl is a ClassificationPartitionedECOC cross-validated ECOC classifier.
% Estimate the generalization error.
disp('Generalization error=')
oosLoss = kfoldLoss(CVMdl)

%% Display the Confusion Matrix

MdlConf = fitcecoc(Inputs,Targets,'Learners',t,...
'ClassNames',{'NORMAL','INTERICTAL','ICTAL'});
svmClass = resubPredict(MdlConf);
[svmResubCM,grpOrder] = confusionmat(Targets,svmClass)

% Calculate the Total Classification Accuracy
TotalAccuracy=(svmResubCM(1,1)+svmResubCM(2,2)+svmResubCM(3,3))/(3*Nofsignal)

disp('Total Classification Accuracy=')
disp(TotalAccuracy)
```

EXAMPLE 5.23 The following MATLAB code is used to extract features from the ECG signals using modified covariance method. Then it classifies these data using SVM. You can download data from the following website: https://www.physionet.org/physiobank/database/mitdb/

```
%% Ch5_Ex45_ECG_MCOV_SVM.m

%The following MATLAB code is used to extract the features from
%the ECG signals using Modified Covariance spectrum.
%Then it classifies ECG Signals Using SVM Classifier
% Cross-Validate ECOC SVM Classifier
clc
clear
%Load Sample ECG Data downloaded from the web site
%https://www.physionet.org/physiobank/database/mitdb/
load MITBIH_ECG.mat
%%
Fs = 320;             % Sampling frequency
T = 1/Fs;             % Sampling period
```

```
segmentLength = 128; % Length of a signal segment
Length=320;% Length of signal
Nofsignal=300; %Number of Signal
order = 34;
%%
% Obtain the Modified Covariance Spectrum of the Normal ECG signal using pmcov.
for i=1:Nofsignal
    [Pxx,F] = pmcov(ECGN(1:Length,i),order,segmentLength,Fs);
    ASData(i,:)=Pxx(:);
end
%%
% Obtain the Modified Covariance Spectrum of the ECG signal with APC using pmcov.
[Pxx,F] = pmcov(ECGAPC(:,1),order,segmentLength,Fs);
for i=Nofsignal+1:2*Nofsignal
    [Pxx,F] = pmcov(ECGAPC(1:Length,i-Nofsignal),order,segmentLength,Fs);
    ASData(i,:)=Pxx(:);
end
%%
% Obtain the Modified Covariance Spectrum of the ECG signal with PVC using pmcov.
for i=2*Nofsignal+1:3*Nofsignal
[Pxx,F] = pmcov(ECGPVC(1:Length,i-2*Nofsignal),order,segmentLength,Fs);
ASData(i,:)=Pxx(:);
end

%%
% Obtain the Modified Covariance Spectrum of the ECG signal with LBBB using pmcov.
for i=3*Nofsignal+1:4*Nofsignal
[Pxx,F] = pmcov(ECGLBBB(1:Length,i-3*Nofsignal),order,segmentLength,Fs);
ASData(i,:)=Pxx(:);
end

%%
% Obtain the Modified Covariance Spectrum of the ECG signal with RBBB using pmcov.
for i=4*Nofsignal+1:5*Nofsignal
[Pxx,F] = pmcov(ECGRBBB(1:Length,i-4*Nofsignal),order,segmentLength,Fs);
ASData(i,:)=Pxx(:);
end

%Create the Data Matrix as Inputs
Inputs=ASData;
%%
%You can create target using excel and the import it from the MATLAB HOME
%menu and Import data and then save it as EEGTargets
%Load Target
load ECGTargets

%% CLASSIFACATION
rng(1); % For reproducibility
% Create an SVM template. It is good practice to standardize the predictors.
t = templateSVM('Standardize',1)
% Train the ECOC classifier. It is good practice to specify the class order.
Mdl = fitcecoc(Inputs,Targets,'Learners',t,...
'ClassNames',{'NORMAL','APC','PVC','LBBB','RBBB'});
% % Mdl is a ClassificationECOC classifier. You can access its properties using dot
% % notation.
% % Cross-validate Mdl using 10-fold cross-validation.
CVMdl = crossval(Mdl);
% CVMdl is a ClassificationPartitionedECOC cross-validated ECOC classifier.
```

```
% Estimate the generalization error.
disp('Generalization error=')
oosLoss = kfoldLoss(CVMdl)

%% Display the Confusion Matrix

MdlConf = fitcecoc(Inputs,Targets,'Learners',t,...
'ClassNames',{'NORMAL','APC','PVC','LBBB','RBBB'});
svmClass = resubPredict(MdlConf);
[svmResubCM,grpOrder] = confusionmat(Targets,svmClass)

% Calculate the Total Classification Accuracy
TotalAccuracy=(svmResubCM(1,1)+svmResubCM(2,2)+svmResubCM(3,3))/(3*Nofsignal)

disp('Total Classification Accuracy=')
disp(TotalAccuracy)
```

EXAMPLE 5.24 The following MATLAB code was used to extract features from the EEG signals using WPD. Then it used statistical values of WPD subbands. Then it classified these data using SVM classifier. You can download data from the following website:
http://epileptologie-bonn.de/cms/front_content.php?idcat = 193&lang = 3&changelang = 3

```
%% Ch5_Ex47_EEG_WPD_SVM.m
% WPD of NORMAL, INTERICTAL and ICTAL EEG signals
%The following MATLAB code is used to Extract WPD features from the EEG signals
%Decompose EEG data using WPD coefficients
%Then uses Statistical features as:

%(1) Mean of the absolute values of the coefficients in each sub-band.
%(2) Standard deviation of the coefficients in each sub-band.
%(3) Skewness of the coefficients in each sub-band.
%(4) Kurtosis of the coefficients in each sub-band.
%(5) RMS power of the wavelet coefficients in each subband.
%(6) Ratio of the mean absolute values of adjacent subbands.

%Then it classifies EEG Signals Using SVM Classifier
% Cross-Validate ECOC SVM Classifier
clc
clear
%Load Sample EEG Data downloaded from the web site
%http://epileptologie-bonn.de/cms/front_content.php?idcat=193&lang=3&changelang=3
load AS_BONN_ALL_EEG_DATA_4096.mat
wname = 'db4';
Length = 4096;    % Length of signal
Nofsignal=100; %Number of Signal
%%
%%%%%%%%%%%%%%%%%%%%%%%%%%%%%% A Z.ZIP EYES OPEN NORMAL SUBJECT

for i=1:Nofsignal;

[wp10,wp11] = dwt(Normal_Eyes_Open(1:Length,i),wname);%WPD Decomposition
[wp20,wp21] = dwt(wp10,wname);
[wp22,wp23] = dwt(wp11,wname);
```

```
[wp30,wp31] = dwt(wp20,wname);
[wp32,wp33] = dwt(wp21,wname);
[wp34,wp35] = dwt(wp22,wname);
[wp36,wp37] = dwt(wp23,wname);

[wp40,wp41] = dwt(wp30,wname);
[wp42,wp43] = dwt(wp31,wname);
[wp44,wp45] = dwt(wp32,wname);
[wp46,wp47] = dwt(wp33,wname);
[wp48,wp49] = dwt(wp34,wname);
[wp4A,wp4B] = dwt(wp35,wname);
[wp4C,wp4D] = dwt(wp36,wname);
[wp4E,wp4F] = dwt(wp37,wname);

ASData(i,1)=mean(abs(wp40));
ASData(i,2)=mean(abs(wp41));
ASData(i,3)=mean(abs(wp42));
ASData(i,4)=mean(abs(wp43));
ASData(i,5)=mean(abs(wp44));
ASData(i,6)=mean(abs(wp45));
ASData(i,7)=mean(abs(wp46));
ASData(i,8)=mean(abs(wp47));
ASData(i,9)=mean(abs(wp48));
ASData(i,10)=mean(abs(wp49));
ASData(i,11)=mean(abs(wp4A));
ASData(i,12)=mean(abs(wp4B));
ASData(i,13)=mean(abs(wp4C));
ASData(i,14)=mean(abs(wp4D));
ASData(i,15)=mean(abs(wp4E));
ASData(i,16)=mean(abs(wp4F));

ASData(i,17)=std(wp40);
ASData(i,18)=std(wp41);
ASData(i,19)=std(wp42);
ASData(i,20)=std(wp43);
ASData(i,21)=std(wp44);
ASData(i,22)=std(wp45);
ASData(i,23)=std(wp46);
ASData(i,24)=std(wp47);
ASData(i,25)=std(wp48);
ASData(i,26)=std(wp49);
ASData(i,27)=std(wp4A);
ASData(i,28)=std(wp4B);
ASData(i,29)=std(wp4C);
ASData(i,30)=std(wp4D);
ASData(i,31)=std(wp4E);
ASData(i,32)=std(wp4F);

ASData(i,33)=skewness(wp40(:));
ASData(i,34)=skewness(wp41(:));
ASData(i,35)=skewness(wp42(:));
ASData(i,36)=skewness(wp43(:));
ASData(i,37)=skewness(wp44(:));
ASData(i,38)=skewness(wp45(:));
ASData(i,39)=skewness(wp46(:));
ASData(i,40)=skewness(wp47(:));
ASData(i,41)=skewness(wp48(:));
ASData(i,42)=skewness(wp49(:));
```

```
ASData(i,43)=skewness(wp4A(:));
ASData(i,44)=skewness(wp4B(:));
ASData(i,45)=skewness(wp4C(:));
ASData(i,46)=skewness(wp4D(:));
ASData(i,47)=skewness(wp4E(:));
ASData(i,48)=skewness(wp4F(:));

ASData(i,49)=kurtosis(wp40);
ASData(i,50)=kurtosis(wp41);
ASData(i,51)=kurtosis(wp42);
ASData(i,52)=kurtosis(wp43);
ASData(i,53)=kurtosis(wp44);
ASData(i,54)=kurtosis(wp45);
ASData(i,55)=kurtosis(wp46);
ASData(i,56)=kurtosis(wp47);
ASData(i,57)=kurtosis(wp48);
ASData(i,58)=kurtosis(wp49);
ASData(i,59)=kurtosis(wp4A);
ASData(i,60)=kurtosis(wp4B);
ASData(i,61)=kurtosis(wp4C);
ASData(i,62)=kurtosis(wp4D);
ASData(i,63)=kurtosis(wp4E);
ASData(i,64)=kurtosis(wp4F);

ASData(i,65)=rms(wp40(:));
ASData(i,66)=rms(wp41(:));
ASData(i,67)=rms(wp42(:));
ASData(i,68)=rms(wp43(:));
ASData(i,69)=rms(wp44(:));
ASData(i,70)=rms(wp45(:));
ASData(i,71)=rms(wp46(:));
ASData(i,72)=rms(wp47(:));
ASData(i,73)=rms(wp48(:));
ASData(i,74)=rms(wp49(:));
ASData(i,75)=rms(wp4A(:));
ASData(i,76)=rms(wp4B(:));
ASData(i,77)=rms(wp4C(:));
ASData(i,78)=rms(wp4D(:));
ASData(i,79)=rms(wp4E(:));
ASData(i,80)=rms(wp4F(:));

ASData(i,81)=mean(abs(wp40))/mean(abs(wp41));
ASData(i,82)=mean(abs(wp41))/mean(abs(wp42));
ASData(i,83)=mean(abs(wp42))/mean(abs(wp43));
ASData(i,84)=mean(abs(wp43))/mean(abs(wp44));
ASData(i,85)=mean(abs(wp44))/mean(abs(wp45));
ASData(i,86)=mean(abs(wp45))/mean(abs(wp46));
ASData(i,87)=mean(abs(wp46))/mean(abs(wp47));
ASData(i,88)=mean(abs(wp47))/mean(abs(wp48));
ASData(i,89)=mean(abs(wp48))/mean(abs(wp49));
ASData(i,90)=mean(abs(wp49))/mean(abs(wp4A));
ASData(i,91)=mean(abs(wp4A))/mean(abs(wp4B));
ASData(i,92)=mean(abs(wp4B))/mean(abs(wp4C));
ASData(i,93)=mean(abs(wp4C))/mean(abs(wp4D));
ASData(i,94)=mean(abs(wp4D))/mean(abs(wp4E));
ASData(i,95)=mean(abs(wp4E))/mean(abs(wp4F));

end

%%
```

```
% D F.ZIP EPILEPTIC SUBJECT (INTERICTAL) were recorded from within
% the epileptogenic zone during seizure free intervals

for i=Nofsignal+1:2*Nofsignal;

[wp10,wp11] = dwt(Interictal(1:Length,i-Nofsignal),wname);%WPD Decomposition
[wp20,wp21] = dwt(wp10,wname);
[wp22,wp23] = dwt(wp11,wname);

[wp30,wp31] = dwt(wp20,wname);
[wp32,wp33] = dwt(wp21,wname);
[wp34,wp35] = dwt(wp22,wname);
[wp36,wp37] = dwt(wp23,wname);

[wp40,wp41] = dwt(wp30,wname);
[wp42,wp43] = dwt(wp31,wname);
[wp44,wp45] = dwt(wp32,wname);
[wp46,wp47] = dwt(wp33,wname);
[wp48,wp49] = dwt(wp34,wname);
[wp4A,wp4B] = dwt(wp35,wname);
[wp4C,wp4D] = dwt(wp36,wname);
[wp4E,wp4F] = dwt(wp37,wname);

ASData(i,1)=mean(abs(wp40));
ASData(i,2)=mean(abs(wp41));
ASData(i,3)=mean(abs(wp42));
ASData(i,4)=mean(abs(wp43));
ASData(i,5)=mean(abs(wp44));
ASData(i,6)=mean(abs(wp45));
ASData(i,7)=mean(abs(wp46));
ASData(i,8)=mean(abs(wp47));
ASData(i,9)=mean(abs(wp48));
ASData(i,10)=mean(abs(wp49));
ASData(i,11)=mean(abs(wp4A));
ASData(i,12)=mean(abs(wp4B));
ASData(i,13)=mean(abs(wp4C));
ASData(i,14)=mean(abs(wp4D));
ASData(i,15)=mean(abs(wp4E));
ASData(i,16)=mean(abs(wp4F));

ASData(i,17)=std(wp40);
ASData(i,18)=std(wp41);
ASData(i,19)=std(wp42);
ASData(i,20)=std(wp43);
ASData(i,21)=std(wp44);
ASData(i,22)=std(wp45);
ASData(i,23)=std(wp46);
ASData(i,24)=std(wp47);
ASData(i,25)=std(wp48);
ASData(i,26)=std(wp49);
ASData(i,27)=std(wp4A);
ASData(i,28)=std(wp4B);
ASData(i,29)=std(wp4C);
ASData(i,30)=std(wp4D);
ASData(i,31)=std(wp4E);
ASData(i,32)=std(wp4F);

ASData(i,33)=skewness(wp40(:));
```

```
ASData(i,34)=skewness(wp41(:));
ASData(i,35)=skewness(wp42(:));
ASData(i,36)=skewness(wp43(:));
ASData(i,37)=skewness(wp44(:));
ASData(i,38)=skewness(wp45(:));
ASData(i,39)=skewness(wp46(:));
ASData(i,40)=skewness(wp47(:));
ASData(i,41)=skewness(wp48(:));
ASData(i,42)=skewness(wp49(:));
ASData(i,43)=skewness(wp4A(:));
ASData(i,44)=skewness(wp4B(:));
ASData(i,45)=skewness(wp4C(:));
ASData(i,46)=skewness(wp4D(:));
ASData(i,47)=skewness(wp4E(:));
ASData(i,48)=skewness(wp4F(:));

ASData(i,49)=kurtosis(wp40);
ASData(i,50)=kurtosis(wp41);
ASData(i,51)=kurtosis(wp42);
ASData(i,52)=kurtosis(wp43);
ASData(i,53)=kurtosis(wp44);
ASData(i,54)=kurtosis(wp45);
ASData(i,55)=kurtosis(wp46);
ASData(i,56)=kurtosis(wp47);
ASData(i,57)=kurtosis(wp48);
ASData(i,58)=kurtosis(wp49);
ASData(i,59)=kurtosis(wp4A);
ASData(i,60)=kurtosis(wp4B);
ASData(i,61)=kurtosis(wp4C);
ASData(i,62)=kurtosis(wp4D);
ASData(i,63)=kurtosis(wp4E);
ASData(i,64)=kurtosis(wp4F);

ASData(i,65)=rms(wp40(:));
ASData(i,66)=rms(wp41(:));
ASData(i,67)=rms(wp42(:));
ASData(i,68)=rms(wp43(:));
ASData(i,69)=rms(wp44(:));
ASData(i,70)=rms(wp45(:));
ASData(i,71)=rms(wp46(:));
ASData(i,72)=rms(wp47(:));
ASData(i,73)=rms(wp48(:));
ASData(i,74)=rms(wp49(:));
ASData(i,75)=rms(wp4A(:));
ASData(i,76)=rms(wp4B(:));
ASData(i,77)=rms(wp4C(:));
ASData(i,78)=rms(wp4D(:));
ASData(i,79)=rms(wp4E(:));
ASData(i,80)=rms(wp4F(:));

ASData(i,81)=mean(abs(wp40))/mean(abs(wp41));
ASData(i,82)=mean(abs(wp41))/mean(abs(wp42));
ASData(i,83)=mean(abs(wp42))/mean(abs(wp43));
ASData(i,84)=mean(abs(wp43))/mean(abs(wp44));
ASData(i,85)=mean(abs(wp44))/mean(abs(wp45));
ASData(i,86)=mean(abs(wp45))/mean(abs(wp46));
ASData(i,87)=mean(abs(wp46))/mean(abs(wp47));
ASData(i,88)=mean(abs(wp47))/mean(abs(wp48));
```

```
ASData(i,89)=mean(abs(wp48))/mean(abs(wp49));
ASData(i,90)=mean(abs(wp49))/mean(abs(wp4A));
ASData(i,91)=mean(abs(wp4A))/mean(abs(wp4B));
ASData(i,92)=mean(abs(wp4B))/mean(abs(wp4C));
ASData(i,93)=mean(abs(wp4C))/mean(abs(wp4D));
ASData(i,94)=mean(abs(wp4D))/mean(abs(wp4E));
ASData(i,95)=mean(abs(wp4E))/mean(abs(wp4F));
end

%% E S.ZIP EPILEPTIC SUBJECT ICTAL DURING SEIZURE %%%%%%%%%%%%%%%%%%%%%%%%%%%%

for i=2*Nofsignal+1:3*Nofsignal;

[wp10,wp11] = dwt(Ictal(1:Length,i-2*Nofsignal),wname);%WPD Decomposition
[wp20,wp21] = dwt(wp10,wname);
[wp22,wp23] = dwt(wp11,wname);

[wp30,wp31] = dwt(wp20,wname);
[wp32,wp33] = dwt(wp21,wname);
[wp34,wp35] = dwt(wp22,wname);
[wp36,wp37] = dwt(wp23,wname);

[wp40,wp41] = dwt(wp30,wname);
[wp42,wp43] = dwt(wp31,wname);
[wp44,wp45] = dwt(wp32,wname);
[wp46,wp47] = dwt(wp33,wname);
[wp48,wp49] = dwt(wp34,wname);
[wp4A,wp4B] = dwt(wp35,wname);
[wp4C,wp4D] = dwt(wp36,wname);
[wp4E,wp4F] = dwt(wp37,wname);

ASData(i,1)=mean(abs(wp40));
ASData(i,2)=mean(abs(wp41));
ASData(i,3)=mean(abs(wp42));
ASData(i,4)=mean(abs(wp43));
ASData(i,5)=mean(abs(wp44));
ASData(i,6)=mean(abs(wp45));
ASData(i,7)=mean(abs(wp46));
ASData(i,8)=mean(abs(wp47));
ASData(i,9)=mean(abs(wp48));
ASData(i,10)=mean(abs(wp49));
ASData(i,11)=mean(abs(wp4A));
ASData(i,12)=mean(abs(wp4B));
ASData(i,13)=mean(abs(wp4C));
ASData(i,14)=mean(abs(wp4D));
ASData(i,15)=mean(abs(wp4E));
ASData(i,16)=mean(abs(wp4F));

ASData(i,17)=std(wp40);
ASData(i,18)=std(wp41);
ASData(i,19)=std(wp42);
ASData(i,20)=std(wp43);
ASData(i,21)=std(wp44);
ASData(i,22)=std(wp45);
ASData(i,23)=std(wp46);
ASData(i,24)=std(wp47);
ASData(i,25)=std(wp48);
ASData(i,26)=std(wp49);
```

```
ASData(i,27)=std(wp4A);
ASData(i,28)=std(wp4B);
ASData(i,29)=std(wp4C);
ASData(i,30)=std(wp4D);
ASData(i,31)=std(wp4E);
ASData(i,32)=std(wp4F);

ASData(i,33)=skewness(wp40(:));
ASData(i,34)=skewness(wp41(:));
ASData(i,35)=skewness(wp42(:));
ASData(i,36)=skewness(wp43(:));
ASData(i,37)=skewness(wp44(:));
ASData(i,38)=skewness(wp45(:));
ASData(i,39)=skewness(wp46(:));
ASData(i,40)=skewness(wp47(:));
ASData(i,41)=skewness(wp48(:));
ASData(i,42)=skewness(wp49(:));
ASData(i,43)=skewness(wp4A(:));
ASData(i,44)=skewness(wp4B(:));
ASData(i,45)=skewness(wp4C(:));
ASData(i,46)=skewness(wp4D(:));
ASData(i,47)=skewness(wp4E(:));
ASData(i,48)=skewness(wp4F(:));

ASData(i,49)=kurtosis(wp40);
ASData(i,50)=kurtosis(wp41);
ASData(i,51)=kurtosis(wp42);
ASData(i,52)=kurtosis(wp43);
ASData(i,53)=kurtosis(wp44);
ASData(i,54)=kurtosis(wp45);
ASData(i,55)=kurtosis(wp46);
ASData(i,56)=kurtosis(wp47);
ASData(i,57)=kurtosis(wp48);
ASData(i,58)=kurtosis(wp49);
ASData(i,59)=kurtosis(wp4A);
ASData(i,60)=kurtosis(wp4B);
ASData(i,61)=kurtosis(wp4C);
ASData(i,62)=kurtosis(wp4D);
ASData(i,63)=kurtosis(wp4E);
ASData(i,64)=kurtosis(wp4F);

ASData(i,65)=rms(wp40(:));
ASData(i,66)=rms(wp41(:));
ASData(i,67)=rms(wp42(:));
ASData(i,68)=rms(wp43(:));
ASData(i,69)=rms(wp44(:));
ASData(i,70)=rms(wp45(:));
ASData(i,71)=rms(wp46(:));
ASData(i,72)=rms(wp47(:));
ASData(i,73)=rms(wp48(:));
ASData(i,74)=rms(wp49(:));
ASData(i,75)=rms(wp4A(:));
ASData(i,76)=rms(wp4B(:));
ASData(i,77)=rms(wp4C(:));
ASData(i,78)=rms(wp4D(:));
ASData(i,79)=rms(wp4E(:));
ASData(i,80)=rms(wp4F(:));
```

```
ASData(i,81)=mean(abs(wp40))/mean(abs(wp41));
ASData(i,82)=mean(abs(wp41))/mean(abs(wp42));
ASData(i,83)=mean(abs(wp42))/mean(abs(wp43));
ASData(i,84)=mean(abs(wp43))/mean(abs(wp44));
ASData(i,85)=mean(abs(wp44))/mean(abs(wp45));
ASData(i,86)=mean(abs(wp45))/mean(abs(wp46));
ASData(i,87)=mean(abs(wp46))/mean(abs(wp47));
ASData(i,88)=mean(abs(wp47))/mean(abs(wp48));
ASData(i,89)=mean(abs(wp48))/mean(abs(wp49));
ASData(i,90)=mean(abs(wp49))/mean(abs(wp4A));
ASData(i,91)=mean(abs(wp4A))/mean(abs(wp4B));
ASData(i,92)=mean(abs(wp4B))/mean(abs(wp4C));
ASData(i,93)=mean(abs(wp4C))/mean(abs(wp4D));
ASData(i,94)=mean(abs(wp4D))/mean(abs(wp4E));
ASData(i,95)=mean(abs(wp4E))/mean(abs(wp4F));
end

%Make input the Data Matrix
Inputs=ASData;

%%
%You can create target using excel and the import it from the MATLAB HOME
%menu and Import data and then save it as EEGTargets
%Load Target
load EEGTargets

%% CLASSIFACATION
rng(1); % For reproducibility
% Create an SVM template. It is good practice to standardize the predictors.
t = templateSVM('Standardize',1)
% Train the ECOC classifier. It is good practice to specify the class order.
Mdl = fitcecoc(Inputs,Targets,'Learners',t,...
'ClassNames',{'NORMAL','INTERICTAL','ICTAL'});
% % Mdl is a ClassificationECOC classifier. You can access its properties using dot
% % notation.
% % Cross-validate Mdl using 10-fold cross-validation.
CVMdl = crossval(Mdl);
% CVMdl is a ClassificationPartitionedECOC cross-validated ECOC classifier.
% Estimate the generalization error.
disp('Estimated generalization error')
oosLoss = kfoldLoss(CVMdl)

%% Display the Confusion Matrix

MdlConf = fitcecoc(Inputs,Targets,'Learners',t,...
'ClassNames',{'NORMAL','INTERICTAL','ICTAL'});
svmClass = resubPredict(MdlConf);
[svmResubCM,grpOrder] = confusionmat(Targets,svmClass)

% Calculate the Total Classification Accuracy
TotalAccuracy=(svmResubCM(1,1)+svmResubCM(2,2)+svmResubCM(3,3))/(3*Nofsignal)

disp('Total Classification Accuracy=')
disp(TotalAccuracy)
```

5.8 DECISION TREE (DT)

A DT is a hierarchical data structure employing the divide-and-conquer approach. It is an effective nonparametric method that can be utilized for both classification and regression. In parametric estimation, a model over the whole input space is defined and learns its parameters from whole training data. Then the same model and the same parameter set can be employed for any test input. In nonparametric estimation, the input space is divided into local regions, defined by a distance measure such as the Euclidean norm, and the related local model formed from the training data in that region is utilized for each input. (Alpaydin, 2014)

A DT is a hierarchical model for supervised learning in which the local region is described in a sequence of recursive splits in a smaller number of stages. A DT consists of internal decision nodes and terminal leaves. Every decision node employs a test function with discrete outcomes labeling the branches. Given an input, a test is employed at each node, and one of the branches is selected depending on the outcome. This procedure begins at the root and is repeated recursively until a leaf node is reached. A DT is a nonparametric model in the sense that any parametric form for the class densities are not assumed, and the tree assembly is not static a priori. The tree grows by adding the branches and leaves during the learning process depending on the complexity of the problem intrinsic in the data. Each leaf node has an output label, the class code in the case of classification, which is a numeric value in regression. A leaf node describes a localized region in the input space in which instances in this region have the same labels in classification, or very similar numeric outputs in regression. The boundaries of the regions are well-defined by the discriminants, which can be coded in the internal nodes on the path from the root to the leaf node. The hierarchical assignment of decisions achieves fast localization of the region covering an input. Furthermore, the DT is interpretable and can be transformed into a set of IF-THEN rules that are easily understandable. Therefore, DTs are widely known and occasionally chosen over more precise but less interpretable approaches. (Alpaydin, 2014)

DT learning algorithms are greedy and, starting at the root with complete training data, the best split should be searched at each step. This splits the training data into two or more, depending on whether the selected attribute is discrete or numeric. We then keep dividing recursively with the related subset until there is no need to split anymore, at which point a leaf node is formed and labeled. In the case of a DT for classification, namely, a classification tree, the goodness of a split is quantified by an impurity measure. A split is pure if, after the split, for all branches, all the instances choosing a branch belong to the same class. One possible way to measure impurity is entropy (Quinlan, 1986). But entropy is not the only possible measure. Hence, for all attributes, discrete or numeric, the impurity must be calculated and the one with the minimum entropy chosen. Then tree construction continues recursively and in parallel for all the branches that are not pure, until all are pure. This is the basis of the classification and regression tree (CART) algorithm (Breiman, Friedman, Olshen, & Stone, 1984), ID3 algorithm (Quinlan, 1986), and its extension C4.5 (Quinlan, 1993; Alpaydin, 2014).

Commonly, a node is not split more if the number of training instances reaching a node is less than a certain percentage of the training set. Stopping tree creation earlier is called prepruning the tree. An alternative way to have simpler trees is postpruning, which works better than prepruning in practice. In postpruning, the tree is grown full until all leaves are pure without any training error. Afterward, subtrees that cause overfitting should be found and pruned. A pruning set should be set aside from the original labeled set that is not employed during training. Each subtree should be replaced with a leaf node labeled with the training instances covered by the subtree. If the leaf node does not achieve worse than the subtree on the pruning set, the subtree must be pruned, and the leaf node must be kept because the additional complexity of the subtree is not acceptable; otherwise, the subtree must be kept. If prepruning is compared with postpruning, prepruning is faster, but postpruning usually produces a more accurate tree (Alpaydin, 2014).

EXAMPLE 5.25 The following MATLAB code was used to extract features from the EEG signals using AR Burg method. Then it classified these data using DT. You can download data from the following website: http://epileptologie-bonn.de/cms/front_content.php?idcat=193&lang=3&changelang=3

```
%% Ch5_Ex51_EEG_ARBURG_TREE.m
%The following MATLAB code is used to extract the features from
%the EEG signals using AR Burg.
%Then it classifies EEG Signals Using Decision Tree Classifier
clc
clear
%Load Sample EEG Data downloaded from the web site
%http://epileptologie-bonn.de/cms/front_content.php?idcat=193&lang=3&changelang=3
```

```
load AS_BONN_ALL_EEG_DATA_4096.mat
%%
Fs = 173.6;          % Sampling frequency
T = 1/Fs;            % Sampling period
segmentLength = 128; % Length of a signal segment
Length=4096;% Length of signal
Nofsignal=100; %Number of Signal
order = 14;

%%
% Obtain the AR Burg Spectrum of the Normal EEG signal using pburg.
for i=1:Nofsignal
    [Pxx,F] = pburg(Normal_Eyes_Open(1:Length,i),order,segmentLength,Fs);
    ASData(i,:)=Pxx(:);
end
%%
% Obtain the AR Burg Spectrum of the Interictal EEG signal using pburg.
for i=Nofsignal+1:2*Nofsignal
    [Pxx,F] = pburg(Interictal(1:Length,i-Nofsignal),order,segmentLength,Fs);
    ASData(i,:)=Pxx(:);
end
%%
% Obtain the AR Burg Spectrum of the Ictal EEG signal using pburg.
for i=2*Nofsignal+1:3*Nofsignal
[Pxx,F] = pburg(Ictal(1:Length,i-2*Nofsignal),order,segmentLength,Fs);
ASData(i,:)=Pxx(:);
end

%Make input the Data Matrix
Inputs=ASData;

%%
%You can create target using excel and the import it from the MATLAB HOME
%menu and Import data and then save it as EEGTargets
%Load Target
load EEGTargets

%% Control Tree Depth
% You can control the depth of the trees using the |MaxNumSplits|,
% |MinLeafSize|, or |MinParentSize| name-value pair parameters. |fitctree|
% grows deep decision trees by default. You can grow shallower
% trees to reduce model complexity or computation time.
%%
% The default values of the tree depth controllers for growing
% classification trees are:
%
% * |n - 1| for |MaxNumSplits|. |n| is the training sample size.
% * |1| for |MinLeafSize|.
% * |10| for |MinParentSize|.
%
% These default values tend to grow deep trees for large training sample
% sizes.
%%
% Train a classification tree using the default values for tree depth control.
% Cross validate the model using 10-fold cross validation.
rng(1); % For reproducibility
MdlDefault = fitctree(Inputs,Targets,'CrossVal','on');
```

```
%%
% view one of the trees.

view(MdlDefault.Trained{1},'Mode','graph')
%%
% The average number of splits is around 15.
%%
% Suppose that you want a classification tree that is not as complex (deep) as
% the ones trained using the default number of splits. Train another
% classification tree, but set the maximum number of splits at 7, which is
% about half the mean number of splits from the default classification tree.
% Cross validate the model using 10-fold cross validation.
Mdl7 = fitctree(Inputs,Targets,'MaxNumSplits',7,'CrossVal','on');
view(Mdl7.Trained{1},'Mode','graph')
%%
% Compare the cross validation classification errors of the models.
classErrorDefault = kfoldLoss(MdlDefault)
classError7 = kfoldLoss(Mdl7)
%%
% |Mdl7| is much less complex and performs only slightly worse than
% |MdlDefault|.

%% Display the Confusion Matrix
treeMdl = fitctree(Inputs,Targets);
%%
% The observations with known class labels are usually called the training data.
% Now compute the resubstitution error, which is the misclassification error
% (the proportion of misclassified observations) on the training set.

treeResubErr = resubLoss(treeMdl)
%%
% You can also compute the confusion matrix on the training set. A
% confusion matrix contains information about known class labels and
% predicted class labels. Generally speaking, the (i,j) element in the
% confusion matrix is the number of samples whose known class label is
% class i and whose predicted class is j. The diagonal elements represent
% correctly classified observations.

treeClass = resubPredict(treeMdl);
[treeResubCM,grpOrder] = confusionmat(Targets,treeClass)

% Calculate the Total Classification Accuracy
TotalAccuracy=(treeResubCM(1,1)+treeResubCM(2,2)+treeResubCM(3,3))/(3*Nofsignal)

disp('Total Classification Accuracy=')
disp(TotalAccuracy)
```

EXAMPLE 5.26 The following MATLAB code is used to extract features from the ECG signals using AR Burg method. Then it classifies these data using DT. You can download data from the following website: https://www.physionet.org/physiobank/database/mitdb/

```
%% Ch5_Ex52_ECG_ARBURG_TREE.m
%The following MATLAB code is used to extract the features from
%the ECG signals using AR Burg spectrum.
%Then it classifies ECG Signals Using Decision Tree Classifier
```

```
clc
clear
%Load Sample ECG Data downloaded from the web site
%https://www.physionet.org/physiobank/database/mitdb/
load MITBIH_ECG.mat
%%
Fs = 320;            % Sampling frequency
T = 1/Fs;            % Sampling period
segmentLength = 128; % Length of a signal segment
Length=320;% Length of signal
Nofsignal=300; %Number of Signal
order = 34;
%%
% Obtain the AR Burg Spectrum of the Normal ECG signal using pburg.
for i=1:Nofsignal
    [Pxx,F] = pburg(ECGN(1:Length,i),order,segmentLength,Fs);
    ASData(i,:)=Pxx(:);
end
%%
% Obtain the AR Burg Spectrum of the ECG signal with APC using pburg.
for i=Nofsignal+1:2*Nofsignal
    [Pxx,F] = pburg(ECGAPC(1:Length,i-Nofsignal),order,segmentLength,Fs);
    ASData(i,:)=Pxx(:);
end
%%
% Obtain the AR Burg Spectrum of the ECG signal with PVC using pburg.
for i=2*Nofsignal+1:3*Nofsignal
[Pxx,F] = pburg(ECGPVC(1:Length,i-2*Nofsignal),order,segmentLength,Fs);
ASData(i,:)=Pxx(:);
end

%%
% Obtain the AR Burg Spectrum of the ECG signal with LBBB using pburg.
for i=3*Nofsignal+1:4*Nofsignal
[Pxx,F] = pburg(ECGLBBB(1:Length,i-3*Nofsignal),order,segmentLength,Fs);
ASData(i,:)=Pxx(:);
end

%%
% Obtain the AR Burg Spectrum of the ECG signal with RBBB using pburg.
for i=4*Nofsignal+1:5*Nofsignal
[Pxx,F] = pburg(ECGRBBB(1:Length,i-4*Nofsignal),order,segmentLength,Fs);
ASData(i,:)=Pxx(:);
end

%Make input the Data Matrix
Inputs=ASData;
%%
%You can create target using excel and the import it from the MATLAB HOME
%menu and Import data and then save it as EEGTargets
%Load Target
load ECGTargets

%% Control Tree Depth
% You can control the depth of the trees using the |MaxNumSplits|,
% |MinLeafSize|, or |MinParentSize| name-value pair parameters. |fitctree|
% grows deep decision trees by default. You can grow shallower
```

```
% trees to reduce model complexity or computation time.
%%
% The default values of the tree depth controllers for growing
% classification trees are:
%
% * |n - 1| for |MaxNumSplits|. |n| is the training sample size.
% * |1| for |MinLeafSize|.
% * |10| for |MinParentSize|.
%
% These default values tend to grow deep trees for large training sample
% sizes.
%%
% Train a classification tree using the default values for tree depth control.
% Cross validate the model using 10-fold cross validation.
rng(1); % For reproducibility
MdlDefault = fitctree(Inputs,Targets,'CrossVal','on');

%%
% view one of the trees.

view(MdlDefault.Trained{1},'Mode','graph')
%%
% The average number of splits is around 15.
%%
% Suppose that you want a classification tree that is not as complex (deep) as
% the ones trained using the default number of splits. Train another
% classification tree, but set the maximum number of splits at 7, which is
% about half the mean number of splits from the default classification tree.
% Cross validate the model using 10-fold cross validation.
Mdl7 = fitctree(Inputs,Targets,'MaxNumSplits',7,'CrossVal','on');
view(Mdl7.Trained{1},'Mode','graph')
%%
% Compare the cross validation classification errors of the models.
classErrorDefault = kfoldLoss(MdlDefault)
classError7 = kfoldLoss(Mdl7)
%%
% |Mdl7| is much less complex and performs only slightly worse than
% |MdlDefault|.

%% Display the Confusion Matrix
treeMdl = fitctree(Inputs,Targets);
%%
% The observations with known class labels are usually called the training data.
% Now compute the resubstitution error, which is the misclassification error
% (the proportion of misclassified observations) on the training set.

treeResubErr = resubLoss(treeMdl)
%%
% You can also compute the confusion matrix on the training set. A
% confusion matrix contains information about known class labels and
% predicted class labels. Generally speaking, the (i,j) element in the
% confusion matrix is the number of samples whose known class label is
% class i and whose predicted class is j. The diagonal elements represent
% correctly classified observations.

treeClass = resubPredict(treeMdl);
[treeResubCM,grpOrder] = confusionmat(Targets,treeClass)
```

```
% Calculate the Total Classification Accuracy
TotalAccuracy=(treeResubCM(1,1)+treeResubCM(2,2)+treeResubCM(3,3))/(3*Nofsignal)

disp('Total Classification Accuracy=')
disp(TotalAccuracy)
```

EXAMPLE 5.27 The following MATLAB code was used to extract features from the EEG signals using WPD. Then it used statistical values of WPD subbands. Then it classified these data using DT classifier. You can download data from the following website:
http://epileptologie-bonn.de/cms/front_content.php?idcat = 193&lang = 3&changelang = 3

```
%% Ch5_Ex57_EEG_WPD_TREE.m
%WPD of NORMAL, INTERICTAL and ICTAL EEG signals
%The following MATLAB code is used to Extract WPD features from the EEG signals
%Decompose EEG data using WPD coefficients
%Then uses Statistical features as:

%(1) Mean of the absolute values of the coefficients in each sub-band.
%(2) Standard deviation of the coefficients in each sub-band.
%(3) Skewness of the coefficients in each sub-band.
%(4) Kurtosis of the coefficients in each sub-band.
%(5) RMS power of the wavelet coefficients in each subband.
%(6) Ratio of the mean absolute values of adjacent subbands.

%Then it classifies EEG Signals Using Decision Tree Classifier
clc
clear
%Load Sample EEG Data downloaded from the web site
%http://epileptologie-bonn.de/cms/front_content.php?idcat=193&lang=3&changelang=3
load AS_BONN_ALL_EEG_DATA_4096.mat
wname = 'db4';
Length = 4096;     % Length of signal
Nofsignal=100; %Number of Signal
%%
%%%%%%%%%%%%%%%%%%%%%%%%%%%%%% A Z.ZIP EYES OPEN NORMAL SUBJECT

for i=1:Nofsignal;

[wp10,wp11] = dwt(Normal_Eyes_Open(1:Length,i),wname);%WPD Decomposition
[wp20,wp21] = dwt(wp10,wname);
[wp22,wp23] = dwt(wp11,wname);

[wp30,wp31] = dwt(wp20,wname);
[wp32,wp33] = dwt(wp21,wname);
[wp34,wp35] = dwt(wp22,wname);
[wp36,wp37] = dwt(wp23,wname);

[wp40,wp41] = dwt(wp30,wname);
[wp42,wp43] = dwt(wp31,wname);
[wp44,wp45] = dwt(wp32,wname);
[wp46,wp47] = dwt(wp33,wname);
[wp48,wp49] = dwt(wp34,wname);
[wp4A,wp4B] = dwt(wp35,wname);
[wp4C,wp4D] = dwt(wp36,wname);
[wp4E,wp4F] = dwt(wp37,wname);
```

```
ASData(i,1)=mean(abs(wp40));
ASData(i,2)=mean(abs(wp41));
ASData(i,3)=mean(abs(wp42));
ASData(i,4)=mean(abs(wp43));
ASData(i,5)=mean(abs(wp44));
ASData(i,6)=mean(abs(wp45));
ASData(i,7)=mean(abs(wp46));
ASData(i,8)=mean(abs(wp47));
ASData(i,9)=mean(abs(wp48));
ASData(i,10)=mean(abs(wp49));
ASData(i,11)=mean(abs(wp4A));
ASData(i,12)=mean(abs(wp4B));
ASData(i,13)=mean(abs(wp4C));
ASData(i,14)=mean(abs(wp4D));
ASData(i,15)=mean(abs(wp4E));
ASData(i,16)=mean(abs(wp4F));

ASData(i,17)=std(wp40);
ASData(i,18)=std(wp41);
ASData(i,19)=std(wp42);
ASData(i,20)=std(wp43);
ASData(i,21)=std(wp44);
ASData(i,22)=std(wp45);
ASData(i,23)=std(wp46);
ASData(i,24)=std(wp47);
ASData(i,25)=std(wp48);
ASData(i,26)=std(wp49);
ASData(i,27)=std(wp4A);
ASData(i,28)=std(wp4B);
ASData(i,29)=std(wp4C);
ASData(i,30)=std(wp4D);
ASData(i,31)=std(wp4E);
ASData(i,32)=std(wp4F);

ASData(i,33)=skewness(wp40(:));
ASData(i,34)=skewness(wp41(:));
ASData(i,35)=skewness(wp42(:));
ASData(i,36)=skewness(wp43(:));
ASData(i,37)=skewness(wp44(:));
ASData(i,38)=skewness(wp45(:));
ASData(i,39)=skewness(wp46(:));
ASData(i,40)=skewness(wp47(:));
ASData(i,41)=skewness(wp48(:));
ASData(i,42)=skewness(wp49(:));
ASData(i,43)=skewness(wp4A(:));
ASData(i,44)=skewness(wp4B(:));
ASData(i,45)=skewness(wp4C(:));
ASData(i,46)=skewness(wp4D(:));
ASData(i,47)=skewness(wp4E(:));
ASData(i,48)=skewness(wp4F(:));

ASData(i,49)=kurtosis(wp40);
ASData(i,50)=kurtosis(wp41);
ASData(i,51)=kurtosis(wp42);
ASData(i,52)=kurtosis(wp43);
ASData(i,53)=kurtosis(wp44);
ASData(i,54)=kurtosis(wp45);
ASData(i,55)=kurtosis(wp46);
ASData(i,56)=kurtosis(wp47);
```

```
ASData(i,57)=kurtosis(wp48);
ASData(i,58)=kurtosis(wp49);
ASData(i,59)=kurtosis(wp4A);
ASData(i,60)=kurtosis(wp4B);
ASData(i,61)=kurtosis(wp4C);
ASData(i,62)=kurtosis(wp4D);
ASData(i,63)=kurtosis(wp4E);
ASData(i,64)=kurtosis(wp4F);

ASData(i,65)=rms(wp40(:));
ASData(i,66)=rms(wp41(:));
ASData(i,67)=rms(wp42(:));
ASData(i,68)=rms(wp43(:));
ASData(i,69)=rms(wp44(:));
ASData(i,70)=rms(wp45(:));
ASData(i,71)=rms(wp46(:));
ASData(i,72)=rms(wp47(:));
ASData(i,73)=rms(wp48(:));
ASData(i,74)=rms(wp49(:));
ASData(i,75)=rms(wp4A(:));
ASData(i,76)=rms(wp4B(:));
ASData(i,77)=rms(wp4C(:));
ASData(i,78)=rms(wp4D(:));
ASData(i,79)=rms(wp4E(:));
ASData(i,80)=rms(wp4F(:));

ASData(i,81)=mean(abs(wp40))/mean(abs(wp41));
ASData(i,82)=mean(abs(wp41))/mean(abs(wp42));
ASData(i,83)=mean(abs(wp42))/mean(abs(wp43));
ASData(i,84)=mean(abs(wp43))/mean(abs(wp44));
ASData(i,85)=mean(abs(wp44))/mean(abs(wp45));
ASData(i,86)=mean(abs(wp45))/mean(abs(wp46));
ASData(i,87)=mean(abs(wp46))/mean(abs(wp47));
ASData(i,88)=mean(abs(wp47))/mean(abs(wp48));
ASData(i,89)=mean(abs(wp48))/mean(abs(wp49));
ASData(i,90)=mean(abs(wp49))/mean(abs(wp4A));
ASData(i,91)=mean(abs(wp4A))/mean(abs(wp4B));
ASData(i,92)=mean(abs(wp4B))/mean(abs(wp4C));
ASData(i,93)=mean(abs(wp4C))/mean(abs(wp4D));
ASData(i,94)=mean(abs(wp4D))/mean(abs(wp4E));
ASData(i,95)=mean(abs(wp4E))/mean(abs(wp4F));

end

%%
% D F.ZIP EPILEPTIC SUBJECT (INTERICTAL) were recorded from within
% the epileptogenic zone during seizure free intervals

for i=Nofsignal+1:2*Nofsignal;

[wp10,wp11] = dwt(Interictal(1:Length,i-Nofsignal),wname);%WPD Decomposition
[wp20,wp21] = dwt(wp10,wname);
[wp22,wp23] = dwt(wp11,wname);

[wp30,wp31] = dwt(wp20,wname);
[wp32,wp33] = dwt(wp21,wname);
[wp34,wp35] = dwt(wp22,wname);
[wp36,wp37] = dwt(wp23,wname);
```

```
[wp40,wp41] = dwt(wp30,wname);
[wp42,wp43] = dwt(wp31,wname);
[wp44,wp45] = dwt(wp32,wname);
[wp46,wp47] = dwt(wp33,wname);
[wp48,wp49] = dwt(wp34,wname);
[wp4A,wp4B] = dwt(wp35,wname);
[wp4C,wp4D] = dwt(wp36,wname);
[wp4E,wp4F] = dwt(wp37,wname);

ASData(i,1)=mean(abs(wp40));
ASData(i,2)=mean(abs(wp41));
ASData(i,3)=mean(abs(wp42));
ASData(i,4)=mean(abs(wp43));
ASData(i,5)=mean(abs(wp44));
ASData(i,6)=mean(abs(wp45));
ASData(i,7)=mean(abs(wp46));
ASData(i,8)=mean(abs(wp47));
ASData(i,9)=mean(abs(wp48));
ASData(i,10)=mean(abs(wp49));
ASData(i,11)=mean(abs(wp4A));
ASData(i,12)=mean(abs(wp4B));
ASData(i,13)=mean(abs(wp4C));
ASData(i,14)=mean(abs(wp4D));
ASData(i,15)=mean(abs(wp4E));
ASData(i,16)=mean(abs(wp4F));

ASData(i,17)=std(wp40);
ASData(i,18)=std(wp41);
ASData(i,19)=std(wp42);
ASData(i,20)=std(wp43);
ASData(i,21)=std(wp44);
ASData(i,22)=std(wp45);
ASData(i,23)=std(wp46);
ASData(i,24)=std(wp47);
ASData(i,25)=std(wp48);
ASData(i,26)=std(wp49);
ASData(i,27)=std(wp4A);
ASData(i,28)=std(wp4B);
ASData(i,29)=std(wp4C);
ASData(i,30)=std(wp4D);
ASData(i,31)=std(wp4E);
ASData(i,32)=std(wp4F);

ASData(i,33)=skewness(wp40(:));
ASData(i,34)=skewness(wp41(:));
ASData(i,35)=skewness(wp42(:));
ASData(i,36)=skewness(wp43(:));
ASData(i,37)=skewness(wp44(:));
ASData(i,38)=skewness(wp45(:));
ASData(i,39)=skewness(wp46(:));
ASData(i,40)=skewness(wp47(:));
ASData(i,41)=skewness(wp48(:));
ASData(i,42)=skewness(wp49(:));
ASData(i,43)=skewness(wp4A(:));
ASData(i,44)=skewness(wp4B(:));
ASData(i,45)=skewness(wp4C(:));
ASData(i,46)=skewness(wp4D(:));
ASData(i,47)=skewness(wp4E(:));
ASData(i,48)=skewness(wp4F(:));
```

```
ASData(i,49)=kurtosis(wp40);
ASData(i,50)=kurtosis(wp41);
ASData(i,51)=kurtosis(wp42);
ASData(i,52)=kurtosis(wp43);
ASData(i,53)=kurtosis(wp44);
ASData(i,54)=kurtosis(wp45);
ASData(i,55)=kurtosis(wp46);
ASData(i,56)=kurtosis(wp47);
ASData(i,57)=kurtosis(wp48);
ASData(i,58)=kurtosis(wp49);
ASData(i,59)=kurtosis(wp4A);
ASData(i,60)=kurtosis(wp4B);
ASData(i,61)=kurtosis(wp4C);
ASData(i,62)=kurtosis(wp4D);
ASData(i,63)=kurtosis(wp4E);
ASData(i,64)=kurtosis(wp4F);

ASData(i,65)=rms(wp40(:));
ASData(i,66)=rms(wp41(:));
ASData(i,67)=rms(wp42(:));
ASData(i,68)=rms(wp43(:));
ASData(i,69)=rms(wp44(:));
ASData(i,70)=rms(wp45(:));
ASData(i,71)=rms(wp46(:));
ASData(i,72)=rms(wp47(:));
ASData(i,73)=rms(wp48(:));
ASData(i,74)=rms(wp49(:));
ASData(i,75)=rms(wp4A(:));
ASData(i,76)=rms(wp4B(:));
ASData(i,77)=rms(wp4C(:));
ASData(i,78)=rms(wp4D(:));
ASData(i,79)=rms(wp4E(:));
ASData(i,80)=rms(wp4F(:));

ASData(i,81)=mean(abs(wp40))/mean(abs(wp41));
ASData(i,82)=mean(abs(wp41))/mean(abs(wp42));
ASData(i,83)=mean(abs(wp42))/mean(abs(wp43));
ASData(i,84)=mean(abs(wp43))/mean(abs(wp44));
ASData(i,85)=mean(abs(wp44))/mean(abs(wp45));
ASData(i,86)=mean(abs(wp45))/mean(abs(wp46));
ASData(i,87)=mean(abs(wp46))/mean(abs(wp47));
ASData(i,88)=mean(abs(wp47))/mean(abs(wp48));
ASData(i,89)=mean(abs(wp48))/mean(abs(wp49));
ASData(i,90)=mean(abs(wp49))/mean(abs(wp4A));
ASData(i,91)=mean(abs(wp4A))/mean(abs(wp4B));
ASData(i,92)=mean(abs(wp4B))/mean(abs(wp4C));
ASData(i,93)=mean(abs(wp4C))/mean(abs(wp4D));
ASData(i,94)=mean(abs(wp4D))/mean(abs(wp4E));
ASData(i,95)=mean(abs(wp4E))/mean(abs(wp4F));
end

%% E S.ZIP EPILEPTIC SUBJECT ICTAL DURING SEIZURE %%%%%%%%%%%%%%%%%%%%%%%%%%%

for i=2*Nofsignal+1:3*Nofsignal;

[wp10,wp11] = dwt(Ictal(1:Length,i-2*Nofsignal),wname);%WPD Decomposition
[wp20,wp21] = dwt(wp10,wname);
[wp22,wp23] = dwt(wp11,wname);
```

```
[wp30,wp31] = dwt(wp20,wname);
[wp32,wp33] = dwt(wp21,wname);
[wp34,wp35] = dwt(wp22,wname);
[wp36,wp37] = dwt(wp23,wname);

[wp40,wp41] = dwt(wp30,wname);
[wp42,wp43] = dwt(wp31,wname);
[wp44,wp45] = dwt(wp32,wname);
[wp46,wp47] = dwt(wp33,wname);
[wp48,wp49] = dwt(wp34,wname);
[wp4A,wp4B] = dwt(wp35,wname);
[wp4C,wp4D] = dwt(wp36,wname);
[wp4E,wp4F] = dwt(wp37,wname);

ASData(i,1)=mean(abs(wp40));
ASData(i,2)=mean(abs(wp41));
ASData(i,3)=mean(abs(wp42));
ASData(i,4)=mean(abs(wp43));
ASData(i,5)=mean(abs(wp44));
ASData(i,6)=mean(abs(wp45));
ASData(i,7)=mean(abs(wp46));
ASData(i,8)=mean(abs(wp47));
ASData(i,9)=mean(abs(wp48));
ASData(i,10)=mean(abs(wp49));
ASData(i,11)=mean(abs(wp4A));
ASData(i,12)=mean(abs(wp4B));
ASData(i,13)=mean(abs(wp4C));
ASData(i,14)=mean(abs(wp4D));
ASData(i,15)=mean(abs(wp4E));
ASData(i,16)=mean(abs(wp4F));

ASData(i,17)=std(wp40);
ASData(i,18)=std(wp41);
ASData(i,19)=std(wp42);
ASData(i,20)=std(wp43);
ASData(i,21)=std(wp44);
ASData(i,22)=std(wp45);
ASData(i,23)=std(wp46);
ASData(i,24)=std(wp47);
ASData(i,25)=std(wp48);
ASData(i,26)=std(wp49);
ASData(i,27)=std(wp4A);
ASData(i,28)=std(wp4B);
ASData(i,29)=std(wp4C);
ASData(i,30)=std(wp4D);
ASData(i,31)=std(wp4E);
ASData(i,32)=std(wp4F);

ASData(i,33)=skewness(wp40(:));
ASData(i,34)=skewness(wp41(:));
ASData(i,35)=skewness(wp42(:));
ASData(i,36)=skewness(wp43(:));
ASData(i,37)=skewness(wp44(:));
ASData(i,38)=skewness(wp45(:));
ASData(i,39)=skewness(wp46(:));
ASData(i,40)=skewness(wp47(:));
ASData(i,41)=skewness(wp48(:));
ASData(i,42)=skewness(wp49(:));
ASData(i,43)=skewness(wp4A(:));
```

```
ASData(i,44)=skewness(wp4B(:));
ASData(i,45)=skewness(wp4C(:));
ASData(i,46)=skewness(wp4D(:));
ASData(i,47)=skewness(wp4E(:));
ASData(i,48)=skewness(wp4F(:));

ASData(i,49)=kurtosis(wp40);
ASData(i,50)=kurtosis(wp41);
ASData(i,51)=kurtosis(wp42);
ASData(i,52)=kurtosis(wp43);
ASData(i,53)=kurtosis(wp44);
ASData(i,54)=kurtosis(wp45);
ASData(i,55)=kurtosis(wp46);
ASData(i,56)=kurtosis(wp47);
ASData(i,57)=kurtosis(wp48);
ASData(i,58)=kurtosis(wp49);
ASData(i,59)=kurtosis(wp4A);
ASData(i,60)=kurtosis(wp4B);
ASData(i,61)=kurtosis(wp4C);
ASData(i,62)=kurtosis(wp4D);
ASData(i,63)=kurtosis(wp4E);
ASData(i,64)=kurtosis(wp4F);

ASData(i,65)=rms(wp40(:));
ASData(i,66)=rms(wp41(:));
ASData(i,67)=rms(wp42(:));
ASData(i,68)=rms(wp43(:));
ASData(i,69)=rms(wp44(:));
ASData(i,70)=rms(wp45(:));
ASData(i,71)=rms(wp46(:));
ASData(i,72)=rms(wp47(:));
ASData(i,73)=rms(wp48(:));
ASData(i,74)=rms(wp49(:));
ASData(i,75)=rms(wp4A(:));
ASData(i,76)=rms(wp4B(:));
ASData(i,77)=rms(wp4C(:));
ASData(i,78)=rms(wp4D(:));
ASData(i,79)=rms(wp4E(:));
ASData(i,80)=rms(wp4F(:));

ASData(i,81)=mean(abs(wp40))/mean(abs(wp41));
ASData(i,82)=mean(abs(wp41))/mean(abs(wp42));
ASData(i,83)=mean(abs(wp42))/mean(abs(wp43));
ASData(i,84)=mean(abs(wp43))/mean(abs(wp44));
ASData(i,85)=mean(abs(wp44))/mean(abs(wp45));
ASData(i,86)=mean(abs(wp45))/mean(abs(wp46));
ASData(i,87)=mean(abs(wp46))/mean(abs(wp47));
ASData(i,88)=mean(abs(wp47))/mean(abs(wp48));
ASData(i,89)=mean(abs(wp48))/mean(abs(wp49));
ASData(i,90)=mean(abs(wp49))/mean(abs(wp4A));
ASData(i,91)=mean(abs(wp4A))/mean(abs(wp4B));
ASData(i,92)=mean(abs(wp4B))/mean(abs(wp4C));
ASData(i,93)=mean(abs(wp4C))/mean(abs(wp4D));
ASData(i,94)=mean(abs(wp4D))/mean(abs(wp4E));
ASData(i,95)=mean(abs(wp4E))/mean(abs(wp4F));
end

%Create Input Data Matrix
Inputs=ASData;
```

```
%%
%You can create target using excel and the import it from the MATLAB HOME
%menu and Import data and then save it as EEGTargets
%Load Target
load EEGTargets

%% Control Tree Depth
% You can control the depth of the trees using the |MaxNumSplits|,
% |MinLeafSize|, or |MinParentSize| name-value pair parameters. |fitctree|
% grows deep decision trees by default. You can grow shallower
% trees to reduce model complexity or computation time.
%%
% The default values of the tree depth controllers for growing
% classification trees are:
%
% * |n - 1| for |MaxNumSplits|. |n| is the training sample size.
% * |1| for |MinLeafSize|.
% * |10| for |MinParentSize|.
%
% These default values tend to grow deep trees for large training sample
% sizes.
%%
% Train a classification tree using the default values for tree depth control.
% Cross validate the model using 10-fold cross validation.
rng(1); % For reproducibility
MdlDefault = fitctree(Inputs,Targets,'CrossVal','on');
%%
% view one of the trees.

view(MdlDefault.Trained{1},'Mode','graph')
%%
% The average number of splits is around 15.
%%
% Suppose that you want a classification tree that is not as complex (deep) as
% the ones trained using the default number of splits. Train another
% classification tree, but set the maximum number of splits at 7, which is
% about half the mean number of splits from the default classification tree.
% Cross validate the model using 10-fold cross validation.
Mdl7 = fitctree(Inputs,Targets,'MaxNumSplits',7,'CrossVal','on');
view(Mdl7.Trained{1},'Mode','graph')
%%
% Compare the cross validation classification errors of the models.
classErrorDefault = kfoldLoss(MdlDefault)
classError7 = kfoldLoss(Mdl7)
%%
% |Mdl7| is much less complex and performs only slightly worse than
% |MdlDefault|.

%% Display the Confusion Matrix
treeMdl = fitctree(Inputs,Targets);

%%
% You can also compute the confusion matrix on the training set. A
% confusion matrix contains information about known class labels and
% predicted class labels. Generally speaking, the (i,j) element in the
% confusion matrix is the number of samples whose known class label is
% class i and whose predicted class is j. The diagonal elements represent
% correctly classified observations.
```

```
treeClass = resubPredict(treeMdl);
[treeResubCM,grpOrder] = confusionmat(Targets,treeClass)

% Calculate the Total Classification Accuracy
TotalAccuracy=(treeResubCM(1,1)+treeResubCM(2,2)+treeResubCM(3,3))/(3*Nofsignal)

disp('Total Classification Accuracy=')
disp(TotalAccuracy)
```

EXAMPLE 5.28 The following MATLAB code was used to extract features from the EEG signals using SWT. Then it used statistical values of SWT subbands. Then it classified these data using DT classifier. You can download data from the following website:
http://epileptologie-bonn.de/cms/front_content.php?idcat = 193&lang = 3&changelang = 3

```
%% Ch5_Ex58_EEG_SWT_TREE.m
%SWT of NORMAL, INTERICTAL and ICTAL EEG signals
%The following MATLAB code is used to Extract SWT features from the EEG signals
%Decompose EEG data using SWT Transfrom
%Then uses Statistical features as:

%(1) Mean of the absolute values of the coefficients in each sub-band.
%(2) Standard deviation of the coefficients in each sub-band.
%(3) Skewness of the coefficients in each sub-band.
%(4) Kurtosis of the coefficients in each sub-band.
%(5) RMS power of the wavelet coefficients in each subband.
%(6) Ratio of the mean absolute values of adjacent subbands.

%Then it classifies EEG Signals Using Decision Tree Classifier
clc
clear
%Load Sample EEG Data downloaded from the web site
%http://epileptologie-bonn.de/cms/front_content.php?idcat=193&lang=3&changelang=3
load AS_BONN_ALL_EEG_DATA_4096.mat

wname = 'db4';
Level = 8;
Length = 4096; % Length of signal
Nofsignal=100; %Number of Signal
%%
%%%%%%%%%%%%%%%%%%%%%%%%%%%%%% A Z.ZIP EYES OPEN NORMAL SUBJECT

for i=1:Nofsignal;

[swa,swd] = swt(Normal_Eyes_Open(1:Length,i),Level,wname);%SWT

ASData(i,1)=mean(abs(swa(1,:)));
ASData(i,2)=mean(abs(swa(2,:)));
ASData(i,3)=mean(abs(swa(3,:)));
ASData(i,4)=mean(abs(swa(4,:)));
ASData(i,5)=mean(abs(swa(5,:)));
ASData(i,6)=mean(abs(swa(6,:)));
ASData(i,7)=mean(abs(swa(7,:)));
ASData(i,8)=mean(abs(swa(8,:)));
ASData(i,9)=mean(abs(swd(1,:)));
ASData(i,10)=mean(abs(swd(2,:)));
ASData(i,11)=mean(abs(swd(3,:)));
```

```
ASData(i,12)=mean(abs(swd(4,:)));
ASData(i,13)=mean(abs(swd(5,:)));
ASData(i,14)=mean(abs(swd(6,:)));
ASData(i,15)=mean(abs(swd(7,:)));
ASData(i,16)=mean(abs(swd(8,:)));

ASData(i,17)=std(swa(1,:));
ASData(i,18)=std(swa(2,:));
ASData(i,19)=std(swa(3,:));
ASData(i,20)=std(swa(4,:));
ASData(i,21)=std(swa(5,:));
ASData(i,22)=std(swa(6,:));
ASData(i,23)=std(swa(7,:));
ASData(i,24)=std(swa(8,:));
ASData(i,25)=std(swd(1,:));
ASData(i,26)=std(swd(2,:));
ASData(i,27)=std(swd(3,:));
ASData(i,28)=std(swd(4,:));
ASData(i,29)=std(swd(5,:));
ASData(i,30)=std(swd(6,:));
ASData(i,31)=std(swd(7,:));
ASData(i,32)=std(swd(8,:));

ASData(i,33)=skewness(swa(1,:));
ASData(i,34)=skewness(swa(2,:));
ASData(i,35)=skewness(swa(3,:));
ASData(i,36)=skewness(swa(4,:));
ASData(i,37)=skewness(swa(5,:));
ASData(i,38)=skewness(swa(6,:));
ASData(i,39)=skewness(swa(7,:));
ASData(i,40)=skewness(swa(8,:));
ASData(i,41)=skewness(swd(1,:));
ASData(i,42)=skewness(swd(2,:));
ASData(i,43)=skewness(swd(3,:));
ASData(i,44)=skewness(swd(4,:));
ASData(i,45)=skewness(swd(5,:));
ASData(i,46)=skewness(swd(6,:));
ASData(i,47)=skewness(swd(7,:));
ASData(i,48)=skewness(swd(8,:));

ASData(i,49)=kurtosis(swa(1,:));
ASData(i,50)=kurtosis(swa(2,:));
ASData(i,51)=kurtosis(swa(3,:));
ASData(i,52)=kurtosis(swa(4,:));
ASData(i,53)=kurtosis(swa(5,:));
ASData(i,54)=kurtosis(swa(6,:));
ASData(i,55)=kurtosis(swa(7,:));
ASData(i,56)=kurtosis(swa(8,:));
ASData(i,57)=kurtosis(swd(1,:));
ASData(i,58)=kurtosis(swd(2,:));
ASData(i,59)=kurtosis(swd(3,:));
ASData(i,60)=kurtosis(swd(4,:));
ASData(i,61)=kurtosis(swd(5,:));
ASData(i,62)=kurtosis(swd(6,:));
ASData(i,63)=kurtosis(swd(7,:));
ASData(i,64)=kurtosis(swd(8,:));

        ASData(i,65)=rms(swa(1,:));
```

```
        ASData(i,66)=rms(swa(2,:));
        ASData(i,67)=rms(swa(3,:));
        ASData(i,68)=rms(swa(4,:));
        ASData(i,69)=rms(swa(5,:));
        ASData(i,70)=rms(swa(6,:));
        ASData(i,71)=rms(swa(7,:));
        ASData(i,72)=rms(swa(8,:));
        ASData(i,73)=rms(swd(1,:));
        ASData(i,74)=rms(swd(2,:));
        ASData(i,75)=rms(swd(3,:));
        ASData(i,76)=rms(swd(4,:));
        ASData(i,77)=rms(swd(5,:));
        ASData(i,78)=rms(swd(6,:));
        ASData(i,79)=rms(swd(7,:));
        ASData(i,80)=rms(swd(8,:));

        ASData(i,81)=mean(abs(swa(1,:)))/mean(abs(swa(2,:)));
        ASData(i,82)=mean(abs(swa(2,:)))/mean(abs(swa(3,:)));
        ASData(i,83)=mean(abs(swa(3,:)))/mean(abs(swa(4,:)));
        ASData(i,84)=mean(abs(swa(4,:)))/mean(abs(swa(5,:)));
        ASData(i,85)=mean(abs(swa(5,:)))/mean(abs(swa(6,:)));
        ASData(i,86)=mean(abs(swa(6,:)))/mean(abs(swa(7,:)));
        ASData(i,87)=mean(abs(swa(7,:)))/mean(abs(swa(8,:)));
        ASData(i,88)=mean(abs(swa(8,:)))/mean(abs(swd(1,:)));
        ASData(i,89)=mean(abs(swd(1,:)))/mean(abs(swd(2,:)));
        ASData(i,90)=mean(abs(swd(2,:)))/mean(abs(swd(3,:)));
        ASData(i,91)=mean(abs(swd(3,:)))/mean(abs(swd(4,:)));
        ASData(i,92)=mean(abs(swd(4,:)))/mean(abs(swd(5,:)));
        ASData(i,93)=mean(abs(swd(5,:)))/mean(abs(swd(6,:)));
        ASData(i,94)=mean(abs(swd(6,:)))/mean(abs(swd(7,:)));
        ASData(i,95)=mean(abs(swd(7,:)))/mean(abs(swd(8,:)));

end

%%
% D F.ZIP EPILEPTIC SUBJECT (INTERICTAL) were recorded from within
% the epileptogenic zone during seizure free intervals

for i=Nofsignal+1:2*Nofsignal;

[swa,swd] = swt(Interictal(1:Length,i-Nofsignal),Level,wname); %SWT

ASData(i,1)=mean(abs(swa(1,:)));
ASData(i,2)=mean(abs(swa(2,:)));
ASData(i,3)=mean(abs(swa(3,:)));
ASData(i,4)=mean(abs(swa(4,:)));
ASData(i,5)=mean(abs(swa(5,:)));
ASData(i,6)=mean(abs(swa(6,:)));
ASData(i,7)=mean(abs(swa(7,:)));
ASData(i,8)=mean(abs(swa(8,:)));
ASData(i,9)=mean(abs(swd(1,:)));
ASData(i,10)=mean(abs(swd(2,:)));
ASData(i,11)=mean(abs(swd(3,:)));
ASData(i,12)=mean(abs(swd(4,:)));
ASData(i,13)=mean(abs(swd(5,:)));
ASData(i,14)=mean(abs(swd(6,:)));
ASData(i,15)=mean(abs(swd(7,:)));
ASData(i,16)=mean(abs(swd(8,:)));
```

```
ASData(i,17)=std(swa(1,:));
ASData(i,18)=std(swa(2,:));
ASData(i,19)=std(swa(3,:));
ASData(i,20)=std(swa(4,:));
ASData(i,21)=std(swa(5,:));
ASData(i,22)=std(swa(6,:));
ASData(i,23)=std(swa(7,:));
ASData(i,24)=std(swa(8,:));
ASData(i,25)=std(swd(1,:));
ASData(i,26)=std(swd(2,:));
ASData(i,27)=std(swd(3,:));
ASData(i,28)=std(swd(4,:));
ASData(i,29)=std(swd(5,:));
ASData(i,30)=std(swd(6,:));
ASData(i,31)=std(swd(7,:));
ASData(i,32)=std(swd(8,:));

ASData(i,33)=skewness(swa(1,:));
ASData(i,34)=skewness(swa(2,:));
ASData(i,35)=skewness(swa(3,:));
ASData(i,36)=skewness(swa(4,:));
ASData(i,37)=skewness(swa(5,:));
ASData(i,38)=skewness(swa(6,:));
ASData(i,39)=skewness(swa(7,:));
ASData(i,40)=skewness(swa(8,:));
ASData(i,41)=skewness(swd(1,:));
ASData(i,42)=skewness(swd(2,:));
ASData(i,43)=skewness(swd(3,:));
ASData(i,44)=skewness(swd(4,:));
ASData(i,45)=skewness(swd(5,:));
ASData(i,46)=skewness(swd(6,:));
ASData(i,47)=skewness(swd(7,:));
ASData(i,48)=skewness(swd(8,:));

ASData(i,49)=kurtosis(swa(1,:));
ASData(i,50)=kurtosis(swa(2,:));
ASData(i,51)=kurtosis(swa(3,:));
ASData(i,52)=kurtosis(swa(4,:));
ASData(i,53)=kurtosis(swa(5,:));
ASData(i,54)=kurtosis(swa(6,:));
ASData(i,55)=kurtosis(swa(7,:));
ASData(i,56)=kurtosis(swa(8,:));
ASData(i,57)=kurtosis(swd(1,:));
ASData(i,58)=kurtosis(swd(2,:));
ASData(i,59)=kurtosis(swd(3,:));
ASData(i,60)=kurtosis(swd(4,:));
ASData(i,61)=kurtosis(swd(5,:));
ASData(i,62)=kurtosis(swd(6,:));
ASData(i,63)=kurtosis(swd(7,:));
ASData(i,64)=kurtosis(swd(8,:));

        ASData(i,65)=rms(swa(1,:));
        ASData(i,66)=rms(swa(2,:));
        ASData(i,67)=rms(swa(3,:));
        ASData(i,68)=rms(swa(4,:));
        ASData(i,69)=rms(swa(5,:));
        ASData(i,70)=rms(swa(6,:));
        ASData(i,71)=rms(swa(7,:));
```

```
        ASData(i,72)=rms(swa(8,:));
        ASData(i,73)=rms(swd(1,:));
        ASData(i,74)=rms(swd(2,:));
        ASData(i,75)=rms(swd(3,:));
        ASData(i,76)=rms(swd(4,:));
        ASData(i,77)=rms(swd(5,:));
        ASData(i,78)=rms(swd(6,:));
        ASData(i,79)=rms(swd(7,:));
        ASData(i,80)=rms(swd(8,:));

        ASData(i,81)=mean(abs(swa(1,:)))/mean(abs(swa(2,:)));
        ASData(i,82)=mean(abs(swa(2,:)))/mean(abs(swa(3,:)));
        ASData(i,83)=mean(abs(swa(3,:)))/mean(abs(swa(4,:)));
        ASData(i,84)=mean(abs(swa(4,:)))/mean(abs(swa(5,:)));
        ASData(i,85)=mean(abs(swa(5,:)))/mean(abs(swa(6,:)));
        ASData(i,86)=mean(abs(swa(6,:)))/mean(abs(swa(7,:)));
        ASData(i,87)=mean(abs(swa(7,:)))/mean(abs(swa(8,:)));
        ASData(i,88)=mean(abs(swa(8,:)))/mean(abs(swd(1,:)));
        ASData(i,89)=mean(abs(swd(1,:)))/mean(abs(swd(2,:)));
        ASData(i,90)=mean(abs(swd(2,:)))/mean(abs(swd(3,:)));
        ASData(i,91)=mean(abs(swd(3,:)))/mean(abs(swd(4,:)));
        ASData(i,92)=mean(abs(swd(4,:)))/mean(abs(swd(5,:)));
        ASData(i,93)=mean(abs(swd(5,:)))/mean(abs(swd(6,:)));
        ASData(i,94)=mean(abs(swd(6,:)))/mean(abs(swd(7,:)));
        ASData(i,95)=mean(abs(swd(7,:)))/mean(abs(swd(8,:)));

end

%% E S.ZIP EPILEPTIC SUBJECT ICTAL DURING SEIZURE
%%%%%%%%%%%%%%%%%%%%%%%%%%%%%%

for i=2*Nofsignal+1:3*Nofsignal;

[swa,swd] = swt(Ictal(1:Length,i-2*Nofsignal),Level,wname);%SWT

ASData(i,1)=mean(abs(swa(1,:)));
ASData(i,2)=mean(abs(swa(2,:)));
ASData(i,3)=mean(abs(swa(3,:)));
ASData(i,4)=mean(abs(swa(4,:)));
ASData(i,5)=mean(abs(swa(5,:)));
ASData(i,6)=mean(abs(swa(6,:)));
ASData(i,7)=mean(abs(swa(7,:)));
ASData(i,8)=mean(abs(swa(8,:)));
ASData(i,9)=mean(abs(swd(1,:)));
ASData(i,10)=mean(abs(swd(2,:)));
ASData(i,11)=mean(abs(swd(3,:)));
ASData(i,12)=mean(abs(swd(4,:)));
ASData(i,13)=mean(abs(swd(5,:)));
ASData(i,14)=mean(abs(swd(6,:)));
ASData(i,15)=mean(abs(swd(7,:)));
ASData(i,16)=mean(abs(swd(8,:)));

ASData(i,17)=std(swa(1,:));
ASData(i,18)=std(swa(2,:));
ASData(i,19)=std(swa(3,:));
ASData(i,20)=std(swa(4,:));
ASData(i,21)=std(swa(5,:));
ASData(i,22)=std(swa(6,:));
```

```
ASData(i,23)=std(swa(7,:));
ASData(i,24)=std(swa(8,:));
ASData(i,25)=std(swd(1,:));
ASData(i,26)=std(swd(2,:));
ASData(i,27)=std(swd(3,:));
ASData(i,28)=std(swd(4,:));
ASData(i,29)=std(swd(5,:));
ASData(i,30)=std(swd(6,:));
ASData(i,31)=std(swd(7,:));
ASData(i,32)=std(swd(8,:));

ASData(i,33)=skewness(swa(1,:));
ASData(i,34)=skewness(swa(2,:));
ASData(i,35)=skewness(swa(3,:));
ASData(i,36)=skewness(swa(4,:));
ASData(i,37)=skewness(swa(5,:));
ASData(i,38)=skewness(swa(6,:));
ASData(i,39)=skewness(swa(7,:));
ASData(i,40)=skewness(swa(8,:));
ASData(i,41)=skewness(swd(1,:));
ASData(i,42)=skewness(swd(2,:));
ASData(i,43)=skewness(swd(3,:));
ASData(i,44)=skewness(swd(4,:));
ASData(i,45)=skewness(swd(5,:));
ASData(i,46)=skewness(swd(6,:));
ASData(i,47)=skewness(swd(7,:));
ASData(i,48)=skewness(swd(8,:));

ASData(i,49)=kurtosis(swa(1,:));
ASData(i,50)=kurtosis(swa(2,:));
ASData(i,51)=kurtosis(swa(3,:));
ASData(i,52)=kurtosis(swa(4,:));
ASData(i,53)=kurtosis(swa(5,:));
ASData(i,54)=kurtosis(swa(6,:));
ASData(i,55)=kurtosis(swa(7,:));
ASData(i,56)=kurtosis(swa(8,:));
ASData(i,57)=kurtosis(swd(1,:));
ASData(i,58)=kurtosis(swd(2,:));
ASData(i,59)=kurtosis(swd(3,:));
ASData(i,60)=kurtosis(swd(4,:));
ASData(i,61)=kurtosis(swd(5,:));
ASData(i,62)=kurtosis(swd(6,:));
ASData(i,63)=kurtosis(swd(7,:));
ASData(i,64)=kurtosis(swd(8,:));

        ASData(i,65)=rms(swa(1,:));
        ASData(i,66)=rms(swa(2,:));
        ASData(i,67)=rms(swa(3,:));
        ASData(i,68)=rms(swa(4,:));
        ASData(i,69)=rms(swa(5,:));
        ASData(i,70)=rms(swa(6,:));
        ASData(i,71)=rms(swa(7,:));
        ASData(i,72)=rms(swa(8,:));
        ASData(i,73)=rms(swd(1,:));
        ASData(i,74)=rms(swd(2,:));
        ASData(i,75)=rms(swd(3,:));
        ASData(i,76)=rms(swd(4,:));
        ASData(i,77)=rms(swd(5,:));
```

```
        ASData(i,78)=rms(swd(6,:));
        ASData(i,79)=rms(swd(7,:));
        ASData(i,80)=rms(swd(8,:));

        ASData(i,81)=mean(abs(swa(1,:)))/mean(abs(swa(2,:)));
        ASData(i,82)=mean(abs(swa(2,:)))/mean(abs(swa(3,:)));
        ASData(i,83)=mean(abs(swa(3,:)))/mean(abs(swa(4,:)));
        ASData(i,84)=mean(abs(swa(4,:)))/mean(abs(swa(5,:)));
        ASData(i,85)=mean(abs(swa(5,:)))/mean(abs(swa(6,:)));
        ASData(i,86)=mean(abs(swa(6,:)))/mean(abs(swa(7,:)));
        ASData(i,87)=mean(abs(swa(7,:)))/mean(abs(swa(8,:)));
        ASData(i,88)=mean(abs(swa(8,:)))/mean(abs(swd(1,:)));
        ASData(i,89)=mean(abs(swd(1,:)))/mean(abs(swd(2,:)));
        ASData(i,90)=mean(abs(swd(2,:)))/mean(abs(swd(3,:)));
        ASData(i,91)=mean(abs(swd(3,:)))/mean(abs(swd(4,:)));
        ASData(i,92)=mean(abs(swd(4,:)))/mean(abs(swd(5,:)));
        ASData(i,93)=mean(abs(swd(5,:)))/mean(abs(swd(6,:)));
        ASData(i,94)=mean(abs(swd(6,:)))/mean(abs(swd(7,:)));
        ASData(i,95)=mean(abs(swd(7,:)))/mean(abs(swd(8,:)));
end

%Create Input Data Matrix
Inputs=ASData;
%%
%You can create target using excel and the import it from the MATLAB HOME
%menu and Import data and then save it as EEGTargets
%Load Target
load EEGTargets

%% Control Tree Depth
% You can control the depth of the trees using the |MaxNumSplits|,
% |MinLeafSize|, or |MinParentSize| name-value pair parameters. |fitctree|
% grows deep decision trees by default. You can grow shallower
% trees to reduce model complexity or computation time.

%%
% The default values of the tree depth controllers for growing
% classification trees are:
%
% * |n - 1| for |MaxNumSplits|. |n| is the training sample size.
% * |1| for |MinLeafSize|.
% * |10| for |MinParentSize|.
%
% These default values tend to grow deep trees for large training sample
% sizes.
%%
% Train a classification tree using the default values for tree depth control.
% Cross validate the model using 10-fold cross validation.
rng(1); % For reproducibility
MdlDefault = fitctree(Inputs,Targets,'CrossVal','on');
%%
% view one of the trees.

view(MdlDefault.Trained{1},'Mode','graph')
%%
% The average number of splits is around 15.
%%
% Suppose that you want a classification tree that is not as complex (deep) as
```

```
% the ones trained using the default number of splits. Train another
% classification tree, but set the maximum number of splits at 7, which is
% about half the mean number of splits from the default classification tree.
% Cross validate the model using 10-fold cross validation.
Mdl7 = fitctree(Inputs,Targets,'MaxNumSplits',7,'CrossVal','on');
view(Mdl7.Trained{1},'Mode','graph')
%%
% Compare the cross validation classification errors of the models.
classErrorDefault = kfoldLoss(MdlDefault)
classError7 = kfoldLoss(Mdl7)
%%
% |Mdl7| is much less complex and performs only slightly worse than
% |MdlDefault|.

%% Display the Confusion Matrix
treeMdl = fitctree(Inputs,Targets);
%%
% The observations with known class labels are usually called the training data.
% Now compute the resubstitution error, which is the misclassification error
% (the proportion of misclassified observations) on the training set.

treeResubErr = resubLoss(treeMdl)
%%
% You can also compute the confusion matrix on the training set. A
% confusion matrix contains information about known class labels and
% predicted class labels. Generally speaking, the (i,j) element in the
% confusion matrix is the number of samples whose known class label is
% class i and whose predicted class is j. The diagonal elements represent
% correctly classified observations.

treeClass = resubPredict(treeMdl);
[treeResubCM,grpOrder] = confusionmat(Targets,treeClass)

% Calculate the Total Classification Accuracy
TotalAccuracy=(treeResubCM(1,1)+treeResubCM(2,2)+treeResubCM(3,3))/(3*Nofsignal)

disp('Total Classification Accuracy=')
disp(TotalAccuracy)
```

5.9 DEEP LEARNING

If a linear model is not sufficient for any learning procedure, new features that are nonlinear functions of the input can be defined, and then a linear model in the space of those features is built. This needs to distinguish what good basis functions are. An alternative way is to employ one of the feature extraction methods such as PCA to study the new space. However, an MLP that extracts such features in its hidden layer is the best method because it has the benefit that the first layer (feature extraction) and the second layer (how those features are combined to predict the output) are learned together in a supervised way. An MLP has inadequate capacity with one hidden layer; on the other hand, an MLP can learn more complex functions of the input with many hidden layers. This is the idea behind deep neural networks (DNNs) in which, starting from a network's raw input, each hidden layer combines the values in its previous layer and learns more complex functions of the input. Another characteristic of DNNs is that consecutive hidden layers are related to more abstract representations until the output layer where the outputs are learned in terms of these intangible notions are involved. In deep learning, the concept is to learn feature levels of increasing abstraction with minimum human involvement (Bengio, 2009) because, in several applications, it is not known what structure exists in the input, and any sort of dependencies must be discovered automatically during training. One key issue with training an MLP with many hidden layers is that, in backpropagating the error to a previous layer, it is necessary to multiply the derivatives in all the layers subsequently, and the gradient becomes extinct. This is also the reason for slow learning of unfolded recurrent neural networks. This situation cannot occur in convolutional neural networks because the fan-in and fan-out of hidden units are naturally insignificant. A DNN is characteristically trained one layer at a time (Hinton & Salakhutdinov, 2006). The objective of each layer is to extract the relevant features in the data that is fed to it, and a technique such as the autoencoder can be utilized for this purpose. Hence, starting from the raw input data, an autoencoder can be trained, and the encoded representation learned in its hidden layer is then employed as an input to train the next autoencoder and so on, until we reach the final layer, which is trained in a supervised manner with the labeled data. After training all the layers one by one, they are all brought together, and the whole network is fine-tuned with the labeled data. If a lot of labeled data and computational power exists, the entire DNN can be trained in a supervised manner, but the compromise is that employing an unsupervised method to initialize the weights works much better than random initialization; as a result, learning can be realized in a faster way with fewer labeled data. Deep learning approaches are prominent mainly as they need less manual interference. The idea of many layers of growing notion that lies beneath deep learning is intuitive. In many applications, the layers of abstraction can be thought that discovering such an abstract representation might be informative and also a better description of the problem (Alpaydin, 2014) (Figs. 5.31–5.35).

EXAMPLE 5.29 The following MATLAB code is used to extract features from the EEG signals using AR Burg method. Then it classifies these data using DNN classifier. You can download data from the following website: http://epileptologie-bonn.de/cms/front_content.php?idcat=193&lang=3&changelang=3

```
%% Ch5_Ex61_EEG_ARBURG_DEEP.m
%The following MATLAB code is used to extract the features from
%the EEG signals using AR Burg.
%Then it classifies EEG Signals Using Deep Neural Network Classifier
% Create a Stacked Network
clc
clear
%Load Sample EEG Data downloaded from the web site
%http://epileptologie-bonn.de/cms/front_content.php?idcat=193&lang=3&changelang=3
load AS_BONN_ALL_EEG_DATA_4096.mat
%%
Fs = 173.6;          % Sampling frequency
T = 1/Fs;            % Sampling period
segmentLength = 128; % Length of a signal segment
Length=4096;% Length of signal
Nofsignal=100; %Number of Signal
order = 14;

%%
% Obtain the AR Burg Spectrum of the Normal EEG signal using pburg.
for i=1:Nofsignal
    [Pxx,F] = pburg(Normal_Eyes_Open(1:Length,i),order,segmentLength,Fs);
    ASData(i,:)=Pxx(:);
end
%%
% Obtain the AR Burg Spectrum of the Interictal EEG signal using pburg.
for i=Nofsignal+1:2*Nofsignal
    [Pxx,F] = pburg(Interictal(1:Length,i-Nofsignal),order,segmentLength,Fs);
    ASData(i,:)=Pxx(:);
end
%%
% Obtain the AR Burg Spectrum of the Ictal EEG signal using pburg.
for i=2*Nofsignal+1:3*Nofsignal
[Pxx,F] = pburg(Ictal(1:Length,i-2*Nofsignal),order,segmentLength,Fs);
ASData(i,:)=Pxx(:);
end
```

```
%Transpose the Data Matrix
Inputs=ASData';
%% TARGET GENERATION
% Your target set must have one label for each sample,
%so it must contains 300 elements. Here is an easy way to build one.
%Start by numbering your classes from 1 to 3.
%Make sure that your X Array is sorted by class,
%meaning all the samples from the first class,
%then all the samples from the second class, and so on. Then:

 % Classes from 1 to 3
 y = 1:3;
 % "Nofsignal" samples per class
 y = repmat(y, Nofsignal, 1);
 % Reshape to obtain a vector
 y = reshape(y, 1, numel(y));
% At this stage, you should have a 1-by-300 vector with numeric class labels from 1 to 3.
%This is what you want if you use the Statistics and Machine Learning Toolbox.
%
% Now, I suppose you intend to use the Neural Network Toolbox.
%In that case the syntax is a bit different: the class labels are in a 3-by-300 matrix,
%each column being a sample, with the row corresponding to the class getting value '1' and the
rest '0'.
%To build it from the previous syntax, type:

Targets = full(ind2vec(y));

%% Train an autoencoder with a hidden layer of size 5 and a linear transfer function for the
% decoder. Set the L2 weight regularizer to 0.001, sparsity regularizer to 4 and sparsity
% proportion to 0.05.
hiddenSize = 5;
autoenc = trainAutoencoder(Inputs, hiddenSize, ...
'L2WeightRegularization', 0.001, ...
'SparsityRegularization', 4, ...
'SparsityProportion', 0.05, ...
'DecoderTransferFunction','purelin');
% Extract the features in the hidden layer.
features = encode(autoenc,Inputs);
% Train a softmax layer for classification using the features .
softnet = trainSoftmaxLayer(features,Targets);
% Stack the encoder and the softmax layer to form a deep network.
stackednet = stack(autoenc,softnet);

deepnet = train(stackednet,Inputs,Targets);
%%
%%%%%%%%%%%%%%%%%%%%TEST

% Estimate the EEG signal types using the deep network, deepnet.
EEG_type = deepnet(Inputs);
% Plot the confusion matrix.
plotconfusion(Targets,EEG_type);

% Plot the ROC.
figure, plotroc(Targets,EEG_type)
```

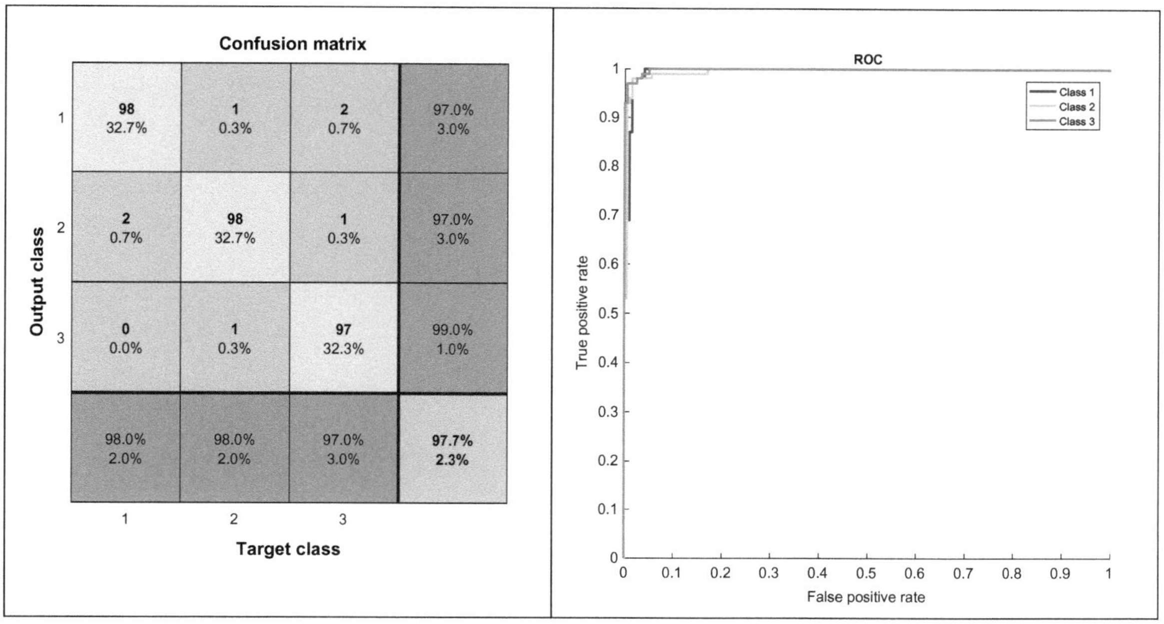

FIG. 5.31 Representation of performance of the DNN classifier for the EEG signals features extracted using AR Burg method.

EXAMPLE 5.30 The following MATLAB code is used to extract features from the ECG signals using AR Burg method. Then it classifies these data using DNN classifier. You can download data from the following website: https://www.physionet.org/physiobank/database/mitdb/

```
%% Ch5_Ex62_ECG_ARBURG_DEEP.m
%The following MATLAB code is used to extract the features from
%the ECG signals using AR Burg spectrum.
%Then it classifies EEG Signals Using Deep Neural Network Classifier
% Create a Stacked Network
clc
clear
%Load Sample ECG Data downloaded from the web site
%https://www.physionet.org/physiobank/database/mitdb/
load MITBIH_ECG.mat
%%
Fs = 320;          % Sampling frequency
T = 1/Fs;            % Sampling period
segmentLength = 128; % Length of a signal segment
Length=320;% Length of signal
Nofsignal=300; %Number of Signal
order = 34;
%%
% Obtain the AR Burg Spectrum of the Normal ECG signal using pburg.
for i=1:Nofsignal
    [Pxx,F] = pburg(ECGN(1:Length,i),order,segmentLength,Fs);
    ASData(i,:)=Pxx(:);
end
%%
% Obtain the AR Burg Spectrum of the ECG signal with APC using pburg.
for i=Nofsignal+1:2*Nofsignal
    [Pxx,F] = pburg(ECGAPC(1:Length,i-Nofsignal),order,segmentLength,Fs);
    ASData(i,:)=Pxx(:);
end
%%
% Obtain the AR Burg Spectrum of the ECG signal with PVC using pburg.
```

```
for i=2*Nofsignal+1:3*Nofsignal
[Pxx,F] = pburg(ECGPVC(1:Length,i-2*Nofsignal),order,segmentLength,Fs);
ASData(i,:)=Pxx(:);
end

%%
% Obtain the AR Burg Spectrum of the ECG signal with LBBB using pburg.
for i=3*Nofsignal+1:4*Nofsignal
[Pxx,F] = pburg(ECGLBBB(1:Length,i-3*Nofsignal),order,segmentLength,Fs);
ASData(i,:)=Pxx(:);
end

%%
% Obtain the AR Burg Spectrum of the ECG signal with RBBB using pburg.
for i=4*Nofsignal+1:5*Nofsignal
[Pxx,F] = pburg(ECGRBBB(1:Length,i-4*Nofsignal),order,segmentLength,Fs);
ASData(i,:)=Pxx(:);
end

%Transpose the Data Matrix
Inputs=ASData';
%% TARGET GENERATION
% Your target set must have one label for each sample,
%so it must contains 300 elements. Here is an easy way to build one.
%Start by numbering your classes from 1 to 3.
%Make sure that your X Array is sorted by class,
%meaning all the samples from the first class,
%then all the samples from the second class, and so on. Then:

 % Classes from 1 to 5
 y = 1:5;
 % Nofsignal samples per class
 y = repmat(y, Nofsignal, 1);
 % Reshape to obtain a vector
 y = reshape(y, 1, numel(y));
% At this stage, you should have a 1-by-1500 vector with numeric class labels from 1 to 3.
%This is what you want if you use the Statistics and Machine Learning Toolbox.
%
% Now, I suppose you intend to use the Neural Network Toolbox.
%In that case the syntax is a bit different: the class labels are in a 3-by-300 matrix,
%each column being a sample, with the row corresponding to the class getting
value '1' and the rest '0'.
%To build it from the previous syntax, type:

Targets = full(ind2vec(y));
%% Train an autoencoder with a hidden layer of size 5 and a linear transfer function for the
% decoder. Set the L2 weight regularizer to 0.001, sparsity regularizer to 4 and sparsity
% proportion to 0.05.
hiddenSize = 5;
autoenc = trainAutoencoder(Inputs, hiddenSize, ...
'L2WeightRegularization', 0.001, ...
'SparsityRegularization', 4, ...
'SparsityProportion', 0.05, ...
'DecoderTransferFunction','purelin');
% Extract the features in the hidden layer.
features = encode(autoenc,Inputs);
% Train a softmax layer for classification using the features .
softnet = trainSoftmaxLayer(features,Targets);
% Stack the encoder and the softmax layer to form a deep network.
```

```
stackednet = stack(autoenc,softnet);

deepnet = train(stackednet,Inputs,Targets);
%%
%%%%%%%%%%%%%%%%%%%TEST

% Estimate the ECG signal types using the deep network, deepnet.

ECG_type = deepnet(Inputs);
% Plot the confusion matrix.
plotconfusion(Targets,ECG_type);

% Plot the ROC.
figure, plotroc(Targets,ECG_type)
```

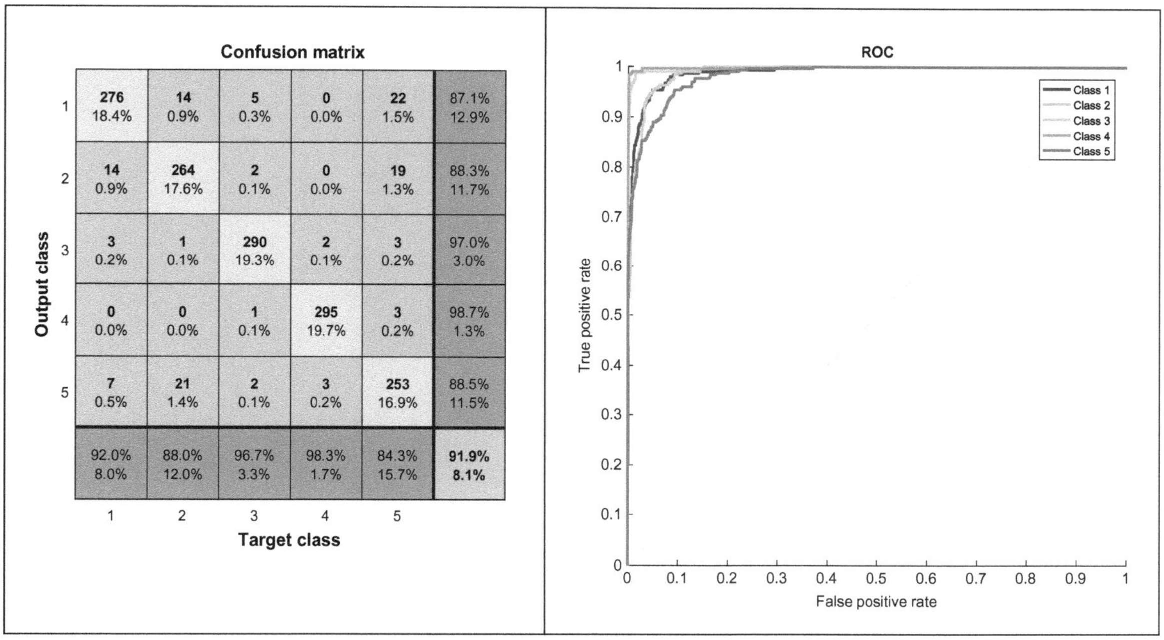

Output class \ Target class	1	2	3	4	5	
1	276 18.4%	14 0.9%	5 0.3%	0 0.0%	22 1.5%	87.1% 12.9%
2	14 0.9%	264 17.6%	2 0.1%	0 0.0%	19 1.3%	88.3% 11.7%
3	3 0.2%	1 0.1%	290 19.3%	2 0.1%	3 0.2%	97.0% 3.0%
4	0 0.0%	0 0.0%	1 0.1%	295 19.7%	3 0.2%	98.7% 1.3%
5	7 0.5%	21 1.4%	2 0.1%	3 0.2%	253 16.9%	88.5% 11.5%
	92.0% 8.0%	88.0% 12.0%	96.7% 3.3%	98.3% 1.7%	84.3% 15.7%	91.9% 8.1%

FIG. 5.32 Representation of performance of the DNN classifier for the ECG signals features extracted using AR Burg method.

EXAMPLE 5.31 The following MATLAB code was used to extract features from the EEG signals using Eigen spectrum method. Then it classified these data using DNN classifier. You can download data from the following website: http://epileptologie-bonn.de/cms/front_content.php?idcat = 193&lang = 3&changelang = 3

```
%% Ch5_Ex63_EEG_EIG_DEEP.m

%The following MATLAB code is used to extract the features from
%the EEG signals using EIGEN Method.
%Then it classifies EEG Signals Using Deep Neural Network Classifier
% Create a Stacked Network
clc
clear
%Load Sample EEG Data downloaded from the web site
%http://epileptologie-bonn.de/cms/front_content.php?idcat=193&lang=3&changelang=3
load AS_BONN_ALL_EEG_DATA_4096.mat
%%
Fs = 173.6;          % Sampling frequency
```

```
T = 1/Fs;           % Sampling period
segmentLength = 128; % Length of a signal segment
Length=4096;% Length of signal
order = 34; %Order of the EIGEN Model
NFFT = 128; %Number of FFT points
Nofsignal=100; %Number of Signal

%% Preparing the Data
% Data for classification problems are set up for a neural network by
% organizing the data into two matrices, the input matrix X and the target
% matrix T.
%
% Each ith column of the input matrix will have 65 elements from the EIGEN
% spectrum
%
% Each corresponding column of the target matrix will have three elements.
%
% Here such the dataset is created.

%%
% Obtain the EIGEN Spectrum of the NORMAL EEG signal using peig.
for i=1:Nofsignal
    [Pxx,F] = peig(Normal_Eyes_Open(1:Length,i),order,NFFT,Fs);
    ASData(i,:)=Pxx(:);
end
%%
% Obtain the EIGEN Spectrum of the INTERICTAL EEG signal using peig.
for i=Nofsignal+1:2*Nofsignal
    [Pxx,F] = peig(Interictal(1:Length,i-Nofsignal),order,NFFT,Fs);
    ASData(i,:)=Pxx(:);
end
%%
% Obtain the EIGEN Spectrum of the ICTAL EEG signal using peig.
for i=2*Nofsignal+1:3*Nofsignal
    [Pxx,F] = peig(Ictal(1:Length,i-2*Nofsignal),order,NFFT,Fs);
    ASData(i,:)=Pxx(:);
end

%Transpose the Data Matrix
Inputs=ASData';

%% TARGET GENERATION
% Your target set must have one label for each sample,
%so it must contains 300 elements. Here is an easy way to build one.
%Start by numbering your classes from 1 to 3.
%Make sure that your X Array is sorted by class,
%meaning all the samples from the first class,
%then all the samples from the second class, and so on. Then:

 % Classes from 1 to 3
 y = 1:3;
 % Nofsignal samples per class
 y = repmat(y, Nofsignal, 1);
 % Reshape to obtain a vector
 y = reshape(y, 1, numel(y));
% At this stage, you should have a 1-by-300 vector with numeric class labels from 1 to 3.
%This is what you want if you use the Statistics and Machine Learning Toolbox.
%
% Now, I suppose you intend to use the Neural Network Toolbox.
```

```
%In that case the syntax is a bit different: the class labels are in a 3-by-300 matrix,
%each column being a sample, with the row corresponding to the class getting value '1' and the
rest '0'.
%To build it from the previous syntax, type:

Targets = full(ind2vec(y));

%% Train an autoencoder with a hidden layer of size 5 and a linear transfer function for the
% decoder. Set the L2 weight regularizer to 0.001, sparsity regularizer to 4 and sparsity
% proportion to 0.05.
hiddenSize = 5;
autoenc = trainAutoencoder(Inputs, hiddenSize, ...
'L2WeightRegularization', 0.001, ...
'SparsityRegularization', 4, ...
'SparsityProportion', 0.05, ...
'DecoderTransferFunction','purelin');
% Extract the features in the hidden layer.
features = encode(autoenc,Inputs);
% Train a softmax layer for classification using the features .
softnet = trainSoftmaxLayer(features,Targets);
% Stack the encoder and the softmax layer to form a deep network.
stackednet = stack(autoenc,softnet);

deepnet = train(stackednet,Inputs,Targets);
%% TEST

% Estimate the EEG signal types using the deep network, deepnet.

EEG_type = deepnet(Inputs);

% Plot the confusion matrix.
plotconfusion(Targets,EEG_type);

% Plot the ROC.
figure, plotroc(Targets,EEG_type)
```

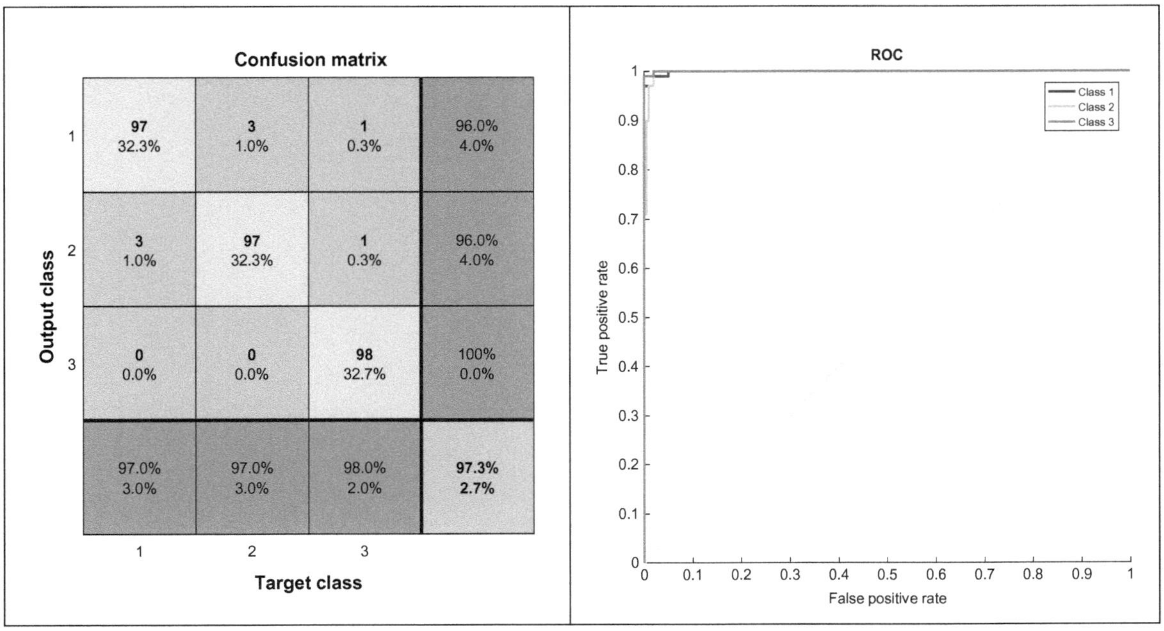

FIG. 5.33 Representation of performance of the DNN classifier for the EEG signals features extracted using Eigen method.

EXAMPLE 5.32 The following MATLAB code was used to extract features from the EEG signals using WPD. Then it used statistical values of WPD subbands. Then it classified these data using DNN classifier. You can download data from the following website:
http://epileptologie-bonn.de/cms/front_content.php?idcat=193&lang=3&changelang=3

```
%% Ch5_Ex67_EEG_WPD_DEEP.m

%WPD of NORMAL, INTERICTAL and ICTAL EEG signals
%The following MATLAB code is used to Extract WPD features from the EEG signals
%Decompose EEG data using WPD coefficients
%Then uses Statistical features as:

%(1) Mean of the absolute values of the coefficients in each sub-band.
%(2) Standard deviation of the coefficients in each sub-band.
%(3) Skewness of the coefficients in each sub-band.
%(4) Kurtosis of the coefficients in each sub-band.
%(5) RMS power of the wavelet coefficients in each subband.
%(6) Ratio of the mean absolute values of adjacent subbands.

%Then it classifies EEG Signals Using Deep Neural Network Classifier
% Create a Stacked Network
clc
clear
%Load Sample EEG Data downloaded from the web site
%http://epileptologie-bonn.de/cms/front_content.php?idcat=193&lang=3&changelang=3
load AS_BONN_ALL_EEG_DATA_4096.mat
wname = 'db4';
Length = 4096; % Length of signal
Nofsignal=100; %Number of Signal

%% Preparing the Data
% Data for classification problems are set up for a neural network by
% organizing the data into two matrices, the input matrix X and the target
% matrix T.
%
% Each ith column of the input matrix will have 95 elements
% representing a statistical features of WPD subbands.
% Here such the dataset is created.

%%
%%%%%%%%%%%%%%%%%%%%%%%%%%%%%% A Z.ZIP EYES OPEN NORMAL SUBJECT

for i=1:Nofsignal;

[wp10,wp11] = dwt(Normal_Eyes_Open(1:Length,i),wname);%WPD Decomposition
[wp20,wp21] = dwt(wp10,wname);
[wp22,wp23] = dwt(wp11,wname);

[wp30,wp31] = dwt(wp20,wname);
[wp32,wp33] = dwt(wp21,wname);
[wp34,wp35] = dwt(wp22,wname);
[wp36,wp37] = dwt(wp23,wname);

[wp40,wp41] = dwt(wp30,wname);
[wp42,wp43] = dwt(wp31,wname);
[wp44,wp45] = dwt(wp32,wname);
[wp46,wp47] = dwt(wp33,wname);
[wp48,wp49] = dwt(wp34,wname);
```

```
[wp4A,wp4B] = dwt(wp35,wname);
[wp4C,wp4D] = dwt(wp36,wname);
[wp4E,wp4F] = dwt(wp37,wname);

ASData(i,1)=mean(abs(wp40));
ASData(i,2)=mean(abs(wp41));
ASData(i,3)=mean(abs(wp42));
ASData(i,4)=mean(abs(wp43));
ASData(i,5)=mean(abs(wp44));
ASData(i,6)=mean(abs(wp45));
ASData(i,7)=mean(abs(wp46));
ASData(i,8)=mean(abs(wp47));
ASData(i,9)=mean(abs(wp48));
ASData(i,10)=mean(abs(wp49));
ASData(i,11)=mean(abs(wp4A));
ASData(i,12)=mean(abs(wp4B));
ASData(i,13)=mean(abs(wp4C));
ASData(i,14)=mean(abs(wp4D));
ASData(i,15)=mean(abs(wp4E));
ASData(i,16)=mean(abs(wp4F));

ASData(i,17)=std(wp40);
ASData(i,18)=std(wp41);
ASData(i,19)=std(wp42);
ASData(i,20)=std(wp43);
ASData(i,21)=std(wp44);
ASData(i,22)=std(wp45);
ASData(i,23)=std(wp46);
ASData(i,24)=std(wp47);
ASData(i,25)=std(wp48);
ASData(i,26)=std(wp49);
ASData(i,27)=std(wp4A);
ASData(i,28)=std(wp4B);
ASData(i,29)=std(wp4C);
ASData(i,30)=std(wp4D);
ASData(i,31)=std(wp4E);
ASData(i,32)=std(wp4F);

ASData(i,33)=skewness(wp40(:));
ASData(i,34)=skewness(wp41(:));
ASData(i,35)=skewness(wp42(:));
ASData(i,36)=skewness(wp43(:));
ASData(i,37)=skewness(wp44(:));
ASData(i,38)=skewness(wp45(:));
ASData(i,39)=skewness(wp46(:));
ASData(i,40)=skewness(wp47(:));
ASData(i,41)=skewness(wp48(:));
ASData(i,42)=skewness(wp49(:));
ASData(i,43)=skewness(wp4A(:));
ASData(i,44)=skewness(wp4B(:));
ASData(i,45)=skewness(wp4C(:));
ASData(i,46)=skewness(wp4D(:));
ASData(i,47)=skewness(wp4E(:));
ASData(i,48)=skewness(wp4F(:));

ASData(i,49)=kurtosis(wp40);
ASData(i,50)=kurtosis(wp41);
ASData(i,51)=kurtosis(wp42);
```

```
ASData(i,52)=kurtosis(wp43);
ASData(i,53)=kurtosis(wp44);
ASData(i,54)=kurtosis(wp45);
ASData(i,55)=kurtosis(wp46);
ASData(i,56)=kurtosis(wp47);
ASData(i,57)=kurtosis(wp48);
ASData(i,58)=kurtosis(wp49);
ASData(i,59)=kurtosis(wp4A);
ASData(i,60)=kurtosis(wp4B);
ASData(i,61)=kurtosis(wp4C);
ASData(i,62)=kurtosis(wp4D);
ASData(i,63)=kurtosis(wp4E);
ASData(i,64)=kurtosis(wp4F);

ASData(i,65)=rms(wp40(:));
ASData(i,66)=rms(wp41(:));
ASData(i,67)=rms(wp42(:));
ASData(i,68)=rms(wp43(:));
ASData(i,69)=rms(wp44(:));
ASData(i,70)=rms(wp45(:));
ASData(i,71)=rms(wp46(:));
ASData(i,72)=rms(wp47(:));
ASData(i,73)=rms(wp48(:));
ASData(i,74)=rms(wp49(:));
ASData(i,75)=rms(wp4A(:));
ASData(i,76)=rms(wp4B(:));
ASData(i,77)=rms(wp4C(:));
ASData(i,78)=rms(wp4D(:));
ASData(i,79)=rms(wp4E(:));
ASData(i,80)=rms(wp4F(:));

ASData(i,81)=mean(abs(wp40))/mean(abs(wp41));
ASData(i,82)=mean(abs(wp41))/mean(abs(wp42));
ASData(i,83)=mean(abs(wp42))/mean(abs(wp43));
ASData(i,84)=mean(abs(wp43))/mean(abs(wp44));
ASData(i,85)=mean(abs(wp44))/mean(abs(wp45));
ASData(i,86)=mean(abs(wp45))/mean(abs(wp46));
ASData(i,87)=mean(abs(wp46))/mean(abs(wp47));
ASData(i,88)=mean(abs(wp47))/mean(abs(wp48));
ASData(i,89)=mean(abs(wp48))/mean(abs(wp49));
ASData(i,90)=mean(abs(wp49))/mean(abs(wp4A));
ASData(i,91)=mean(abs(wp4A))/mean(abs(wp4B));
ASData(i,92)=mean(abs(wp4B))/mean(abs(wp4C));
ASData(i,93)=mean(abs(wp4C))/mean(abs(wp4D));
ASData(i,94)=mean(abs(wp4D))/mean(abs(wp4E));
ASData(i,95)=mean(abs(wp4E))/mean(abs(wp4F));

end

%%
% D F.ZIP EPILEPTIC SUBJECT (INTERICTAL) were recorded from within
% the epileptogenic zone during seizure free intervals

for i=Nofsignal+1:2*Nofsignal;

[wp10,wp11] = dwt(Interictal(1:Length,i-Nofsignal),wname);%WPD Decomposition
[wp20,wp21] = dwt(wp10,wname);
[wp22,wp23] = dwt(wp11,wname);
```

```
[wp30,wp31] = dwt(wp20,wname);
[wp32,wp33] = dwt(wp21,wname);
[wp34,wp35] = dwt(wp22,wname);
[wp36,wp37] = dwt(wp23,wname);

[wp40,wp41] = dwt(wp30,wname);
[wp42,wp43] = dwt(wp31,wname);
[wp44,wp45] = dwt(wp32,wname);
[wp46,wp47] = dwt(wp33,wname);
[wp48,wp49] = dwt(wp34,wname);
[wp4A,wp4B] = dwt(wp35,wname);
[wp4C,wp4D] = dwt(wp36,wname);
[wp4E,wp4F] = dwt(wp37,wname);

ASData(i,1)=mean(abs(wp40));
ASData(i,2)=mean(abs(wp41));
ASData(i,3)=mean(abs(wp42));
ASData(i,4)=mean(abs(wp43));
ASData(i,5)=mean(abs(wp44));
ASData(i,6)=mean(abs(wp45));
ASData(i,7)=mean(abs(wp46));
ASData(i,8)=mean(abs(wp47));
ASData(i,9)=mean(abs(wp48));
ASData(i,10)=mean(abs(wp49));
ASData(i,11)=mean(abs(wp4A));
ASData(i,12)=mean(abs(wp4B));
ASData(i,13)=mean(abs(wp4C));
ASData(i,14)=mean(abs(wp4D));
ASData(i,15)=mean(abs(wp4E));
ASData(i,16)=mean(abs(wp4F));

ASData(i,17)=std(wp40);
ASData(i,18)=std(wp41);
ASData(i,19)=std(wp42);
ASData(i,20)=std(wp43);
ASData(i,21)=std(wp44);
ASData(i,22)=std(wp45);
ASData(i,23)=std(wp46);
ASData(i,24)=std(wp47);
ASData(i,25)=std(wp48);
ASData(i,26)=std(wp49);
ASData(i,27)=std(wp4A);
ASData(i,28)=std(wp4B);
ASData(i,29)=std(wp4C);
ASData(i,30)=std(wp4D);
ASData(i,31)=std(wp4E);
ASData(i,32)=std(wp4F);

ASData(i,33)=skewness(wp40(:));
ASData(i,34)=skewness(wp41(:));
ASData(i,35)=skewness(wp42(:));
ASData(i,36)=skewness(wp43(:));
ASData(i,37)=skewness(wp44(:));
ASData(i,38)=skewness(wp45(:));
ASData(i,39)=skewness(wp46(:));
ASData(i,40)=skewness(wp47(:));
ASData(i,41)=skewness(wp48(:));
ASData(i,42)=skewness(wp49(:));
```

```
ASData(i,43)=skewness(wp4A(:));
ASData(i,44)=skewness(wp4B(:));
ASData(i,45)=skewness(wp4C(:));
ASData(i,46)=skewness(wp4D(:));
ASData(i,47)=skewness(wp4E(:));
ASData(i,48)=skewness(wp4F(:));

ASData(i,49)=kurtosis(wp40);
ASData(i,50)=kurtosis(wp41);
ASData(i,51)=kurtosis(wp42);
ASData(i,52)=kurtosis(wp43);
ASData(i,53)=kurtosis(wp44);
ASData(i,54)=kurtosis(wp45);
ASData(i,55)=kurtosis(wp46);
ASData(i,56)=kurtosis(wp47);
ASData(i,57)=kurtosis(wp48);
ASData(i,58)=kurtosis(wp49);
ASData(i,59)=kurtosis(wp4A);
ASData(i,60)=kurtosis(wp4B);
ASData(i,61)=kurtosis(wp4C);
ASData(i,62)=kurtosis(wp4D);
ASData(i,63)=kurtosis(wp4E);
ASData(i,64)=kurtosis(wp4F);

ASData(i,65)=rms(wp40(:));
ASData(i,66)=rms(wp41(:));
ASData(i,67)=rms(wp42(:));
ASData(i,68)=rms(wp43(:));
ASData(i,69)=rms(wp44(:));
ASData(i,70)=rms(wp45(:));
ASData(i,71)=rms(wp46(:));
ASData(i,72)=rms(wp47(:));
ASData(i,73)=rms(wp48(:));
ASData(i,74)=rms(wp49(:));
ASData(i,75)=rms(wp4A(:));
ASData(i,76)=rms(wp4B(:));
ASData(i,77)=rms(wp4C(:));
ASData(i,78)=rms(wp4D(:));
ASData(i,79)=rms(wp4E(:));
ASData(i,80)=rms(wp4F(:));

ASData(i,81)=mean(abs(wp40))/mean(abs(wp41));
ASData(i,82)=mean(abs(wp41))/mean(abs(wp42));
ASData(i,83)=mean(abs(wp42))/mean(abs(wp43));
ASData(i,84)=mean(abs(wp43))/mean(abs(wp44));
ASData(i,85)=mean(abs(wp44))/mean(abs(wp45));
ASData(i,86)=mean(abs(wp45))/mean(abs(wp46));
ASData(i,87)=mean(abs(wp46))/mean(abs(wp47));
ASData(i,88)=mean(abs(wp47))/mean(abs(wp48));
ASData(i,89)=mean(abs(wp48))/mean(abs(wp49));
ASData(i,90)=mean(abs(wp49))/mean(abs(wp4A));
ASData(i,91)=mean(abs(wp4A))/mean(abs(wp4B));
ASData(i,92)=mean(abs(wp4B))/mean(abs(wp4C));
ASData(i,93)=mean(abs(wp4C))/mean(abs(wp4D));
ASData(i,94)=mean(abs(wp4D))/mean(abs(wp4E));
ASData(i,95)=mean(abs(wp4E))/mean(abs(wp4F));
end
```

```
%% E S.ZIP EPILEPTIC SUBJECT ICTAL DURING SEIZURE %%%%%%%%%%%%%%%%%%%%%%%%%%%%%%

for i=2*Nofsignal+1:3*Nofsignal;

[wp10,wp11] = dwt(Ictal(1:Length,i-2*Nofsignal),wname);%WPD Decomposition
[wp20,wp21] = dwt(wp10,wname);
[wp22,wp23] = dwt(wp11,wname);

[wp30,wp31] = dwt(wp20,wname);
[wp32,wp33] = dwt(wp21,wname);
[wp34,wp35] = dwt(wp22,wname);
[wp36,wp37] = dwt(wp23,wname);

[wp40,wp41] = dwt(wp30,wname);
[wp42,wp43] = dwt(wp31,wname);
[wp44,wp45] = dwt(wp32,wname);
[wp46,wp47] = dwt(wp33,wname);
[wp48,wp49] = dwt(wp34,wname);
[wp4A,wp4B] = dwt(wp35,wname);
[wp4C,wp4D] = dwt(wp36,wname);
[wp4E,wp4F] = dwt(wp37,wname);

ASData(i,1)=mean(abs(wp40));
ASData(i,2)=mean(abs(wp41));
ASData(i,3)=mean(abs(wp42));
ASData(i,4)=mean(abs(wp43));
ASData(i,5)=mean(abs(wp44));
ASData(i,6)=mean(abs(wp45));
ASData(i,7)=mean(abs(wp46));
ASData(i,8)=mean(abs(wp47));
ASData(i,9)=mean(abs(wp48));
ASData(i,10)=mean(abs(wp49));
ASData(i,11)=mean(abs(wp4A));
ASData(i,12)=mean(abs(wp4B));
ASData(i,13)=mean(abs(wp4C));
ASData(i,14)=mean(abs(wp4D));
ASData(i,15)=mean(abs(wp4E));
ASData(i,16)=mean(abs(wp4F));

ASData(i,17)=std(wp40);
ASData(i,18)=std(wp41);
ASData(i,19)=std(wp42);
ASData(i,20)=std(wp43);
ASData(i,21)=std(wp44);
ASData(i,22)=std(wp45);
ASData(i,23)=std(wp46);
ASData(i,24)=std(wp47);
ASData(i,25)=std(wp48);
ASData(i,26)=std(wp49);
ASData(i,27)=std(wp4A);
ASData(i,28)=std(wp4B);
ASData(i,29)=std(wp4C);
ASData(i,30)=std(wp4D);
ASData(i,31)=std(wp4E);
ASData(i,32)=std(wp4F);

ASData(i,33)=skewness(wp40(:));
ASData(i,34)=skewness(wp41(:));
```

```
ASData(i,35)=skewness(wp42(:));
ASData(i,36)=skewness(wp43(:));
ASData(i,37)=skewness(wp44(:));
ASData(i,38)=skewness(wp45(:));
ASData(i,39)=skewness(wp46(:));
ASData(i,40)=skewness(wp47(:));
ASData(i,41)=skewness(wp48(:));
ASData(i,42)=skewness(wp49(:));
ASData(i,43)=skewness(wp4A(:));
ASData(i,44)=skewness(wp4B(:));
ASData(i,45)=skewness(wp4C(:));
ASData(i,46)=skewness(wp4D(:));
ASData(i,47)=skewness(wp4E(:));
ASData(i,48)=skewness(wp4F(:));

ASData(i,49)=kurtosis(wp40);
ASData(i,50)=kurtosis(wp41);
ASData(i,51)=kurtosis(wp42);
ASData(i,52)=kurtosis(wp43);
ASData(i,53)=kurtosis(wp44);
ASData(i,54)=kurtosis(wp45);
ASData(i,55)=kurtosis(wp46);
ASData(i,56)=kurtosis(wp47);
ASData(i,57)=kurtosis(wp48);
ASData(i,58)=kurtosis(wp49);
ASData(i,59)=kurtosis(wp4A);
ASData(i,60)=kurtosis(wp4B);
ASData(i,61)=kurtosis(wp4C);
ASData(i,62)=kurtosis(wp4D);
ASData(i,63)=kurtosis(wp4E);
ASData(i,64)=kurtosis(wp4F);

ASData(i,65)=rms(wp40(:));
ASData(i,66)=rms(wp41(:));
ASData(i,67)=rms(wp42(:));
ASData(i,68)=rms(wp43(:));
ASData(i,69)=rms(wp44(:));
ASData(i,70)=rms(wp45(:));
ASData(i,71)=rms(wp46(:));
ASData(i,72)=rms(wp47(:));
ASData(i,73)=rms(wp48(:));
ASData(i,74)=rms(wp49(:));
ASData(i,75)=rms(wp4A(:));
ASData(i,76)=rms(wp4B(:));
ASData(i,77)=rms(wp4C(:));
ASData(i,78)=rms(wp4D(:));
ASData(i,79)=rms(wp4E(:));
ASData(i,80)=rms(wp4F(:));

ASData(i,81)=mean(abs(wp40))/mean(abs(wp41));
ASData(i,82)=mean(abs(wp41))/mean(abs(wp42));
ASData(i,83)=mean(abs(wp42))/mean(abs(wp43));
ASData(i,84)=mean(abs(wp43))/mean(abs(wp44));
ASData(i,85)=mean(abs(wp44))/mean(abs(wp45));
ASData(i,86)=mean(abs(wp45))/mean(abs(wp46));
ASData(i,87)=mean(abs(wp46))/mean(abs(wp47));
ASData(i,88)=mean(abs(wp47))/mean(abs(wp48));
ASData(i,89)=mean(abs(wp48))/mean(abs(wp49));
```

```
ASData(i,90)=mean(abs(wp49))/mean(abs(wp4A));
ASData(i,91)=mean(abs(wp4A))/mean(abs(wp4B));
ASData(i,92)=mean(abs(wp4B))/mean(abs(wp4C));
ASData(i,93)=mean(abs(wp4C))/mean(abs(wp4D));
ASData(i,94)=mean(abs(wp4D))/mean(abs(wp4E));
ASData(i,95)=mean(abs(wp4E))/mean(abs(wp4F));
end

%Transpose the Data Matrix
Inputs=ASData';

%% TARGET GENERATION
%
% Your target set must have one label for each sample,
%so it must contains 300 elements. Here is an easy way to build one.
%Start by numbering your classes from 1 to 3.
%Make sure that your X Array is sorted by class,
%meaning all the samples from the first class,
%then all the samples from the second class, and so on. Then:

 % Classes from 1 to 3
 y = 1:3;
 % Nofsignal samples per class
 y = repmat(y, Nofsignal, 1);
 % Reshape to obtain a vector
 y = reshape(y, 1, numel(y));
% At this stage, you should have a 1-by-300 vector with numeric class labels from 1 to 3.
%This is what you want if you use the Statistics and Machine Learning Toolbox.
%
% Now, I suppose you intend to use the Neural Network Toolbox.
%In that case the syntax is a bit different: the class labels are in a 3-by-300 matrix,
%each column being a sample, with the row corresponding to the class getting value '1' and the
rest '0'.
%To build it from the previous syntax, type:

Targets = full(ind2vec(y));

%% Train an autoencoder with a hidden layer of size 5 and a linear transfer function for the
% decoder. Set the L2 weight regularizer to 0.001, sparsity regularizer to 4 and sparsity
% proportion to 0.05.
hiddenSize = 5;
autoenc = trainAutoencoder(Inputs, hiddenSize, ...
'L2WeightRegularization', 0.001, ...
'SparsityRegularization', 4, ...
'SparsityProportion', 0.05, ...
'DecoderTransferFunction','purelin');
% Extract the features in the hidden layer.
features = encode(autoenc,Inputs);
% Train a softmax layer for classification using the features .
softnet = trainSoftmaxLayer(features,Targets);
% Stack the encoder and the softmax layer to form a deep network.
stackednet = stack(autoenc,softnet);

deepnet = train(stackednet,Inputs,Targets);
%% TEST

% Estimate the EEG signal types using the deep network, deepnet.
```

```
EEG_type = deepnet(Inputs);

% Plot the confusion matrix.
plotconfusion(Targets,EEG_type);

% Plot the ROC.
figure, plotroc(Targets,EEG_type)
```

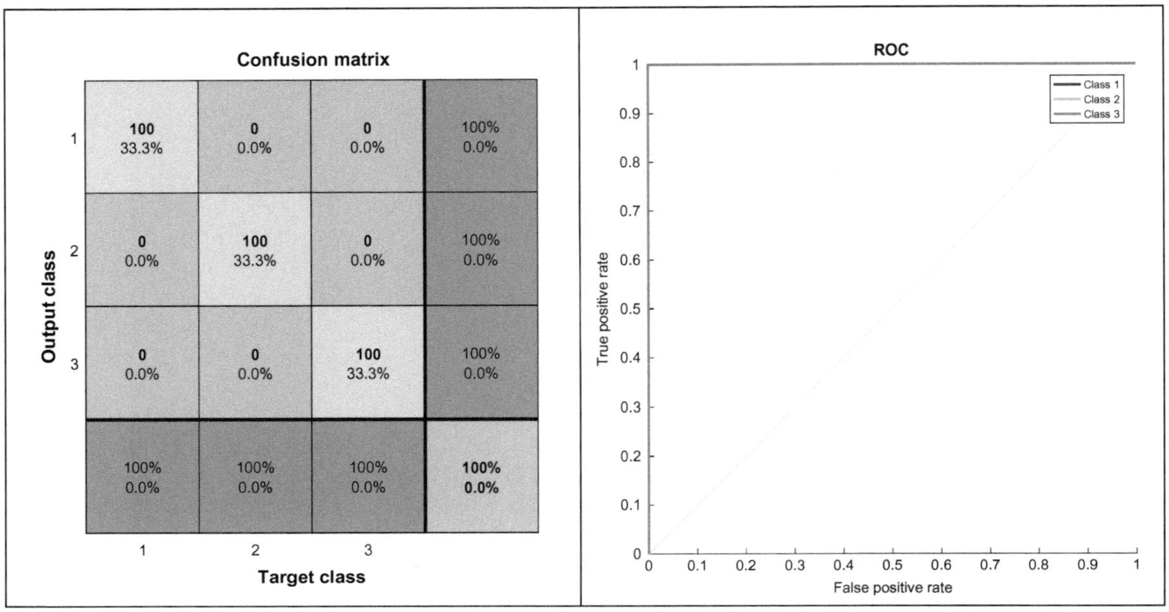

FIG. 5.34 Representation of performance of the DNN classifier for the EEG signals features extracted using WPD.

EXAMPLE 5.33 The following MATLAB code was used to extract features from the EEG signals using SWT. Then it used statistical values of SWT subbands. Then it classified these data using DNN classifier. You can download data from the following website:
http://epileptologie-bonn.de/cms/front_content.php?idcat=193&lang=3&changelang=3

```
%% Ch5_Ex68_EEG_SWT_DEEP.m

%SWT of NORMAL, INTERICTAL and ICTAL EEG signals
%The following MATLAB code is used to Extract SWT features from the EEG signals
%Decompose EEG data using SWT Transfrom
%Then uses Statistical features as:

%(1) Mean of the absolute values of the coefficients in each sub-band.
%(2) Standard deviation of the coefficients in each sub-band.
%(3) Skewness of the coefficients in each sub-band.
%(4) Kurtosis of the coefficients in each sub-band.
%(5) RMS power of the wavelet coefficients in each subband.
%(6) Ratio of the mean absolute values of adjacent subbands.

%Then it classifies EEG Signals Using Deep Neural Network Classifier
% Create a Stacked Network
clc
```

```
clear
%Load Sample EEG Data downloaded from the web site
%http://epileptologie-bonn.de/cms/front_content.php?idcat=193&lang=3&changelang=3
load AS_BONN_ALL_EEG_DATA_4096.mat

wname = 'db4';
Level = 8;
Length = 4096; % Length of signal
Nofsignal=100; %Number of Signal
%%
%%%%%%%%%%%%%%%%%%%%%%%%%%%% A Z.ZIP EYES OPEN NORMAL SUBJECT

for i=1:Nofsignal;

[swa,swd] = swt(Normal_Eyes_Open(1:Length,i),Level,wname);%SWT

ASData(i,1)=mean(abs(swa(1,:)));
ASData(i,2)=mean(abs(swa(2,:)));
ASData(i,3)=mean(abs(swa(3,:)));
ASData(i,4)=mean(abs(swa(4,:)));
ASData(i,5)=mean(abs(swa(5,:)));
ASData(i,6)=mean(abs(swa(6,:)));
ASData(i,7)=mean(abs(swa(7,:)));
ASData(i,8)=mean(abs(swa(8,:)));
ASData(i,9)=mean(abs(swd(1,:)));
ASData(i,10)=mean(abs(swd(2,:)));
ASData(i,11)=mean(abs(swd(3,:)));
ASData(i,12)=mean(abs(swd(4,:)));
ASData(i,13)=mean(abs(swd(5,:)));
ASData(i,14)=mean(abs(swd(6,:)));
ASData(i,15)=mean(abs(swd(7,:)));
ASData(i,16)=mean(abs(swd(8,:)));

ASData(i,17)=std(swa(1,:));
ASData(i,18)=std(swa(2,:));
ASData(i,19)=std(swa(3,:));
ASData(i,20)=std(swa(4,:));
ASData(i,21)=std(swa(5,:));
ASData(i,22)=std(swa(6,:));
ASData(i,23)=std(swa(7,:));
ASData(i,24)=std(swa(8,:));
ASData(i,25)=std(swd(1,:));
ASData(i,26)=std(swd(2,:));
ASData(i,27)=std(swd(3,:));
ASData(i,28)=std(swd(4,:));
ASData(i,29)=std(swd(5,:));
ASData(i,30)=std(swd(6,:));
ASData(i,31)=std(swd(7,:));
ASData(i,32)=std(swd(8,:));

ASData(i,33)=skewness(swa(1,:));
ASData(i,34)=skewness(swa(2,:));
ASData(i,35)=skewness(swa(3,:));
ASData(i,36)=skewness(swa(4,:));
ASData(i,37)=skewness(swa(5,:));
ASData(i,38)=skewness(swa(6,:));
ASData(i,39)=skewness(swa(7,:));
```

```
ASData(i,40)=skewness(swa(8,:));
ASData(i,41)=skewness(swd(1,:));
ASData(i,42)=skewness(swd(2,:));
ASData(i,43)=skewness(swd(3,:));
ASData(i,44)=skewness(swd(4,:));
ASData(i,45)=skewness(swd(5,:));
ASData(i,46)=skewness(swd(6,:));
ASData(i,47)=skewness(swd(7,:));
ASData(i,48)=skewness(swd(8,:));

ASData(i,49)=kurtosis(swa(1,:));
ASData(i,50)=kurtosis(swa(2,:));
ASData(i,51)=kurtosis(swa(3,:));
ASData(i,52)=kurtosis(swa(4,:));
ASData(i,53)=kurtosis(swa(5,:));
ASData(i,54)=kurtosis(swa(6,:));
ASData(i,55)=kurtosis(swa(7,:));
ASData(i,56)=kurtosis(swa(8,:));
ASData(i,57)=kurtosis(swd(1,:));
ASData(i,58)=kurtosis(swd(2,:));
ASData(i,59)=kurtosis(swd(3,:));
ASData(i,60)=kurtosis(swd(4,:));
ASData(i,61)=kurtosis(swd(5,:));
ASData(i,62)=kurtosis(swd(6,:));
ASData(i,63)=kurtosis(swd(7,:));
ASData(i,64)=kurtosis(swd(8,:));

        ASData(i,65)=rms(swa(1,:));
        ASData(i,66)=rms(swa(2,:));
        ASData(i,67)=rms(swa(3,:));
        ASData(i,68)=rms(swa(4,:));
        ASData(i,69)=rms(swa(5,:));
        ASData(i,70)=rms(swa(6,:));
        ASData(i,71)=rms(swa(7,:));
        ASData(i,72)=rms(swa(8,:));
        ASData(i,73)=rms(swd(1,:));
        ASData(i,74)=rms(swd(2,:));
        ASData(i,75)=rms(swd(3,:));
        ASData(i,76)=rms(swd(4,:));
        ASData(i,77)=rms(swd(5,:));
        ASData(i,78)=rms(swd(6,:));
        ASData(i,79)=rms(swd(7,:));
        ASData(i,80)=rms(swd(8,:));

        ASData(i,81)=mean(abs(swa(1,:)))/mean(abs(swa(2,:)));
        ASData(i,82)=mean(abs(swa(2,:)))/mean(abs(swa(3,:)));
        ASData(i,83)=mean(abs(swa(3,:)))/mean(abs(swa(4,:)));
        ASData(i,84)=mean(abs(swa(4,:)))/mean(abs(swa(5,:)));
        ASData(i,85)=mean(abs(swa(5,:)))/mean(abs(swa(6,:)));
        ASData(i,86)=mean(abs(swa(6,:)))/mean(abs(swa(7,:)));
        ASData(i,87)=mean(abs(swa(7,:)))/mean(abs(swa(8,:)));
        ASData(i,88)=mean(abs(swa(8,:)))/mean(abs(swd(1,:)));
        ASData(i,89)=mean(abs(swd(1,:)))/mean(abs(swd(2,:)));
        ASData(i,90)=mean(abs(swd(2,:)))/mean(abs(swd(3,:)));
        ASData(i,91)=mean(abs(swd(3,:)))/mean(abs(swd(4,:)));
        ASData(i,92)=mean(abs(swd(4,:)))/mean(abs(swd(5,:)));
        ASData(i,93)=mean(abs(swd(5,:)))/mean(abs(swd(6,:)));
```

```
        ASData(i,94)=mean(abs(swd(6,:)))/mean(abs(swd(7,:)));
        ASData(i,95)=mean(abs(swd(7,:)))/mean(abs(swd(8,:)));

end

%%
% D F.ZIP EPILEPTIC SUBJECT (INTERICTAL) were recorded from within
% the epileptogenic zone during seizure free intervals

for i=Nofsignal+1:2*Nofsignal;

[swa,swd] = swt(Interictal(1:Length,i-Nofsignal),Level,wname); %SWT

ASData(i,1)=mean(abs(swa(1,:)));
ASData(i,2)=mean(abs(swa(2,:)));
ASData(i,3)=mean(abs(swa(3,:)));
ASData(i,4)=mean(abs(swa(4,:)));
ASData(i,5)=mean(abs(swa(5,:)));
ASData(i,6)=mean(abs(swa(6,:)));
ASData(i,7)=mean(abs(swa(7,:)));
ASData(i,8)=mean(abs(swa(8,:)));
ASData(i,9)=mean(abs(swd(1,:)));
ASData(i,10)=mean(abs(swd(2,:)));
ASData(i,11)=mean(abs(swd(3,:)));
ASData(i,12)=mean(abs(swd(4,:)));
ASData(i,13)=mean(abs(swd(5,:)));
ASData(i,14)=mean(abs(swd(6,:)));
ASData(i,15)=mean(abs(swd(7,:)));
ASData(i,16)=mean(abs(swd(8,:)));

ASData(i,17)=std(swa(1,:));
ASData(i,18)=std(swa(2,:));
ASData(i,19)=std(swa(3,:));
ASData(i,20)=std(swa(4,:));
ASData(i,21)=std(swa(5,:));
ASData(i,22)=std(swa(6,:));
ASData(i,23)=std(swa(7,:));
ASData(i,24)=std(swa(8,:));
ASData(i,25)=std(swd(1,:));
ASData(i,26)=std(swd(2,:));
ASData(i,27)=std(swd(3,:));
ASData(i,28)=std(swd(4,:));
ASData(i,29)=std(swd(5,:));
ASData(i,30)=std(swd(6,:));
ASData(i,31)=std(swd(7,:));
ASData(i,32)=std(swd(8,:));

ASData(i,33)=skewness(swa(1,:));
ASData(i,34)=skewness(swa(2,:));
ASData(i,35)=skewness(swa(3,:));
ASData(i,36)=skewness(swa(4,:));
ASData(i,37)=skewness(swa(5,:));
ASData(i,38)=skewness(swa(6,:));
ASData(i,39)=skewness(swa(7,:));
ASData(i,40)=skewness(swa(8,:));
ASData(i,41)=skewness(swd(1,:));
ASData(i,42)=skewness(swd(2,:));
```

```
ASData(i,43)=skewness(swd(3,:));
ASData(i,44)=skewness(swd(4,:));
ASData(i,45)=skewness(swd(5,:));
ASData(i,46)=skewness(swd(6,:));
ASData(i,47)=skewness(swd(7,:));
ASData(i,48)=skewness(swd(8,:));

ASData(i,49)=kurtosis(swa(1,:));
ASData(i,50)=kurtosis(swa(2,:));
ASData(i,51)=kurtosis(swa(3,:));
ASData(i,52)=kurtosis(swa(4,:));
ASData(i,53)=kurtosis(swa(5,:));
ASData(i,54)=kurtosis(swa(6,:));
ASData(i,55)=kurtosis(swa(7,:));
ASData(i,56)=kurtosis(swa(8,:));
ASData(i,57)=kurtosis(swd(1,:));
ASData(i,58)=kurtosis(swd(2,:));
ASData(i,59)=kurtosis(swd(3,:));
ASData(i,60)=kurtosis(swd(4,:));
ASData(i,61)=kurtosis(swd(5,:));
ASData(i,62)=kurtosis(swd(6,:));
ASData(i,63)=kurtosis(swd(7,:));
ASData(i,64)=kurtosis(swd(8,:));

          ASData(i,65)=rms(swa(1,:));
          ASData(i,66)=rms(swa(2,:));
          ASData(i,67)=rms(swa(3,:));
          ASData(i,68)=rms(swa(4,:));
          ASData(i,69)=rms(swa(5,:));
          ASData(i,70)=rms(swa(6,:));
          ASData(i,71)=rms(swa(7,:));
          ASData(i,72)=rms(swa(8,:));
          ASData(i,73)=rms(swd(1,:));
          ASData(i,74)=rms(swd(2,:));
          ASData(i,75)=rms(swd(3,:));
          ASData(i,76)=rms(swd(4,:));
          ASData(i,77)=rms(swd(5,:));
          ASData(i,78)=rms(swd(6,:));
          ASData(i,79)=rms(swd(7,:));
          ASData(i,80)=rms(swd(8,:));

          ASData(i,81)=mean(abs(swa(1,:)))/mean(abs(swa(2,:)));
          ASData(i,82)=mean(abs(swa(2,:)))/mean(abs(swa(3,:)));
          ASData(i,83)=mean(abs(swa(3,:)))/mean(abs(swa(4,:)));
          ASData(i,84)=mean(abs(swa(4,:)))/mean(abs(swa(5,:)));
          ASData(i,85)=mean(abs(swa(5,:)))/mean(abs(swa(6,:)));
          ASData(i,86)=mean(abs(swa(6,:)))/mean(abs(swa(7,:)));
          ASData(i,87)=mean(abs(swa(7,:)))/mean(abs(swa(8,:)));
          ASData(i,88)=mean(abs(swa(8,:)))/mean(abs(swd(1,:)));
          ASData(i,89)=mean(abs(swd(1,:)))/mean(abs(swd(2,:)));
          ASData(i,90)=mean(abs(swd(2,:)))/mean(abs(swd(3,:)));
          ASData(i,91)=mean(abs(swd(3,:)))/mean(abs(swd(4,:)));
          ASData(i,92)=mean(abs(swd(4,:)))/mean(abs(swd(5,:)));
          ASData(i,93)=mean(abs(swd(5,:)))/mean(abs(swd(6,:)));
          ASData(i,94)=mean(abs(swd(6,:)))/mean(abs(swd(7,:)));
          ASData(i,95)=mean(abs(swd(7,:)))/mean(abs(swd(8,:)));

end
```

```
%% E S.ZIP EPILEPTIC SUBJECT ICTAL DURING SEIZURE
%%%%%%%%%%%%%%%%%%%%%%%%%%%%%%

for i=2*Nofsignal+1:3*Nofsignal;

[swa,swd] = swt(Ictal(1:Length,i-2*Nofsignal),Level,wname);%SWT

ASData(i,1)=mean(abs(swa(1,:)));
ASData(i,2)=mean(abs(swa(2,:)));
ASData(i,3)=mean(abs(swa(3,:)));
ASData(i,4)=mean(abs(swa(4,:)));
ASData(i,5)=mean(abs(swa(5,:)));
ASData(i,6)=mean(abs(swa(6,:)));
ASData(i,7)=mean(abs(swa(7,:)));
ASData(i,8)=mean(abs(swa(8,:)));
ASData(i,9)=mean(abs(swd(1,:)));
ASData(i,10)=mean(abs(swd(2,:)));
ASData(i,11)=mean(abs(swd(3,:)));
ASData(i,12)=mean(abs(swd(4,:)));
ASData(i,13)=mean(abs(swd(5,:)));
ASData(i,14)=mean(abs(swd(6,:)));
ASData(i,15)=mean(abs(swd(7,:)));
ASData(i,16)=mean(abs(swd(8,:)));

ASData(i,17)=std(swa(1,:));
ASData(i,18)=std(swa(2,:));
ASData(i,19)=std(swa(3,:));
ASData(i,20)=std(swa(4,:));
ASData(i,21)=std(swa(5,:));
ASData(i,22)=std(swa(6,:));
ASData(i,23)=std(swa(7,:));
ASData(i,24)=std(swa(8,:));
ASData(i,25)=std(swd(1,:));
ASData(i,26)=std(swd(2,:));
ASData(i,27)=std(swd(3,:));
ASData(i,28)=std(swd(4,:));
ASData(i,29)=std(swd(5,:));
ASData(i,30)=std(swd(6,:));
ASData(i,31)=std(swd(7,:));
ASData(i,32)=std(swd(8,:));

ASData(i,33)=skewness(swa(1,:));
ASData(i,34)=skewness(swa(2,:));
ASData(i,35)=skewness(swa(3,:));
ASData(i,36)=skewness(swa(4,:));
ASData(i,37)=skewness(swa(5,:));
ASData(i,38)=skewness(swa(6,:));
ASData(i,39)=skewness(swa(7,:));
ASData(i,40)=skewness(swa(8,:));
ASData(i,41)=skewness(swd(1,:));
ASData(i,42)=skewness(swd(2,:));
ASData(i,43)=skewness(swd(3,:));
ASData(i,44)=skewness(swd(4,:));
ASData(i,45)=skewness(swd(5,:));
ASData(i,46)=skewness(swd(6,:));
ASData(i,47)=skewness(swd(7,:));
ASData(i,48)=skewness(swd(8,:));
```

```
ASData(i,49)=kurtosis(swa(1,:));
ASData(i,50)=kurtosis(swa(2,:));
ASData(i,51)=kurtosis(swa(3,:));
ASData(i,52)=kurtosis(swa(4,:));
ASData(i,53)=kurtosis(swa(5,:));
ASData(i,54)=kurtosis(swa(6,:));
ASData(i,55)=kurtosis(swa(7,:));
ASData(i,56)=kurtosis(swa(8,:));
ASData(i,57)=kurtosis(swd(1,:));
ASData(i,58)=kurtosis(swd(2,:));
ASData(i,59)=kurtosis(swd(3,:));
ASData(i,60)=kurtosis(swd(4,:));
ASData(i,61)=kurtosis(swd(5,:));
ASData(i,62)=kurtosis(swd(6,:));
ASData(i,63)=kurtosis(swd(7,:));
ASData(i,64)=kurtosis(swd(8,:));

        ASData(i,65)=rms(swa(1,:));
        ASData(i,66)=rms(swa(2,:));
        ASData(i,67)=rms(swa(3,:));
        ASData(i,68)=rms(swa(4,:));
        ASData(i,69)=rms(swa(5,:));
        ASData(i,70)=rms(swa(6,:));
        ASData(i,71)=rms(swa(7,:));
        ASData(i,72)=rms(swa(8,:));
        ASData(i,73)=rms(swd(1,:));
        ASData(i,74)=rms(swd(2,:));
        ASData(i,75)=rms(swd(3,:));
        ASData(i,76)=rms(swd(4,:));
        ASData(i,77)=rms(swd(5,:));
        ASData(i,78)=rms(swd(6,:));
        ASData(i,79)=rms(swd(7,:));
        ASData(i,80)=rms(swd(8,:));

        ASData(i,81)=mean(abs(swa(1,:)))/mean(abs(swa(2,:)));
        ASData(i,82)=mean(abs(swa(2,:)))/mean(abs(swa(3,:)));
        ASData(i,83)=mean(abs(swa(3,:)))/mean(abs(swa(4,:)));
        ASData(i,84)=mean(abs(swa(4,:)))/mean(abs(swa(5,:)));
        ASData(i,85)=mean(abs(swa(5,:)))/mean(abs(swa(6,:)));
        ASData(i,86)=mean(abs(swa(6,:)))/mean(abs(swa(7,:)));
        ASData(i,87)=mean(abs(swa(7,:)))/mean(abs(swa(8,:)));
        ASData(i,88)=mean(abs(swa(8,:)))/mean(abs(swd(1,:)));
        ASData(i,89)=mean(abs(swd(1,:)))/mean(abs(swd(2,:)));
        ASData(i,90)=mean(abs(swd(2,:)))/mean(abs(swd(3,:)));
        ASData(i,91)=mean(abs(swd(3,:)))/mean(abs(swd(4,:)));
        ASData(i,92)=mean(abs(swd(4,:)))/mean(abs(swd(5,:)));
        ASData(i,93)=mean(abs(swd(5,:)))/mean(abs(swd(6,:)));
        ASData(i,94)=mean(abs(swd(6,:)))/mean(abs(swd(7,:)));
        ASData(i,95)=mean(abs(swd(7,:)))/mean(abs(swd(8,:)));
end

%Transpose the Data Matrix
Inputs=ASData';

%% TARGET GENERATION
%
% Your target set must have one label for each sample,
```

```
%so it must contains 300 elements. Here is an easy way to build one.
%Start by numbering your classes from 1 to 3.
%Make sure that your X Array is sorted by class,
%meaning all the samples from the first class,
%then all the samples from the second class, and so on. Then:

 % Classes from 1 to 3
 y = 1:3;
 % Nofsignal samples per class
 y = repmat(y, Nofsignal, 1);
 % Reshape to obtain a vector
 y = reshape(y, 1, numel(y));
% At this stage, you should have a 1-by-300 vector with numeric class labels from 1 to 3.
%This is what you want if you use the Statistics and Machine Learning Toolbox.
%
% Now, I suppose you intend to use the Neural Network Toolbox.
%In that case the syntax is a bit different: the class labels are in a 3-by-300 matrix,
%each column being a sample, with the row corresponding to the class getting value '1' and the
rest '0'.
%To build it from the previous syntax, type:

Targets = full(ind2vec(y));

%% Train an autoencoder with a hidden layer of size 5 and a linear transfer function for the
% decoder. Set the L2 weight regularizer to 0.001, sparsity regularizer to 4 and sparsity
% proportion to 0.05.
hiddenSize = 5;
autoenc = trainAutoencoder(Inputs, hiddenSize, ...
'L2WeightRegularization', 0.001, ...
'SparsityRegularization', 4, ...
'SparsityProportion', 0.05, ...
'DecoderTransferFunction','purelin');
% Extract the features in the hidden layer.
features = encode(autoenc,Inputs);
% Train a softmax layer for classification using the features .
softnet = trainSoftmaxLayer(features,Targets);
% Stack the encoder and the softmax layer to form a deep network.
stackednet = stack(autoenc,softnet);

deepnet = train(stackednet,Inputs,Targets);
%%
%%%%%%%%%%%%%%%%%%%%TEST

% Estimate the EEG signal types using the deep network, deepnet.

EEG_type = deepnet(Inputs);

% Plot the confusion matrix.
plotconfusion(Targets,EEG_type);

% Plot the ROC.
figure, plotroc(Targets,EEG_type)
```

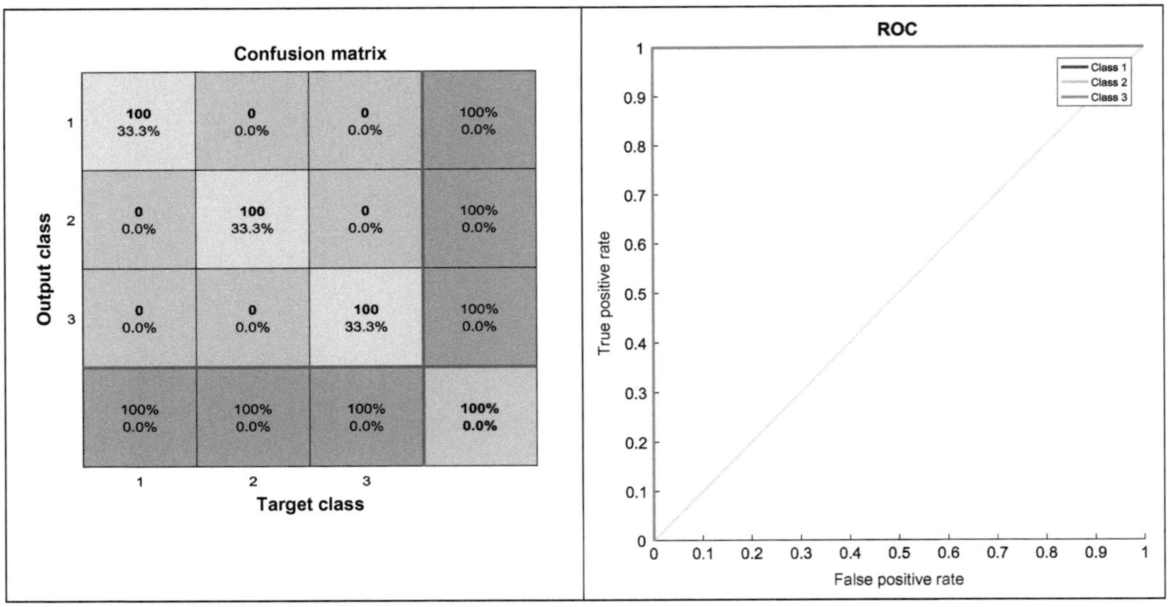

FIG. 5.35 Representation of performance of the DNN classifier for the EEG signals features extracted using SWT.

REFERENCES

Alpaydin, E. (2014). *Introduction to machine learning*. MIT Press.

Begg, R., Lai, D. T., & Palaniswami, M. (2007). *Computational intelligence in biomedical engineering*. CRC Press.

Bengio, Y. (2009). Learning deep architectures for AI. *Foundations and Trends® in Machine Learning*, *2*(1), 1–127.

Bishop, C. M. (2007). *Pattern recognition and machine learning (information science and statistics)*. New York: Springer.

Breiman, L., Friedman, J. H., Olshen, R. A., & Stone, C. J. (1984). *Classification and Regression Trees*. Boca Raton: Hall/CRC.

Cichosz, P. (2014). *Data mining algorithms: Explained using R*. John Wiley & Sons.

Cohen, J. (1960). A coefficient of agreement for nominal scales. *Educational and Psychological Measurement*, *20*(1), 37–46.

Guo, G., Wang, H., Bell, D., Bi, Y., & Greer, K. (2003). KNN model-based approach in classification. In *Presented at the OTM confederated international conferences "On the Move to Meaningful Internet Systems"* (pp. 986–996). Springer.

Hall, M., Witten, I., & Frank, E. (2011). *Data mining: Practical machine learning tools and techniques*. Burlington: Kaufmann.

Han, J., Pei, J., & Kamber, M. (2011). *Data mining: Concepts and techniques*. Elsevier.

Hand, D. J., Mannila, H., & Smyth, P. (2001). *Principles of data mining (adaptive computation and machine learning)*. Cambridge, MA: MIT Press.

Hinton, G. E., & Salakhutdinov, R. R. (2006). Reducing the dimensionality of data with neural networks. *Science*, *313*(5786), 504–507.

Mitchell, T. M. (1997). *Machine learning. Vol. 45(37)* (pp. 870–877). Burr Ridge, IL: McGraw Hill.

Quinlan, J. R. (1993). Improved use of continuous attributes in C4.5. *J Artif. Intell. Res. 5*, 77–90.

Quinlan, J. R. (1986). Induction of decision trees. *Machine Learning*, *1*(1), 81–106.

Rumelhart, D. E., Hinton, G. E., & Williams, R. J. (1986). Learning representations by back-propagating errors. *Nature*, *323*(6088), 533.

Sanei, S. (2013). *Adaptive processing of brain signals*. John Wiley & Sons.

Siuly, S., Li, Y., & Zhang, Y. (2016). *EEG signal analysis and classification*. Springer.

Sokolova, M., Japkowicz, N., & Szpakowicz, S. (2006). Beyond accuracy, F-score and ROC: a family of discriminant measures for performance evaluation. In *Presented at the Australasian joint conference on artificial intelligence* (pp. 1015–1021): Springer.

Wu, X., Kumar, V., Quinlan, J. R., Ghosh, J., Yang, Q., Motoda, H., … Philip YS, Y. (2008). Top 10 algorithms in data mining. *Knowledge and Information Systems*, *14*(1), 1–37.

Yang, Z., & Zhou, M. (2015). Kappa statistic for clustered physician-patients polytomous data. *Computational Statistics and Data Analysis*, *87*, 1–17.

Index

Note: Page numbers followed by *f* indicate figures, *t* indicate tables, and *b* indicate boxes.

D

E

F

N

O

P

Q

R

S

T

U

V

W